T0181631

The series Lecture Notes in Computer Science (LNCS), including its subseries Lecture Notes in Artificial Intelligence (LNAI) and Lecture Notes in Bioinformatics (LNBI), has established itself as a medium for the publication of new developments in computer science and information technology research, teaching, and education.

LNCS enjoys close cooperation with the computer science R & D community, the series counts many renowned academics among its volume editors and paper authors, and collaborates with prestigious societies. Its mission is to serve this international community by providing an invaluable service, mainly focused on the publication of conference and workshop proceedings and postproceedings. LNCS commenced publication in 1973.

Jiří Mikyška · Clélia de Mulatier ·
Maciej Paszynski · Valeria V. Krzhizhanovskaya ·
Jack J. Dongarra · Peter M. A. Sloot
Editors

Computational Science – ICCS 2023

23rd International Conference Prague, Czech Republic, July 3–5, 2023 Proceedings, Part IV

Editors
Jiří Mikyška
Czech Technical University in Prague
Prague, Czech Republic

Maciej Paszynski
AGH University of Science and Technology
Krakow, Poland

Jack J. Dongarra
University of Tennessee at Knoxville
Knoxville, TN, USA

Clélia de Mulatier
University of Amsterdam
Amsterdam, The Netherlands

Valeria V. Krzhizhanovskaya
University of Amsterdam
Amsterdam, The Netherlands

Peter M. A. Sloot
University of Amsterdam
Amsterdam, The Netherlands

ISSN 0302-9743 ISSN 1611-3349 (electronic)
Lecture Notes in Computer Science
ISBN 978-3-031-36026-8 ISBN 978-3-031-36027-5 (eBook)
https://doi.org/10.1007/978-3-031-36027-5

This Springer imprint is published by the registered company Springer Nature Switzerland AG
The registered company address is: Gewerbestrasse 11, 6330 Cham, Switzerland

Preface

Welcome to the 23rd annual International Conference on Computational Science (ICCS - https://www.iccs-meeting.org/iccs2023/), held on July 3–5, 2023 at the Czech Technical University in Prague, Czechia.

In keeping with the new normal of our times, ICCS featured both in-person and online sessions. Although the challenges of such a hybrid format are manifold, we have always tried our best to keep the ICCS community as dynamic, creative, and productive as possible. We are proud to present the proceedings you are reading as a result.

ICCS 2023 was jointly organized by the Czech Technical University in Prague, the University of Amsterdam, NTU Singapore, and the University of Tennessee.

Standing on the Vltava River, Prague is central Europe's political, cultural, and economic hub.

The Czech Technical University in Prague (CTU) is one of Europe's largest and oldest technical universities and the highest-rated in the group of Czech technical universities. CTU offers 350 accredited study programs, 100 of which are taught in a foreign language. Close to 19,000 students are studying at CTU in 2022/2023. The Faculty of Nuclear Sciences and Physical Engineering (FNSPE), located along the river bank in Prague's beautiful Old Town (Staré Mesto) and host to ICCS 2023, is the only one in Czechia to offer studies in a broad range of fields related to Nuclear Physics and Engineering. The Faculty operates both fission (VR-1) and fusion (GOLEM Tokamak) reactors and hosts several cutting-edge research projects, collaborating with a number of international research centers (CERN, ITER, BNL-STAR, ELI).

The International Conference on Computational Science is an annual conference that brings together researchers and scientists from mathematics and computer science as basic computing disciplines, as well as researchers from various application areas who are pioneering computational methods in sciences such as physics, chemistry, life sciences, engineering, arts, and humanitarian fields, to discuss problems and solutions in the area, identify new issues, and shape future directions for research.

Since its inception in 2001, ICCS has attracted increasingly higher-quality attendees and papers, and this year is not an exception, with over 300 participants. The proceedings series have become a primary intellectual resource for computational science researchers, defining and advancing the state of the art in this field.

The theme for 2023, "**Computation at the Cutting Edge of Science**", highlights the role of Computational Science in assisting multidisciplinary research. This conference was a unique event focusing on recent developments in scalable scientific algorithms; advanced software tools; computational grids; advanced numerical methods; and novel application areas. These innovative novel models, algorithms, and tools drive new science through efficient application in physical systems, computational and systems biology, environmental systems, finance, and others.

ICCS is well known for its excellent lineup of keynote speakers. The keynotes for 2023 were:

- **Helen Brooks**, United Kingdom Atomic Energy Authority (UKAEA), UK
- **Jack Dongarra**, University of Tennessee, USA
- **Derek Groen**, Brunel University London, UK
- **Anders Dam Jensen**, European High Performance Computing Joint Undertaking (EuroHPC JU), Luxembourg
- **Jakub Šístek**, Institute of Mathematics of the Czech Academy of Sciences & Czech Technical University in Prague, Czechia

This year we had 531 submissions (176 to the main track and 355 to the thematic tracks). In the main track, 54 full papers were accepted (30.7%); in the thematic tracks, 134 full papers (37.7%). A higher acceptance rate in the thematic tracks is explained by the nature of these, where track organizers personally invite many experts in a particular field to participate in their sessions. Each submission received at least 2 single-blind reviews (2.9 reviews per paper on average).

ICCS relies strongly on our thematic track organizers' vital contributions to attract high-quality papers in many subject areas. We would like to thank all committee members from the main and thematic tracks for their contribution to ensuring a high standard for the accepted papers. We would also like to thank *Springer, Elsevier,* and *Intellegibilis* for their support. Finally, we appreciate all the local organizing committee members for their hard work in preparing for this conference.

We are proud to note that ICCS is an A-rank conference in the CORE classification.

We hope you enjoyed the conference, whether virtually or in person.

July 2023

Jiří Mikyška
Clélia de Mulatier
Maciej Paszynski
Valeria V. Krzhizhanovskaya
Jack J. Dongarra
Peter M. A. Sloot

Organization

The Conference Chairs

General Chair

Valeria Krzhizhanovskaya — University of Amsterdam, The Netherlands

Main Track Chair

Clélia de Mulatier — University of Amsterdam, The Netherlands

Thematic Tracks Chair

Maciej Paszynski — AGH University of Science and Technology, Poland

Scientific Chairs

Peter M. A. Sloot — University of Amsterdam, The Netherlands | Complexity Institute NTU, Singapore

Jack Dongarra — University of Tennessee, USA

Local Organizing Committee

LOC Chair

Jiří Mikyška — Czech Technical University in Prague, Czechia

LOC Members

Pavel Eichler — Czech Technical University in Prague, Czechia
Radek Fučík — Czech Technical University in Prague, Czechia
Jakub Klinkovský — Czech Technical University in Prague, Czechia
Tomáš Oberhuber — Czech Technical University in Prague, Czechia
Pavel Strachota — Czech Technical University in Prague, Czechia

Thematic Tracks and Organizers

Advances in High-Performance Computational Earth Sciences: Applications and Frameworks – IHPCES

Takashi Shimokawabe, Kohei Fujita, Dominik Bartuschat

Artificial Intelligence and High-Performance Computing for Advanced Simulations – AIHPC4AS

Maciej Paszynski, Robert Schaefer, Victor Calo, David Pardo, Quanling Deng

Biomedical and Bioinformatics Challenges for Computer Science – BBC

Mario Cannataro, Giuseppe Agapito, Mauro Castelli, Riccardo Dondi, Rodrigo Weber dos Santos, Italo Zoppis

Computational Collective Intelligence – CCI

Marcin Maleszka, Ngoc Thanh Nguyen

Computational Diplomacy and Policy – CoDiP

Michael Lees, Brian Castellani, Bastien Chopard

Computational Health – CompHealth

Sergey Kovalchuk, Georgiy Bobashev, Anastasia Angelopoulou, Jude Hemanth

Computational Modelling of Cellular Mechanics – CMCM

Gabor Zavodszky, Igor Pivkin

Computational Optimization, Modelling, and Simulation – COMS

Xin-She Yang, Slawomir Koziel, Leifur Leifsson

Computational Social Complexity – CSCx

Vítor V. Vasconcelos. Debraj Roy, Elisabeth Krüger, Flávio Pinheiro, Alexander J. Stewart, Victoria Garibay, Andreia Sofia Teixeira, Yan Leng, Gabor Zavodszky

Computer Graphics, Image Processing, and Artificial Intelligence – CGIPAI

Andres Iglesias, Lihua You, Akemi Galvez-Tomida

Machine Learning and Data Assimilation for Dynamical Systems – MLDADS

Rossella Arcucci, Cesar Quilodran-Casas

MeshFree Methods and Radial Basis Functions in Computational Sciences – MESHFREE

Vaclav Skala, Samsul Ariffin Abdul Karim

Multiscale Modelling and Simulation – MMS

Derek Groen, Diana Suleimenova

Network Models and Analysis: From Foundations to Complex Systems – NMA

Marianna Milano, Pietro Cinaglia, Giuseppe Agapito

Quantum Computing – QCW

Katarzyna Rycerz, Marian Bubak

Simulations of Flow and Transport: Modeling, Algorithms, and Computation – SOFTMAC

Shuyu Sun, Jingfa Li, James Liu

Smart Systems: Bringing Together Computer Vision, Sensor Networks and Machine Learning – SmartSys

Pedro Cardoso, Roberto Lam, Jânio Monteiro, João Rodrigues

Solving Problems with Uncertainties – SPU

Vassil Alexandrov, Aneta Karaivanova

Teaching Computational Science – WTCS

Angela Shiflet, Nia Alexandrov

Reviewers

Zeeshan Abbas
Samsul Ariffin Abdul Karim
Tesfamariam Mulugeta Abuhay
Giuseppe Agapito
Elisabete Alberdi
Vassil Alexandrov
Nia Alexandrov
Alexander Alexeev
Nuno Alpalhão
Julen Alvarez-Aramberri
Domingos Alves
Sergey Alyaev
Anastasia Anagnostou
Anastasia Angelopoulou
Fabio Anselmi
Hideo Aochi
Rossella Arcucci
Konstantinos Asteriou
Emanouil Atanassov
Costin Badica
Daniel Balouek-Thomert
Krzysztof Banaś
Dariusz Barbucha
Luca Barillaro
João Barroso
Dominik Bartuschat
Pouria Behnodfaur
Jörn Behrens
Adrian Bekasiewicz
Gebrail Bekdas
Mehmet Belen
Stefano Beretta
Benjamin Berkels
Daniel Berrar
Piotr Biskupski
Georgiy Bobashev
Tomasz Boiński
Alessandra Bonfanti
Carlos Bordons
Bartosz Bosak
Lorella Bottino
Roland Bouffanais
Lars Braubach
Marian Bubak
Jérémy Buisson
Aleksander Byrski
Cristiano Cabrita
Xing Cai
Barbara Calabrese
Nurullah Çalık
Victor Calo
Jesús Cámara
Almudena Campuzano
Cristian Candia
Mario Cannataro
Pedro Cardoso
Eddy Caron
Alberto Carrassi
Alfonso Carriazo
Stefano Casarin
Manuel Castañón-Puga
Brian Castellani
Mauro Castelli
Nicholas Chancellor
Ehtzaz Chaudhry
Théophile Chaumont-Frelet
Thierry Chaussalet
Sibo Cheng
Siew Ann Cheong
Lock-Yue Chew
Su-Fong Chien
Marta Chinnici
Bastien Chopard
Svetlana Chuprina
Ivan Cimrak
Pietro Cinaglia

Noélia Correia
Adriano Cortes
Ana Cortes
Anna Cortes
Enrique Costa-Montenegro
David Coster
Carlos Cotta
Peter Coveney
Daan Crommelin
Attila Csikasz-Nagy
Javier Cuenca
António Cunha
Luigi D'Alfonso
Alberto d'Onofrio
Lisandro Dalcin
Ming Dao
Bhaskar Dasgupta
Clélia de Mulatier
Pasquale Deluca
Yusuf Demiroglu
Quanling Deng
Eric Dignum
Abhijnan Dikshit
Tiziana Di Matteo
Jacek Długopolski
Anh Khoa Doan
Sagar Dolas
Riccardo Dondi
Rafal Drezewski
Hans du Buf
Vitor Duarte
Rob E. Loke
Amir Ebrahimi Fard
Wouter Edeling
Nadaniela Egidi
Kareem Elsafty
Nahid Emad
Christian Engelmann
August Ernstsson
Roberto R. Expósito
Fangxin Fang
Giuseppe Fedele
Antonino Fiannaca
Christos Filelis-Papadopoulos
Piotr Frąckiewicz
Alberto Freitas
Ruy Freitas Reis
Zhuojia Fu
Kohei Fujita
Takeshi Fukaya
Wlodzimierz Funika
Takashi Furumura
Ernst Fusch
Marco Gallieri
Teresa Galvão Dias
Akemi Galvez-Tomida
Luis Garcia-Castillo
Bartłomiej Gardas
Victoria Garibay
Frédéric Gava
Piotr Gawron
Bernhard Geiger
Alex Gerbessiotis
Josephin Giacomini
Konstantinos Giannoutakis
Alfonso Gijón
Nigel Gilbert
Adam Glos
Alexandrino Gonçalves
Jorge González-Domínguez
Yuriy Gorbachev
Pawel Gorecki
Markus Götz
Michael Gowanlock
George Gravvanis
Derek Groen
Lutz Gross
Tobias Guggemos
Serge Guillas
Xiaohu Guo
Manish Gupta
Piotr Gurgul
Zulfiqar Habib
Yue Hao
Habibollah Haron
Mohammad Khatim Hasan
Ali Hashemian
Claire Heaney
Alexander Heinecke
Jude Hemanth

Pedro Medeiros
Wen Mei
Wagner Meira Jr.
Roderick Melnik
Pedro Mendes Guerreiro
Yan Meng
Isaak Mengesha
Ivan Merelli
Tomasz Michalak
Lyudmila Mihaylova
Marianna Milano
Jaroslaw Miszczak
Dhruv Mittal
Miguel Molina-Solana
Fernando Monteiro
Jânio Monteiro
Andrew Moore
Anabela Moreira Bernardino
Eugénia Moreira Bernardino
Peter Mueller
Khan Muhammad
Daichi Mukunoki
Judit Munoz-Matute
Hiromichi Nagao
Kengo Nakajima
Grzegorz J. Nalepa
I. Michael Navon
Vittorio Nespeca
Philipp Neumann
James Nevin
Ngoc-Thanh Nguyen
Nancy Nichols
Marcin Niemiec
Sinan Melih Nigdeli
Hitoshi Nishizawa
Algirdas Noreika
Manuel Núñez
Joe O'Connor
Frederike Oetker
Lidia Ogiela
Ángel Javier Omella
Kenji Ono
Eneko Osaba
Rongjiang Pan
Nikela Papadopoulou
Marcin Paprzycki
David Pardo
Anna Paszynska
Maciej Paszynski
Łukasz Pawela
Giulia Pederzani
Ebo Peerbooms
Alberto Pérez de Alba Ortíz
Sara Perez-Carabaza
Dana Petcu
Serge Petiton
Beata Petrovski
Toby Phillips
Frank Phillipson
Eugenio Piasini
Juan C. Pichel
Anna Pietrenko-Dabrowska
Gustavo Pilatti
Flávio Pinheiro
Armando Pinho
Catalina Pino Muñoz
Pietro Pinoli
Yuri Pirola
Igor Pivkin
Robert Platt
Dirk Pleiter
Marcin Płodzień
Cristina Portales
Simon Portegies Zwart
Roland Potthast
Małgorzata Przybyła-Kasperek
Ela Pustulka-Hunt
Vladimir Puzyrev
Ubaid Qadri
Rick Quax
Cesar Quilodran-Casas
Issam Rais
Andrianirina Rakotoharisoa
Célia Ramos
Vishwas H. V. S. Rao
Robin Richardson
Heike Riel
Sophie Robert
João Rodrigues
Daniel Rodriguez

Marcin Rogowski
Sergio Rojas
Diego Romano
Albert Romkes
Debraj Roy
Adam Rycerz
Katarzyna Rycerz
Mahdi Saeedipour
Arindam Saha
Ozlem Salehi
Alberto Sanchez
Ayşin Sancı
Gabriele Santin
Vinicius Santos Silva
Allah Bux Sargano
Azali Saudi
Ileana Scarpino
Robert Schaefer
Ulf D. Schiller
Bertil Schmidt
Martin Schreiber
Gabriela Schütz
Jan Šembera
Paulina Sepúlveda-Salas
Ovidiu Serban
Franciszek Seredynski
Marzia Settino
Mostafa Shahriari
Vivek Sheraton
Angela Shiflet
Takashi Shimokawabe
Alexander Shukhman
Marcin Sieniek
Joaquim Silva
Mateusz Sitko
Haozhen Situ
Leszek Siwik
Vaclav Skala
Renata Słota
Oskar Slowik
Grażyna Ślusarczyk
Sucha Smanchat
Alexander Smirnovsky
Maciej Smołka
Thiago Sobral
Isabel Sofia
Piotr Sowiński
Christian Spieker
Michał Staniszewski
Robert Staszewski
Alexander J. Stewart
Magdalena Stobinska
Tomasz Stopa
Achim Streit
Barbara Strug
Dante Suarez
Patricia Suarez
Diana Suleimenova
Shuyu Sun
Martin Swain
Edward Szczerbicki
Tadeusz Szuba
Ryszard Tadeusiewicz
Daisuke Takahashi
Osamu Tatebe
Carlos Tavares Calafate
Andrey Tchernykh
Andreia Sofia Teixeira
Kasim Terzic
Jannis Teunissen
Sue Thorne
Ed Threlfall
Alfredo Tirado-Ramos
Pawel Topa
Paolo Trunfio
Hassan Ugail
Carlos Uriarte
Rosarina Vallelunga
Eirik Valseth
Tom van den Bosch
Ana Varbanescu
Vítor V. Vasconcelos
Alexandra Vatyan
Patrick Vega
Francesc Verdugo
Gytis Vilutis
Jackel Chew Vui Lung
Shuangbu Wang
Jianwu Wang
Peng Wang

Katarzyna Wasielewska
Jarosław Wątróbski
Rodrigo Weber dos Santos
Marie Weiel
Didier Wernli
Lars Wienbrandt
Iza Wierzbowska
Maciej Woźniak
Dunhui Xiao
Huilin Xing
Yani Xue
Abuzer Yakaryilmaz
Alexey Yakovlev
Xin-She Yang
Dongwei Ye
Vehpi Yildirim
Lihua You
Drago Žagar
Sebastian Zając
Constantin-Bala Zamfirescu
Gabor Zavodszky
Justyna Zawalska
Pavel Zemcik
Wenbin Zhang
Yao Zhang
Helen Zhang
Jian-Jun Zhang
Jinghui Zhong
Sotirios Ziavras
Zoltan Zimboras
Italo Zoppis
Chiara Zucco
Pavel Zun
Karol Życzkowski

Contents – Part IV

Multiscale Modelling and Simulation

Network Models and Analysis: From Foundations to Complex Systems

Computational Social Complexity

The Social Graph Based on Real Data

Tomasz M. Gwizdała(✉) and Aleksandra Piecuch

Faculty of Physics and Applied Informatics, University of Lodz, Lodz, Poland
tomasz.gwizdalla@uni.lodz.pl

Abstract. In this paper, we propose a model enabling the creation of a social graph corresponding to real society. The procedure uses data describing the real social relations in the community, like marital status or number of kids. Results show the power-law behavior of the distribution of links and, typical for small worlds, the independence of the clustering coefficient on the size of the graph.

Keywords: social network · scale-free network · small-world

1 Introduction and Model

The computational study of different social processes requires knowledge of the structure of the community studied. Among the seminal papers devoted to study of graph representations of communities we have especially to mention the Pastor-Satorras paper [10] where different properties of graphs are studied in the context of epidemic. Now, a lot of problems can be studied with the use of social graphs. We can mention here the processes which can be generally called social contagion [1]. Particularly, such topics like disease spreading [4–6], transfer of information [9], opinion formation [8,11] or innovation emergence [12] are studied with the use of particular graphs. It is important to emphasize that speaking about social network in the context of studied problem we have take into account the real community which structure does not necessarily correspond to the structures obtained by studying, recently very popular, online networks. The technique which have recently become popular for such networks is the study of homophily effect, [13,14]. In our paper, we will present a new approach to creating a society graph, based on the real statistics describing it. The special attention is devoted to study the real social links like families, acquaintances. Due to ease of access, we use the data provided for Polish society, mainly our city - Lodz.

The population is typically presented as an undirected, connected graph, where nodes correspond to individuals and edges to relations between them. Both features mentioned above are essential for the understanding of the model. When assuming the graph's undirected character, we consider the mutual type of interaction. It means that compared to well-known social networks, we consider rather a connection (like the most popular way of using linkedin) than observation (like twitter). When assuming the connected graph, we ensure that, in

J. Mikyška et al. (Eds.): ICCS 2023, LNCS 14076, pp. 3–9, 2023.
https://doi.org/10.1007/978-3-031-36027-5_1

society, we do not have individuals that are completely excluded. We will come back to this issue when discussing the levels of connections.

We propose to use initially four levels of connection between individuals/ nodes

- **I.** Household members and inmates. In this class, we collect the people who have permanent contact - members of families, cohabitants, or people who rent apartments together.
- **II.** In this class, several types of connections are considered. The crucial property of people in these groups is that they are strongly connected but outside households. We include small groups created in the school or work environment and further family contacts here. These people meet themselves every day or almost every day, and small social distances characterize their meetings.
- **III.** Acquaitances. In this class, we consider loosely related individuals but having the possibility of frequent contacts - kids in schools or employees in companies.
- **IV.** Accidental but possible contacts - here we sample mainly the people living in the common areas - villages, city districts. We can meet them accidentally on the street, in the shop, but these meetings are strongly random.

In the paper, we focus on the first two categories. However, the creation of a third class will also be described because the second one is based on it. We assume that at least one individual is always considered as a family member (1st class) or caregiver (2nd class).

To ensure the correctness of the data, we use only the ones which follow some basic criteria.

- The source of data can be considered reliable. This criterion, indeed, led to the strong restriction that all data come from the polish GUS (Statistics Poland - https://stat.gov.pl)
- The studies were conducted in a similar period. It is tough to find all data sampled simultaneously during the same study because they cover different aspects of social or economic statistics.
- The studies were conducted on a similar statistical population. Sometimes, the easily accessible data concern different communities, e.g., big cities and village populations. In such a situation, we do rather consider the global data for the whole country (like the employment vs. size of company statistics).

In the paper, we model the closed community corresponding rather to a big city than to scattered areas of villages and small towns. Such a choice is made for several reasons. The regulations for such environments are more regular, without many exceptions, like e.g., the size of schools. The modeling of the distribution of individuals, for example, between districts, is also easier than creating a large number of small communities, where additional conditions should be taken into account (like e.g., the general possibilities of connecting between such communities). The data presented below differ significantly for various communities. In the calculations, we mainly use the data for our city - Lodz (Łódź), or for the whole of Poland.

The successive steps of the proposed approach is as follows:

- We start by selecting the correct number of population members according to the age pyramid. Here we use the data for Lodz, and the borders of intervals for which data are known are given by $\{2, 6, 12, 15, 18, 24, 34, 44, 54, 64, 75, 95\}$. For a relatively small percentage of people in the last interval (above 95), we merge the last two intervals.
- The marital status of men is sampled. We use the data concerning the five possible states: bachelor, married, cohabitant, divorced, and a widower for different age intervals (different from the age pyramid). We start from the men subset because the total number of men is smaller than the number of women.
- The marriages and cohabitation relations are created as the first level connection. We use the distribution of differences between men's and women's ages in relations and sample the partners from the women's subset. For the remaining women, we attach the status in the same manner as for men. The correctness of the final marital distribution is the test for the correctness of the whole procedure, but we do not present the results confirming this effect here.
- The kids are attached as the first-level connection. Having the distribution of kids typical for a particular age and the type of family (pair, single men, single woman), we attach the number of kids of an age different from the age of the woman (or man) by a value from interval [18,40].
- The kids and youth are attached to schools as the 3rd-level connections. We know the average size of a school of a particular type, so we can, taking into account the number of individuals in the corresponding age, determine the number of schools. Then we create cliques, attaching every individual under 18 (in accordance with Polish law) to the one of the established schools.
- People working in the same companies are connected as the 3rd-level connections. The basis for assigning people to jobs is the distribution of a number of employees in one of the four groups specified by polish statistics (micro, small, average, big - sometimes also the group of large companies is described, but we omit it). After assigning the type of company to every individual of the right age (before the retirement age, different for men and women), we create cliques of size sampled uniformly from the interval characteristic for particular group.
- The selector of narrower groups as the 2nd-level connections. We choose the smaller cliques among the 3rd-level connections. We can choose them arbitrarily, but for the presented calculations, we use the Poisson distribution with an average equal to 3 as the size of these cliques.
- Finalization. After the steps described above, there can still be some unconnected nodes. It mainly concerns retired people, who are not in pairs at the first level, are not in a household with children and do not work professionally. For these people, we attach the person, younger by 20–40 years, who can be considered as a separately living child or a caregiver.

The procedure described above allows the creation of the connected graph when the first and second-level connections are considered. Intentionally, we do not describe the procedure of creating connections belonging to the 4th-level because it is not analyzed in the paper.

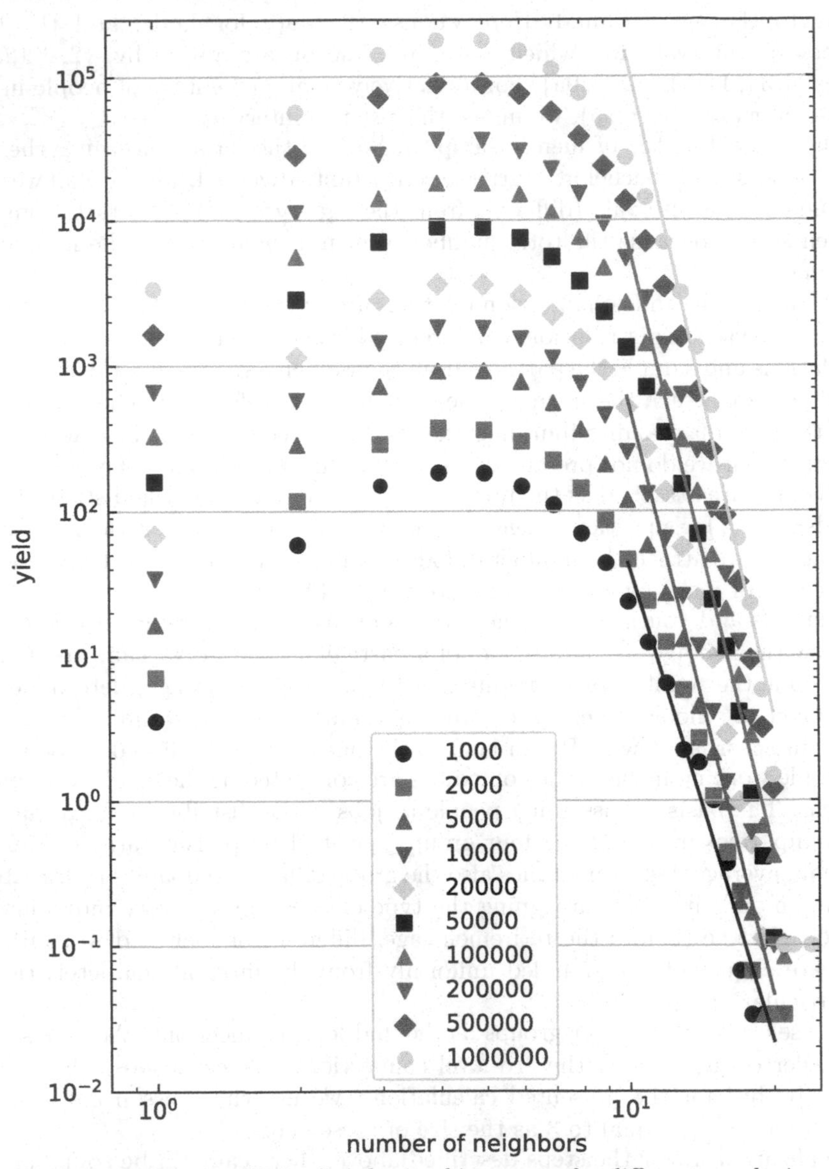

Fig. 1. The comparison of links distribution histograms for different graph sizes. The solid lines, created for selected set of graph sizes (10^3, 10^4, 10^5, 10^6) correspond the power-law dependencies and the colors correspond to the color of marker for particular size of community.

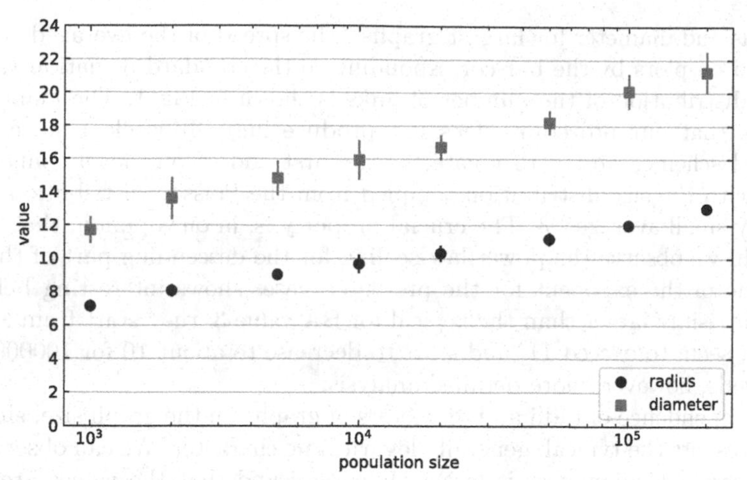

Fig. 2. The dependence of radii and diameters on the size of community for graphs created according to the presented procedure.

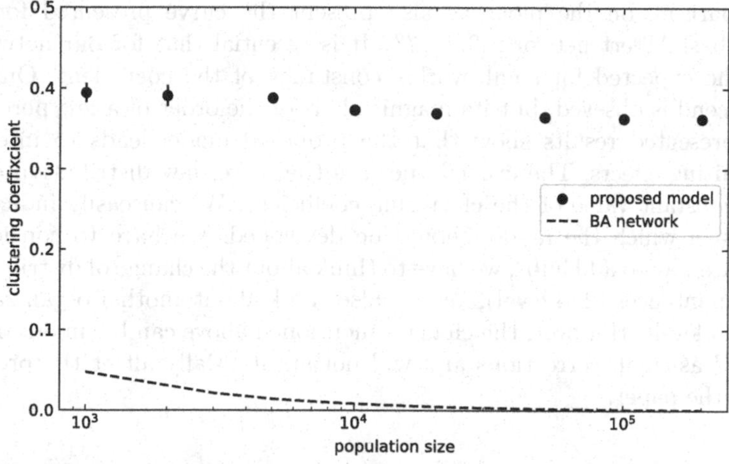

Fig. 3. The dependence of clustering coefficient on the size of community for graphs created according to the presented procedure. The values were obtained in the same wat as those from Fig. 2.

2 Results and Conclusions

We concentrate our attention on four parameters that are typically used to describe the properties of a graph as the form representing the community. They are the number of neighbors/links distribution, radius and diameter, and clustering coefficient. The results are presented for graph sizes from 1000 to 1000000, distributed uniformly in the log scale. The calculations for every number are performed several times, and the presented values are averaged over independent runs. The number of repetitions varies from 30 for the smaller sizes to 5 for greater ones. Due to large computational complexity we do not present results for radius and diameter for largest graphs. The spread of the averaged values is presented on plots by the bar corresponding to the standard deviation value.

The distribution of the number of links is shown in Fig. 1. The main observation is that our procedure does not produce hubs. It is clear since in the presented scheme, we try to create, as the first and second level connections, cliques with the size distribution sampled from the Poisson distribution with a relatively small average - 3. The crucial property is, in our opinion, that for our approach, we observe the power-law scaling for the descending part of the plot. The value of the exponent for the presented cases shows interesting behavior. They are visibly larger than the typical for BA value 3; they start from about 8 for 1000, seem to exceed 11, and start to decrease to about 10 for 100000. This effect needs, however, more detailed analysis.

The dependence of radii and diameters of graphs on the graph size, shown in Fig. 2, presents the typical, generally logarithmic character. We can observe that the dispersion of diameters is larger than radii and that the values are larger than the predicted for scale-free Barabasi-Albert network [3].

A significant effect can be observed in the plot of clustering coefficient 3. For comparison, on the plot, we also present the curve presented for scale-free Barabasi-Albert network [2,7]. ??? It is essential that for our network we obtain, the expected for small-worlds constancy of the coefficient. Only very small descend is obseved, but its magnitude is of the order of a few percent.

The presented results show that the proposed model leads to interesting and promising effects. The crucial ones are the power-law distribution of links and the constant value of the clustering coefficient. We can easily indicate the directions in which the model should be developed. We have to consider the appropriate way to add hubs; we have to think about the change of distribution of close acquaintances (2nd level), and we also think about another organization of connection levels. However, the changes mentioned above can be, in our opinion, considered as small corrections and will not substantially affect the properties shown in the paper.

References

1. Ferraz de Arruda, G., Petri, G., Rodriguez, P.M., Moreno, Y.: Multistability, intermittency, and hybrid transitions in social contagion models on hypergraphs. Nat. Commun. **14**(1), 1375 (2023)
2. Bollobas, B., Riordan, O.: Mathematical results on scale-free random graphs. In: Bormholdt, S., Schuster, A.G. (eds.) Handbook of Graphs and Networks, pp. 1–34. Wiley (2003)
3. Bollobas, B., Riordan, O.: The diameter of a scale-free random graph. Combinatorica **24**, 5–34 (2004)
4. Brattig Correia, R., Barrat, A., Rocha, L.M.: Contact networks have small metric backbones that maintain community structure and are primary transmission subgraphs. PLOS Comput. Biol. **19**(2), 1–23 (02 2023). https://doi.org/10.1371/journal.pcbi.1010854
5. Gwizdałła, T.M.: Viral disease spreading in grouped population. Comput. Meth. Prog. Biomed. **197**, 105715 (2020)

6. Karaivanov, A.: A social network model of covid-19. Plos One **15**(10), 1–33 (2020). https://doi.org/10.1371/journal.pone.0240878
7. Klemm, K., Eguíluz, V.M.: Growing scale-free networks with small-world behavior. Phys. Rev. E **65**, 057102 (2002)
8. Kozitsin, I.V.: A general framework to link theory and empirics in opinion formation models. Sci. Rep. **12**, 5543 (2022)
9. Kumar, S., Saini, M., Goel, M., Panda, B.: Modeling information diffusion in online social networks using a modified forest-fire model. J. Intell. Inf. Syst. **56**, 355–377 (2021)
10. Pastor-Satorras, R., Castellano, C., Van Mieghem, P., Vespignani, A.: Epidemic processes in complex networks. Rev. Mod. Phys. **87**, 925–979 (2015)
11. Ramaciotti Morales, P., Cointet, J.P., Muñoz Zolotoochin, G., Fernández Peralta, A., Iñiguez, G., Pournaki, A.: Inferring attitudinal spaces in social networks. Soc. Netw. Anal. Min. **13**(1), 1–18 (2023)
12. Sziklai, B.R., Lengyel, B.: Finding early adopters of innovation in social networks. Soc. Netw. Anal. Mining **13**(1), 4 (2023)
13. Talaga, S., Nowak, A.: Homophily as a process generating social networks: Insights from social distance attachment mode. JASSS **23**(2) (2020)
14. Zhao, T., Hu, J., He, P., Fan, H., Lyu, M., King, I.: Exploiting homophily-based implicit social network to improve recommendation performance. In: 2014 International Joint Conference on Neural Networks (IJCNN), pp. 2539–2547 (2014)

Longitudinal Analysis of the Topology of Criminal Networks Using a Simple Cost-Benefit Agent-Based Model

Louis Félix Weyland[2(✉)], Ana Isabel Barros[1,2], and Koen van der Zwet[2]

[1] Police Academy of The Netherlands, Apeldoorn, The Netherlands
[2] TNO Defense, Safety and Security, The Hague, The Netherlands
louis.weyland@tno.nl

Abstract. Recently, efforts have been made in computational criminology to study the dynamics of criminal organisations and improve law enforcement measures. To understand the evolution of a criminal network, current literature uses social network analysis and agent-based modelling as research tools. However, these studies only explain the short-term adaptation of a criminal network with a simplified mechanism for introducing new actors. Moreover, most studies do not consider the spatial factor, i.e. the underlying social network of a criminal network and the social environment in which it is active. This paper presents a computational modelling approach to address this literature gap by combining an agent-based model with an explicit social network to simulate the long-term evolution of a criminal organisation. To analyse the dynamics of a criminal organisation in a population, different social networks were modelled. A comparison of the evolution between the different networks was carried out, including a topological analysis (secrecy, flow of information and size of largest component). This paper demonstrates that the underlying structure of the network does make a difference in its development. In particular, with a preferentially structured population, the prevalence of criminal behaviour is very pronounced. Moreover, the preferential structure provides criminal organisations a certain efficiency in terms of secrecy and flow of information.

Keywords: criminal networks · agent-based modelling · social network analysis · social opportunity

1 Introduction

To adequately respond to a persistent and rampant issue of illicit activities, such as human trafficking and drug production [26,27], the latest advancements in mathematics and computational power are used as a decision-driven tool to increase the effectiveness of the intervention methods against organised crime

J. Mikyška et al. (Eds.): ICCS 2023, LNCS 14076, pp. 10–24, 2023.
https://doi.org/10.1007/978-3-031-36027-5_2

[11,12,14]. The objective of this research was to explore and analyse the possible evolution of any criminal network through an agent-based model while addressing the gap in the literature. Based on the data provided by the intelligence services or constructed from publicly available data, a network can be reconstructed upon which a social network analysis is applied to determine its characteristics and define the key players[1] [1,4,6,9–12,34]. Given the covert nature of criminal networks and the difficulty of updating information about them, current literature is often forced to base its research on static portrayals of criminal ties[2] [11,30]. However, criminal organisations have been found to have a fluid structure where re-structuring is a constant process to adapt to endogenous and exogenous factors such as gang rivalries or police interventions [35]. Thus, the resilience of a network is defined by how well the organisation can mitigate damage and maintain its illegal activity. It has been found that interventions can have unintended effects in the long term, such as becoming relatively ineffective or forcing the dispersion of criminal activities across an area [12,17,24]. To increase the efficiency of dismantling interventions, it is imperative to understand the dynamic of the criminal organisation (CO) and the possible repercussions an intervention can have. As noted earlier, most studies focus on static networks and attempt to provide information about their characteristics when interventions are simulated. Yet, static snapshots of criminal networks contain only partial information and thus longitudinal data is preferred[3]. To tackle the lack of longitudinal data about criminal networks and provide insights into the criminal network resilience behaviour, various approaches have been adopted.

On the one hand, there is a large volume of published studies describing the evolution of a criminal network after introducing a fictitious police intervention [4,12,14]. In a study investigating the dynamics of a CO, Behzadan et al. [4] simulated the evolution of a CO from a game-theoretic perspective. To mimic the dynamics, information about the actors is needed, such as: their nationality, languages spoken, or function within the organisation. However, most data sets do not contain an exhaustive list of variables. Duxbury and Haynie [14] performed a similar study, using a combination of agent-based modelling with social network

[1] In this context, key players are regarded as important actors who ensure the proper functioning of the organisation. In their absence, the organisation would break into considerably smaller fragments leading to a reduction in productive capacity.

[2] A static criminal network is defined as a network whose links between nodes are assumed to be unchanged. Thereby, in social network analysis, nodes (vertices) correspond to actors and links (edges) to relations. The term "relation" depends on the context of the data in question. It can represent a simple acquaintance between two persons, defined by work relation, friendship, kingship or membership of the association [9], or it can also represent a medium by which resources are shared between actors [33]. Moreover, a network can be shaped by various types of ties, including strong ties which denote close and trusted connections, weak ties which refer to more distant and casual connections with lower intimacy and less frequent interaction, and latent ties which represent potential connections that are not actively utilised [13].

[3] The term longitudinal refers to the succession of consecutive snap-shots of the criminal networks over a certain period, resulting in a time-varying network.

analysis to provide some insights into the mechanism behind the resilience by introducing the concept of trustworthiness: a liaison between actors also depends on trust. Distinguishing between profit-oriented and security-oriented networks, the formation of new ties will depend on the CO's motive. In security-oriented networks, the flow of information is restricted, thus liaisons between actors with low degree centrality are more likely to be created. In a profit-oriented network, a high interconnection between actors is the driving force of the simulated dynamic. However, the models in [4,14] assume rational behaviour of the actors and perfect knowledge of the network; the actors know each other's attributes. In contrast, access to information in a security-oriented organisation–such as the terrorist network studied in the aforementioned article–is severely restricted so as not to jeopardise its activities [6,14]. Duijn et al. [12] analysed the evolution of a CO by introducing the economical concept of the human capital approach; new liaisons are created between actors holding either a similar role or interrelated roles. Calderoni et al. [9] tackled the question of the dynamics of a criminal network from a recruitment point of view using various policies as an exogenous factor. Amongst the policies tested, insights into the influence of family members belonging to the mafia and the influence of social support to families at risk were highlighted. One of the limitations of the model is that the research was applied to a case study, which does not necessarily have a broader application. Furthermore, the impact of imprisonment on an agent's decision to re-enter the organisation is neglected. Having access to a longitudinal criminal network, some studies compared various mechanisms such as preferential attachment, triadic closure, or role-similarity to assess their goodness of fit of the simulated dynamics [6,7]. However, the authors faced the challenge of missing data. These simulations explain the short-term adaptation of a criminal network. After an intervention, most COs are able to fill the role of the neutralised criminal and resume their activity after a few days to weeks [19,24]. Additionally, the emergence of new actors is either not explained or not taken into account in these models. On the other hand, there are publications that concentrate on adopting a game-theoretical approach to model the evolution of criminal behaviour within a population [5,18,21]. Martinez-Vaquero et al. [18] and Perc et al. [21] attempted to show a more general analysis of the emergence of criminal behaviour by considering the effect of policies, such as jurisdiction penalty, social exclusion, and clerical self-justice. During the simulation, actors decide to act either as honest citizens or criminals and collect the respective remuneration, referred to as fitness points. The policies act as control variables influencing the presence of the different roles within the population. Berenji et al. [5] further add to the policy-focused model approach by tackling the line of inquiry by introducing the aspect of rehabilitation. Despite the promising results, these studies do not account for the possible influence of a network-like structure on the model. It is believed that when studying the spread of criminal behaviour within a population, it is important to consider its underlying social ties. Based on the aforementioned studies, a literature gap can be asserted concerning the publications focusing on short-term evolutions of a CO [4,9,12,14] and those focusing on long-term developments of criminal behaviour in a population but neglecting the influence of a social network structure in their simulations [18,21].

To bridge the identified literature gap, this paper aims at formulating a model that simulates the long-term evolution of a CO by combining a developed recruitment mechanism with network-explicit configurations. As noted in [9,18], it is believed that understanding the recruitment mechanism of a CO will add to the longevity of the dynamic evolution and give an adequate answer to how new actors emerge in a CO. This is achieved by using a different adaptation of Martinez-Vaquero et al. [18]'s model, where social ties are taken into account: this is the network-explicit configuration that is modelled in this paper. By developing this model, using the network-explicit configuration, the research will provide (i) social relevance insofar as it could further aid the understanding of criminal network growth, and (ii) practical relevance in that it could further the ability of authorities to predict this evolution and develop appropriate interventions.

2 Methodology

2.1 Agent-Based Modelling

The following agent-based model is extended based on the conceptual framework proposed by [18]. As stated by Martinez-Vaquero et al. [18], to model the growth or decline of a criminal network it is important to define the conditions leading a person to join such an organisation. Consequently, that is most adequately done using three different types of agents: honest, member of a criminal organisation (MCO), and lone actors. The lone actors are deemed important since they account for cases where one does not join an organisation but prefers to act independently. Through a cost-benefit game, the entities interact with each other and decide to adopt the best strategy, which involves either becoming honest, or MCO, or a lone actor. The driving force of the model is a punishment/reward system which influences the decision-making of the agents. For honest citizens, MCOs and lone actors, the reward reflects the success of their activities. Based on the theory of social opportunity, an agent's success can influence his/her entourage to imitate him/her [12,18]. The punishment system is exerted through the introduction of three meta-agents: the criminal court of justice, social control, and the pressure exerted by a CO. The meta-agents do not correspond to a specific node but merely represent an abstract agent which oversees the game. The mechanism of the agent-based model, including the acting, investigation and evolutionary stage is based on the original paper [18]. For the purpose of conciseness, the reader is directed to consult the original paper for further details. The distinction between this model and the original lies in the fact that the interactions between agents are governed by an underlying network. Thus, the groups, as defined in the original literature, are composed of the selected player and all its immediate neighbours who share a connection. As a result, the computation of reward and damage points can be carried out directly, rather than using the mean-filed approach. The following pseudo-code gives an overview of the agent-based model's (ABM) mechanism. Alternatively, a more detailed explanation including the ODD+ protocol can be studied. The definition and default values of the parameters are indicated in Table 1. The table serves merely as a reference point for subsequent analysis in which one or more parameters are subject to changes. The default combinations of parameters have been chosen such that the reward points are the same for MCOs and lone actors, and the penalty systems are equal. Furthermore, the influence of the CO on lone actors as well as the penalty share between MCOs was set to a relatively low value.

Algorithm 1: Modified Agent-based model based on Martinez-Vaquero et al.

```
input  : A network with nodes having a fitness and status attribute
output: The evolution of status of the population and the topological change of the criminal network
for i ← 1 to n_rounds do
    victimiser = select_random_person(network); // select random victimiser
    // Acting stage
    if victimiser.status == criminal then
        inflict_damage(neighbours,c_c); // inflict damage to all non-MCO neighbours
        victimiser.fitness +(r_c × c_c)/n_c;
        criminal_neighbours.fitness+(r_c × c_c)/n_c // all the MCO neighbours get the same benefit
    else if victimiser.status == wolf then
        if U(0,1) ≥ 1 − δ(1 − n_c) then
            inflict_damage(neighbours,c_w); // inflict damage to all the neighbours
            victimiser.fitness +(c_w × r_w);
            criminal_neighbour.fitness+τ(r_w × c_w)/n_c
    // Investigation stage
    // Investigation is successful if victimiser has been found
    // If select_random_person(network) == victimiser
    if state_investigation == successful then
        victimiser.fitness -β_s;
    if honest_investigation == successful then
        victimiser.fitness -(β_h × n_h);
    if criminal_investigation == successful AND victimiser.status == wolf then
        victimiser.fitness -(β_c × n_c);
    if honest_investigation/state_investigation == successful AND victimiser.status ==criminal then
        // All the criminal neighbours get also a penalty
        // If an investigation stage is unsuccessful, the respective term (β_s/β_h)is 0
        criminal_neighbours.fitness -γ × (β_s + β_h × n_h);
    // Evolutionary stage
    random_person == select_random_person(network);
    if U(0,1) ≤ mutation_prob then
        random_person.status = random_select(["c","h","w"]); // randomly switch stage
    else
        fermi_function(random_person,random_neighbour); // adopt random_neighbour status based
            on Fermi function
```

2.2 Network Initialisation

To introduce a network-explicit agent-based model, [18]'s model has been built upon by firstly taking an initial criminal network as defined in the literature, and constructing it with the inclusion of "honest" and "lone actors" nodes using different attachment methods. The attachments are: random, preferential, and small-world. These attachment methods have helped explain various networks observed in nature [3,15,31]. The preferential attachment has proven to be interesting, especially as it can explain the properties of social networks such as the scale-free property [2]. The random network, also called Erdos-Rényi network,

Table 1. Overview of the parameters used for the Agent-based model. The reward, damage, and punishment parameters are based on [18].

Parameters	Default value	Explanation
δ	0.7	Influence factor of an independent actor to act
τ	0.1	Influence factor of an independent actor's action on CO
γ	0.1	Punishment sharing factor for members of a CO
β_s	1	State punishment value
β_h	1	Civil punishment value
β_c	1	Criminal punishment value
c_w	1	Damage caused by a lone actor
c_c	1	Damage caused by a member of a CO
r_w	1	Reward factor for a lone actor
r_c	1	Reward factor for a member of a CO
T	10	Temperature factor for the Fermi function
$mutation_{prob}$	0.0001	Probability of undergoing random mutation

accounts for the small-world theory where any two persons are separated by a chain of social acquaintances of a maximum length of six. In contrast, the small-world network, also referred to as the Watts-Strogatz network, generates a random network which can predict a precise clustering coefficient and average path length [2]. In other words, honest citizens and lone actors have been added to a criminal network using the different attachment methods. Thereby, the initial links amongst the criminal network were not modified. The intention was to preserve the structure and thus the properties of a criminal network. The attachment methods were based on the pseudo-algorithm presented in [22] which are stochastic. Visualisation of the attachment methods can be accessed by this link. Thus the rationale behind using these different attachment methods is to provide different simulations and analyse how the configuration of ties will influence the outcome of the model. For all the attachment methods, the ratio of honest, lone actors, and MCOs were set based on the data collected by the United Nations Office on Drugs and Crime [25] concerning the average ratio of incarceration within a population. Obtaining accurate data on the number of criminals within a population is challenging due to, among others, underreporting and a lack of standardised reporting across law enforcement agencies and jurisdictions. Therefore, for simplicity, it was assumed that the number of convicts reflects approximately the number of criminals within a population. For this thesis, data on the United States was used, where approximately 1%[4] of the population is unlawful [28]. Yet, it was not possible to define out of those unlawful individuals within the population how many are part of a CO. Nevertheless, the triangular phase diagrams presented in [18] indicate that the ratio between

[4] The United Nations Office on Drugs and Crime (UNODC) presents the data as persons per 100,000 population, taking into account the entire demographic population. Thus, the interest is in the active population, defined as persons aged 18–65 years, who make up 60% of the population according to the U.S. Census Bureau. Thus on average 600 out of 100000 are incarcerated in the U.S.A., which reduced to the active population, gives 600/(0.6*100000)=1%.

lone actors and MCOs does not impact the convergence to the same equilibrium, merely that the convergence rate is expected to be subject to changes.

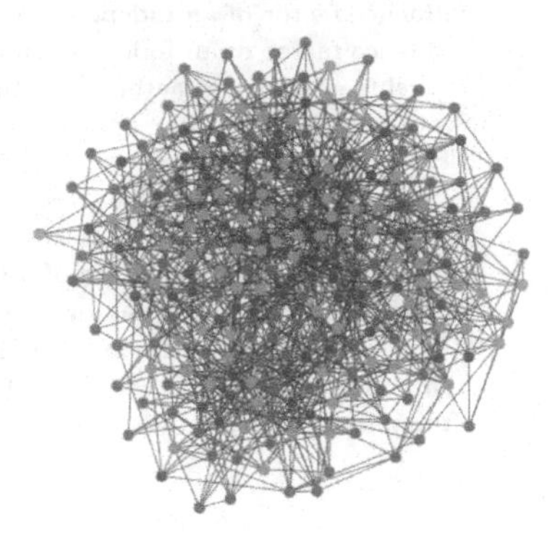

Fig. 1. Visualisation of the population created with the small-world algorithm, including criminals (red), lone actors (blue) and honest citizens (green). (Color figure online)

To achieve a fair comparison of the results between the populations of different structures (preferential/random/small-world), it was important that they had common characteristics (see Table 2). It was decided that the different structures should have the same density δ which corresponds to having approximately the same amount of nodes and edges. This density was set to the one observed in the initial criminal network[5]. To make sure that the small-world network does not become a random network by setting the rewiring probability $prob$ too high, the proximity-ratio test was performed [29]. Furthermore, the average clustering coefficient is higher in small-world networks than in random networks despite having similar average path length, which confirms the literature [22].

2.3 Determine Topological Changes

The focus of this research is on the development of the criminal network within the population. As a result, only the nodes with a criminal state were retained for the topological measurements. Throughout each round, some measures were collected, which gives an indication of the structure of the network and how it evolves. The measurements include some standard metrics, such as the size of the largest components, the flow of information, and the density of the network [10,11,14,34]. [11] defined the size of the largest component as a measure of interest, based on the reasoning that it represents a "self-organised criminal phase", a threat to national security. The flow of information defines how well

[5] The density has been chosen to be the same as the initial criminal network in order to avoid forcing the criminal network to evolve into a less dense network. By doing so, the criminal network has the option to either keep the same density or evolve towards a less dense configuration.

Table 2. Overview of the properties of the different population structures. *prob* corresponds to the probability of rewiring an existing link. δ corresponds to the density. $\langle k \rangle$ corresponds to the average degree of a node. $\langle p \rangle$ corresponds to the average degree path length. $\pm$ values represent the standard deviation for continuous distributions, respective for discrete distributions.

	Preferential	Random	Small-world
percentage of honests	99%	99%	99%
percentage of actors	0.1%	0.1%	0.1%
percentage of MCOs	0.9%	0.9%	0.9%
prob	–	–	0.2
nodes	10555 ± 0	10555 ± 0	10555 ± 0
edges	1.50e6 $\pm$ 1066	1.54e6 $\pm$ 260	1.50e6 $\pm$ 200
δ	0.027 $\pm$ 2e–05	0.028 $\pm$ 4e–06	0.027 $\pm$ 3e–06
clustering coefficient	0.05 $\pm$ 8e–04	0.01 $\pm$ 4e-05	0.4 $\pm$ 1e–04
$\langle k \rangle$	285 ± 0.2	291 ± 0.05	283 ± 0.04
$\langle p \rangle$	2.08 $\pm$ 3e–03	2.12 $\pm$ 1e–04	2.47 $\pm$ 1e–03
# of components	1 ± 0	1 ± 0	1 ± 0

the exchange of information and goods circulates within the largest component G and is defined by the following equation within the largest component G [11,12]:

$$\eta(G) = \frac{1}{N(N-1)} \sum_{i<j \in G} \frac{1}{d_{ij}} \tag{1}$$

where N is the total number of nodes in G and d_{ij} is the distance between node i and j. The aim is to reduce the communication flow of a network, which will lead to a decrease in its efficiency. As a result, the CO will become less successful and less attractive to be a part of. Additionally, Duijn et al. [12] formulated the secrecy metric:

$$Secrecy(G) = \frac{N(N-1)}{2E} \tag{2}$$

where E is the number of edges and N is the number of nodes in the largest component G. This corresponds to the inverse of the density metric, which expresses the ratio of actual relations over the number of possible relations. Thus, the metric indicates how exposed the network is. In other words, the more connections there are in a network, the more direct neighbours are possibly exposed by an actor questioned by the police. The difference between Eq. 1 and Eq. 2 is that the efficiency metric η captures the underlying structure of the network, while the secrecy metric does not take into account how the nodes are connected.

2.4 Data

For this research, the data from Cavallaro et al. [10] was selected. These data represents the social network of the Mistretta family and the Batanesi clan, a sub-branch of the Sicilian Mafia mainly involved in a cartel of construction companies. The Sicilian Mafia is described as a reactive organisation that efficiently adapts to external stimuli to pursue its economic and social goals. In this paper we have used the data corresponding to the recorded phone calls to construct the network used in our simulation model. The characteristics of the resulting network (Table 3) do not show anything exceptional or peculiar.

Table 3. Overview of the criminal network. N corresponds to the number of nodes, E corresponds to the number of edges, D corresponds to the diameter, $\langle k \rangle$ corresponds to the average degree, δ corresponds to the density, η corresponds to the flow of information in the network [11,12].

Montagna Phone Calls			
N = 95	E = 120	$\langle k \rangle = 2.526$	giant comp. = 84
D = 14	$\eta = 0.173$	$\delta = 0.027$	

2.5 Measurement

To respond to the research question raised, the topological changes were measured for the different population structures; preferential, random, and small-world. Thereby, two different scenarios were analysed: growth and steady size of the criminal network. Due to the stochastic nature of the model and the structure of the population, the measurements were done multiple times. Each time, a simulation was done on a newly generated population of the same attachment method. To have a fair comparison between the different attachment methods, the same parameter values were used for the different case scenarios. To achieve meaningful results, each node should converge to its local equilibrium: each node should undergo approximately 100 evolutionary stages during the simulation. The following results show the evolution of the criminal network in the different populations for two scenarios: (i) flat evolution of a CO and (ii) increase of a CO within the population. For this experiment, the reward r_c and c_c values were modified. An extensive analysis of the implications of the other parameters cited in Table 1 can be found in [32]. Thereby, the evolution of the CO in different population structures was compared to one-other[6].

3 Results and Discussion

Figure 2 (left) shows the evolution of a CO with a slight rise in the size of the CO at around 0.015%. One can also notice that with the evolution the standard error increased lightly. In the case of a random structured network, the CO evolved around 0.01%. For a small-world network, the COs were slightly lower with a value of approximately 0.0075%. Using the default parameters, it was possible to trigger an evolution with no particular trend, where the size of the criminal network is approximately constant. Figure 2 (right) presents the evolution of the

[6] The entirety of the code can be found in the following Github repository.

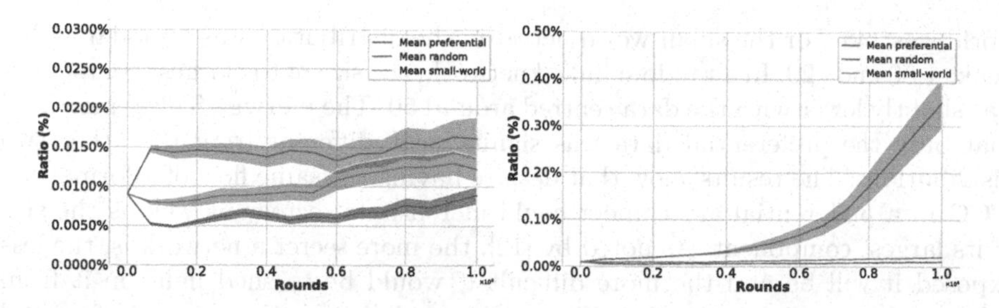

Fig. 2. Evolution of the CO within a population based on the different attachment methods (preferential, random and small-world) using the default parameters (left) as outlined in Table 1. For the simulations on the right, the default parameters from Table 1 were used with $r_c = c_c = 100$. The darker line corresponds to the mean value of 50 games and respective lighter colours correspond to the standard error. (Color figure online)

criminal network in the case where the reward factor r_c and damage factor c_c were set to 100. What stands out in this figure is the rapid increase of the CO in a preferential structured network with exponential growth. In comparison, the increase of a CO in a random or small-world network was minimal, despite having the same initial parameters. It is worth noting that the three curves and their standard error did not overlap, which is a strong indication that the trends were significantly different. Criminal behaviour seemed to spread differently across populations. It is all the more interesting to see that criminal behaviour quickly proliferated in a preferential network. The aforementioned results showed that the evolution of the population differed based on its structure. The next step was to examine how the topological structure of the criminal network differs within the different population structures. Figure 3 shows the topological properties of the flat evolution of a criminal network in the different population structures. The three sub-plots visualise the distribution of the properties (secrecy, flow of information, and size of largest component) using a kernel density estimate plot. In the left sub-figure, the secrecy plot shows different distributions between the social network structures. In the case of a preferential network, a uni-modal distribution centring around 45 can be noticed. In the case of a small-world population, a multi-modal distribution with a mean secrecy value of 35 can be seen. In a random world, the secrecy distribution took a very slim peak with a mean value of around 33. A pairwise Tukey's Honestly Significant Differences test [16] was performed with $\alpha = 0.05$ on the distributions indicating that the random and small-world distributions had identical mean. On the other hand, the flow of information plot shows almost perfectly overlapping distributions. Thereby, the distributions resembled a Gaussian distribution with a centre of approximately 0.165. With a p-value$>$ 0.05, it can be assumed that three distributions have the same distribution. The right sub-figure reveals the distribution of the size of the largest component of a criminal network. With a perfect Gaussian curve, the mean size of the largest intertwined criminal network in a preferential structured world was 145. For the small-world network, the distribution was bimodal with a peak at 80 and 120. In a random linked network, the size of the largest component was slightly lower with the data centred around 50. The pairwise Tukey indicated that only the preferential data was significantly different from the other two distributions. The results show that despite having the same flow of information, a CO in a preferential attachment could increase its secrecy as well as the size of its largest component. As noted by [12], the more secret a network is, the less exposed it will be and the more difficult it would be to shed light on it from a law enforcement perspective. Thus, the results show that in a preferential population, a criminal network can take an optimal configuration, allowing for a high flow of information without compromising its secrecy.

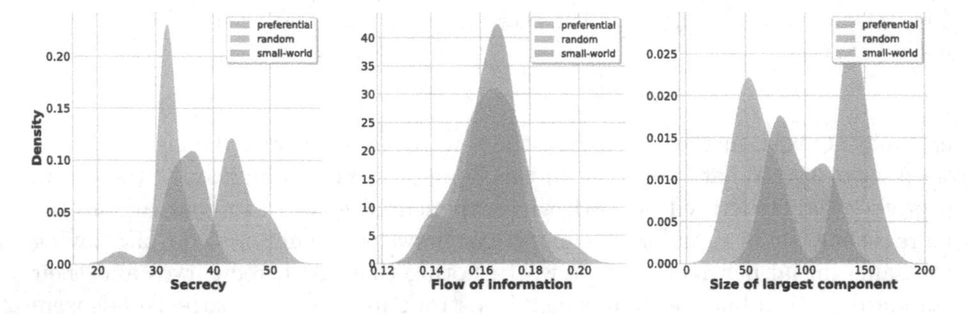

Fig. 3. Measurement of the topological evolution of the criminal network in the different populations (preferential/random/small-world). For each simulation, the games were repeated 50 times with each game played 1e6 rounds. For a fair comparison, the area under the curve was normalised to 1.

Figure 4 presents the results obtained in case the CO grows within the population. Looking at the secrecy plot over the rounds, one notes a rapid increase in secrecy for the preferential attachment in comparison to the random and small-world. A slight nadir can be noticed in the case of a small-world population. However, after that, a slightly linear increase along the simulation can be observed in the three populations. The flow of information in the middle sub-figure 4 illuminated a more complex pattern where the flow rapidly decreased in the case of the small-world population to stabilise around 0.15. The CO underwent a quick reduction in the flow of information in a preferential attachment before having an almost linear increase along the simulation. For a CO evolving in a random structured network, the flow of information seemed to evolve around the initial value of 0.17. The right sub-figure demonstrates a clear pattern where the preferential data increased exponentially from 95 to around 3500 interconnected MCO in only 1e6 rounds. In contrast, for the random and small-world network, the evolution of the largest component was minimal. Overall, in each sub-figure, the curves seemed to follow unique trajectories. Furthermore, there was little to no overlap in the standard errors in the secrecy and size of the largest component, which could be an indication of significant differences between the

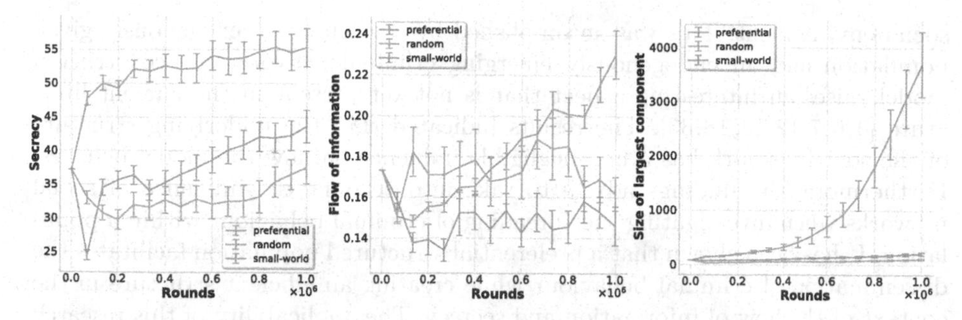

Fig. 4. Measurement of the topological evolution of the criminal network in the different populations (preferential/random/small-world). The left sub-figure shows the evolution of secrecy within a CO. The middle figure shows the evolution of the flow of information in a CO. The right sub-figure depicts the size of the largest connected component of a CO within a population. For this simulation, the default parameters (Table 1) were used with $r_c = c_c = 100$. The line corresponds to the mean value of 50 games with the respective standard error.

curves. For the flow of information figure, it is more difficult to assume no overlap. Based on the authoritative reference from [23], a Monte Carlo bootstrapping test was performed to assess the statistical significance. Since it is assumed that the points are correlated for each simulation, block bootstrapping was applied. The hypothesis test stated that for each metric, the distributions statistically differ. For preferential attachment, the size of the largest component increased faster than the other structures. A possible explanation for this might be that, as pointed out by [3], in a preferential configuration the existence of so-called hubs–highly connected nodes–might ensure the high connectivity between nodes and so easily create giant components. The giant component is judged as being a threat to national security [11]. Moreover, contrary to [12] claims, the results seemed to agree with [8,20] who stated that the presence of hubs increases the efficiency of the CO. While increasing the size of the largest component, the CO was able to maintain or even increase the flow of information/efficiency. Thus, if the preferential configuration is indeed the surrounding structure of a CO, then particular attention should be brought to its evolution. Also, in comparison, the small-world exerted a certain degree of local clustering compared to the random population, which might explain why, in a small-world, the spread of criminal behaviour was slightly more favoured. However, one needs to account for the limitation of the used metrics. As pointed out earlier, the secrecy Eq. 2 does take into account only the number of edges and vertices but does not capture the cohesion between these nodes and edges. Therefore a combination of metrics, including the flow of information, was necessary.

4 Conclusion

In this paper, we introduce a modelling approach to increase the understanding of the evolution of criminal networks given the underlying social network and the social environment. This was accomplished by combining an exogenously given population and an endogenously emerging criminal network: the constructed model raises an interesting aspect that is not yet present in the current literature [4,6,7,12,14,18,33]. The results indicated that the underlying structure of the social network has a non-negligible influence on how the model behaves. Furthermore, the findings advise to take into account the influence of social networks when investigating the spreading of criminal behaviour within a population. It has been shown that a preferential structured population facilitates the dissemination of criminal behaviour while creating an efficient structure in the context of the flow of information and secrecy. The applicability of this research in practice can be ensured by calibrating the presented model against longitudinal data and fine-tuning the model's parameters. The data could correspond, for example, to the social network of a population collected by a national statistical office combined with longitudinal data of criminal networks defined by criminal investigations. The calibration would allow for a close-to-reality simulation, where specific scenarios can be analysed. In this manner, law enforcement can use this simulation to mimic police interventions and examine the resilience of a criminal network and consequently adapt their strategies. The research can be extended by introducing the concept of strong/weak ties to better model the evolving social ties of the surroundings of a criminal network. Additionally, as discussed in Sect. 3, it is presumed that local clustering in the network could be a potent force for the spreading of a criminal network. Hence, it might be interesting to elaborate on this claim. Local clustering is often characterised by local communities. Thus, an analysis could be conducted on the different communities to ascertain how criminal behaviour spreads within them. This research is theoretical at the moment and aims to provide a scientific basis for further research to build on it. One important criterion is the creation of the population. A limitation of the design is the exogenously imposed ties within the population. Furthermore, the different parameters used for the simulation add uncertainty to it. It is important to bear in mind that the presented findings may be somewhat limited by the huge degree of freedom the simulation entails and the enormous search space generated. In this research specific cases have been analysed and thus cannot be easily translated to more general situations. Nonetheless, the results have shown the importance of taking into account network explicit structures which were not taken into account in [5,12,14,18,21].

References

1. Agreste, S., Catanese, S., Meo, P.D., Ferrara, E., Fiumara, G.: Network structure and resilience of mafia syndicates. Inf. Sci. **351**, 30–47 (2016). https://doi.org/10.1016/j.ins.2016.02.027
2. Barabasi, A.L.: Network Science. Cambridge University Pr. (2016). https://www.ebook.de/de/product/24312547/albert_laszlo_barabasi_network_science.html. Accessed 29 June 2022
3. Barabási, A.L., Albert, R.: Emergence of scaling in random networks. Science **286**(5439), 509–512 (1999)

4. Behzadan, V., Nourmohammadi, A., Gunes, M., Yuksel, M.: On fighting fire with fire. In: Proceedings of the 2017 IEEE/ACM International Conference on Advances in Social Networks Analysis and Mining 2017. ACM (2017). https://doi.org/10.1145/3110025.3119404
5. Berenji, B., Chou, T., D'Orsogna, M.R.: Recidivism and rehabilitation of criminal offenders: a carrot and stick evolutionary game. PLoS One **9**(1), e85531 (2014)
6. Berlusconi, G.: Come at the king, you best not miss: criminal network adaptation after law enforcement targeting of key players. Global Crime, 1–21 (2021). https://doi.org/10.1080/17440572.2021.2012460
7. Bright, D., Koskinen, J., Malm, A.: Illicit network dynamics: the formation and evolution of a drug trafficking network. J. Quant. Criminol. **35**(2), 237–258 (2018). https://doi.org/10.1007/s10940-018-9379-8
8. Bright, D.A., Delaney, J.J.: Evolution of a drug trafficking network: mapping changes in network structure and function across time. Global Crime **14**(2–3), 238–260 (2013). https://doi.org/10.1080/17440572.2013.787927
9. Calderoni, F., Campedelli, G.M., Szekely, A., Paolucci, M., Andrighetto, G.: Recruitment into organized crime: an agent-based approach testing the impact of different policies. CrimRxiv (2022). https://doi.org/10.21428/cb6ab371.d3cb86db
10. Cavallaro, L., et al.: Disrupting resilient criminal networks through data analysis: the case of sicilian mafia. Plos One **15**(8), e0236476 (2020). https://doi.org/10.1371/journal.pone.0236476
11. da Cunha, B.R., Gonçalves, S.: Topology, robustness, and structural controllability of the Brazilian Federal Police criminal intelligence network. Appl. Netw. Sci. **3**(1), 1–20 (2018). https://doi.org/10.1007/s41109-018-0092-1
12. Duijn, P.A.C., Kashirin, V., Sloot, P.M.A.: The relative ineffectiveness of criminal network disruption. Sci. Rep. **4**(1), 1–15 (2014). https://doi.org/10.1038/srep04238
13. Duijn, P.A.C.: Detecting and disrupting criminal networks: a data driven approach. University of Amsterdam (2016)
14. Duxbury, S.W., Haynie, D.L.: Criminal network security: an agent-based approach to evaluating network resilience. Criminology **57**(2), 314–342 (2019). https://doi.org/10.1111/1745-9125.12203
15. Erdos, P., Rényi, A., et al.: On the evolution of random graphs. Publ. Math. Inst. Hung. Acad. Sci **5**(1), 17–60 (1960)
16. Haynes, W.: Tukey's test. In: Encyclopedia of Systems Biology, pp. 2303–2304. Springer, New York (2013). https://doi.org/10.1007/978-1-4419-9863-7_1212
17. Magliocca, N.R., et al.: Modeling cocaine traffickers and counterdrug interdiction forces as a complex adaptive system. In: Proceedings of the National Academy of Sciences, vol. 116, no. 16, pp. 7784–7792 (2019). https://doi.org/10.1073/pnas.1812459116
18. Martinez-Vaquero, L.A., Dolci, V., Trianni, V.: Evolutionary dynamics of organised crime and terrorist networks. Sci. Rep. **9**(1), 9727 (2019). https://doi.org/10.1038/s41598-019-46141-8
19. Morselli, C.: Inside Criminal Networks. Springer, New York (2009). https://doi.org/10.1007/978-0-387-09526-4
20. Morselli, C., Giguère, C., Petit, K.: The efficiency/security trade-off in criminal networks. Soc. Netw. **29**(1), 143–153 (2007). https://doi.org/10.1016/j.socnet.2006.05.001
21. Perc, M., Donnay, K., Helbing, D.: Understanding recurrent crime as system-immanent collective behavior. PloS one **8**(10), e76063 (2013)

22. Prettejohn, B.J., Berryman, M.J., McDonnell, M.D.: Methods for generating complex networks with selected structural properties for simulations: a review and tutorial for neuroscientists. Front. Comput. Neurosci. **5**, 11 (2011)
23. Sanchez, J.: Comparing 2 sets of longitudinal data. Cross Validated (2015). https://stats.stackexchange.com/q/156896, https://stats.stackexchange.com/users/48943/jason-sanchez. Accessed 26 Sept 2022
24. Spapens, T.: Interaction between criminal groups and law enforcement: the case of ecstasy in the Netherlands. Global Crime **12**(1), 19–40 (2011). https://doi.org/10.1080/17440572.2011.548955
25. United Nations Office on Drugs and Crime: Prison population, regional and global estimates — dataunodc. United Nations (2019). https://dataunodc.un.org/content/prison-population-regional-and-global-estimates. Accessed 29 June 2022
26. United Nations Office on Drugs and Crime: Global Report on Trafficking in Persons 2020. UN, Office on Drugs and Crime (2021). https://doi.org/10.18356/9789210051958
27. United Nations Office on Drugs and Crime: World drug report 2021. UN, Office on Drugs and Crime, Vienna : 2021–06 (2021). http://digitallibrary.un.org/record/3931425. Accessed 29 June 2022. Includes bibliographical references
28. Wagner, P., Betram, W.: What percent of the U.S. is incarcerated? (and other ways to measure mass incarceration). Prison Policy Initiative (2020). https://www.prisonpolicy.org/blog/2020/01/16/percent-incarcerated/. Accessed 20 Sept 2022
29. Walsh, T., et al.: Search in a small world. In: Ijcai. vol. 99, pp. 1172–1177. Citeseer (1999)
30. Wandelt, S., Sun, X., Feng, D., Zanin, M., Havlin, S.: A comparative analysis of approaches to network-dismantling. Sci. Rep. **8**(1), 1–5 (2018). https://doi.org/10.1038/s41598-018-31902-8
31. Watts, D.J., Strogatz, S.H.: Collective dynamics of 'small-world'networks. Nature **393**(6684), 440–442 (1998)
32. Weyland, L.F.: Longitudinal Analysis of the Topology of Criminal Networks using a Simple Cost-Benefit Agent-Based Model. Master's thesis, Universiteit van Amsterdam (2022)
33. Will, M., Groeneveld, J., Frank, K., Müller, B.: Combining social network analysis and agent-based modelling to explore dynamics of human interaction: a review. Socio-Environ. Syst. Model. **2**, 16325 (2020). https://doi.org/10.18174/sesmo.2020a16325
34. Wood, G.: The structure and vulnerability of a drug trafficking collaboration network. Soc. Netw. **48**, 1–9 (2017). https://doi.org/10.1016/j.socnet.2016.07.001
35. van der Zwet, K., Barros, A.I., van Engers, T.M., van der Vecht, B.: An agent-based model for emergent opponent behavior. In: Rodrigues, J.M.F. (ed.) ICCS 2019. LNCS, vol. 11537, pp. 290–303. Springer, Cham (2019). https://doi.org/10.1007/978-3-030-22741-8_21

Manifold Analysis for High-Dimensional Socio-Environmental Surveys

Charles Dupont(✉) and Debraj Roy

Faculty of Science, Informatics Institute, University of Amsterdam, Science Park 904, 1090 XH Amsterdam, The Netherlands
{c.a.dupont,d.roy}@uva.nl

Abstract. Recent studies on anthropogenic climate change demonstrate a disproportionate effect on agriculture in the Global South and North. Questionnaires have become a common tool to capture the impact of climatic shocks on household agricultural income and consequently on farmers' adaptation strategies. These questionnaires are high-dimensional and contain data on several aspects of an individual (household) such as spatial and demographic characteristics, socio-economic conditions, farming practices, adaptation choices, and constraints. The extraction of insights from these high-dimensional datasets is far from trivial. Standard tools such as Principal Component Analysis, Factor Analysis, and Regression models are routinely used in such analysis, but they either rely on a pairwise correlation matrix, assume specific (conditional) probability distributions, or assume that the survey data lies in a linear subspace. Recent advances in manifold learning techniques have demonstrated better detection of different behavioural regimes from surveys. This paper uses Bangladesh Climate Change Adaptation Survey data to compare three non-linear manifold techniques: Fisher Information Non-Parametric Embedding (FINE), Diffusion Maps and t-SNE. Using a simulation framework, we show that FINE appears to consistently outperform the other methods except for questionnaires with high multi-partite information. Although not limited by the need to impose a grouping scheme on data, t-SNE and Diffusion Maps require hyperparameter tuning and thus more computational effort, unlike FINE which is non-parametric. Finally, we demonstrate FINE's ability to detect adaptation regimes and corresponding key drivers from high-dimensional data.

Keywords: Survey Analysis · Climate Change Adaptation · Fisher Information · t-SNE · Diffusion Maps

1 Introduction

Climate change is one of the significant global challenges of the 21st century and floods are the costliest climate-induced hazard. Rapid urbanization and climate change exacerbate flood risks worldwide, undermining humanity's aspirations to achieve sustainable development goals (SDG) [12]. Current global warming

J. Mikyška et al. (Eds.): ICCS 2023, LNCS 14076, pp. 25–39, 2023.
https://doi.org/10.1007/978-3-031-36027-5_3

trends and their adverse impacts such as floods represent a complex problem, which cannot be understood independently of its socioeconomic, political, and cultural contexts. In particular, the impact of climate change on farmers and their livelihoods is at a critical juncture and adaptation is key for embracing best practices as new technologies and pathways to sustainability emerge. As the amount of available data pertaining to farmers' adaptation strategies has increased, so has the need for robust computational methods to improve the facility with which we can extract insights from high-dimensional survey data. Although standard methods such as Principal Component Analysis, Factor Analysis or Regression models are routinely used and have been effective to some degree, they either rely on a pairwise correlation matrix, assume specific (conditional) probability distributions or that the high-dimensional survey data lies in a linear subspace [6]. Recent advances in manifold learning techniques have shown great promise in terms of improved detection of behavioural regimes and other key non-linear features from survey data [2].

In this paper, we compare three non-linear manifold learning techniques: t-SNE [14], Diffusion Maps [4], and Fisher Information Non Parametric Embedding (FINE) [1]. We start by extending prior work [8] done with a simulation framework which allows for the generation of synthetic questionnaires. Because the underlying one-dimensional statistical manifolds are known, we are able to quantify how well each algorithm is able to recover the structure of the simulated data. Next, we apply the various methods to the Bangladesh Climate Change Adaptation Survey [10], which contains rich data regarding aspects of individual households such as spatial information, socio-economic and demographic indicators, farming practices, adaptation choices and constraints to adaptations. This allows us to investigate whether behavioural regimes of adaptation can be extracted, and more broadly to better understand each method's utility and relative trade-offs for the analysis of high-dimensional, real world questionnaires. Although all three methods yield comparable results, we uncover key differences and relative advantages that are important to take into consideration. By virtue of being non-parametric, FINE benefits from decreased computational efforts in contrast to t-SNE and Diffusion Maps which require hyperparameter tuning. Although FINE typically outperforms the other two methods, its performance degrades when there is high interdependence between survey items, and we identify a cutoff point beyond which adding more features does not result in increasing the differential entropy of pairwise distances between observations in the resulting embedding. FINE requires the researcher to impose a grouping scheme on observations, which may not always be intuitive, while t-SNE and Diffusion Maps allow clusters (groups of similar observations) to emerge more naturally since no prior structure is assumed. Nonetheless, FINE is shown to be particularly successful in the extraction of adaptation regimes. Lastly, FINE allows one to use as much data as possible since missing feature values can simply be ignored, whereas t-SNE and Diffusion Maps can be significantly impacted by imputed or missing values, which may require removing incomplete observations.

The structure of the rest of the paper is as follows. Section 2 provides an overview of the algorithms studied in this work, as well as the simulation framework, climate change adaptation questionnaire, and experiments that are carried out. Section 3 presents key results obtained for the various experiments. Lastly, Sect. 4 discusses our findings as well as future directions of research.

2 Methods

2.1 Dimension Reduction Algorithms

t-SNE. t-Distributed Stochastic Neighbour Embedding (t-SNE) was first introduced by Laurens van der Maaten and Geoffrey Hinton [14], and is based on prior work on Stochastic Neighbour Embedding [9]. Key steps are presented and summarized in Supporting Information (SI), Algorithm 1. First, a probability distribution over pairs of data points in the original feature space is constructed such that the similarity of some data point $\mathbf{x}_j$ to data point $\mathbf{x}_i$ is defined as

$$p_{j|i} = \frac{\exp\left(-||\mathbf{x}_i - \mathbf{x}_j||^2/2\sigma_i^2\right)}{\sum_{k\neq i}\exp\left(-||\mathbf{x}_i - \mathbf{x}_k||^2/2\sigma_i^2\right)}.$$

We can interpret this quantity as the conditional probability of selecting $\mathbf{x}_j$ as a neighbour of $\mathbf{x}_i$. $p_{i|i} = 0$ since a data point cannot be its own neighbour. σ_i denotes the variance of the Gaussian distribution centered around $\mathbf{x}_i$. It is tuned for each $\mathbf{x}_i$ separately such that the resulting conditional probability distribution P_i over all other datapoints $\mathbf{x}_{j\neq i}$ yields a perplexity value specified by the user, calculated as perplexity$(P_i) = 2^{H(P_i)}$, where $H(P_i)$ is the Shannon entropy [14]. Because typically $p_{j|i} \neq p_{i|j}$, we define the joint distribution $p_{ij} = \frac{p_{j|i}+p_{i|j}}{2N}$, where N denotes the total number of observations in the dataset. The next step is to construct another probability distribution over the data in a lower-dimensional space with the aim of minimizing the Kullback-Leibler (KL) divergence between the previous probability distribution and this newly constructed one, thus preserving similarities between data points in the original space. The joint probabilities for data points in this lower dimensional map are given by

$$q_{ij} = \frac{\left(1 + ||\mathbf{y}_i - \mathbf{y}_j||^2\right)^{-1}}{\sum_{k\neq l}\left(1 + ||\mathbf{y}_k - \mathbf{y}_l||^2\right)^{-1}},$$

which is a heavy-tailed Student t-distribution [14]. KL divergence is minimized iteratively using gradient descent by updating vectors $\mathbf{y}_i$ at each step.

Diffusion Maps. Diffusion Maps is a method introduced by Coifman and Lafron which takes inspiration from the processes of heat diffusion and random walks [4]. Intuitively, if we were to take a random walk over observations in a dataset, starting at some random point, we would be more likely to travel to a nearby, similar point than to one that is much further away. The Diffusion Maps algorithm leverages this idea in order to estimate the connectivity k between pairs of data points using a Gaussian kernel as follows:

$k(\mathbf{x}_i, \mathbf{x}_j) = \exp\left(-\frac{||\mathbf{x}_i - \mathbf{x}_j||^2}{\epsilon}\right)$, where ϵ is some normalization parameter. Subsequently, a diffusion process is constructed using the transition matrix of a Markov Chain M on the data set, which allows us to map diffusion distances to a lower-dimensional space. Key parameters are t, used to construct the t-step transition matrix M^t, and additional normalization parameter α. SI Algorithm 2 summarizes the important steps of this process.

FINE. The Fisher Information Non Parametric Embedding (FINE) algorithm was developed by Carter et al. and works by constructing a statistical manifold upon which lives a family of probability distributions (estimated from some dataset) for which we can compute inter-distances [1]. This algorithm was further developed and applied to questionnaire data by Har-Shemesh et al. [8]. Algorithm 1 summarizes key steps. First, respondents to the questionnaire are divided into K groups. For each of these groups, a probability distribution is constructed over the set of all possible responses I (each element being a string, e.g. "ABDC"). By considering the square roots of these probabilities, we can regard each probability distribution as a point on the unit hypersphere and compute distances between these points using the arc length. With this distance matrix, non-linear dimension reduction is achieved by applying classical Multidimensional Scaling (MDS), which is another non-linear technique for visualizing similarities between observations in a dataset [11]. Questionnaire items are assumed to be independent such that probabilities may be factorized.

Algorithm 1: FINE (for questionnaire data)

Data: $D = (\mathbf{x}_1, \mathbf{x}_2, \ldots, \mathbf{x}_n)$
Input: dimension d, choice of grouping scheme
Result: E, a lower-dimensional representation of the data
begin
  Divide observations into K groups using some grouping scheme;
  for $k = 1, 2, \ldots, K$ **do**
    Estimate (square root) probabilities $\xi^{(k)}$ for responses in k^{th} group;
  end
  for $j, k = 1, 2, \ldots, K$ **do**
    Compute $M_{jk} = \cos^{-1}\left(\sum_I \xi_I^{(j)} \xi_I^{(k)}\right) \rightarrow$ arc length on the unit hypersphere;
  end
  Construct embedding $E = \mathrm{MDS}(M, d)$;
end

2.2 Simulation Framework

Framework Description. The authors of [8] propose a simulation framework for generating questionnaire responses in a controlled way. This allows us to compare the embeddings generated by the three algorithms to a "ground-truth" embedding. This is done by parameterising angles ϕ_i as

$$\phi_i^\kappa(t) = \begin{cases} \frac{\pi}{2}\sin^2(m\pi t), & i = \kappa \\ \frac{\pi}{2}t, & i \neq \kappa \end{cases}, \tag{1}$$

where κ allows us to choose which angle is proportional to the squared sine term, and $t \in [0, 1]$ is the unique parameter of this family of probability distributions. Furthermore, m controls the non-linearity of the family. There are $N - 1$ angles for a questionnaire with N possible distinct responses, and we compute the square root probabilities as follows:

$$\begin{aligned} \xi_1 &= \cos(\phi_1) \\ \xi_2 &= \sin(\phi_1)\cos(\phi_2) \\ \xi_3 &= \sin(\phi_1)\sin(\phi_2)\cos(\phi_3) \\ &\vdots \\ \xi_{N-1} &= \sin(\phi_1)\ldots\sin(\phi_{N-2})\cos(\phi_{N-1}) \\ \xi_N &= \sin(\phi_1)\ldots\sin(\phi_{N-2})\sin(\phi_{N-1}) \end{aligned} \tag{2}$$

For some choice K, which denotes the total number of groups (see Algorithm 1), we draw K values uniformly on the curve given by Eq. (1). Then, we compute probabilities $p_I^{(k)} = (\xi_I^{(k)})^2$ and randomly generate a number of questionnaire responses for each group $k = 1, 2, \ldots, K$ using these probabilities.

Experiments. We wish to compare the embeddings generated by t-SNE, Diffusion Maps, and FINE for various simulated questionnaire responses. Similarly to [8], we generate responses for $\kappa \in \{1, 2, N-1\}$, and $K \in \{20, 50\}$. We keep $m = 3$ fixed as well as the number of questions ($N_Q = 8$) and the number of possible answers for each question ($N_A = 3$), yielding $N = 3^8 = 6561$. For each of the 6 possible combinations of parameters κ and K, there is a unique theoretical embedding and 30 questionnaires are simulated. When $K = 20$ we generate 25 responses per group, and when $K = 50$ we generate 50 responses per group. Then, for each set of 30 questionnaires, we apply all three non-linear dimension reduction algorithms.

In order to evaluate the quality of the generated embeddings, we apply the Procrustes algorithm [7], which can stretch, rotate or reflect the generated embeddings so that they match up with the theoretical embedding as closely as possible. Once this is done, we compute the Pearson correlation coefficient between the coordinates of each generated embedding and those of the theoretical embedding. Note that the theoretical embedding is determined via application of the MDS algorithm using arc length distances between the exact probability distributions calculated using Eq. (2).

Parameter Tuning. FINE does not require any parameterization, although a grouping scheme must be provided, which in this case is defined by the simulation framework. On the other hand, both t-SNE and Diffusion Maps require some parameter tuning. For t-SNE, we perform a grid search over the following parameters: perplexity $\in \{1, 2, 5, 10\}$, learning rate $\eta \in \{10, 50, 100, 200\}$, distance metric $\in$ {weighted hamming (with/without one-hot encoding), cosine (with one-hot encoding)}. The maximum number of steps T and momentum $\alpha(t)$ are fixed at 1000 and 0.8 respectively. For Diffusion Maps, we perform a grid search over: $\epsilon \in \{0.5, 1.0, 1.5, 2.0\}$, $t \in \{0, 0.5, 1, 5\}$, distance metric $\in$ {weighted hamming (with/without one-hot encoding), cosine (with one-hot encoding)}. We fix $\alpha = \frac{1}{2}$. See [3] for a review of one-hot encoding, and note that the weighted hamming distance is simply the number of positions that two strings differ, each positional contribution (1 if different, 0 if identical) being weighted by the reciprocal of the number of possible values at that position.

2.3 Bangladesh Climate Change Adaptation Survey

The non-linear manifold learning algorithms of interest are applied to a questionnaire dataset pertaining to the economics of adaptation to climate change in Bangladesh with the aim of identifying different regimes of behaviour and adaptation in response to climate change.

Dataset Description. Data collection was carried out in 2012 amongst 827 households in Bangladesh in 40 different communities [10]. This survey is a follow-up to a first round of data collection, which was studied in detail in [5]. Some households have frequently missing response fields, so we retain 805 households having responded to at least 30% of survey questions. Each of the 40 distinct communities has a unique combination of district, "upazila", and union codes, where upazilas are sub-units of districts, and unions are even smaller administrative units. Households additionally possess one of 7 distinct codes corresponding to different agro-ecological zones.

Handpicked Features. We construct a set of handpicked features that we expect to be important for detecting adaptation strategies based on existing literature. Specifically, we keep track of household income and expenditure, what occupations are held by household members, total monetary loss due to climatic and personal shocks, what actions were taken in response, social capital, collective action, constraints to adaptations, what adaptations were implemented, and finally what community groups household members are a part of as well as associated benefits. This set of 95 features is summarized in SI Table 8. Summary statistics for various features are also provided in other SI tables. Note that we discretise continuous features into at most 5 bins using the Bayesian Blocks dynamic programming method, first introduced by Scargle [13]. Additional details regarding this method are available in the SI document.

Experiments. We apply FINE to the set of handpicked features, using communities as our grouping scheme for individual household observations. Additionally, we examine the impact of how much a particular feature varies across communities on the embedding produced by FINE as follows. For each handpicked feature, we compute the KL divergence of that feature's values for each pair of communities. We record the median, and after producing an embedding using FINE, we compute the differential entropy of the distribution of pairwise distances between communities. Finally, we apply t-SNE and Diffusion Maps to the set of handpicked features in order to see if any clusters naturally emerge.

3 Results

3.1 Simulation Framework Results

Table 1 displays the best hyperparameter combinations for t-SNE and Diffusion Maps. For t-SNE, lower perplexity values typically perform better, as does the weighted hamming distance. For Diffusion Maps, using the cosine distance metric (with one-hot encoding) yields optimal performance for all (κ, K) pairs. However, choices for ϵ and t seem to be more delicate and dependent on the κ, K values. Overall, for both t-SNE and Diffusion Maps, not all hyperparameter combinations are found to yield good performance, emphasizing the importance of hyperparameter tuning.

Table 1. Summary of best hyperparameters for t-SNE and Diffusion Maps

(κ, K)	t-SNE	Diffusion Maps
$(\kappa = 1, K = 20)$	perplexity $= 2$, $\eta = 50$	$\epsilon = 1.5$, $t = 0.5$
$(\kappa = 1, K = 50)$	perplexity $= 5$, $\eta = 200$	$\epsilon = 1.5$, $t = 0.5$
$(\kappa = 2, K = 20)$	perplexity $= 2$, $\eta = 50$	$\epsilon = 0.5$, $t = 0.5$
$(\kappa = 2, K = 50)$	perplexity $= 10$, $\eta = 200$	$\epsilon = 0.5$, $t = 0.5$
$(\kappa = 6560, K = 20)$	perplexity $= 2$, $\eta = 50$	$\epsilon = 2.0$, $t = 0.5$
$(\kappa = 6560, K = 50)$	perplexity $= 1$, $\eta = 50$	$\epsilon = 2.0$, $t = 0.0$

Figure 1 displays the distribution of performance (correlation with theoretical embedding) of each algorithm for all 30 questionnaires and each (κ, K) combination using best-performing hyperparameters. FINE significantly outperforms the other algorithms in all cases and with lower variance in performance across the 30 questionnaires, except when $\kappa = 1$ and $K = 20$ where Diffusion Maps performs similarly. t-SNE consistently performs worse, and is significantly more sensitive to which of the 30 questionnaires is being analyzed (as evidenced by the high variance in performance). Overall, all algorithms achieve a mean correlation with the theoretical embedding of at least 0.87 (at a 95% confidence level).

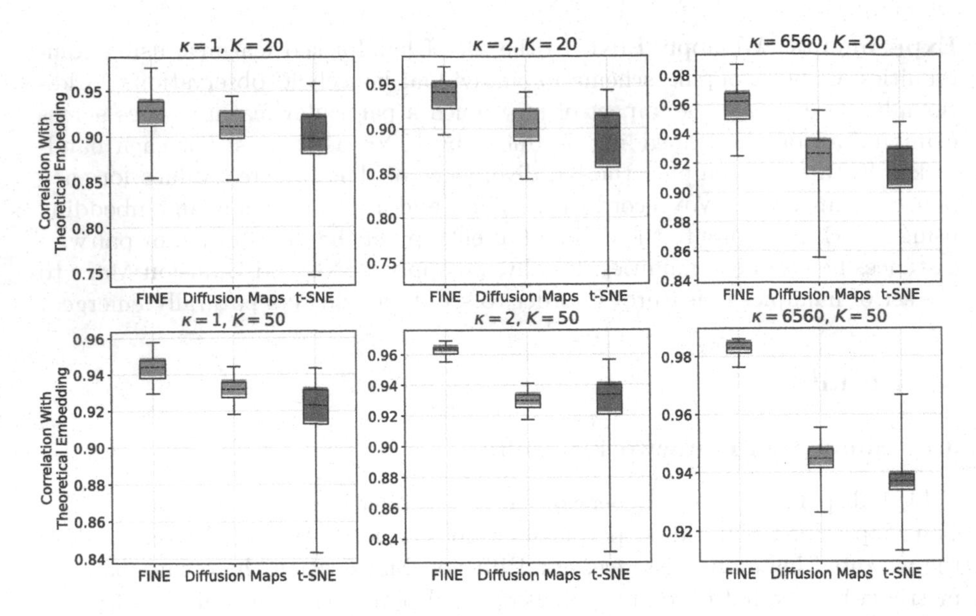

Fig. 1. Distributions of correlation coefficients with respect to the theoretical embedding for each algorithm and (κ, K) pair using best hyperparameters. Dashed lines and whiskers denote the mean, maximum and minimum values.

Figure 2 displays the theoretical embedding for each (κ, K) pair, along with embeddings obtained using t-SNE, Diffusion Maps, and FINE for one sample questionnaire. Overall the embeddings produced by FINE most closely resemble the theoretical embeddings out of all three algorithms. Additionally, the underlying structure of the data is better recovered with a larger number of groups K and responses in all cases except for t-SNE when comparing $(\kappa = 1, K = 20)$ and $(\kappa = 1, K = 50)$. Multi-partite information, which measures the amount of dependence between the questions of the simulated questionnaires, is also displayed and is defined as

$$\text{MI} \equiv \sum_{I} p_I(q_1, q_2, \ldots, q_{N_Q}) \ln \frac{p_I(q_1, q_2, \ldots, q_{N_Q})}{p_I(q_1)p_I(q_2)\ldots p_I(q_{N_Q})}. \tag{3}$$

In order to more closely investigate the dependence of the various algorithms' performance on multi-partite information, we generate an additional 300 questionnaires (each one with its own theoretical embedding and distribution of multi-partite information values), using $N_Q = 7$, $N_A = 3$, and 30 uniformly spaced κ values between 1 and $N - 1$ as well as $m \in \{1, 2, \ldots, 10\}$. We fix the number of groups at $K = 20$, and generate 50 responses per group. As always, FINE does not require any parameter tuning. For t-SNE, using Table 1 as a guide, we use perplexity $= 2$, $\eta = 50$ and a weighted hamming distance metric after one-hot encoding. For Diffusion Maps, relying on Table 1, we select $\epsilon = 0.5$ and $t = 0.5$, and use the cosine distance metric after one-hot encoding.

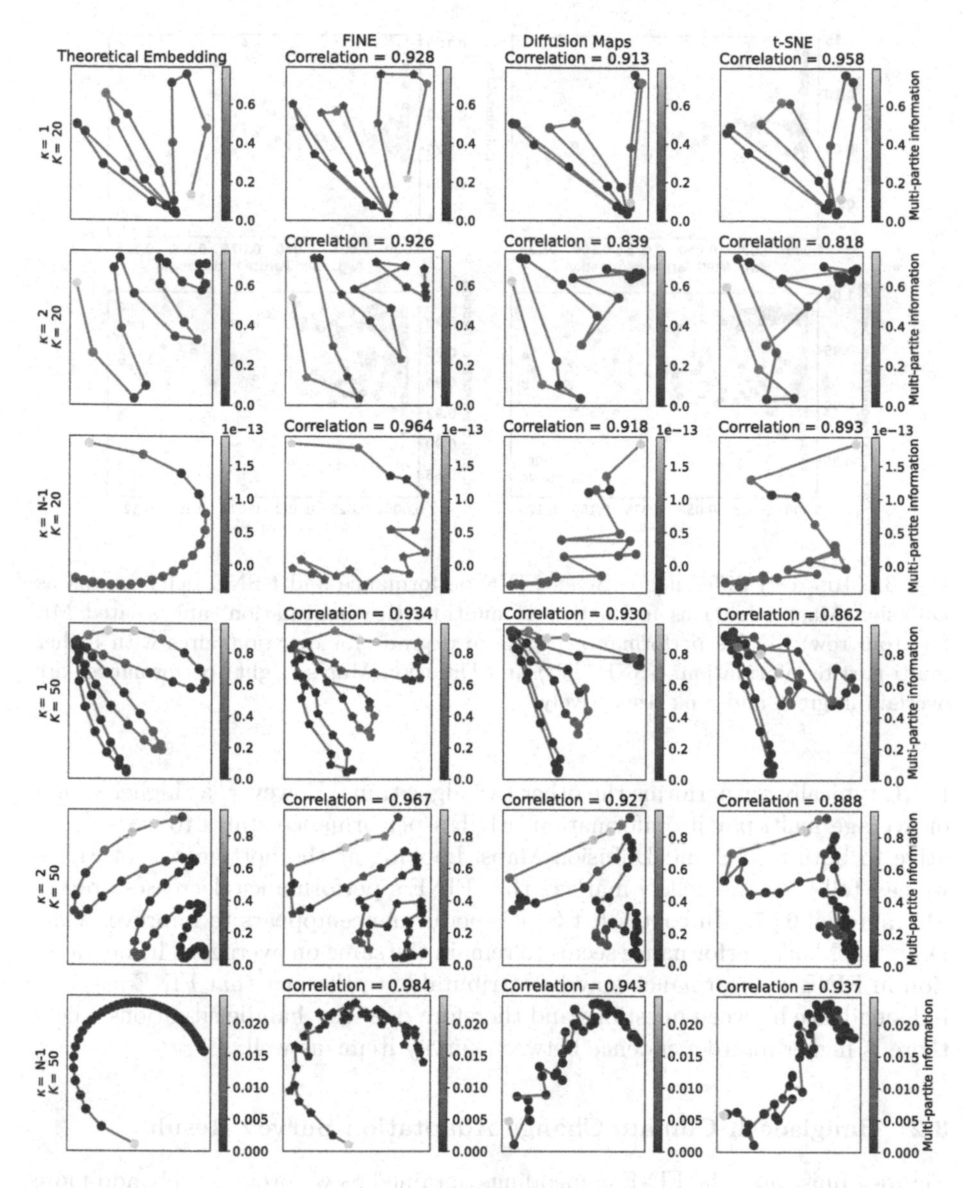

Fig. 2. Comparison of embeddings obtained with t-SNE, Diffusion Maps, and FINE with respect to the theoretical embedding using the simulation framework. t-SNE achieves comparable performance to FINE due to hyperparameter tuning, which is a departure from prior results presented in [8].

The top row of Fig. 3 displays the differences in correlation with respect to the theoretical embedding between FINE and t-SNE as well as FINE and Diffusion Maps for 300 different values of (averaged) multi-partite information. In agreement with Fig. 1, the differences are almost always positive, indicating that

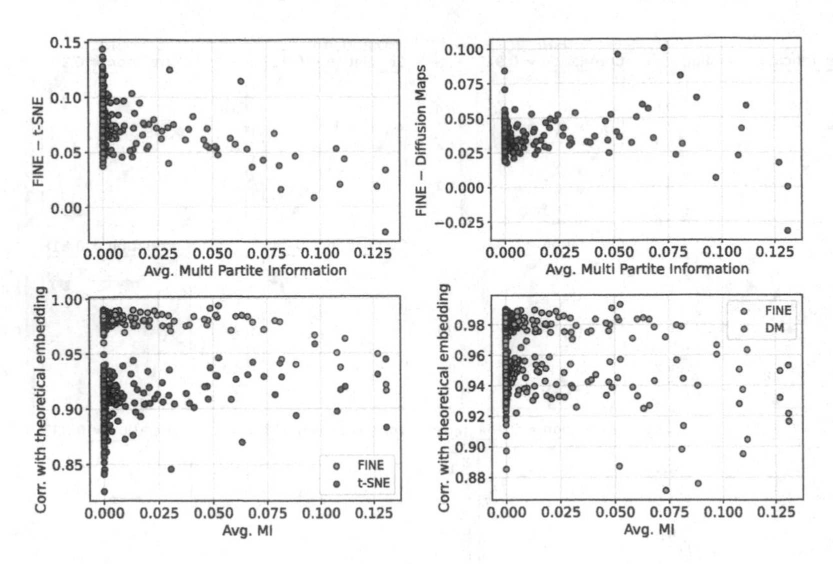

Fig. 3. (Top row) difference between FINE performance and t-SNE (left) as well as Diffusion Maps (right) as a function of multi-partite information, abbreviated MI. (Bottom row) FINE's performance begins to degrade for questionnaires with higher multi-partite information. t-SNE (left) and Diffusion Map's (right) performances are overlaid in green and pink respectively.

FINE typically outperforms the other two algorithms. However, at higher values of average multi-partite information, FINE's performance starts to worsen relative to both t-SNE and Diffusion Maps. Looking at the bottom row of Fig. 3, we can tell from the yellow markers that FINE's performance decreases around MI values of 0.075. In contrast, t-SNE's performance appears to improve, while Diffusion Map's performance seems to remain the same on average. The degradation in FINE's performance may be attributable to the fact that FINE assumes independence between questions and therefore does not handle situations where there is higher interdependence between survey items as well.

3.2 Bangladesh Climate Change Adaptation Survey Results

Figure 4 illustrates the FINE embeddings obtained as we progressively add more handpicked features, starting with ones with lower median KL divergence. The top left embedding includes a single feature corresponding to monetary loss due to sea level rise with median KL divergence close to zero, which signifies that almost all communities have the same distribution for this feature. This results in a distribution of pairwise distances with very low differential entropy – nearly all communities collapse to the same coordinate, except for community 6 which appears as an outlier due to being the only one containing a household having suffered damages due to sea level rise. Community 21 also appears as a clear outlier in subsequent embeddings. Upon investigation, we found that 65.5%

of households in this community reported having to migrate due to suffering heavy losses as a result of soil and river erosion, which is significantly more than households in any other community.

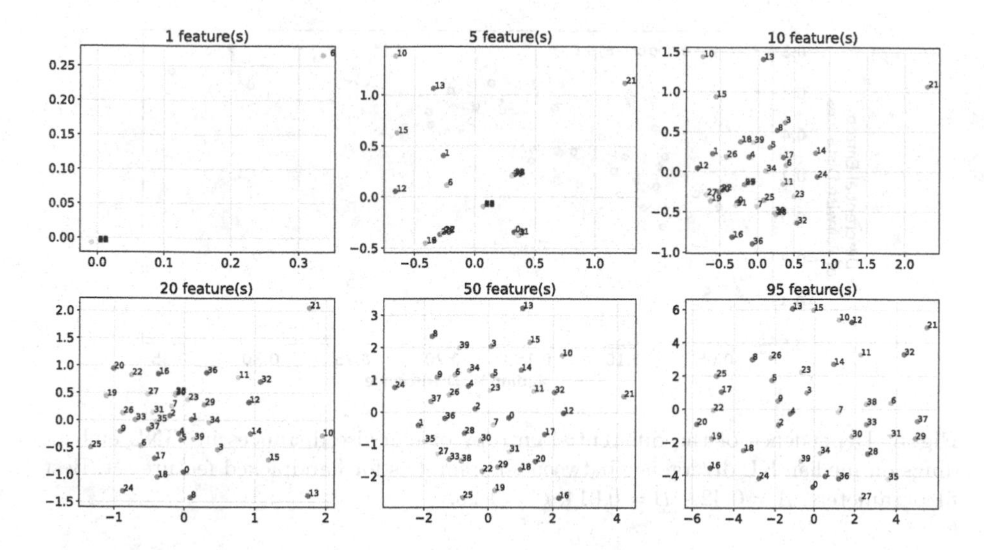

Fig. 4. FINE embeddings using an increasing number of handpicked features, added in order from lowest to highest median KL divergence.

The bottom right subplot includes all 95 handpicked features. Despite not including the agro-ecological zone in the set of handpicked features, we notice the influence of spatial characteristics on adaptation regimes quite clearly in some cases. For example, communities 29, 30, 31, and 33 all appear close together in the embedding and in fact are all located in the same agro-ecological zone. Since agro-ecological zones are defined as regions with similar climate conditions, it is perhaps unsurprising that communities in the same geographical areas would be similarly impacted by climatic shocks as well as respond in a similar fashion. However, such proximity is certainly not the only driver of adaptation. Communities 10, 12, 13, and 15 appear close together at the top of the embedding, but belong to three different agro-ecological zones. In fact, these are the only communities in which at least two households needed to sell assets in response to salinity increases. Furthermore, communities 10 and 12 had over 30% of households with at least one member needing to seek off-farm employment, which could explain their appearing especially close together. Only communities 21 and 32 also have this property, and as a result they appear quite isolated in the embedding as well (especially 21 for reasons mentioned earlier). As a last example, despite being spread over four different agro-ecological zones, at least 65% of households in communities 16–20, 22, 24 and 25 decided to change their planting dates, and notice that these communities form an elongated vertical cluster in the bottom left of the embedding. The only other community satisfying this property is community 15, which differs in other, more pronounced regards (as described earlier). Additional embeddings were generated using FINE for various feature sets, which may be found in the SI document.

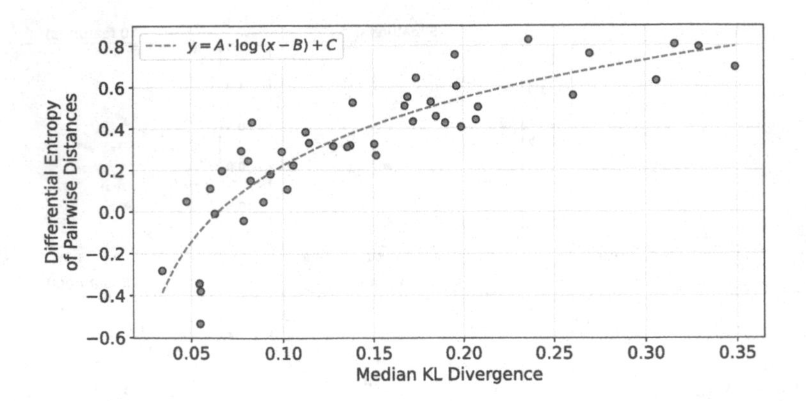

Fig. 5. Dependence of the differential entropy of pairwise distances in FINE embeddings on median KL divergence between communities for handpicked feature set. Best fit parameters: $A = 0.422$, $B = 0.014$, $C = 1.26$.

Figure 5 displays how the differential entropy of pairwise distances between communities behaves as a function of the median KL divergence for embeddings produced with one handpicked feature at a time. We observe a logarithmic trend, which seems to imply that past a certain threshold, a feature containing more information and richer differences between communities does not necessarily yield a distribution of pairwise distances with higher entropy.

We now turn our attention to the top row of Fig. 6, which shows the embeddings obtained by t-SNE and Diffusion Maps for the set of handpicked features after one-hot encoding and removing any households with missing values for any of the features, leaving a total of 256 households. t-SNE uses a weighted hamming distance while Diffusion Maps relies on a cosine distance metric. Households appear closely packed together with no discernible clusters, which we find to be consistent across different runs of t-SNE and Diffusion Maps. In the bottom row, we plot community barycenters by collapsing households in the same community to their mean coordinates. The cluster of communities 16–20, 22, 24 and 25 emerges somewhat for both algorithms. However, the cluster of communities 10, 12, 13, and 15 is not clear-cut for t-SNE, and Diffusion Maps does not highlight community 21 as an outlier.

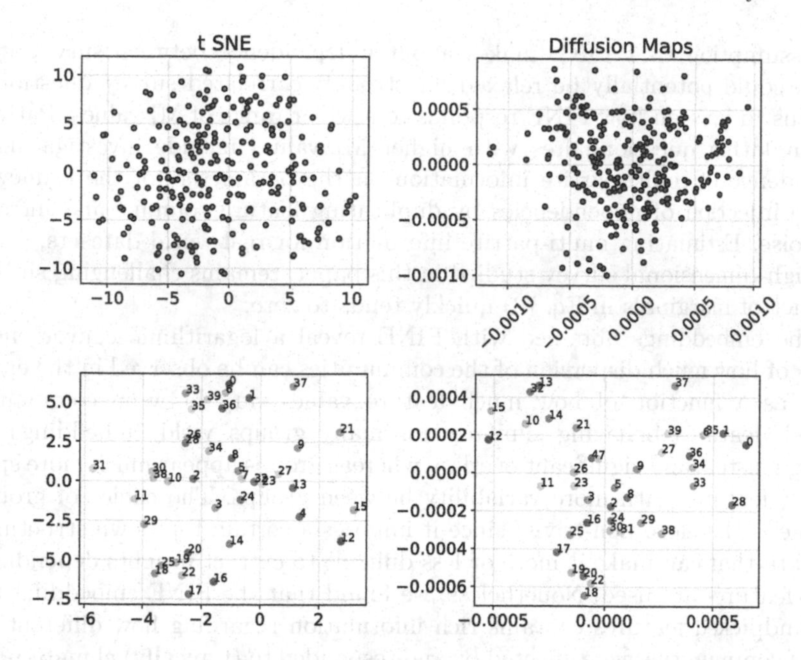

Fig. 6. (Top left) t-SNE embedding with handpicked features after one-hot encoding, using a weighted hamming distance metric. (Top right) Diffusion Maps embedding for same feature set after one-hot encoding, using cosine distance metric. (Bottom row) community barycenters for t-SNE and Diffusion maps.

Lastly, we compare pairwise distances between community coordinates for each pair of algorithms using handpicked features. Overall, we find that there is agreement across all three methods regarding the arrangement of the communities in relation to one another. Pearson correlation coefficients are found to be: 0.569 (t-SNE and Diffusion Maps), 0.514 (FINE and Diffusion Maps), and 0.318 (FINE and t-SNE). Perhaps unsurprisingly, correlation is highest between t-SNE and Diffusion Maps since the community grouping scheme was not applied to these two methods and many households were omitted due to missing feature values. An additional visualization of these correlations may be found in SI Fig. 3.

4 Discussion

Experiments carried out with a simulation framework reveal that all three methods achieve comparable performance in terms of recovering the general structure of the underlying one-dimensional manifolds. This is a departure from the previous study using this framework, which underestimated the performance of t-SNE in particular due to a lack of hyperparameter tuning. FINE appears to consistently outperform the other two methods except for questionnaires with high multi-partite information. This reduction in performance may be due to the assumption that FINE makes about independence between survey items, which could potentially be relaxed for strongly-correlated survey questions. It remains to be seen how FINE responds to a wider range of MI values. Pathways to simulating questionnaires with higher MI values include investigating the dependence of multi-partite information on the parameters of the framework, or the injection of dependencies by duplicating feature columns and introducing noise. Estimating multi-partite information for real-world datasets, such as the high-dimensional survey studied in this paper, remains challenging since the product of marginals in Eq. (3) quickly tends to zero.

The embeddings obtained with FINE reveal a logarithmic convergence in terms of how much dispersion of the communities can be observed in the embeddings as a function of how much feature values vary between communities. Indeed, features behaving similarly for many groups yield embeddings with strong clusters and significant overlap, whereas groups appear much more spread out for features with more variability between groups. The choice of grouping scheme is therefore non-trivial since it imposes a certain top-down structure on the data that can make it more or less difficult to extract insights depending on what features are used. Nonetheless, we found that the FINE embedding using all handpicked features contains rich information regarding how different clusters of communities were affected by and responded to (typically) climate-related shocks. This enabled us to identify key drivers to explain why certain communities were clustered together and to identify underlying human behavioural patterns. Applying FINE to other real-world datasets would be highly instructive regarding its capabilities and limitations. While not being limited by the need to impose a grouping scheme on data, t-SNE and Diffusion Maps require hyperparameter tuning and thus more computational effort, unlike FINE which is non-parametric. Another drawback is that t-SNE and Diffusion Maps do not seem to handle missing feature values well, which caused us to remove a significant portion of households in order to generate the embeddings displayed in Fig. 6. FINE on the other hand can simply use all available values to estimate group probability mass functions. Nonetheless, t-SNE and Diffusion Maps allow clusters to emerge in a more bottom-up way, which can be desirable when a natural grouping of observations is not clear.

The choice of algorithm ultimately depends on the researcher's goals. t-SNE and Diffusion Maps may be more suitable for exploratory data analysis and for discovering whether data contains any intrinsic clusters. On the other hand, when a grouping scheme is obvious or supported by existing literature, then FINE seems to be a more suitable and straightforward choice. Of course, using a combination of these approaches is possible, and in fact can help to extract greater insight from data, as well as ensure that results are robust across different methods.

Acknowledgements. The authors acknowledge the support from the Netherlands eScience project of the Dutch NWO, under contract 27020G08, titled "Computing societal dynamics of climate change adaptation in cities".

Data and Code Availability. Data from the Bangladesh Climate Change Adaptation Survey is available at: https://dataverse.harvard.edu/dataset.xhtml?persistentId=doi:10.7910/DVN/27883. All code used in this paper can be found at: https://github.com/charlesaugdupont/cca-manifold-learning

Supporting Information. Supporting tables and figures can be found at: https://github.com/charlesaugdupont/cca-manifold-learning/blob/main/SI.pdf

References

1. Carter, K.M., Raich, R., Finn, W.G., Hero, A.O., III.: Fine: fisher information nonparametric embedding. IEEE Trans. Pattern Anal. Mach. Intell. **31**(11), 2093–2098 (2009). https://doi.org/10.1109/TPAMI.2009.67
2. Cayton, L.: Algorithms for manifold learning. Univ. California San Diego Tech. Rep. **12**(1–17), 1 (2005)
3. Cerda, P., Varoquaux, G., Kégl, B.: Similarity encoding for learning with dirty categorical variables. Mach. Learn. **107**(8), 1477–1494 (2018)
4. Coifman, R.R., Lafon, S.: Diffusion maps. Appl. Computat. Harmonic Anal. **21**(1), 5–30 (2006). https://doi.org/10.1016/j.acha.2006.04.006
5. Delaporte, I., Maurel, M.: Adaptation to climate change in Bangladesh. Clim. Policy **18**(1), 49–62 (2018)
6. Fodor, I.K.: A survey of dimension reduction techniques. Tech. rep., Lawrence Livermore National Lab., CA (US) (2002)
7. Gower, J.C.: Generalized procrustes analysis. Psychometrika **40**(1), 33–51 (1975). https://doi.org/10.1007/BF02291478
8. Har-Shemesh, O., Quax, R., Lansing, J.S., Sloot, P.M.A.: Questionnaire data analysis using information geometry. Sci. Rep. **10**(1), 8633 (2020). https://doi.org/10.1038/s41598-020-63760-8
9. Hinton, G., Roweis, S.: Stochastic neighbor embedding. Adv. Neural Inf. Process. Syst. **15**, 833–840 (2003). http://citeseerx.ist.psu.edu/viewdoc/download?=10.1.1.13.7959&rep=rep1&type=pdf
10. (IFPRI), I.F.P.R.I.: Bangladesh Climate Change Adaptation Survey (BCCAS), Round II (2014). https://doi.org/10.7910/DVN/27883
11. Kruskal, J.: Multidimensional scaling by optimizing goodness of fit to a nonmetric hypothesis. Psychometrika **29**(1), 1–27 (1964)
12. Reckien, D., et al.: Climate change, equity and the sustainable development goals: an urban perspective. Environ. Urban. **29**(1), 159–182 (2017). https://doi.org/10.1177/0956247816677778
13. Scargle, J.D.: Studies in astronomical time series analysis. v. Bayesian blocks, a new method to analyze structure in photon counting data. Astrophys. J. **504**(1), 405–418 (1998). https://doi.org/10.1086/306064
14. van der Maaten, L., Hinton, G.: Visualizing high-dimensional data using t-SNE. J. Mach. Learn. Res. **9**, 2579–2605 (2008)

Toxicity in Evolving Twitter Topics

Marcel Geller[1(✉)], Vítor V. Vasconcelos[2,3], and Flávio L. Pinheiro[1]

[1] NOVA IMS – Universidade Nova de Lisboa, Lisboa, Portugal
fpinheiro@novaims.unl.pt

[2] Computational Science Lab, Informatics Institute, University of Amsterdam, Amsterdam, The Netherlands
v.v.vasconcelos@uva.nl

[3] Institute for Advanced Study, University of Amsterdam, Amsterdam, The Netherlands

Abstract. Tracking the evolution of discussions on online social spaces is essential to assess populations' main tendencies and concerns worldwide. This paper investigates the relationship between topic evolution and speech toxicity on Twitter. We construct a Dynamic Topic Evolution Model (DyTEM) based on a corpus of collected tweets. To build DyTEM, we leverage a combination of traditional static Topic Modelling approaches and sentence embeddings using sBERT, a state-of-the-art sentence transformer. The DyTEM is represented as a directed graph. Then, we propose a hashtag-based method to validate the consistency of the DyTEM and provide guidance for the hyperparameter selection. Our study identifies five evolutionary steps or Topic Transition Types: Topic Stagnation, Topic Merge, Topic Split, Topic Disappearance, and Topic Emergence. We utilize a speech toxicity classification model to analyze toxicity dynamics in topic evolution, comparing the Topic Transition Types in terms of their toxicity. Our results reveal a positive correlation between the popularity of a topic and its toxicity, with no statistically significant difference in the presence of inflammatory speech among the different transition types. These findings, along with the methods introduced in this paper, have broader implications for understanding and monitoring the impact of topic evolution on the online discourse, which can potentially inform interventions and policy-making in addressing toxic behavior in digital communities.

Keywords: Social Media Platforms · Twitter · Topic Modelling · Topic Evolution · Discourse Toxicity

1 Introduction

The study of how topics in collections of documents evolve is not new [8,18]. It follows naturally from the problem of automatically identifying topics [10] and then considering the time at which each document was produced and their lineage [2], i.e., when they emerge or collapse and their parent-child relationships. Such a description can provide insights into trends across many areas – such as in scientific literature [21,27,41], the web [6,12], media [3,26,47], news [5,36,44,49]

J. Mikyška et al. (Eds.): ICCS 2023, LNCS 14076, pp. 40–54, 2023.
https://doi.org/10.1007/978-3-031-36027-5_4

– and the determinants of why some topic lineages extend longer while others fall short.

Online social networks – such as Twitter, Facebook, or Reddit – provide a valuable resource to study human behavior at large [40], constituting a rich source of observational data of individuals' actions and interactions over time. Text-based corpora from discussions on OSN can also be studied from a topic-level description and benefit from considering their temporal evolution. Contrary to collections of published documents – manuscripts or books – we often look into speech to better understand the intricacies of social dynamics and human behavior. The dynamics of topics emergence, merging, branching, persistence, or decline comes then as a consequence of our choices on which discussions we engage in and which not. In other words, our choices operate as a selective force that defines which topics prevail and which fade away from collective memory. Relevant in such dynamics are the language used within a topic, their efficiency in carrying information, and the resulting perception actors have of speech.

OSN have been used not only to revisit old theories but also document new phenomena such as social polarization and influence [16], information diffusion [42], the spread of disinformation [35] and information virality [22]. While OSNs provide access to large datasets, they come at the expense of requiring pre-processing and feature engineering to be studied [1,15], of underlying biases that need to be accounted for [45], and experiments that need to be designed [43]. Many techniques have become popularly adopted to address such challenges: text-mining and machine learning methods have been used to estimate the Sentiment [32,34,37], Morality [4,24,28], or Toxicity [17] load in speech; network analysis [19] is often used to study patterns of information diffusion, connectivity, and community structure within Twitter.

Given this background, it is pertinent to ask how speech and associated features can modulate the evolutionary dynamics of topics over time. In this paper, we look at a large corpus of geolocated Tweets from New York (USA) to study the extent to which the evolution of topics is modulated by the toxicity of the embedded discourse. We use Topic Modelling methods and clustering techniques to track the emergence, branching, merging, persistence, and disappearance of topics from the social discussion. Specifically, we focus on discourse toxicity and its impact on topic evolution. Toxicity, in the context of online communication, refers to the presence of harmful, offensive, or aggressive language within a text. It encompasses a wide range of negative behaviors and expressions, such as hate speech, profanity, targeted harassment, personal attacks, threats, discriminatory language, and other forms of abusive or derogatory communication. Toxicity can manifest in various degrees, from mildly offensive remarks to extreme cases of online harassment and cyberbullying.

Our goal is to analyze if the Toxicity of topics tends to drift into higher/lower levels throughout their evolution and if there is any association between toxicity level and the topic's popularity. We contribute to better understand and effectively detect toxicity in online discourse, which is crucial for evaluating the health of digital communication ecosystems and informing policy-making and advancements in the domain of computational social science.

2 Related Work

OSNs have become a valuable source of observational data to study human behavior and social dynamics [7,31]. Among various OSNs, Twitter, a microblogging online social network, has attracted significant attention from academic researchers. Its unique features include short posts limited to 280 characters, high frequency of posting, real-time accessibility to a global audience, and the provision of a free API allowing researchers to extract unfiltered and filtered content randomly or targeted from specific users or geolocations.

Although topic modeling is usually associated with the study and categorization of extensive collections of documents, it can also be used to identify sets of Tweets that share a common vocabulary. Naturally, dealing with short documents or tweets brings along several challenges, such as sparse co-occurrence of words across documents, informal language, high content variability, and noisy and irrelevant data. These challenges can affect the accuracy and reliability of topic models, so pre-processing techniques, such as text normalization, removal of stop words, and feature selection, are often applied to improve the quality of the input data [48].

Most studies approach Topic Modelling as a static process, neglecting the temporal dimension. However, when using Twitter data, we can track the evolution of discourse through the timestamp of each Tweet. In the particular case of topic evolution, several authors looked into dynamically modeling the public discourse. Malik et al. [33], propose a visual framework to study complex trends in public discussions over time. Abulaish and Fazil [2], studied topic evolution as an m-bipartite Graph and applied it to analyze the Evolution of tweets from Barack Obama, Donald Trump, and a Twitter Socialbot.

In both cases, the tweet corpus was divided into time bins, and Latent Dirichlet Allocation (LDA) Topic Modelling was utilized to compute topics of each time bin [9], which were then compared to subsequent time bins based on the cosine similarity of constructed topic embeddings. However, these approaches were tested on relatively small datasets of 16,199 and 3,200 Tweets, respectively, and a Topic Evolution Model has not yet been applied to studying the dynamics of speech toxicity.

In Alam et al. [3], the temporal evolution of hashtag distributions for trending keywords is studied, the authors argue that combining hashtag information allows for a more suitable breakdown of topics underlying trending keywords while providing a context that offers better interpretability.

In addition to semantics, several speech characteristics can be inferred, which might be relevant in understanding both the intention of the writer but also how a reader perceives it. In that sense, perhaps the most popular text-based metric used is the sentiment of a text, which attempts to capture whether a person was expressing positive or negative thoughts through their speech [25]. However, the sentiment is somewhat narrow and lacks the nuance of the depth and variety of emotions. Other metrics that can be inferred from speech include the moral load of a document based on the Moral Foundations Theory or the embedded

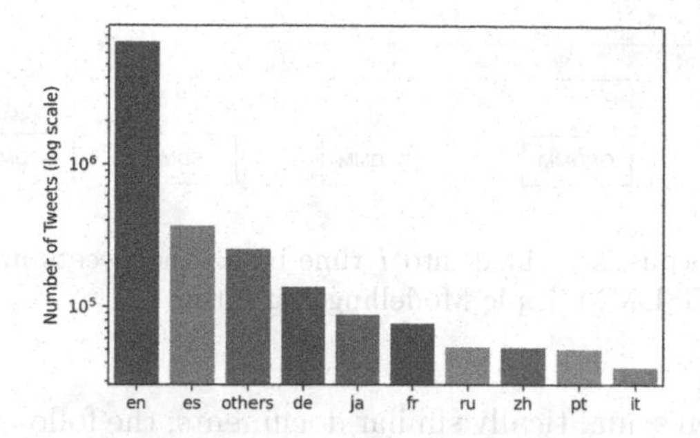

Fig. 1. Language of collected Tweets as classified by the fasttext language detector

toxicity of a text, which attempts to capture the existence of hateful or harmful language.

There are many proposed Text Mining approaches for detecting properties like toxicity or the moral load of a document. Most are based on a corpus of human-labeled documents [23]. Each document is transformed into a feature vector that embeds the relevant properties of the document. Then Machine Learning [39] is applied to train a Model on the feature vectors using the human-generated annotations as the ground truth.

The original contribution of this work lies in the construction of a dynamic topic evolution model to analyze the relationship between topic evolution and speech toxicity on geolocated tweets. By combining traditional static topic modeling approaches and sBERT sentence embeddings, we represent the topic evolution model as a directed graph and introduce a hashtag-based method for validation and hyperparameter selection. Our study expands on the existing literature by addressing the gap in understanding the dynamics of speech toxicity in topic evolution and its implications for online discourse and digital communities.

3 Data and Methods

We use a corpus of approximately 8 Million Tweets published between 2020/06/02 and 2020/11/03 and geo-located around the city of New York, USA. These cover a particularly interesting period of recent US history, which was marked by, for instance, the George Floyd protests, the strengthening of the Black Lives Matter movement and the 2020 presidential election race. The dataset was obtained using the free Twitter Academic API, with the only restriction being the geolocation of the tweets. Contrary to other studies, we have not focused on a particular topic or account. Instead, we resort to the collection in bulk of random tweets from the Twitter timeline.

Tweets are uniformly distributed in time, and on average we have 53,859 tweets per day (154 days in total). The dataset contains 364,918 unique hashtags with every hashtag occurring on average 4.81 times in the corpus. In order to further reduce the sparsity of the dataset and increase the number of token co-

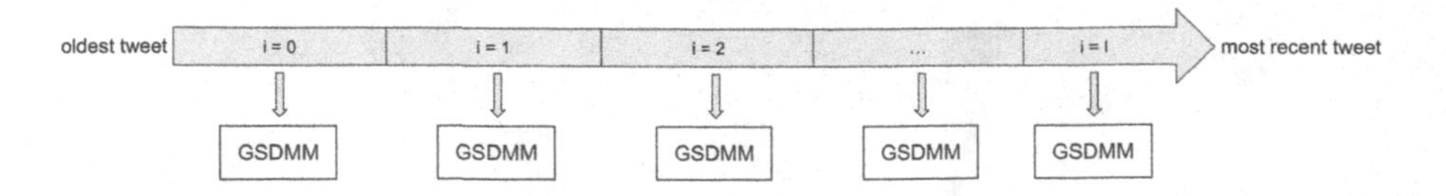

Fig. 2. The tweet corpus is divided into I time bins. The tweets in each bin are then clustered using the GSDMM Topic Modelling Algorithm.

occurrences between semantically similar documents, the following preprocessing steps were applied to the entire tweet corpus.

- Lowercasing of alphabetical characters.
- Removal of all emojis from the dataset.
- Deduplication of Tweets.
- Removal of all non-alphanumeric characters.
- Discarding of tokens containing less than 2 characters.
- Filtering out Tweets with less than 20 characters.
- Application of the fastText language detector [29,30] to remove tweets not written in English.
- Lemmatization.

After applying these preprocessing steps the number of tweets in the corpus reduces by 37% to 5,197,172 Tweets.

Since the goal is to study topic evolution, we start by splitting the tweet corpus into non-overlapping time intervals of 10 days. Which leads to approximately 350.000 tweets per time interval, and eleven time intervals. The choice of non-overlapping time intervals over a sliding window method was done in order to reduce the number of time intervals to analyze and subsequently lower computational costs.

3.1 Topic Modelling and DAG Lineage

We use the Generalized Scalable Dirichlet Multinomial Mixture Model (GSDMM) to identify existing topics in the corpus [46]. We run GSDMM independently for each of the time bins, see Fig. 2, and subsequently, we perform a topic linkage between the different time windows in order to obtain an approximation of topic evolution throughout the studied period, and as such their lineage.

The GSDMM is a variant of the popular LDA model, that is particularly useful for handling sparse data. GSDMM requires two hyperparameters, one that is used to represent the relative weight given to each cluster of words (α) and a second to represent the significance of each word in determining a document's topic distribution (β). Following Yin and Wang [46], we consider $\alpha = \beta = 0.1$ and we run GSDMM for 20 iterations for each time bin. GSDMM is able to automatically infer the number of topics, therefore it is simply necessary to initialize the number of topics, which we set to 120.

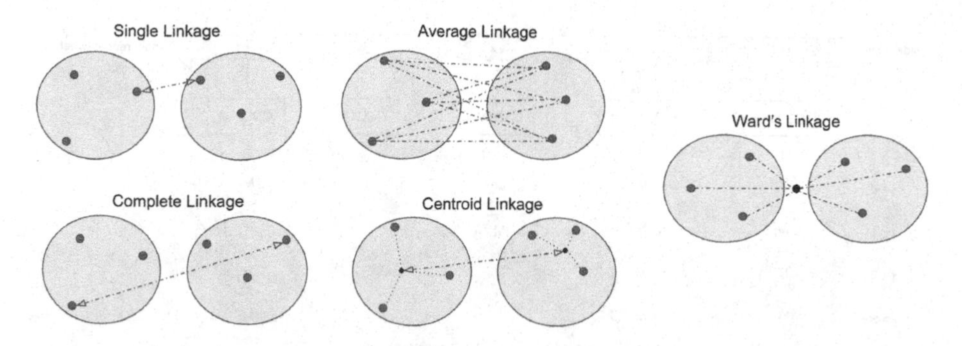

Fig. 3. Collection of Linkage Methods that can be used to quantify the distance between two Topics. Please note that it is required to have an embedding on the document level. In our case, we utilize sBERT to obtain a numerical vector for each Tweet.

At this stage, all time bins are considered independently from one another and topics are disconnected. Each tweet can be associated with a specific time bin and topic. To maintain consistent notations throughout the manuscript, let us define that M_j^i refers to a set of tweets that are associated with topic j at time i. Moving to the temporal evolution of Topics, we represent Topics and their parent-child relationship by means of an M-partite Directed Acyclic Graph (DAG). Each partition corresponds to all topics (j) of a given time bin (i), and edges connecting topics are an ordered pair of nodes (M_j^i, M_j^{i+1}) in two adjacent time windows.

In order to identify relationships between topics, we measure the semantic similarity between all topics from adjacent partitions. To that end, we use SentenceBERT (short sBERT), a pre-trained transformer-based model that can be used to encode each tweet into a high-dimensional feature vector [38]. The cosine distance between data points in this vector space is a metric to quantify the inverse semantic similarity of documents. A topic is a set of data points in the vector space that in order to compute the proximity of two topics needs to be compared to another set of data points in the same vector space created by sBERT. When computing the distance of two sets of data points we perform a Centroid Linkage approach [13]. We choose this approach over others due to its relative robustness against outliers when compared to, for instance, single or complete linkage methods and for being more computationally efficient than Ward linkage.

Finally, we use a threshold-based approach to select which candidate edges should be considered, in that sense we only consider edges that represent similarities (measured by the cosine similarity) between topics that are greater or equal to a threshold parameter ε.

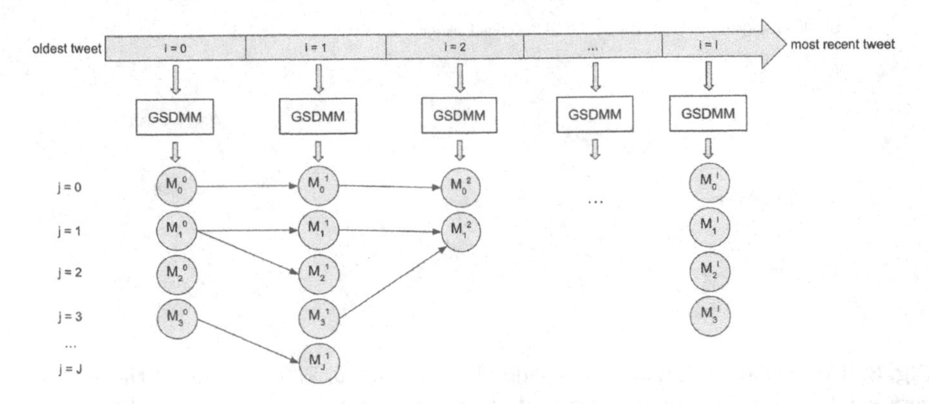

Fig. 4. Static Topic results are used to build a graph representation of the Topic Evolution. Topics of adjacent Time Bins are connected if a proximity threshold ε is exceeded.

3.2 Transition Types

By definition, a topic M_j^i can be connected to multiple topics from time bin $i+1$. Since multiple topics from time bin $i+1$ can have a similarity with M_j^i that exceeds ε, a parent topic node can have more than one child topic. This is also true the other way around. A child topic M_j^i can have more than one parent topic, i.e., if multiple topics from time bin $i+1$ are similar enough to M_j^i to exceed the proximity threshold ε. Hence, between time bin i and $i+1$ the following scenarios are possible to observe:

1. A Topic M_j^i splits into multiple Topics if the outdegree of M_j^i exceeds 1.
2. Multiple Topics merge into one Topic M_j^{i+1} if the indegree of M_j^{i+1} exceeds 1.
3. A Topic M_j^i stagnates if it has a single child topic M_j^{i+1} that has a single parent Topic. In other words, M_j^i needs to have an outdegree of 1 while its child Topic needs to have an indegree of 1.
4. M_j^i disappears if it has an outdegree of 0 (in other words: it has no child topic).
5. M_j^{i+1} emerges if it has an indegree of 0 (in other words: it has no parent topic).

These five scenarios are referred to as transition types throughout the rest of this paper. Moreover, by definition, topics in the first time bin do not emerge and topics in the last time bin do not disappear.

Figure 5 shows an illustrative example of the topic evolution. Taking topics at Bin 1 that are associated with the Black-Lives-Matter (BLM) movement, we follow their topic lineage for time 4 iterations. Each box corresponds to a topic and shows the five most popular hashtags. Topics that have BLM-related hashtags among the most popular hashtags are mostly connected to topics that

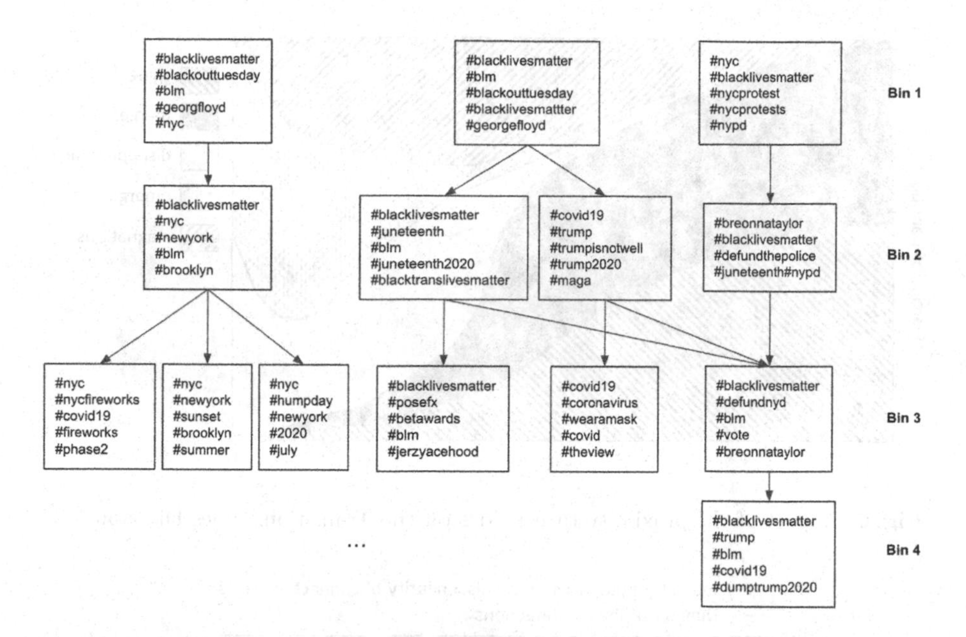

Fig. 5. Sample Topic Trajectories. Topics labeled with the 5 most frequent hashtags. A proximity threshold $\varepsilon = 0.8$ was chosen.

also contain BLM-related hashtags frequently. In this spot check a topic split was detected between Time Bin 1 and Time Bin 2, where BLM seems to split into two topics: one continues to be around the BLM events, and the second gears towards a more political topic involving Donald Trump.

As discussed above, the threshold ε is the only hyperparameter in the Topic Evolution approach presented in this paper. However, the selection of ε has a strong effect on the structure of the resulting DAG. To give an overview of the sensitivity of ε, Fig. 6 shows the proportion of each transition type in different values of ε. When selecting a value of ε close to its maximum value, only topics are connected that are almost identical between time bins. In this scenario, almost all transitions would be of type emergence and disappearance, because child or parent topics are due to the strict proximity requirement barely detected. In a scenario where ε is chosen to be a value close to its minimum value, almost all topics are connected to all topics from adjacent time bins. Therefore, in this case most topics are splitting and merging between time bins.

In order to validate the dynamic topic evolution DAG, we analyse whether the linkage done based on the BERT text embeddings is reflected by the distribution of hashtags across topics. We quantify the similarity of hashtags using the Jensen-Shannon-Divergence and compute it for each edge of the DAG. Starting from ε close to 1 only very similar topics are connected. The number of connections is therefore small while these few connected topics have a similar distribution of hashtags. As we decrease ε less similar topics are connected, which is why

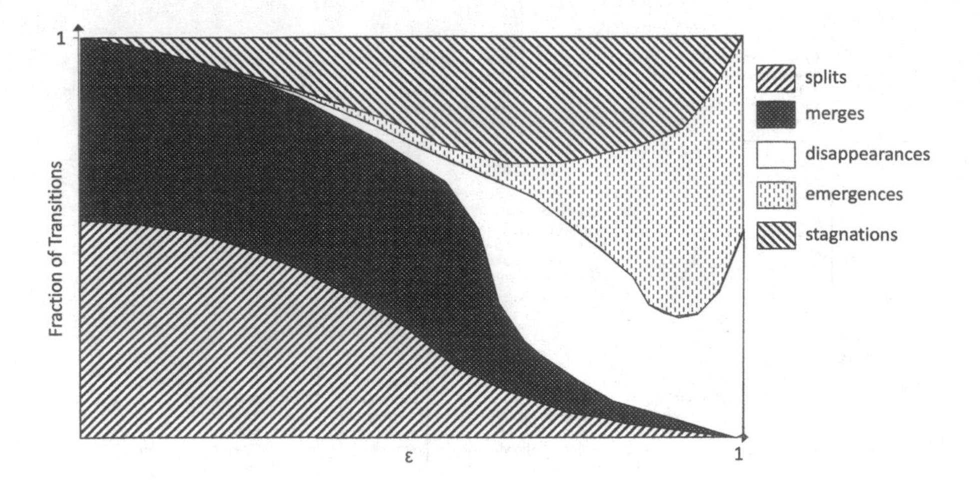

Fig. 6. Impact of the proximity threshold ε on the Transition Types Distribution

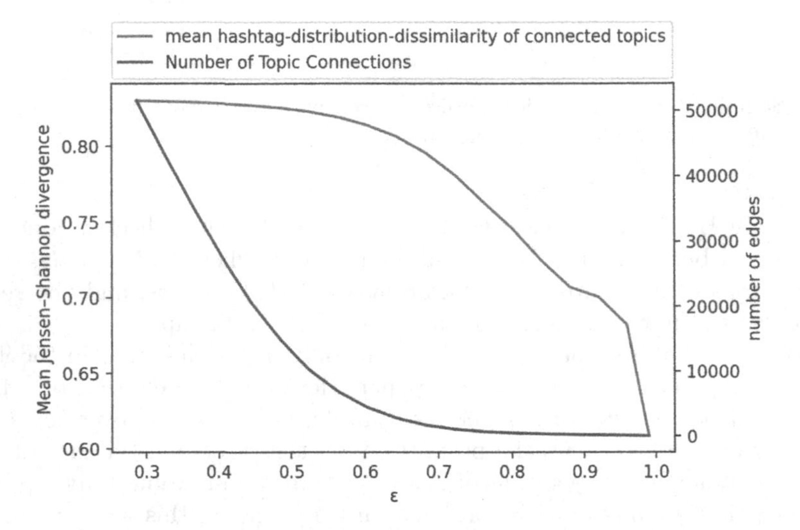

Fig. 7. Similarity of the distribution of hashtags across connected Topics in different value for ε

the number of edges in the DAG increases and the distribution of hashtags becomes less similar across connected topics. The fact that the hashtag similarity of connected topics is increasing in ε confirms the consistency of the model. Hashtags can arguably serve as a user-generated ground truth label for the Topic of a Tweet and in the given case support the linkage of topics discussed above.

4 Studying Toxicity in Topic Evolution

When analyzing the language used in a text corpus, toxicity means the existence of harmful or inflammatory language. Laura Hanu [20] proposed a machine learning model, Detoxify, that can estimate the level of embedded toxicity in a text. Although Detoxify is not trained on tweets but on a set of human-labeled Wikipedia comments, it is assumed that the model generalizes well enough to provide a good estimate for the level of toxicity in tweets. While Detoxify is a multi-label model that provides estimates for multiple speech characteristics, in the following analysis solely the toxicity of a document is used. While Detoxify returns a continuous level of toxicity between 0 and 1, the bimodal distribution suggests that it is reasonable to map the values to a binary variable. Binarizing the toxicity distribution is in line with the model's training on discrete human-annotated labels, which helps maintain consistency in the analysis and makes interpreting the results in relation to the original training data more straightforward. Using a cut-off threshold of 0.5, approximately 11% of the tweets are identified as toxic. This value is in line with other studies. Studies focusing on more controversial topics typically have a higher percentage of toxic tweets. Broad studies which focus on stronger definitions of toxicity, like hate speech and hateful content, have lower. For instance, a study by Davidson et al. (2017) [11] analyzed a dataset of over 25,000 tweets and found that around 5% of them contained hate speech. Another study by Founta et al. (2018) [14] analyzed a dataset of 80,000 tweets and found around 4% of abusive and hateful tweets in their random sample. Since we are focusing on the emergence of topics a higher fraction can help keep track of more nuanced topics.

The binarized toxicity can be included in the Topic Evolution DAG in the following manner. Each topic node is assigned with an attribute indicating the percentage of toxic tweets detected in the topic. Each edge of the Graph is assigned with an attribute called *Δtoxicity* that is calculated as follows, indicating the relative change in the percentage of toxic tweets.

$$\Delta Toxicity = \frac{Toxicity(childTopic)}{Toxicity(parentTopic)} - 1 \qquad (1)$$

4.1 Toxicity per Transition Type

Each edge of the DAG belongs to at least one of the 5 transition types defined in Sect. 3. Each merge and split transition, by definition, has multiple $\Delta Toxicity$ values assigned. Do topics tend to become more/less toxic when they merge/split/emerge/disappear/stagnate? The goal is to find out whether or not certain transition types have significantly different average $\Delta Toxicity$. The distribution of $\Delta Toxicity$ per transition type is presented in Fig. 8.

The results indicate no significant difference between the distributions of $\Delta Toxicity$ across transition types. The mean $\Delta Toxicity$ is near zero for all distributions, meaning that none of the transition types is correlated with a significant change in the percentage of inflammatory speech.

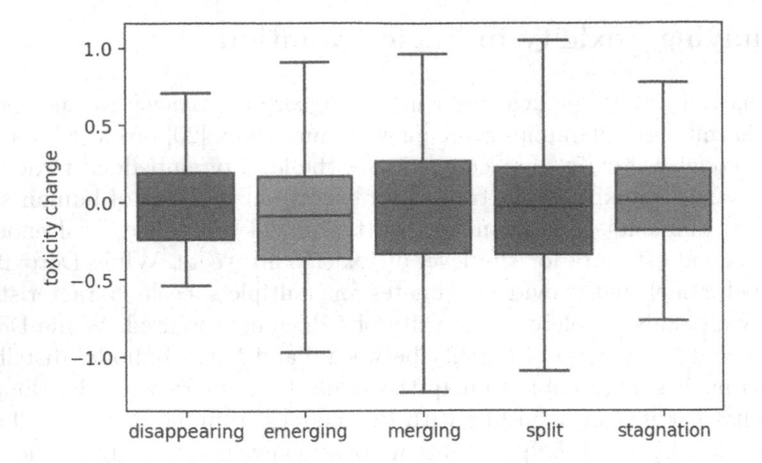

Fig. 8. Distribution of $\Delta Toxicity$ broken down by transition type. A proximity threshold of $\varepsilon = 0.8$ was chosen. Topics with a Popularity <30 were discarded.

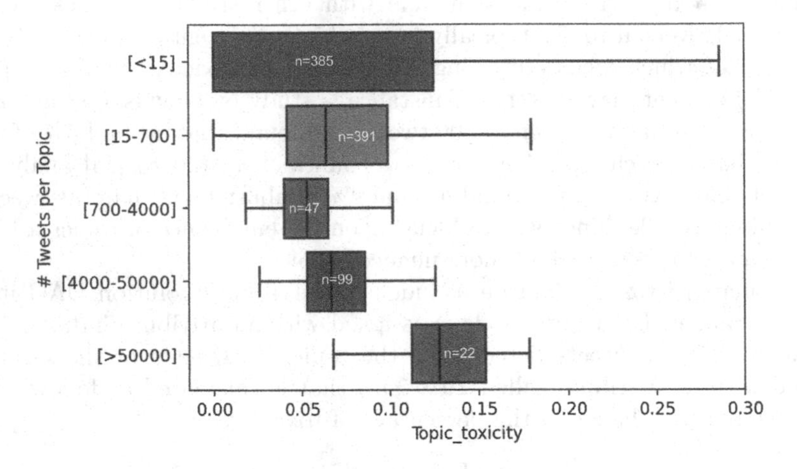

Fig. 9. Relationship between the Topic Toxicity and Topic Popularity (measured in Tweets per Topic)

4.2 Relationship Topic Popularity - Toxicity

We can measure topic popularity by assessing the number of tweets in each topic. We now analyse the relationship between Topic Popularity and Topic Toxicity. Do popular topics tend to be more or less toxic than niche Topics? As presented in Fig. 10 and 9, our results indicate a parabola-shaped distribution of topics in the popularity-toxicity space. Ignoring micro topics with a popularity of less than 15 tweets, topic popularity is positively correlated with topic toxicity. Because of its right-skewed distribution, the topic popularity was log-transformed.

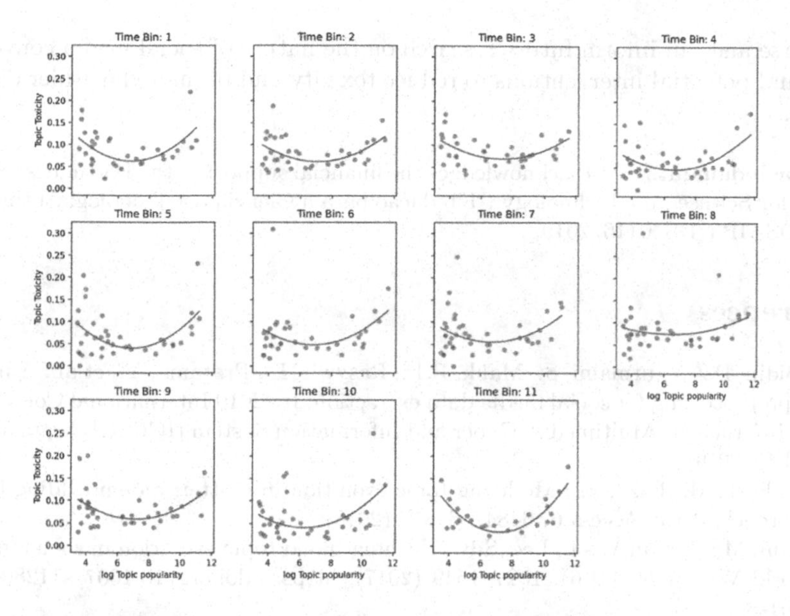

Fig. 10. Relationship between the Topic Toxicity and Topic Popularity grouped by time bin. Because of its skewed distribution the Topic Popularity was log transformed. Topics with an Popularity <30 were discarded

5 Conclusion

In conclusion, this study introduces a dynamic Topic Evolution Modeling approach that represents topic trajectories in an M-partite-DAG. We analyzed the impact of the proximity threshold selection on the graph structure and validated the model's consistency using hashtags present in the tweets. Both manual spot checks of sample topic trajectories and quantification of hashtag similarity for all connected topics in the graph confirmed the model's consistency. Our findings reveal a positive correlation between topic popularity and toxicity, suggesting that viral topics tend to contain more inflammatory speech characteristics than niche topics. Moreover, we examined whether the level of speech toxicity evolves equally across all transition types. Our results indicate no significant difference in toxicity changes across the transition types, with changes being close to zero for all types. Despite the insights provided by our study, future research should aim to validate these results on other datasets with larger sample sizes and denser data. Additionally, while our current model focuses on topics emerging in adjacent time bins, it could be extended to consider the re-emergence of topics by comparing topics from non-adjacent time bins. This extension would remove the M-partite property of the graph but enable a more profound understanding of topic evolution.

In summary, our dynamic Topic Evolution Modeling approach offers a valuable tool for analyzing the relationship between topic dynamics and toxicity in Twitter discussions. This work contributes to the existing literature on online discourse and can inform future research on the nature of social media conversations and potential interventions to reduce toxicity and promote healthier digital spaces.

Acknowledgments. FLP acknowledges the financial support of the Portuguese Foundation for Science and Technology ("Fundação para a Ciência e a Tecnologia") through grant DSAIPA/DS/0116/2019.

References

1. Abidin, D.Z., Nurmaini, S., Malik, R.F., Rasywir, E., Pratama, Y., et al.: A model of preprocessing for social media data extraction. In: 2019 International Conference on Informatics, Multimedia, Cyber and Information System (ICIMCIS), pp. 67–72. IEEE (2019)
2. Abulaish, M., Fazil, M.: Modeling topic evolution in twitter: an embedding-based approach. IEEE Access **6**, 64847–64857 (2018)
3. Alam, M.H., Ryu, W.-J., Lee, S.K.: Hashtag-based topic evolution in social media. World Wide Web **20**(6), 1527–1549 (2017). https://doi.org/10.1007/s11280-017-0451-3
4. Araque, O., Gatti, L., Kalimeri, K.: Moralstrength: exploiting a moral lexicon and embedding similarity for moral foundations prediction. Knowl.-Based Syst. **191**, 105184 (2020)
5. Bai, Y., Jia, S., Chen, L.: Topic evolution analysis of covid-19 news articles. J. Phys. Conf. Series **1601**, 052009 (2020)
6. Bar-Ilan, J., Peritz, B.C.: A method for measuring the evolution of a topic on the web: the case of informetrics. J. Am. Soc. Inf. Sci. Technol. **60**(9), 1730–1740 (2009)
7. Bello-Orgaz, G., Jung, J.J., Camacho, D.: Social big data: recent achievements and new challenges. Inf. Fusion **28**, 45–59 (2016)
8. Blei, D.M., Lafferty, J.D.: Dynamic topic models. In: Proceedings of the 23rd International Conference on Machine Learning, pp. 113–120 (2006)
9. Blei, D.M., Ng, A.Y., Jordan, M.I.: Latent dirichlet allocation. J. Mach. Learn. Res. **3**, 993–1022 (2003)
10. Boyd-Graber, J., Hu, Y., Mimno, D., et al.: Applications of topic models. Found. Trends® Inf. Retrieval **11**(2–3), 143–296 (2017)
11. Davidson, T., Warmsley, D., Macy, M., Weber, I.: Automated hate speech detection and the problem of offensive language. In: Proceedings of the International AAAI Conference on Web and Social Media, vol. 11, pp. 512–515 (2017)
12. Derntl, M., Günnemann, N., Tillmann, A., Klamma, R., Jarke, M.: Building and exploring dynamic topic models on the web. In: Proceedings of the 23rd ACM International Conference on Conference on Information and Knowledge Management, pp. 2012–2014 (2014)
13. El-Hamdouchi, A., Willett, P.: Comparison of hierarchic agglomerative clustering methods for document retrieval. Comput. J. **32**(3), 220–227 (1989)
14. Founta, A., et al.: Large scale crowdsourcing and characterization of twitter abusive behavior. In: Proceedings of the International AAAI Conference on Web and Social Media, vol. 12 (2018)
15. Gani, R., Chalaguine, L.: Feature engineering vs bert on twitter data. arXiv preprint arXiv:2210.16168 (2022)

16. Garimella, V.R.K., Weber, I.: A long-term analysis of polarization on twitter. In: Eleventh International AAAI Conference on Web and Social Media, vol. 11, no. 1, pp. 528–531 (2017)
17. Georgakopoulos, S.V., Tasoulis, S.K., Vrahatis, A.G., Plagianakos, V.P.: Convolutional neural networks for toxic comment classification. In: Proceedings of the 10th Hellenic Conference on Artificial Intelligence, pp. 1–6 (2018)
18. Gohr, A., Hinneburg, A., Schult, R., Spiliopoulou, M.: Topic evolution in a stream of documents. In: Proceedings of the 2009 SIAM International Conference on Data Mining, pp. 859–870. SIAM (2009)
19. Grandjean, M.: A social network analysis of twitter: mapping the digital humanities community. Cogent Arts Hum. **3**(1), 1171458 (2016)
20. Hanu, L.: Unitary team: Detoxify. Github (2020). https://github.com/unitaryai/detoxify
21. He, Q., Chen, B., Pei, J., Qiu, B., Mitra, P., Giles, L.: Detecting topic evolution in scientific literature: how can citations help? In: Proceedings of the 18th ACM Conference on Information and Knowledge Management, pp. 957–966 (2009)
22. Hoang, T.-A., Lim, E.-P., Achananuparp, P., Jiang, J., Zhu, F.: On modeling virality of twitter content. In: Xing, C., Crestani, F., Rauber, A. (eds.) ICADL 2011. LNCS, vol. 7008, pp. 212–221. Springer, Heidelberg (2011). https://doi.org/10.1007/978-3-642-24826-9_27
23. Hoover, J., et al.: Moral foundations twitter corpus: a collection of 35k tweets annotated for moral sentiment. Soc. Psychol. Pers. Sci. **11**(8), 1057–1071 (2020)
24. Hopp, F.R., Fisher, J.T., Cornell, D., Huskey, R., Weber, R.: The extended moral foundations dictionary (EMFD): Development and applications of a crowd-sourced approach to extracting moral intuitions from text. Behav. Res. Meth. **53**, 232–246 (2021)
25. Hu, R., Rui, L., Zeng, P., Chen, L., Fan, X.: Text sentiment analysis: a review. In: 2018 IEEE 4th International Conference on Computer and Communications (ICCC), pp. 2283–2288. IEEE (2018)
26. Hu, Y., Xu, X., Li, L.: Analyzing topic-sentiment and topic evolution over time from social media. In: Lehner, F., Fteimi, N. (eds.) KSEM 2016. LNCS (LNAI), vol. 9983, pp. 97–109. Springer, Cham (2016). https://doi.org/10.1007/978-3-319-47650-6_8
27. Jo, Y., Hopcroft, J.E., Lagoze, C.: The web of topics: discovering the topology of topic evolution in a corpus. In: Proceedings of the 20th International Conference on World Wide Web, pp. 257–266 (2011)
28. Johnson, K., Goldwasser, D.: Classification of moral foundations in microblog political discourse. In: Proceedings of the 56th Annual Meeting of the Association for Computational Linguistics (volume 1: long papers), pp. 720–730 (2018)
29. Joulin, A., Grave, E., Bojanowski, P., Douze, M., Jégou, H., Mikolov, T.: Fasttext.zip: Compressing text classification models. arXiv preprint arXiv:1612.03651 (2016)
30. Joulin, A., Grave, E., Bojanowski, P., Mikolov, T.: Bag of tricks for efficient text classification. arXiv preprint arXiv:1607.01759 (2016)
31. Lazer, D., et al.: Computational social science. Science **323**(5915), 721–723 (2009)
32. Li, N., Wu, D.D.: Using text mining and sentiment analysis for online forums hotspot detection and forecast. Decis. Supp. Syst. **48**(2), 354–368 (2010)
33. Malik, S., et al.: Topicflow: visualizing topic alignment of twitter data over time. In: Proceedings of the 2013 IEEE/ACM International Conference on Advances in Social Networks Analysis and Mining, pp. 720–726 (2013)

34. Medhat, W., Hassan, A., Korashy, H.: Sentiment analysis algorithms and applications: a survey. Ain Shams Eng. J. **5**(4), 1093–1113 (2014)
35. Murayama, T., Wakamiya, S., Aramaki, E., Kobayashi, R.: Modeling the spread of fake news on twitter. Plos One **16**(4), e0250419 (2021)
36. Neo, S.Y., Ran, Y., Goh, H.K., Zheng, Y., Chua, T.S., Li, J.: The use of topic evolution to help users browse and find answers in news video corpus. In: Proceedings of the 15th ACM International Conference on Multimedia, pp. 198–207 (2007)
37. Redhu, S., Srivastava, S., Bansal, B., Gupta, G.: Sentiment analysis using text mining: a review. Int. J. Data Sci. Technol. **4**(2), 49–53 (2018)
38. Reimers, N., Gurevych, I.: Sentence-bert: sentence embeddings using siamese bert-networks. In: Proceedings of the 2019 Conference on Empirical Methods in Natural Language Processing and the 9th International Joint Conference on Natural Language Processing (EMNLP-IJCNLP), pp. 3982–3992 (2019)
39. Roy, S., Pacheco, M.L., Goldwasser, D.: Identifying morality frames in political tweets using relational learning. arXiv preprint arXiv:2109.04535 (2021)
40. Salganik, M.J.: Bit by Bit: Social Research in the Digital Age. Princeton University Press (2019)
41. Song, M., Heo, G.E., Kim, S.Y.: Analyzing topic evolution in bioinformatics: investigation of dynamics of the field with conference data in DBLP. Scientometrics **101**, 397–428 (2014)
42. Stai, E., Milaiou, E., Karyotis, V., Papavassiliou, S.: Temporal dynamics of information diffusion in twitter: modeling and experimentation. IEEE Trans. Comput. Soc. Syst. **5**(1), 256–264 (2018)
43. Tan, C., Lee, L., Pang, B.: The effect of wording on message propagation: Topic-and author-controlled natural experiments on twitter. arXiv preprint arXiv:1405.1438 (2014)
44. Viermetz, M., Skubacz, M., Ziegler, C.N., Seipel, D.: Tracking topic evolution in news environments. In: 2008 10th IEEE Conference on E-Commerce Technology and the Fifth IEEE Conference on Enterprise Computing, E-Commerce and E-Services, pp. 215–220. IEEE (2008)
45. Yang, K.C., Hui, P.M., Menczer, F.: How twitter data sampling biases us voter behavior characterizations. Peer J. Comput. Sci. **8**, e1025 (2022)
46. Yin, J., Wang, J.: A dirichlet multinomial mixture model-based approach for short text clustering. Proceedings of the 20th ACM SIGKDD International Conference on Knowledge Discovery and Data Mining, pp. 233–242 (2014)
47. Zhang, Y., Mao, W., Lin, J.: Modeling topic evolution in social media short texts. In: 2017 IEEE International Conference on Big Knowledge (ICBK), pp. 315–319. IEEE (2017)
48. Zhao, W.X., et al.: Comparing twitter and traditional media using topic models. In: Clough, P., et al. (eds.) ECIR 2011. LNCS, vol. 6611, pp. 338–349. Springer, Heidelberg (2011). https://doi.org/10.1007/978-3-642-20161-5_34
49. Zhou, H., Yu, H., Hu, R., Hu, J.: A survey on trends of cross-media topic evolution map. Knowl.-Based Syst. **124**, 164–175 (2017)

Structural Validation of Synthetic Power Distribution Networks Using the Multiscale Flat Norm

Rounak Meyur[1(✉)], Kostiantyn Lyman[2], Bala Krishnamoorthy[2], and Mahantesh Halappanavar[1]

[1] Pacific Northwest National Lab, Richland, USA
{rounak.meyur,hala}@pnnl.gov

[2] Department of Mathematics and Statistics, Washington State University, Washington, USA
{kostiantyn.lyman,kbala}@wsu.edu

Abstract. We study the problem of comparing a pair of geometric networks that may not be similarly defined, i.e., when they do not have one-to-one correspondences between their nodes and edges. Our motivating application is to compare power distribution networks of a region. Due to the lack of openly available power network datasets, researchers synthesize realistic networks resembling their actual counterparts. But the synthetic digital twins may vary significantly from one another and from actual networks due to varying underlying assumptions and approaches. Hence the user wants to evaluate the quality of networks in terms of their structural similarity to actual power networks. But the lack of correspondence between the networks renders most standard approaches, e.g., subgraph isomorphism and edit distance, unsuitable.

We propose an approach based on the *multiscale flat norm*, a notion of distance between objects defined in the field of geometric measure theory, to compute the distance between a pair of planar geometric networks. Using a triangulation of the domain containing the input networks, the flat norm distance between two networks at a given scale can be computed by solving a linear program. In addition, this computation automatically identifies the 2D regions (patches) that capture where the two networks are different. We demonstrate our approach on a set of actual power networks from a county in the USA. Our approach can be extended to validate synthetic networks created for multiple infrastructures such as transportation, communication, water, and gas networks.

Keywords: synthetic networks · multiscale flat norm · network validation

---

The research is supported in part by the U.S. DOE Exascale Computing Project's (ECP) (17-SC-20-SC) ExaGraph codesign center and Laboratory Directed Research and Development Program at Pacific Northwest National Laboratory (PNNL).

J. Mikyška et al. (Eds.): ICCS 2023, LNCS 14076, pp. 55–69, 2023.
https://doi.org/10.1007/978-3-031-36027-5_5

1 Introduction

The power grid is the most vital infrastructure that provides crucial support for the delivery of basic services to most segments of society. Once considered a passive entity in power grid planning and operation, the power distribution system poses significant challenges in the present day. The increased adoption of rooftop solar photovoltaics (PVs) and electric vehicles (EVs) augmented with residential charging units has altered the energy consumption profile of an average consumer. Access to extensive datasets pertaining to power distribution networks and residential consumer demand is vital for public policy researchers and power system engineers alike. However, the proprietary nature of power distribution system data hinders their public availability. This has led researchers to develop frameworks that synthesize realistic datasets pertaining to the power distribution system [4,13,16,17,27,28]. These frameworks create digital replicates similar to the actual power distribution networks in terms of their structure and function. Hence the created networks can be used as *digital duplicates* in simulation studies of policies and methods before implementation in real systems.

The algorithms associated with these frameworks vary widely—ranging from first principles based approaches [17,27] to learning statistical distributions of network attributes [28] to using deep learning models such as generative adversarial neural networks [14]. Validating the synthetic power distribution networks with respect to their physical counterpart is vital for assessing the suitability of their use as effective digital duplicates. Since the underlying assumptions and algorithms of each framework are distinct from each other, some of them may excel compared to others in reproducing digital replicates with better precision for selective regions. To this end, we require well-defined metrics to rank the frameworks and judge their strengths and weaknesses in generating digital duplicates of power distribution networks for a particular geographic region.

The literature pertaining to frameworks for synthetic distribution network creation include certain validation results that compare the generated networks to the actual counterpart [4,12,28]. But the validation results are mostly limited to comparing the statistical network attributes such as degree and hop distributions and power engineering operational attributes such as node voltages and edge power flows. Since power distribution networks represent real physical systems, the created digital replicates have associated geographic embedding. Therefore, a structural comparison of synthetic network graphs to their actual counterpart becomes pertinent for power distribution networks with geographic embedding. Consider an example where a digital twin is used to analyze impact of a weather event [26]. Severe weather events such as hurricanes, earthquakes and wild fires occur in specific geographic trajectories, affecting only portions of societal infrastructures. In order to correctly identify them during simulations, the digital twin should structurally resemble the actual infrastructure.

Problem Statement. In recent years, the problem of evaluating quality of reconstructed networks has been studied for street maps. Certain metrics were defined to compare outputs of frameworks that use GPS trajectory data to reconstruct street map graphs [1,2]. The abstract problem can be stated as follows: *compute the similarity between a given pair of embedded planar graphs.* This is similar to the well known subgraph isomorphism problem [7] wherein we look for isomorphic subgraphs in a pair of given graphs. A major precursor to this problem is that we require a one-to-one mapping between nodes and edges of the two graphs. While such mappings are well-defined for street networks, the same cannot be inferred for power distribution networks. Since power network datasets are proprietary, the node and edge labels are redacted from the network before it is shared. The actual network is obtained as a set of "drawings" with associated geographic embeddings. Each drawing can be considered as a collection of line segments termed a *geometry.* Hence the problem of comparing a set of power distribution networks with geographic embedding can be stated as the following: *compute the similarity between a given pair of geometries lying on a geographic plane.*

Our Contributions. We propose a new distance measure to compare a pair of geometries using the *flat norm*, a notion of distance between generalized objects studied in geometric measure theory [9,20]. This distance combines the difference in length of the geometries with the area of the patches contained between them. The area of patches in between the pair of geometries accounts for the lateral displacement between them. We employ a *multiscale* version of the flat norm [21] that uses a scale parameter $\lambda \geq 0$ to combine the length and area components (for the sake of brevity, we refer to the multiscale flat norm simply as the flat norm). Intuitively, a smaller value of λ captures larger patches of area between the geometries while a large value of λ captures more of the (differences in) lengths of the geometries. Computing the flat norm over a range of values of λ allows us to compare the geometries at multiple scales. For computation, we use a discretized version of the flat norm defined on simplicial complexes [10], which are triangulations in our case. A lack of one-to-one correspondence between edges and nodes in the pair of networks prevents us from performing one-to-one comparison of edges. Instead we can sample random regions in the area of interest and compare the pair of geometries within each region. For performing such local comparisons, we define a *normalized flat norm* where we normalize the flat norm distance between the parts of the two geometries by the sum of the lengths of the two parts in the region. Such comparison enables us to characterize the quality of the digital duplicate for the sampled region. Further, such comparisons over a sequence of sampled regions allows us to characterize the suitability of using the entire synthetic network as a duplicate of the actual network.

Our main **contributions** are the following: (i) we propose a distance measure for comparing a pair of geometries embedded in the same plane using the flat norm that accounts for deviation in length and lateral displacement between the geometries; and (ii) we perform a region-based characterization of synthetic networks by sampling random regions and comparing the pair of geometries contained within the sampled region. The proposed distance allows us to perform a global as well as local comparison between a pair of network geometries.

Related Work. Several well defined graph structure comparison metrics such as subgraph isomorphism and edit distance have been proposed in the literature along with algorithms to compute them efficiently. Tantardini et al. [30] compare graph network structures for the entire graph (global comparison) as well as for small portions of the graph known as motifs (local comparison). Other researchers have proposed methodologies to identify structural similarities in embedded graphs [3,22]. However, all these methods depend on one-to-one correspondence of graph nodes and edges rather than considering the node and edge geometries of the graphs. The edit distance, i.e., the minimum number of edit operations to transform one network to the other, has been widely used to compare networks having structural properties [24,25,31]. Riba et al. [25] used the Hausdorff distance between nodes in the network to compare network geometries. Majhi et al. [15] modified the traditional definition of graph edit distance to be applicable in the context of "geometric graphs" embedded in a Euclidean space. Along with the usual insertion and deletion operations, the authors have proposed a cost for translation in computing the geometric edit distance between the graphs. However, the authors also show that the problem of computing this metric is $\mathcal{NP}$-hard.

Meyur et al. [18] compared network geometries using the Hausdorff distance after partitioning the geographic region into small rectangular grids and comparing the geometries for each grid. However, the Hausdorff metric is sensitive to outliers as it focuses only on the maximum possible distance between the pair of geometries. When the geometries coincide almost entirely except in a few small portions, the Hausdorff metric still records the discrepancy in those small portions without accounting for the similarity over the majority of portions. The similar approach used by Brovelli et al. [5] to compare a pair of road networks in a geographic region suffers from the same drawback. This necessitates a well-defined distance metric between networks with geographic embedding [2].

Several comparison methods have been proposed in the context of planar graphs embedded in a Euclidean space [6,19]. They include local and global metrics to compare road networks. The local metrics characterize the networks based on cliques and motifs, while the global metrics involve computing the *efficiency* of constructing the infrastructure network. The most efficient network is assumed to be the one with only straight line geometries connecting node pairs. Albeit useful to characterize network structures, these methods are not suitable for a numeric comparison of network geometries.

2 Methods

Following Mahji and Wenk [15], we use the term *geometric graph* to define network graphs embedded in a Euclidean space. Next, we define what we mean by structurally similar geometric graphs.

Definition 1 (Geometric graph). *A graph $\mathcal{G}(\mathcal{V}, \mathcal{E})$ with node set $\mathcal{V}$ and edge set $\mathcal{E}$ is said to be a geometric graph of $\mathbb{R}^d$ if the set of nodes $\mathcal{V} \subset \mathbb{R}^d$ and the edges are Euclidean straight line segments $\{\overline{uv} \mid e := (u, v) \in \mathcal{E}\}$ which intersect (possibly) at their endpoints.*

Definition 2 (Structurally similar geometric graphs). *Two geometric graphs* $\mathcal{G}_0(\mathcal{V}_0, \mathcal{E}_0)$ *and* $\mathcal{G}_1(\mathcal{V}_1, \mathcal{E}_1)$ *are said to be* structurally similar *at the level of* $\delta \geq 0$, *termed* δ*-similar, if* $\text{dist}(\mathcal{G}_0, \mathcal{G}_1) \leq \delta$ *for the distance function* dist *between the two graphs.*

We could consider a given network as a set of edge geometries. Hence we could consider the problem of comparing geometric graphs $\mathcal{G}_0$ and $\mathcal{G}_1$ as that of comparing the set of edge geometries $\mathcal{E}_0$ and $\mathcal{E}_1$. In this paper, we propose a suitable distance that allows us to compare between a pair of geometric graphs or a pair of geometries. We use the multiscale flat norm, which has been well explored in the field of geometric measure theory, to define such a distance between the geometries.

The other aspect of this paper is to identify a suitable threshold δ for inferring the structural similarity of a pair of geometric graphs. But there is no general method to choose the threshold. Here, we perform a statistical analysis for our particular case of comparing power distribution networks. We validate the comparison results with visual inspection to conclude that the proposed metric serves its purpose to identify structurally similar geometric graphs.

2.1 Multiscale Flat Norm

We use the multiscale simplicial flat norm proposed by Ibrahim et al. [10] to compute the distance between two networks. We now introduce some background for this computation. A d-dimensional *current* T (referred to as a d-current) is a generalized d-dimensional geometric object with orientations (or direction) and multiplicities (or magnitude). An example of a 2-current is a surface with finite area (multiplicity) and a specific orientation (clockwise or counterclockwise). The boundary of T, denoted by ∂T, is a $(d-1)$-current. The *multiscale flat norm* of a d-current T at scale $\lambda \geq 0$ is defined as

$$\mathbb{F}_\lambda(T) = \min_S \{V_d(T - \partial S) + \lambda V_{d+1}(S)\}, \quad (1)$$

where the minimum is taken over all $(d+1)$-currents S, and V_d denotes the d-dimensional *volume*, e.g., length in 1D or area in 2D. Computing the flat norm of a 1-current (curve) T identifies the optimal 2-current (area patches) S that minimizes the sum of the length of current $T - \partial S$ and the area of patch(es) S. Figure 1 shows the flat norm computation for a generic 1D current T (blue). The 2D area patches S (magenta) are computed such that the expression in Eq. (1) is minimized for the chosen value of λ that ends up using most of the patch under the sharper spike on the left but only a small portion of the patch under the wider bump to the right.

The scale parameter λ can be intuitively understood as follows. Rolling a ball of radius $1/\lambda$ on the 1-current T traces the output current $T - \partial S$ and the untraced regions constitute the patches S. Hence we observe that for a large λ, the radius of the ball is very small and hence it traces major features while smoothing out (i.e., missing) only minor features (wiggles) of the input current. But for a small λ, the ball with a large radius smoothes out larger scale features

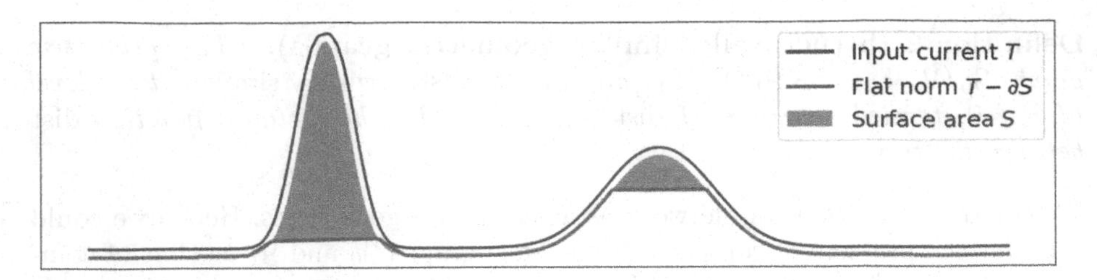

Fig. 1. Multiscale flat norm of a 1D current T (blue). The flat norm is the sum of length of the resulting 1D current $T - \partial S$ (green) and the area of 2D patches S (magenta). We show $T - \partial S$ slightly separated for easy visualization. (Color figure online)

(bumps) in the current. Note that for smaller λ, the cost of area patches is smaller in the minimization function and hence more patches are used for computing the flat norm. We can use the flat norm to define a natural distance between a pair of 1-currents T_1 and T_2 as follows [10].

$$\mathbb{F}_\lambda (T_1, T_2) = \mathbb{F}_\lambda (T_1 - T_2) \tag{2}$$

We compute the flat norm distance between a pair of input geometries (synthetic and actual) as the flat norm of the current $T = T_1 - T_2$ where T_1 and T_2 are the currents corresponding to individual geometries. Let Σ denote the set of all line segments in the input current T. We perform a constrained triangulation of Σ to obtain a 2-dimensional finite oriented simplicial complex K. A constrained triangulation ensures that each line segment $\sigma_i \in \Sigma$ is an edge in K, and that T is an oriented 1-dimensional subcomplex of K.

Let m and n denote the numbers of edges and triangles in K. We can denote the input current T as a 1-chain $\sum_{i=1}^{m} t_i \sigma_i$ where σ_i denotes an edge in K and t_i is the corresponding multiplicity. Note that $t_i = -1$ indicates that orientation of σ_i and T are opposite, $t_i = 0$ denotes that σ_i is not contained in T, and $t_i = 1$ implies that σ_i is oriented the same way as T. Similarly, we define the set S to be the 2-chain of K and denote it by $\sum_{i=1}^{m} s_i \omega_i$ where ω_i denotes a 2-simplex in K and s_i is the corresponding multiplicity.

The boundary matrix $[\partial] \in \mathbb{Z}^{m \times n}$ captures the intersection of the 1 and 2-simplices of K. The entries of the boundary matrix $[\partial]_{ij} \in \{-1, 0, 1\}$. If edge σ_i is a face of triangle ω_j, then $[\partial]_{ij}$ is nonzero and it is zero otherwise. The entry is -1 if the orientations of σ_i and ω_j are opposite and it is $+1$ if the orientations agree.

We can respectively stack the t_i's and s_i's in m and n-length vectors $\mathbf{t} \in \mathbb{Z}^m$ and $\mathbf{s} \in \mathbb{Z}^n$. The 1-chain representing $T - \partial S$ is denoted by $\mathbf{x} \in \mathbb{Z}^m$ and is given as$\mathbf{x} = \mathbf{t} - [\partial]\, \mathbf{s}$. The multiscale flat norm defined in Eq. (1) can be computed by solving the following optimization problem:

$$\begin{aligned} \mathbb{F}_\lambda (T) = \min_{\mathbf{s} \in \mathbb{Z}^n} \sum_{i=1}^{m} w_i \left| x_i \right| + \lambda \left(\sum_{j=1}^{n} v_j \left| s_j \right| \right) \\ \text{s.t.} \quad \mathbf{x} = \mathbf{t} - [\partial]\, \mathbf{s}, \quad \mathbf{x} \in \mathbb{Z}^m, \end{aligned} \tag{3}$$

where $V_d(\tau)$ in Eq. (1) denotes the volume of the d-dimensional simplex τ. We denote volume of the edge σ_i as $V_1(\sigma_i) = w_i$ and set it to be the Euclidean length, and volume of a triangle τ_j as $V_2(\tau_j) = v_j$ and set it to be the area of the triangle.

In this work, we consider geometric graphs embedded on the geographic plane and are associated with longitude and latitude coordinates. We compute the Euclidean length of edge σ_i as $w_i = R\Delta\phi_i$ where $\Delta\phi_i$ is the Euclidean normed distance between the geographic coordinates of the terminals of σ_i and R is the radius of the earth. Similarly, the area of triangle τ_j is computed as $v_j = R^2\Delta\Omega_j$ where $\Delta\Omega_j$ is the solid angle subtended by the geographic coordinates of the vertices of τ_j.

Using the fact that the objective function is piecewise linear in $\mathbf{x}$ and $\mathbf{s}$, the minimization problem can be reformulated as an integer linear program (ILP) as follows:

$$\mathbb{F}_\lambda(T) = \min \sum_{i=1}^{m} w_i\left(x_i^+ + x_i^-\right) + \lambda\left(\sum_{j=1}^{n} v_j\left(s_j^+ + s_j^-\right)\right) \tag{4a}$$

$$\text{s.t.} \quad \mathbf{x}^+ - \mathbf{x}^- = \mathbf{t} - [\partial]\left(\mathbf{s}^+ - \mathbf{s}^-\right) \tag{4b}$$

$$\mathbf{x}^+, \mathbf{x}^- \geq 0, \quad \mathbf{s}^+, \mathbf{s}^- \geq 0 \tag{4c}$$

$$\mathbf{x}^+, \mathbf{x}^- \in \mathbb{Z}^m, \quad \mathbf{s}^+, \mathbf{s}^- \in \mathbb{Z}^n \tag{4d}$$

The linear programming relaxation of the ILP in Eq. (4) is obtained by ignoring the integer constraints Eq. (4d). We refer to this relaxed linear program (LP) as the ***flat norm LP***. Ibrahim et al. [10] showed that the boundary matrix $[\partial]$ is totally unimodular for our application setting. Hence the flat norm LP will solve the ILP, and hence the flat norm can be computed in polynomial time.

2.2 Proposed Algorithm

Algorithm 1 describes how we compute the distance between a pair of geometries with the associated embedding on a metric space $\mathcal{M}$. We assume that the geometries (networks) $\mathcal{G}_1(\mathcal{V}_1, \mathcal{E}_1)$ and $\mathcal{G}_2(\mathcal{V}_2, \mathcal{E}_2)$ with respective node sets $\mathcal{V}_1, \mathcal{V}_2$ and edge sets $\mathcal{E}_1, \mathcal{E}_2$ have no one-to-one correspondence between the $\mathcal{V}_i$'s or $\mathcal{E}_i$'s. Note that each vertex $v \in \mathcal{V}_1, \mathcal{V}_2$ is a point and each edge $e \in \mathcal{E}_1, \mathcal{E}_2$ is a straight line segment in $\mathcal{M}$. We consider the collection of edges $\mathcal{E}_1, \mathcal{E}_2$ as input to our algorithm. First, we orient the edge geometries in a particular direction (left to right in our case) to define the currents T_1 and T_2, which have both magnitude and direction. Next, we consider the bounding rectangle $\mathcal{E}_{\text{bound}}$ for the edge geometries and define the set Σ to be triangulated as the set of all edges in either geometry and the bounding rectangle. We perform a constrained Delaunay triangulation [29] on the set Σ to construct the 2-dimensional simplicial complex K. The constrained triangulation ensures that the set of edges in Σ is included in the simplicial complex K. Then we define the currents T_1 and T_2 corresponding to the respective edge geometries $\mathcal{E}_1$ and $\mathcal{E}_2$ as 1-chains in K. Finally, the flat norm LP is solved to compute the simplicial flat norm.

Algorithm 1: Distance between a pair of geometries

Input: Geometries $\mathcal{E}_1, \mathcal{E}_2$

Parameter: Scale λ

1: Orient each edge in the edge sets from left to right:
$\tilde{\mathcal{E}}_1 := \mathsf{Orient}(\mathcal{E}_1)$; $\tilde{\mathcal{E}}_2 := \mathsf{Orient}(\mathcal{E}_2)$.

2: Find bounding rectangle for the pair of geometries: $\mathcal{E}_{\text{bound}} = \mathsf{rect}\left(\tilde{\mathcal{E}}_1, \tilde{\mathcal{E}}_2\right)$.

3: Define the set of line segments to be triangulated: $\Sigma = \tilde{\mathcal{E}}_1 \cup \tilde{\mathcal{E}}_2 \cup \mathcal{E}_{\text{bound}}$.

4: Perform constrained triangulation on set Σ to construct 2-dimensional simplicial complex K.

5: Define the currents T_1, T_2 as 1-chains of oriented edges $\tilde{\mathcal{E}}_1$ and $\tilde{\mathcal{E}}_2$ in K.

6: Solve the flat norm LP to compute flat norm $\mathbb{F}_\lambda(T_1 - T_2)$.

Output: Flat norm distance $\mathbb{F}_\lambda(T_1 - T_2)$.

2.3 Normalized Flat Norm

Recall that in our context of synthetic power distribution networks, the primary goal of comparing a synthetic network to its actual counterpart is to infer the quality of the replica or the *digital duplicate* synthesized by the framework. The proposed approach using the flat norm for structural comparison of a pair of geometries provides us a method to perform global as well as local comparison. While we can produce a global comparison by computing the flat norm distance between the two networks, it may not provide us with complete information on the quality of the synthetic replicate. On the other hand, a local comparison can provide us details about the framework generating the synthetic networks. For example, a synthetic network generation framework might produce higher quality digital replicates of actual power distribution networks for urban regions as compared to rural areas. A local comparison highlights this attribute and identifies potential use case scenarios of a given synthetic network generation framework.

Furthermore, availability of actual power distribution network data is sparse due to its proprietary nature. We may not be able to produce a global comparison between two networks due to unavailability of network data from one of the sources. Hence, we want to restrict our comparison to only the portions in the region where data from either network is available, which also necessitates a local comparison between the networks.

For a local comparison, we consider uniform sized regions and compute the flat norm distance between the pair of geometries within the region. However, the computed flat norm is dependent on the length of edges present within the region from either network. Hence we define the *normalized* multiscale flat norm, denoted by $\widetilde{\mathbb{F}}_\lambda$, for a given region as

$$\widetilde{\mathbb{F}}_\lambda(T_1 - T_2) = \frac{\mathbb{F}_\lambda(T_1 - T_2)}{|T_1| + |T_2|}. \tag{5}$$

For a given parameter ϵ, a local region is defined as a square of size $2\epsilon \times 2\epsilon$ steradians. Let $T_{1,\epsilon}$ and $T_{2,\epsilon}$ denote the currents representing the input geometries inside the local region characterized by ϵ. Note that the "amount" or the total length of network geometries within a square region varies depending on the location of the local region. In this case, the lengths of the network geometries are respectively $|T_{1,\epsilon}|$ and $|T_{2,\epsilon}|$. Therefore, we use the ratio of the total length of network geometries inside a square region to the parameter ϵ to characterize this "amount" and denote it by $|T|/\epsilon$ where

$$|T|/\epsilon = \frac{|T_{1,\epsilon}| + |T_{2,\epsilon}|}{\epsilon}. \tag{6}$$

Note that while performing a comparison between a pair of network geometries in a local region using the multiscale flat norm, we need to ensure that comparison is performed for similar length of the networks inside similar regions. Therefore, the ratio $|T|/\epsilon$, which indicates the length of networks inside a region scaled to the size of the region, becomes an important aspect of characterization while performing the flat norm based comparison.

3 Results and Discussion

We use the proposed multiscale flat norm to compare a pair of network geometries from power distribution networks for a region in a county in USA. The two networks considered are the actual power distribution network for the region and the synthetic network generated using the methodology proposed by Meyur et al. [17]. We provide a brief overview of these networks.

Actual Network. The actual power distribution network was obtained from the power company serving the location. Due to its proprietary nature, node and edge labels were redacted from the shared data. Further, the networks were shared as a set of handmade drawings, many of which had not been drawn to a well-defined scale. We digitized the drawings by overlaying them on OpenStreetMaps [23] and georeferencing to particular points of interest [8]. Geometries corresponding to the actual network edges are obtained as shape files.

Synthetic Network. The synthetic power distribution network is generated using a framework with the underlying assumption that the network follows the road network infrastructure to a significant extent [17]. To this end, the residences are connected to local pole top transformers located along the road network to construct the low voltage (LV) secondary distribution network. The local transformers are then connected to the power substation following the road network leading to the medium voltage (MV) primary distribution network. That is, the primary network edges are chosen from the underlying road infrastructure network such that the structural and power engineering constraints are satisfied.

3.1 Comparing Network Geometries

The primary goal of computing the flat norm is to compare the pair of input geometries. As mentioned earlier, the flat norm provides an accurate measure of

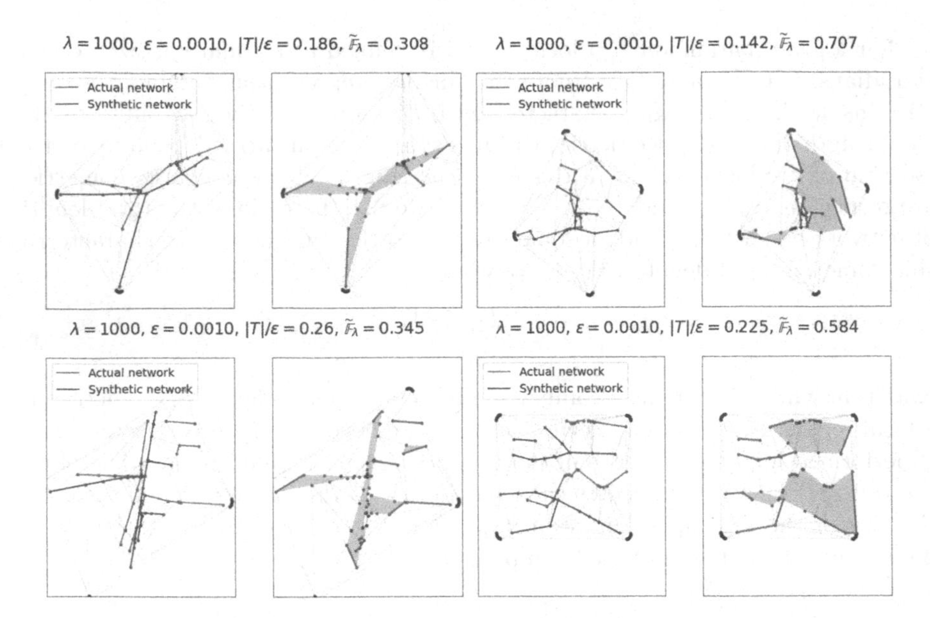

Fig. 2. Normalized flat norm (with scale $\lambda = 1000$) distances for pairs of regions in the network of same size ($\epsilon = 0.001$) with similar $|T|/\epsilon$ ratios (two pairs each in the top and bottom rows). The pairs of geometries for the first plot (on left) are quite similar, which is reflected in the low flat norm distances between them. The network geometries on the right plots are more dissimilar and hence the flat norm distances are high.

difference between the geometries by considering both the length deviation and area patches in between the geometries. Further, we normalize the computed flat norm to the total length of the geometries. In this section, we show examples where we computed the normalized flat norm for the pair of network geometries (actual and synthetic) for a few regions.

The top two plots in Fig. 2 show two regions characterized by $\epsilon = 0.001$ and almost similar $|T|/\epsilon$ ratios. This indicates that the length of network scaled to the region size is almost equal for the two regions. From a mere visual perspective, we can conclude that the first pair of network geometries resemble each other where as the second pair are fairly different. This is further validated from the results of the flat norm distance between the network geometries computed with the scale $\lambda = 1000$, since the first case produces a smaller flat norm distance compared to the latter. The bottom two plots show another example of two regions with almost similar $|T|/\epsilon$ ratios and enable us to infer similar conclusions. The results strengthens our case of using flat norm as an appropriate measure to perform a local comparison of network geometries.

3.2 Comparison of Flat Norm and Hausdorff Distance Metrics

In this section, we compare the proposed flat norm metric for structural comparison with the Hausdorff distance metric which has been extensively used in the literature for similar purposes. The Hausdorff distance is considered to be a *stable* metric since minor perturbations to the geometries do not affect the metric. While this property is advantageous when we are dealing with noisy data, this fails to capture structural differences unless they are significantly large.

The comparison metric is said to be "stable" if the computed normalized flat norm for the pair of perturbed network geometries is close to the normalized flat norm of the unperturbed geometries. A perturbed network is similar to the original network with only the geographic embeddings of the nodes perturbed. To this end, we consider a circular region around each node in the network by defining a perturbation radius (in meters). We then uniformly sample a point in each circular region and use them as the perturbed embeddings of the nodes. We compare the stability of the flat norm metric with the Hausdorff metric—both empirically on our sample networks and using a simple theoretical example.

We consider two simple curves T_1, T_2 in the plane whose end points are the same (see Fig. 3). We perturb T_1 within a small neighborhood of each point on it while keeping T_2 fixed and the end points of both curves also fixed. Hence we consider perturbed versions $\widetilde{T}_1$ that lie within an ε-tube of T_1. We could have cases where $\widetilde{T}_1$ lies mostly at the upper envelope of this ε-tube, or mostly at the lower envelope. In both cases, one would expect the distance between $\widetilde{T}_1$ and T_2 to also change significantly (from that between T_1 and T_2). The flat norm distance accurately captures all such changes (to keep the example simple, we consider the default flat norm distance and not the normalized version). At the same time, both such variations could have the same Hausdorff distance H from T_2 as T_1, which completely misses all the changes to T_1 in either case.

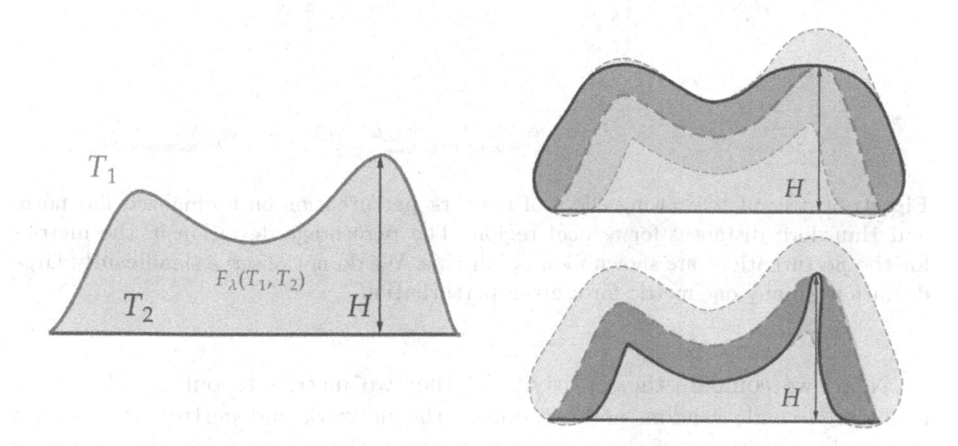

Fig. 3. Left: Curves T_1 and T_2 with shared end points, and their flat norm distance $\mathbb{F}_\lambda(T_1, T_2)$ and Hausdorff distance H. **Right:** Two perturbed versions $\widetilde{T}_1$ (solid blue) of T_1 (now in dashed blue) that lie within an ε-neighborhood of T_1. The Hausdorff distance between $\widetilde{T}_1$ and T_2 remains same, i.e., H. But the flat norm distance increased in the first case (Right, top) as captured by the green patch, and it decreased in the second case (Right, Bottom) by the area shown in pink.

A modification of this example can illustrate the other extreme case—when Hausdorff distance changes by a lot but the flat norm distance does not change much at all. Consider moving *only* the highest point on T_1 further up so that Hausdorff distance becomes $2H$ (this is the point on T_1 in the left figure in Fig. 3 at the arrow indicating the Hausdorff distance H). We keep T_1 still a connected curve, thus creating a sharp spike in it. While the Hausdorff distance between the curves has doubled, the flat norm distance sees only a minute increase as measured by the tiny area under this spike. Once again, the flat norm distance accurately captures the intuition that the curves have *not* changed much when just a single point moves away while the rest of the curve stays the same. Hence the flat norm is a more robust metric that always captures significant changes while maintaining stability to small perturbations.

We observe similar behavior to those illustrated by the theoretical example (Fig. 3) in our computational experiments. Figure 4 shows scatter plots denoting empirical distribution of percentage deviation of the two metrics from the original values $\left(\%\Delta\mathbb{D}_{\text{Haus}}, \%\Delta\widetilde{\mathbb{F}}_{\lambda}\right)$ for a local region. The perturbations are considered for three different radii shown in separate plots. We note that the percentage deviations in the two metrics are comparable in most cases. In other words, neither metric behaves abnormally for a small perturbation in one of the networks.

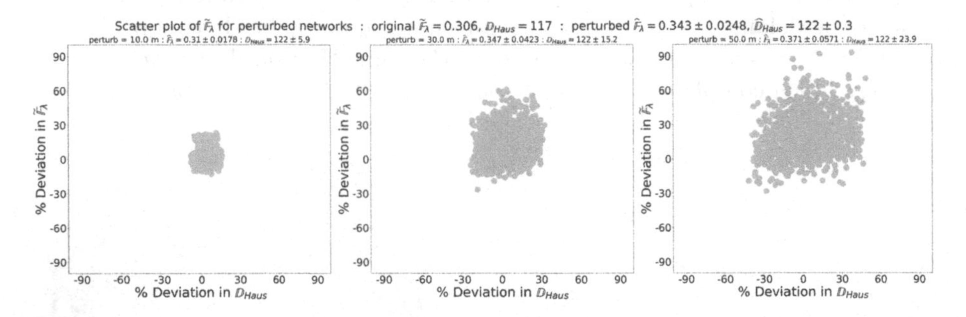

Fig. 4. Scatter plots showing effect of network perturbation on normalized flat norm and Hausdorff distances for a local region. The percentage deviation in the metrics for the perturbations are shown along each axis. We do not observe significantly large deviations in any one metric for a given perturbation.

Next, we compare the sensitivity of the two metrics to outliers. Here, we consider a single random node in one of the network and perturb it. Figure 5 shows the sensitivity of the metrics to these outliers. The original normalized flat norm and Hausdorff distance metrics are shown by the horizontal and vertical dashed lines respectively. The points along the horizontal dashed line denote the cases where Hausdorff distance metric is more sensitive to the outliers, while the normalized flat norm metric remains the same. These cases occur when the perturbed random node determines the Hausdorff distance, similar to the theoretical case (where Hausdorff distance went from H to $2H$). On the flip

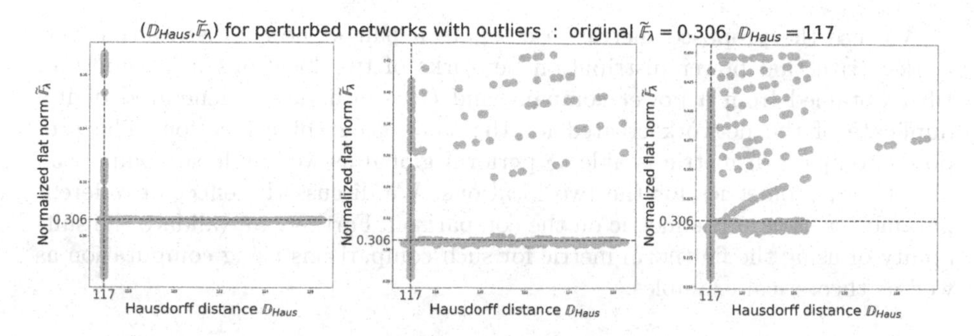

Fig. 5. Scatter plots showing effect of few outliers on normalized flat norm and Hausdorff distance for a local region. The original normalized flat norm and Hausdorff distance are highlighted by the dashed horizontal and vertical lines. We observe multiple cases where the Hausdorff distance is more sensitive to outliers compared to the proposed normalized flat norm metric.

side, the points along the vertical dashed line denote the Hausdorff distance remaining unchanged while the normalized flat norm metric shows variation. Just as in the theoretical example (Fig. 3), such variation in the normalized flat norm metric implies a variation in the network structure. However, such variation is not captured by the Hausdorff distance metric. Hence, our proposed metric is capable of identifying structural differences due to perturbations, while remaining stable when widely separated nodes (which are involved in Hausdorff distance computation) are perturbed. The other points which are neither on the horizontal nor on the vertical dashed lines indicate that either metric is able to identify the structural variation due to the perturbation.

4 Conclusions

We have proposed a fairly general metric to compare a pair of network geometries embedded on the same plane. Unlike standard approaches that map the geometries to points in a possibly simpler space and then measuring distance between those points [11], or comparing "signatures" for the geometries, our metric works directly in the input space and hence allows us to capture all details in the input. The metric uses the multiscale flat norm from geometric measure theory, and can be used in more general settings as long as we can triangulate the region containing the two geometries. It is impossible to derive *standard* stability results for this distance measure that imply only small changes in the flat norm metric when the inputs change by small amount—there is no alternative metric to measure the *small change in the input*. For instance, our theoretical example (in Fig. 3) shows that the commonly used Hausdorff metric cannot be used for this purpose. At the same time, we do get natural stability results for our distance following the properties of the flat norm—small changes in the input geometries lead to only small changes in the flat norm distance between them [9,20].

We use the proposed metric to compare a pair of power distribution networks: (i) actual power distribution networks of two locations in a county of USA obtained from a power company and (ii) synthetically generated digital duplicate of the network created for the same geographic location. The proposed comparison metric is able to perform global as well as local comparison of network geometries for the two locations. We discuss the effect of different parameters used in the metric on the comparison. Further, we validate the suitability of using the flat norm metric for such comparisons using computation as well as theoretical examples.

References

1. Ahmed, M., Fasy, B.T., Hickmann, K.S., Wenk, C.: A path-based distance for street map comparison. ACM Trans. Spat. Algor. Syst. **1**(1) (2015)
2. Ahmed, M., Fasy, B.T., Wenk, C.: Local persistent homology based distance between maps. In: Proceedings of the 22nd ACM SIGSPATIAL International Conference on Advances in Geographic Information Systems, pp. 43–52. SIGSPATIAL 2014. Association for Computing Machinery, New York, NY, USA (2014)
3. Bai, Y., Ding, H., Bian, S., Chen, T., Sun, Y., Wang, W.: Simgnn: a neural network approach to fast graph similarity computation. https://arxiv.org/abs/1808.05689 (2018). Accessed 26 Sept 2022
4. Bidel, A., Schelo, T., Hamacher, T.: Synthetic distribution grid generation based on high resolution spatial data. In: 2021 IEEE International Conference on Environment and Electrical Engineering and 2021 IEEE Industrial and Commercial Power Systems Europe (EEEIC/ICPS Europe), pp. 1–6. IEEE, Bari, Italy (2021)
5. Brovelli, M.A., Minghini, M., Molinari, M., Mooney, P.: Towards an automated comparison of openstreetmap with authoritative road datasets. Trans. GIS **21**(2), 191–206 (2017)
6. Cardillo, A., Scellato, S., Latora, V., Porta, S.: Structural properties of planar graphs of urban street patterns. Phys. Rev. E **73**, 066107 (2006)
7. Eppstein, D.: Subgraph isomorphism in planar graphs and related problems. In: Proceedings of the Sixth Annual ACM-SIAM Symposium on Discrete Algorithms, pp. 632–640. SODA 1995, Society for Industrial and Applied Mathematics, USA (1995)
8. ESRI: Georeferencing a raster to a vector (2022). Accessed 26 Sept 2022
9. Federer, H.: Geometric Measure Theory. Die Grundlehren der mathematischen Wissenschaften, Band 153. Springer-Verlag, New York (1969)
10. Ibrahim, S., Krishnamoorthy, B., Vixie, K.R.: Simplicial flat norm with scale. J. Comput. Geometry **4**(1), 133–159 (2013)
11. Kendall, D.G., Barden, D.M., Carne, T., Le, H.: Shape and Shape Theory. Wiley Series in Probability and Statistics, Wiley, Hoboken, NJ, USA (2009)
12. Krishnan, V., et al.: Validation of synthetic U.S. electric power distribution system data sets. IEEE Trans. Smart Grid **11**(5), 4477–4489 (2020)
13. Li, H., et al.: Building highly detailed synthetic electric grid data sets for combined transmission and distribution systems. IEEE Open Access J. Power Energy **7**, 478–488 (2020)
14. Liang, M., Meng, Y., Wang, J., Lubkeman, D.L., Lu, N.: Feedergan: synthetic feeder generation via deep graph adversarial nets. IEEE Trans. Smart Grid **12**(2), 1163–1173 (2021)

15. Majhi, S., Wenk, C.: Distance measures for geometric graphs. https://arxiv.org/abs/2209.12869 (2022). Accessed 26 Sept 2022
16. Mateo, C., et al.: Building large-scale u.s. synthetic electric distribution system models. IEEE Trans. Smart Grid **11**(6), 5301–5313 (2020)
17. Meyur, R., et al.: Creating realistic power distribution networks using interdependent road infrastructure. In: 2020 IEEE International Conference on Big Data (Big Data), pp. 1226–1235. IEEE, Atlanta, GA, USA (2020)
18. Meyur, R., et al.: Ensembles of realistic power distribution networks. Proc. Nat. Acad. Sci. **119**(26), e2123355119 (2022)
19. Morer, I., Cardillo, A., Díaz-Guilera, A., Prignano, L., Lozano, S.: Comparing spatial networks: a one-size-fits-all efficiency-driven approach. Phys. Rev. E **101**, 042301 (2020)
20. Morgan, F.: Geometric Measure Theory: A Beginner's Guide. Academic Press, Cambridge, MA, USA, fourth edn (2008)
21. Morgan, S.P., Vixie, K.R.: L^1TV computes the flat norm for boundaries. Abstr. Appl. Anal. 45153, 14 (2007)
22. Ok, S.: A graph similarity for deep learning. In: Larochelle, H., Ranzato, M., Hadsell, R., Balcan, M., Lin, H. (eds.) Advances in Neural Information Processing Systems, vol. 33, pp. 1–12. Curran Associates Inc, Vancouver, BC, Canada (2020)
23. Open Street Map Foundation: Open Street Maps (2022). Accessed 20 Aug 2022
24. Paaβen, B.: Revisiting the tree edit distance and its backtracing: A tutorial. https://arxiv.org/abs/1805.06869 (2022). Accessed 26 Sept 2022
25. Riba, P., Fischer, A., Lladós, J., Fornés, A.: Learning graph edit distance by graph neural networks. Pattern Recogn. **120**, 108132 (2021)
26. Roy, K.C., Hasan, S., Culotta, A., Eluru, N.: Predicting traffic demand during hurricane evacuation using real-time data from transportation systems and social media. Transp. Res. Part C Emerg. Technol. **131**, 103339 (2021)
27. Saha, S.S., Schweitzer, E., Scaglione, A., Johnson, N.G.: A framework for generating synthetic distribution feeders using openstreetmap. In: 2019 North American Power Symposium (NAPS), pp. 1–6. IEEE, Wichita, KS, USA (2019)
28. Schweitzer, E., Scaglione, A., Monti, A., Pagani, G.A.: Automated generation algorithm for synthetic medium voltage radial distribution systems. IEEE J. Emerg. Select. Top. Circ. Syst. **7**(2), 271–284 (2017)
29. Si, H.: Constrained delaunay tetrahedral mesh generation and refinement. Finite Elem. Anal. Des. **46**, 33–46 (2010)
30. Tantardini, M., Ieva, F., Tajoli, L., Piccardi, C.: Comparing methods for comparing networks. Sci. Rep. **9**(1), 17557 (2019)
31. Xu, H.: An algorithm for comparing similarity between two trees. https://arxiv.org/abs/1508.03381 (2015). Accessed 26 Sept 2022

OptICS-EV: A Data-Driven Model for Optimal Installation of Charging Stations for Electric Vehicles

Kazi Ashik Islam(✉), Rounak Meyur, Aparna Kishore, Swapna Thorve, Da Qi Chen, and Madhav Marathe

Biocomplexity Institute, University of Virginia, Charlottesville, USA
{ki5hd,rm5nz,ak8mj,st6ua,wny7gj,marathe}@virginia.edu

Abstract. As the demand for electric vehicles continues to surge worldwide, it becomes increasingly imperative for the government to plan and anticipate its practical impact on society. In particular, any city/state needs to guarantee sufficient and proper placement of charging stations to service all current/future electric vehicle adopters. Furthermore, it needs to consider the inevitable additional strain these charging stations put on the existing power grid. In this paper, we use data-driven models to address these issues by providing an algorithm that finds optimal placement and connections of electric vehicle charging stations in the state of Virginia. Specifically, we found it suffices to build 10,733 additional charging stations to cover 75% of the population within 0.33 miles (and everyone within 2.5 miles). We also show optimally connecting the stations to the power grid significantly improves the stability of the network. Additionally, we study 1) the trade-off between the average distance a driver needs to travel to their nearest charging station versus the number of stations to build, and 2) the impact on the grid under various adoption rates. These studies provide further insight into various tools policymakers can use to prepare for the evolving future.

Keywords: Electric Vehicle · Charging Station Placement · Power Grid

1 Introduction

The transportation sector is responsible for 17% of the total GHG emissions, of which 41% of emissions come from passenger cars[1]. Thus, reducing carbon footprint has become a critical goal in the transportation domain. Electric vehicles

[1] https://www.statista.com/statistics/1185535/transport-carbon-dioxide-emissions-breakdown/.

This work was partially supported by University of Virginia Strategic Investment Fund SIE160, NSF EAGER CMMI-1745207, NSF Grant OAC-1916805, NSF BIGDATA INS-1633028, and NSF CINES-1916805.

K. A. Islam and R. Meyur—These authors contributed equally to this work.

J. Mikyška et al. (Eds.): ICCS 2023, LNCS 14076, pp. 70–85, 2023.
https://doi.org/10.1007/978-3-031-36027-5_6

(EV) are a robust solution to addressing this problem given their eco-friendly characteristics. In recent years, the U.S. has witnessed widespread adoption of EVs in the residential sector[2]. Multiple incentives, policy changes, rising fuel prices, and improvements in the range of EVs are some of the influential factors for a rapid increase in EV adoption. Thus, many entities such as urban planners, government, and utilities are increasingly interested in finding optimal placement of EV charging stations (EVCS). Existing research works have focused on optimal EVCS placement with the goal of minimizing total construction cost [20], transportation and substation energy loss [17], or considering driver preferences [21]. Two common goals are (i) supporting consumer charging demand and, (ii) maintaining power grid reliability.

In general, users prefer to be within a reasonable distance of a charging station [7]. In reality, however, it might be impossible to accomplish such a guarantee for everyone, especially in less densely populated areas. Thus, one may impose a much larger upper bound on the maximum distance while minimizing the average across all users. To ensure grid reliability, the new charging stations should be connected to the power network in a way so that the voltages at nodes across the network are within acceptable engineering standard (e.g. 0.9–1.1 per unit (p.u.) for rated voltage of 1 p.u. [3] implying a maximum allowable voltage deviation of 0.1 p.u.). One can accomplish this by connecting new charging stations directly to the closest substation but this incurs additional costs due to long connecting lines. Hence, another realistic goal is to minimize the connection cost (or equivalently, connection distance) of new charging stations to the power grid while ensuring that the node voltages adhere to engineering standards of a reliable power grid. Formally, the problem can be stated as follows:

Problem 1. Given the locations of EV users and the associated power grid, let $d_{max}, d_{avg}, \Delta v$ be constants. Find locations to build and connect EVCS to the grid that minimizes the total connection cost while ensuring that all EV users are within d_{max} to some station, the overall average distance for clients are within d_{avg} and, the voltage deviation at any node in the network is within Δv.

1.1 Our Contributions

1. We present a scalable two-part algorithm that tackles this multi-objective problem in stages. The first part efficiently computes the best placements of charging station to cover the population within the shortest distance possible by iteratively solving an integer program (Sect. 3.1).
2. The second part aims to ease the potential strain to the power grid after building new charging stations. We formulate an integer problem (7) to find the optimal way to connect the stations to the power grid. This provides essential factors to consider for policymakers when preparing for the surge in power consumption (Sect. 3.2).
3. We show that, in the first part of the algorithm, we can efficiently find a solution, consisting of 10,733 new charging stations, that covers 75% of the

[2] https://afdc.energy.gov/data/10962.

population within 0.33 miles and guarantees everyone is within 2.5 miles of a charging station. We also demonstrate a trade-off between the average distance of a user to its nearest EVCS and the number of EVCS.
4. For the second part, we consider synthetically created power networks of Virginia [25] and focus on networks in the Montgomery County to show the effect of different adoption scenarios. In particular, it reveals the reduced reliability of the network at high adoption rates. However, an optimal routing algorithm to connect the newly constructed charging stations to the power grid can ensure higher level of reliability.

2 Related Work

Existing works in the literature have addressed the problem of optimal placement of EV charging stations. A detailed review of such works has been done in [2].

In general, flow-capturing models [9,13,18,33] are popular in literature. They model the charging demand as a directed flow along consumer routes of travel in the transportation network and optimally place stations along the routes to cover the demands. Another class of models is the set covering model [8,31]. They model demand locations as polygons or aggregated points and optimally place EVCS facilities to cover the demand locations (located within a threshold distance). Vehicle movement simulation models [7,16] develop activity simulation frameworks. This class of models evaluate the feasibility of daily travel activities of EV owners, given a selection of EVCS locations in the region.

The impact on the voltage profile, phase imbalance, and power quality due to residential EV charging on the power distribution network is studied in [26]. In another work, Gupta et al. [21] pose the optimal EVCS location problem in the context of an oligopolistic market instead of an urban system planner. The work considers locational marginal prices (LMPs, which are the wholesale electricity rate) and uses a penalty function for introducing grid instabilities. A methodology to compute an optimal EV charging network that maximizes profit and satisfies grid constraints, space limitations, and considers time-varying charging demand is proposed in [34]. A post-processing algorithm known as *Removing and Merging Possible Locations algorithm* is proposed to improve the total profit by excluding and merging some of the initial choices made. The problem of simultaneous allocation of EVCS location is considered in [22] from the perspective of a social or urban planner (minimizing social costs, maximizing environmental benefits, and minimizing power losses). The distribution of EV arrivals is estimated from the distribution of vehicle parking times at different parking lots and on different days of the week. The problem of optimal allocation of EVCSs in a balanced [11] and unbalanced [27] radial distribution grid is done where the loss in the power grid and voltage deviation is minimized.

3 Methodology

We tackle the complex Problem 1 by splitting it into two natural components: (*i*) placing charging stations to cover existing and future EV users, and (*ii*) connecting the charging stations to the power-grid (Sect. 3.1, 3.2 respectively).

3.1 EV Charging Station Placement

We formally define the EV charging station placement problem as follows:

Problem 2. Let C be a set of existing and potential EV users, S_{cur} a set of existing EVCS, S_{cand} a set of candidate locations for placing new EVCS, and distance thresholds d_{max} and d_{avg}. Find the smallest set $S_{new} \subseteq S_{cand}$ such that every user $u \in C$ has a charging station within d_{max} and the average distance between users and their nearest charging station is at most d_{avg}.

Problem 2 is a variation of the classic well-known NP-hard facility location problem. Even though approximation schemes exist [4,10,19], it is not clear if they are computationally feasible in practice. Furthermore, constant approximation might not be desirable either; for example, one may be willing to walk half a mile but not one mile to charge their car and, the cost of building 10,000 and 20,000 charging stations may differ significantly. Thus, we propose an alternative problem. Instead of ensuring the average distance is below some threshold, we introduce multiple thresholds and attempt to cover as much of the population as possible within the lowest threshold possible. For example, due to geographic and personal preferences, we discover three categories of distances, $< 0.25, < 0.5, < 2.5$ miles, the population are willing to travel to charge their car. First, we attempt to cover as many people as possible within 0.25 miles, then 0.5 miles and lastly 2.5 miles. This in turn also ensures a small average distance. To formally define our problem, we first introduce the following definition:

Definition 1. *Let S denote a set of charging stations, and $D = \{d_1, d_2, ..., d_k\}$ be a set of distances. We say, coverage$(u, S, D) = d_i$ if $d_i \in D$ is the smallest distance such that there exists a charging station $s \in S$ where distance$(u, s) \leq d_i$.*

Given these thresholds D, a natural constraint to impose is to ensure that if an user can be covered within distance d_i, it must be covered within distance d_i by the final solution as well. Then, our problem is the following:

Problem 3 (EV Charging Station Placement Problem (EVCSPP)). Let C denote a set of existing and potential EV users, S_{cur} a set of existing charging stations, S_{cand} a set of candidate locations for placing new charging stations, and $D = \{d_1, d_2, ..., d_k\}$ a set of distances where $d_1 < d_2 < ... < d_k$. Find the smallest set $S_{new} \subseteq S_{cand}$ so that $coverage(u, S_{cur} \cup S_{new}, D) = coverage(u, S_{cur} \cup S_{cand}, D)$ for every EV user $u \in C$.

A special case of Problem 3 is when we have a single threshold value d_{th} within D, i.e. $D = \{d_{th}\}$. To ensure a feasible solution exists in this special case, we may assume that for every EV user $u \in C$, there exists a station $s \in S_{cur} \cup S_{cand}$ such that $distance(u, s) \leq d_{th}$. This problem is known to be NP-hard to approximate to a small factor (1.46) [10]. We formulate it as an Integer Program (IP) and use known solvers to obtain a good solution. Consider the following IP (notations within the program are described in Table 1).

$$\min_{x,y} \sum_{j \in S} y_j \tag{1}$$

$$s.t. \quad \sum_{j \in S_i} x_{i,j} \geq 1 \qquad \forall i \in C \tag{2}$$

$$x_{i,j} \leq y_j \qquad \forall i, j \tag{3}$$

Table 1. IP (1–3) notations and descriptions of the objective and constraints.

Notation	Description
S	$S_{cur} \cup S_{cand}$
S_i	Set of existing/candidate charging stations within distance d_{th} of user i
y_j	1 if charging station j is built
$x_{i,j}$	1 if user i will be serviced by charging station j
Objective 1	Minimize the number of charging stations constructed
Constraint 2	Every user must be serviced by at least one charging station that is within d_{th} distance of the user
Constraint 3	User i can use charging station j, only if station j is built

Algorithm 1: Single Threshold Placement: STP($C, S_{cur}, S_{cand}, d_{th}$)

Input: $C, S_{cur}, S_{cand}, d_{th}$
1 Construct IP (1–3) from the inputs.
2 Set $y_j = 1, \forall j \in S_{cur}$.
3 Solve the IP. Let, S_{new} be the set of newly built stations.
4 **return** S_{new}

Before solving this IP, we set $y_j = 1, \forall j \in S_{cur}$. The IP can be solved by existing solvers such as Gurobi [12]. The entire process for solving the special case problem is shown in Algorithm 1, we call this method *Single Threshold Placement* (STP).

Algorithm 2 describes our method *Multi-Threshold Placement* (MTP) for solving Problem 3. The main idea here is that, we go in increasing order of the

Algorithm 2: Multi-threshold Placement: MTP(C, S_{cur}, S_{cand}, D)

Input: $C, S_{cur}, S_{cand}, D = \{d_1, d_2, ..., d_k\}$
1 Set of newly built stations: $S_{new} \leftarrow \{\}$
2 $C_{covered} \leftarrow \{\}$
3 **for** $i = 1$ *to* k **do**
4 | $C_i \leftarrow$ set of users $u \in C$ who have a station $s \in S_{cur} \cup S_{cand}$ within distance d_i.
5 | $S^i_{new} \leftarrow \text{STP}(C_i, S_{cur}, S_{cand}, d_i)$
6 | $S_{cur} \leftarrow S_{cur} \cup S^i_{new}$
7 | $S_{new} \leftarrow S_{new} \cup S^i_{new}$
8 | $C_{covered} \leftarrow C_{covered} \cup C_i$
9 $C' \leftarrow C \backslash C_{covered}$
10 **if** C' *is not empty* **then**
11 | $S^c_{new} \leftarrow \text{STP}(C', \{\}, C', d_k)$
12 **return** $S_{new} \cup S^c_{new}$

distance thresholds in D, and cover all the users who can be covered within the current distance threshold. We ensure that at each distance threshold the number of newly built stations is minimized by applying STP (line 5). Note that, stations that are built at threshold d_i are considered as already built when processing the threshold d_{i+1} (line 6). Also, some users in C might not have any location $s \in S_{cur} \cup S_{cand}$ within the largest distance threshold d_k. To cover such users within distance d_k, we consider each of these user locations as candidate locations to build charging stations. We then use STP to determine the minimum number of stations required to cover them with threshold d_k (line 11).

3.2 Connecting EV Charging Stations

The existing power distribution network is a *tree* $\mathcal{G}(\mathcal{V}, \mathcal{E})$ comprising of $N+1$ nodes (also called *buses*) collected in the set $\mathcal{V} := \{0, 1, 2, \cdots, N\}$. The tree is rooted at substation node $\{0\}$ and consists of primary and secondary distribution lines collected in the edge set $\mathcal{E}$. Secondary distribution lines connect residences to local pole top transformers, which are fed by the distribution substation through the primary distribution lines. We denote the branch bus incidence matrix $\mathbf{E} \in \mathbb{R}^{N \times (N+1)}$ with its element along row l and column k

$$\mathbf{E}(l, k) := \quad 1 \text{ if } k = i, \quad -1 \text{ if } k = j, \;\; 0 \text{ otherwise}, \;\; \forall l = (i, j) \in \mathcal{E} \tag{4}$$

We define $\mathbf{E} = \left[\mathbf{e}_0 \; \mathbf{E}_{\text{red}}\right]$, where $\mathbf{E}_{\text{red}}$ is the reduced branch bus matrix obtained after removing the column corresponding to the substation (root) node.

In this section, we consider the problem of identifying the optimal connection points for the new EVCSs contained in set $\mathcal{P}$ to the existing distribution network $\mathcal{G}(\mathcal{V}, \mathcal{E})$. Albeit routing power delivery to these new nodes by connecting them to the nearest distribution network node can result in reduced investment for construction, it can lead to power grid reliability issues where node voltages and edge power flows violate prescribed engineering standards. We can formalize this problem as follows.

Problem 4 (Optimal Routing Problem). Given a set of EVCS locations $\mathcal{P}$, find the set of connecting edges $\mathcal{E}_{\text{new}} = \{(p, v)|\ p \in \mathcal{P}, v \in \mathcal{V}\}$ to the existing power distribution network $\mathcal{G}(\mathcal{V}, \mathcal{E})$ such that each new node is connected to exactly one node in the network and the power grid reliability is maintained.

To this end, we start by considering a set of candidate edges for each EVCS location $p \in \mathcal{P}$. In this paper, we consider all nodes $v \in \mathcal{V}$ within ϵ-radius of p and define $\mathcal{E}_D = \{(p, v) \forall p \in \mathcal{P} \mid v \in \mathcal{V},\ \text{dist}(p, v) \leq \epsilon\}$ as the candidate set of edges. Then we define a integer optimization program to identify the optimal set of edges $\mathcal{E}_{\text{new}} \subseteq \mathcal{E}_D$ from these candidate edges which minimizes the cost of constructing new distribution line connection to the EVCSs as well as adhere to the power grid reliability standards. Table 2 lists the vectors and matrices used in the optimization problem. A bold character denotes vector/matrix and scalar values are denoted by non-bold subscripted symbols.

Table 2. Vectors and matrices for optimization problem

Var.	Description	Var.	Description
$\mathbf{E}$	branch bus incidence matrix	$\mathbf{E}_{\text{red}}$	reduced branch bus incidence matrix
$\mathbf{p}$	vector of power demand at all nodes	$\mathbf{v}$	vector of node voltage magnitudes
$\mathbf{f}$	vector of power flow through all edges	$\mathbf{c}$	vector of cost of edges
$\mathbf{R}$	diagonal matrix of edge resistances	$\underline{v}, \overline{v}$	lower and upper voltage limits
$\overline{\mathbf{S}}$	diagonal matrix of edge thermal limits		

Node Variables. Each node $i \in \mathcal{V} \cup \mathcal{P}$ has associated voltage magnitude v_i which can be stacked into a $(|\mathcal{V}| + |\mathcal{P}|)$-length vector $\mathbf{v}$. Each node in the network has an associated power demand consumption denoted by p_i. The residential load demands are obtained from [30] and the EVCS loads are estimated by considering average number of customers arriving at the location. The power demands can be stacked to vector $\mathbf{p}$.

Edge Variables. We define binary variable x_e for each edge $e \in \mathcal{E} \cup \mathcal{E}_D$. $x_e = 1$ denotes that edge e is included in the optimal network, while $x_e = 0$ implies otherwise. Note that $x_e = 1$ for $e \in \mathcal{E}$ since the existing distribution network topology is not altered. We also define f_e to be the power flowing through edge e and cost of each edge to be c_e. The edge variables x_e, f_e and c_e can be respectively stacked to $(|\mathcal{E}| + |\mathcal{E}_D|)$-length vectors $\mathbf{x}$, $\mathbf{f}$ and $\mathbf{c}$.

Radiality Constraints. The resulting power distribution network after new edges are added has to maintain a radial or tree structure. This is ensured by ensuring that the number of edges is equal to number of non-root nodes. After the EVCSs in $\mathcal{P}$ are connected, the number of non-root nodes is given by $N + |\mathcal{P}|$. We use the following linear equality constraint: $\sum_{e \in \mathcal{E} \cup \mathcal{E}_D} x_e = N + |\mathcal{P}|$.

Power Flow Constraints. The power flowing through the edges $\mathbf{f}$ is linearly related to power consumption $\mathbf{p}$ at nodes in the network through the branch bus incidence matrix $\mathbf{E}$. The power flow equations for a network relate node voltages to the power flowing through edges in the network. The standard power flow constraints are quadratic equality constraints which make them non-convex. However, following assumptions of small line impedance values, we arrive at an approximate linear relation between node voltages at edge terminals and power flowing through the edge. This approximation is also known as *Linearized Distribution Flow* (LDF) model [6]. Note that such approximation holds true for edges where $x_e = 1$. Therefore, the LDF model for our case is given as:

$$x_e \left(v_i - v_j - r_e f_e\right) = 0 \quad \forall e := (i, j) \tag{5}$$

where r_e is resistance of edge e. (5) is non-convex because it has bi-linear terms in the equality. In order to deal with this non-convexity, McCormick relaxation has been used widely in several previous works [28,29]. In general, McCormick relaxation replaces the non-convex equality constraint with its convex envelope [24]. However, in the case of bi-linear variables with at least one binary variable, this relaxation becomes exact. The convex relaxed version of (5) is:

$$-\left(1 - x_e\right) M \le v_i - v_j - r_e f_e \le \left(1 - x_e\right) M \quad \forall e := (i, j) \in \mathcal{E} \cup \mathcal{E}_D \tag{6}$$

Here M is a sufficiently large number such that the inequality turns into a strict equality for $x_e = 1$, while it remains irrelevant when $x_e = 0$. We can construct the diagonal matrix $\mathbf{R}$ with resistance of edges as the entries. An important aspect of the optimization problem is the consideration of power grid reliability constraints. This includes constraints which force node voltages to remain within engineering standards $\underline{v}, \overline{v}$ and edge power flows to be limited by the respective thermal constraints $\overline{s}_e$. We can diagonalize the edge thermal constraints to the diagonal matrix $\overline{\mathbf{S}}$.

The optimization problem aims to minimize overall investment of constructing new power lines required to connect the EVCSs to existing power network. Meanwhile, the voltage at all nodes need to be as close to the rated voltage as possible. This ensures that all consumers have high quality of power delivered to them. In engineering practice, voltage is expressed in per unit (p.u.) which is the ratio of actual voltage to the rated value. Thus, we can minimize the deviation of node voltages to voltage of 1 p.u. We use hyperparameter λ to scale these two separate expressions in the objective function and obtain the following:

$$\min_{\mathbf{x},\mathbf{v},\mathbf{f}} \quad \mathbf{c}^T\mathbf{x} + \lambda||\mathbf{v} - \mathbf{1}||^2 \tag{7a}$$

$$\text{s. to.} \quad \mathbf{E}_{\text{red}}^T\mathbf{f} = -\mathbf{p}, \qquad \overline{\mathbf{S}}\mathbf{x} \ge \mathbf{f} \ge -\overline{\mathbf{S}}\mathbf{x} \tag{7b}$$

$$\left(\mathbf{1} - \mathbf{x}\right) M \ge \mathbf{E}\mathbf{v} - \mathbf{R}\mathbf{f} \ge -\left(\mathbf{1} - \mathbf{x}\right) M, \qquad \underline{v}\mathbf{1} \le \mathbf{v} \le \overline{v}\mathbf{1} \tag{7c}$$

$$x_e = 1,\ \forall e \in \mathcal{E};\ x_e \in \{0, 1\}\ \forall e \in \mathcal{E}_D, \qquad \mathbf{1}^T\mathbf{x} = N + |\mathcal{P}| \tag{7d}$$

4 Experimental Results

For our experiments, we use the state of Virginia as our study area. To construct a problem instance for this area, we first collected the home locations within the state from a synthetic population data [1], and existing EV charging station locations from US Department of Energy[3]. We also collected locations of 16 different types of POIs (e.g. gas station, train station, airport) from HERE maps [15]. We consider these as candidate locations for building new charging stations. We have collected ~3.1 million home locations, 1090 existing EV charging stations, and ~1.47 million POIs. Finally, we use the synthetically created power distribution networks [25] for Montgomery County of Virginia, USA to consider the implications on the power grid.

4.1 EV Charging Station Placement

As mentioned earlier, the number of homes to be covered and the number of candidate locations to build charging stations are quite large. To make the problem more tractable, we do the following: we construct a road network of our study area (using data from HERE maps) and then map each home location, existing EV charging station, and POI to the nearest node in the road network. By doing this, the home locations were mapped to 161,324; the POIs to 25,044; and the existing EV stations to 863 unique nodes. We denote the set of these nodes by C, S_{cand}, S_{cur} respectively. For the set of distance threshold values D, we have chosen $D = \{0.25, 0.5, 2.5\}$ where the unit is miles (experiment with different values of D performed later). C, S_{cand}, S_{cur}, and D together represents our problem instance for Virginia. A visualization of C and S_{cur} is provided in left plot of Fig. 1.

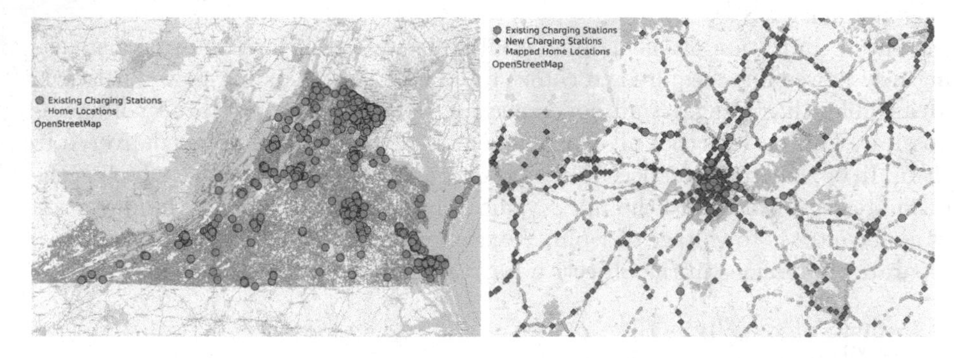

Fig. 1. Plots showing uneven distribution of existing charging stations in Virginia, USA (**left**) and an equitable solution provided by the proposed algorithm for city of Charlottesville (**right**). New stations are built to ensure availability of charging options for all residences.

[3] https://afdc.energy.gov/fuels/electricity_locations.html#/analyze?fuel=ELEC.

We applied MTP on our problem instance to find a solution. We ran this experiment on a high-performance computing cluster, with 256 GB RAM and 24 CPU cores allocated to our task. MTP terminated with a runtime of ~15 min. Our solution suggests 10,733 new charging stations needs to be built (11,596 stations including the existing ones). Following are some of our observations from this experiment:

(i) Within MTP, STP solves the special case single threshold problem to optimality in every iteration (Algorithm 2 line 5).
(ii) Figure 2 **(left)** shows the distribution of the distances between homes and their nearest charging station, in MTP solution, and when considering only the existing charging stations. Note that, the vertical scale of the two plots are different. With only existing stations, the average distance is 3.64 miles. In MTP solution, 75% of the homes have a charging station within 0.33 miles; the average distance is 0.31 miles.
(iii) The right plot in Fig. 1 shows a visualization of the MTP solution in Charlottesville city, Virginia. We see that new charging stations are built to ensure that homes that are not covered by existing charging stations, are now covered by the new ones.

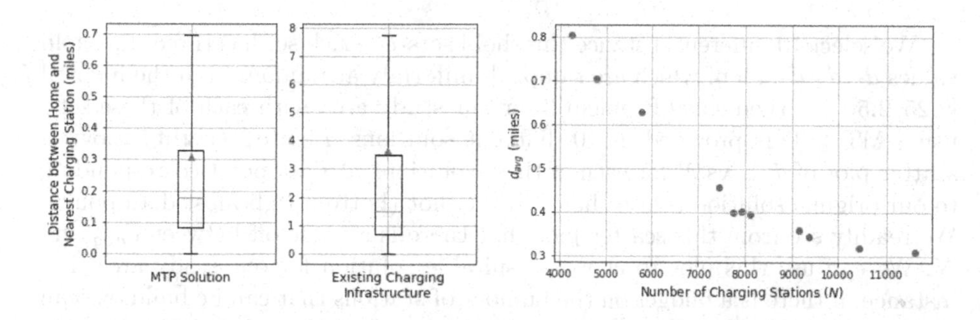

Fig. 2. (left) Box-plots showing the distribution of distances between homes and their nearest charging stations, in MTP solution, and when considering only the existing charging stations (scales of the two box-plots are different, outliers are not shown). In MTP solution, 75% of the homes have a charging station within 0.33 miles. **(right)** Scatter plot showing average distance between homes and their nearest charging station (d_{avg}) vs number of charging stations (N). The average distance decreases when we build more stations.

Trade-Off Between Number of Charging Stations and Average Distance Between Homes and Their Nearest Charging Stations. In our previous experiment, we have used the distance threshold values $D = \{0.25, 0.5, 2.5\}$ (miles). Intuitively, if we choose smaller threshold values then we will need more charging stations to cover the homes within the smaller distance. On the other hand, if we choose, larger threshold values, then we can cover more homes with fewer number of charging stations. Therefore, we expect a trade-off between the number of charging stations (N) and the average distance between homes and their nearest charging stations (d_{avg}). Now, we investigate this experimentally.

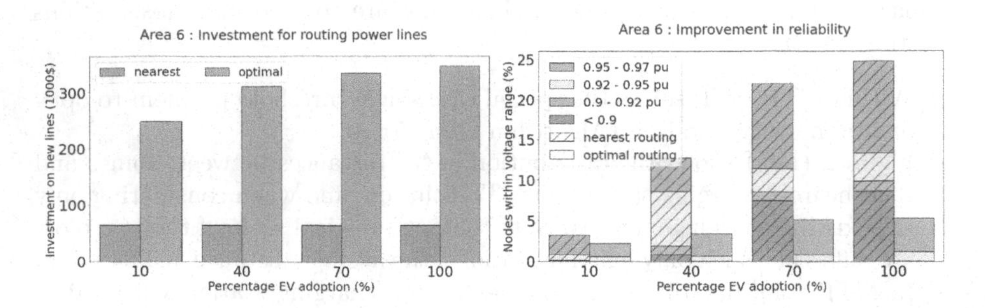

Fig. 3. Plots comparing two different routing algorithms for connecting EV charging stations to existing power distribution network: connecting charging station to the nearest node requires smaller lines as opposed to optimally routing stations. Connecting to nearest available power network node can reduce the additional investment of constructing new power lines (**left**), while it hampers the power grid reliability since we observe multiple undervoltage nodes (**right**) for higher EV adoption rates.

We select 10 different distance threshold sets D. Each set has three threshold values d_1, d_2, d_3, all of which are sampled uniformly at random from the interval $[0.25, 2.5]$. We then solve Problem 3 for our study area with each of these sets, using MTP. This provided us 10 different solutions. Figure 2 (**right**) shows a scatter plot of d_{avg} vs N for each of these solutions. A data point corresponding to our original solution is also shown in the plot (bottom-rightmost data point). We readily see from this scatter plot that there is a trade-off between d_{avg} and N. We can use this plot to choose a suitable solution for our study area. For instance, if there is a budget on the number of stations that can be build, we can filter out the solutions where we go over budget and then choose the solution with the minimum average distance.

4.2 Optimal Routing Problem

In this section we compare the proposed optimal routing algorithm to the scenario where each EVCS is connected to the nearest available node in the power distribution network. We term this alternate algorithm as the *nearest* routing algorithm. Since EVCSs are connected to the nearest possible node, the new distribution line connections are minimum length edges, which ensures less investment on upgrading existing power infrastructure. However, this comes at a cost of reduced reliability. A *reliable* power grid is considered to be one, which has adequate generation to support the consumer load demand and is operated without violating standard power engineering constraints [5]. In our case, we assume that adequate generation is available to supply the increased demand of EVCSs. We consider the power network to be *reliable* when the line flows (edge flows) are within their rated capacities and the node voltages are within acceptable engineering standards of 0.9–1.1 p.u. [3]. Figure 3 compares the two routing algorithms for different levels of EV adoption. The *nearest* routing algorithm requires minimum investment to be made on installing new lines, while we observe a significant fraction of the nodes in the network having undervoltage issues (less than 0.9 pu) for higher levels of EV adoption. The undervoltage problem disappears when we implement the *optimal* routing algorithm which strictly imposes the voltage limit constraints, but the investment on new line construction increases. We performed our experiments on one of the synthetic networks from Montgomery County in Virginia, USA and identify the region as 'Area 6' in the plots.

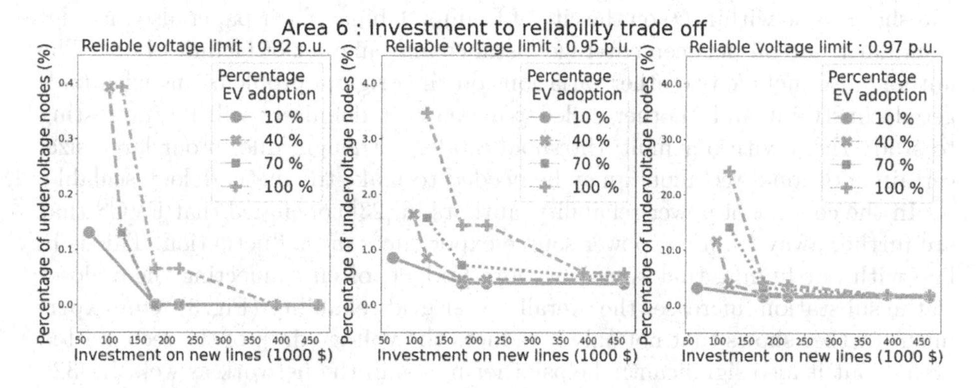

Fig. 4. Plots showing trade off between investment on additional distribution lines and reliability of the power network (vertical scales differ across the three panels). With an increased investment in longer lines, the EVCSs are connected to optimal nodes in the distribution network such that less number of nodes experience undervoltage. For smaller investment, the node voltages are acceptable by standard engineering practices (>0.9 pu), yet they are far from the rated voltage of 1 pu, making the power grid unreliable.

The optimal routing algorithm ensures that EVCSs are connected to the power distribution network in a way such that all node voltages are within the accepted engineering standards (greater than 0.9 p.u.). However, this does not ensure that the node voltages are close to the rated voltage (1 p.u.). To this end, we have used the parameter λ in the optimization problem in (7). A higher value of λ ensures that the node voltages are closer to rated voltage of 1 p.u. Note that λ has been used as a weight in the objective function – this means that for high values of λ, the aspect of having node voltages closer to the rated voltage is given more importance than minimizing investment on new lines. This trade off is shown in Fig. 4 (vertical scales differ across panels). We plot the investment on new line construction and the number of undervoltage nodes for different values of parameter λ. Since voltages above 0.9 p.u cannot be considered as 'undervoltage' as per engineering standards, we define the voltage limits as 0.92 p.u., 0.95 p.u. and 0.97 p.u. for the three plots and consider node voltages less than this limit as 'undervoltage'.

5 Discussions and Conclusion

Comparison to Related Works: Many current works separately study the optimal placement of EVCS and the effect of EVCS on the power grid. To the best of our knowledge, this work is the first to provide a methodology that combines both into consideration. Individually, our experiment also provides similar findings as some of the previous works.

To build EVCS in order to cover the need of a population, authors in [14] uses the maximum coverage problem (equivalent to our Problem 3 with a single threshold value within D) on the city of Beijing, China. Their paper also includes a similar trade-off between coverage distance and amount of facilities built. The authors also include two other variations on the original problem, one with budgeted constraint and another called p-median. It definitely will be interesting to study them with our multi-threshold model; although, due to our large size, certain additional techniques may be needed to make the methodology scalable.

In the context of power reliability, authors in [23] concluded that EVCS that are further away from the power source experiences more fluctuation. This is in line with our finding that spending more budget, often connecting them closer to the substation, increases the overall power grid reliability (Fig. 3). Our experiment further shows that not only it reduces the voltage drop of the reconnected nodes, but it also significantly helps other nodes in the network as well. In [32], the authors used simulations to show a drastic increase in power consumption (109%) even at 60% EV penetration. This is consistent with our findings that even at a 40% adoption rate, the number of nodes experiencing undervoltage exceeds double the number observed for 10% adoption rate. Even with optimal connection (Fig. 4), across different undervoltage thresholds, similar trends exist where higher adoption with a limiting budget necessarily induces more number of undervoltage nodes (decreased reliability). Although our demand at each charging station is based on the population data and directly correlated with the size and density of the region, unlike [32], we do not propose methods for smoothing the charging demand (e.g. via tariffs). Further studies can be done with these considerations.

Policy Suggestions: By combining real population data and synthetic models of the power grid, we provide a useful analytical tool for policymakers when planning for EVCS. For example, given the chosen threshold of 0.25, 0.5, and 2.5 miles, we see that 10,733 additional EVCS are required. Furthermore, greedily connecting them to the closest point in the existing power grid, even with a 40% adoption rate, imposes a significant decrease in power grid reliability. However, by connecting intelligently, the strain on power grid can be almost entirely eliminated. Policymakers may assess each region independently and decide if optimal connections are warranted. From our experiment, for example, there does not seem to be much difference between the cost in a scenario with 40% adoption versus a 100% adoption, suggesting that if sufficient budget exists, it is worthwhile to prepare for the worst-case scenario.

Future Directions: There are many directions to further extend our model. For example, we may impose limits on how many users may access a particular station due to capacity/space constraints. Our estimation of the demand can also be refined. By using traffic flows or Migration data, one may be able to better predict when and where a person will use an EVCS. This time-refined analysis will provide a better estimate of fluctuation since electricity usage varies throughout the day. We can also generalize our optimal connection to allow the rerouting of existing power lines. It is conceivable that altering existing infrastructure might improve the grid reliability regardless of adoption rates.

References

1. Adiga, A., et al.: Generating a synthetic population of the United States. Technical report NDSSL 15-009, Network Dynamics and Simulation Science Laboratory (2015)
2. Ahmad, F., Iqbal, A., Ashraf, I., Marzband, M., Khan, I.: Optimal location of electric vehicle charging station and its impact on distribution network: a review. Energy Rep. **8**, 2314–2333 (2022)
3. American National Standards Institute. ANSI C84.1-2020: Electric Power Systems Voltage Ratings (60 Hz) (2020)
4. Bādoiu, M., Har-Peled, S., Indyk, P.: Approximate clustering via core-sets. In: Proceedings of the Thiry-Fourth Annual ACM symposium on Theory of computing, pp. 250–257 (2002)
5. Billinton, R., Li, W.: Basic concepts of power system reliability evaluation. In: Billinton, R., Li, W. (eds.) Reliability Assessment of Electric Power Systems Using Monte Carlo Methods, pp. 9–31. Springer, Boston (1994). https://doi.org/10.1007/978-1-4899-1346-3_2
6. Bolognani, S., Dörfler, F.: Fast power system analysis via implicit linearization of the power flow manifold. In: 53rd Annual Allerton Conference on Communication, Control, and Computing (Allerton), pp. 402–409 (2015)
7. Dong, J., Liu, C., Lin, Z.: Charging infrastructure planning for promoting battery electric vehicles: an activity-based approach using multiday travel data. Transp. Res. Part C: Emerg. Technol. **38**, 44–55 (2014)
8. Frade, I., Ribeiro, A., Goncalves, G., Antunes, A.: Optimal location of charging stations for electric vehicles in a neighborhood in Lisbon, Portugal. Transp. Res. Rec. **2252**, 91–98 (2011)
9. Fredriksson, H., Dahl, M., Holmgren, J.: Optimal placement of charging stations for electric vehicles in large-scale transportation networks. Procedia Comput. Sci. **160**(2018), 77–84 (2019)
10. Guha, S., Khuller, S.: Greedy strikes back: improved facility location algorithms. J. Algorithms **31**(1), 228–248 (1999)
11. Gupta, K., Narayanankutty, R.A., Sundaramoorthy, K., Sankar, A.: Optimal location identification for aggregated charging of electric vehicles in solar photovoltaic powered microgrids with reduced distribution losses. Energy Sour. Part A: Recovery Util. Environ. Effects 1–16 (2020)

12. Gurobi Optimization, LLC. Gurobi Optimizer Reference Manual (2021)
13. He, J., Yang, H., Tang, T.Q., Huang, H.J.: An optimal charging station location model with the consideration of electric vehicle's driving range. Transp. Res. Part C: Emerg. Technol. **86**, 641–654 (2018)
14. He, S.Y., Kuo, Y.-H., Wu, D.: Incorporating institutional and spatial factors in the selection of the optimal locations of public electric vehicle charging facilities: A case study of Beijing, China. Trans. Res. Part C: Emerg. Technol. **67**, 131–148 (2016)
15. HERE Premium Streets Data set for the U.S. (2020)
16. Hess, A., Malandrino, F., Reinhardt, M.B., Casetti, C., Hummel, K.A., Barceló-Ordinas, J.M.: Optimal deployment of charging stations for electric vehicular networks. In: Proceedings of the First Workshop on Urban Networking, UrbaNe 2012, New York, NY, USA, pp. 1–6. Association for Computing Machinery (2012)
17. Islam, M.M., Shareef, H., Mohamed, A.: Improved approach for electric vehicle rapid charging station placement and sizing using google maps and binary lightning search algorithm. PLOS ONE **12**(12), 1–20 (2017)
18. Kuby, M., Lim, S.: Location of alternative-fuel stations using the flow-refueling location model and dispersion of candidate sites on arcs. Netw. Spat. Econ. **7**, 129–152 (2007)
19. Kumar, P., Kumar, P.: Almost optimal solutions to k-clustering problems. Int. J. Comput. Geom. Appl. **20**(04), 431–447 (2010)
20. Lam, A.Y., Leung, Y.-W., Chu, X.: Electric vehicle charging station placement. In: 2013 IEEE International Conference on Smart Grid Communications (SmartGridComm), pp. 510–515 (2013)
21. Luo, C., Huang, Y.-F., Gupta, V.: Placement of EV charging stations-balancing benefits among multiple entities. IEEE Trans. Smart Grid **8**(2), 759–768 (2017)
22. Luo, L., Gu, W., Wu, Z., Zhou, S.: Joint planning of distributed generation and electric vehicle charging stations considering real-time charging navigation. Appl. Energy **242**, 1274–1284 (2019)
23. Ma, G., Jiang, L., Chen, Y., Dai, C., Ju, R.: Study on the impact of electric vehicle charging load on nodal voltage deviation. Arch. Electr. Eng. **66**(3), 495–505 (2017)
24. McCormick, G.P.: Computability of global solutions to factorable nonconvex programs: part I - convex underestimating problems. Math. Program. **10**(1), 147–175 (1976)
25. Meyur, R., et al.: Ensembles of realistic power distribution networks. Proc. Natl. Acad. Sci. **119**(26), e2123355119 (2022)
26. Putrus, G.A., Suwanapingkarl, P., Johnston, D., Bentley, E.C., Narayana, M.: Impact of electric vehicles on power distribution networks. In: 2009 IEEE Vehicle Power and Propulsion Conference, pp. 827–831 (2009)
27. Reddy, M.S.K., Selvajyothi, K.: Optimal placement of electric vehicle charging station for unbalanced radial distribution systems. Energy Sour. Part A: Recovery Util. Environ. Effects (2020)
28. Singh, M.K., Kekatos, V., Liu, C.C.: Optimal distribution system restoration with microgrids and distributed generators. In: 2019 IEEE Power & Energy Society General Meeting (PESGM), Atlanta, GA, USA, pp. 1–5. IEEE (2019)
29. Singh, M.K., Taheri, S., Kekatos, V., Schneider, K.P., Liu, C.C.: Joint grid topology reconfiguration and design of Watt-VAR curves for DERs. In: IEEE Power & Energy Society General Meeting, Denver, CO, USA. IEEE (2022)
30. Thorve, S., et al.: High-resolution synthetic residential energy use profiles for the United States. Sci. Data **10**(1), 76 (2023)

31. Tu, W., Li, Q., Fang, Z., Shaw, S.l., Zhou, B., Chang, X.: Optimizing the locations of electric taxi charging stations: a spatial-temporal demand coverage approach. Transp. Res. Part C: Emerg. Technol. **65**(3688), 172–189 (2016)
32. Ul-Haq, A., Cecati, C., Strunz, K., Abbasi, E.: Impact of electric vehicle charging on voltage unbalance in an urban distribution network. Intell. Industr. Syst. **1**, 51–60 (2015)
33. Wang, Y.W., Lin, C.C.: Locating multiple types of recharging stations for battery-powered electric vehicle transport. Transp. Res. Part E: Logist. Transp. Rev. **58**, 76–87 (2013)
34. Zhang, Y., Chen, J., Cai, L., Pan, J.: EV charging network design with transportation and power grid constraints. In: IEEE INFOCOM 2018 - IEEE Conference on Computer Communications, pp. 2492–2500 (2018)

Computer Graphics, Image Processing and Artificial Intelligence

Radial Basis Function Neural Network with a Centers Training Stage for Prediction Based on Dispersed Image Data

Kwabena Frimpong Marfo and Małgorzata Przybyła-Kasperek(✉)

Institute of Computer Science, University of Silesia in Katowice, Będzińska 39, 41-200 Sosnowiec, Poland
{kwabena.marfo,malgorzata.przybyla-kasperek}@us.edu.pl

Abstract. Neural networks perform very well on difficult problems such as image or speech recognition as well as machine text translation. Classification based on fragmented and dispersed data representing certain properties of images or computer's vision is a complex problem. Here, the suitability of a Radial Basis Function (RBF) neural network was evaluated using fragmented data in the problem of recognizing objects in images. The great difficulty of the considered problem is, there is not images data as such but only data on some properties of images stored in a dispersed form. More specifically, it was demonstrated that applying a $k-$nearest neighbors classifier in the first step to generate predictions based on fragmented data, and then using a RBF neural network to learn how to correctly recognize the systems of generated predictions for making a final classification is a good approach for recognizing objects in images. An additional step of training the weights (centers) between the input and hidden layers of a RBF network was proposed. In general, this investigation demonstrates that adding this step significantly improves the correctness of recognizing objects in images.

Keywords: Radial Basis Function Neural Network · Dispersed Data · Image Data Processing

1 Introduction

Object detection in images is very important in today's world and many applications such as surveillance systems can be found. Examples of important uses include: recognizing types of vehicles in road traffic [18], recognizing types of objects in satellite images [12], or even recognizing components on a production line [2,20]. In literature, we can find numerous applications of neural networks for processing images and recognizing objects in images [1,19,21]. These are mainly applications of convolutional neural networks. A very interesting approach where neural networks were used for recognizing handwriting can be found in [14]. In [5], the generative adversarial network (GAN) was used to generate images.

J. Mikyška et al. (Eds.): ICCS 2023, LNCS 14076, pp. 89–103, 2023.
https://doi.org/10.1007/978-3-031-36027-5_7

Paper [15] provided an overview of very interesting approaches that used neural networks to generate artistic patterns.

It is very rare to find studies that deal with recognizing images that are available in fragmentary form. This problem can be considered as a set of images obtained from the perspective of several cameras, each of which observes the object from different angles [16]. In such a case – when the image data is fragmented, recognizing the object in the photo is more difficult. In the paper [7], the approach of assembling fragments of photos and matching parts to merge into one can be found. This approach required that the fragmented data do not overlap, however, in this study, fragmentation concerns the dispersion of information about the recognized object with shared fragments and does not mean disjointed information. We also assume that we do not have images as such, nor its fragments, but only the characteristics extracted from these images. These may be the average values recorded in a certain area of pixels or certain shape characteristics such as the length and width depicted in the image. We also assume that these data are available in tabular form – a set of decision tables. However, the data may overlap, i.e. there may be common conditional attributes in several decision tables. In addition, in the training set which comprises a set of local tables, there may appear fragmentary characteristics for the same physical object or for other objects belonging to the same decision class. The focus of this study was not on image fusion but on object recognition, assigning the correct decision class for an object that is seen in a fragmented way.

Other approaches used to recognize fragmented images can also be found in literature. The paper [22] proposed the use of hidden Markov models for gesture recognition based on fragmentary vision. In the paper [3], fuzzy rules were used for fragmented handwritten digit recognition.

This paper proposes the use of a Radial Basis Function (RBF) neural network with a centers training stage in combination with the k−nearest neighbors algorithm to classify objects observed in images based on fragmented characteristics – data stored as a set of local decision tables. For this purpose, three problems were considered: classification of car type based on photo's characteristics, classification of land type based on satellite images, and classification of bean type based on fragmentary computer vision. To the best of our knowledge, such an issue has not been studied before in literature. In this study research results, comparisons and statistical tests are presented. It has been justified that the proposed approach gives much better results than the RBF networks without a centers training stage as well as the baseline approach that uses a heterogeneous ensemble of classifiers.

The paper is organized as follows. In Sect. 2, the proposed classification model using a RBF neural network is described. The algorithm's description and the discussion about the key features of the proposed approach are given. Section 3 addresses the datasets that were used and presents the conducted experiments and discussion on obtained results. Section 4 is on conclusions and future research plans.

2 Model and Methods

When data is stored in dispersed form, aggregating local tables into a table becomes a difficult task due to data inconsistencies. To overcome this, we consider local data separately, look for patterns in them and find a way to combine the dependencies already discovered into a single piece. More formally, we assume that some characteristics of images are available in dispersed form – in the form of a set of local tables. We assume that a set of decision tables $D_i = (U_i, A_i, d)$, $i \in \{1, \ldots, n\}$ is available, where U_i is the universe comprising a set of objects – images; A_i is a set of conditional attributes – some features that describe the image; d is a decision attribute – object shown in the image. Objects and attributes in local tables can be different, however, some objects may be common among local tables as we do not impose any restrictions here.

This is a case where different sensors capture features of an image based on the object identified in the image. In a real-life situation, it could be pictures taken at different angles of the same object by different cameras. Thus, local tables will be generated (one local table for each camera) with different sets of conditional attributes (but some may be shared). Another instance could be cameras set up in different locations in a city where each camera takes pictures of vehicles on the road and based on the features identified, the type of vehicle is determined. In this situation, same object could be recognized by different cameras, but most of the objects identified will be different.

Since aggregation of these local tables into a single table is not immediately possible, one possible approach that can be used is to generate classifiers based on each local table separately. For this purpose, the $k-$nearest neighbors classifier is chosen, as it has low computational complexity and is a suitable method for image classification based on local features [8]. We expect objects of one type in images to have similar features. To calculate the distances between the test objects and the objects in the local tables, the Gower measure is used [13]. This measure allows to compute the distance even when there are attributes of different types (quantitative, qualitative, binary) and from different ranges in the decision table (it does not require normalization or standardization). In addition, it should be noted that for the test objects, values of all attributes occurring in the local tables should be known. Each base classifier makes its classification using a subset of all attributes – more strictly, a classifier i that is built based on a decision table D_i uses the set of attributes A_i. A modification of the $k-$nearest neighbors classifier is proposed, i.e. instead of generating a prediction from the abstract level (a decision class most frequent in the neighborhood), a prediction from the measurement level is generated. That is, a classifier i generates a probability vector over decision classes for a test object x (denoted by $\mu_i(x)$). The dimension of vector $\mu_i(x) = [\mu_{i,1}(x), \ldots, \mu_{i,c}(x)]$ is equal to the number of decision classes $c = card\{V^d\}$, where V^d is a set of decision attribute values (a set of the types of objects in the image), $card\{V^d\}$ is the cardinality of this set. Each coefficient $\mu_{i,j}(x)$ is determined using the k-nearest neighbors of the test object x belonging to a given decision class j and decision table D_i. In this way, the information we get from the base classifiers is more complete. These are average similarities determined from fragmented data for each decision class (i.e. objects that may appear in the image).

To summarize the previous step, the base classifiers for the test object x generates n prediction vectors $\mu_i(x)$, $i \in \{1,\ldots,n\}$, each vector is c dimensional, i.e. a composite of similarities to decision classes obtained based on fragmented data. The task of recognizing the correct decision class based on such vectors – identifying the object in the image – is a difficult task. For this purpose the Radial Basis Function (RBF) neural network is used.

We make use of a RBF neural network because they have the property of separating spaces that are not linearly separable. This is achieved by transforming the input vectors into a new feature space, where classes are linearly separable. For this to be possible, there must be more neurons in the hidden layer than in the input layer. In the considered problem, the input vector is formed through the prediction vectors generated by the base classifiers. So, more formally, a vector will be created $[\mu_1(x),\ldots,\mu_n(x)]$ for the test object x. Its dimension is equal to $n \cdot c$ because we have n base classifiers, each generating a c dimensional vector. Thus, we have $n \cdot c$ neurons in the input layer. RBF neural networks always contain only one hidden layer, thus, there are no problems with determining the appropriate number of hidden layers for a given problem. Each neuron in the hidden layer has a parameter called center. Formally, the k–th neuron in the hidden layer has the center c_k. The center is interpreted as a representative of a certain group of objects. RBF networks are similar to the k–nearest neighbors approach. However, instead of eliminating distant objects from the classification process (as it is done in the k–nearest neighbors classifier), this time we simply reduce the influence of distant objects on the output of the neural network, but still use them in the classification process. We obtain this property by using the Gaussian function as the activation function for each of the neurons in the hidden layer. The more the input vector is similar to the center of the neuron in the hidden layer, the greater its influence on the output of the neural network. Let us denote the vector given as the input of the network in the input layer by y. Then for the k–th neuron of the hidden layer the Gaussian function is as follows

$$\Phi_k(y) = \exp\left[-\frac{\| y - c_k \|}{2\sigma_k^2}\right], \tag{1}$$

where c_k is the center of the i–th neuron, σ_k is the k–th neuron's bandwidth and $\| \cdot \|$ is the Euclidean norm. The Gaussian width σ_k of the k–th neuron in the hidden layer was estimated as $\sigma_k = \frac{\rho_{\max}}{2 \cdot n_H}$ where $\rho_{\max}$ is the maximum distance between the chosen centers and n_H is the number of neurons in the hidden layer. The weights and biases assigned to the connections of neurons from the hidden layer to the output layer in the RBF network are trained using the back-propagation algorithm.

A very important issue in RBF networks is the appropriate determination of the center of neurons in the hidden layer. This is usually done by a clustering algorithm realized on the training set (this is implemented before the training of the network). Usually the k–means clustering algorithm is used and the centroids determined by this algorithm are used as centers connecting the input layer to the hidden layer neurons. The Lloyd's $k-$means algorithm – a modification of the $k-$means algorithm is also very often used. In the study, one of the approaches analyzed is the Lloyd's $k-$means algorithm for determining centers. But more interesting is the proposed approach in which instead of designating centers only once (determined by the Lloyd's $k-$means algorithm), a step of training these centers is used. This is implemented as follows.

- In the first stage, the Lloyd's $k-$means algorithm is used to determine the centers which serves as connections between the input layer and the hidden layer.
- After, random values are assigned as weights and biases, which serve as connections between the hidden layer and the output layer.
- Then, training the RBF network begins, however, here the propagation of the error is extended up to the centers. This way, the centers are trained iteratively together with the weights and biases using the back-propagation method.

The pseudo-code for the proposed RBF neural network is given below in Listing 2.

RBF Neural Network with a Centers Training Stage

```
# X: matrix of prediction vectors over all local tables
# centers: array of values determined by the Lloyd's algorithm
# neurons: number of neurons in the hidden layer
# sigma: Gaussian width

def RBFModel(X:array, centers:array, neurons:int, sigma:float):
  model = Sequential()
  model.add(Flatten(input_shape=(X.shape[1],)))
  model.add(RBFLayer(units= neurons, gamma=sigma))
  model.add(Dense(classes, activation='softmax'))
  model.layers[1].set_weights([centers])
  model.compile(loss='categorical_crossentropy',
  optimizer='adam',metrics=['accuracy'])
  return model
```

The above describes how to build and train the RBF network. An important issue still is the set used for training, the number of epochs used and the optimal batch-size. In the problem presented here, local tables were used to generate prediction vectors for objects. In addition, the local tables contain only fragmented data, and to train the network we need objects that have values for all attributes that are present in the local tables (in order to designate vectors for all tables). So, to train the RBF network, a stratified 10-fold cross-validation method on the test set was used. That is, the test set was divided into 10 folds with equal number of objects and proportional shares of decision classes. At each iteration, the RBF network was trained using 9 folds (to be very accurate – prediction vectors generated for test objects from these 9 folds), and the quality of classification of the model was evaluated on the last independent fold. This procedure was repeated three times and the results was averaged to determine the classification error level of the model. To determine the number of epochs and batch-size, different values were experimented to determine the optimal values for each dataset. After numerous trials, the optimal (epoch, batch-size) for Vehicle Silhouettes, Landsat Satellite and Dry Bean datasets were (400, 200), (400, 500), (400,15) respectively. RBF networks trains considerable fast so the whole process was fairly quick. A big advantage of the RBF networks was how interpretable the results obtained from it were. This was achieved thanks to properly selected/trained centers and their reduced influence in the case of large distances.

3 Datasets and Results

The system proposed in the paper for the classification of objects in images based on characteristics stored in fragmented form – a set of local decision tables – was tested on three datasets. These datasets in their original form were retrieved from the UC Irvine Machine Learning Repository:

- Vehicle Silhouettes: eighteen quantitative conditional attributes, four decision classes, 846 objects – 592 training, 254 test set [17]. The goal of the data was to classify a given silhouette as one of four vehicle types, using a set of characteristics extracted from the silhouette. The vehicle can be viewed from one of many different angles. The images were acquired by a camera looking down at the vehicle model from a fixed elevation angle. The vehicles were rotated and their orientation angle was measured using a radial grid placed under the vehicle.
- Landsat Satellite: thirty-six quantitative conditional attributes, six decision classes, 6435 objects – 4435 training, 1000 test set [6]. Multispectral pixel values in a 3×3 neighborhood in a satellite image and a classification associated with the central pixel in each neighborhood. Each row of data corresponds to a neighborhood of pixels in a 3×3 square completely contained within an 82×100 sub-area. Each row contains the pixel values in the four spectral bands of each of the 9 pixels in the 3×3 neighborhood and a number indicating the classification label (earth type: red soil, cotton crop, grey soil, damp grey soil, soil with vegetation stubble, very damp grey soil) of the center pixel.
- Dry Bean: seventeen quantitative conditional attributes, seven decision classes, 13611 objects – 9527 training, 4084 test set [9]. The goal of the data was to classify type of beans based on the characteristics obtained from the image. Images of 7 different registered dry beans were taken with a high-resolution camera. Bean images obtained by computer vision systems were subjected to segmentation and feature extraction stages, and a total of 16 features – 12 dimensions and 4 shape forms were obtained from the grains.

Each of the datasets were originally available as a single decision table. However, the proposed system explored the possibilities of classification based on fragmentary data. Therefore, the data was preprocessed – randomly dispersed. Different numbers of local tables were considered during dispersion. The data was divided into 3, 5, 7, 9 and 11 local tables. Each table contained a reduced set of conditional attributes and all objects from the original table. Some attributes were common between local tables. It should also be noted that, the more local tables, the fewer attributes in each local table. The data was imbalanced – the number of objects in individual decision classes, both in the training and test sets, varied strongly. The specific cardinality are presented in Fig. 1. Two variants for each of the datasets were considered in the study. Experiments were performed both on dispersed imbalanced data and on data that had been modified by applying one of the known methods for imbalanced data. The Synthetic Minority Over-sampling Technique (SMOTE) method was used in the paper [4]. This is an over-sampling method which adds artificially created objects to a dataset. The objects were created based on randomly selected minority class objects. For this purpose, the $k-$nearest neighbors algorithm was used. On the line connecting the selected object with its closest neighbors, a new object from the minority class was created. The implementation of this algorithm available in WEKA [11] software was used. Each local table was balanced separately. Each decision class except the most numerous one, was changed using SMOTE method in such a way that all decision classes had the same number of objects after balancing. Thus, in the end, we obtained 30 dispersed datasets: Vehicle Silhouettes, Vehicle Silhouettes balanced, Landsat Satellite, Landsat Satellite balanced, Dry Bean, Dry Bean balanced; and for each dataset five versions of the dispersion:

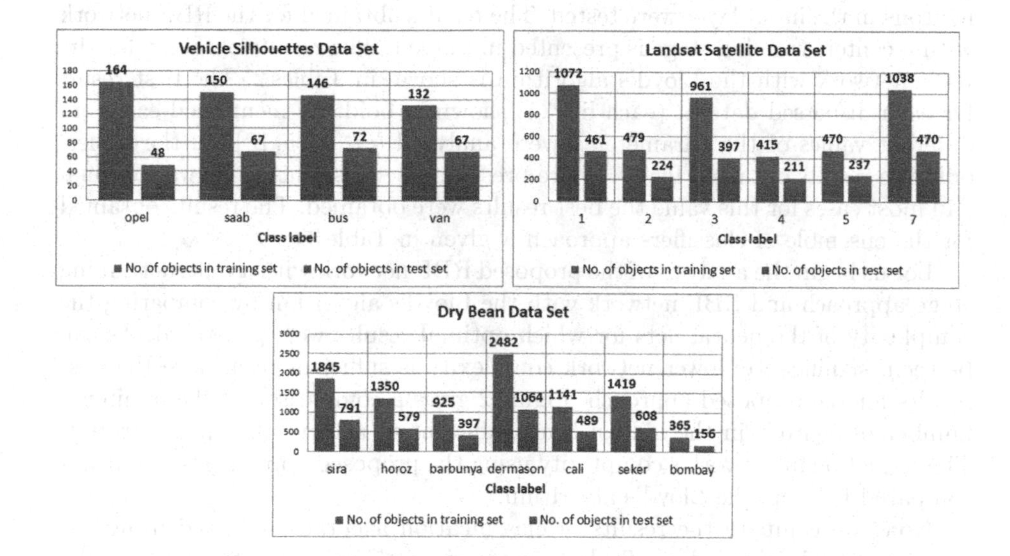

Fig. 1. Imbalance of data – cardinality of decision classes in training and test sets.

3, 5, 7, 9, 11 local tables. The quality of classification was evaluated based on the test set. A classification accuracy measure (*acc*) was used for this purpose. That is, a fraction of the total number of objects in the test set that were classified correctly. As was mentioned before, a 10-fold cross-validation was used on the test set, i.e., the neural network was trained 10 times with 9 folds and tested on one remaining fold. In addition, each test was performed three times to ensure that the results were reliable and not distorted by the influence of randomness. The results for the neural network approach that are given below is the average of the obtained results.

Three approaches were tested and the results are presented below. The two tested approaches use the RBF networks described above. The first RBF approach used a centers training stage by means of back-propagating the error up to the centers during training whiles the second RBF approach used only the Lloyd's $k-$means algorithm to determine the centers. As a baseline approach, the approach proposed in [10] was used. This ensemble of classifiers method consists of creating three base classifiers: $k-$nearest neighbors, decision tree and naive bayes classifier (KNN, DT, NB) based on each local table. The approach was implemented in python programming language using implementations available in the sklearn library. Based on each local table, the three classifiers were built. The final decision was made using soft voting for all classifiers. Different parameter values were tested for approaches based on RBF networks. For the generation of prediction vectors using the $k-$nearest neighbors classifier, $k \in \{1, 5, 10\}$ parameters were studied. Due to the limited space of this paper, only the results for the optimal k value is presented. Different numbers of neurons in the hidden layer were tested for RBF neural networks. The following values $\{0.25, 0.5, 0.75, 1, 1.5, 1.75, 2, 2.5, 2.75, 3, 3.5, 3.75, 4, 4.5, 4.75, 5\}\times$ the number of neurons in the input layer were tested. The results obtained for the RBF network with a centers training stage is presented in Table 1. The results obtained for the RBF network with the Lloyd's algorithm is shown in Tables 2. The best result for each dispersed dataset (each line) is shown in bold. As mentioned early on, different values of the parameter k were analyzed $k \in \{1, 5, 10\}$. In the tables, only the results obtained for $k = 5$ are presented because this value was optimal – in most cases for this value the best results were obtained. The results obtained for the ensemble of classifiers approach is given in Table 3.

Let us begin the analysis of the proposed RBF network with a centers training stage approach and RBF network with the Lloyd's algorithm by comparing the complexity of the neural nets for which optimal results were generated. As can be seen, significantly lower network complexity is sufficient to achieve the best results for the proposed approach. Figure 2 gives a comparison of the minimum number of neurons in the hidden layer sufficient to achieve the optimal result. The reduction in network's complexity using the proposed approach is significant compared to using the Lloyd's algorithm.

Now, we compare the results of classification accuracy obtained using the three analyzed approaches. Table 3 summarizes all results – the best results obtained for the proposed approach, the best results using the RBF networks

Table 1. Results of classification accuracy *acc* for the RBF network with a centers training stage. Designation I is used for the number of neurons in the input layer.

Dataset	No.	No. of neurons in hidden layer															
optimal k parameter	tables	0.25 × I	0.5 × I	0.75 × I	1 × I	1.5 × I	1.75 × I	2 × I	2.5 × I	2.75 × I	3 × I	3.5 × I	3.75 × I	4 × I	4.5 × I	4.75 × I	5 × I
Vehicle	3	0.483	0.608	0.642	0.661	0.702	0.719	0.725	0.735	0.734	0.759	0.756	0.763	0.752	**0.769**	0.762	0.765
imbalanced	5	0.568	0.651	0.667	0.683	0.716	0.724	0.733	0.75	0.754	0.758	0.754	0.743	0.75	0.749	0.749	**0.762**
k = 5	7	0.567	0.639	0.671	0.668	0.685	0.702	**0.709**	0.7	0.701	0.694	0.685	0.682	0.69	0.68	0.682	0.684
	9	0.634	0.664	0.7	0.721	**0.73**	0.707	0.706	0.709	0.691	0.7	0.681	0.672	0.671	0.65	0.643	0.637
	11	0.6	0.654	0.69	0.707	0.711	0.702	0.707	0.705	**0.718**	0.703	0.714	0.714	0.707	0.712	0.707	0.702
Vehicle	3	0.721	0.735	0.741	0.735	0.73	0.744	0.752	0.752	0.748	0.759	0.756	0.744	0.76	0.755	0.755	**0.764**
balanced	5	0.709	0.72	0.755	0.757	0.754	**0.759**	0.749	0.745	0.755	0.745	0.745	0.734	0.73	0.725	0.719	0.724
k = 5	7	0.726	0.727	**0.734**	0.714	0.722	0.706	0.707	0.7	0.684	0.684	0.667	0.658	0.653	0.629	0.625	0.62
	9	0.723	**0.732**	0.718	0.701	0.68	0.685	0.681	0.672	0.654	0.637	0.626	0.628	0.606	0.609	0.593	0.57
	11	**0.727**	**0.727**	0.72	0.722	0.702	0.698	0.693	0.668	0.668	0.659	0.647	0.631	0.597	0.593	0.58	0.588
Satellite	3	0.788	0.834	0.849	0.854	0.874	0.88	0.885	0.891	0.892	0.896	0.894	0.896	0.896	**0.897**	0.894	0.896
imbalanced	5	0.827	0.845	0.855	0.866	0.879	0.883	0.885	0.887	0.888	0.888	0.886	0.891	0.888	0.886	**0.892**	**0.892**
k = 5	7	0.835	0.855	0.87	0.877	0.884	0.883	0.886	**0.887**	0.884	0.883	0.884	**0.887**	0.884	0.883	0.882	0.882
	9	0.845	0.857	0.87	0.877	0.884	**0.889**	0.883	0.886	0.883	0.885	0.883	0.88	0.882	0.882	0.881	0.882
	11	0.843	0.856	0.864	0.87	0.875	0.877	0.877	0.874	**0.878**	0.874	0.874	0.876	0.872	0.87	0.865	0.868
Satellite	3	0.659	0.775	0.813	0.832	0.852	0.856	0.858	0.864	0.864	0.867	0.867	0.868	0.872	0.873	**0.875**	0.874
balanced	5	0.753	0.808	0.839	0.849	0.85	0.855	0.854	0.858	0.858	0.863	0.862	0.864	**0.866**	0.865	**0.866**	**0.866**
k = 5	7	0.78	0.834	0.852	0.851	0.85	0.856	0.859	0.861	**0.862**	0.861	0.861	0.86	0.861	0.86	0.859	0.859
	9	0.801	0.843	0.847	0.849	0.857	0.859	0.861	0.86	0.859	0.861	**0.863**	0.86	0.86	0.861	0.855	0.855
	11	0.815	0.843	0.846	0.845	0.851	0.851	0.855	0.859	**0.861**	0.859	0.857	0.854	0.85	0.849	0.848	0.846
Dry Bean	3	0.786	0.857	0.883	0.895	0.907	0.907	0.911	0.914	0.916	0.916	0.917	0.916	0.918	0.919	**0.92**	**0.92**
imbalanced	5	0.851	0.893	0.904	0.912	0.915	0.916	0.917	**0.918**	0.917	**0.918**	0.917	**0.918**	0.917	0.917	0.917	0.917
k = 5	7	0.871	0.903	0.913	0.914	0.916	0.916	**0.917**	**0.917**	0.916	**0.917**	**0.917**	**0.917**	**0.917**	0.916	0.916	0.915
	9	0.881	0.905	0.912	0.915	0.917	0.917	0.917	0.918	**0.919**	0.918	0.918	0.918	0.917	0.918	0.918	0.918
	11	0.888	0.908	0.912	0.915	**0.916**	0.914	**0.916**	0.915	**0.916**	0.915	0.915	**0.916**	**0.916**	0.914	0.914	0.915
Dry Bean	3	0.789	0.855	0.882	0.892	0.902	0.908	0.91	0.913	0.916	0.916	**0.918**	**0.918**	**0.918**	**0.918**	0.917	**0.918**
balanced	5	0.852	0.892	0.905	0.91	0.915	0.916	**0.918**	**0.918**	0.917	**0.918**	0.917	0.916	0.917	0.916	0.916	0.917
k = 5	7	0.865	0.901	0.91	0.914	**0.917**	**0.917**	0.915	0.916	0.916	**0.917**	0.916	0.916	**0.917**	**0.917**	0.914	0.915
	9	0.882	0.906	0.911	0.916	0.918	**0.919**	0.918	0.918	**0.919**	0.917	0.918	0.917	0.918	0.915	0.917	0.917
	11	0.886	0.906	0.911	0.915	0.917	**0.918**	0.916	0.916	0.917	0.916	0.917	0.917	0.916	0.914	0.915	0.916

Table 2. Results of classification accuracy *acc* for the RBF network with the Lloyd's algorithm. Designation I is used for the number of neurons in the input layer.

Data set	No.	No. of neurons in hidden layer															
optimal k parameter	tables	$0.25 \times I$	$0.5 \times I$	$0.75 \times I$	$1 \times I$	$1.5 \times I$	$1.75 \times I$	$2 \times I$	$2.5 \times I$	$2.75 \times I$	$3 \times I$	$3.5 \times I$	$3.75 \times I$	$4 \times I$	$4.5 \times I$	$4.75 \times I$	$5 \times I$
Vehicle	3	0.268	0.351	0.427	0.476	0.54	0.55	0.571	0.588	0.613	0.629	0.631	0.639	0.642	0.641	0.65	**0.656**
imbalanced	5	0.289	0.429	0.487	0.529	0.57	0.574	0.612	0.625	0.638	0.631	0.657	0.657	0.671	0.67	0.688	**0.692**
k = 5	7	0.341	0.424	0.479	0.523	0.566	0.597	0.603	0.633	0.638	0.638	0.656	0.661	0.669	0.671	0.672	**0.676**
	9	0.363	0.496	0.545	0.582	0.616	0.623	0.637	0.659	0.655	0.665	0.686	0.688	0.716	0.719	0.719	**0.727**
	11	0.359	0.469	0.537	0.575	0.601	0.61	0.623	0.651	0.647	0.659	0.674	0.683	0.687	0.687	0.686	**0.698**
Vehicle	3	0.556	0.692	0.689	0.691	0.722	0.722	0.742	0.738	0.746	0.731	0.741	0.746	0.75	0.751	0.75	**0.761**
balanced	5	0.509	0.618	0.643	0.659	0.692	0.719	0.71	0.718	0.731	0.735	0.744	**0.757**	0.751	0.753	0.748	0.747
k = 5	7	0.683	0.695	0.702	0.71	0.735	0.717	0.739	0.731	0.733	0.725	0.731	**0.741**	0.739	0.734	0.738	0.738
	9	0.503	0.671	0.674	0.687	0.707	0.709	0.717	0.72	0.721	**0.729**	0.728	0.72	0.724	0.725	0.725	0.72
	11	0.561	0.611	0.637	0.691	0.71	0.715	0.713	**0.722**	**0.722**	0.715	0.705	0.709	0.701	0.701	0.694	0.694
Satellite	3	0.729	0.789	0.817	0.823	0.83	0.833	0.834	0.841	0.846	0.846	0.85	0.848	0.851	0.851	0.852	**0.853**
imbalanced	5	0.759	0.809	0.828	0.83	0.838	0.84	0.845	0.849	0.85	0.852	0.852	0.852	0.855	0.855	**0.859**	**0.859**
k = 5	7	0.773	0.825	0.834	0.833	0.841	0.846	0.848	0.851	0.853	0.853	0.854	0.858	0.858	0.859	0.859	**0.863**
	9	0.814	0.832	0.832	0.839	0.841	0.845	0.843	0.849	0.849	0.851	0.857	0.857	0.859	0.861	0.861	**0.862**
	11	0.81	0.831	0.841	0.841	0.844	0.847	0.848	0.85	0.851	0.851	0.854	0.855	0.856	0.857	**0.86**	0.859
Satellite	3	0.56	0.674	0.72	0.757	0.791	0.804	0.81	0.818	0.825	0.829	0.837	0.84	0.843	0.846	0.848	**0.849**
balanced	5	0.629	0.714	0.758	0.787	0.815	0.819	0.822	0.839	0.841	0.842	0.843	0.845	0.848	**0.85**	0.847	0.847
k = 5	7	0.644	0.75	0.799	0.809	0.823	0.829	0.838	0.843	0.843	0.843	0.845	0.846	0.849	0.85	**0.851**	0.85
	9	0.669	0.774	0.8	0.818	0.828	0.832	0.843	0.843	0.844	0.841	0.846	**0.847**	0.846	0.843	0.844	0.845
	11	0.714	0.789	0.814	0.82	0.834	0.837	0.837	0.841	0.844	0.846	0.849	0.851	0.848	**0.851**	0.85	0.849
Dry Bean	3	0.591	0.777	0.828	0.842	0.864	0.869	0.871	0.882	0.886	0.887	0.891	0.892	0.895	0.897	**0.9**	0.898
imbalanced	5	0.762	0.833	0.858	0.874	0.89	0.893	0.894	0.897	0.899	0.9	0.902	0.904	0.905	0.905	0.904	**0.906**
k = 5	7	0.796	0.847	0.873	0.883	0.893	0.896	0.9	0.901	0.904	0.903	0.905	0.905	0.906	0.907	0.908	**0.909**
	9	0.817	0.856	0.877	0.885	0.892	0.894	0.897	0.899	0.9	0.901	0.901	0.901	0.904	0.905	**0.906**	**0.906**
	11	0.827	0.867	0.886	0.891	0.898	0.9	0.9	0.901	0.901	0.902	0.901	0.902	0.902	0.903	0.904	**0.905**
Dry Bean	3	0.58	0.774	0.818	0.841	0.859	0.873	0.876	0.884	0.885	0.889	0.891	0.892	0.893	0.893	**0.896**	0.894
balanced	5	0.766	0.835	0.854	0.87	0.886	0.891	0.892	0.895	0.898	0.9	0.899	0.903	0.902	0.901	0.903	**0.905**
k = 5	7	0.798	0.852	0.869	0.881	0.891	0.894	0.898	0.9	0.901	0.901	0.902	0.905	0.906	0.907	0.908	**0.91**
	9	0.817	0.859	0.873	0.885	0.894	0.896	0.897	0.9	0.9	0.901	0.902	0.903	0.905	0.906	**0.907**	0.905
	11	0.828	0.868	0.882	0.889	0.896	0.897	0.9	0.901	0.902	0.902	0.901	0.902	0.903	**0.904**	**0.904**	**0.904**

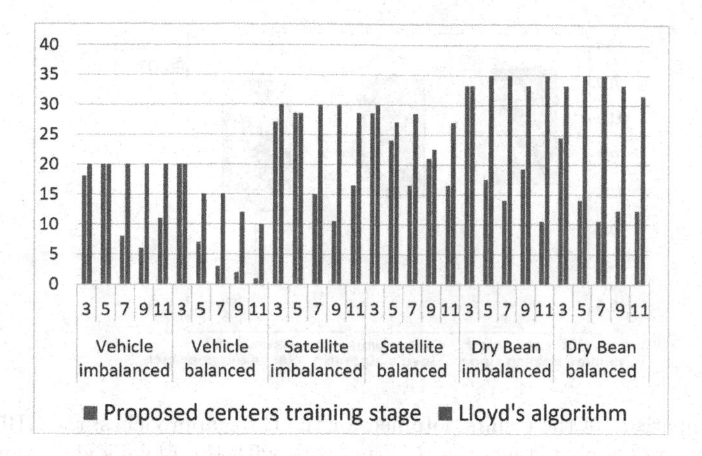

Fig. 2. The minimum number of neurons in the hidden layer sufficient to achieve the optimal result – comparison of RBF networks with a centers training stage approach and with the Lloyd's algorithm.

with the Lloyd's algorithm and the results for the ensemble of classifiers approach. The best result for each dispersed dataset (each line) is shown in bold. As can be seen, in all cases (except one) the best results were generated using the proposed approach. Statistical tests were performed in order to confirm significant differences in the obtained results *acc*. The received classification accuracy were divided into three dependent data samples, results from Table 3. The Friedman test was used to detect differences in multiple test samples. There was a statistically significant difference in the results obtained for the three different approaches being considered, $\chi^2(29, 2) = 39.467, p = 0.000001$. Additionally, comparative box-whiskers charts for the results with three approaches were created (Fig. 3). As can be observed, the values of the classification accuracy for the proposed approach – RBF network with a centers training stage is the best (much better than the others approaches). In the next step, the Wilcoxon each-pair test was used. This test confirmed that the differences in the classification accuracy were significant between the RBF network with a centers training stage and both the RBF network with the Lloyd's algorithm and the ensemble of classifiers approach. There in no statistically significant difference in the classification accuracy between the RBF network with the Lloyd's algorithm and the ensemble of classifiers approach.

For the proposed approach, a comparison of the results obtained by using balanced and imbalanced datasets were also made. Figure 4 shows the comparison of the results in a bar chart and box-whiskers charts. As can be seen for the Dry Bean set, balancing the dataset had no effect on the results. For the Satellite set, better results were obtained for the imbalanced dataset. On the other hand, for the Vehicle dataset, for a smaller number of local tables (3 and 5 tables) we got better results for the imbalanced data. Also, the box-whiskers charts confirmed that there was no difference between the results for balanced and imbalanced

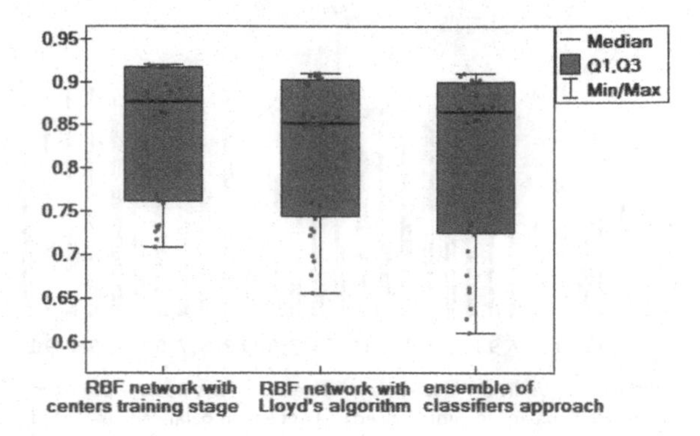

Fig. 3. Comparison of the results obtained for the three approaches: the RBF network with the centers training stage, the RBF network with the Lloyd's algorithm and the ensemble of classifiers approach.

datasets. The Wilcoxon test confirmed that there was no statistically significant difference in the average classification accuracy obtained for balanced and imbalanced sets for the proposed approach. Thus, it can be concluded that the proposed approach performs very well for imbalanced data.

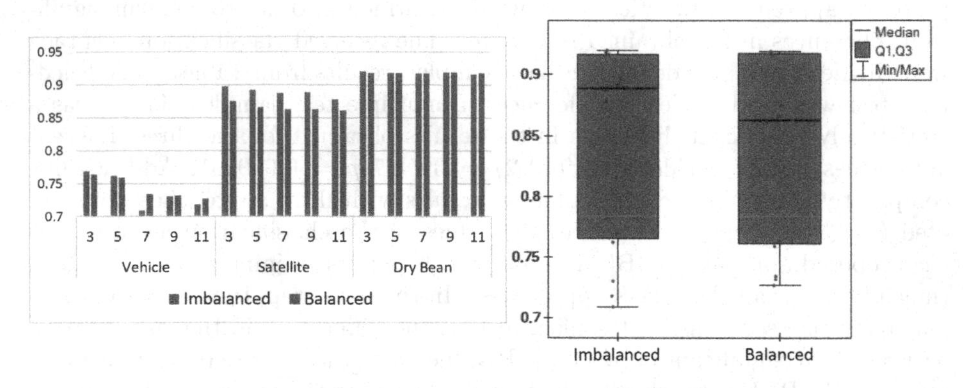

Fig. 4. Comparison of the results obtained for the RBF network with a centers training stage and imbalanced versus balanced datasets.

For the proposed approach, a comparison of the results obtained for different versions of dispersion was made. Figure 5 shows the comparison of the results in a bar chart and box-whiskers charts. As can be seen for the Dry Bean dataset, degree of dispersion of the dataset had no effect on the results. For the Satellite and the Vehicle datasets, better results were obtained for a smaller number of local tables – a smaller degree of dispersion, but the differences were not large. The Wilcoxon test confirmed that there was statistically significant differences in the average classification accuracy only between pairs of dispersion: 3 and

Table 3. Comparison of classification accuracy *acc* obtained for approaches: the RBF network with a centers training stage, the RBF network with the Lloyd's algorithm and the ensemble of classifiers approach from paper [10].

Dataset	No. tables	RBF network with a centers training stage	RBF network with Lloyd's algorithm	ensemble of classifiers from paper [10]
Vehicle	3	**0.769**	0.656	0.657
imbalanced	5	**0.762**	0.692	0.638
k=5	7	**0.709**	0.676	0.626
	9	**0.73**	0.727	0.661
	11	**0.718**	0.698	0.61
Vehicle	3	**0.764**	0.761	0.728
balanced	5	**0.759**	0.757	0.724
k=5	7	0.734	**0.741**	0.736
	9	**0.732**	0.729	0.705
	11	**0.727**	0.722	0.677
Satellite	3	**0.897**	0.853	0.885
imbalanced	5	**0.892**	0.859	0.872
k=5	7	**0.887**	0.863	0.868
	9	**0.889**	0.862	0.868
	11	**0.878**	0.86	0.863
Satellite	3	**0.875**	0.849	0.87
balanced	5	**0.866**	0.85	0.864
k=5	7	**0.862**	0.851	0.856
	9	**0.863**	0.847	0.856
	11	**0.861**	0.851	0.854
Dry Bean	3	**0.92**	0.9	0.906
imbalanced	5	**0.918**	0.906	0.902
k=5	7	**0.917**	0.909	0.899
	9	**0.919**	0.906	0.894
	11	**0.916**	0.905	0.9
Dry Bean	3	**0.918**	0.896	0.909
balanced	5	**0.918**	0.905	0.899
k=5	7	**0.917**	0.91	0.9
	9	**0.919**	0.907	0.898
	11	**0.918**	0.904	0.903

7 local tables; 5 and 7 local tables; 9 and 11 local tables. The conclusion of this comparison is that the proposed approach handles both small and large data dispersion quite well which is a very important property because often in real situations we have to deal with large dispersion – many units providing independent datasets. The proposed method requires optimization of several parameters in both the k-nearest neighbors classifier and the neural network, which can be considered a drawback of the method.

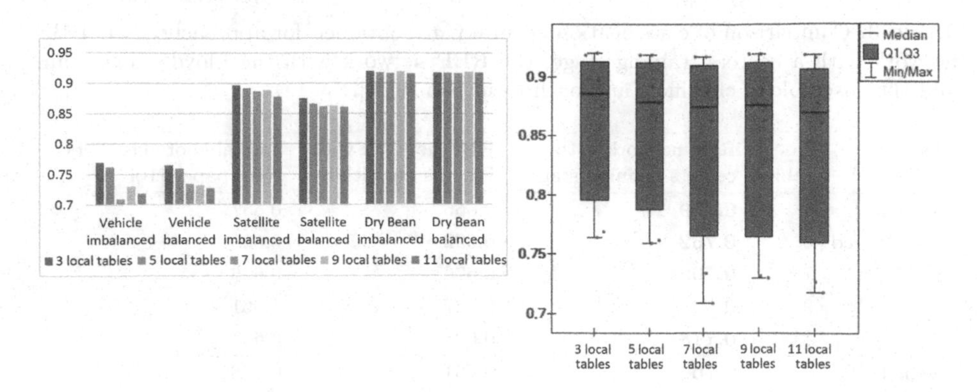

Fig. 5. Comparison of the results obtained for the RBF network with a centers training stage and different dispersion versions of datasets – $3, 5, 7, 9$ and 11 local tables.

4 Conclusion

The study concerned analysis of the classification problem of an object presented in an image based on the characteristics of the object stored in a fragmentary form – a set of local decision tables. For this purpose, a RBF network model with a centers training stage was proposed. The paper shows that this approach gives much better results than the RBF network model with the Lloyd's algorithm. In addition, the network's structure is much simpler for the proposed approach. Moreover, the proposed approach gives better results than the ensemble of classifiers approach. It was also shown that the proposed model copes very well with imbalanced datasets and that the degree of dispersion does not have a large impact on classification accuracy. In future works, it is planned to use neural networks to define predictions based on local tables. The possibility of using a global learning stage after building the RBF network that combines predictions is also being considered. This stage would be implemented by using some artificially generated data.

References

1. Basha, S.S., Dubey, S.R., Pulabaigari, V., Mukherjee, S.: Impact of fully connected layers on performance of convolutional neural networks for image classification. Neurocomputing **378**, 112–119 (2020)
2. Caggiano, A., Zhang, J., Alfieri, V., Caiazzo, F., Gao, R., Teti, R.: Machine learning-based image processing for on-line defect recognition in additive manufacturing. CIRP Ann. **68**(1), 451–454 (2019)
3. Chaki, J., Dey, N.: Fragmented handwritten digit recognition using grading scheme and fuzzy rules. Sādhanā **45**(1), 1–23 (2020). https://doi.org/10.1007/s12046-020-01410-5
4. Chawla, N.V., Bowyer, K.W., Hall, L.O., Kegelmeyer, W.P.: SMOTE: synthetic minority over-sampling technique. J. Artif. Intell. Res. **16**, 321–357 (2002)

5. Chen, J., Liu, G., Chen, X.: AnimeGAN: a novel lightweight GAN for photo animation. In: International Symposium on Intelligence Computation and Applications, pp. 242–256. Springer, Singapore (2020)
6. Dua, D., Graff, C.: UCI Machine Learning Repository [http://archive.ics.uci.edu/ml]. Irvine, CA: University of California, School of Information and Computer Science (2019)
7. Fornasier, M., Toniolo, D.: Fast, robust and efficient 2D pattern recognition for re-assembling fragmented images. Pattern Recogn. **38**(11), 2074–2087 (2005)
8. Giuseppe, A., Falchi, F.: *k*-NN based image classification relying on local feature similarity. In: International Conference on Similarity Search and Applications, SISAP 2010, Istanbul, Turkey, pp. 101–108 (2010)
9. Koklu, M., Ozkan, I.A.: Multiclass classification of dry beans using computer vision and machine learning techniques. Comput. Electron. Agri. **174**, 105507 (2020)
10. Kurian, R.A., Lakshmi, K.: An ensemble classifier for the prediction of heart disease. Int. J. Sci. Res. Comput. Sci. **3**, 25–31 (2018)
11. Markov, Z., Russell, I.: An introduction to the WEKA data mining system. SIGCSE Bull. **38**(3), 367–368 (2006). https://doi.org/10.1145/1140123.1140127
12. Mozgovoy, D.K., Hnatushenko, V.V., Vasyliev, V.V.: Automated recognition of vegetation and water bodies on the territory of megacities in satellite images of visible and IR bands. ISPRS Ann. Photogrammetry Remote Sens. Spat. Inf. Sci. **4**(3), 167–172 (2018)
13. Przybyła-Kasperek, M., Wakulicz-Deja, A.: Global decision-making system with dynamically generated clusters. Inform. Sci. **270**, 172–191 (2014)
14. Ptucha, R., Such, F.P., Pillai, S., Brockler, F., Singh, V., Hutkowski, P.: Intelligent character recognition using fully convolutional neural networks. Pattern Recogn. **88**, 604–613 (2019)
15. Santos, I., Castro, L., Rodriguez-Fernandez, N., Torrente-Patino, A., Carballal, A.: Artificial neural networks and deep learning in the visual arts: a review. Neural Comput. Appl. **33**(1), 121–157 (2021)
16. Shelepin, Y.E., Chikhman, V.N., Foreman, N.: Analysis of the studies of the perception of fragmented images: global description and perception using local features. Neurosci. Behav. Physiol. **39**(6), 569–580 (2009)
17. Siebert, J.P.: Vehicle Recognition Using Rule Based Methods, Turing Institute Research Memorandum TIRM-87-0.18, March 1987
18. Sochor, J., Špaňhel, J., Herout, A.: Boxcars: improving fine-grained recognition of vehicles using 3-D bounding boxes in traffic surveillance. IEEE Trans. Intell. Trans. Syst. **20**(1), 97–108 (2018)
19. Traore, B.B., Kamsu-Foguem, B., Tangara, F.: Deep convolution neural network for image recognition. Ecol. Inform. **48**, 257–268 (2018)
20. Wan, S., Goudos, S.: Faster R-CNN for multi-class fruit detection using a robotic vision system. Comput. Netw. **168**, 107036 (2020)
21. Yadav, S.S., Jadhav, S.M.: Deep convolutional neural network based medical image classification for disease diagnosis. J. Big Data **6**(1), 1–18 (2019). https://doi.org/10.1186/s40537-019-0276-2
22. Yang, R., Sarkar, S.: Gesture recognition using hidden markov models from fragmented observations. In: 2006 IEEE Computer Society Conference on Computer Vision and Pattern Recognition (CVPR 2006), vol. 1, 766–773. IEEE (2006)

Database of Fragments of Medieval Codices of the 11th–12th Centuries – The Uniqueness of Requirements and Data

Jakub Leszek Pach[1,2](✉)

[1] Warsaw University of Life Sciences – SGGW, Nowoursynowska 166, 02-787 Warsaw, Poland
jakub_pach@sggw.edu.pl
[2] The National Library of Poland, al. Niepodległości 213, 02-086 Warsaw, Poland

Abstract. This paper presents a new offline dataset called the Fragments of Medieval Codices (FOMC). It contains medieval Latin handwritings coming from 11th–12th century and can be used to evaluate the performance of offline writer identification and to find the handwriting similarity between the writers, or to test the handwritten optical character recognition systems. It consists of 117 fragments of handwritten documents of medieval codices and contains in total over two thousand very high quality images. The collection was assembled using the IIIF standard. We describe the collecting and processing steps performed to develop the dataset and define several evaluation tasks regarding the use of this dataset.

Keywords: Latin manuscripts database · offline writer identification · optical character recognition

1 Introduction

Databases of handwritten texts in image processing have a dual purpose, which are the identification of writers and the text recognition. For the first purpose, correct classification is possible without the need to recognize letters and, consequently, the content. In the case of the second purpose, it is necessary to correctly identify each character - a letter - to recognize the content of the document. OCR and writer identification can work together, because this makes it possible to analyze the similarity of specific characters. Both purposes already have many dedicated methods that perform the task with very high efficiency for modern writing.

However, the identification of Latin manuscripts is still a major challenge today, partly because of how medieval writing was done and what it was written on. Writing materials, e.g. parchment, were incomparably more expensive than paper used today is, and this resulted in a much tighter written text, in order to save every fragment of such a valuable material. The letters of the text often

J. Mikyška et al. (Eds.): ICCS 2023, LNCS 14076, pp. 104–112, 2023.
https://doi.org/10.1007/978-3-031-36027-5_8

overlapped, the margins contained comments by copyists and descendants, and some of the content was written in complicated abbreviations. The ability to decipher the content of such documents is now only within the reach of specialist historians. Hence, in order for such writer identification systems to implement the discussed topic, interdisciplinary cooperation of highly qualified specialists in the discipline of history and computer science is necessary. The second reason why there is still so much to be done in the discussed topic is the lack of databases on which these systems could be taught and refined, while for modern writings such databases are easier to find. Therefore, in this paper a database of Latin manuscripts from the 11th–12th centuries, which is supposed to help in this matter, is presented. The database is accessible at [15].

The remaining part of this paper is organized as follows. In the second section, an overview of existing databases in the field of writer recognition is provided. In Sect. 3, a detailed description of the presented database and its analysis is given. Section 4 contains conclusions and discussion on further work with the database.

2 Existing Databases

The criterion of selection of the databases described here is their high frequency of use. The author searched databases of papers in the field of author identification and handwriting recognition coming from the last two decades, with the largest scientific value measured by the number of citations, and selected those which are most often used to validate systems of writer identification. If a base is no longer available or it is very rarely cited by other authors, such base is omitted. Due to that the Latin scripts are the object of studies of the author, the use of the Latin alphabet script was the key to the selection of reference databases. Non-alphabetic script appears only as a reference.

The description of the databases is made in alphabetical order.

The BFL Database [4] was made in the first decade of 21th century in Latin America that has 315 writers, three samples per writer, in Portuguese language.

The CERUG-MIXED dataset [17] established in University of Groningen also in 21th century contains handwritten documents collected from 105 writers, four samples per writer that wrote in two different languages - Chinese and English.

The next database which is the second most commonly used database written in modern English and German is CVL [8]. It contains 311 different writers where one writer wrote between five and seven samples, mainly in English but one sample is in German. This database was created in 21th century as well.

On the last year of the second millennium, the Firemaker [1] database was created that contains 1000 images of scanned handwritten text, containing pages of text written by 250 writers, four pages per writer, in Dutch.

The GDRS dataset [11] is next database generated by a research group in the Computational Intelligence Laboratory at the National Center for Scientific Research “Demokritos”, Greece, in 21th century. 26 writers made eight sample texts written in four languages, that is, English, German, Greek and French.

The following database IAM [14] is probably the most commonly used dataset in writer identification and contains 657 writers, where writers wrote samples between one to almost sixty. The part of this database which is used mainly is known as MIAM (Modified IAM).

Another popular historic database is IAM-HistDB [3]. It contains three medieval manuscripts from 9th, 13th and 18th century and has 127 pages.

The next benchmarking dataset ICDAR2013 [10] was created in 21th century with the help of 250 writers who wrote four parts of text in two languages: English and Greek. This base contains the whole ICDAR2011.

The largest historical database written in Latin is ICDAR2017 [2] known also as HistoricalWI-2017. It contains 3600 handwritten pages originating from 13th to 20th century. It contains manuscripts from 720 different writers where each writer contributed five pages.

The IFN/ENIT database [16] contains Arabic handwriting from 411 writers, five samples each.

Database JEITA-HP [9] was prepared by Hewlett-Packard Japan and distributed by Japan Electronics and Information Technology Industries Association, in Japanese. It consists of two datasets: Dataset A and Dataset B, which store handwritten character patterns from 480 writers and 100 writers, respectively.

KHATT database [13] is a database of unconstrained handwritten Arabic Text written by 1000 different writers, four samples per writer.

The next offline dataset [12] is called the Qatar University Writer Identification dataset (QUWI). This dataset contains both Arabic and English handwritings. It consists of handwritten documents of 1017 writers, four samples each. The last database is RIMES [5] and was created to evaluate automatic systems of recognition and indexing of handwritten letters. The collection was a success with more than 1,300 people who have participated to the RIMES database creation by writing up to five mails (handwritten correspondence) in French and Bengali.

The information an all databases have been compiled in Table 1 for easier comparison. Let us analyze those of the databases with have a large historical component.

The IAM-HistDB database is composed of three codes that were created in different historical eras, styles and languages. This is profitable for a machine learning because there is a transcript of content and it is also possible to train a neural network to read content from the same age in other manuscripts. However, when it comes to author identification, three writers are not sufficient to properly validate the effectiveness of writer identification systems.

In addition, from the perspective of historical science, it is necessary to take into account the fact that two scientific disciplines, paleography and neography, deal with handwriting in this period of time. Paleography deals with handwriting until the creation of the first incunabula - the first printed books (until the 15th century), and since then, neography has been dealing with modern handwriting. The fact that the separation of two separate scientific disciplines was necessary

Table 1. Comparison between datasets.

Database acronym	Language	No. of writers	No. of images	Samples per writer	Type of a sample	Rounded width of a sample	Age of samples (century)
BFL [4]	Portuguese	315	945	3	page	2500	21
CERUG-MIXED [17]	Chinese; English	105	420	4	page	–	21
CVL [8]	English; German	310	1604	5–7	page	1800	21
Firemaker [1]	Dutch	250	1000	4	page	2500	20
GRDS [11]	English; German; Greek; French	26	208	8	page	–	21
IAM [14]	English	657	1539	1–57	page	–	21
IAM-HistDB [3]	Latin; Medieval German; English	3	127	60; 47; 20	page	3300	9; 13; 18
ICDAR2013 [10]	Greek	250	1000	5	page	2500	21
ICDAR2017 [2]	Latin	720	3600	5	page	<1000; 1500;	13–20
IFN/ENIT [16]	Arabic	411	2200	5	page; words	2600	21
JEITA-HP [9]	Japanese	480; 100	–	3306	character	–	21
KHATT [13]	Arabic	1000	4000	4	page	2000	21
QUWI [12]	Arabic; English; French	1017	5085	4	page	–	21
RIMES [5]	French; Bengali	1300	12093	9	line	<1000	21
FOMC [15]	Latin	117	2040	11–39	page	>2000	11–12

for the analysis of handwriting indicates that the changes in the handwriting itself and its evolution were so large that it is not justified to compare them together, because they are so different.

It is important that the oldest document from this database was written on parchment, which accounted for a very significant percentage of the cost of creating the codex, in contrast to the last one written on paper.

To emphasize how expensive parchment was in the Middle Ages, there was the practice of scraping off the contents of a document no longer in use and writing it down again (palimpsest). As writing has developed over time, writing materials have evolved. The replacement of parchment from use by its much cheaper equivalent – paper – has had profound consequences. One of them, which is important in the aspect of the analysis of manuscripts in the discipline of computer science, is the reduction of the text density in relation to the document format. Therefore, the second historical database ICDAR2017 contains 720 writers, each represented by five samples. It was created to validate the correct identification of the author of the manuscript. However, the problem of the huge time interval over which the samples were written carries the same problems as the previous database.

The oldest documents from this database are from the 13th century and are written on parchment, in a completely different style and according to different rules than the last ones written in the 20th century on paper. For example, two manuscripts from this database are shown, which are treated as equal in relation to each other (see Fig. 1). Figure (a) shows a manuscript that contains over 3 million pixels, and figure (b) contains only 700 000. If the amount of writing material between these two samples is taken into account, a significant disproportion appears, and it is not an isolated case in this database. Consider that some writers are represented with only five samples. For a classifier, one writer will be underrepresented and the other will be overrepresented, in the extreme case. Of all the databases included in Table 1, only ICDAR2017 and IAM-HistDB are databases that have samples of Latin writing from the medieval period, while the rest are built from modern writing in modern languages. Therefore, they cannot be considered in the validation of manuscript identification systems dedicated to the Latin manuscripts, and can only be used as reference research.

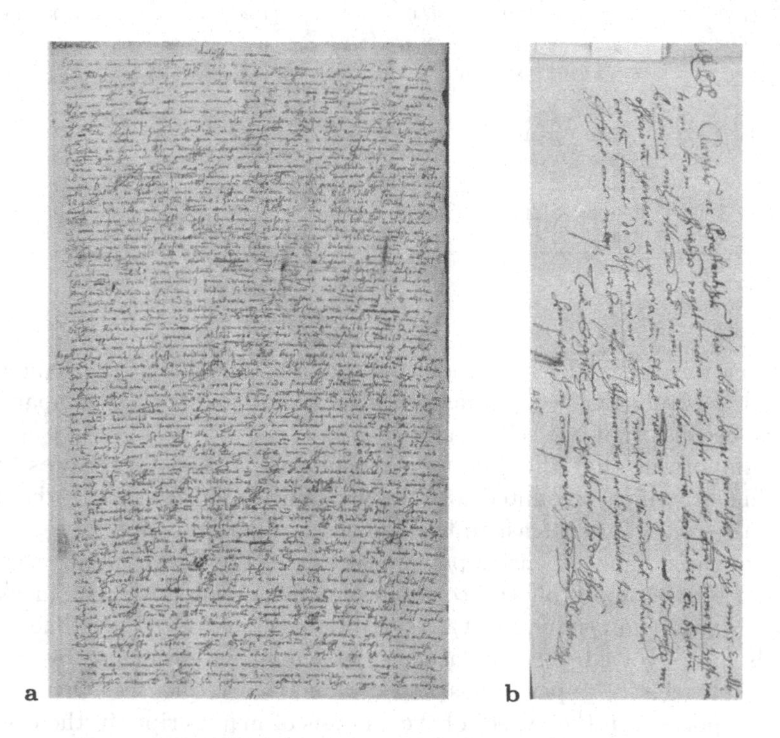

Fig. 1. Two samples of manuscripts from the ICDAR2017.

3 Dataset Description and Analysis

In the 21st century, the most significant libraries in the world holding the heritage of medieval manuscripts make their materials available in a digitized form in excellent quality. In order to highlight the order of magnitude of the size of the collections we are talking about, let us highlight that the digital repository of the National Library in Poland alone has over three thousand digitized sources in Latin of the period belonging only to the Middle Ages, and some of them are divided into over a hundred manuscripts. In addition, materials are being digitized all the time and this number is growing.

The digital representation of such a codex inherits all description features from its analogue counterpart. By this the identifier of such a source, called the signature, assigned according to the rules prevailing in a specific library, is meant.

This is where the first problem arises when using sources to try to create a larger database of manuscripts for research purposes from more than one library. In order for the source to be clearly identifiable, we need to know the information from which library it comes, and its signature, which forms a two-level key. In the IT aspect, it is necessary to create a single-level key, and in the case of access to sources by a normal user, this poses a specific difficulty if the user does not have knowledge in the field of historian's workshop.

Another problem is the standard of shared content in digital form - should the manuscripts be made available as raster images or as a PDF containing the entire codex? The images are of huge resolutions and consequently they take up huge amounts of space. The transfer of these data can significantly burden the network traffic of such a digital repository. In addition, downloading a specific image from the code means that the user has to leave the page being viewed and use the image viewing application. For a user/historian who is used to turning pages and analyzing content in a specific context, it is a huge impediment and makes it very difficult to become familiar with the content of the codex under examination.

Hence, a standard interface for scalar data access using the IIIF [6,18] standard was developed. The discussed digital library repositories, using the created Application Programming Interface (API), made it possible to view these documents in a web browser and download individual documents if necessary, and otherwise it enabled viewing the code as a whole. Thanks to this, it did not require the highest resolution of viewed documents and ultimately saved server resources. However, a manifest file has been developed for serial (scientific) data retrieval, but additional software is required to use it, which requires expertise in the discipline of computer science. Here, by solving one problem, we create a new one that makes it difficult to collect large amounts of data.

The problem of two-level access to sources is not a new thing in this field, it found its solution in a very interesting way.

Most of the medieval content in Latin is the liturgical content of the Catholic Church, which is related to the issues of singing, hymns, etc. Therefore, instead of using the internal library signatures that hold the sources for internal purposes, the RISM (Répertoire International des Sources Musicales) [7] is used as signatures. Hence, when using the discussed IIIF standard and its API, it was decided to collect digitized manuscripts available in excellent quality from digital libraries located in Bamberg, Cambridge, Düsseldorf-Gerresheim, Heidelberg, Köln, Lisbon, London, München, Oxford, Paris, Schaffhausen, Vendôme and Warsaw. The double ID problem was solved by using RISM directory IDs. Those identifiers that did not exist in the database have been added.

It should be remembered that only a small subset of the available codices from the digital repositories of the above-mentioned libraries was selected to create the database. Then, from the entire codes, handwriting samples (raster images) were carefully selected for the best preservation of the ink, the least amount of noise, absence of comments from other users of the codes (didascals), lack of miniatures and decorations. Another important aspect was the selection of samples that were digitized with lines of text kept horizontal, without any rotation.

The database currently has 117 fragments of codices with the number of samples between 11 and 39, with an average of 17, which gives 2040 writing samples. The resolution of the width of the manuscripts varies between two thousand pixels and almost nine thousand pixels (see Fig. 2). As it can be seen, even within two centuries, the script differs significantly, let alone such a wide period of time of sources as in IAM-HistDB or ICDAR2017. Hence, in the future, a careful standardization of this database is possible along with its development. this is not easy, because the codes differ significantly in their sizes, and the distance between the digital camera and the code is also different, so the DPI in the file properties does not reflect the actual values. Therefore, it is necessary in the future to analyze, among others, the resolution, text density and font (duct) thickness.

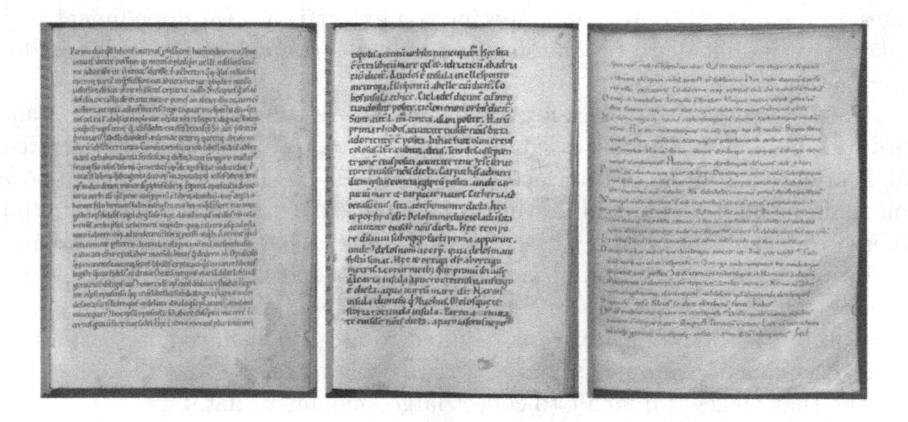

Fig. 2. Three sample manuscripts from the discussed database.

4 Conclusion and Future Research

The database contains fragments from 117 medieval codices in Latin from the 11th–12th centuries. The manuscripts were collected using the IIIF standard, and the source identifier is the MISM signature. When selecting handwriting samples, special care was taken to ensure that the documents were free of imperfections, comments, miniatures and decorations that may hinder the analysis of the database in question. The database is dedicated to verifying the effectiveness of author identification systems dedicated to Latin script or to universal use and is resistant to specific data, because systems dedicated to, for example, English script, were also validated with JEITA-HP databases with Japanese script or CERUG with Chinese script. The second use of the database may be not to identify the hand that wrote the manuscript, but the style in which the handwriting sample was written down, which could be used to show similarity between them. As a third application, the database can be used to teach neural networks to recognize content after obtaining a text layer. This content can be obtained from a specialist historian who can read the content, or the existing character recognition systems can be used to try to read the content.

In the future, it may be helpful to analyze the database from the IT aspect, i.e. to standardize the resolution of images in this database. Most databases discussed in Table 1 show that 2000 pixels is sufficient for correct handwriting recognition of A4 sheets at 300 DPI. Here the writing is very different from the modern one, and the sources are not of equal size. In addition, the writing, unlike the modern one, is stylistic, careful and the thickness of the writing (duct) may require greater resolution. Only a thorough analysis of the duct should approximate the resolution of the base standardization.

Historically, it may be helpful to analyze samples within a single codex to see if they were written by the same person. An analysis of the manuscript's contributions can help a lot with this. Another way to solve the same problem may be to validate multiple writer identification systems and analyze them against each other. On top of that, adding content transcription can help to use this database to train OCR handwritten models.

In the future, it is also possible to expand the database with new codices available in digital library repositories.

References

1. Bulacu, M., Schomaker, L., Vuurpijl, L.: Writer identification using edge-based directional features. In: Proceedings of the 7th International Conference on Document Analysis and Recognition ICDAR, pp. 937–941. IEEE (2003). https://doi.org/10.1109/ICDAR.2003.1227797
2. Fiel, S., et al.: ICDAR2017 competition on historical document writer identification (Historical-WI). In: Proceedings of the 14th International Conference on Document Analysis and Recognition ICDAR, vol. 01, pp. 1377–1382. IEEE (2017). https://doi.org/10.1109/ICDAR.2017.225
3. Fischer, A., Indermühle, E., Bunke, H., Viehhauser, G., Stolz, M.: Ground truth creation for handwriting recognition in historical documents. In: Proceedings of

the 9th IAPR International Workshop on Document Analysis Systems DAS, pp. 3–10 (2010). https://doi.org/10.1145/1815330.1815331
4. Freitas, C., Oliveira, L.S., Sabourin, R., Bortolozzi, F.: Brazilian forensic letter database. In: 11th International Workshop on Frontiers on Handwriting Recognition, Montreal, Canada (2008)
5. Grosicki, E., Carre, M., Brodin, J.M., Geoffrois, E.: Rimes evaluation campaign for handwritten mail processing. In: Proceedings of the ICFHR 2008: 11th International Conference on Frontiers in Handwriting Recognition (2008)
6. International Image Interoperability Framework: Gain richer access to the world's image and audio/visual files. https://iiif.io. Accessed Feb 2023
7. Keil, K.: Repertoire international des sources musicales (RISM). Fontes Artis Musicae **59**(4), 343–346 (2012)
8. Kleber, F., Fiel, S., Diem, M., Sablatnig, R.: CVL-database: an off-line database for writer retrieval, writer identification and word spotting. In: Proceedings of the 12th International Conference on Document Analysis and Recognition ICDAR, pp. 560–564. IEEE (2013). https://doi.org/10.1109/ICDAR.2013.117
9. Liu, C.L., Sako, H., Fujisawa, H.: Handwritten Chinese character recognition: alternatives to nonlinear normalization. In: Proceedings of the 7th International Conference on Document Analysis and Recognition ICDAR, vol. 3, pp. 524–528. IEEE (2003). https://doi.org/10.1109/ICDAR.2003.1227720
10. Louloudis, G., Gatos, B., Stamatopoulos, N., Papandreou, A.: ICDAR 2013 competition on writer identification. In: Proceedings of the 12th International Conference on Document Analysis and Recognition ICDAR, pp. 1397–1401 (2013). https://doi.org/10.1109/ICDAR.2013.282
11. Louloudis, G., Stamatopoulos, N., Gatos, B.: Writer identification. In: Document Analysis and Text Recognition: Benchmarking State-of-the Art Systems, pp. 121–154 (2018)
12. Maadeed, S.A., Ayouby, W., Hassaïne, A., Aljaam, J.M.: QUWI: an Arabic and English handwriting dataset for offline writer identification. In: 2012 International Conference on Frontiers in Handwriting Recognition, pp. 746–751 (2012). https://doi.org/10.1109/ICFHR.2012.256
13. Mahmoud, S.A., et al.: KHATT: an open Arabic offline handwritten text database. Pattern Recogn. **47**(3), 1096–1112 (2014). https://doi.org/10.1016/j.patcog.2013.08.009
14. Marti, U.V., Bunke, H.: The IAM-database: an English sentence database for offline handwriting recognition. Int. J. Doc. Anal. Recogn. **5**(1), 39–46 (2002). https://doi.org/10.1007/s100320200071
15. Pach, J.L.: New database of fragments of medieval codices of the 11th–12th centuries (2023). https://doi.org/10.5281/zenodo.7569006
16. Pechwitz, M., Maddouri, S.S., Märgner, V., Ellouze, N., et al.: IFN/ENIT-database of handwritten Arabic words. In: Proceedings of the 7th Colloque International Francophone sur l'Ecrit et le Document (CIFED), vol. 2, pp. 127–136 (2002). http://ifnenit.com/download/CIFED_02_ifn-enit-database.pdf
17. Semma, A., Hannad, Y., Siddiqi, I., Lazrak, S., Elkettani, Y.: Feature learning and encoding for multi-script writer identification. Int. J. Doc. Anal. Recogn. **25**, 79–93 (2022). https://doi.org/10.1007/s10032-022-00394-8
18. Snydman, S., Sanderson, R., Cramer, T.: The international image interoperability framework (IIIF): a community & technology approach for web-based images. In: Proceedings of the IS&T Archiving 2015, pp. 16–21 (2015). https://library.imaging.org/archiving/articles/12/1/art00005. Article ID: art00005

Global Optimisation for Improved Volume Tracking of Time-Varying Meshes

Jan Dvořák[✉], Filip Hácha, and Libor Váša

Department of Computer Science and Engineering, University of West Bohemia, Pilsen, Czech Republic
{jdvorak,hachaf,lvasa}@kiv.zcu.cz

Abstract. Processing of deforming shapes represented by sequences of triangle meshes with connectivity varying in time is difficult, because of the lack of temporal correspondence information, which makes it hard to exploit the temporal coherence. Establishing surface correspondence is not an easy task either, especially since some surface patches may have no corresponding counterpart in some frames, due to self-contact. Previously, it has been shown that establishing sparse correspondence via tracking volume elements might be feasible, however, previous methods suffer from severe drawbacks, which lead to tracking artifacts that compromise the applicability of the results. In this paper, we propose a new, temporally global optimisation step, which allows to improve the intermediate results obtained via forward tracking. Together with an improved formulation of volume element affinity and a robust means of identifying and removing tracking irregularities, the procedure yields a substantially better model of temporal volume correspondence.

Keywords: Time-varying mesh · model · animation · tracking · analysis · surface

1 Introduction

Sequences of triangle meshes are becoming more common in computer graphics due to a recent boom in both image acquisition hardware and reconstruction techniques aimed at estimating a static shape from a set of views. Since the most common data source is a reconstruction from video sequences, which treats each frame as an independent reconstruction problem, the most common type of resulting mesh sequences is the *Time-Varying Mesh (TVM)*, i.e. a sequence, where both the geometry (vertex coordinates) and the connectivity (triangles/polygons) are different in each frame. Effort has been put previously into converting this data into a more convenient form, e.g. a *dynamic mesh*, where the connectivity is shared by all the frames and implicitly captures *inter-frame surface correspondence*. Not only is such a representation more efficient for storage and transmission, because of the shared connectivity, it is also much more convenient for processing, since common procedures, such as texturing,

J. Mikyška et al. (Eds.): ICCS 2023, LNCS 14076, pp. 113–127, 2023.
https://doi.org/10.1007/978-3-031-36027-5_9

editing or movement analysis can exploit the known surface correspondence. On the other hand, current state of the art approaches to converting a TVM into a dynamic mesh suffer from many problems and work robustly only under rather constraining conditions.

Working directly with TVMs provides a much greater versatility, however, for many tasks, such as compression, time-consistent editing and others, it is necessary to build an *auxiliary model* that captures the temporal correspondence that is present in the data. It has been observed previously that it is often difficult to establish correspondence of surface elements, since such correspondence loses bijectivity even in the common case of self-contact of objects in the input data. Volume correspondence, on the other hand, is bijective for a wider variety of possible inputs, limited by the requirement of approximately constant overall volume. Therefore, some recent methods have focused on establishing correspondence of volume elements by means of tracking, achieving partial success and applicability in certain scenarios.

One of the drawbacks of the current approaches is that they rely on a frame-by-frame processing procedure, gathering information about the nature of the objects captured in the data in chronological order. This approach prevents information from frames that appear later in the sequence to influence the tracking results (and the induced correspondence information) of preceding frames. This leads to certain artifacts in the tracking results, which in turn hinder application of the tracking in scenarios that are sensitive to tracking errors, such as compression or time-consistent editing.

In particular, when tracking volume elements through time, it is necessary to capture information on which elements are tightly bound together - in previous works, this binding has been termed **affinity**. When processing the sequence forwards in time, the affinity gets constantly updated, as parts of the objects separate or come into contact. The updating may in turn lead to volume elements transitioning between separate components (we refer to such centers as *irregular*), due to temporary self contact, as illustrated in Fig. 1. Also, even when the qualitative change of separating a volume element from its component does not occur, the continuously updated affinity constantly lags behind the actual shape changes, leading to sub-optimal tracking results.

In this paper, we address these issues by proposing an improved tracking procedure, which efficiently eliminates most of the problems encountered previously. Our main contributions are:

- an improved approach to evaluating volume element affinity, which eliminates reconnecting of previously separated components caused by the infinite impulse response (IIR) filter used in the state of the art,
- a robust measure capable of identifying incorrectly tracked volume elements, which allows removing them from the intermediate result,
- a new post-processing phase that optimises tracking criteria globally, taking the whole sequence into account, allowing temporal propagation of information in both directions.

Fig. 1. Example of irregularly tracked centers. Consistent colouring of legs of the subject reveals two yellow coloured centers transitioning to the red coloured part. (Color figure online)

The global optimisation step allows for correcting the tracking imperfections caused by the removal of incorrectly tracked elements. We demonstrate the superiority of the proposed tracking strategy both quantitatively and qualitatively.

2 Related Work

The most popular way of obtaining a temporal model to represent TVMs is surface tracking. Usually, a certain template surface is sequentially aligned to all the frames using non-rigid registration [13,15]. The simplest methods rely on the template surface given prior and an assumption that it reflects the ground-truth topological information. Such methods usually fail in the presence of frequent erroneous self-contact in the input. This issue might be mitigated to some extent by subdividing the surfaces into patches [5,11], or by identifying frames with significant change in appearance and working with subsequences between such frames [6,9,14,17]. Bojsen-Hansen et al. [2] were able to detect topology changes and adjust the template shape accordingly, however, they incorrectly assume surface correspondences to be bijective outside of the adjusted parts. Budd et al. [4] build a shape similarity tree, which allows alignment of the more similar rather than subsequent frames.

In the presence of self-contacts, the volume correspondences are more likely bijective than the surface correspondences. This has been already utilised by Huang et al. [10,12], who proposed non-rigid registration of centroidal Voronoi tessellations (CVTs). However, their approach does not consider that volume elements move coherently together. Dvořák et al. [7] proposed to track volume elements called *centers*, which are not inherently CVTs, but instead are regularised to achieve coherence of the movement, considering a spatial neighbourhood of each center. This approach was improved [8] by introducing a more appropriate notion of center neighbourhood based on similarity of motion in already tracked frames and motion regularisation, which works better for rigidly moving parts.

Since our proposed approach builds on this method, it will be described in more detail in Sect. 3.

Recently, machine learning models became popular for representing temporal sequences. These are especially successful on sparse data (e.g., single-view RGBD video). Relevant to our work is, for example, OccupancyFlow [16], a learned occupancy function deformed by a neural vector field, as well as the work of Božič et al. [3] who train a neural deformation graph. The main limitation of neural models is, however, that working with them is less intuitive and thus their application is limited for example in time-consistent editing.

The rest of the paper is organised as follows. First, we will describe the key concepts of volume tracking relevant to this paper. In Sect. 4, we formulate improved affinity weights based on the maximum distances of centers and motion dissimilarities encountered in the already processed frames. These weights are a drop-in replacement for the original IIR filter based affinity. While significantly reducing the tracking error, these weights still do not prevent the occurrence of irregular centers. To this end, in Sect. 5 we discuss an irregularity measure based on the distance to the trajectory of the nearest center, which allows identifying centers to be removed from the tracking results. In Sect. 6, we discuss a post-processing of the centers to attenuate the influence of irregular center removal.

3 As-Rigid-as-Possible Volume Tracking

In this section, we briefly review the principles of a state-of-the-art method for volume tracking of TVMs, focusing on parts that are relevant to this paper. For a full description of the method, we refer the reader to the original paper [8]. The input to the method is a sequence of triangle meshes, denoted frames, with no assumption on the coherence of their connectivity. The method finds a fixed set C of N points (denoted centers), each representing a small volume surrounding it, whose positions vary in time. Each center follows a certain trajectory $\mathbf{c}_i = \left[\mathbf{c}_i^{(0)}, \mathbf{c}_i^{(1)}, \ldots, \mathbf{c}_i^{(F-1)}\right] \in \mathbb{R}^{3F}$, where F is the number of frames and $\mathbf{c}_i^{(f)}$ is the position of the i-th center in the f-th frame. The method aims to uniformly distribute the centers inside the enclosed volume of each frame, while ensuring that each center moves coherently with its neighbouring centers.

First, each frame is converted into a dense regular square voxel grid by sampling the indicator function $IF(\mathbf{x})$, which returns 1 in the interior and 0 otherwise. Alternatively, the method can also accept the sequence of voxel grids directly as input, which means that it can be applied to any sequence of shapes for which it is possible to determine the inside/outside information with acceptable amount of certainty (e.g., implicit representations, point clouds, etc.).

Center positions in the first frame are obtained by sampling n random occupied voxels and uniform distribution of centers is achieved by the Lloyd's algorithm: For each center, its Voronoi cell V_i^0 of occupied voxel positions is iteratively evaluated, such as

$$V_i^{(f)} = \left\{ \mathbf{x} : IF(\mathbf{x}) = 1 \wedge \|\mathbf{x} - \mathbf{c}_i^{(f)}\| \leq \|\mathbf{x} - \mathbf{c}_j^{(f)}\| \right\},$$

for every j, and the center is moved to the centroid $\bar{\mathbf{x}}_i^{(0)}$ of such cell.

For each subsequent frame, the method first obtains an initial distribution of the centers by linear extrapolation from the previous frame. Then, the positions are adjusted in an optimisation process, in which a tracking energy $E = E_s + \beta E_u$ is minimised, where β is a weighting constant ($\beta = 1$ by default). The uniformity energy term E_u enforces uniform distribution of the centers inside the volume. It is formulated as a sum of squared distances between the centers and the centroids of their corresponding Voronoi cells:

$$E_u = \frac{1}{2} \sum_{\mathbf{c}_i \in C} \|\mathbf{c}_i^{(f)} - \bar{\mathbf{x}}_i^{(f)}\|^2.$$

The smoothness of the movement measured in E_s is evaluated as

$$E_s = \frac{1}{2} \sum_{\mathbf{c}_i \in C} \|\mathbf{c}_i^{(f)} - \mathbf{p}_i^{(f)}\|^2,$$

where $\mathbf{p}_i^{(f)}$ is a prediction of the center position obtained using rigid transformation estimated from the movement of neighbouring centers and affinity weights from the previous frame $w^{(f-1)}$:

$$\mathbf{p}_i^{(f)} = \mathcal{A}_{i|w^{(f-1)}}^{(f)}(\mathbf{c}_i^{(f-1)}) = \mathbf{R}_{i|w^{(f-1)}}^{(f)} \mathbf{c}_i^{(f-1)} + \mathbf{t}_{i|w^{(f-1)}}^{(f)}.$$

Considering center positions fixed, the rigid transformation $\mathcal{A}_{i|w}^{(f)} = (\mathbf{R}_{i|w}^{(f)}, \mathbf{t}_{i|w}^{(f)})$ at a frame f given a certain set of weights w can be found minimising

$$(\mathbf{R}_{i|w}^{(f)}, \mathbf{t}_{i|w}^{(f)}) = \underset{\mathbf{R} \in \mathrm{SO}(3), \mathbf{t} \in \mathbb{R}^3}{\arg\min} \sum_{w(i,j) \geq \mu} w(i,j) \left\| \mathbf{c}_j^{(f)} - (\mathbf{R}\mathbf{c}_j^{(f-1)} + \mathbf{t}) \right\|^2,$$

where $\mu = 0.001$ is a threshold parameter to speedup the computation process by considering only relevant weights. Such transformation can be found in closed form using singular-value decomposition [18].

To optimise the energy E, the method interleaves between calculating the predictions $\bar{\mathbf{x}}_i^{(f)}$ and $\mathbf{p}_i^{(f)}$ with fixed positions of centers and then updating the positions with fixed predictions:

$$\mathbf{c}_i^{(f)} = \frac{\mathbf{p}_i^{(f)} + \beta \bar{\mathbf{x}}_i^{(f)}}{1 + \beta}.$$

The optimisation process is terminated when the change in $\mathbf{c}_i^{(f)}$ is sufficiently small or a fixed number of iterations has been reached.

Once the final positions in the current frame are obtained, the affinity weights are updated, so that they reflect the observed changes in the relations between centers in the currently processed frame. The method considers center relations based on spatial proximity in a single frame

$$a_p^{(f)}(i,j) = \exp(-\sigma_p \cdot \left\| \mathbf{c}_i^{(f)} - \mathbf{c}_j^{(f)} \right\|^2),$$

where σ_p is a parameter controlling the width of the Gaussian function.

The method also attempts to separate topologically distant parts by combining the spatial proximity with motion dissimilarity that is measured as:

$$a_m^{(f)}(i,j) = \exp(-\sigma_m \cdot d_i^{(f)}(\mathcal{A}_{i|w^{(f-1)}}^{(f)}, \mathcal{A}_{j|w^{(f-1)}}^{(f)})^2),$$

where σ_m is another Gaussian width parameter and $d_i^{(f)}(\mathcal{A}, \mathcal{B})$ is the distance between two rigid transformations that is evaluated by measuring distances of points around $\mathbf{c}_i^{(f)}$ (voxel positions in $V_i^{(f)}$ in our case) transformed by both $\mathcal{A}$ and $\mathcal{B}$

$$d_i^{(f)}(\mathcal{A}, \mathcal{B}) = \frac{1}{|V_i^{(f)}|} \sum_{\mathbf{v}_k \in V_i^{(f)}} \|\mathcal{A}(\mathbf{v}_k) - \mathcal{B}(\mathbf{v}_k)\| .$$

Instead of setting the Gaussian width parameters σ directly, they are calculated from parameters ρ with clearer geometric meaning: $\sigma = -\ln(0.5)/\rho^2$, which determine at which distance the Gaussian function drops to 0.5.

To propagate the already observed information throughout the sequence, an IIR filter with falloff parameter α is applied on motion dissimilarity:

$$a_{\mathrm{IIR}}^{(f)}(i,j) = \alpha a_m^{(f)}(i,j) + (1-\alpha) a_{\mathrm{IIR}}^{(f-1)}(i,j).$$

Finally, the spatial proximity $a_p^{(f)}(i,j)$ is combined with the IIR filtered motion dissimilarity $a_{\mathrm{IIR}}^{(f)}(i,j)$ to form the weights $w^{(f)}(i,j)$ that will be used to optimise the positions in the next frame:

$$w^{(f)}(i,j) = a_p^{(f)}(i,j) \cdot a_{\mathrm{IIR}}^{(f)}(i,j). \tag{1}$$

With this knowledge, we can proceed with discussing the contributions of this paper.

4 Maximum Distance Based Affinity

Similarly to the original weight formulation in Eq. 1, the new weight is also computed as a product of spatial proximity and motion dissimilarity:

$$\tilde{w}^{(f)}(i,j) = \tilde{a}_p^{(f)}(i,j) \cdot \tilde{a}_m^{(f)}(i,j).$$

The difference is how the center proximity $\tilde{a}_p^{(f)}(i,j)$ and the motion dissimilarity $\tilde{a}_p^{(f)}(i,j)$ are formulated.

The previous method measured the spatial center proximity using the Euclidean distance of centers in a single frame, ignoring the information from previous frames. The main limitation of such approach is the fact that as two topologically distant or separated parts come to near proximity, the affinity between their centers increases. Assuming the tracked sequence represents piecewise rigid objects, it could be more appropriate to use geodesic distance inside the volume, which, ideally, should be roughly constant throughout the sequence. However, due to self contact present in real-world data, the topological information in a given frame might be incorrect, resulting in introduction of erroneous decrease in such a measure. Additionally, geodesic distance is computationally expensive to evaluate at the frequency required by the tracking pipeline. We instead propose to approximate this quantity by the largest Euclidean distance encountered in all frames up to and including the current frame:

$$\tilde{a}_p^{(f)}(i,j) = \exp\left(-\sigma_p \cdot \max_{0 \leq k \leq f} \left\| \mathbf{c}_i^{(f)} - \mathbf{c}_j^{(f)} \right\|^2\right).$$

When examining relative positions of a certain pair of centers in time, we observe that the Euclidean distance between them fluctuates, but it is never larger than their geodesic distance. Note that our goal is not to evaluate this quantity precisely, but to correctly differentiate between the true connected neighbours of a center and the topologically distant centers in near proximity (see Fig. 2).

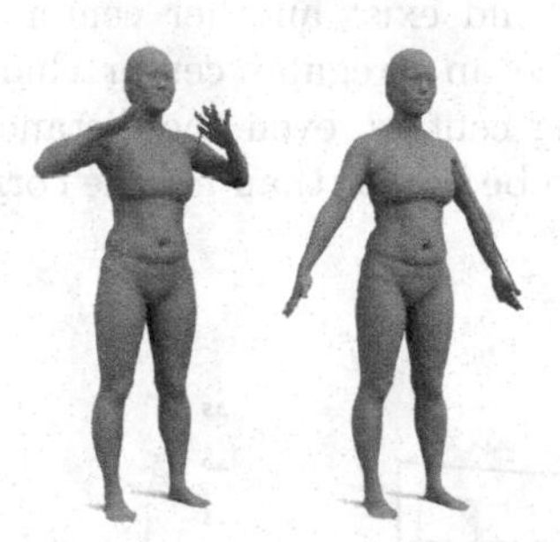

Fig. 2. Spatial proximity in a single frame might not reflect the underlying topology of the represented object. Left: Two topologically distant points in near proximity. Right: Examining a different frame reveals that they should not be considered as neighbouring/affine.

An analogous observations can be made about the similarity of the movement. If a pair of centers moved significantly differently in the past, then they cannot both belong to the same rigid part, even when the movement has been almost identical in several previous frames. Instead of an IIR filter, we thus propose to also use the maximum dissimilarity over all already processed frames:

$$\tilde{a}_m^{(f)}(i,j) = \begin{cases} 1, & f = 0 \\ \exp\left(-\sigma_m \cdot \max_{1 \leq k \leq f} d_i^{(f)}(\mathcal{A}_{i|\tilde{w}^{(f-1)}}^{(f)}, \mathcal{A}_{j|\tilde{w}^{(f-1)}}^{(f)})^2\right), & \text{otherwise} \end{cases}.$$

In the first frame, we have no information about the movement, therefore we assume there is no difference and instead solely rely on $\tilde{a}_p^{(0)}(i,j)$ when computing the affinity weights.

Setting Gaussian widths σ_m and σ_p (resp. ρ_m and ρ_p) to obtain satisfactory results is a task specific to the scale of the data and complexity of the motion. In our experiments, we have obtained best results with $\rho_p = \rho_m = 0.125$ for human performance capture. For synthetic datasets, where the bounding box was significantly larger and the motion was mainly rigid, we have determined that best results were obtained with $\rho_p = 0.2$ and $\rho_m = 0.05$.

The previous IIR-filter-based affinity depends on a falloff parameter α, which controls how the affinity reacts to occurring changes. Setting α too low results in slow reactions. On the other hand, too high α means that the affinity forgets faster the separation that occurred in the past. Our new formulation of affinity reacts more dynamically to changes and also reflects every observed separation.

5 Irregular Center Detection

Once the frame-by-frame tracking is finished (regardless of the affinity weights used), we can analyse the achieved results and detect the *irregular centers*. To this end, we evaluate an irregularity measure $I_i = \min_j \|\mathbf{c}_i - \mathbf{c}_j\|_2^2$, where $\mathbf{c}_i$ is the center trajectory and $\|\cdot\|_2^2$ is the squared Euclidean norm. If a center is correctly tracked, there should exist another center with a similar trajectory in the near proximity. Since an irregular center changes suddenly its relative position to its neighbouring centers, even the distance to the closest center to its trajectory is expected to be higher than for the correctly tracked centers (see Fig. 3).

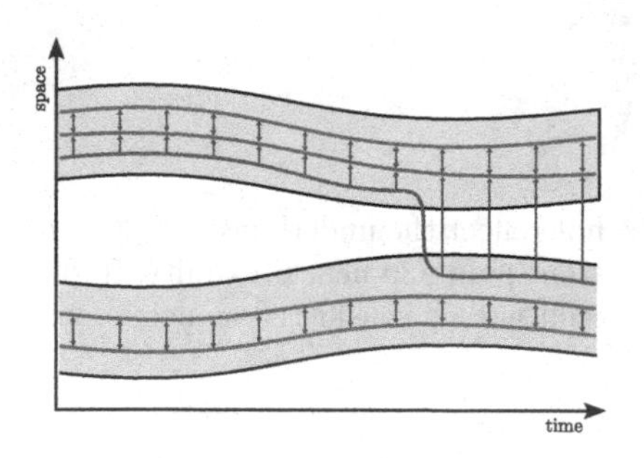

Fig. 3. Irregular center detection using distance to closest trajectory. Arrows indicate distances that contributed to the computation. Red trajectory has a much higher I_i and is correctly detected as irregular. (Color figure online)

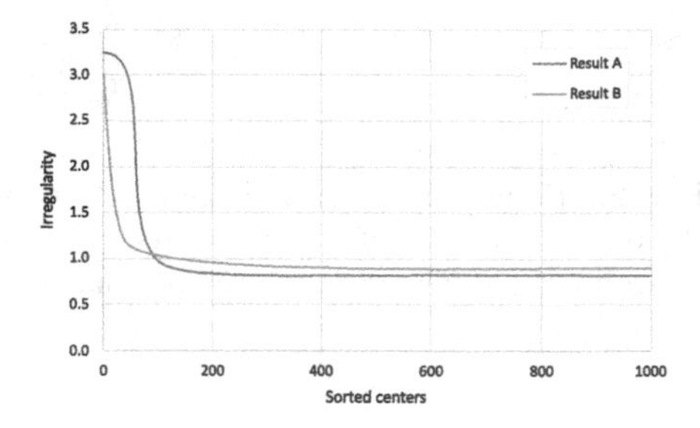

Fig. 4. Example comparison of two irregularity curves. The Result B outperforms the Result A as its curve drops faster to satisfying values of I_i.

The value of I_i must be considered in the context of the values of all centers, as it depends on various factors, e.g., center count, scale of the data and the dynamics of the movement. We can also use this measure to quantify the success of tracking in terms of the presence of irregular centers, by sorting all the values in descending order and plotting them as a curve. By comparing the curves resulting from different tracking methods, we can determine which results are less influenced by the presence of irregular centers (see Fig. 4), as long as the results were tracked in the same input sequence and the center count is similar (although not necessarily equal). When attempting to improve the tracking results, one of our goals is to narrow or eliminate the part of the curve with I_i significantly higher than the correctly tracked centers, while not significantly increasing the irregularity of such centers.

6 Global Optimisation

Simply removing a certain number of centers with the highest I_i actually does not lead to an improvement in terms of flattening the irregularity curve. The uniformity of the centers distribution is violated, since removed centers leave an uncovered volume. Re-running the frame-by-frame tracking with the irregular centers removed might still not prevent new irregular centers from appearing, even when the final affinity weights obtained in the initial tracking are utilised. Instead, we propose to follow the irregular center removal with adjustment of the remaining tracked trajectories of centers in a global optimisation process.

6.1 Global Tracking Energy

The objectives of the global optimisation are identical to the frame-by-frame tracking. We optimise a global energy $\hat{E}$ consisting of uniformity and motion smoothness energy terms $\hat{E} = \hat{E}_s + \hat{\beta}\hat{E}_u$.

The uniformity term is the same as in the frame-by-frame tracking, except for that it is evaluated for all the frames in the sequence at once:

$$\hat{E}_u = \frac{1}{2} \sum_{\mathbf{c}_i \in C} \sum_{f=0}^{F-1} \|\mathbf{c}_i^{(f)} - \bar{\mathbf{x}}_i^{(f)}\|^2.$$

If we consider the centroids $\bar{\mathbf{x}}_i^{(f)}$ fixed, we can approximate the gradient by these partial derivatives:

$$\frac{\partial \hat{E}_u}{\partial \mathbf{c}_i^{(f)}} \approx \mathbf{c}_i^{(f)} - \bar{\mathbf{x}}_i^{(f)}.$$

The global smoothness energy is evaluated as

$$\hat{E}_s = \frac{1}{2}\left(\sum_{\mathbf{c}_i \in C}\sum_{f=1}^{F-1}\left\|\mathbf{c}_i^{(f)} - \mathbf{p}_i^{(f)}\right\|^2 + \sum_{\mathbf{c}_i \in C}\sum_{f=0}^{F-2}\left\|\mathbf{c}_i^{(f)} - \mathbf{q}_i^{(f)}\right\|^2\right),$$

$$\mathbf{p}_i^{(f)} = \mathcal{A}_{i|\omega}^{(f)}\left(\mathbf{c}_i^{(f-1)}\right),$$

$$\mathbf{q}_i^{(f)} = \mathcal{A}_{i|\omega}^{(f+1)^{-1}}\left(\mathbf{c}_i^{(f+1)}\right),$$

where $\mathbf{p}_i^{(f)}$ and $\mathbf{q}_i^{(f)}$ are forward and backward rigid motion predictions of the center position at frame f, using rigid transformations estimated given overall movement-based affinity weights ω (see Eq. 2). Considering such predictions fixed, the partial derivatives, which form the approximated gradient, are as follows:

$$\frac{\partial \hat{E}_s}{\partial \mathbf{c}_i^{(f)}} \approx \begin{cases} \mathbf{c}_i^{(f)} - \mathbf{q}_i^{(f)}, & f = 0 \\ \mathbf{c}_i^{(f)} - \mathbf{p}_i^{(f)} - \mathbf{q}_i^{(f)}, & 1 \leq f \leq F-2 \\ \mathbf{c}_i^{(f)} - \mathbf{p}_i^{(f)}, & f = F-1 \end{cases}$$

6.2 Optimisation Strategy

The optimisation process is iterative, working with a set of trajectories C, whose initial values are given by the original tracking results with irregular centers removed. In each iteration, we evaluate the energy $\hat{E}(C)$ and the approximated gradient $\nabla\hat{E}$, and construct a candidate set of trajectories $\bar{C}$, where the center positions are calculated as

$$\bar{\mathbf{c}}_i^{(f)} = \mathbf{c}_i^{(f)} - \lambda\left(\frac{\partial \hat{E}_s}{\partial \mathbf{c}_i^{(f)}} + \hat{\beta}\frac{\partial \hat{E}_u}{\partial \mathbf{c}_i^{(f)}}\right).$$

First, lambda is set to $\lambda = 0.1$ and then it is iteratively scaled by $\frac{1}{2}$ until $\hat{E}(\bar{C})$ is smaller than $\hat{E}(C)$, or a specified number of attempts has been reached. If an improvement in terms of energy is achieved, we set $C = \bar{C}$ and continue to the next iteration. Otherwise, the process is terminated and C is the resulting set of trajectories. The optimisation process can also be terminated after a specified number of iterations (20 in our experiments).

Such an optimisation strategy does not necessarily converge to a global optimum. If an irregular center was left in the initial set C, the local steps in the gradient direction will not straighten its trajectory in order to eliminate the transition between disconnected components. The locality of the changes is, however, also an advantage, since the local trajectory adjustments ensure that the objectives are met, while not introducing any large sudden changes, and therefore no new irregular centers can appear.

6.3 Global Movement-Based Affinity

The affinity utilised in the global optimisation process directly considers only the dissimilarity of motion:

$$\omega(i,j) = \exp\left(-\sigma_{\text{gm}} \cdot \max_{0 \le f < F} d_i^{(f)} \left(\mathcal{A}_{i|\omega_{\max}}^{(f)}, \mathcal{A}_{j|\omega_{\max}}^{(f)}\right)^2\right), \tag{2}$$

where σ_{gm} is a parameter controlling the tolerance of the affinity to dissimilar motions. The overall spatial proximity $\omega_{\max}(i,j)$ of centers is also considered, but only for estimating the transformations $\mathcal{A}_{i|\omega_{\max}}^{(f)}$:

$$\omega_{\max}(i,j) = \exp\left(-\sigma_{\text{gp}} \cdot \max_{0 \le f < F} \left\| \mathbf{c}_i^{(f)} - \mathbf{c}_j^{(f)} \right\|^2\right).$$

The motivation to mainly rely on the motion information instead of using both motion and proximity combined is the following: If a center belongs to a large rigidly moving part of the object, we want it to be influenced by the whole part rather than only a certain small neighbourhood around it. Having all the positions in each frame at hand, we can confidently rely solely on this information without worrying about the rigid part suddenly splitting in a later frame.

7 Experimental Results

7.1 Influence of the Proposed Affinity

To evaluate how the previous forward tracking pipeline benefits from the proposed maximum distance based affinity, we have compared the new tracking results with those reported previously [8] using default configurations for both methods. The comparison included all the previously studied datasets (including selected sequences from D-FAUST dataset [1]) except for *pentagonal_prism* and *collision* datasets, which we believe were already tracked correctly with the previous method. The tracking quality was evaluated using the metrics PCAC, which measures the complexity of the tracked trajectories (lower is assumed better, although lowering below a certain threshold given by the true complexity of the movement is not desirable) and DFU, which measures relative standard deviation of Voronoi cell sizes (a lower value indicates a more uniform covering). For details on the measures, see the original paper [8]. The results are shown in Table 1.

Incorporating the newly proposed affinity results in a considerable improvement over the original affinity on all the sequences. Visually, the results contain less irregular centers, which can be seen when assigning each center a consistent color by interpreting the first three PCA coefficients of its trajectory as RGB values, and the centers achieve better coverage over problematic parts (see Fig. 5a). This is also reflected by the irregularity curves (see Fig. 5b).

Table 1. Comparison of forward-tracked results using proposed and original (IIR) affinity. Highlighted are the best results for a given dataset.

	F	Proposed		IIR	
		PCAC	DFU	PCAC	DFU
gears	60	**1.785**	**0.070**	1.816	0.071
casual_man	545	**10.587**	**0.127**	12.289	0.206
samba	175	**5.060**	**0.215**	5.547	0.262
DF_50020_knees	515	**5.117**	**0.111**	6.764	0.179
DF_50009_chicken_wings	212	**2.855**	**0.116**	3.583	0.155
DF_50004_jumping_jacks	360	**6.040**	**0.130**	7.826	0.175

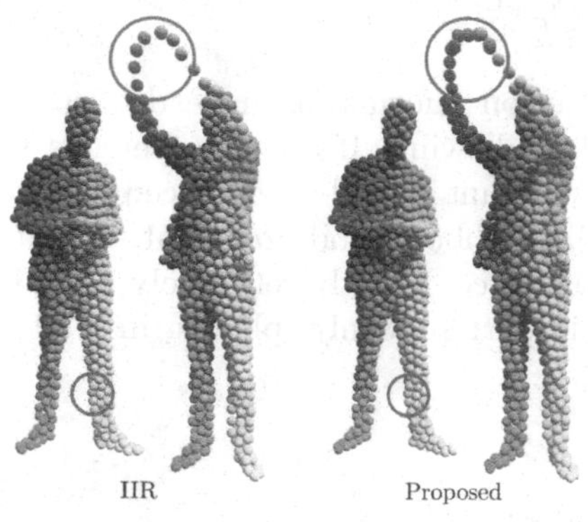

(a) Visual comparison.

(b) Largest 100 values of I_i.

Fig. 5. Tracking results for *casual_man* dataset.

7.2 Irregular Center Removal

In this experiment, we have studied the effects of various strategies for removing 20 irregular centers from tracking results obtained through forward tracking with our proposed affinity on the *casual_man* dataset from Sect. 7.1. The strategies differed in the number of global optimisations performed n_{go} and in number of removed centers each optimisation n_{rem}. Parameters of the global optimisation were as follows: $\rho_{\text{gp}} = 0.125, \rho_{\text{gm}} = 0.03$ and $\hat{\beta} = 0.5$. Note that these values were selected empirically and slightly different values yield similar results. Table 2 shows the measured PCAC and DFU values and the irregularity curves are shown in Fig. 6. For comparison, we also include results for center removal without global optimisation.

With growing n_{go}, the improvement process achieves better coverage of volume, which is reflected in the DFU measure. However, we can also see a negative trend in terms of irregularity. Best values of PCAC were achieved with $n_{\text{go}} = 5$. This is also reflected in a visual inspection of results (see Fig. 7). Figure 7a shows centers that were detected as irregular. It can be seen that the detected centers

Table 2. Comparison of PCAC and DFU measures for various irregular center removal strategies on *casual_ man* dataset forward tracked using our proposed affinity.

n_{rem}	20	20	4	1
n_{go}	—	1	5	20
PCAC	10.154	8.760	**8.578**	8.869
DFU	0.160	0.149	0.144	**0.134**

indeed travel across different body parts. The increased irregularity with growing n_{go} is reflected by certain number of centers oscillating to cover a larger volume, which is unfortunately visible only when the tracked centers are animated.

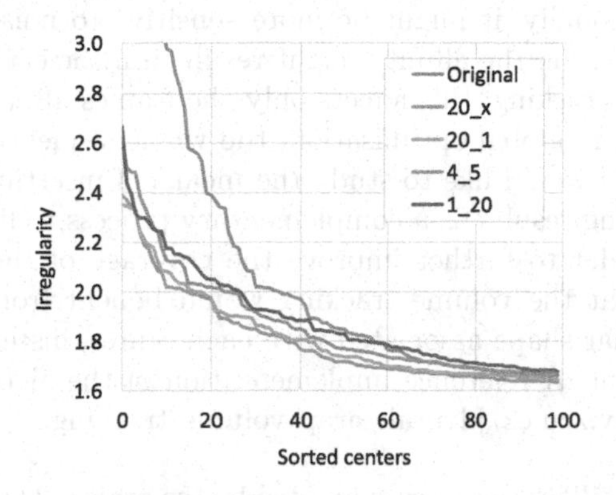

Fig. 6. Comparison of first 100 I_i after center removal.

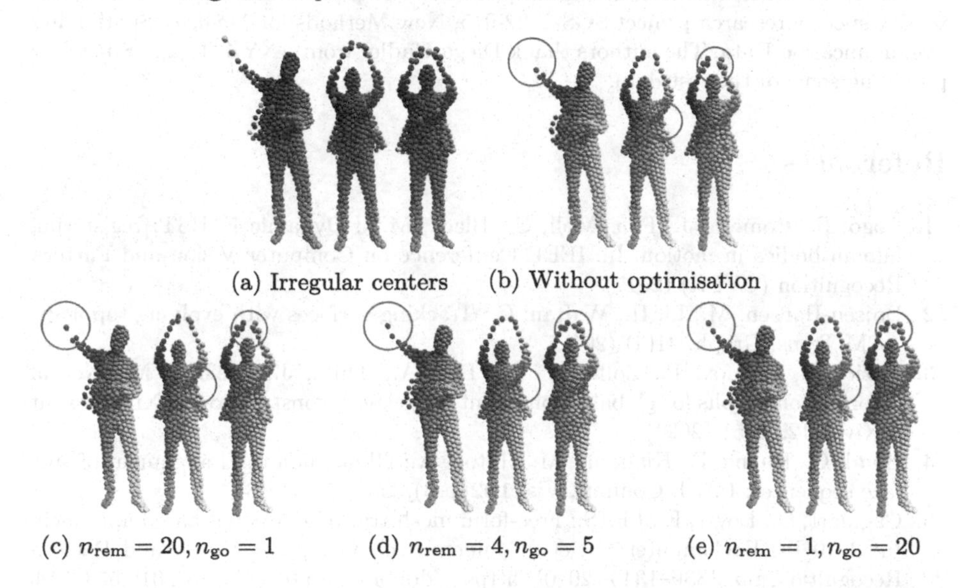

Fig. 7. Results for various irregular center removal strategies for *casual_ man* dataset with target of 20 centers to be removed. Highlighted are the areas with the most notable differences.

8 Conclusions

In this paper, we have shown that the forward volume tracking results can be further considerably improved by incorporating volume element filtering followed by a global optimisation, which assures volume coverage without introducing new irregularly tracked centers. The experiments show that it is beneficial to remove centers in small batches, rather than all at once or each center separately. The novel affinity for the forward tracking method, which we also proposed, considerably reduces tracking imperfections, which is reflected in all considered tracking quality metrics.

Our approach shares the limitations of the previous volume tracking methods, as it also cannot handle sequences without sufficient notion of inside/outside information. Additionally, it might be more sensitive to noise, since any error introduced in computing the affinity might result in incorrect split of two affine centers. In forward tracking, this affects only the frames after the occurrence of the error. However, in global optimisation, the whole sequence is influenced.

In the future, we would like to study the means of inserting centers into the intermediate tracking results as a complementary process to filtering and global optimisation, in order to further improve the coverage of the tracked volume. We also believe that the volume tracking would benefit from incorporating a notion of a particular shape associated with each center, instead of representing it as a discrete point. A reference implementation of the algorithm is available at https://gitlab.kiv.zcu.cz/jdvorak/arap-volume-tracking.

Acknowledgement. This work was supported by the project 20-02154S of the Czech Science Foundation. Jan Dvořák and Filip Hácha were partially supported by the University specific research project SGS-2022-015, New Methods for Medical, Spatial and Communication Data. The authors thank Diego Gadler from AXYZ Design, S.R.L. for providing some of the test data.

References

1. Bogo, F., Romero, J., Pons-Moll, G., Black, M.J.: Dynamic FAUST: registering human bodies in motion. In: IEEE Conference on Computer Vision and Pattern Recognition (CVPR) (2017)
2. Bojsen-Hansen, M., Li, H., Wojtan, C.: Tracking surfaces with evolving topology. ACM Trans. Graph. 31(4) (2012)
3. Božič, A., Palafox, P., Zollhöfer, M., Dai, A., Thies, J., Nießner, M.: Neural deformation graphs for globally-consistent non-rigid reconstruction. arXiv preprint arXiv:2012.01451 (2020)
4. Budd, C., Huang, P., Klaudiny, M., Hilton, A.: Global non-rigid alignment of surface sequences. Int. J. Comput. Vis. **102**(1–3), 256–270 (2013)
5. Cagniart, C., Boyer, E., Ilic, S.: Free-form mesh tracking: a patch-based approach. In: 2010 IEEE Computer Society Conference on Computer Vision and Pattern Recognition, pp. 1339–1346 (2010). https://doi.org/10.1109/CVPR.2010.5539814
6. Collet, A., et al.: High-quality streamable free-viewpoint video. ACM Trans. Graph. **34**(4), 1–13 (2015). https://doi.org/10.1145/2766945

7. Dvořák, J., Vaněček, P., Váša, L.: Towards understanding time varying triangle meshes. In: Paszynski, M., Kranzlmüller, D., Krzhizhanovskaya, V.V., Dongarra, J.J., Sloot, P.M.A. (eds.) ICCS 2021. LNCS, vol. 12746, pp. 45–58. Springer, Cham (2021). https://doi.org/10.1007/978-3-030-77977-1_4
8. Dvořák, J., Káčereková, Z., Vaněček, P., Hruda, L., Váša, L.: As-rigid-as-possible volume tracking for time-varying surfaces. Comput. Graph. **102**, 329–338 (2022). https://doi.org/10.1016/j.cag.2021.10.015
9. Guo, K., Xu, F., Wang, Y., Liu, Y., Dai, Q.: Robust non-rigid motion tracking and surface reconstruction using l0 regularization. In: 2015 IEEE International Conference on Computer Vision (ICCV), pp. 3083–3091 (2015). https://doi.org/10.1109/ICCV.2015.353
10. Huang, C.H., Allain, B., Franco, J.S., Navab, N., Ilic, S., Boyer, E.: Volumetric 3d tracking by detection. In: Proceedings of the IEEE Conference on Computer Vision and Pattern Recognition (CVPR) (2016)
11. Huang, C.H., Boyer, E., Ilic, S.: Robust human body shape and pose tracking. In: 2013 International Conference on 3D Vision - 3DV 2013, pp. 287–294 (2013). https://doi.org/10.1109/3DV.2013.45
12. Huang, C.H.P., et al.: Tracking-by-detection of 3d human shapes: from surfaces to volumes. IEEE Trans. Pattern Anal. Mach. Intell. **40**(8), 1994–2008 (2018). https://doi.org/10.1109/TPAMI.2017.2740308
13. Li, H., Adams, B., Guibas, L.J., Pauly, M.: Robust single-view geometry and motion reconstruction. ACM Trans. Graph. **28**(5), 1–10 (2009). https://doi.org/10.1145/1618452.1618521
14. Moynihan, M., Ruano, S., Pages, R., Smolic, A.: Autonomous tracking for volumetric video sequences. In: Proceedings of the IEEE/CVF Winter Conference on Applications of Computer Vision (WACV), pp. 1660–1669 (2021)
15. Myronenko, A., Song, X.: Point set registration: coherent point drift. IEEE Trans. Pattern Anal. Mach. Intell. **32**(12), 2262–2275 (2010)
16. Niemeyer, M., Mescheder, L., Oechsle, M., Geiger, A.: Occupancy flow: 4d reconstruction by learning particle dynamics. In: Proceedings of the IEEE/CVF International Conference on Computer Vision (ICCV) (2019)
17. Prada, F., Kazhdan, M., Chuang, M., Collet, A., Hoppe, H.: Spatiotemporal atlas parameterization for evolving meshes. ACM Trans. Graph. **36**(4), 1–12 (2017). https://doi.org/10.1145/3072959.3073679
18. Sorkine-Hornung, O., Rabinovich, M.: Least-squares rigid motion using svd (2016). Technical note

Detection of Objects Dangerous for the Operation of Mining Machines

Jakub Szymkowiak, Marek Bazan(✉), Krzysztof Halawa, and Tomasz Janiczek

Wrocław University of Science and Technology, 27 Wybrzeże Wyspiańskiego st., 50-370 Wrocław, Poland
252868@student.pwr.edu.pl, {marek.bazan, krzysztof.halawa,tomasz.janiczek}@pwr.edu.pl
https://wit.pwr.edu.pl/en/

Abstract. Deep learning was used to detect boulders that can damage excavators in opencast mines. Different network architectures were applied, i.e., modern YOLOv5, RetinaNet and Mask-RCNN. Studies were carried out in which the results obtained using a few networks were compared. The abovementioned neural networks were exploited in a framework for detection of oversized boulders on a conveyor belt operating in an opencast coal mine. The method is based on the analysis of a certain number of consecutive frames of the film from an industrial camera. The novelty relies on checking the detection of a boulder within subsequent frames and allowing the skipping of a prescribed small number of neighboring frames with false negative detections. This allows one to make a decision about stopping a conveyor belt after detecting a boulder in consecutive frames even when they are interleaved with frames that contained a boulder missed by a detector due to misleading environmental conditions such as shadows or sand. The method was tested on recordings from an opencast mine in Poland. The proposed method can help prevent the failure of expensive equipment.

Keywords: deep learning · YOLOv5 · object detection · digital image processing · opencast mining

1 Introduction

Currently, there are many opencast mines around the world, where various raw materials are extracted. The vast majority of mineral resources is developed by opencast mining [22]. Often in such mines there are giant excavators [6], the height of which is similar to buildings that are several storeys. There is a high risk of damage if there is a big rock on the conveyor belt. Failures caused by unnoticed large stones are very costly, complicated to repair and cause mining to stop. Therefore, it is very important to develop methods that will reduce the risk of the excavator operator not noticing such dangerous cases. For the

J. Mikyška et al. (Eds.): ICCS 2023, LNCS 14076, pp. 128–139, 2023.
https://doi.org/10.1007/978-3-031-36027-5_10

automatic detection of dangerous boulders, this article proposes a method using deep neural networks. Among other things, modern and fast YOLOv5 networks were used.

This work was created in connection with a practical problem that occurred in an open-pit mine and can occur in other situations concerning stones recognition against a difficult background [8,20,23,25].

During the transfer of spoil in the form of small gravel or sand, a large stone appeared on the conveyor belt from time to time, which caused significant danger to mining machines. To obtain models operating in different lighting conditions, both video streams taken in good daylight and movies shot under artificial lighting in the middle of the night were used. A camera with image stabilization hardware was mounted on the excavator above the conveyor belt. An example view from the camera is shown in Fig. 1. A stone is visible on the conveyor belt. Even such a relatively small object should stop the transport of the excavated material. In many cases, much larger stones have been mined, which, if overlooked, can cause severe damage (Fig. 2).

Fig. 1. Camera view with a visible stone on the conveyor belt

2 Preparing a Dataset

The 27 recordings over several days were used for a training process. Additional 10 recordings were excluded from training and put aside for testing the final framework after the training of the models is finished. After dividing 27 videos into individual frames, over 28,000 images were obtained. Then, appropriate labels were assigned to images in which at least one large rock was observed. A Python script was created to make labeling faster and easier. At least one stone was visible in 2399 images. Sometimes not only one stone may appear on a frame. However, such situations occur very rarely. Large stones were marked in the images with bounding boxes. For this purpose, the makesense.ai platform was

Fig. 2. Camera view with a visible stone on the conveyor belt (artificial light)

used, which enables the preparation of datasets with selected objects. Then, the dataset with ready-made annotations was processed on the roboflow.ai platform [18]. This platform allows users to download annotations in various formats. It was also used to divide the data into training, validation, and test sets. To inspect the stability of the training process the models were trained in two configurations of a data split:

1. the training set – 80% of the frames and the test and validation sets – 10% of the frames each,
2. the training set – 70% of the frames, and the test and validation sets – 15% of the frames each.

The sizes of images to which frames have to be scaled in datasets used for different object detection models are depicted in Sect. 3 when describing the models.

One has to note that all frames in each set contained a boulder. It is a requirement for object detection models training that images without positive detection are not provided. It means that not all frames from videos are included in the datasets, but only those containing a positive detection. In the datasets, frames are shuffled. One has to note that validation and test sets in the context of detection model building are different from validation and test sets that are formed from ten videos not used in the training. The validation set of the later datasets is used for fine-tuning of hyper-parameters and the test set for testing the whole framework.

3 Applied and Tested Models of Neural Networks

Experiments were carried out in which the results of the use of the following neural networks were checked: RetinaNet, Mask RCNN and YOLOv5. A brief comparison of these architectures is e.g., in [5,16].

3.1 RetinaNet

RetinaNet is a one-shot object detection model that utilizes anchor boxes and focal loss to handle the class imbalance problem in object detection [12]. RetinaNet has achieved state-of-the-art results on several benchmark datasets and is widely used for object detection tasks in real-world applications. Because of these advantages, it was also decided to test its effectiveness in detecting stones. This network is built of one backbone and two other subnets that are responsible for different tasks. The backbone creates a convolutional feature map by analyzing the entire input image. One subnet performs convolutional object classification on the output of the backbone subnet. The latter subnet performs convolutional regression, creating bounding boxes [12]. The images processed by this architecture were scaled to a rectangular image size between 800 × 800 and 1333 × 1333 pixels. The architecture of RetinaNet is presented in Fig. 3. Fizyr's implementation of RetinaNet was used in the conducted experiments [7].

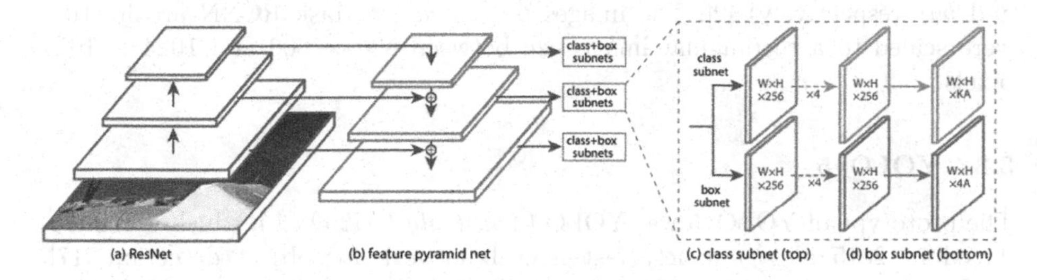

Fig. 3. The RetinaNet architecture [12].

3.2 Mask RCNN

The Mask R-CNN (Mask Region-based Convolutional Neural Network) extends the popular Faster R-CNN and, besides detecting objects, it also performs instance segmentation [4,9]. An interesting case of use in the context of our problems was presented in [8,15]. The Mask R-CNN first generates object proposals and then classifies the proposals and generates masks in the second stage by applying a fully convolutional network. The network which is based on Matterport's implementation but fully compatible with TensorFlow 2 [19] were applied. This implementation is based on the Feature Pyramid Network (FPN) and ResNet101 [1,11]. Combining the FPN and the ResNet allowed us to increase both the quality and the speed of data processing. The comparison between the application of FPN and ResNet is shown in Fig. 4, where the numbers signify the spatial resolution and channels. The arrows denote convolutional, deconvolutional, and fully-connected layers.

In Fig. 4 it can be seen that the head of Mask RCNN is built of two branches - one of them is responsible for generating masks and the other one for classifying and calculating bounding boxes. The latter consists of two parts: one that is for

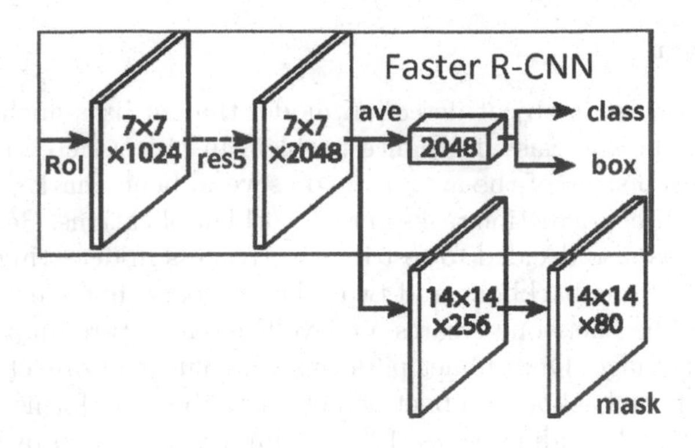

Fig. 4. The head architecture of the Mask RCNN [11].

classification and the other is responsible for detection (they are marked as *class* and *box* respectively) [9]. The images processed by Mask RCNN architecture were scaled to a rectangular image size between 800 × 800 and 1024 × 1024 pixels.

3.3 YOLOv5

The prototype of YOLOv5 was YOLO (*You Only Look Once*), which was introduced in 2015 as a uniquely fast neural network for object detection [17]. YOLOv5 was created in 2020 as a significantly improved version of the first YOLO architecture [21]. YOLOv5 was implemented using PyTorch framework. The first YOLO network is built of 24 convolutional layers and two fully-connected layers [17]. The architecture of this network is shown in Fig. 5. YOLO is a one-shot object detection model with CSP (Cross-Stage Partial connections), which allows for better feature reuse and reduces computation time [24].

YOLOv5 is built of three parts:

- New CSP-Darknet53 (*backbone*)—contains 29 convolutional layers.
- SPPF, New CSP-PAN (Cross-Stage Partial-connection & Path Aggregation Network) (*neck*) [2,10]—SPPF is built of three `MaxPooling2D` layers, two `ConvBNSiLU` layers, and one `Concat` layer. `ConvBNSiLU` is built of one convolutional layer followed by a batch-normalization layer and Sigmoid Linear Unit [16]. `Concat` is a layer that combines outputs of the preceding layers. The application of a `CSP` (*Cross-Stage-Partial connection*) allows us to improve the quality of object detection. It is a technique derived from CSPNet [14].
- YOLOv3 Head (*head*)—Three subnets built of convolutional layers that were used since YOLOv3 [13].

The images processed by the YOLOv5 architecture were scaled to the image size of 416 × 416 pixels. This is a standard image size for output from roboflow.ai [18]. Scaling images in other models to that size did not improve the results.

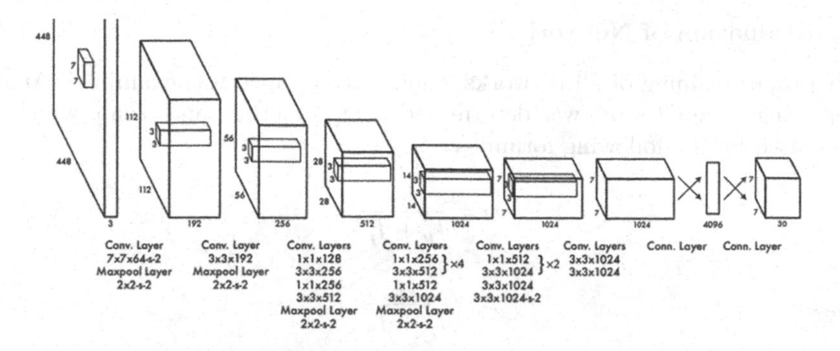

Fig. 5. The architecture of the first YOLO network [17]

4 Detection System

The presented detection system is designed as follows. On its input, the frames received from the camera are delivered in real-time. After detecting a stone, the system can send a signal to stop the conveyor belt automatically. Subsequent frames are analyzed on an ongoing basis. In order to minimize false alarms, the presence of a boulder is signaled when the neural network detects a boulder in in most frames out of 30 consecutive frames. Although there is a chance that the one stone is observable, it may be difficult to detect even by a well-trained model. Such situations have occurred, such, as dark images, blurry images, boulders covered in sand, etc. Such cases were taken it into account, and it was decided that there should be a possible break in the detection process. This break lasts no more than a small number of frames. For example, if a stone was detected in 10 frames in a row, then undetected in 4 frames in a row, and subsequently again detected in 20 frames in a row, it would be considered that a large stone had really appeared on the line. The optimal number of consecutive frames that may constitute a break in the sequence of frames where true detection occurred, is established using a validation set by maximizing the number of proper signals to stop a conveyor belt. For such a method a high-quality stone detector is required to minimize false negatives due to the model but not due to the environmental conditions. To our knowledge, such a method has not been published before.

Additionally, the created program prints information about a number of frames in which a stone appeared (not counting the sets of frames that fulfilled the conditions of sending a signal). The total time of analysis and analyzed frames per second is also shown. It should be noted that the system works in the same way for each network but the implementations differ in detail.

5 Results

This section first presents the results of the network operation on single images and then performance of the created detectors based on 30 consecutive frames from recordings. There was one stone in each recording.

5.1 Evaluation of Networks

After proper training of all networks, their performance was examined. Among other things, the F1-score was determined using precision and recall, which are represented by the following formulas:

$$p = \frac{t_p}{t_p + f_p} \tag{1}$$

$$r = \frac{t_p}{t_p + f_n} \tag{2}$$

where t_p is true positives, f_p denotes false positives, f_n is false negatives.

The formula for F1-score is the harmonic mean of precision and recall:

$$F_1 = \frac{2pr}{p + r} \tag{3}$$

where p is precision, r denotes recall.

Moreover, mean average precision (mAP) was calculated. This metric is commonly used to assess the quality of object detection models. mAP takes into account the trade-off between precision and recall, and accounts for both false positives and false negatives. This property makes mAP such a relevant measure. The method of calculating mAP is described, among others, in [3] Each network was evaluated on the validation data set. The results are shown in Table 1.

Table 1. Comparison of the networks for the following data split: training set 80%, validation set 10%, test set 10%

model	F1	precision	recall	mAP
YOLOv5 based	0.991	0.992	0.991	0.992
Mask RCNN based	0.611	0.588	0.637	0.884
RetinaNet based	0.328	0.199	0.950	0.994

Based on the analysis of the data in Table 1, it can be seen that the best results were obtained using the YOLO network. In order to ensure that the obtained results do not significantly depend on the choice of data for the test and validation sets, the experiments were repeated for the following data division: training set 70%, validation set 15%, test set 15%. The obtained results are presented in Table 2 (Fig. 6).

5.2 Evaluation of Created Detectors

The detection system were tested only on recordings that were not used during the training. The results are shown in the Tables 3–4. For clarification, here is the explanation what each column means:

Table 2. The YOLOv5-based model quality measures for a data split: training set 70%, validation set 15%, test set 15%.

model	F1	precision	recall	mAP
YOLOv5 based	0.989	0.988	0.992	0.995

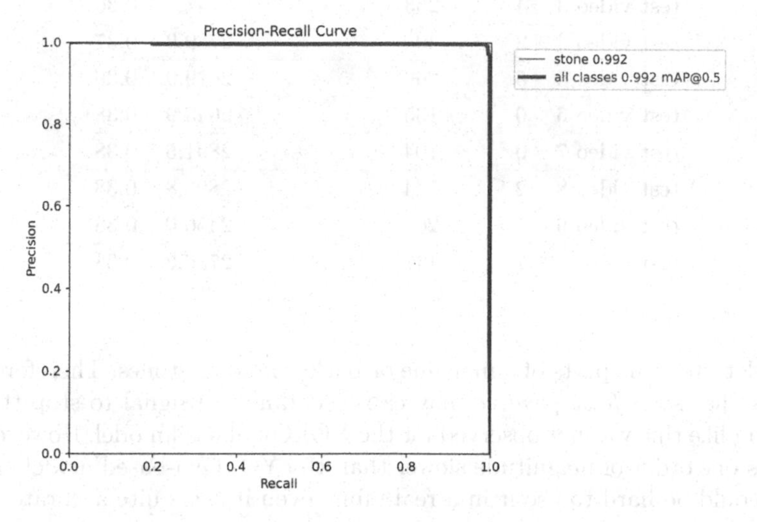

Fig. 6. Precision-recall curve for YOLOv5 for data split: training set 80%, validation set 10%, test set 10%

- video—input/output AVI file
- signals—a number of such series of detection that lasted at least 30 frames with possible breaks of 5 frames maximum,
- other detections—a number of other frames on which at least one stone was detected,
- time—total time of analyzing the recording,
- FPS—frames per second.

The following three detectors, which use 30 consecutive frames, were created: Retinanet based model, Mask RCNN based model and YOLOv5-based model.

Detector with Mask R-CNN. For two video the model could not detect stones in a sufficient number of frames in a row. It should be noticed that one of these recordings is the same one where the stone was not well detected also by the YOLOv5-based model. It can also be noticed that, on the one hand, this model detects stones more frequently than the YOLOv5-based model. On the other hand, after analyzing the output files, one can see that it detects more *true positives* and also more *false positives*. In addition, the confidence is set to 0.9, so the possible amount of *false positives* had already been decreased. Although there is a threat that this model may generate a false signal - it sometimes happens

Table 3. The results for Mask RCNN-based detector. Results for 5 skip frames.

video	signals	other detections	time [s]	FPS
test video 1	2	36	2796.0	0.37
test video 2	1	295	3103.1	0.34
test video 3	4	233	2741.4	0.36
test video 4	2	49	2789.9	0.37
test video 5	5	200	2679.9	0.36
test video 6	0	135	2643.9	0.38
test video 7	0	104	2891.5	0.38
test video 8	2	134	2806,8	0.38
test video 9	3	26	2456.0	0.38
test video 10	7	147	2736.2	0.38

that it detects some parts of a machine or background as stones. Therefore there is a risk that such *false positive* may cause sending the signal to stop the line. Anything like this was not observed for the YOLOv5-based model. However, this model is one order of magnitude slower than the YOLOv5-based model. That is why it could be hard to use it in a real mine, even if it is quite accurate.

Detector with RetinaNet. The RetinaNet-based model was able to detect a sufficient number of stones only for four recordings. It can be said it is less accurate than the other described models. In this case, there is a similar risk to the case of Mask RCNN-based model—it sometimes detects parts of a machine or background as stones, which can lead to sending false signals. This model is about twice as fast as the Mask RCNN-based model. It could not be used in a real mine because it is not accurate enough and still much slower than the YOLOv5-based model.

Detector with YOLOv5. For the YOLOv5-based detector we performed a fine-tuning of a number of skip frames. The result of this process in shown in Table 5. For this purpose, from the test set five videos are treated as validation videos on which the optimal number of skip frames is obtained by maximising the number of positive signals and retaining as low a number as possible of false positives or true positives that did not form a 30 successive frames sequence described by a column "other detections". The optimum on the validation set was attained for 7 skip frames. The performance on the remaining 5 test videos (not used in any stage of the training) is shown in Table 6.

Table 4. The results for RetinaNet-based detector. Results were obtained for 5 skip frames.

video	signals	other detections	time [s]	FPS
test video 1	2	68	1339.7	0.77
test video 2	0	85	1363.9	0.77
test video 3	0	69	1299.3	0.77
test video 4	0	88	1337.6	0.77
test video 5	1	225	1263.0	0.77
test video 6	0	58	1298.0	0.77
test video 7	0	108	1414.2	0.77
test video 8	0	55	1373.5	0.77
test video 9	2	73	1210.7	0.77
test video 10	3	130	1340.0	0.77

Table 5. The results for the YOLOv5-based detector for a varying number of skip frames on a validation set consisting of 5 videos. Fine-tuning a number of skip frames relies on maximisation of positive signals, as well as keeping the number of other detections frames as low as possible on videos from the validation set. The best results are bolded.

video	1 skip frame		3 skip frames		5 skip frames		7 skip frames		9 skip frames	
	signals	other detections	signals	other detections	signals	other detections	signals	other detections	signals	other detections
val. video 1	2	30	2	30	2	30	**2**	**30**	2	30
val. video 2	0	42	0	42	0	42	**0**	**42**	0	42
val. video 3	0	69	0	69	1	39	**1**	**39**	0	69
val. video 4	1	53	1	53	1	53	**2**	**23**	2	23
val. video 5	5	105	6	75	7	56	**7**	**56**	7	56

One can see that for two videos (one in the validation set and one in the test set), the model was unable to detect a stone for a sufficient number of frames in a row in order to generate a signal. One of these recordings is really dark, and in the other one, the stone is barely visible, which is caused by sand covering it. It can be said that for most cases, the YOLOv5-based model detects big stones well. It should also be noted that this model is very fast in comparison to the others. It processes 12–14 frames per second, which makes it unmatched if it comes to operation time. In conclusion, the YOLOv5-based detector would be right to use in real mines because the results would be delivered with low delay.

Table 6. The results for the YOLOv5-based detector for seven skip frames on a test set of five videos not seen by the model at any stage of the training.

video	signals	other detections	time [s]	FPS
test video 1	2	14	76.0	13.1
test video 2	1	136	86.0	12.7
test video 3	6	67	80.4	13.1
test video 4	0	22	80.0	13.1
test video 5	3	9	78.9	13.1

6 Conclusion

The use of deep models can be very desirable for detecting dangerous objects on a conveyor belt. Due to the monotony and lack of focus, there is a high risk of overlooking the boulders by the operator. The use of 30 consecutive frames allows to significantly improve the correctness of an operation.

The YOLOv5-based detector is the most suitable to use in a real mine because it is fast and accurate enough to detect dangerous objects in a video stream quite effectively. The results on test recordings are satisfying in terms of accuracy and processing speed. The Mask RCNN-based detector is slower but also accurate enough. Although it cannot match the YOLOv5-based detector in terms of processing speed, the RetinaNet-based model can detect objects only in some cases. It is twice as fast as the Mask RCNN-based detector but still one order of magnitude slower than the YOLOv5-based model. It is possible to try optimizing each of these detector models for a specific mine by manipulating the training hyperparameters. If the detection system is not equipped with a powerful GPU and detection time is crucial, it is recommended to use the YOLOv5-based model.

Acknowledgments. The authors would like to thank Produs S.A. and The Bełchatów coal mine for their cooperation and research data and to members of the CyberTech circle.

References

1. Abdulla, W.: Mask RCNN for Object Detection and Segmentation. https://github.com/matterport/Mask_RCNN. Accessed 14 Nov 2022
2. Bochkovskiy, A., Wang, C.Y., Liao, H.Y.M.: Yolov4: optimal speed and accuracy of object detection. arXiv preprint arXiv:2004.10934 (2020)
3. Cao, Y., Chen, K., Loy, C.C., Lin, D.: Prime sample attention in object detection. In: Proceedings of the IEEE/CVF Conference on Computer Vision and Pattern Recognition, pp. 11583–11591 (2020)
4. Cheng, R.: A survey: comparison between convolutional neural network and yolo in image identification. In: Journal of Physics: Conference Series, vol. 1453, p. 012139. IOP Publishing (2020)

5. Córdova, M., et al.: Litter detection with deep learning: a comparative study. Sensors **22**(2), 548 (2022)
6. Dhillon, B.S.: Mining Equipment Reliability. Springer, London (2008)
7. Fizyr: Keras-RetinaNet. https://github.com/fizyr/keras-retinanet. Accessed 10 Nov 2022
8. Fujita, H., Itagaki, M., Ichikawa, K., Hooi, Y.K., Kawano, K., Yamamoto, R.: Fine-tuned pre-trained mask R-CNN models for surface object detection. arXiv preprint arXiv:2010.11464 (2020)
9. He, K., Gkioxari, G., Dollár, P., Girshick, R.: Mask r-CNN. In: Proceedings of the IEEE International Conference on Computer Vision, pp. 2961–2969 (2017)
10. He, K., Zhang, X., Ren, S., Sun, J.: Spatial pyramid pooling in deep convolutional networks for visual recognition. IEEE Trans. Pattern Anal. Mach. Intell. **37**(9), 1904–1916 (2015)
11. He, K., Zhang, X., Ren, S., Sun, J.: Deep residual learning for image recognition. In: Proceedings of the IEEE Conference on Computer Vision and Pattern Recognition, pp. 770–778 (2016)
12. Lin, T.Y., Goyal, P., Girshick, R., He, K., Dollár, P.: Focal loss for dense object detection. In: Proceedings of the IEEE International Conference on Computer Vision, pp. 2980–2988 (2017)
13. Nepal, U., Eslamiat, H.: Comparing yolov3, yolov4 and yolov5 for autonomous landing spot detection in faulty UAVs. Sensors **22**(2), 464 (2022)
14. Parico, A.I.B., Ahamed, T.: Real time pear fruit detection and counting using yolov4 models and deep sort. Sensors **21**(14), 4803 (2021)
15. Qu, X., Wang, J., Wang, X., Hu, Y., Zeng, T., Tan, T.: Gravelly soil uniformity identification based on the optimized mask r-CNN model. Expert Syst. Appl. **212**, 118837 (2023)
16. Rain Juhl, E.F.: Real-time object detection and classification for ASL alphabet. http://cs231n.stanford.edu/reports/2022/pdfs/147.pdf. Accessed 11 Dec 2022
17. Redmon, J., Divvala, S., Girshick, R., Farhadi, A.: You only look once: unified, real-time object detection. In: Proceedings of the IEEE Conference on Computer Vision and Pattern Recognition, pp. 779–788 (2016)
18. Roboflow.ai (2023). https://roboflow.com/ Accessed 20 Apr 2023
19. Sciancalepore, M.: Mask R-CNN for object detection and segmentation (working with tf 2.4.1). https://github.com/masc-it/Mask-RCNN Accessed 14 Nov 2022
20. Suresh, M., Abhishek, M.: Kidney stone detection using digital image processing techniques. In: 2021 Third International Conference on Inventive Research in Computing Applications (ICIRCA), pp. 556–561. IEEE (2021)
21. Thuan, D.: Evolution of yolo algorithm and yolov5: The state-of-the-art object detention algorithm (2021)
22. Velikanov, V., Kozyr, A., Dyorina, N.: Engineering implementation of view objectives in mine excavator design. Procedia Engineering **206**, 1592–1596 (2017)
23. Vishmitha, D., Yoshika, K., Sivalakshmi, P., Chowdary, V., Shanthi, K., Yamini, M., et al.: Kidney stone detection using deep learning and transfer learning. In: 2022 4th International Conference on Inventive Research in Computing Applications (ICIRCA), pp. 987–992. IEEE (2022)
24. Wang, C.Y., Liao, H.Y.M., Wu, Y.H., Chen, P.Y., Hsieh, J.W., Yeh, I.H.: CSP-NET: a new backbone that can enhance learning capability of CNN. In: Proceedings of the IEEE/CVF Conference on Computer Vision and Pattern Recognition Workshops, pp. 390–391 (2020)
25. Zhang, H.: Image processing for the oil sands mining industry [in the spotlight]. IEEE Signal Process. Mag. **25**(6), 200–198 (2008)

A Novel DAAM-DCNNs Hybrid Approach to Facial Expression Recognition to Enhance Learning Experience

Rayner Alfred[1(✉)], Rayner Henry Pailus[1], Joe Henry Obit[1], Yuto Lim[2], and Haviluddin Sukirno[3]

[1] Creative Advanced Machine Intelligence Research Centre, Faculty of Computing and Informatics, Universiti Malaysia Sabah, Kota Kinabalu, Malaysia
{ralfred,joehenry}@ums.edu.my

[2] School of Information Science, Japan Advanced Institute of Science and Technology, Access 1-1 Asahidai, Nomi, Ishikawa 923-1292, Japan
ylim@jaist.ac.jp

[3] Department of Informatics, Universitas Mulawarman, Samarinda, Kalimantan Timur, Indonesia
haviluddin@unmul.ac.id

Abstract. Many machine learning models are applied on facial expression classification and there are three main issues affecting the performance of any algorithms in classifying emotions based on facial expressions, and these issues include image illumination, image quality and partial features recognition. Many approaches have been proposed to handle these issues. Unfortunately, one of the main challenges in detecting and classifying facial expression process is minimal differences of features between different types of emotions that can be used to differentiate these different types of emotions. Thus, there is a need to enrich each type of emotion with more relevant extracted features by having a more effective approach to extract features that can be used to represent each type of emotions more effectively and efficiently. This work addresses the issue of improving the emotion recognition accuracy by introducing a novel hybrid approach that combines the Depth Active Appearance Model (DAAM) and Deep Convolutional Neural Networks (DCNNs). The proposed DAAM and DCNNs model can be used to assist one in identifying emotions and classify learner involvement and interest in the topic which are plotted as feedback to the instructor to improve learner experience. The proposed method is evaluated on two publicly available datasets namely, JAFFE and CK+ and the results are compared to the state-of-the-art results. The empirical study showed that the proposed DAMM-CNNs hybrid method managed to perform the face expression recognition with 97.4% for the JAFFE dataset and 96.9% for the CK+ dataset.

Supported by Universiti Malaysia Sabah under Grant No: SDN0057-2019.

J. Mikyška et al. (Eds.): ICCS 2023, LNCS 14076, pp. 140–154, 2023.
https://doi.org/10.1007/978-3-031-36027-5_11

Keywords: Emotions Classification · Facial Expression Recognition · Depth Active Appearance Model · Deep Convolutional Neural Networks · Deep Learning · Feature Extractions

1 Introduction

As early as the twentieth century, Ekman and Friesen [1] defined six basic emotions based on cross-culture study [2], which indicated that humans perceive certain basic emotions in the sameway regardless of culture [3]. These prototypical facial expressions are anger, disgust, fear, happiness, sadness, and surprise. Motivations are closely related to emotions. An emotion is a mental and physiological feeling state that directs our attention and guides our behaviour. There is a need to detect and classify one's emotion in classroom with more effectively and efficiently as emotions affect learning and education. This is due to the fact that students' emotional experiences can impact on their ability to learn, their engagement in school, and their career choices. In fact, facial expression is the most frequently used nonverbal communication mode by the students in the virtual classroom. In addition to that the facial expressions of the students are significantly correlated to their emotions which helps to recognize their comprehension towards the lecture [4].

Emotion detection and classification can be performed based on neurophysiological measurements, behaviour patterns, speech [6] and facial expressions [7]. Electroencephalography (EEG) received considerable attention from researchers, since it can provide a simple, cheap, portable, and ease-to-use solution for identifying emotions [8]. However, this approach requires intervention during the process of learning and it does not provide a seamless approach to detect and recognize emotions. In a classroom, body movements and gestures can also be used to detect and classify students' emotions. Analyzing body movements and gestures also helps in emotion detection with the help of machine learning [9]. The body movements, posture, and gestures change significantly with changes in emotions. This is the reason why we can generally guess a person's basic mood with a combination of his hand/arm gestures and body movements. However, body movements and gestures are seldom used to assess emotion in classroom due to the limitation of body movements and gestures during the process of learning [10]. In speech emotion recognition, several different classifiers and different methods for features extraction can be developed to analyze extracted features that include Mel-frequency cepstrum coefficients (MFCC) and modulation spectral (MS) features [5]. A recurrent neural network (RNN), multivariate linear regression (MLR) and support vector machines (SVM) techniques are widely used in the field of emotion recognition for spoken audio signals.

Facial Recognition is a useful emotion detection technique in which through the identification of facial features using machine learning (e.g., deep learning), features of important facial regions are extracted and analyzed to classify facial expressions [11]. Major facial features used in emotion detection through machine learning and these features include eyes, nose, lips, jaw, eyebrows, mouth (open/close), and more. By having an efficient and effective feature extraction method, more enriched facial representation can be obtained for effective emotion detection through identifying facial features. One of the main challenges in detecting and classifying facial expression process is the limitation in extracting relevant features that can be used to differentiate different types of emotions. In addition to that, insufficient datasets used for training, data bias and inconsistent annotations are very common among different facial expression datasets due to different collecting conditions and the subjectiveness of annotating. In fact, it remains challenging for most researchers to design and implement algorithms to classify facial expressions under different light conditions, poses, and backgrounds and across people of different ages, genders, and ethnicities [12].

Although pure expression recognition based on only visible face images can achieve promising results, incorporating with other models into a high-level framework can provide complementary information and further enhance the robustness although this will add more complexities in designing and analyzing the results produced by multi modals [13]. Nevertheless, the fusion of other modalities, such as infrared images, depth information from 3D face models and physiological data, is becoming a promising research direction due to the large complementarity for facial expressions.

Thus, there is a need to enrich each type of emotion with more relevant extracted features by having a more effective approach to extract features that can be used to represent each type of emotions more effectively and efficiently. This work addresses the issue of improving the emotion recognition accuracy by introducing a novel hybrid approach that combines the Depth Active Appearance Model (DAAM) [15] and Deep Convolutional Neural Networks (DCNNs). DAAM is a well-known statistical model that is used to detect face model by training the image and matching with new image through shape and appearance of the image which efficiency in detecting face position. DAAM has better performance than other methods in eliminations of the influence of different facial region size, head pose and lighting condition and thus can effectively increase the recognition accuracy [16]. DCNNs is well known a deep learning algorithm that is used in computer vision task such as face recognition or object detection. As it has been shown that the method combining Active Appearance Model (AAM) and deep learning achieves significant segmentation accuracy in Prostate segmentation on 3D MR images [17], it is expected that the proposed DAAM-DCNNs hybrid approach to emotion classification is able to perform better than the stand alone CNNs methods. The proposed DAAM and DCNNs model can further be used to assist us in identifying emotions and classify learner involvement and interest in the topic which are plotted as feedback to the instructor to improve learner experience [18]. In this paper, the effects of using different values of the parameters used in the DCNNs are also investigated. These parameters include the number of epochs, the percentages of dropout and validation dataset used in this work.

The remainder of this paper is organized as follows. Section 2 discusses related works in emotion detections and classifications. Section 3 presents the formalization of the novel hybrid approach that combines the Depth Active Appearance Model (DAAM) and Deep Convolutional Neural Networks (DCNNs) in detecting and classifying emotions. Section 4 describes the experimental setup and discusses the obtained results related to emotion detections and classifications. Finally, the paper is concluded in Sect. 5 by drawing a conclusion and providing some future works.

2 Related Works

CNN has been extensively used in diverse computer vision applications, including Face Expression Recognition [19], increase agriculture productivity [20] and diseases detection [21]. This section presents several related works according to three categories; total solution for face expression recognition using CNN, feature extractions using CNN and finally applying CNN as the main classifiers.

2.1 Complete Solutions for Face Expression Recognition

There are several well-known CNN architectures used in diverse computer vision applications. For face expression recognition, AlexNet [22,23], VGGNet [24,25] and Inception [26] are well-known CNN architectures.

AlexNet contains eight layers with weights; the first five are convolutional and the remaining three are fully connected. [22] In facial expression recognition, AlexNet network can be improved by the aid of Batch Normalization (BN) layer to the existing network. [27] In contrast, VGG16 contains sixteen layers with weights in which the first thirteen are convolutional and the remaining three are fully connected. Meanwhile, the VGG19 contain nineteen layers with weights in which the first sixteen are convolutional and the remaining three are fully connected. [25] With VGGNet deep convolutional neural network, having a deeper network architecture and a 3×3 small convolution kernel and a 2×2 small pool kernel, the recognition rate is significantly improved, and the number of parameters is only slightly larger than that of the shallow layer. [24] Inception V3 has 22 layers with weights deep network, the first twenty one are convolutional and the remaining one is a fully connected. [28] Agrawal and Mittal investigated the effects of CNN parameters namely kernel size and number of filters on the classification accuracy using the FER-2013 dataset [19].

A Faster R-CNN (Faster Regions with Convolutional Neural Network Features) [29] for facial expression recognition has been proposed that utilizes Region Proposal Networks (RPN) [30]. Firstly, the facial expression image is normalized and the implicit features are extracted by using the trainable convolution kernel. Then, the maximum pooling is used to reduce the dimensions of the extracted implicit features. After that, RPNs [30] is used to generate high-quality region proposals, which are used by Faster R-CNN for detection. Finally, the Softmax classifier and regression layer is used to classify the facial expressions and predict boundary box of the test sample, respectively. It was reported that the mean Average Precision (mAP) is around 0.82.

Facial expression recognition using Efficient Local Binary Pattern (LBP) for feature extraction and General Regression Neural Network (GRNN) [31] for classification has also been presented. GRNN trains the network faster and does not require iterative training procedure. The proposed algorithm with the optimum window sizes of 64×64 improves the recognition rate.

Real-time convolutional neural networks for emotion and gender classification has also been studied [32]. In this work, two models were proposed in which they are evaluated in accordance to their test accuracy and number of parameters. The first model relies on the idea of eliminating completely the fully connected layers. The second architecture combines the deletion of the fully connected layer and the inclusion of the combined depth-wise separable convolutions and residual modules. They reported accuracies of 96% in the IMDB gender dataset and 66% in the FER-2013 emotion dataset.

A Video-based Emotion Recognition Using Deeply-Supervised CNN (DSN) architecture has been proposed to recognition emotion in Video [33]. The proposed DSB takes the multi-level and multi-scale features extracted from different convolutional layers to provide a more advanced representation of emotion recognition. CNN and LSTM based facial expression analysis model for a humanoid robot was also proposed and studied. First, a convolutional neural network (CNN) is used to extract visual features by learning on a large number of static images. Second, a long short-term memory (LSTM) recurrent neural network is used to determine the relationship between the transformation of facial expressions in image sequences and the six basic emotions. Third, CNN and LSTM are combined to exploit their advantages in the proposed model [34].

2.2 Applying Image Pre-processing Prior to CNNs Classification

A simple solution for facial expression recognition that uses a combination of Convolutional Neural Network and specific image pre-processing steps has also been proposed [31]. In this work, several pre-processing techniques are applied to extract only expression specific features from a face image and explore the presentation order of the samples during training. The proposed method achieved competitive results when compared with other facial expression recognition methods with 96.76% of accuracy in the CK+ database [35].

Automatic facial expression recognition based on a deep convolutional-neural-network structure is also introduced in which a face detection based on Haar-Like Feature is applied before feeding them into the CNNs for facial expression recognition [36]. The best recognition result is obtained when the learning rate, η, is 0.5 in JAFFE [37], and 0.7 in CK+ [35].

A 3D facial expression recognition (FER) algorithm using convolutional neural networks (CNNs) and landmark features and masks was introduced by Huiyuan and Lijun [38], which is invariant to pose and illumination variations due to the solely use of 3D geometric facial models without any texture information. The results show that the CNN model benefits from the masking, and the combination of landmark and CNN features can further improve the 3D FER accuracy.

2.3 Features Extraction Using CNNs Coupled with Other Machine Learning Classifiers

Several works have been conducted in which the deep neural network (particularly a CNN) is used as a feature extraction tool and then apply additional independent classifiers such as Random Forest (RF), Support Vector Machine (SVM) and k Nearest Neighbour (k-NN).

For instance, a novel method for automatically recognizing facial expressions using Deep Convolutional Neural Network(DCNN) features and SVM is proposed [39]. In this work, automatic facial expression recognition using DCNN features is investigated. Two publicly available datasets CK+ [35] and JAFFE [37] are used to carry out the experiment. Pre-processing step involves detecting and cropping the frontal faces using OpenCV2. Then facial features are extracted using the DCNN framework and the facial expression classification is performed using a SVM.

Another facial expression recognition method has also been introduced that makes full use of CNNs to detect face features globally and locally and then combines these global and local generic features for improving accuracy in facial expression recognition using the Support Vector Machine (SVM) classifier.

3 DAAM-DCNNs Hybrid Approach to Facial Expression Recognition

The methodology of the paper consists of three parts (i.e., face detection, face alignment, feature refinement, classification) which is in Fig. 1 described below.

Before applying Active Appearance Model, firstly we collect enough face images with various shapes as training set. The faces are automatically detected using viola Jones algorithm, which is one of the most used algorithms for face detection, the system is trained with face and non-face images, this technique can easily detect single and multiple faces from images and video. Next, an affine transformation (rotation, scaling and translation) is used to normalize the face geometric position [40]. This transformation allows creating perspective distortion. The affine transformation is used for scaling, skewing and rotation. At the end of this process, all faces across an entire dataset will:

1. Be centered in the image
2. Be rotated that such the eyes lie on a horizontal line (i.e., the face is rotated such that the eyes lie along the same y-coordinates)
3. Be scaled such that the size of the faces are approximately identical

Finally, these captured faces are then passed to the DAAM algorithm for feature refinement in order to obtain the mean shape of all the faces in order to construct shape model for facial expression classification. In feature refinement, the Depth Active Appearance Model (DAAM) [41] method is incorporated for both shape and texture information from facial images which has shown strong potential in a variety of facial recognition technologies. AAM allows accurate,

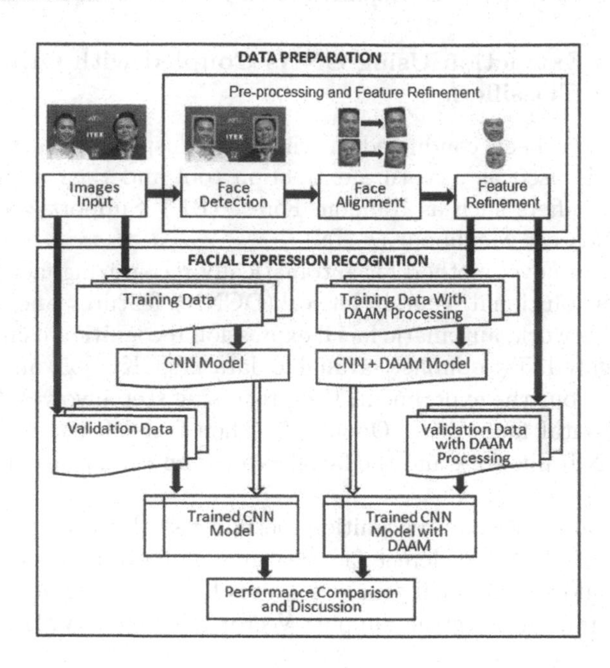

Fig. 1. DAAM-DCNNs Hybrid framework for Facial Expression Recognition system

real-time tracking of human faces in 2D and can be extended to track faces in 3D by constraining its fitting with a linear 3D morphable model. It then was extended to Depth Active Appearance Model (DAAM) [15], where a new constraint is introduced into AAM fitting that uses depth data from a commodity RGBD camera (Kinect). AAD consists of two combined models: Shape and Texture models. The shape model represents the shape of each model image and the texture model warps each image so that its landmark points match the mean shape. The PCA is applied to the normalized values and the model of texture.

3.1 Convolutional Neural Network

Figure 2 displays the architecture of the CNN model. The CNN architecture comprises 3 layers: two consecutives convolutional-pooling layers and a fully-connected classification layer. The two convolutional pooling layers use the same fixed 5 × 5 convolutional kernel and 2 × 2 pooling kernel, but have 64 and 128 neurons, respectively. The last layer has 256 neurons, which are all connected to the final six neurons for all siz types of facial expression classification.

In this work we apply a deep learning algorithm (e.g., Convolutional Neural Network) in implementing the facial expression recognition. The input image was an RGB image of 224×224 pixels. So, input size = $224 \times 224 \times 3$. The **Convolution layer** applies a 2D convolution of the input feature maps and a convolution kernel. The first hidden layer is a convolutional layer that applies 64 filters with size 5×5 each and stride 1×1. Then, the second hidden layer is

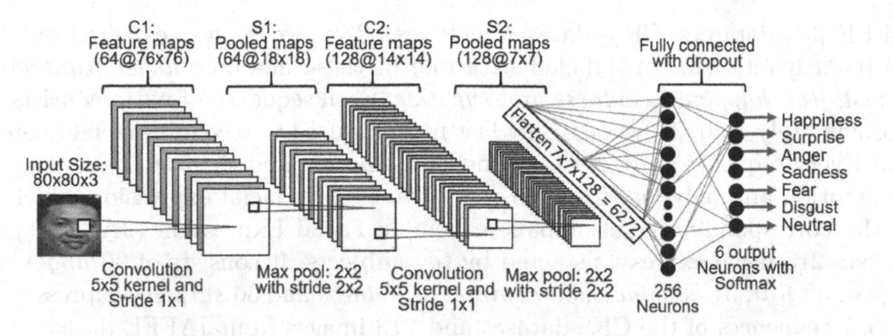

Fig. 2. CNN architecture comprises 3 layers: two consecutives convolutional-pooling layers and a fully-connected classification layer.

the **Pooling layer** that applies a Max pooling function over a spatial window without overlapping (pooling kernel) per each output feature map. This helps to reduce processing time and also helps to reduce overfitting. Each Max Pooling layer has configuration - windows size $= 2 \times 2$ and stride $= 2 \times 2$. Thus, we half the size of the image at every Pooling layer. The third hidden layer is a **Convolutional layer** that uses 128 filters with size 5×5 each and stride 1×1. Then the fourth hidden layer will be the **Pooling layer** that applies a Max pooling function. Each Max Pooling layer has configuration - windows size $=$ 2×2 and stride $= 2 \times 2$. Thus, we half the size of the image at every Pooling layer.

Then next is a **Flatten layer** that converts the 2D matrix data $(7 \times 7) \times 128$ kernels to a 1D vector with 6272 parameters before building the fully connected layers. After that we will use a **Fully connected layer** with $17 \times 17 \times 128 =$ 6272 neurons and *Relu* activation function. Then, we will use a regularization layer called Dropout. It is configured to randomly exclude 20% of neurons in the layer in order to reduce overfitting. Finally, the output layer which has 7 neurons for the 7 classes and a *Softmax* activation function to output probability-like predictions for each class that includes happiness, surprise, anger, sadness, fear, disgust, and neutral.

4 Experimental Setup and Results

In this work, the effects of CNN parameters namely the number of epoch, the percentage of dropout for the fully connected network (see Fig. 2) and the percentages of validation datasets are investigated using the CK+ [35] and JAFFE [37] datasets.

4.1 Datasets

In this work, two publicly available facial expression datasets were used to evaluate the proposed DAAM-DCNN hybrid method, namely CK+ [35] and

JAFFE [37] datasets. CK+ dataset includes 327 video sequences acted out by 118 participants which is labeled with one of these motions: *anger*, *contempt*, *disgust*, *fear*, *happiness*, *sadness* and *surprise*. Each sequence of video consists of approximately 15 to 35 frames, only last frame is used to recognize facial expression. Every sequence starts with the neutral emotion and the last frame depicts the emotion and only last frame is used to recognize facial expression which is for the corresponding label. Japanese Female Facial Expression (JAFFE) [37] consists 213 facial expressions acted by ten subjects. It consists of 30 *anger*, 29 *disgust*, 32 *fear*, 31 *happiness*, 30 *neutral*, 31 *sadness* and 30 *surprise* expressions. All 327 sequences of the CK+dataset and 213 images from JAFFE dataset are used for evaluating the proposed model.

4.2 Investigated Parameters

In this experiment, we vary the number of epochs in order to observe the performance of the proposed DAAM-DCNN hybrid method. We also vary the percentage of dropout and also the percentage number of samples used to validate the performance of the proposed DAAM-DCNN hybrid method. All the values for all the parameters used in this experiment are listed in Table 2 (Table 1).

Table 1. Values of parameters Investigated in this work.

Data Sets	Parameters		
	Epoch	% of Dropout	% of Validation Dataset Used
CK+	15, 30, 50, 100	5, 10, 25, 50	10, 20, 30, 40
JAFFE	15, 30, 50, 100	5, 10, 25, 50	10, 20, 30, 40

In the DAAM-DCNNs method, the initial values for the number of epochs (**E**), the percentage of dropout (**D**) and validation dataset (**V**) are set to 15, 25 and 20 respectively. Then, the best value for **E** is determined by varying its values (e.g., 13, 30, 50 and 100) and at the same time maintaining the values for **D** and **V**. Once, the best value for **E** has been determined, this value is then used to determine the best value for **D** by varying its values (e.g., 5, 10, 25 and 50) and at the same time maintaining the values for **V**. Finally, once the best value for **D** has been determined, these **E**'s and **D**'s values are then used to determine the best value for **V** by varying its values (e.g., 10, 20, 30 and 40) and at the same time maintaining the values for **E** and **D**.

Once all the parameters' values are determined that will produce the optimise recognition rate for the two publicly available datasets, JAFFE and CK+, the same setting will be used to determine the recognition rate for CNN alone without the DAAM preprocessing steps. THe recognition rate obtained will then be compared to the one obtained from the DAAM-DCNNs framework. The results obtained on this work will also be compared to the all results that were published previously [42,43].

4.3 Results and Discussion

Table 2. Comparison of recognition rate with different values of epochs, dropouts and validation percentages.

Dataset	V (Validation), D (Dropout), E (Epoch)	Recognition Rate	
		Values	Acc (%)
CK+	D = 25%, V = 20%, Epoch	15	66.8
		30	75.0
		50	76.5
		100	**77.5**
	E = 100, V = 20%, Dropout (%)	5	**86.3**
		10	85.1
		25	80.2
		50	75.6
	E = 100, D = 5%, Validation (%)	10	87.2
		20	91.3
		30	93.9
		40	**96.9**
JAFFA	D = 25%, V = 20%, Epoch	15	80.7
		30	78.9
		50	83.2
		100	**85.8**
	E = 100, V = 20%, Dropout (%)	5	**92.8**
		10	89.3
		25	83.2
		50	75.1
	E = 100, D = 5%, Validation (%)	10	93.6
		20	92.8
		30	**97.4**
		40	96.4

Table 2 tabulates all the results obtained in work. In this work, the effects of varying the values for the number of epoch, the percentage of dropout for the fully connected network (see Fig. 2) and the percentages of validation datasets are investigated. As shown in Table 2, **V** stands for the percentage of *validation* dataset, **D** stands for the percentages of *dropout* and the **E** stands for the number of *epoch* used in the experiments.

JAFFE Dataset. As for the JAFFE dataset, the best recognition rate was 97.4% in which the values for epoch, dropout percentage and percentage of samples used for validation were 100, 5% and 30% respectively for dataset JAFFE. It can be observed that the number of epochs has positive relationship with the recognition rate. The higher is the epoch number, the better is the recognition rate as shown in Table 2. Based on Table 2, the best recognition rate is 85.8% having epoch's, dropout's and validation's values of 100, 25% and 20%. Since epoch of 100 produced highest recognition rate, this epoch's value is maintained and we vary other parameters' values (e.g., dropout's and validation's percentage). Based on the results obtained, a higher recognition rate can be obtained when the dropout percentage value is lowered to 5% as shown in Table 2. It can be observed that, the lower the dropout's value, the better is the recognition rate of the proposed DAAM-DCNN hybrid method. At this point the best recognition rate is 92.8%.

Finally, as the epoch's and dropout's values are maintained as 100 and 5%, the value of the percentage of validation dataset is varied and based on the results, it showed that the best recognition rate can be achieved is 97.4%, when the percentage of validation dataset is set to 30% as shown in Table 2.

CK+ Dataset. As for the CK+dataset, the best recognition rate was 97.4% in which the values for epoch, dropout percentage and percentage of samples used for validation were 100, 5% and 30% respectively. When observing the patterns of recognition rate with respect to the number of epochs, percentage of dropout and also the percentage of samples used for validations, similar patterns can be observed in which the number of epoch has positive relationship with the recognition rate. The higher is the epoch number, the better is the recognition rate as shown in Table 2. Based on Table 2, the best recognition rate for CK+ dataset is 77.5% having epoch's, dropout's and validation's values of 100, 25% and 20%.

Similarly, the epoch's value of 100 is maintained in order to produce better recognition rate, other parameters' values are varied (e.g., dropout's and validation's percentage). Based on the results obtained, a higher recognition rate can be obtained when the dropout percentage value is lowered to 5% as shown in Table 2. It also can be observed that, the lower the dropout's value, the better is the recognition rate of the proposed DAAM-DCNN hybrid method. At this point the best recognition rate is 86.3%. Finally, as the epoch's and dropout's values are maintained as 100 and 5%, the value of the percentage of validation dataset is varied and based on the results, it showed that the best recognition rate can be achieved is 96.9%, when the percentage of validation dataset is set to 40% as shown in Table 2. Table 3 shows the recognition rate of the CNN alone without the DAAM preprocessing stage with predefined parameters'values for the number of epochs, the percentages of dropouts and percentages of validation dataset used. Based on Table 3, the recognition rates obtained for the CK+ and JAFFA datasets are 87.3% and 92.1% respectively. Finally, Table 4 shows the comparison of recognition rate obtained by the proposed DAAM-DCNNS model with state-of-art literature results.

Table 3. Recognition rate using CNN alone with predefined values of epoch, dropout and validation percentages.

Data Sets	Parameters			Recognition Rate (%)
	Epoch	Dropout (%)	Validation (%)	
CK+	100	5	40	87.3
JAFFE	100	5	30	92.1

Table 4. Comparison of recognition rate obtained by the proposed DAAM DCNN model with state-of-art literature.

Data Sets	Recognition Rate	
	Methods	Recognition Rate (%)
CK+	**DCNN**	**87.3**
	DAAM-DCNN	**96.9**
	Shan *et al.* [46](2009)	89.1
	Jeni *et al.* [47](2011)	96.0
	Kahou *et al.* [48](2015)	91.3
JAFFA	**DCNN**	**92.1**
	DAAM-DCNN	**97.4**
	Lyon *et al.* [42](1999)	92.0
	Zhao *et al.* [43](2011)	81.6
	Zhang *et al.* [44](2011)	92.9
	Mlakar *et al.* [45](2015)	87.8

5 Conclusion

Facial expressions convey the emotional state of an individual to the observers. An efficient and effective method to recognize facial expressions has been proposed in this paper. The hybrid approach that combines the DAAM and DCNNs can be used to effectively extract relevant features to be used for facial expression recognition. The proposed method is evaluated on two publicly available datasets name, JAFFE and CK+ and the results are compared to the state-of-the-art results. The empirical study showed that the proposed DAMM-CNNs hybrid method managed to perform the face expression recognition with 97.4% for the JAFFE dataset and 96.9% for the CK+ dataset. The proposed model can be adopted to any generic facial expressions recognition dataset that either involves recognition in static images or video sequences. No retraining or extensive pre-processing techniques are required to adopt the proposed method for facial feature extraction. The proposed DAAM and DCNNs model can be used to assist us in identifying emotions and classify learner involvement and interest in the topic which are plotted as feedback to the instructor to improve learner experience. However, the performance of any CNNs architecture can be tuned by changing the size of the filters, the number of strides, the number of convolutional layers, the number of layers for the fully connected network, the number of neurons for each layer and the learning and activation functions used in learning any data. Thus, future work involves exploring ways or approaches used to automatically opitmize the CNNs parameters in order to learn any data presented to the deep learning algorithms.

Acknowledgements. This work was supported in part by the Universiti Malaysia Sabah internal grant no. SDN0057-2019 (Biometric Patient Authentication System Using Face Recognition Approach for Smart Hospital).

References

1. Ekman, P., Friesen, W.V.: Constants across cultures in the face and emotion. J. Pers. Soc. Psychol. **17**(2), 124 (1971)
2. Huang, G., Cui, J., Alam, M., Wong, K.H.: Experimental analysis of the facial expression recognition of male and female. In: Proceedings of the 3rd International Conference on Computer Science and Application Engineering, pp. 1–5, October 2019
3. Tyng, C.M., Amin, H.U., Saad, M.N., Malik, A.S.: The influences of emotion on learning and memory. Front. Psychol. **1454** (2017)
4. Sathik, M., Jonathan, S.G.: Effect of facial expressions on student's comprehension recognition in virtual educational environments. Springerplus **2**, 1–9 (2013)
5. Kerkeni, L., Serrestou, Y., Mbarki, M., Raoof, K., Mahjoub, M.A., Cleder, C.: Automatic speech emotion recognition using machine learning (2019)
6. Sun, L., Zou, B., Fu, S., Chen, J., Wang, F.: Speech emotion recognition based on DNN-decision tree SVM model. Speech Commun. **115**, 29–37 (2019)
7. Marechal, C., et al.: Survey on AI-based multimodal methods for emotion detection. High-Perform. Model. Simul. Big Data Appl. **11400**, 307–324 (2019)
8. Alarcao, S.M., Fonseca, M.J.: Emotions recognition using EEG signals: a survey. IEEE Trans. Affect. Comput. **10**(3), 374–393 (2017)
9. Shen, Z., Cheng, J., Hu, X., Dong, Q.: Emotion recognition based on multi-view body gestures. In: 2019 IEEE International Conference on Image Processing (ICIP), pp. 3317–3321. IEEE, September 2019
10. Nafisi, J.S.A.: Gesture and body-movement as teaching and learning tools in western classical singing (Doctoral dissertation, Monash University) (2013)
11. Ko, B.C.: A brief review of facial emotion recognition based on visual information. Sensors **18**(2), 401 (2018)
12. Valstar, M.F., Mehu, M., Jiang, B., Pantic, M., Scherer, K.: Meta-analysis of the first facial expression recognition challenge. IEEE Trans. Syst. Man Cybern. Part B (Cybernetics) **42**(4), 966–979 (2012)
13. Ringeval, F., et al.: Avec 2017: real-life depression, and affect recognition workshop and challenge. In: Proceedings of the 7th Annual Workshop on Audio/Visual Emotion Challenge, pp. 3–9, October 2017
14. Valstar, M., Gratch, J., Schuller, B., Ringeval, F., Cowie, R., Pantic, M.: Summary for AVEC 2016: depression, mood, and emotion recognition workshop and challenge. In: Proceedings of the 24th ACM International Conference on Multimedia, pp. 1483–1484, October 2016
15. Smolyanskiy, N., Huitema, C., Liang, L., Anderson, S.E.: Real-time 3D face tracking based on active appearance model constrained by depth data. Image Vis. Comput. **32**(11), 860–869 (2014)

16. Wang, L., Li, R., Wang, K.: A novel automatic facial expression recognition method based on AAM. J. Comput. **9**(3), 608–617 (2014)
17. Cheng, R., et al.: Active appearance model and deep learning for more accurate prostate segmentation on MRI. In: Medical Imaging 2016: Image Processing, vol. 9784, pp. 678–686. SPIE (2016)
18. Krithika, L.B., GG, L.P.: Student emotion recognition system (SERS) for e-learning improvement based on learner concentration metric. Procedia Comput. Sci. **85**, 767–776 (2016)
19. Agrawal, A., Mittal, N.: Using CNN for facial expression recognition: a study of the effects of kernel size and number of filters on accuracy. Vis. Comput. **36**(2), 405–412 (2020)
20. Alfred, R., Obit, J.H., Chin, C.P.-Y., Haviluddin, H., Lim, Y.: Towards paddy rice smart farming: a review on big data, machine learning, and rice production tasks. IEEE Access **9**, art. no. 9389541, 50358–50380 (2021). https://doi.org/10.1109/ACCESS.2021.3069449
21. Alfred, R., Obit, J.H.: The roles of machine learning methods in limiting the spread of deadly diseases: a systematic review. Heliyon **7**(6), art. no. e07371 (2021). https://doi.org/10.1016/j.heliyon.2021.e07371
22. Pedraza, A., Gallego, J., Lopez, S., Gonzalez, L., Laurinavicius, A., Bueno, G.: Glomerulus classification with convolutional neural networks. In: Valdés Hernández, M., González-Castro, V. (eds.) MIUA 2017. CCIS, vol. 723, pp. 839–849. Springer, Cham (2017). https://doi.org/10.1007/978-3-319-60964-5_73
23. Krizhevsky, A., Sutskever, I., Hinton, G.E.: Imagenet classification with deep convolutional neural networks. Commun. ACM **60**(6), 84–90 (2017)
24. Jun, H., Shuai, L., Jinming, S., Yue, L., Jingwei, W., Peng, J.: Facial expression recognition based on VGGNet convolutional neural network. In: 2018 Chinese Automation Congress (CAC), pp. 4146–4151. IEEE, November 2018
25. Gopalakrishnan, K., Khaitan, S.K., Choudhary, A., Agrawal, A.: Deep convolutional neural networks with transfer learning for computer vision-based data-driven pavement distress detection. Constr. Build. Mater. **157**, 322–330 (2017)
26. Ning, C., Zhou, H., Song, Y., Tang, J.: Inception single shot multibox detector for object detection. In: 2017 IEEE International Conference on Multimedia & Expo Workshops (ICMEW), pp. 549–554. IEEE (2017)
27. Chen, X., Yang, X., Wang, M., Zou, J.: Convolution neural network for automatic facial expression recognition. In: 2017 International Conference on Applied System Innovation (ICASI), pp. 814–817. IEEE (2017)
28. Szegedy, C., et al.: Going deeper with convolutions. In: Proceedings of the IEEE Conference on Computer Vision and Pattern Recognition, pp. 1–9 (2015)
29. Li, J., et al.: Facial expression recognition with faster R-CNN. Procedia Comput. Sci. **107**, 135–140 (2017)
30. Ren, S., He, K., Girshick, R., Sun, J.: Faster R-CNN: towards real-time object detection with region proposal networks. Adv. Neural Inf. Process. Syst. **28** (2015)
31. Talele, K., Shirsat, A., Uplenchwar, T., Tuckley, K.: Facial expression recognition using general regression neural network. In: 2016 IEEE Bombay Section Symposium (IBSS), pp. 1–6. IEEE (2016)
32. Arriaga, O., Valdenegro-Toro, M., Plöger, P.: Real-time convolutional neural networks for emotion and gender classification. arXiv preprint arXiv:1710.07557. (2017)
33. Fan, Y., Lam, J.C., Li, V.O.: Video-based emotion recognition using deeply-supervised neural networks. In: Proceedings of the 20th ACM International Conference on Multimodal Interaction, pp. 584–588 (2018)

34. Li, T.H.S., Kuo, P.H., Tsai, T.N., Luan, P.C.: CNN and LSTM based facial expression analysis model for a humanoid robot. IEEE Access **7**, 93998–94011 (2019)
35. Lucey, P., Cohn, J.F., Kanade, T., Saragih, J., Ambadar, Z., Matthews, I.: The extended cohn-kanade dataset (ck+): S complete dataset for action unit and emotion-specified expression. In: 2010 IEEE Computer Society Conference on Computer Vision and Pattern Recognition-Workshops, pp. 94–101. IEEE (2010)
36. Shan, K., Guo, J., You, W., Lu, D., Bie, R.: Automatic facial expression recognition based on a deep convolutional-neural-network structure. In: 2017 IEEE 15th International Conference on Software Engineering Research, Management and Applications (SERA), pp. 123–128. IEEE (2017)
37. Lyons, M., Akamatsu, S., Kamachi, M., Gyoba, J.: Coding facial expressions with gabor wavelets. In: Proceedings Third IEEE International Conference on Automatic Face and Gesture Recognition, pp. 200–205. IEEE (1998)
38. Yang, H., Yin, L.: CNN based 3D facial expression recognition using masking and landmark features. In: 2017 Seventh International Conference on Affective Computing and Intelligent Interaction (ACII), pp. 556–560. IEEE (2017)
39. Mayya, V., Pai, R.M., Pai, M.M.: Automatic facial expression recognition using DCNN. Procedia Comput. Sci. **93**, 453–461 (2016)
40. Jain, A.K.: Fundamentals of Digital Image Processing. Prentice-Hall, Inc., Hoboken (1989)
41. Cootes, T.F., Edwards, G.J., Taylor, C.J.: Active appearance models. In: Burkhardt, H., Neumann, B. (eds.) ECCV 1998. LNCS, vol. 1407, pp. 484–498. Springer, Heidelberg (1998). https://doi.org/10.1007/BFb0054760
42. Lyons, M.J., Budynek, J., Akamatsu, S.: Automatic classification of single facial images. IEEE Trans. Pattern Anal. Mach. Intell. **21**(12), 1357–1362 (1999)
43. Zhao, X., Zhang, S.: Facial expression recognition based on local binary patterns and kernel discriminant isomap. Sensors **11**(10), 9573–9588 (2011)
44. Zhang, L., Tjondronegoro, D.: Facial expression recognition using facial movement features. IEEE Trans. Affect. Comput. **2**(4), 219–229 (2011)
45. Mlakar, U., Potočnik, B.: Automated facial expression recognition based on histograms of oriented gradient feature vector differences. SIViP **9**(1), 245–253 (2015). https://doi.org/10.1007/s11760-015-0810-4
46. Shan, C., Gong, S., McOwan, P.W.: Facial expression recognition based on local binary patterns: a comprehensive study. Image Vis. Comput. **27**(6), 803–816 (2009)
47. Jeni, L.A., Takacs, D., Lorincz, A.: High quality facial expression recognition in video streams using shape related information only. In: 2011 IEEE International Conference on Computer Vision Workshops (ICCV Workshops), pp. 2168–2174. IEEE (2011)
48. Kahou, S.E., Froumenty, P., Pal, C.J.: Facial expression analysis based on high dimensional binary features. In: ECCV Workshops (2), pp. 135–147 (2014)

Champion Recommendation in League of Legends Using Machine Learning

Kinga Wiktoria Błaszczyk and Dominik Szajerman(✉)

Institute of Information Technology, Lodz University of Technology, Łódź, Poland
dominik.szajerman@p.lodz.pl

Abstract. League of Legends (LoL) is a Multiplayer Online Battle Arena (MOBA) game with over 160 champions and a competitive esports scene. The ban-and-pick system allows players to choose champions before a match, which can greatly impact the outcome, especially in professional leagues. This article presents an overview of the development of a champion recommendation system for League of Legends, with a focus on the evaluation and comparison of multiple machine learning models. Our system offers real-time recommendations during the pick and ban phase, utilizing data from professional and high-level games. The accuracy and performance of the various models are analyzed and presented, providing insights into the strengths and weaknesses of each approach in solving the task of champion recommendation. Results show that the player's statistics on a champion are the biggest determinants of a game's outcome.

Keywords: League of Legends · Champion recommendation · Machine Learning

1 Introduction

League of Legends (Lol) is a worldwide famous multiplayer online battle arena (MOBA) game highly known for its competitive esports scene. The game released in 2009 by Riot Games has been consistently ranked as one of the most played games in the world, with an active player base of over 100 million people.

The game consists of two teams of five players competing against each other in a strategic battle to destroy the enemy's base. Before the match starts, the player must choose a unique champion to play, due to the ban and pick system used by the game code. With 162 champions to choose from, each with their own set of abilities and play style, the choice of the champions to play as, becomes a tough one. When choosing a champion, players must carefully weigh their strengths and weaknesses, as well as how they will complement their team's strategy and counter their opponents' chosen champions. This decision can have a significant impact on the game's outcome, especially in professional leagues.

To help players make a more informed decision, we developed a real-time recommendation system. The system uses machine learning algorithms to suggest champions that may be effective in the current game based on factors such as the player's past performance.

J. Mikyška et al. (Eds.): ICCS 2023, LNCS 14076, pp. 155–170, 2023.
https://doi.org/10.1007/978-3-031-36027-5_12

1.1 Gameplay Overview

League of Legends (LoL) is a game where two teams of five players compete against each other. Players control a champion with unique abilities and battle it out on a map called the Rift. The goal is to destroy the enemy's nexus, a structure located in their base, while defending your own.

Each game starts with champions picking their loadout and heading to their lanes, where they will fight against the enemy team and minions (computer-controlled creatures). The objective is to earn gold and experience by killing minions and enemy champions, which will allow players to level up and purchase powerful items to enhance their champion's abilities.

Victory in LoL requires not only individual skill but also teamwork and strategy. Champions have unique abilities that can be combined with others to execute powerful combinations and take down the enemy. Teams must communicate and coordinate their movements and abilities to secure objectives and push towards the enemy nexus.

Overall, League of Legends is a fast-paced, action-packed game that rewards strategic thinking, teamwork, and quick reflexes. With over 160 champions to choose from, there is always a new challenge to tackle and a new way to play.

1.2 Pick and Ban Phase

In MOBA games, the drafting stage is of paramount importance. The selection of characters with strong synergy can be the deciding factor in a team's victory. Games like League of Legends have implemented the pick and ban system, which holds great power in shaping the outcome of the match. The pick and ban system is a way for teams to strategically select the champions they will play as, while also preventing the enemy team from using certain champions that may be particularly strong against them. Each team takes turns selecting and banning champions, with the first team selecting two champions, the second team selecting two champions and banning one champion, and the first team banning one champion. This process is repeated until all champions have been selected and banned.

2 Related Work

Recently, there have been limited publications on various strategies for developing effective Champion Recommendation systems in MOBA games. Costa et al. [5] investigated a method of creating team compositions in League of Legends using genetic algorithms. Their research found that the use of genetic algorithms guided by appropriate fitness functions significantly enhanced team composition, with generated teams showing a quality of between 76% and 95%. Hong et al. [9] found that a neural network classifier was more effective than a random forest classifier in their study of a champion recommendation system for League of Legends. In contrast, Porokhnenko et al. [11] found that linear regression was the fastest model when considering different machine learning methods for predicting match results in Dota 2. Harikumar et al. [2] found that their models achieved high accuracy (between 95% and 99%) for predicting match outcomes in League of Legends using feature selection and ensemble methods. Hanke et al. [7] also evaluated a recommender system for hero line-ups in MOBA games and found that the neural network they trained was capable of predicting the winning team with an accuracy of 88.63 In a study of a predictor for the MOBA game Dota 2, Kinkade et al. [10] used Logistic Regression and Random Forest Classifier models to achieve an accuracy of 73% at the beginning of the match. Chen et al. [3] treated the drafting process as a combinatorial game and confirmed that Monte Carlo Tree Search-based recommendations could lead to stronger hero line-ups compared to other baselines. Yu et al. [6] found that a SVD-based collaborative filtering approach was suitable for recommending champions in League of Legends, while Conley et al. [4] achieved promising results for a recommendation engine for picking heroes in Dota 2 using logistic regression and K-nearest neighbor models. More recent work has focused on predicting the results of the games rather than building larger systems on this basis [8,12]. Overall, these studies demonstrate that it is possible to make reliable predictions of match outcomes and recommend effective hero line-ups in MOBA games using various machine learning techniques.

3 Methodology

3.1 Formulation of Machine Learning Problem

The objective of this research is to develop an efficient system for champion recommendation in League of Legends. To achieve this goal, the first step is to construct an algorithm for predicting match outcomes by using the League of Legends champions chosen by players or other relevant factors. This problem is a binary classification problem with two output classes, representing a victory for the "blue" or "red" team.

3.2 Datasets

Pre-made Datasets. The datasets were sourced from Kaggle[1], each featuring a unique set of characteristics for describing professional League of Legends matches. The datasets include statistics from 2,840 matches that took place between January 1st, 2021, and March 23rd, 2021. Data for these datasets was specifically gathered from Oracle's Elixir[2], a reputable source for professional gaming information since 2015. A general explanation of the datasets is presented below:

[1] https://www.kaggle.com/datasets/tekpixo/leagueoflegendsprematch2021.
[2] oracleselixir.com/about.

- Players Statistics (**PS**) dataset: This dataset contains pre-game statistics for each player with their picked champion, including win rate percentage (WR), games played (GP), and the ratio of kills plus assists over deaths (KDA) for that champion in previous matches. The dataset includes a total of 31 features;
- Picked Champions and Players Statistics (**PC+PS**) dataset: This dataset merges data from the Picked Champions and Players Statistics datasets, resulting in a total of 41 features;
- Complete (**C**) dataset: This dataset includes all the information from the previous datasets, resulting in a total of 51 features.
- Picked Champions (**PC**) dataset: This dataset includes 11 features that provide detailed information about the champions picked by both teams in each match and the outcome of the game;
- Banned Champions (**BC**) dataset: The dataset includes 11 features that provide detailed information about the champions banned by both teams in each match, as well as the outcome of the match;

Table 1 provides the structure for PS dataset and a preview of data. The columns contain: btgp – blue top games played, btwr – blue top win ratio, btkda – blue top kill/death/assists ratio etc. for the remaining positions and side. The [...] columns include analogous statistics for red. Result: "1" – blue team won, "0" – it lost.

Table 1. Pre-made Player Statistics dataset preview.

game	btgp	btwr	btkda	bjgp	bjwr	bjkda	bmgp	bmwr	bmkda	...	result
6911–9185	31	0.32	2.4	17	0.47	4.3	6	0.67	2.7	...	0
6911–9186	10	0.6	3.4	21	0.33	3.6	6	0.17	1.9	...	1
6912–9187	35	0.77	2.8	11	0.82	8.2	84	0.64	5.3	...	0

Tables 2and 3 provide a similar preview of the pre-made dataset's categorization PC and BC.

Table 2. Pre-made – Picked Champions dataset preview. Names: btop, bjg, bmid etc. mean champion id picked from the blue side. Names: rtop, rjg, rmid etc. mean champion id banned from the red side. Result: "1" – blue team won, "0" – it lost.

game	btop	bjg	bmid	badc	bsupp	rtop	rjg	rmid	radc	rsupp	result
6911–9185	58	113	134	135	12	106	80	142	145	57	0
6911–9186	79	104	134	523	412	39	2	61	145	875	1
6912–9187	150	76	61	21	3	58	104	112	202	89	0

As previously mentioned, data were obtained from Oracle's Elixir portal. The platform provides a daily updated CSV file containing all professional matches of the year. The dataset includes the performance history of the 10 players involved in each match with the champions they selected for the game.

Table 3. Pre-made – Banned Champions dataset preview. Names: bban1, bban2, bban3 etc. mean champion id banned from the blue side. Names: rban1, rban2 etc. mean champion id banned from the red side. Result: "1" – blue team won, "0" – it lost.

game	bban1	bban2	bban3	bban4	bban5	rban1	rban2	rban3	rban4	rban5	result
6911–9185	84	777	236	516	54	2	104	3	266	164	0
6911–9186	3	58	266	164	142	84	135	236	777	516	1
6912–9187	163	4	135	516	106	84	134	12	429	777	0

Riot API Generated Datasets. The datasets were created utilizing the API provided by Riot Games, the developer of League of Legends. The datasets include data from 2277 games, collected between July 1st, 2022, and October 30th, 2022. These datasets were modeled after the pre-made datasets to expand the available data for the study. The data satisfies the following requirements:

- The game mode is RANKED SOLO/DUO, which features a ban and pick system. This system allows for the collection of valuable data on banned and picked champions for the study.
- The skill level of the players is "very high", which corresponds to roughly the top 0.011% of players. It is believed that the performance of available champions will be optimal in the hands of these highly skilled players.
- The game duration is set at over 20 min, this criteria is established to ensure that the games are as fair as possible (LoL has a surrender system, which allows players to end the game from the 15th minute, it requires 100% agreement from the whole team of 5 players. After the 20th minute of the game, the surrender requires 80% agreement).

Similar to the pre-made datasets, a crawler was developed to collect performance history data on the 10 players involved in each match with the champions they selected for the game. The crawler gathered data from the 20 previously played games for each individual player from the 10 in question. This data were used to create Players Statistics datasets.

Both of these dataset groups pertain to binary classification, with the target feature in the final column indicating the game result for the blue team, represented by "1" for a win or "0" for a loss. The target feature is consistent across all datasets, with a class distribution of:

- Pre-made datasets: 46.37% wins for the blue team and 53.63% wins for the red team.
- Riot API generated datasets: 51.12% wins for the blue team and 48.88% wins for the red team.

It is noteworthy that the pre-made datasets included data for 155 champions available in League of Legends, whereas the datasets generated through the Riot API included data for 161 champions.

3.3 Machine Learning Models for Solving the Problem

This research employed various models to determine the likelihood of victory for the blue team. These models were grouped into seven categories:

1. Ensemble Models, which combine multiple base models to create a stronger overall model. These models are often used when the base models are weak learners, but when combined, they can increase the overall performance.
   - Gradient Boosting (GB) [1]: a method that iteratively trains base models and adjusts the weights of the training instances to focus on the misclassified instances in the previous iteration.
   - Extreme Gradient Boosting (XGB) [1]: an optimized version of Gradient Boosting that uses histogram-based algorithms for faster and more accurate training.
   - Bagging Classifier (BC) [1]: a method that trains multiple instances of a base model on different subsets of data and combines their predictions.
   - AdaBoost (AdaB) [1]: an algorithm that combines weak learners to improve classification accuracy. It adjusts instance weights in each iteration to focus on misclassifications.
   - Random Forest Classifier (RFC) [1]: a method that trains multiple decision trees and combines their predictions.
   - Random Forest Regressor (RFR) [1]: a method that trains multiple decision trees for regression tasks and combines their predictions.
2. Neural Networks (NN), which are inspired by the structure and function of the human brain. They are widely used in tasks such as image recognition, speech recognition, and natural language processing, and are characterized by multiple layers and high adaptability.
   - Neural Network MLP (MLP) [1]: a simple feedforward NN that has multiple layers of neurons, it s used for supervised learning tasks.
   - Recurrent Neural Network (RNN) [1]: a type of neural network that is particularly well-suited for sequential data, it uses feedback connections to process sequences of variable length.
   - Fully Connected Neural Network (FCNN) [1]: a type of neural network that is commonly used in image recognition tasks, it is a neural network where all the neurons in a layer are fully connected to the neurons of the previous and next layers.
3. Bayesian Models, which are based on Bayes' theorem. They are characterized by their ability to handle uncertainty and complexity in data, and are widely used in tasks such as classification and regression.
   - Gaussian Process Classifier (GPC) [1]: a probabilistic model that can be used for classification tasks, it makes predictions based on the probability distribution of the input.
   - Gaussian Process Regressor (GPR) [1]: a probabilistic model that can be used for regression tasks, it makes predictions based on the probability distribution of the input.

   - Quadratic Discriminant Analysis (QDA) [1]: a type of Bayesian model that is used for classification, it is characterized by its ability to model quadratic decision boundaries.
   - Gaussian Naive Bayes (GNB) [1]: a probabilistic model that uses Bayes' theorem to classify data based on the probability of a given class given a set of features.
4. Linear Models, which are based on linear equations to make predictions. They are simple yet powerful models and are used for prediction and forecasting purposes.
   - Logistic Regression (LogR) [1]: a linear model that is used for binary classification, it is characterized by the logistic function that maps the input to a probability value between 0 and 1.
   - Linear Support Vector Classification (LSVC) [1]: a type of SVM model that separates the data with a linear boundary.
   - Ridge Classifier (RC) [1]: a linear model that is used for classification, it uses L2 regularization method to prevent overfitting.
   - Linear Regression (LR) [1]: a technique used to model the relationship between a scalar dependent variable and one or more independent variables represented by a vector, it is a simple yet powerful model that is widely used for prediction and forecasting purposes.
5. Decision Tree Models, which create a tree-like structure to make predictions. These models are characterized by their ability to handle non-linear relationships and categorical variables.
   - Decision Tree Classifier (DTC) [1]: a model that creates a tree-like structure to make predictions, it is characterized by its ability to handle non-linear relationships and categorical variables.
   - Cat Boost Classifier (CBC) [1]: an optimization of the decision tree algorithm, it is particularly useful for dealing with categorical variables.
6. Support Vector Machines (SVMs), which are based on the concept of maximizing the margin between different classes. These models are widely used in tasks such as classification and regression.
   - SVM linear (SVM-L) [1]: a type of SVM model that separates the data with a linear boundary.
   - SVM radial (SVM-R) [1]: as above with a non-linear boundary.
7. k-Nearest Neighbors (k-NN), which is a non-parametric, instance-based learning algorithm. It does not have a model equation like linear or logistic regression models. Instead, it makes predictions based on the similarity between the input instance and the instances in the training set.
   - k-Nearest Neighbors (kNN) [1]: a non-parametric, instance-based learning algorithm, it makes predictions based on the similarity between the input instance and the instances in the training set.

All data were divided into training and test sets. The data were split into 70% for training, and 30% for testing. After that, they were submitted to the k-fold cross-validation method with 10 folds and 3 repetitions. Followed by parameter optimization with the GridSearchCV algorithm. Table 4 lists the models that contain hyperparameters.

Table 4. Enumerated values of hyperparameters of machine learning models.

Model	Hyperparameter and Values
RFC	n-estimators: [100, 300, 500, 800, 1200]
	max-depth: [5, 10, 15, 25, 30]
	min-samples-split: [2, 5, 10, 15, 100]
	min-samples-leaf: [1, 2, 5, 10]
SVM-R	C: np.arange(0.1, 1, 0.1)
	kernel: ['rbf']
SVM-L	C: np.arange(0.1, 1, 0.1)
	kernel: ['linear']
DTC	max-depth: range(1, 16, 1)
	min-samples-split: range(2, 10, 1)
	min-samples-leaf: range(1, 21, 1)
	criterion: ['gini', 'entropy', 'log-loss']
GNB	var-smoothing: np.logspace(0, -10, num=100)
kNN	leaf-size: np.arange(1, 50, 1)
	n-neighbors: np.arange(10, 30, 1), 'p: [1, 2]
AdaB	n-estimators: range(10, 200, 10)
GB	n-estimators: range(10, 200, 10)
	max-depth: range(1, 16, 1)
	min-samples-split: range(200, 1001, 200)
	min-samples-leaf: range(30, 71, 10)
	max-features: range(7, 20, 2)
	subsample: [0.6, 0.7, 0.75, 0.8, 0.85, 0.9, 1]
XGB	n-estimators: range(10, 200, 10)
	max-depth: range(1, 16, 1)
MLP	hidden-layer-sizes: [(sp-randint.rvs(100, 600, 1), sp-randint.rvs(100, 600, 1),), (sp-randint.rvs(100, 600, 1),)]
	activation: ['identity', 'logistic', 'tanh', 'relu']
	solver: ['sgd', 'adam', 'lbfgs']
	alpha: np.arange(0.0001, 0.01, 0.001)
	learning-rate: ['constant', 'adaptive', 'invscaling']
LogR	penalty: ['l1', 'l2', 'elasticnet', 'none']
	C: np.logspace(−4, 4, 20)
	solver: ['lbfgs', 'newton-cg', 'liblinear', 'sag', 'saga']
	max-iter: [100, 1000, 2500, 5000, 10000]
GPC	kernel: [1*RBF(), 1*DotProduct(), 1*Matern(), 1*RationalQuadratic(), 1*WhiteKernel()]
GPR	'kernel: [1*RBF(), 1*DotProduct(), 1*Matern(), 1*RationalQuadratic(), 1*WhiteKernel()]
	alpha: [1e-2, 1e-3]
QDA	reg-param: np.arange(1e-04, 1e-01, 1e-04)

(*continued*)

Table 4. (*continued*)

Model	Hyperparameter and Values
LSVC	penalty: ['l1', 'l2']
	C: np.logspace(1, 4, 20)
	tol: np.arange(1e-6, 1e-5, 1e-6)
	loss: ['hinge', 'squared-hinge']
	multi-class: ['ovr', 'crammer-singer']
	class-weight: ['None', 'balanced']
CBC	learning-rate: [0.001, 0.01, 0.1]
	eval-metric: ['RMSE', 'AUC', 'MultiClass']
	loss-function: ['Logloss', 'CrossEntropy']
	iterations: np.arange(100, 200, 10)
	depth: np.arange(4, 10, 1)
	l2-leaf-reg: np.arange(2, 10, 1)
	random-strength: np.arange(0, 10, 1)
BC	n-estimators: np.arange(500, 1000, 100)
	max-samples: np.arange(0.1, 1, 0.1)
	max-features: np.arange(1, 20, 1)
RC	alpha: np.logspace(0.1, 1, 20)

4 Results and Discussion

4.1 Pre-made Datasets

Table 5 represents the accuracy of the models for different datasets from Pre-made set. Observing the data behavior, it is noticeable that models have the highest accuracy for the Players Statistics (**PS**) dataset, whereas that accuracy drops slightly for the Picked Champions and Players Statistics (**PC+PS**) and Complete (**C**) datasets and then drastically for the Picked Champions (**PC**) and Banned Champions (**BC**) datasets. The difference that causes the model precision to drop between the Players Statistics and Picked Champions and Players Statistics is that the Picked Champions and Players Statistics dataset has an additional feature (Picked Champion ID), which is correlated with other features in the Players Statistics dataset so including it in Picked Champions and Players Statistics dataset increases the multicollinearity and decrease the model's performance. A similar thing is happening with the Complete dataset, which also has this extra feature (Picked Champion ID) as well as more additional features which are Banned Champions ID. Not only does the correlated feature lower the accuracy but so do the other features, because they are deemed not relevant to the task at hand. Information about the banned champions does not impact the outcome of the game. It leads us to the two last datasets, PC and BC. Models performance was the worst for those datasets simply because the data about champions included and excluded in the game does not provide enough information to greatly affect the outcome of the game. The best-performing models are Gradient Boosting, Linear Support Vector Classification, Random Forest Regressor, and Cat Boost Classifier. It's noticeable that some of the models struggle to perform as well as others on all of the available datasets. Them being: Support Vector Machines radial, k-Nearest Neighbors, Neural Network MLP, Recurrent Neural Networks, and Fully Connected Neural Networks. Support Vector Machines radial and Neural Network MLP are both complex models that have many parameters to adjust, which can make them more prone to overfitting if not properly regularized. They also require more data to train properly. k-Nearest Neighbors is a non-parametric method that is sensitive to the scale and distribution of the features. It is also sensitive to the presence of outliers. Recurrent Neural Networks and Fully Connected Neural Networks are both deep learning models that require large amounts of data to train and can be sensitive to the quality and quantity of data. They are also sensitive to the architecture of the network, which can be difficult to optimize.

Table 5. Experimental Results for Machine Learning Models on Pre-made Datasets: Model Name, Accuracy. The bold text in the table represents the best-scoring model for a specific dataset, while the grey background indicates the worst-scoring model overall.

Model	**PS**	**PC+PS**	**C**	**PC**	**BC**
Linear Regression	91.67%	86.97%	86.97%	53.05%	51.76%
Random Forest Regressor	**92.02%**	88.26%	87.,21%	55.75%	51.29%
Random Forest Classifier	87.44%	80.28%	81.10%	55.87%	53.76%
Support Vector Machines radial	78.99%	52.81%	52.93%	51.06%	51.53%
Support Vector Machines linear	91.31%	87.56%	87.44%	53.52%	53.87%
Decision Tree Classifier	83.92%	80.05%	78.52%	50.82%	53.40%
Gaussian Naive Bayes radial	88.50%	83.10%	82.75%	50.94%	50.59%
k-Nearest Neighbors	69.95%	52.46%	51.41%	51.41%	51.88%
AdaBoost	91.78%	86.74%	86.97%	56.81%	51.76%
Gradient Boosting	**92.37%**	86.62%	86.74%	55.99%	51.41%
Extreme Gradient Boosting	92.14%	88.26%	87.09%	55.40%	51.88%
Neural Network MLP	85.56%	53.87%	53.87%	53.87%	53.87%
Logistic Regression	90.85%	81.46%	81.92%	52.46%	51.99%
Recurrent Neural Network	91.31%	52.82%	52.58%	52.11%	52.58%
Fully Connected Neural Network	78.64%	46.13%	46.24%	46.13%	46.13%
Gaussian Process Classifier	90.96%	87.44%	87.32%	53.52%	53.52%
Gaussian Process Regressor	91.67%	86.97%	86.97%	52.93%	51.88%
Quadratic Discriminant Analysis	88.26%	82.04%	82.75%	51.06%	49.53%
Linear Support Vector Classification	**92.25%**	87.68%	87.56%	53.05%	51.76%
Cat Boost Classifier	**92.02%**	87.44%	87.91%	55.16%	51.53%
Bagging Classifier	89.79%	85.92%	84.51%	52.82%	51.88%
Ridge Classifier	91.78%	87.44%	87.21%	53.05%	51.76%

Table 6 shows the improved accuracy of each model when evaluated on the same datasets as in Table 5. These models were optimized using optimal parameters obtained from the model tuning process. However, some models did not improve at all, with accuracy scores remaining the same as in Table 5. These include Linear Regression, Random Forest Regressor, Support Vector Machines (linear), Recurrent Neural Network, Fully Connected Neural Network, Gaussian Process Regressor, and Linear Support Vector Classification. This lack of improvement suggests that either the models have reached their maximum potential or that the chosen parameters were not suitable for the task at hand. The greatest improvement in accuracy was seen in the Neural Network MLP model, with an increase of 30.75% for the "Picked Champions and Players Statistics" dataset, 24.06% for the "Complete" dataset, and 5.99% for the "Player Statistics" dataset. In the "Player Statistics" dataset, the models with the best improvement were k-Nearest Neighbors (9.51%), Neural Network MLP (5.99%), and Support Vector Machines radial (4.70%). Two models, Extreme Gradient

Table 6. Experimental Results for Hyperparameterized Machine Learning Models on Pre-made Datasets: Model Name, Accuracy. The bold text in the table represents the best-scoring model for a specific dataset, while the grey background indicates the worst-scoring model overall.

Model	PS	PC+PS	C	PC	BC
Linear Regression	91.67%	86.97%	86.97%	53.05%	51.76%
Random Forest Regressor	92.02%	88.26%	87.21%	55.75%	51.29%
Random Forest Classifier	92.14%	87.68%	87.44%	56.92%	52.00%
Support Vector Machines radial	79.23%	53.40%	54.11%	53.05%	53.87%
Support Vector Machines linear	91.31%	87.56%	87.44%	53.52%	53.87%
Decision Tree Classifier	85.56%	77.11%	80.40%	53.87%	52.82%
Gaussian Naive Bayes radial	88.50%	83.10%	82.75%	53.52%	52.82%
k-Nearest Neighbors	79.46%	56.10%	50.94%	54.11%	52.35%
AdaBoost	**92.49%**	87.79%	87.79%	54.58%	54.58%
Gradient Boosting	92.02%	89.08%	88.62%	53.76%	52.93%
Extreme Gradient Boosting	**92.61%**	88.73%	88.50%	54.23%	52.70%
Neural Network MLP	91.55%	84.62%	77.93%	54.34%	53.87%
Logistic Regression	92.14%	86.03%	86.85%	53.87%	52.93%
Recurrent Neural Network	91.31%	52.82%	52.58%	52.11%	52.58%
Fully Connected Neural Network	78.64%	46.13%	46.24%	46.13%	46.13%
Gaussian Process Classifier	92.14%	87.68%	87.32%	53.52%	52.35%
Gaussian Process Regressor	91.67%	86.97%	86.97%	52.93%	51.88%
Quadratic Discriminant Analysis	88.26%	82.86%	82.04%	51.17%	49.77%
Linear Support Vector Classification	92.25%	87.68%	87.56%	53.05%	51.76%
Cat Boost Classifier	91.55%	88.62%	87.68%	55.05%	50.70%
Bagging Classifier	91.90%	88.50%	88.38%	58.22%	54.93%
Ridge Classifier	91.78%	87.56%	87.44%	53.05%	51.76%

Boosting (92.61%) and AdaBoost (92.49%), even outperformed the highest-performing model from the previous experiment (92.37%).

4.2 Riot API Datasets

Table 7 presents the accuracy scores of various models evaluated on datasets sourced from the Riot API. These models were optimized using optimal parameters determined through the model-tuning process. It's apparent that the results for this data set, especially for "Players Statistics", "Picked Champions and Players Statistics", and "Complete" datasets, surpass those from the previous experiments. The reason behind that first is the accumulation of the data gathered in order to calculate the "Player Statistics" on related champions. There's no information provided on the pre-made set of data, about how many matches were used to determine the player statistics whereas, in this experiment, a number of 20 last played games was used. It's assumed that a pre-made set of data used a smaller volume of matches to compute the player statistics. The best-performing models are Random Forest Classifier, Bagging Classifier, and Gradient Boosting. Random Forest Classifier and Bagging Classifier are both Ensemble methods that build multiple decision trees and combine their predictions. These methods are robust to overfitting and outliers and handle well with high dimensionality and correlated features. Gradient Boosting is also an ensemble method that combines multiple decision trees. Gradient Boosting is a powerful technique that builds new trees which complement the already-built trees. It is also robust to overfitting and outliers and is able to capture complex interactions among the

Table 7. Experimental Results for Hyperparameterized Machine Learning Models on Riot API Datasets: Model Name, Accuracy, Execution Time for PS. The bold text in the table represents the best-scoring model for a specific dataset. The best models for the task at hand, in terms of accuracy and execution time, are indicated by an underline.

Model	PS	PC+PS	C	PC	BC	Ex Time [s] (for PS)
LR	98.68%	98.83%	98.83%	49.93%	48.17%	0.0180
RFR	97.95%	98.10%	97.80%	52.12%	49.19%	1.1571
RFC	**98.98%**	98.98%	98.83%	52.56%	51.83%	2.5695
SVM-R	89.46%	51.24%	53.00%	51.24%	49.05%	0.3335
SMV-L	98.54%	98.68%	98.24%	52.56%	49.19%	0.0922
DTC	95.90%	95.02%	94.58%	48.61%	51.98%	0.0336
GNB	97.51%	97.36%	97.51%	50.51%	49.78%	0.0200
kNN	89.31%	51.83%	53.59%	49.78%	52.12%	0.1225
AdaB	98.68%	98.68%	98.68%	49.93%	52.42%	1.1007
GB	**98.83%**	98.54%	98.54%	49.93%	48.02%	0.2319

(*continued*)

Table 7. (*continued*)

Model	PS	PC+PS	C	PC	BC	Ex Time [s] (for PS)
XGB	98.39%	98.68%	98.98%	51.39%	51.24%	0.3107
MLP	97.36%	95.75%	94.29%	51.39%	49.78%	0.1389
LogR	98.68%	98.54%	98.24%	47.88%	49.78%	0.1075
RNN	97.22%	96.34%	56.52%	48.17%	47.73%	16.5964
FCNN	87.99%	51.98%	52.12%	52.27%	49.78%	5.4875
GPC	98.68%	98.54%	97.95%	50.95%	53.73%	21.3553
GPR	98.68%	98.83%	98.83%	49.93%	48.17%	6.3128
QDA	97.07%	97.07%	97.07%	49.63%	48.76%	0.0580
LSVC	97.95%	88.29%	83.75%	52.27%	51.54%	0.0470
CBC	98.54%	98.39%	98.68%	50.80%	51.10%	3 .6650
BC	**98.96%**	98.83%	98.83%	49.49%	50.37%	5.0829
RC	98.68%	98.83%	98.68%	49.93%	48.17%	0.0520

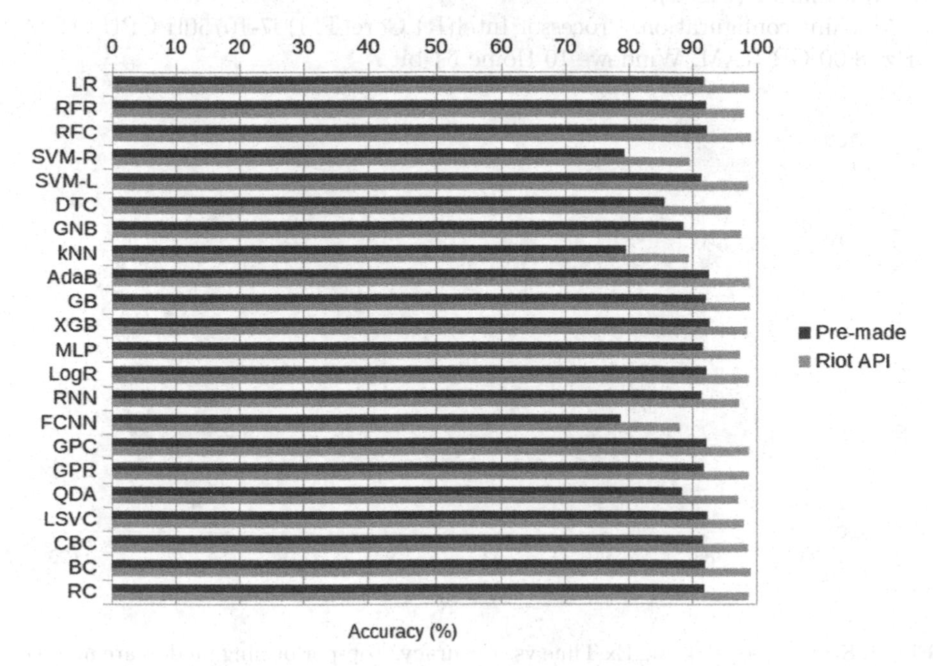

Fig. 1. Performance of tested models for PS and both datasets.

features. This dataset has the right characteristics to train these models effectively, they are well-suited to the task at hand.

The 98% accuracy achieved is slightly higher than the scores reported by Harikumar et al. [2] and Costa et al. [5].

Overall, all models achieved greater accuracy for the Riot API dataset. Figure 1 shows the differences in accuracy for individual models processing data from both datasets: Pre-made and Riot API.

4.3 Execution Time

Table 7 including the column "Ex Time" presents juxtaposition of models accuracy and their execution time for the Riot API – "Player Statistics" dataset. It is important to note that the execution time of a model is essential in a recommendation system because it is performing in real time. Therefore a rapid and accurate model is more valued than just an accurate but slow one. It is observable that between the three highest-scoring models and the three fastest models, there is no correlation. That is because more accurate models need more time to perform above others. In such a situation, it is better to forsake the best scoring models and focus on faster-executing ones. This being the case, the best models for the given task are Linear Regression, Ridge Classifier, and Support Vector Machines linear (Fig. 2).

Machine configuration: Processor Intel(R) Core(TM) i7-10750H CPU @ 2.60 GHz, 8.00 GB RAM, Windows 10 Home 64-bit.

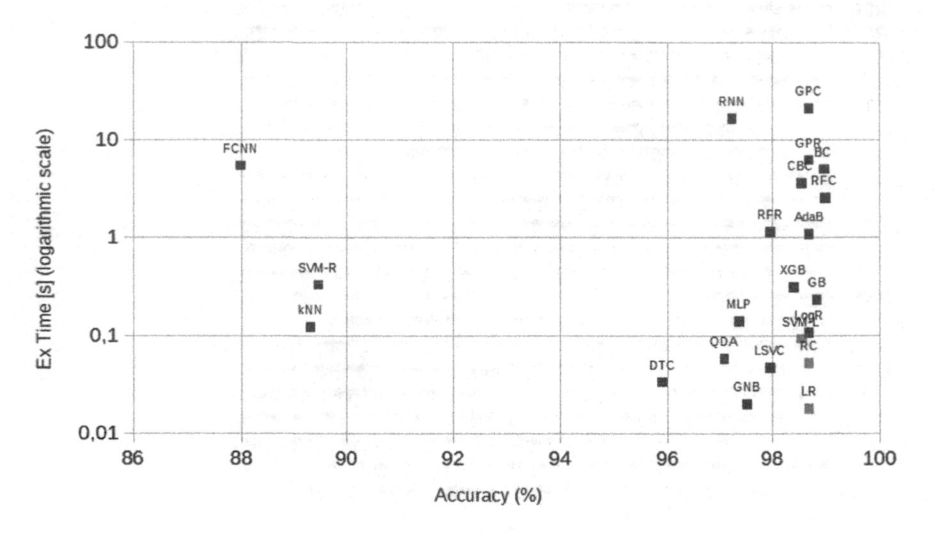

Fig. 2. Scatter plot showing Ex Time vs. Accuracy. Top-performing models are marked green. (Color figure online)

5 Conclusions

In conclusion, the research results show that the models have the highest accuracy for the "Players Statistics" dataset, but accuracy drops slightly for datasets with additional correlated features. The highest performing models are Extreme Gradient Boosting and Random Forest Classifier. The 98% accuracy achieved in this study represents a potentially improved outcome compared to previous findings in the literature. However, when considering accuracy as well as execution time, the optimal models for the task at hand are Linear Regression, Support Vector Machines linear, and Ridge Classifier. Some models, such as Support Vector Machines radial, k-Nearest Neighbors, Neural Network MLP, Recurrent Neural Networks, and Fully Connected Neural Networks, struggle to perform as well on all datasets due to factors such as overfitting, sensitivity to scale and distribution of features, and the need for large amounts of data. Additionally, the results from using datasets sourced from the Riot API show improved performance, likely due to the accumulation of data used to calculate player statistics. It is worth noting that some of the models may have longer execution time compared to others, it is important to consider the trade-off between accuracy and execution time when selecting a model for recommendation system.

References

1. Supervised learning. https://scikit-learn.org/stable/supervised_learning.html
2. Ani, R., Harikumar, V., Devan, A.K., Deepa, O.: Victory prediction in league of legends using feature selection and ensemble methods (2019). https://doi.org/10.1109/ICCS45141.2019.9065758
3. Chen, Z., et al.: The art of drafting: A team-oriented hero recommendation system for multiplayer online battle arena games (2018). https://doi.org/10.48550/arXiv.1806.10130
4. Conley, K., Perry, D.: How does he saw me ? a recommendation engine for picking heroes in dota 2 (2013)
5. Costa, L.M., Souza, A.C.C., Souza, F.C.M.: An approach for team composition in league of legends using genetic algorithm (2019). https://doi.org/10.1109/sbgames.2019.00018
6. Do, T.D., Yu, D.S., Anwer, S., Wang, S.I.: Using collaborative filtering to recommend champions in league of legends (2020). https://doi.org/10.1109/CoG47356.2020.9231735
7. Hanke, L., Chaimowicz, L.: A recommender system for hero line-ups in MOBA games. In: Proceedings of the AAAI Conference on Artificial Intelligence and Interactive Digital Entertainment, vol. 13, no. 1, pp. 43–49 (2021). https://doi.org/10.1609/aiide.v13i1.12938
8. Hitar-Garcia, J.A., Moran-Fernandez, L., Bolon-Canedo, V.: Machine learning methods for predicting league of legends game outcome. IEEE Trans. Games (2022). https://doi.org/10.1109/tg.2022.3153086
9. Hong, S.J., Lee, S.K., Yang, S.I.: Champion recommendation system of league of legends (2020). https://doi.org/10.1109/ICTC49870.2020.9289546
10. Kinkade, N.: Dota 2 win prediction (2015)

11. Porokhnenko, I., Polezhaev, P., Shukhman, A.: Machine learning approaches to choose heroes in dota 2 (2019). https://doi.org/10.23919/FRUCT.2019.8711985
12. Shen, Q.: A machine learning approach to predict the result of league of legends. In: 2022 International Conference on Machine Learning and Knowledge Engineering (MLKE). IEEE (2022). https://doi.org/10.1109/mlke55170.2022.00013

Classification Performance of Extreme Learning Machine Radial Basis Function with K-means, K-medoids and Mean Shift Clustering Algorithms

Aleksandra Konopka[1], Karol Struniawski[1(✉)], and Ryszard Kozera[1,2]

[1] Institute of Information Technology, Warsaw University of Life Sciences - SGGW, ul. Nowoursynowska 159, 02-776 Warsaw, Poland
{aleksandra_konopka,karol_struniawski,ryszard_kozera}@sggw.edu.pl

[2] School of Physics, Mathematics and Computing, The University of Western Australia, 35 Stirling Highway, Crawley, Perth, WA 6009, Australia
ryszard.kozera@uwa.edu.au

Abstract. Extreme Learning Machine (ELM) is a feed-forward neural network with one hidden layer. In its modification called ELM Radial Basis Function the input data is a priori clustered into a number of sets represented by their centroids. The matrix of distances between each sample and centroid is calculated and applied as input data to the neural network. This work conducts a comparison study of the ELM Radial Basis Function classification performance upon applying either k-means, k-medoids or mean shift clustering methods. Generated results are obtained from two datasets i.e. Wine Quality-White and Ionosphere. The computations are based on full datasets or on the same both sets reduced by a feature selection algorithm. The parameters of the classifiers such as number of neurons in hidden layer, the value of k in k-means and k-medoids, the value of radius in mean shift are optimized through an iterative procedure upon maximizing an accuracy or minimizing Mean Square Error and computation time. The different distance metrics for k-means and k-medoids, and mean shift with Gaussian or flat kernel function are also compared. The results obtained with Softplus and linear activation function (applied in most of the computations in this work) are juxtaposed with the results generated by other activation functions.

Keywords: Neural Networks · Machine Learning · Extreme Learning Machine · Radial Basis Function · Clustering Algorithms

1 Introduction

The Backpropagation Algorithm (BA) introduced in 1986 by Rumelhart et al. [21] represents an important component of machine learning. The main problem of BA stems from the fact that it usually yields local minima of associated network's residual error function. In addition, BA computational cost and training

J. Mikyška et al. (Eds.): ICCS 2023, LNCS 14076, pp. 171–186, 2023.
https://doi.org/10.1007/978-3-031-36027-5_13

time, especially for the large datasets, may preclude its practical application. The new concept of neural network called Extreme Learning Machine (ELM) was introduced by Huang et al. in 2004 [7]. ELM converges much faster than traditional learning schemes as relieved from the time-consuming iterations and is also more likely to reach a global optimum [10]. ELM is successfully adapted to various machine learning applications such as classification and regression in medicine, chemistry, transportation, economy, agriculture, robotics etc. It also outperforms other methods in training time and approximation ability [9,18]. ELM is characterized by yielding extremely fast training time in comparison to other machine learning methods like e.g. Multilayer Perceptron trained with BA [11]. Currently, ELM still evolves to further improve generalization capacity in case of special applications. One of its variants called Extreme Learning Machine Radial Basis Function (ELM-RBF), weaving core principles of ELM with feature space mapping using RBF kernels, yields comparable results to BA with considerably faster computation time [2]. In the field of ELM-RBF, the most commonly used clustering method is k-means, although application of k-medoids is also found in the literature.

In this paper, the performance of ELM-RBF combined with k-means, k-medoids or mean shift clustering methods is thoroughly investigated. The comparison involves manipulating with multiple variables, such as parameter values of clustering methods, number of neurons or different activation functions in ELM-RBF. In order to obtain the most significant results a comparison is conducted on two datasets - Wine Quality-White and Ionosphere [3]. The characteristics of the selected benchmark sets allow to obtain significant results and compare different algorithms, such as those used for solving k-medoids problem, as their application relies on the size of the input dataset.

2 Extreme Learning Machine

Extreme Learning Machine consists of input, hidden and output layer aimed to solve classification and regression tasks by supervised learning [7]. Assume input data is described as pairs of values $\eta = \{(x_i, t_i)\}_{i=1}^N$, where $x_i = \{x_{ij}\}_{j=1}^d$ forms matrix $X_{d \times N}$ of d features. Here t_i is recognized as affiliation to the given class establishing T. For the classification need $t_i \in [0, M] \subseteq \mathbb{N}$, whereas for regression the target value $t_i \in \mathbb{R}^M$. Here M represents either the number of classes or dimension of target values. The number of neurons in input, output and hidden layer is assumed here to be equal to d, M and L, respectively. Here L is given a priori as there is no universal method for L optimization. Neurons in ELM are McCulloch-Pitts neurons [14] with identity function on input and output layers and any activation function $f : \mathbb{R} \to \mathbb{R}$ on hidden layer units. Matrix of weights $W_{d \times L}$ with coefficients $w_{ij} \in (-1, 1)$ connecting input neurons with L hidden layer neurons and bias values b_i of $b = \{b_i\}_{i=1}^N$ are randomly selected with the aid of uniform distribution function, where $i = 1, \ldots, d$ and $j = 1, \ldots, L$. Thus, $H = f(X^T W + b)$ is obtained as an output of the hidden layer. Wages β between hidden and output layer are computed once using algebraic transformations. To calculate β the following matrix equation $H\beta = T$ should be solved. The matrix H is non-invertible and therefore solving $H\beta = T$ can be reformulated into the optimization task estimating $\hat{\beta}$ based on minimization of a Mean Square Error (MSE) between H and T [7]. The corresponding solution reads as $\hat{\beta} = H^{\dagger}T$, where $H^{\dagger}$ is the Moore-Penrose pseudo-inverse operation [20]. Various methods for $H^{\dagger}$ evaluation can be applied including e.g. Cholesky factorization of a singular matrix [12].

3 Extreme Learning Machine Radial Basis Function

Extreme Learning Machine Radial Basis Function is a method with training similar to its archetype which is based on random generation of W, b and calculation of β using generalized inverse of matrix H. The extra component involves input data transformation with the aid of Radial Basis Function [15]. Specifically, let $X_{d \times N}$ be a set of d features for N observations. First, a vector quantization technique is applied as a clustering algorithm. The aim of the latter is to partition N samples into a certain number (given a priori or designated automatically during algorithm's run) of k clusters using e.g. k-means clustering algorithm. Each sample x_i is assigned to exactly one cluster $\{c_j\}_{j=1}^{k}$, which in turn is recognized as the closest centroid to x_i in terms of a considered metric. Next, x_i is transformed to the new feature space based on a chosen kernel function. Note here that when $k > d$ then X is mapped to a higher dimension. The definition of the RBF kernel for ELM-RBF is given as $K(x_i, c_j) = exp\{\frac{-\|x_i - c_j\|^2}{2\sigma^2}\}$. The matrix $K_{N \times k}$ is computed as a measure of distance between each x_i and c_j. The σ_j value is determined as a $\sigma_j = \frac{max\{d_j\}}{\sqrt{2k}}$. Finally, K is treated as an input matrix to the typical ELM network.

4 Clustering Methods

Clustering methods divide samples into disjoint groups. In k-means, k-medoids and mean shift each cluster is represented by a centroid which is calculated through an iterative procedure.

4.1 Mean Shift

Mean shift is an unsupervised learning algorithm [5] commonly applied in clustering, tracking and smoothing. This algorithm locates maxima of a density function with the aid of an iterative procedure upon updating candidates for centroids. Mean shift requires to specify the bandwidth of a window, which is shifted until the algorithm converges. The number of clusters is a priori unknown and depends on the density of input data samples. The points are assigned to the corresponding local maxima computed by the algorithm.

The input data of the algorithm is a set of n data points $\mathcal{Q} = \{q_i\}_{i=1}^{n}$, where $q_i = (q_{i_1}, q_{i_2}, \ldots, q_{i_d}) \in \mathbb{R}^d$. The value of bandwidth b is specified arbitrarily. Let $C = \{\{c_{ij}\}_{i=1}^{n}\}_{j=1}^{s_i}$ be the set of all locations of the window shifted in the algorithm, where $c_{ij} = (c_{ij_1}, c_{ij_2}, \ldots, c_{ij_d})$, $i = 1, \ldots, n$ is a number of point that is currently considered and $j = 1, \ldots, s_i$ is a number of the shift for i'th point. Then first point $q_1 \in \mathcal{Q}$ is selected and it is set as the first location of window's center ($c_{11} = q_1$). Then Euclidean distances $d(c_{ij}, q_i) = \sqrt{(c_{ij_1} - q_{ij_1})^2 + \cdots + (c_{ij_d} - q_{ij_d})^2}$ between current c_{ij} and each q_i are computed. All q_i with $d(c_{ij}, q_i) \leq b$ (i.e. inside the window centered at c_{ij}) are selected to the set $\hat{\mathcal{Q}}_{ij} = \{\hat{q}_m\}_{m=1}^{w_{ij}}$. Subsequently, the new location of the window is calculated as a mean value of all $\hat{\mathcal{Q}}_{ij}$ for each of their d dimensions $c_{i(j+1)} = (\frac{1}{w_{ij}} \sum_{m=1}^{w_{ij}} \hat{q}_{m_1}, \ldots, \frac{1}{w_{ij}} \sum_{m=1}^{w_{ij}} \hat{q}_{m_d})$. Such procedure is repeated for all $\mathcal{Q}$ until they all converge to their corresponding local maxima. An output of this method is the set of points assigned to each of the generated disjoint clusters.

The above procedure iterates over each of the points from the dataset resulting in high computation time for large input data. Despite the fact that shifting a selected point does not influence other iterations and the process can be parallelized it is still very slow. The elapsing computation time is highly correlated with the size of input matrix of data making this approach inefficient. The mean shift algorithm can be optimized to render the results with less computation. The Bart Finkston's implementation (available on Mathworks [4]) presents an approach which highly reduces the computational complexity of the algorithm. The idea is to mark the points which were inside a shifting window (in any of s_i iterations) as visited to disregard them in upcoming iterations when new starting points c_{i1} are to be selected. We note how many times each of the points is positioned inside a bandwidth considering all s_i iterations for i'th starting point until q_i converges to c_{is_i}. The procedure is repeated selecting random points from $\mathcal{Q}$ not so-far visited. Subsequently, the matrix with votes for all generated centroids is used to attach $\mathcal{Q}$ to the appropriate clusters applying majority voting.

4.2 K-means

K-means is a common clustering method for which each of the samples is assigned to one of k disjoint clusters [19]. The number of k is selected arbitrarily. Each of the clusters is represented by a centroid which is equal to mean value of all observations within a group. The samples in a given iteration are assigned to the closest centroid in accordance with a distance metric i.e Euclidean distance.

Let a set of n data points $\mathcal{Q} = \{q_i\}_{i=1}^{n}$, where $q_i = (q_{i_1}, q_{i_2}, \ldots, q_{i_d}) \in \mathbb{R}^d$ be given. Assume $C = \{\{c_{lj}\}_{l=1}^{k}\}_{j=1}^{m}$ is the set of all locations of k centroids-to-be in the algorithm, where $c_{lj} = (c_{lj_1}, c_{lj_2}, \ldots, c_{lj_d})$, $l = 1, \ldots, k$ and $j = 1, \ldots, m$ is a number of the algorithm's iteration. Let $C_m = \{c_{lm}\}_{l=1}^{k}$ be the set of centroids yielded by the algorithm after reaching termination condition (after m iterations). The value of m is set a posteriori once the stopping conditions are met. We start by selecting starting locations of all the k centroids $C_1 = \{c_{l1}\}_{l=1}^{k}$, they can be set randomly or upon applying an optimization algorithm (e.g. k-means++ [1]). Then, the samples are assigned to the closest centroid by means of selected distance metric. In case of Euclidean distance ρ_E, the respective values are equal to $d_{i,l,j} = \rho_E(q_i, c_{lj})$. Subsequently, new locations of the centroids $C_{j+1} = \{c_{l(j+1)}\}_{l=1}^{k}$ are computed, where the new coordinates of each $c_{l(j+1)}$ are mean values of all the respective coordinates for data points q_i assigned to a given l'th centroid in j'th iteration. Here $c_{l(j+1)} = (\bar{c}_{lj_1}, \bar{c}_{lj_2}, \ldots, \bar{c}_{lj_d})$ with $\bar{c}_{lj_{dm}} = (1/a_{lj}) \sum_{i=1}^{a_{lj}} \hat{q}_{lj_i}$, where $dm = 1, \ldots, d$, a_{lj} is a number of points $\hat{q}_{lj_i}$ linked to l'th centroid in j'th iteration. The algorithm iterates until one of the stopping conditions is met i.e. either when computed centroids no longer switch their location or the distance between c_{lj} and $c_{l(j+1)}$ is smaller than prescribed ε or lastly, if the preselected maximal number of iterations is exceeded.

4.3 K-medoids

K-medoids is a method applied for clustering [8]. The data is assigned to one of k medoids, where each medoid is a specific point from the dataset and represents its cluster. The value of k and the dissimilarity measure are arbitrally selected. In the first step of the algorithm, k points are chosen as starting medoids. The latter is achieved either in a random manner or upon applying a specific method (e.g. k-means++ [1]) to solve an optimization problem leading to a reduction of computational time of the algorithm. The k-medoids problem can be solved with numerous algorithms such as: Partitioning Around Medoids (PAM), Voronoi-iteration k-medoids, Reynolds' improvements, FastPAM, FasterPAM algorithms, CLARA, CLARANS, FastCLARA and FastCLARANS, Lloyd's iterations [8,17].

In Partitioning Around Medoids algorithm, once a starting set of medoids is selected the values of a cost function are evaluated for all possible swaps of a medoid in a given cluster with another point belonging to the same cluster. When all the combinations are computed we finally apply only this one which has the minimal value of evaluated function. Subsequently, the affiliation of all points to points being set as current medoids is recalculated. This procedure is repeated for all the medoids as long as the value of the cost function decreases, otherwise the algorithm terminates. PAM algorithm has a high computation complexity as it calculates all possible swaps of all the medoids. In such setting, this algorithm is in practice predominantly applicable to a small amount of input data. For more complex computations one can apply variants of PAM (such as FastPAM or FasterPAM).

5 Experiments and Results

The computations were conducted on two sample datasets, namely Wine Quality-White and Ionosphere [3]. Wine Quality-White comprises of 4898 wine samples described by eleven features. Each of the wine samples was categorized by experts to the quality measure (ranging within $C = \{3, 4, \ldots, 9\}$). In order to perform computations on this set of features, a Mean Square Error is calculated to verify how a prediction deviates from an actual class as wine quality is represented by scale measure from C. We remark here that a misclassification of a wine from the class 3 to the predicted class 4 is a minor concern as opposite to assigning it to class 9. The Ionosphere is a dataset describing signals that pass through the ionosphere. These signals are classified into two disjoint groups: "good" signals having evidence of some type of structure and "bad" signals deprived of such a feature. This dataset consists of 351 samples described by 34 features. In this binary classification task, the rendered results are compared calculating accuracy (ACC) representing the percentage of correctly classified samples in the whole classification process. These datasets are used here for classification with the aid of ELM-RBF applying a selected clustering method: k-means, k-medoids or mean shift. All computations described in this work are performed in Matlab.

In preliminary computations, which were carried on Wine Quality-White dataset, some of the parameters were selected and they are fixed in this work. The estimation of generalized inverse for ELM-RBF is based on Cholesky factorization of a singular matrix [12] yielding fast computation time and low MSE. To compare k-means, k-medoids and mean shift, different activation functions are chosen to obtain the best clustering results. In doing so, a linear activation function $f_L(x) = ax$ is used as it yields prominent classification results while applying k-means and k-medoids. On the other hand, Softplus activation function $f_{SP}(x) = log(1+\exp(x))$ is also applied as it renders the best categorization results for mean shift.

Our computations exploit Matlab implementations of k-means, k-medoids and mean shift. Most of the conducted experiments admit in these methods default parameters unless specified otherwise. More specifically, the k-means and k-medoids clustering algorithms in Matlab have default distance metric set to Squared Euclidean distance. In addition, a default method for choosing initial cluster centroid positions is k-means++ algorithm [1]. In case of k-medoids, an algorithm to find medoids is available in three variants which are applied by default depending on number of rows (samples) in the input data. More specifically, for the number of rows less than 3000 PAM algorithm is applied. For the number of rows between 3000 and 10000 a variant of the Lloyd's iterations is selected. In case of larger datasets, a default algorithm is CLARA [8]. Thus, the algorithm applied for Wine Quality-White and Ionosphere datasets are Lloyd's iterations and PAM, respectively. In this work, the applied mean shift algorithm [6] allows the implementation of Gaussian or flat kernel for distance calculations. The classification results were generated for both datasets applying 10 times a 20% cross-validation. Note that the computation results obtained for same values of parameters can still vary as all three clustering methods rely on randomness. Such difference is still noticeable even upon applying multiple cross-validations. The computations are conducted on three computers: K1 - Ryzen 5600G CPU, 16 GB DDR4 3600 MHz RAM, K2 - Ryzen 3900X CPU, 64 GB DDR4 3600 MHz and K3 - Dell 7750 Xeon W10885M CPU, 128 GB DDR4 2933 MHz MHz.

A classification task is performed on Wine Quality-White dataset for ELM-RBF with k-means and k-medoids for k ranging from 10 to 100 (with step-size 10) (see Fig. 1). These computations are applied on computer K1. The number of hidden-layer neurons n varies from 100 to 1000 (with step-size 100). Lastly, a linear or Softplus activation function is applied. For both k-means and k-medoids MSE is the highest for $k \in \{10, 20\}$. The lower value of k gets, the worse classification result is rendered. Indeed, in the extreme case MSE attains the value 4.780 for k-medoids with $k = 10$ and $n = 500$ combined with Softplus activation function. For $k > 20$ the value of MSE for k-means and k-medoids stabilizes within the interval [0.619,0.793]. The best MSE result equal to 0.619 is achieved by k-medoids with linear activation function, $k = 100$ and $n = 700$ rendering the computation time equal to 409 s. Slightly worse result MSE = 0.628 is achieved for k-medoids with $k = 40$, $n = 100$ and linear activation function. Nevertheless, selecting the last set of parameters reduces computation time to 78 s. The best result obtained applying k-means yields MSE = 0.620 for $k = 100$, $n = 700$ and linear activation function, for which the computation time is equal to 402 s.

The mean shift clustering method is combined with ELM-RBF and the results obtained for Wine Quality-White dataset are based on the following choice of parameters: the values of the boundary width (radius) r are attuned from 0.3 to 1.3 (with step-size 0.1 - as it generated amounts of clusters from around 2 to 614) with the number of neurons n varying from 100 to 1000 (with step-size 100) combined with a linear or Softplus activation functions and Gaussian or flat kernel (see Fig. 2). The best result achieved for ELM-RBF with mean shift for the tested parameters is MSE = 0.686 for Softplus activation function, $n = 700$, $r = 0.6$ (which renders an average of 22 clusters in 10 cross-validations) with flat kernel function applied. This result is worse then the best result for k-medoids (and k-means) for this dataset for same tested number of neurons. The computation time is equal to 323 s. The best MSE (the lowest) are achieved for $r \in \{0.6, 0.7\}$ rendering between 10 and 22 clusters. ELM-RBF with mean shift obtains the worst MSE results for $r \leq 0.5$ (when $r = 0.5$ around 52 clusteres are rendered) applying linear kernel function and reaches even 2.02.

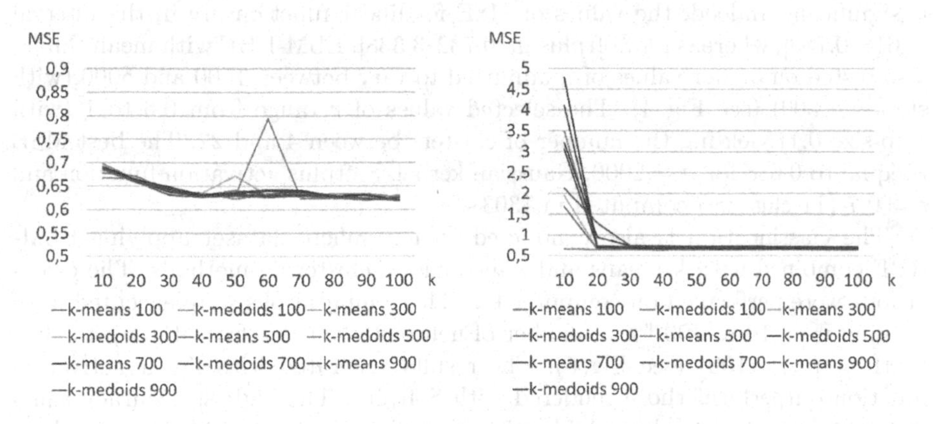

Fig. 1. MSE calculated on classification result for Wine Quality-White dataset for ELM-RBF with k-means or k-medoids , k varying between 10 and 100, 100–900 neurons applying linear (left) or Softplus activation function (right).

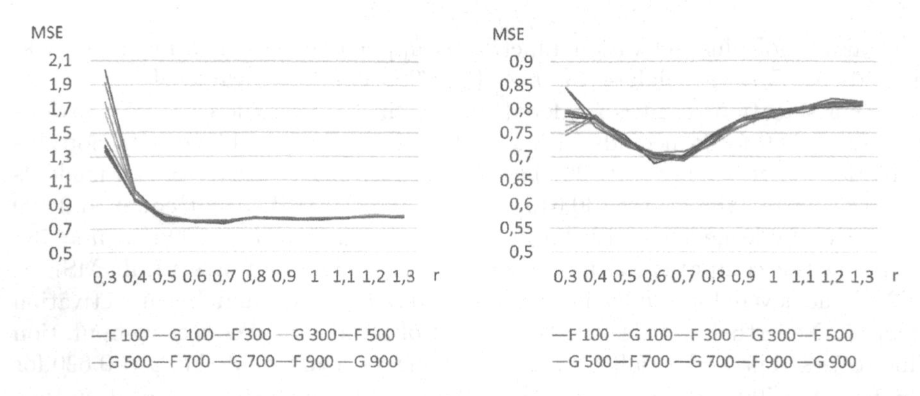

Fig. 2. MSE calculated on classification result for Wine Quality-White dataset for ELM-RBF with mean shift for r varying between 0.3 and 1.3, 100–900 neurons applying linear (left) or Softplus activation function (right), and Gaussian (G) or flat kernel (F).

The results generated by ELM-RBF on Wine Quality-White dataset are also analyzed on higher number of neurons n ranging from 1500 to 5000 (with step-size 500), for k varying from 10 to 100 (with step-size 10) in k-means and k-medoids applying linear or Softplus activation function (see Fig. 3). These computations are performed on the computer K2. Top three results are attained for k-medoids with linear activation function for $n = 2000$, $k = 100$, MSE $= 0.618$, for $n = 2000$, $k = 70$, MSE $= 0.618$ and for $n = 2500$, $k = 70$, MSE $= 0.619$. These results are equal to the best result for n ranging between 100 and 1000, but their computation time is much longer amounting to 1241, 1239 and 1676 s, respectively. Even if these computations were performed on faster computer, a huge increase in time is noticeable due to the enlarged number of neurons. The difference in results generated by Softplus and linear activation function is significant. Indeed, the values of MSE for linear function are in the interval $[0.618, 0.788]$, whereas for Softplus in $[0.732, 3.338]$. ELM-RBF with mean shift is also tested on higher values of n admitted to vary between 1500 and 5000 (with step-size 500) (see Fig. 4). The selected values of r range from 0.6 to 1 (with step-size 0.1) yielding the number of clusters between 4 and 22. The best MSE is equal to 0.688 for $n = 2000$, Gaussian kernel, Softplus activation function and $r = 0.7$ (11 clusters) computed in 1303 s.

The classification is also conducted on Ionosphere dataset applying ELM-RBF combined with k-means and k-medoids as clustering methods. The calculations were performed on computer K1. The computations were executed for k ranging from 10 to 100 for a number of neurons n varying from 100 up to 1000 (with step-size 100) (see Fig. 5). The results generated with linear activation function outperform those rendered with Softplus. The highest accuracy equal to 0.946 is reached for k-medoids with linear function, $n = 100$, $k = 96$ taking 13 s of execution time. The best result for k-means ACC $= 0.941$ is obtained for linear function, $n = 200$, $k = 80$ in 9 s. The accuracy for linear activation function rapidly increases with k running over $k = \{1, 2, \ldots, 10\}$. Once $k > 10$

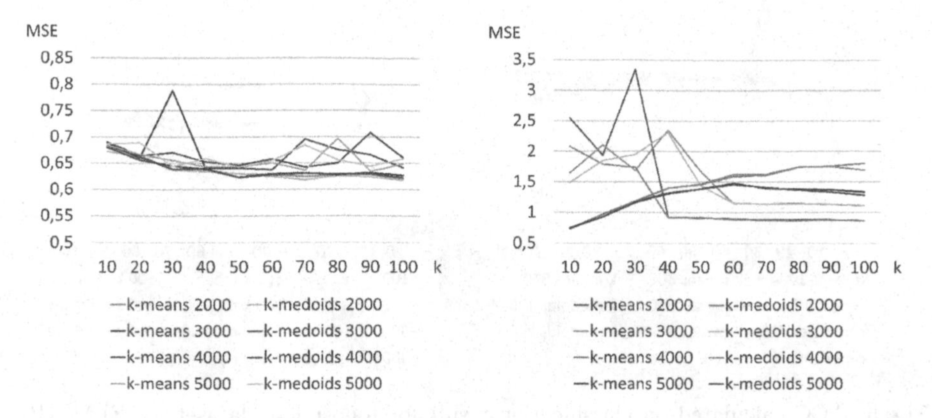

Fig. 3. MSE calculated on classification result for Wine Quality-White dataset for ELM-RBF with k-means or k-medoids for k varying between 10 and 100, 2000–5000 neurons applying linear (left) or Softplus activation function (right).

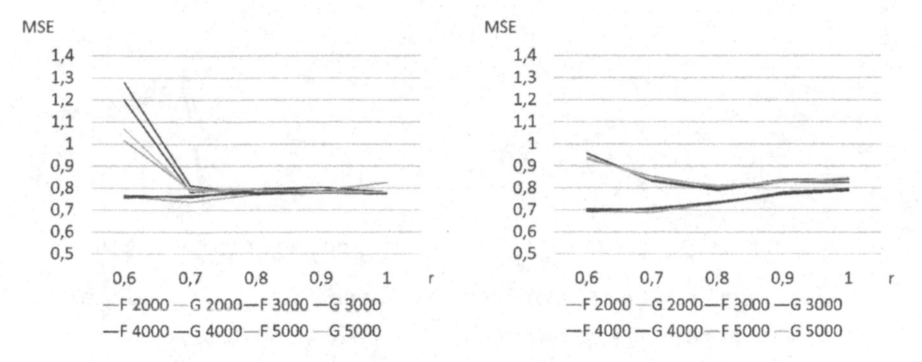

Fig. 4. MSE calculated on classification result for Wine Quality-White dataset for ELM-RBF with mean shift for r varying between 0.6 and 1, 2000–5000 neurons applying linear (left) or Softplus activation function (right), and Gaussian (G) or flat kernel (F).

the rate of improvement in classification results decelerates. The best 39 results are generated for $n \leq 200$. The lowest ACC for linear function equal to 0.599 is attained for k-medoids $k = 3$, $n = 900$ and the worst overall result for the considered methods and parameters reads as ACC $= 0.587$, and is achieved for k-medoids with Softplus activation function combined with $n = 200$ and $k = 11$. ELM-RBF with mean shift is also used as a classification method on Ionosphere dataset. The considered parameters' values are: n ranging from 100 to 1000 (with step-size 100), r from 0.5 to 7 (step-size 0.5) rendering from 1 to 216 clusters (see Fig. 6). The highest ACC $= 0.808$ is registered for Softplus activation function, $n = 100$, $r = 4.5$ rendering around 14 clusters. The 40 best results calculated with Softplus function are generated for $r = 4.5$ or $r = 5$ yielding around 14 and 4 clusters, respectively.

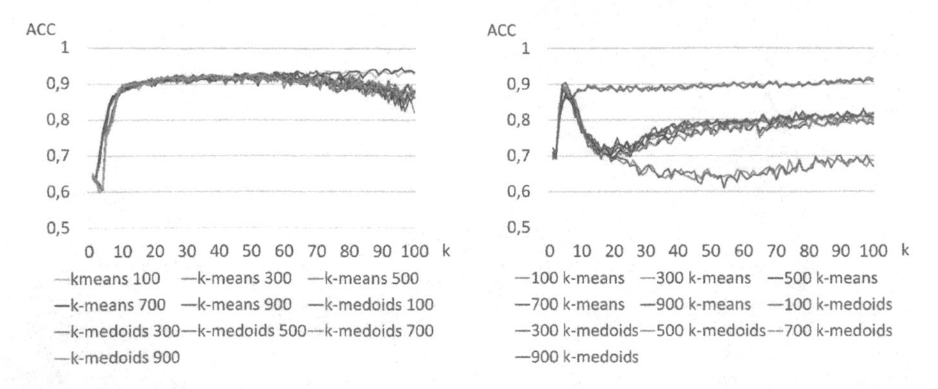

Fig. 5. ACC calculated on classification result for Ionosphere dataset for ELM-RBF with k-means or k-medoids for k varying between 1 and 100, 100–900 neurons applying linear (left) or Softplus activation function (right).

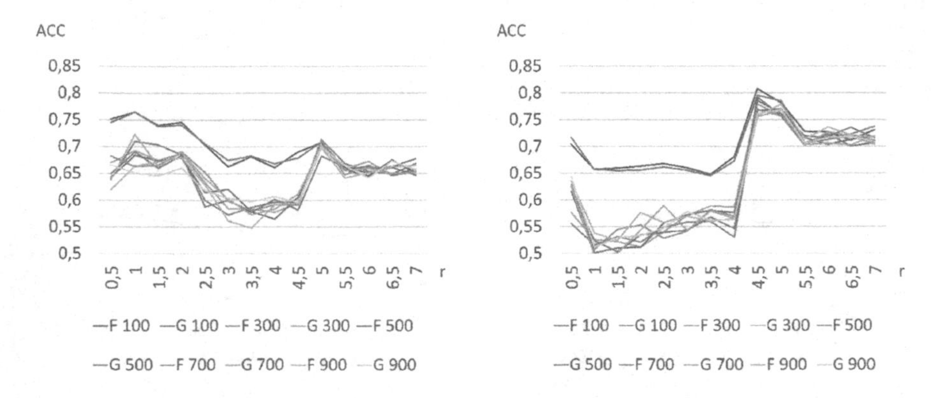

Fig. 6. ACC calculated on classification result for Ionosphere dataset for ELM-RBF with mean shift for r varying between 0.5 and 7, 100–900 neurons applying linear (left) or Softplus activation function (right), and Gaussian (G) or flat kernel (F).

K-means and k-medoids applied to ELM-RBF for clustering are tested on Ionosphere for larger number of neurons in the hidden layer. Tested parameters are n ranging from 1500 to 5000 (with step-size 500) and k from 1 to 100 (see Fig. 7). These computations are performed on the computer K3. The best ACC $= 0.937$ is attained for k-means with linear function, $n = 1500$ and $k = 49$. In contrast, the worst ACC equal to 0.602 is obtained for k-medoids with linear function, $n = 3000$ and $k = 4$. The best result for $1500 \leq n \leq 5000$ equal to 0.937 is lower then the best result for $100 \leq n \leq 1000$ reading as ACC $= 0.946$.

The classification process is performed on Ionosphere with ELM-RBF combined with mean shift testing the values of n ranging from 1500 to 5000 (with step-size 500) (see Fig. 8). The admitted values of r are taken from 0.5 to 7 (with step-size 0.5). The top 32 ACC, which are higher than 0.74, are obtained for $r = 4.5$ or $r = 5$ (around 13 or 4 centroids, respectively). The best ACC $= 0.792$ is achieved with Softplus function, $n = 3500$, $r = 4.5$ and Gaussian kernel. This result is worse than the best one calculated for $100 \leq n \leq 1000$ - ACC $= 0.808$.

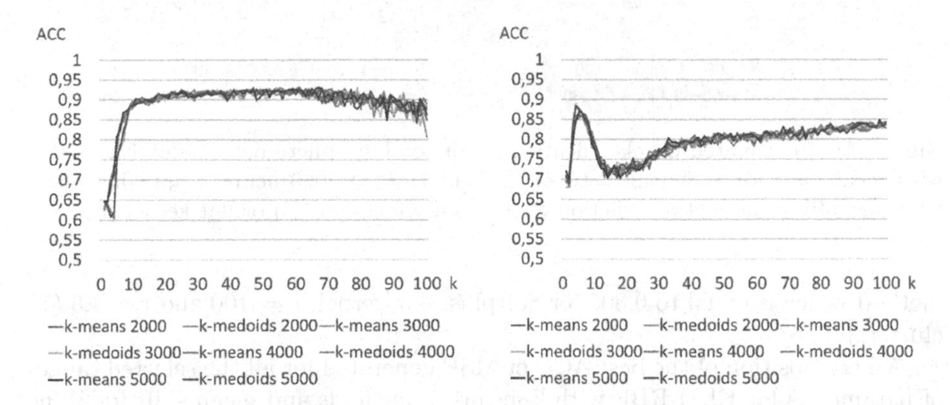

Fig. 7. ACC calculated on classification result for Ionosphere dataset for ELM-RBF with k-means or k-medoids for k varying between 1 and 100, 2000–5000 neurons applying linear (left) or Softplus activation function (right).

Feature selection method - Fast Correlation Based Filter (FCBF) [22] - is applied on Ionosphere to reduce the set of features leaving those that are highly correlated with affiliation to the class and their correlation between other features is low. The initial set of features is reduced from 34 to 4 represented by numbers: 5, 6, 28 and 33. The real aim of applying feature selection filtering is the hope to improve classification result and to reduce computation time. In the next step, ELM-RBF combined with k-means and k-medoids is tested on the set of features selected from Ionosphere. These computations are performed on the computer K2. The parameters involved are n varying between 100 and 1000 (with step-size 100) and k ranging from 1 to 100 (with step-size 1). The best ACC equal to 0.910 is obtained for k-means, linear activation function, for $n = 200$ and $k = 34$ and is computed in 11 s. This result is worse then the best result for the whole set of features which is equal to 0.946. The classification is also conducted on the selected Ionosphere features applying ELM-RBF with mean shift clustering method for n running from 100 to 1000 (with step-size 100), $r \in [0.01, 1.5]$ (with step-size 0.01) rendering from 1 up to 255 clusters. The best achieved result ACC $= 0.892$ is obtained for linear activation function (on the whole set of features Softplus activation function gave the best results for mean shift), flat kernel, $n = 100$, $r = 0.51$ rendering 40 clusters and is calculated in 6 s outperforming the best result obtained for the whole set of features applying mean shift clustering

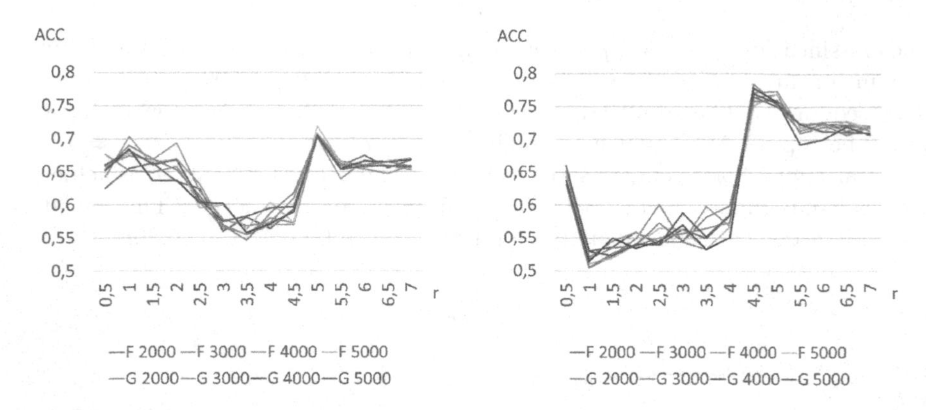

Fig. 8. ACC calculated on classification result for Ionosphere dataset for ELM-RBF with mean shift for r varying between 0.5 and 7, 2000–5000 neurons applying linear (left) or Softplus activation function (right), and Gaussian (G) or flat kernel (F).

method which is equal to 0.808 for Softplus, flat kernel, $n = 100$ and $r = 4.5$ (14 clusters).

A juxtaposition of the best ACC or MSE generated for all the selected ranges of parameters for ELM-RBF with k-means, k-medoids and mean shift for Wine Quality-White and Ionosphere is presented in Table 1 and Table 2.

In previous computations, the default distances for k-means and k-medoids were applied and the activation functions (Softplus and linear) were selected a priori as they gave the best classification results for Wine Quality-White dataset. Subsequently, the parameters that yielded the best ACC for the whole set of features in Ionosphere for the three considered clustering methods were selected and other distance functions for k-means and k-medoids were considered and combined with various activation functions tested for all three clustering methods. In case of k-means, the tested distances are: *cityblock*, *correlation*, *cosine* and *sqeuclidean*. In case of k-medoids: *chebychev*, *cityblock*, *correlation*, *cosine*, *euclidean*,

Table 1. The best classification results (measured with MSE) on Wine Quality-White dataset applying ELM-RBF for each of the considered clusterization methods (k-means, k-medoids and mean shift) on analyzed ranges of parameters. *Act fun* column stands here for the activation function, t for computation time in seconds and *rng* for range.

n rng	k rng	r rng	Method	act fun	n	r	k	Kernel	MSE	t
100–1000	10–100	-	k-medoids	linear	700	-	100.0	-	0.6188	409
100–1000	10–100	-	k-means	linear	700	-	100.0	-	0.6202	402
100–1000	-	0.3–1.3	mean shift	Softplus	700	0.6	21.5	flat	0.6856	323
1500–5000	10–100	-	k-means	Softplus	4000	-	10.0	-	0.7315	2357
1500–5000	10–100	-	k-medoids	Softplus	4500	-	10.0	-	0.7368	2566
1500–5000	-	0.6–1	mean shift	Softplus	2000	0.7	10.8	Gaussian	0.6882	1303

Table 2. The best classification results (measured with ACC) on Ionosphere dataset applying ELM-RBF for each of the considered clustering methods (k-means, k-medoids and mean shift) on analyzed ranges of parameters. *Act fun* stands here for the activation function, *rng* for range, *krn* for kernel (f - flat or g - Gaussian) and *t* for computation time in seconds. In features column there is an information about whether the computations were performed on the whole set of features or only on ones selected by FCBF.

n rng	k rng	r rng	features	Method	act fun	n	r	k	krn	ACC	t
100–1000	1–100	-	all	k-medoids	linear	100	-	96	-	0.946	13
100–1000	1–100	-	all	k-means	linear	200	-	80	-	0.941	9
100–1000	-	0.5–7	all	mean shift	Softplus	100	4.50	14	f	0.808	5
1500–5000	1–100	-	all	k-means	linear	1500	-	49	-	0.938	60
1500–5000	1–100	-	all	k-medoids	linear	4500	-	48	-	0.933	169
1500–5000	-	0.5–7	all	mean shift	Softplus	3500	4.50	13	g	0.792	155
100–1000	1–100	-	selected	k-means	linear	200	-	34	-	0.910	11
100–1000	1–100	-	selected	k-medoids	linear	200	-	41	-	0.907	14
100–1000	-	0.01–1.5	selected	mean shift	linear	100	0.51	40	f	0.892	6

hamming, *jaccard*, *minkowski*, *spearman* and *sqeuclidean* were analyzed [13]. For k-means, k-medoids and mean shift the tested activation functions for ELM-RBF are: sigmoid, tanh, relu, rbf, linear, swish, ELiSH, HardTanH, TanhRe, ELUs, Softplus, LReLU and BinaryStep [16]. The best five results obtained for k-means and k-medoids are presented in Table 3. For both k-means and k-medoids the best classification results are rendered for linear activation function. The highest ACC is obtained for *cityblock* (ACC = 0.94) and *sqeuclidean* distance (ACC = 0.93). The best five results for mean shift with $n = 100$, $r = 4.5$ (rendering 14

Table 3. The top 5 ACC on the Ionosphere dataset for k-means, $n = 100$, $k = 96$ and k-medoids, $n = 200$, $k = 80$ analyzed for different distances and activation functions.

Distance	Method	activation function	n	k	ACC	time
cityblock	k-medoids	linear	100	96	0.940	11
sqeuclidean	k-medoids	linear	100	96	0.933	11
cityblock	k-medoids	TanhRe	100	96	0.929	13
cityblock	k-medoids	ELUs	100	96	0.927	13
cityblock	k-medoids	Softplus	100	96	0.927	12
cityblock	k-means	linear	200	80	0.942	9
sqeuclidean	k-means	linear	200	80	0.927	9
sqeuclidean	k-means	relu	200	80	0.850	10
sqeuclidean	k-means	HardTanH	200	80	0.833	11
cityblock	k-means	TanhRe	200	80	0.828	11

or 13 clusters) and flat kernel, which are computed in $t \in [4.1, 5.1]$ seconds are rendered for: Softplus, tanh, swish, TanhRe and sigmoid activation functions, and their ACC are equal to: 0.794, 0.783, 0.780, 0.769 and 0.763, respectively.

6 Conclusions

The best classification results for Wine Quality-White with ELM-RBF are obtained for k-medoids and k-means. The latter is attained for $n \in N_{1000} = \{100, 200, \ldots, 1000\}$ with MSE $= 0.62$ which is 0.07 better than the best result applying mean shift. As it turns out, further enlargement of n (from 1500 to 5000) does not improve the MSE results. The best classification result is achieved for $n = 700$. Similarly, the best classification results for Ionosphere dataset are derived for k-means and k-medoids with $n \in N_{1000}$ for which ACC $\in [0.94, 0.95]$. Again, admitting a higher number of neurons on this dataset does not ameliorate the classification results - a fenomenon also manifested on Wine Quality-White data. The best classification result is achieved for $n = 100$, which is the lower bound of the analyzed numbers of neurons. For both datasets and lower values of n the best results for k-means and k-medoids are attained for high values of $k \in \{80, 81, \ldots, 100\}$. Again, admitting a higher number of neurons n the best results are obtained for lower values of k, i.e. $k = 10$ for Wine Quality-White and $k \in \{48, 49\}$ for Ionosphere. Furthermore, classification conducted on features selected by FCBF from Ionosphere dataset does not improve the overall best classification result. Nevertheless, the ACC obtained by ELM-RBF with mean shift on reduced set of data increases accuracy rate from 0.81 to 0.89 ACC. For k-means and k-medoids the best results are obtained with the aid of linear activation function (and *cityblock* or *sqeuclidean* distance metrics). Mean shift rendered in most computations the best results on Softplus activation function; however, the best outcome achieved with this clustering method on reduced set of Ionosphere features is attained with linear activation function. In further research, one should verify results rendered applying other parameters especially when the best classification in this work is observed on their boundary values as it is expected to obtain in those cases better results. The *cityblock* metric should be further analyzed for k-means and k-medoids in ELM-RBF.

References

1. Arthur, D., Vassilvitskii, S.: K-means++: the advantages of careful seeding. In: Proceedings of the Annual ACM-SIAM Symposium on Discrete Algorithms, pp. 1027–1035 (2007)
2. Dhini, A., Surjandari, I., Kusumoputro, B., Kusiak, A.: Extreme learning machine - radial basis function (ELM-RBF) networks for diagnosing faults in a steam turbine. J. Ind. Prod. Eng. **39**(7), 572–580 (2022). https://doi.org/10.1080/21681015.2021.1887948
3. Dua, D., Graff, C.: UCI machine learning repository (2017). http://archive.ics.uci.edu/ml

4. Finkston, B.: Mean shift clustering (2023). http://bit.ly/3wVVngu
5. Fukunaga, K., Hostetler, L.: The estimation of the gradient of a density function, with applications in pattern recognition. IEEE Trans. Inf. Theory **21**(1), 32–40 (1975). https://doi.org/10.1109/TIT.1975.1055330
6. Gong, H.: An open-source implementation of meanshift clustering for matlab/octave (2015). https://github.com/hangong/meanshift_matlab
7. Huang, G.B., Zhu, Q.Y., Siew, C.K.: Extreme learning machine: a new learning scheme of feedforward neural networks. In: IEEE Proceedings of International Joint Conference on Neural Networks, vol. 2, pp. 985–990 (2004). https://doi.org/10.1109/IJCNN.2004.1380068
8. Kaufman, L., Rousseeuw, P.: Finding Groups in Data: An Introduction to Cluster Analysis. Wiley, Hoboken (1990). https://doi.org/10.2307/2532178
9. Konopka, A., et al.: Classification of soil bacteria based on machine learning and image processing. In: Groen, D., de Mulatier, C., Paszynski, M., Krzhizhanovskaya, V.V., Dongarra, J.J., Sloot, P.M.A. (eds.) ICCS 2022. LNCS, vol. 13352, pp. 263–277. Springer, Cham (2022). https://doi.org/10.1007/978-3-031-08757-8_23
10. Leung, H.C., Leung, C.S., Wong, E.W.M.: Fault and noise tolerance in the incremental extreme learning machine. IEEE Access **7**, 155171–155183 (2019). https://doi.org/10.1109/ACCESS.2019.2948059
11. Li, H.T., Chou, C.Y., Chen, Y.T., Wang, S.H., Wu, A.Y.: Robust and lightweight ensemble extreme learning machine engine based on eigenspace domain for compressed learning. IEEE TCAS-I **66**(12), 4699–4712 (2019). https://doi.org/10.1109/TCSI.2019.2940642
12. Lu, S., Wang, X., Zhang, G., Zhou, X.: Effective algorithms of the Moore-Penrose inverse matrices for extreme learning machine. Intell. Data Anal. **19**(4), 743–760 (2015). https://doi.org/10.3233/IDA-150743
13. MathWorks: k-medoids clustering - Matlab k-medoids (2023). http://bit.ly/3RwXlNR
14. McCulloch, W.S., Pitts, W.: A logical calculus of the ideas immanent in nervous activity. Bull. Math. Biophys. **5**(4), 115–133 (1943). https://doi.org/10.1007/BF02478259
15. Mojrian, S., et al.: Hybrid machine learning model of extreme learning machine radial basis function for breast cancer detection and diagnosis; a multilayer fuzzy expert system. In: RIVF, pp. 1–7 (2020). https://doi.org/10.1109/RIVF48685.2020.9140744
16. Nader, A., Azar, D.: Evolution of activation functions: an empirical investigation. ACM TELO **1**(2), 1–36 (2021). https://doi.org/10.1145/3464384
17. Park, H.S., Jun, C.H.: A simple and fast algorithm for k-medoids clustering. Expert Syst. Appl. **36**(2), 3336–3341 (2009). https://doi.org/10.1016/j.eswa.2008.01.039
18. Peng, X., Lin, P., Zhang, T., Wang, J.: Extreme learning machine-based classification of ADHD using brain structural MRI data. PLoS ONE **8**(11), 1–12 (2013). https://doi.org/10.1371/journal.pone.0079476
19. Pérez-Ortega, J., Almanza-Ortega, N.N., Vega-Villalobos, A., Pazos-Rangel, R., Zavala-Díaz, C., Martínez-Rebollar, A.: The k-means algorithm evolution. In: Introduction to Data Science and Machine Learning, chap. 5. IntechOpen, Rijeka (2019). https://doi.org/10.5772/intechopen.85447
20. Rao, C.R., Mitra, S.K.: Generalized Inverse of Matrices and its Applications. Wiley, Hoboken (1971)

21. Rumelhart, D.E., Hinton, G.E., Williams, R.J.: Learning representations by back-propagating errors. Nature **323**(6088), 533–536 (1986). https://doi.org/10.1038/323533a0
22. Yu, L., Liu, H.: Feature selection for high-dimensional data: a fast correlation-based filter solution. In: ICML, pp. 856–863 (2003)

Impact of Text Pre-processing on Classification Accuracy in Polish

Urszula Gumińska, Aneta Poniszewska-Marańda(✉), and Joanna Ochelska-Mierzejewska

Institute of Information Technology, Lodz University of Technology, Lodz, Poland
urszula.krzeszewska@dokt.p.lodz.pl,
{aneta.poniszewska-maranda,joanna.ochelska-mierzejewska}@p.lodz.pl

Abstract. Natural language processing (NLP) like other Machine Learning (ML), Deep Learning (DL) and data processing tasks, requires a large amount of data to be effective. Thus, one of the most significant challenges confronting ML/DL tasks, including NLP, is a lack of data. This is especially noticeable in the case of text data for niche languages like Polish. Manual collection and labelling of text data is the primary method for obtaining language-specific data. However, this is a lengthy and labour-intensive process. As a result, researchers use a variety of other solutions, such as machine translation from another, more developed language in the field, to obtain data more quickly and affordably. For these reasons the suitable experiments, described in the paper, were carried out, resulting in a machine translation model that translates texts from English into Polish. Its results were compared with those of a pre-trained model and translations were subjected to human testing. Moreover, the paper presents the influence of different pre-processing stages on the final result of text classification in Polish in terms of one of six emotions: anger, fear, joy, love, sadness, surprise.

Keywords: Natural language processing · Machine translation task · Polish emotion classification · Text pre-processing

1 Introduction

Natural Language Processing (NLP) as an important sub-field of Artificial Intelligence (AI) is used to analyse, understand and generate language, it studies the interaction between human and computer through natural language – used by humans in everyday communication [1]. Natural language processing, understood as a field of computer science and linguistics, can encompass areas outside of both machine learning and deep learning, as well as take advantage of the strengths of both.

The concept of noise in Natural Language Processing is remain unclear and frequently described as difficult to define. Text pre-processing is regarded as a standard procedure in automatic text analysis [2,3]. It is also the first step conducted during most of the language specific tasks. However, during this process

J. Mikyška et al. (Eds.): ICCS 2023, LNCS 14076, pp. 187–201, 2023.
https://doi.org/10.1007/978-3-031-36027-5_14

both redundant and relevant information can be removed. It is possible to distinguish stages such as: removal of special characters, alignment of case, removal of punctuation marks and inverted commas, removal of possessive pronouns, lemmatization, stop words removal. There are three basic stages in automatic text processing:

1. Text preprocessing – the scope of text preprocessing includes case uniformity, removal of special characters that will not be relevant for further analysis (e.g. newline character).
2. Vectorization – the scope of vectorization includes the presentation of previously prepared texts in a form understandable for a computer, placing in specific data structures.
3. The final processing of the text – in our case, it is text classification.

Natural language processing, as well as other Machine Learning (ML), Deep Learning (DL) and data processing tasks, need a large amount of data to accomplish their task effectively [4]. Thus, one of the biggest challenges facing ML/DL tasks, including NLP, is the lack of data. In language processing, this is particularly evident in the case of text data for niche languages, such as Polish.

The primary method for obtaining language-specific data is manual collection and labelling of text data. However, this is a very time-consuming and labour-intensive process [1]. As a result, researchers use a variety of other solutions to obtain data more quickly and less expensively, including machine translation from another, more developed language in the field [5]. Most often, this is English, as the universal language used by researchers around the world and developed in the field of NLP. Also, all the operations that modify the underlying dataset that are part of pre-processing in inflectional languages, as in Polish, can have a significant impact on the effective use of such difficult-to-acquire data.

The work presented in this paper contributes in creation of machine translation model for English-Polish translation and its comparison with existing pre-trained model. What is more, checking the impact of noise reduction methods from text pre-processing task on classification accuracy for texts with six basic emotions specifically in Polish language is described. Moreover, the conducted experiments provide a basis for further research for better automatic text analysis.

The paper is structured as follows: Sect. 2 presents the related works in area of text pre-processing on classification accuracy. Section 3 describes the process of machine translation tasks for text pre-processing while Sect. 4 deals with the influence of preprocessing on classification in Polish. Both sections contain described task-specific datasets, experiments and analysis of their results.

2 Related Works in Text Pre-processing on Classification Accuracy

In Natural Language Processing a concept of noise is not well understood and often described as hard to define. The text pre-processing itself is considered a standard process in automatic text analysis. This often includes such procedures as stop-words removal, lemmatization, character size uniformity [2].

Research works often focus on the role of using such methods [3], treating this process as necessary for successful text analysis. Given the huge amount of data available for processing text, especially in the earlier stages of NLP development, a reduction of data was necessary to make it more manageable for existing algorithms. It results in the creation of many tools that allow automatic pre-processing of text data [6]. In addition, experiments were performed on texts domain, such as technical texts using common pre-processing techniques [7], or texts in different languages like Arabic [8] or Chinese [9].

However, few of the published works have dealt with the fact that some of the information removed during text preprocessing can be a significant factor affecting the final interpretation of the text. One such paper is [10], which distinguishes between useful and harmful noise. Another paper focusing on categorizing noise [11] divides it into 7 categories: Orthography, Grammatical Errors, Disfluencies in Human Data, Internet Jargon, URLs, Links and Markup, Repetition of Punctuation and Code-switching. Additionally, it states that in various tasks, different elements of texts can be both useful and harmful.

Most work on NLP focuses on English. It is a fairly easy and structured language and, in addition to being acknowledged as a scientific language, is known and understood by the scientific-research community worldwide. Despite the development of NLP research in national languages other than English, there is still a lack of data for specific tasks in these languages [12]. In response to this problem, many researchers are using automatic methods to translate collections of text data from English into national languages – using machine translation.

Currently, most work on machine translation focuses on neural machine translation approaches using deep machine learning. One of the approaches used for this task is the use of Recurrent Neural Networks (RNN). One RNN model is used as a message encoder, while the other is used as a decoder [13]. This is the most intuitive approach within the sequence to sequence translation. Another approach is the use of Convolutional Naural Networks (CNN). Although this solution was not initially popular in machine translation, thanks to the attentional mechanism [15], it has succeeded in being introduced on a large scale. One more approach is the use of the Transformers architecture [15] – a simpler approach than the RNN usage. The Transformer model's self-attention layers learn the dependencies between words in a sequence by examining links between all of the words in the paired sequences and directly modeling those relationships.

3 Machine Translation Model for English-Polish Translation

One of the steps needed to test the impact of noise on the classification of texts in Polish was to prepare a dataset of texts in Polish. Due to the lack of a large dataset containing all 6 basic emotions in Polish, it was decided to create such a dataset by applying machine translation on such a dataset in English.

This section outlines the process of creating and verifying an English-to-Polish machine translation model. It describes in detail the prepared network architecture, the data needed to learn and verify the performance of the model, and discusses the results by comparing them with another pretrained model.

3.1 Chosen Model Architecture

The prepared model for English-Polish translation is based on the Encoder Decoder architecture [14] with an attention layer [15]. An example of how such a model works is shown in Fig. 1. It is one of the architectures that fits naturally with the machine translation task.

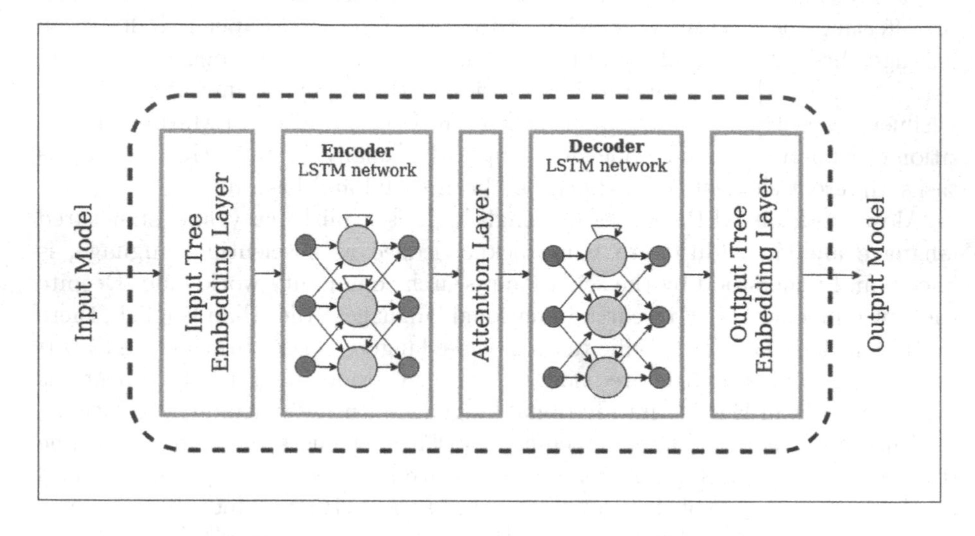

Fig. 1. Encoder Decoder architecture for English-Polish translation.

This model consists of five basic layers:

1. Input embedding layer – responsible for converting input texts into embeddings. At the input of this layer, natural language texts are provided, in the prepared implementation English texts. Its output, on the other hand, receives vectors of texts of a specified length.
2. Encoder layer – its task is to encode the data using the chosen network architecture. For the prepared implementation, the Long-Short Term Memory (LSTM) model architecture was used.
3. Attention layer – uses the attention mechanism which is a technique that is meant to mimic cognitive attention, helps a neural network in memorizing the large sequences of data. This mechanism in machine translation is responsible for both align and translate. Alignment means finding which parts of the input sequence are relevant to each word in the output, whereas translation is the process of using the relevant information to select the appropriate output.

4. Decoder layer – decodes the information using the selected network architecture. For the prepared implementation, the Long-Short Term Memory (LSTM) model architecture was used.
5. Output embbedding layer – responsible for converting the embeddings obtained from the decoder into natural language texts. At the input of this layer, text vectors of a specific length are given. On its output, there are natural language texts, in the prepared implementation, texts in Polish. The result of this layer is the translated texts.

Machine translation task uses LSTM networks [16] as layers in the Encoder and Decoder. The exact appearance of the prepared model along with the parameters is shown in Fig. 2. This diagram was generated from a created Python application. It reflects the theoretical Encoder-Decoder architecture presented above together with the connections between specific layers.

Attention layer was implemented according to [5]. The Tensorflow library's built-in layers, i.e. LSTM, Input, Dense, Embedding, Concatenate, were used to create the other layers. *Adam optimizer* was applied to the model thus prepared. The model was trained for 25 epochs, with 80 iterations used in each epoch.

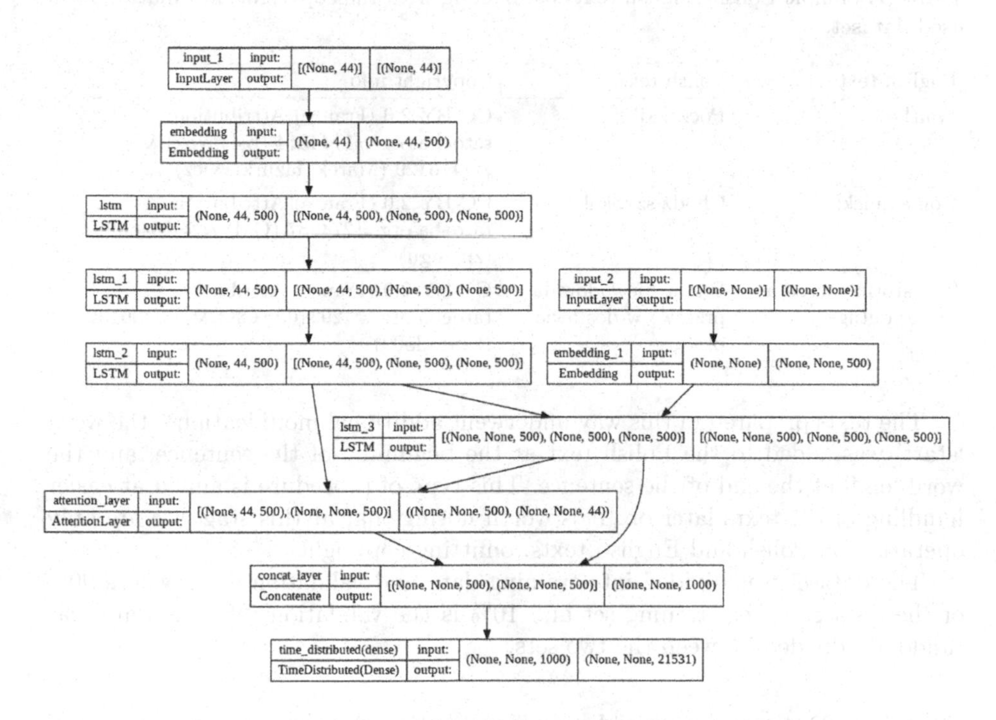

Fig. 2. Prepared model architecture with all parameters.

3.2 The Dataset Used for Machine Translation Task

To conduct the study, it was necessary to select two datasets:

1. The first set serves as a training set – it is on its basis that a DL model was created to allow translations; it is the set that contains texts in both Polish and English.
2. The second set, on the other hand, is the set that we care about translating – it is the collection in English along with the corresponding labels that characterize the emotion assigned to the texts.

Training Dataset. Is the training dataset, the Tab-delimited Bilingual Sentence Pairs for Polish-English pairs [17] was used. The dataset contains about 46.5k Polish-English pairs in a single text file separated by tabs. Each pair additionally contains copyrights information assigned to it. The sentences in the database are of varying difficulty, ordered from shortest to longest. A sample texts pairs are shown in Table 1.

Table 1. Sample Polish-English texts pairs along with the copyright information from used dataset.

English text	Polish text	Copyright info
Wait!	Poczekaj!	CC-BY 2.0 (France) Attribution: tatoeba.org #1744314 (belgavox) & #4476129 (MarekMazurkiewicz)
Come quick!	Chodź szybko!	CC-BY 2.0 (France) Attribution: tatoeba.org #274037 (CM) & #354669 (zipangu)
The storm caused a power outage	Burza spowodowała przerwę w dostawie prądu	CC-BY 2.0 (France) Attribution: tatoeba.org #1293100 (CK) & #3430132 (konrad509)

The texts prepared in this way underwent additional modification – the word 'start' was added to the Polish text at the beginning of the sentence, and the word 'end' at the end of the sentence. This type of procedure is aimed at easier handling of the texts later on. It is worth noting that at this stage we are only operating on Polish and English texts, omitting copyright.

The dataset was divided into training data and validation data, where 90% of the dataset is the training set and 10% is the validation set. The data was randomly divided between the two sets.

Emotions Dataset to Translate. The Emotion Dataset for Emotion Recognition Tasks [18] was used to create a database under emotion dataset in Polish. It is a dataset containing 2,000 texts in English along with one of 6 emotions [19] assigned to them:

- anger,

- fear,
- joy,
- love,
- sadness
- and surprise.

A dataset containing tweets exhibiting six different emotions. The dataset comes in the form of a *.csv file containing English text in the first column and a digital designation of one of the 6 emotions in the second column. Each of the emotions in the text is equidistant. The texts are already pre-processed and therefore contain only lowercase letters and no punctuation, allowing the texts to be used directly in ML models. Sample texts along with assigned emotions are shown in Table 2.

Table 2. Sample texts along with assigned emotions in tweets dataset.

tweet_id	sentiment	content
1956967666	sadness	Layin n bed with a headache ughhhh...waitin on your call...
1956972270	worry	I ate Something I don't know what it is... Why do I keep Telling things about food
1957088179	happiness	@mrssunshine96 big now!!! Vanessa is going to be 3 in September, its going by so fast! its hard cuz Im workin so much, I miss out on alot

It is this dataset that, once translated, is to serve as a text base to test the impact of noise reduction on classification in Polish. This means that all texts from this dataset were translated using both the pre-trained model and the pre-trained Fairseq model.

3.3 Machine Translation Task Results

The prepared model was initially tested on the validation set. The model was trained for 25 epochs, each of which had 82 iterations. During this time, the values such as loss function and accuracy were measured for both training and validation datasets. As the loss results for the validation set were quite high, especially compared to the test set (Fig. 3). It can be seen that the loss function for the training set decreased its value very rapidly in the first 2 epochs and in the later epochs its decrease was smaller, but it still managed to reach values of around 0.15, while for the test set, the function had a much smaller variation and reached a minimum value of about 0.53.

The obtained results are likely to suggest that the model will not perform particularly well with translations. Therefore it was decided to manually check how the sample translations look like for the prepared model. A few of the translations are shown in Table 3. The first column contains three randomly selected sentences in their original version, the second column contains the corresponding Polish translation included in the translation dataset, while the third column contains the translation obtained from the prepared model.

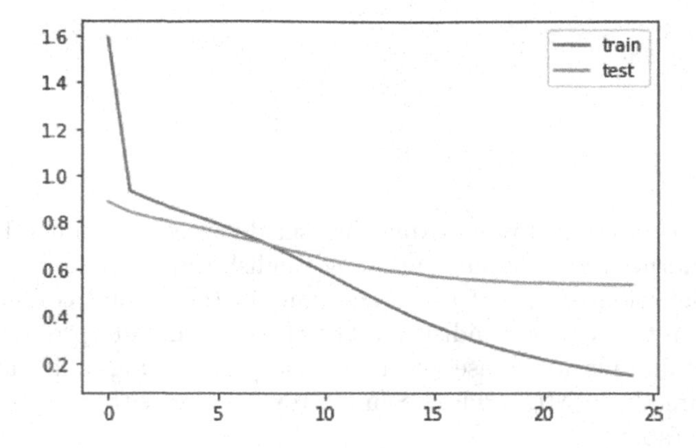

Fig. 3. Loss function over the epochs for train and test dataset.

Table 3. Results of translation in comparison to original text.

Sentence	Original translation	Our translation
tom couldnt stand to be in the same room with mary	tom nie mogl zniesc przebywania razem z mary w jednym pokoju	tom nie potrafil sie przyznac ze mary mine mary
this house belongs to my uncle	to jest dom mojego wujka	dom jest tutaj przez dom
is the hotel far from here	czy hotel jest daleko stad	czy jest tu daleko stad

The results of the manual check were not satisfactory. It can be concluded that the model does not manage longer sentences. At this stage, it was known that it would not be possible to apply the model to create a database of emotionally tinged texts in Polish. Nonetheless, it was decided to use the model to translate the target texts with emotional overtones in order to compare the results to those obtained from the Fairseq model. A summary of the translations along with the original text is shown in Table 4.

Table 4. Comparison of emotion texts translation between prepared model and pre-trained model.

Sentence	Our model translation	Fairseq model translation
i never make her separate from me because i don t ever want her to feel like i m ashamed with her	lubie nauczyc sie jak robilem jak sie nauczyc sie nauczyc jak sie poprawil jak to sie spodoba sie jak ty i uslyszalem jak sie tam	Nigdy nie odrywam jej od siebie, bo nie chce, zeby czula sie zawstydzona
i start to feel emotional	spalem jak noc	Zaczynam czuc sie emocjonalnie
i like to have the same breathless feeling as a reader eager to see what will happen next	nigdy nie mialem okazji pomagac ale musze go znalezc od ciebie ale nie przeczytam wszystkiego	Lubie miec to samo uczucie, gdy czytelnik chce zobaczyc, co bedzie dalej

It is worth noting that translations using the pre-trained model are inaccurate and seem to be completely unrelated to the original sentence. Translations from Fairseq's pre-trained model are much more accurate, plus the model added capital letters, punctuation, etc. making the sentences grammatically and inflectionally correct.

3.4 Summary of Machine Translation Task

Although no metric was used at this stage to measure the quality of the translations, issue a clear difference in quality between the self-trained model and the pre-trained one. What is worth noting is that the model prepared as part of the task did not produce translations that would allow the preparation of an emotion dataset in Polish. Possible reasons for this result are:

- too small teaching set for the model to master the relevant language relationships,
- the general difficulty of translating from English into inflectional languages, which is the Polish language (many words do not have their direct equivalents, and there is a variety by persons, cases, etc.),
- choosing too simple DL model to create effective translations.

It is worth noting that the pre-trained English-Polish convolutional model for Fairseq predicts translations at a high level. When analysed manually, the translations are of high quality and appear to be sufficient to create a database of emotionally tinged texts in Polish.

In the future, it would be worthwhile to additionally use an independent metric to check the quality of the translation, or to conduct human tests with an evaluation of the quality of resulting translations.

The lack of data in individual languages is still a very big problem, and regardless of the fact that tools are being developed to effectively map a dataset from English to other languages, none of these tools will be as reliable as data manually collected and evaluated by humans. English translations may become the only viable option for researchers working on niche languages, but at this point they do not capture the dynamics associated with a language, and these models need to be refined all the time.

All of this leads to the next part of the paper responsible for testing the impact of noise reduction in the data, using a pre-trained model to obtain data containing emotions.

4 Text Pre-processing Impact on Text

The second process was to evaluate the impact of noise reduction on the final emotion classification result for Polish texts. To accomplish this, a suitable dataset containing texts in Polish with the six basic emotions assigned was prepared, the necessary steps were designed and test cases were proposed using the further pre-processing steps.

4.1 Polish Sentences Dataset Used in Classification

Due to the very poor translation performance of the prepared model, a dataset translated pre-trained Fairseq model was used to test the effect of pre-processing methods on the effectiveness of multi-class classification for emotions in Polish. This dataset was further expanded by 1000 texts in Polish containing some emotional tinge and information about the formality of the text. All six basic emotions mentioned before were available. The texts were collected as part of a project carried out by students of the Lodz University of Technology, majority in computer science. The data was collected from private correspondence of college mailboxes. They were then anonymized.

In the same way, a value indicating the formality of the text was assigned, where "0" meant informal text, "1" partially formal and "2" very formal. However, for the purposes of this study, we will skip the formality aspect of the texts and focus only on the basic emotions.

From the values assigned to the successive emotions, the highest one was selected and the emotion assigned to this highest value became the text label. A sample texts along with assigned values for specific emotions and its label are shown in Table 5.

Table 5. Sample text along with assigned values for specific emotions and its label.

id	text	formal	joy	sadness	fear	disgust	surprise	love	label
1	W zalaczeniu pismo od dyrekcji w sprawie dni wolnych od pracy w 2021 r (i jeden dzien juz podany na 2022). Pozdrawiam Jan Kowalski (ang. Attached is a letter from the management regarding public holidays in 2021 (and one day already given for 2022). Best regards Jan Kowalski)	2	0	1	0	0	0	0	sadness
2	Dzien dobry, bardzo, bardzo, bardzo dziekuje. Wszystko jest juz w jak najlepszym porzadku.Pozdrawiam, Ania. (ang. Good morning, thank you very, very, very much. Everything is now in the best possible order.Regards, Ania.)	0.5	1.5	1	0	0	0.5	1.5	happy

4.2 Development Tools Used for Performing Experiments

The Jupiter Notebook was prepared as part of the experiments. It contains all the experiments as well as downloading the necessary libraries, database reading, data preparation and summary of experiment results.

To facilitate the development process the Google Colab environment was used, as it already contains the basic Python libraries installed, and also allows the use of computing resources necessary for the machine learning processes.

The following libraries were used in this phase:

- *stop-words* [20] – library was used to obtain a publicly available list of stop words for the Polish language (for this purpose the list available within the *nltk* library is often used, unfortunately it does not contain a list of words for Polish language),
- *morfeusz2* [21] – library contains functions based on Polish language dictionary that allow basic operations related to vocabulary and sentence syntax; it has been used for lemmatization of single words,
- *pandas* [22] – the basic library used to work on the database,
- *gensim* [23] – library containing the used *doc2vec* text vectorization method.

4.3 Experiments Verifying the Impact of Noise Removal

Within the framework of noise removal task, the six experiments corresponding to different types of pre-processing were carried out. The basic steps for which the experiments were conducted include:

1. texts without any processing (1),
2. texts with commonly available stop-words removed (2),
3. texts with deleted stop-words selected manually (3),
4. texts after stemming (4),
5. texts after deleting commonly available stop words and stemming (5),
6. texts after deleting selected manually stop words and stemming (6).

To conduct the experiments on such a small dataset, the 5-fold cross validation [24] on datasets divided 4:1 was decided to used, where 4 is the training set and 1 is the validation set. In this way, the best result obtained on the validation set was considered as the result of the experiment.

The following activities-functions were performed to carry out the following steps:

1. Removal of publicly available stop words.
2. Removal of stop words selected manually (The stop words list differs only from the publicly available one by removing the word "no"). The list of used stop words for the Polish language is shown in Fig. 4
3. Lemmatization.
4. Validation.

For all texts in the performed experiments, the appropriate combination of presented functions was chosen, and before validation, the model was trained on each of the pre-processed datasets. The doc2vec model architecture was used.

The validation process implementation is shown in Fig. 5.

This process takes into account the comparison of expected attributed emotion with result of doc2vec model [25]. All positive and negative results were counted and then efficacy was calculated from these, comparing with the whole data set. No additional analyses were performed to check whether any of the six basic emotions are classified better or worse. This was not necessary to obtain the results of effect of noise removal on the classification result, which should be independent of the specific emotion.

```
my_stop_words = ['ach', 'aj', 'albo', 'bardzo', 'bez', 'bo', 'być', 'ci', 'cię',
'ciebie', 'co', 'czy', 'daleko', 'dla', 'dlaczego', 'dlatego', 'do', 'dobrze', 'dokąd',
'dość', 'dużo', 'dwa', 'dwaj', 'dwie', 'dwoje', 'dziś', 'dzisiaj', 'gdyby', 'gdzie', 'go',
'ich', 'ile', 'im', 'inny', 'ja', 'ją', 'jak', 'jakby', 'jaki', 'je', 'jeden', 'jedna', 'jedno',
'jego', 'jej', 'jemu', 'jeśli', 'jest', 'jestem', 'jeżeli', 'już', 'każdy', 'kiedy',
'kierunku', 'kto', 'ku', 'lub', 'ma', 'mają', 'mam', 'mi', 'mną', 'mnie', 'moi',
'mój', 'moja', 'moje', 'może', 'mu', 'my', 'na', 'nam', 'nami', 'nas', 'nasi', 'nasz',
'nasza', 'nasze', 'natychmiast', 'nią', 'nic', 'nich', 'niego', 'niej', 'niemu', 'nigdy',
'nim', 'nimi', 'niż', 'obok', 'od', 'około', 'on', 'ona', 'one', 'oni', 'ono', 'owszem',
'po', 'pod', 'ponieważ', 'przed', 'przedtem', 'są', 'sam', 'sama', 'się', 'skąd', 'tak',
'taki', 'tam', 'ten', 'to', 'tobą', 'tobie', 'tu', 'tutaj', 'twoi', 'twój', 'twoja', 'twoje',
'ty', 'wam', 'wami', 'was', 'wasi', 'wasz', 'wasza', 'wasze', 'we', 'więc', 'wszystko',
'wtedy', 'wy', 'żaden', 'zawsze', 'że']
```

Fig. 4. List of stop words used in the experiments of noise removal.

In this paper, the main focus of the analyses is on the influence of noise reduction on the final task of prepared text vectors, namely, to understand them. Therefore, it was decided to use the classification as a determinant of the effectiveness of the used methods. What is more, the simplest of the methods was used for classification to avoid the influence of choice of the method itself on the final result.

Each model was learned for 200 epochs, where the text vector consisted of 100 elements.

```
def validate(test_doc, model1):
  true = 0
  false = 0
  total = 0
  for test in test_doc:
    vector = model1.infer_vector(test.words)
    total += 1
    if model1.docvecs.most_similar([vector])[0][0] == test.tags[0]:
      true += 1
    else:
      false += 1

  print("true: " + str(true))
  print("false: " + str(false))
  print("total: " + str(total))
```

Fig. 5. Validation process implementation for noise removal.

To summarize – the process of verification the impact of noise reduction has 3 main steps as in most text analysis tasks:

1. Text pre-processing - which includes in separate experiments all noise reduction scenarios.

2. Tokenization and vectorization - using doc2vec model.
3. The final processing - in this case, classification, this step is also the validation process, as the classification accuracy was chosen as the impact determinant for noise reduction.

4.4 Results of the Noise Reduction Experiments

The results of noise reduction experiments on texts in Polish language are shown in Table 6. The best results were obtained for texts that had not undergone any pre-processing. The worst results were obtained by texts for which stop words were removed.

Both the result for commonly available and chosen manually stop words is very low. If we classify 6 classes, as in this case (we classify 6 emotions), with random assignment of classes, we will get a result close to 17%. Those obtained by removing the stop word are not much better. Such a low score may result from the fact that the texts are very short, usually containing only a few single sentences. Therefore, when we take away the stop words from the texts, they do not carry much information, which makes it much more difficult to classify them well.

Table 6. Results of noise reduction experiments on texts in Polish.

Type of pre-processing on data	Max accuracy
Raw data	34%
Generally available stop words removed	18%
Selected by myself stop words removed	19%
Lemmatized texts	31%
Lemmatized texts with generally available stop words removed	30%
Lemmatized texts with selected by myself stop words removed	32%

Nevertheless, it is worth noting that the same tendency, which is present in the texts in which only stop words are removed, persists also in the texts subjected to lemmatization with the removal of stop words. Namely, the list of stop words, appropriately selected for the needs of language processing, allows to obtain higher classification efficiency.

As for the results from the model with lemmatized texts, it is not the worst, compared to the others. In this case, words from the same word family will have exactly the same representation, unlike in the case of non-lemmatized texts, where the words are treated as completely different but similar.

5 Conclusions

Text pre-processing has an impact on the final text classification accuracy – regardless of whether the final classification result will be higher or lower than with raw data, one cannot help but notice the difference in performance of models trained on differently processed data. Such an impact is particularly noticeable in Polish, which, as an inflected language, carries a large part of the information specifically in the elements that are lost during preprocessing.

Using deep learning modelling methods, it is worth leaving the whole text unprocessed (more information for model to learn) – the deep learning method used within the study that created text vectors performed better with raw data. This is likely due to the fact that the texts were short and feeding them whole allowed the model to learn additional dependencies available in the texts [4,25,26]. If it is possible, it is worth leaving the texts unchanged, especially with usage of deep learning methods. Such texts contain the most information. Unfortunately, this is not always possible due to limitations in memory and computational resources. In addition, for the Polish language, retaining information on variety, punctuation or word order further increases the time cost of the learning process.

Choosing the right list of stop words to remove depending on the text classification task has an impact on model accuracy – most of the generally available stop word lists were prepared for classification of the subject of the utterance into not emotion or sentiment of the text, so when we want to get information about the emotion in the text, it is important to choose a stop word list for this need. Simply leaving the word 'not' in the texts allowed us to get better results of classification efficiency in Polish. The reason may be that the Polish language allows for double negation and the structure in this language is not defined so strictly.

In summary, it is worthwhile for the research to consider whether the preprocessing method used affects the final classification result too much. Especially when we perform classification in a language other than English like Polish, for which most of the pre-processing methods have been prepared.

References

1. Hedderich, M.A., et al.: A Survey on Recent Approaches for Natural Language Processing in Low-Resource Scenarios (2021). (https://aclanthology.org/2021.naacl-main.201). NAACL
2. Tabassum, A., Patil, R.R.: A survey on text pre-processing & feature extraction techniques in natural language processing. Int. Res. J. Eng. Technol. **7**, 4864–4867 (2020)
3. Haddi, E., Xiaohui, L., Yong, S.: The role of text pre-processing in sentiment analysis. In: International Conference on Information Technology and Quantitative Management, vol. 15, pp. 26–32 (2013)
4. Goyal, P., Pandey, S., Jain, K.: Deep Learning for Natural Language Processing: Creating Neural Networks with Python, Apress, 1st ed. Edition (2018)
5. Bahdanau, D., Cho, K., Bengio, Y.: Neural Machine Translation by Jointly Learning to Align and Translate (2014). ArXiv. 1409
6. Khyani, D., Siddhartha, B.S.: An interpretation of lemmatization and stemming in natural language processing. Shanghai Ligong Daxue Xuebao/J. Univ. Shanghai Sci. Technol. **22**, 350–357 (2021)

7. Sarica, S., Luo, J.: Stopwords in technical language processing. PLoS ONE **16**(8), e0254937 (2020)
8. Abu, E.K., Khair, I.: Effects of stop words elimination for Arabic information retrieval: a comparative study. Int. J. Comput. Inf. Sci. **4**, p119-133 (2006)
9. Feng, Z., Fu, L.W., Xiaotie, D., Song, H.: Automatic identification of Chinese stop words. Res. Compt. Sci. **18**, 51–162 (2006)
10. Camacho-Collados, J., Pilevar, M.T.: On the role of text preprocessing in neural network architectures: an evaluation study on text categorization and sentiment analysis. In: Proceedings of 2018 EMNLP Workshop BlackboxNLP: Analyzing and Interpreting Neural Networks for NLP, pp. 40–46 (2018)
11. Al Sharou, K., Li, Z., Specia, L.: Towards a better understanding of noise in natural language processing. In: Proceedings of the International Conference on Recent Advances in Natural Language Processing (RANLP 2021), pp. 53–62, Held Online. INCOMA Ltd. (2021)
12. Przybyla, O.: Issues of polish question answering. In: Proceedings of 1st Conference Information Technologies: Research and their Interdisciplinary Applications (ITRIA 2012), pp. 96–102 (2012)
13. Dzmitry, B., Cho K., Bengio, I.: Neural machine translation by jointly learning to align and translate. In: Proceedings of 3rd International Conference on Learning Representations, ICLR 2015, San Diego (2015)
14. Aitken, K., Ramasesh, V., Cao, Y., Maheswaranathan, N.: Understanding how encoder-decoder architectures attend. In: proceedings of 35th Conference on Neural Information Processing Systems (NeurIPS 2021) (2021)
15. Vaswani, A., et al.: Attention is all you need. In: Proceedings of 31st Conference on Neural Information Processing Systems (NIPS 2017), USA (2017)
16. Hochreiter, S., Schmidhuber, J.: Long short-term memory. In: Neural Computation, vol. 9, pp. 1735–80 (1998). https://doi.org/10.1162/neco.1997.9.8.1735
17. Tab-delimited Bilingual Sentence Pairs dataset page. http://www.manythings.org/anki/. Accessed 1 Feb 2023
18. Emotion Dataset for Emotion Recognition Tasks. https://www.kaggle.com/datasets/parulpandey/emotion-dataset. Accessed 1 Feb 2023
19. Piorkowska, M., Wrobel, M.: Basic emotions. In: Encyclopedia of Personality and Individual Differences, Editors: Zeigler-Hill, V. and Shackelford, T.K (2017). https://doi.org/10.1007/978-3-319-28099-8-495-1
20. Stop words documentation. https://pypi.org/project/stop-words/. Accessed 1 Feb 2023
21. Kieras, W., Wolinski, M.: (In Polish) Morfeusz 2 - analizator i generator fleksyjny dla jezyka polskiego. Jezyk Polski, XCVI **I**(1), 75–83 (2017)
22. Pandas documentation. https://pandas.pydata.org/. Accessed 1 Feb 2023
23. Gensim documentation. https://pypi.org/project/gensim/. Accessed 1 Feb 2023
24. Yadav, S., Shukla, S.: Analysis of k-fold cross-validation over hold-out validation on colossal datasets for quality classification. In: Proceedings of IEEE 6th International Conference on Advanced Computing (IACC), pp. 78–83 (2016)
25. Le, Q., Mikolov, T.: Distributed representations of sentences and documents. In: Proceedings of 31st International Conference on Machine Learning, ICML (2014)
26. Mikolov, T., Chen, K., Corrado, G., Dean, J.: Efficient estimation of word representations in vector space. In: Proceedings of International Conference on Learning Representations (2013)

A Method of Social Context Enhanced User Preferences for Conversational Recommender Systems

Zhanchao Gao[1,2], Lingwei Wei[1,2], Wei Zhou[1,2], Meng Lin[1](✉), and Songlin Hu[1,2]

[1] Institute of Information Engineering, Chinese Academy of Sciences, Beijing, China
{gaozhanchao,weilingwei,zhouwei,linmeng,husonglin}@iie.ac.cn
[2] School of Cyber Security, University of Chinese Academy of Sciences, Beijing, China

Abstract. Conversational recommender systems (CRS) can dynamically capture user fine-grained preference by directly asking whether a user likes an attribute or not. However, like traditional recommender systems, accurately comprehending users' preferences remains a critical challenge for CRS to make effective conversation policy decisions. While there have been various efforts made to improve the performance of CRS, they have neglected the impact of the users' social context, which has been proved to be valuable in modeling user preferences and enhancing the performance of recommender systems. In this paper, we propose a social-enhanced user preference estimation model (SocialCRS) to leverage the social context of users to better learn user embedding representation. Specifically, we construct a user-item-attribute heterogeneous graph and apply a graph convolution network (GCN) to learn the embeddings of users, items, and attributes. Another GCN is used on the user social context graph to learn the social embedding of users. To estimate better user preference, the attention mechanism is adopted to aggregate the embedding of the user's friends. By aggregating these users' embeddings, we obtain social-enhanced user preferences. Through extensive experiments on two public benchmark datasets in a multi-round conversational recommendation scenario, we demonstrate the effectiveness of our model, which significantly outperforms the state-of-the-art CRS methods.

Keywords: conversational recommender systems · user preference · social context

1 Introduction

With the advent of intelligent assistants such as Siri (Apple), Alexa (Amazon), and Google Assistant, research on conversational information systems has become increasingly significant.

Conversational recommender systems (CRS) aim to capture dynamic and fine-grained user preferences through interactive conversations with the user [18,19].

J. Mikyška et al. (Eds.): ICCS 2023, LNCS 14076, pp. 202–216, 2023.
https://doi.org/10.1007/978-3-031-36027-5_15

In a multi-round conversational recommendation (MCR) scenario, conversational recommendation typically involves two components: recommender component (RC) and conversational component (CC) [17,28]. The recommender component is responsible for estimating the user's preference for items and attributes, while the conversational component interacts with the user based on the results of the RC and the historical conversational state. Through multi-round conversations with the user, the system can collect rich user feedback, which can clarify the user preference and lead to better recommendation quality.

CRS combines the interactive form of the conversation with recommender systems, thus it can directly ask a user whether he/she likes an item/attribute or not [19] to explicitly obtain the exact preferences of users. Previous studies, such as EAR [17] and SCPR [11] adopt Factorization Machine (FM) [26] as their recommender components to estimate user preference and do not utilize the social context of users. Considering the relation between attribute-level and item-level feedback signals, Xu et al. [32] proposes a user preference estimation model called FPAN in multi-round conversational recommender systems, which does not consider the social context of users either. Among these studies, they only consider the interaction between users and the system and elicit users' current preferences through conversations, ignoring the social context of users, which is known to be helpful for modeling users' potential preferences.

In this study, we aim to address the research gap by investigating the impact of social context on conversational recommender systems. Inspired by social correlation theories such as social influence [23], a user's preference is similar to or influenced by his/her socially connected friends [29]. For example, we always like to share our preferences for movies, music, or book with our friends in reality. Analogous to the fact that users like to spread their preferences with their friends and users' interests are influenced by their friends. Since connected users tend to share similar preferences, we think a user's preferences can not only be estimated from the items he interacted with but also can be inferred from his social context. With the assumption of users' interests are influenced by their friends, we believe that leveraging the social context of users can help CRS better understand their users' preferences and thus provide more accurate recommendations.

In this paper, we focus on the recommender component of conversational recommender systems and propose a novel social-enhanced user preference estimation model (SocialCRS) for multi-round conversational recommendation to better estimate user preferences. Specifically, we leverage the social context of users and user-item interaction history to construct a user-item-attribute heterogeneous graph. And then apply Graph Convolution Network (GCN) [15] on the heterogeneous graph to learn the embeddings of users, items, and attributes. Besides, GCN is also used on the user social context graph to learn the social embedding of users. Then attention mechanism is adopted to aggregate users' social context to estimate better user preference. With the social context information, the CRS can conduct more accurate and personalized recommendations.

In summary, the main contributions of this work are as follows.

- We propose a social-enhanced user preference estimation model (SocialCRS) to leverage the social context information of users. By extracting the social context sub-graph in the recommender component of conversational recommender systems and learning graph embedding of users, we improved the performance of conversational recommender systems.
- We use the attention mechanism to integrate users' social information to estimate a better representation of user preferences.
- We conduct experiments on two public datasets. Extensive results show that our model significantly outperforms the state-of-the-art methods.

2 Related Work

2.1 Recommender Systems \Conversational Recommender Systems

Traditional recommender systems have achieved much commercial success and are becoming increasingly popular in the era of big data. They often assume a one-shot inter-action paradigm [14], which makes use of historical user-item interactions to estimate user preference on items and work statically. The most representative methods are Matrix Factorization (MF) [16] and Factorization Machine (FM) [26]. With the development of deep neural networks, Neural FM [12] and DeepFM [9] were developed to enhance FM's representation ability by modeling higher-order and non-linear feature interactions. He et al. [12] presented a Neural Collaborative Filtering (NCF) framework by modeling user-item interactions and estimate users' preferences with deep learning methods.

By combing conversational techniques with recommender systems, conversational recommender systems (CRS) were proposed. Therefore, CRS has the natural advantage of obtaining dynamic and fined-grained users' preferences through user online feedback, having become one of the trending research topics for recommender systems. A variety of conversational recommendation task formulations have been proposed.

Interactive recommender systems [2,11,30,34] and critiquing-based recommender systems [3,21,25] utilize real-time user feedback on previously recommended items to improve online recommendation strategy, which can be seen as early forms of CRS [7]. Zhang et al. [33] proposed a System Ask-User Respond paradigm for conversation search and recommendation, which made the system can actively ask appropriate questions to understand the user needs. Li et al. [20] developed a conditional generative model of natural language recommendation conversations and make recommendations of movies in the cold-start setting. Christakopoulou et al. [4] presented a large-scale learned interactive recommendation system that asked the users question about the topic and gives item recommendations. CRM [28] integrated recommender systems and dialogue system technologies and used reinforcement learning (RL) to find the policy to interact with the user.

Item and attribute-based conversational recommender systems, such as EAR [17] and SCPR [11], both of them adopted Factorization Machine (FM) [26] as their recommender components to estimate user preference and based on a dynamic weighted graph, Deng et al. [5] proposed an adaptive RL framework and model a unified policy learning method to make decisions in CRS. Closely related to our work, Xu et al. [32] proposed a user preference estimation model called FPAN in multi-round conversational recommender systems to capture the relation between users' feedback information.

Same as traditional recommender systems, one of the key tasks of conversational recommender systems is still to correctly understand the users' preferences. Only by correctly understanding the users' preferences, CRS can make better conversational policy decisions and make better recommendations.

2.2 Social Context Information

Extensive research has demonstrated that the user's social context provides additional information that improves the understanding of user behavior in recommender systems and has been widely used in recommender systems. SoRec [22] exploited social context information by decomposing both the user-item rating matrix and the user-social matrix. Jamali et al. [13] proposed the SocialMF model, which introduces a social trust propagation mechanism in the matrix decomposition, making users' preferences indicate users close to their trust. TrustSVD [8] introduced social trust relationships in SVD++ (Singular Value Decomposition), which can be seen as a combination of both SVD++ and SoRec. With the development of graph neural networks, various graph convolution techniques have also been used to model users' social relationships and obtain better representations of user preferences. For example, Fan et al. [6] used graph neural networks and attention mechanisms to capture information between users and users and users and users and items. Wu et al. [31] used graph neural networks to model the social influence propagation process, which enables better representation of users and items.

Although the use of additional information provided by social context has been widely used in traditional recommender systems, the modeling of user preferences in conversational recommender systems does not consider the use of users' social information to enhance user preference representation and thus improve the performance of conversational recommender systems.

3 Preliminaries

3.1 Problem Formulation

In this section, we introduce the notation used to formalize our task. Formally, let $u \in \mathcal{U}$ denotes a user u from the user set $\mathcal{U}$, $i \in \mathcal{I}$ denotes an item i from the item set $\mathcal{I}$, and $a \in \mathcal{A}$ denotes an attribute a from the attribute set $\mathcal{A}$. Each item i is associated with a set of attributes $\mathcal{A}_i \subseteq \mathcal{A}$, which describe its properties.

In an MCR setting, the session starts with a preferred attribute specified by the user. At each turn, the system's recommender component evaluates the user's preference for items and attributes. According to the results of the RC and the historical conversational state, the conversational component decides whether to ask or recommend. When the system chooses to ask the user whether he likes a given attribute, the user replies with binary feedback, either accepting or rejecting the asked attribute. When the system decides to make recommendations to the user, the user gives explicit feedback, indicating whether he likes or dislikes the recommended list of items. The session ends when the user accepts the recommendation or the conversational process reaches the maximum number of turn T.

In this paper, we focus on the recommender component that estimates the users' attribute preference and item preference, which supports the action decision of the conversational component. By utilizing users' social context, the recommender component can better estimate user preference.

4 Methodology

4.1 Model Overview

Figure 1 illustrates the architecture of the proposed Social-enhanced user preference estimation model for Conversational Recommender Systems (SocialCRS). It consists of an offline representation learning module in which the initial embeddings of all the users, items, and attributes are generated by a graph convolution network. In addition to the online feedback module realized by Xu et al. [32], we have added two additional modules to take advantage of users' social context information. One of the two modules is a graph convolution network on the user

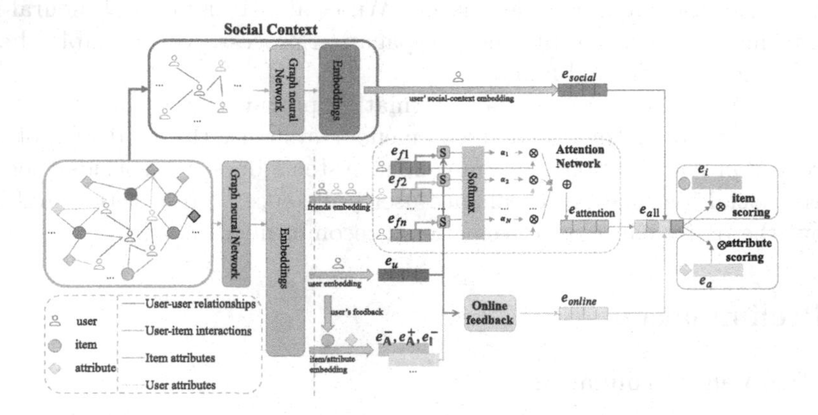

Fig. 1. The overall architecture of the proposed SocialCRS. First, we use GCN on the user-item-attribute heterogeneous graph to pre-train the embeddings of users, items, and attributes. Then, GCN is used on the user social context graph to learn the social embedding of users. The attention mechanism is adopted to aggregate the embedding of users' friends to estimate better user preference. Finally, the final social-enhanced user preference representation is obtained by aggregating three different representations: e_{social}, $e_{attention}$ and e_{online}

social context graph, the other is an attention network to aggregate users' social context. By combining these three modules, we can get social-enhanced user preference representation e_{all}. Finally, user preference on items and attributes are respectively estimated by modeling interactions between social-enhanced user preference representation with the item embedding and attribute embedding.

4.2 Representation Learning

Different from FPAN, SocialCRS not only utilizes the historical user-item interaction data and the relations between items and attributes but also leverages the social connections between users to learn the initial representations of users, items, and attributes. Specifically, a user-item-attribute heterogeneous graph is constructed.

Formally, let $\mathcal{G} = (\mathcal{V}, \mathcal{E})$ denotes the constructed graph, where the set of nodes is denoted as $\mathcal{V} = \mathcal{U} \cup \mathcal{I} \cup \mathcal{A}$, and the set of edges $\mathcal{E}$ consists of four types of edges: the user-item edge (u, i) means the user u interacted with item i, the user-attributes edge (u, a) means the user u prefers the attribute a, and the item-attribute edge (i, a) means that the item i contains the attribute a, and the user-user edge (u, u_f) means the relationship between the user u and his friend u_f.

Following previous work [32], GraphSAGE [10] is adopted to learn the node representations. For the k_{th} layer representation of node $v \in \mathcal{V}$, GraphSAGE generates node embedding:

$$h_v^k = \sigma(W^k \cdot CONCAT(h_v^{k-1}, h_{N(v)}^k)), \tag{1}$$

where σ means the activate function, W^k is trainable parameters, and h_v^{k-1} is the $k - 1_{th}$ layer representation of node v, $N(v)$ denotes the set of node $v's$ neighbors.

Then, to avoid the over-smoothed embedding at the last layer and capture different semantics at different layers, we aggregate the representations generated at different layers and get the final representation of the nodes:

$$e_v = \frac{1}{L+1} \sum_{j=0}^{L} h_v^j, \tag{2}$$

where $v \in \mathcal{V}$. Since $\mathcal{V} = \mathcal{U} \cup \mathcal{I} \cup \mathcal{A}$, we use e_u, e_i, e_a to denote the embeddings of the user, item, and attribute respectively.

4.3 Social-Enhanced User Preference Estimation

Social Context Graph Convolution. We have applied GCN on the user-item-attribute heterogeneous graph to learn the node representations. To better learn the user's embedded representation with social context information, we extract the social context sub-graph $\mathcal{G}_{\mathcal{U}}$ from the above heterogeneous graph $\mathcal{G}$. The social context sub-graph $\mathcal{G}_{\mathcal{U}}$ only contains user nodes and edges that consist of social connections between users. GCN can utilize users' social relations naturally and aggregate multi-hop neighbor nodes to learn users' social embedding. Similarly, we adopt GraphSAGE on $\mathcal{G}_{\mathcal{U}}$ to learn users' social embedding as rich context information:

$$e_{social} = \frac{1}{L+1}\sum_{j=0}^{L} h_u^j, \tag{3}$$

where h_u^j denotes the j_{th} layer representation of node $u \in \mathcal{U}$.

Attention Mechanism. The attention mechanism has been shown effective in many machine learning tasks. It simulates human recognition by focusing on some selective parts of the whole image or the whole sentence while ignoring some other informative parts [1]. Depending on how close a user is to his or her friends, those friends have different effects on a user's preferences. By regarding the user's friends as an image or a sentence, an attention mechanism is applied to learn the weights of the user's friends' influence on the user's preference. Therefore we can further fuse social influence among users. To be specific, we first calculate the similarity between the user's representation e_u and his friend's representation e_{f_i}:

$$S_i = f(e_u, e_{f_i}) = e_u W^Q \cdot e_{f_i} W^K, \tag{4}$$

where W^Q and W^K are trainable parameters. Then, we use the softmax function to get the weights of the user's friends' influence:

$$\alpha_i = softmax(S_i) = \frac{exp(S_i)}{\sum_i exp(S_i)}. \tag{5}$$

Finally, we get user representation that aggregates the user's social context information:

$$e_{attention} = \textstyle\sum_i \alpha_i e_{f_i}. \tag{6}$$

Online Feedback. Inspired by FPAN [32], with the user online feedback, including $\mathcal{A}_u^+, \mathcal{A}_u^-, \mathcal{I}_u^-$, we derive the online user's preference representation by aggregating different kinds of feedback signals:

$$e_{online} = e_u - e_{\mathcal{I}}^- + e_{\mathcal{A}}^+ - e_{\mathcal{A}}^-, \tag{7}$$

where $e_{\mathcal{A}}^+$ denotes the embeddings of the user's positive attribute feedback $\mathcal{A}_u^+$, $e_{\mathcal{A}}^-$ denotes the embeddings of the user's negative attribute feedback $\mathcal{A}_u^-$, and $e_{\mathcal{I}}^-$ denotes the embeddings of the user's negative item feedback $\mathcal{I}_u^-$. More details can be seen from [32].

Social-Enhanced User Preference. Now, the final social-enhanced user preference representation is obtained by aggregating above three different representations:

$$e_{all} = \alpha e_{social} + \beta e_{attention} + \gamma e_{online}, \tag{8}$$

where α, β, γ are hyper parameters, and $\alpha + \beta + \gamma = 1$.

4.4 Item and Attribute Scoring

Since we have gotten the social-enhanced user preference representation e_{all}, SocialCRS next needs to score items and attributes, deciding which items to recommend and which attribute to ask.

Item Scoring. Given an arbitrary item $i \in \mathcal{I}$, we predict how likely u will like i in the conversation session by the dot product between item embedding e_i and the aggregated user preference representation e_{all}:

$$f(i, u) = e_{all} \cdot e_i. \tag{9}$$

Attribute Scoring. Similarly, given an arbitrary attribute $a \in \mathcal{A}$, the affinity score between the user u and attribute a can be estimated as the dot product between the attribute embedding e_a and e_{all}:

$$g(a, u) = e_{all} \cdot e_a. \tag{10}$$

4.5 Model Training

Since the scoring of items and attributes are independent, we formulate the task of the goal of accurate item scoring and attribute scoring separately. Following previous works [17,32], the training objective consists of two loss functions: $\mathcal{L}_{item}$ and $\mathcal{L}_{attr}$.

Item Scoring Loss. To make accurate item scoring, we optimize the pairwise Bayesian Personalized Ranking (BPR) [27] loss. We use two types of negative samples $\mathcal{D}_1$ and $\mathcal{D}_2$ tailored for MCR:

$$\begin{aligned} \mathcal{L}_{item} = & \sum_{(u,i^+,i^-)\in\mathcal{D}_1} -ln\sigma(f(i^+,u) - f(i^-,u)) + \\ & \sum_{(u,i^+,i^-)\in\mathcal{D}_2} -ln\sigma(f(i^+,u) - f(i^-,u)) + \lambda_\Theta \|\Theta\|^2, \end{aligned} \tag{11}$$

where

$$\mathcal{D}_1 := \{(u, i^+, i^-) | i^- \in \mathcal{I} \setminus \mathcal{I}_u\}$$

and

$$\mathcal{D}_2 := \{(u, i^+, i^-) | i^- \in \mathcal{I}_{cand} \setminus (\mathcal{I}_u \cup \mathcal{I}_u^-\}$$

i^+ denotes the user's target item in a conversation session. $\mathcal{I}_u$ is the set of items historically interacted by user u. $\mathcal{I}_{cand}$ is the candidate item set containing items that satisfy the user's attribute requirements in the current conversation session. And the first loss learns u's general preference, the second loss learns u's specific preference given the current candidates. λ_Θ is the regularization parameter to prevent overfitting.

Table 1. Dataset Statistics of LastFM and Yelp*

		LastFM	Yelp*
User-Item Interaction	#Users	1801	27675
	#Items	7432	70311
	#Attributes	33	590
	#Interactions	76693	1368606
	#Avg. friend relations per user	13.303	24.868

Attribute Scoring Loss. For attribute scoring, we also employ BPR loss, and assume that the attributes of the ground truth item i^+ should be ranked higher than other attributes:

$$\mathcal{L}_{attr} = \sum_{(u,a^+,a^-)\in\mathcal{D}_3} -ln\sigma(g(a^+,u) - g(a^-,u)) + \lambda_\Theta \|\Theta\|^2, \tag{12}$$

where the pairwise training data $\mathcal{D}_3$ is defined as:

$$\mathcal{D}_3 := \left\{(u,a^+,a^-)|a^+ \in \mathcal{A}_{i^+} \setminus \mathcal{A}_u^+, a^- \in \mathcal{A} \setminus (\mathcal{A}_{i^+} \cup \mathcal{A}_u^-\right\}.$$

Multi-task Training. We perform joint training on the two tasks of item scoring and attribute scoring, which has the potential of mutual benefits since their parameters are shared. The multi-task training objective is:

$$\mathcal{L} = \mathcal{L}_{item} + \mathcal{L}_{attr}. \tag{13}$$

5 Experiments Setups

5.1 Datasets

For better comparison, we follow [5,17,19,32] to conduct experiments on two publicly available datasets: (1) **LastFM**[1] for music artist recommendation and (2) **Yelp**[2] for business recommendation.

Specifically, LastFM contains 1,801 users and 7,432 items, and 76,693 interactions. **Yelp** contains 27,675 users and 70311 items and 1,368,606 interactions. Following the practices in [17], the users that have less than 10 reviews are pruned to reduce the data sparsity. Each of the datasets is split in the ratio of 7:2:1 for training, validation and testing. For the item attributes, Lei et al. [17] preprocess the original attributes of LastFM by manually merging relevant attributes into 33 coarse-grained attributes, and constructing a two-level taxonomy with 29 first-level categories and 590 s-level attributes for Yelp. The Yelp dataset is

[1] https://grouplens.org/datasets/hetrec-2011/.
[2] https://www.yelp.com/dataset/.

slightly different from [17,32] and the same as the dataset used in [5,19] called **Yelp***, using 590 s-level attributes instead of 29 first-level categories, which can help us reduce heavy manual work and make it more practical. The statistics of the two datasets are presented in Table 1.

5.2 Evaluation Metrics

Following previous work on conversational recommender systems, we use the success rate (SR@T) [28] to evaluate the ratio of successfully recommend the ground truth item by turn T. Besides, we use the average number of turns (AT) to measure the efficiency of conversation. The larger the SR we get, the better recommendation performance we have. And a smaller AT denotes more efficient conversation.

5.3 Baselines

We compare SocialCRS with the following state-of-the-art baselines.

- **Max Entropy** [17]**:** This is a ruled-based method to decide whether to ask or recommend. Each turn it asks the attribute having the maximum entropy among the candidate items or chooses to recommend the top-ranked items with a certain probability. Details can be found at [17].
- **Abs Greedy** [17]**:** This method only has a recommendation component. It only recommends items and updates itself, until it makes a successful recommendation or fails after reaching the maximum number of turns.
- **CRM** [28]**:** This approach is originally designed for the single-round conversational recommendation. It integrates recommender systems and dialogue system technologies and uses reinforcement learning (RL) to find the policy to interact with the user. To adapt this method into the MCR scenario, we follow the description of [17].
- **EAR** [17]**:** The EAR framework consists of three stages, including estimation, action, and reflection, to better converse with users. This method is based on the multi-round conversational recommendation scenario and enhances the interaction between the conversation and recommendation components with a RL framework. FM [26] with attribute-aware BPR [27] is adopted as its recommendation component.
- **SCPR** [19]**:** This method is based on the multi-round conversational recommendation setting, which models conversational recommendation as an interactive path reasoning problem on the graph. It adopts the DQN [24] framework to determine when to ask an attribute or to recommend items, intending to achieve successful recommendations in the fewest turns.
- **UNICORN** [5]**:** This method treats three separated decision-making processes in CRS, including when to ask or recommend, what to ask and which to recommend as a unified policy learning problem. Based on a dynamic weighted graph, UNICORN proposed an adaptive RL framework.

- **FPAN** [32]: This work concerns user preference estimation in multi-round conversational recommender systems. It makes use of GNN to learn the offline representations and two gating modules to aggregate the online feedback information, achieving more accurate user preference estimation.

5.4 Implementation Details

To maintain a fair comparison, we follow [32] and set the size of the recommendation list as 10, and the maximum turn T as 15. The same reward settings are used in our experiments. The embedding size is set to 64. The number of GraphSAGE layers is set to 2. We optimize the model with Adam optimizer. The L_2 norm regularization is set to be 1e-4. The learning rate is set to 0.005 and 0.001 for the item prediction task and attribute prediction task, respectively.

6 Results and Discussion

6.1 Performance Comparison for Multi-round CRS

Table 2 presents the statistics of the model's performances in terms of SR@15 and AT on two datasets. The results are encouraging. That is, the proposed SocialCRS achieves significantly higher SR and less AT than state-of-the-art baselines, demonstrating the superior performance of SocialCRS. From the reported results, we have the following observations. (1) FPAN and SocialCRS which estimate user preference on both attributes and items, achieve better performance than other baselines. (2) Compared with FPAN, SocialCRS obtains better performance on both datasets. Take the dataset LastFM as an example, SocialCRS gains 9.90% SR@15 improvements and decrease 3.85% AT against FPAN. This

Table 2. Performance comparison of all methods on two datasets by SR@15 and AT, where the best performance is bold-faced. SR@15 is the higher the better, while AT is the lower the better. The number in bold denotes that the improvement of SocialCRS over other methods is statistically significant for $p < 0.05$.

	LastFM		Yelp*	
	SR@15↑	AT↓	SR@15↑	AT↓
Max Entropy	0.283	13.91	0.398	13.42
Abs Greedy [17]	0.222	13.48	0.189	13.43
CRM [28]	0.325	13.75	0.177	13.69
EAR [17]	0.429	12.88	0.182	13.63
SCPR [19]	0.465	12.86	0.489	12.62
UNICORN [5]	0.535	11.82	0.520	11.31
FPAN [32]	0.667	10.14	0.642	10.16
SocialCRS	**0.733(+9.90%)**	**9.75(-3.85%)**	**0.717(+11.68%)**	**9.73(-4.23%)**

validates our social-enhanced user preference estimation in the recommender component helps the conversational component to make better decisions, and improves confidence when making recommendations.

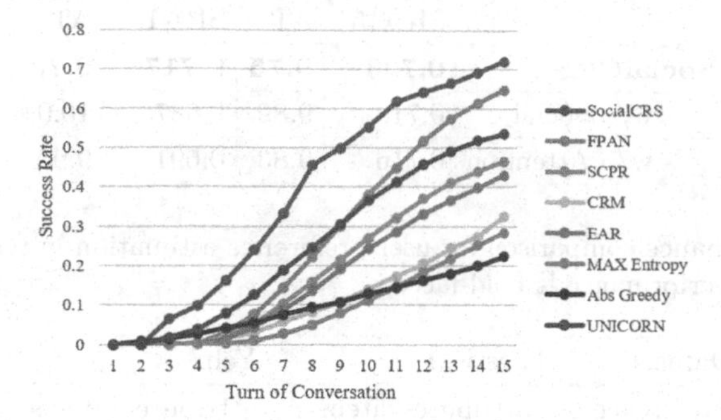

Fig. 2. Success rate of compared methods at different turns on the LastFM dataset.

6.2 Performance Comparison at Different Conversation Turns

In this part, we report the performance comparison of the success rate at each turn (SR@t) in Fig. 2.

As can be seen, our SocialCRS model significantly outperforms all baselines in various settings. At almost every turn, SocialCRS has higher success rate. Furthermore, the success rate of SocialCRS starts to promote on the first few turns, earlier than other methods, which indicates the effectiveness of our proposed model.

6.3 Ablation Study

We further conducted an ablation study to show the contributions of different types of integrating the social context of the users on the two datasets. By setting $\alpha = 0$, $\beta = 0$ respectively in (8), we investigate the influence of social context graph convolution (**Social**) and the attention mechanism (**Attention**) on conversational recommender systems.

Table 3 shows the results of the ablation study. From the reported results, we can see that removing any type of integration of the social context of the users results in a performance drop. This indicates that combining both social context graph convolution and the attention mechanism is important to learn user preferences with their social context. Besides, both of them outperform the state-of-the-art baselines, which confirms the effectiveness of introducing social context to estimate user preference for CRS.

6.4 Performance Comparison for User Preference Estimation

In this part, we compare the proposed model with FPAN for user preference estimation of the attribute prediction and item prediction w.r.t AUC score.

Table 3. Ablation Study on the two dataset. SR@15 is the higher the better, while AT is the lower the better.

Model	LastFM		Yelp*	
	SR@15↑	AT↓	SR@15↑	AT↓
SocialCRS	**0.733**	**9.75**	**0.717**	**9.73**
- w/o Social	0.710	9.89	0.687	10.04
- w/o Attention	0.716	9.83	0.691	9.90

Table 4. Performance comparison for user preference estimation in terms of the AUC score. The best performance is bold-faced.

Dataset	LastFM		Yelp*	
Preference on	attributes	items	attributes	items
FPAN [32]	0.7852	0.6258	0.9731	0.7771
SocialCRS	**0.7904**	**0.6575**	**0.9837**	**0.7915**

Table 4 reports the performance of the attribute prediction and item prediction. From the reported results, we can see that SocialCRS achieves a better performance of preference estimation on both attributes and items, which indicates the model learns better user preference with social context information. Specifically, compared to FPAN, on the dataset—LastFM, the AUC of SocialCRS's item prediction increased to 0.6575 while the attribute prediction increased to 0.7904. The same improvement can be observed in the Yelp* dataset. These improvements demonstrate introducing the social context can improve the prediction accuracy not only for item prediction but also for attribute prediction.

7 Conclusion

In this paper, we have investigated the social context to improve the performance of conversational recommender systems. We present a novel social-enhanced model named SocialCRS for integrating user social context to better estimate user preference. Specifically, SocialCRS applies Graph Convolution Network (GCN) to learn the embeddings of users, items, and attributes on a user-item-attributes heterogeneous graph. Additionally, GCN is used on the users' social context graph to learn the social embeddings of users. Furthermore, an attention mechanism is adopted to aggregate users' social context. By aggregating the above three different representations, the proposed model successfully learns better user preferences with the guidance of social context. Extensive experiments show that SocialCRS is a simple yet powerful method to leverage users' social context to estimate better user preferences and provide accurate and personalized recommendations for users in conversational recommender systems.

References

1. Cao, D., He, X., Miao, L., Xiao, G., Chen, H., Xu, J.: Social-enhanced attentive group recommendation. IEEE Trans. Knowl. Data Eng. **33**(3), 1195–1209 (2021)
2. Chen, H., et al.: Large-scale interactive recommendation with tree-structured policy gradient. In: Proceedings of the AAAI Conference on Artificial Intelligence, pp. 3312–3320 (2019)
3. Chen, L., Pu, P.: Critiquing-based recommenders: survey and emerging trends. User Model. User-Adap. Inter. **22**(1), 125–150 (2012)
4. Christakopoulou, K., Beutel, A., Li, R., Jain, S., Chi, E.H.: Q&R: a two-stage approach toward interactive recommendation. In: Proceedings of the 24th ACM SIGKDD International Conference on Knowledge Discovery & Data Mining, pp. 139–148 (2018)
5. Deng, Y., Li, Y., Sun, F., Ding, B., Lam, W.: Unified conversational recommendation policy learning via graph-based reinforcement learning. In: Proceedings of the 44th International ACM SIGIR Conference on Research and Development in Information Retrieval, pp. 1431–1441 (2021)
6. Fan, W., et al.: Graph neural networks for social recommendation. In: The World Wide Web Conference, pp. 417–426 (2019)
7. Gao, C., Lei, W., He, X., de Rijke, M., Chua, T.S.: Advances and challenges in conversational recommender systems: a survey. AI Open **2**, 100–126 (2021)
8. Guo, G., Zhang, J., Yorke-Smith, N.: TrustSVD: collaborative filtering with both the explicit and implicit influence of user trust and of item ratings. In: Proceedings of the 29th AAAI Conference on Artificial Intelligence, pp. 123–129 (2015)
9. Guo, H., Tang, R., Ye, Y., Li, Z., He, X.: DeepFM: a factorization-machine based neural network for CTR prediction. In: Proceedings of the 26th International Joint Conference on Artificial Intelligence, pp. 1725–1731 (2017)
10. Hamilton, W.L., Ying, R., Leskovec, J.: Inductive representation learning on large graphs. In: Proceedings of the 31st International Conference on Neural Information Processing Systems, pp. 1025–1035 (2017)
11. He, C., Parra, D., Verbert, K.: Interactive recommender systems: a survey of the state of the art and future research challenges and opportunities. Expert Syst. Appl. **56**, 9–27 (2016)
12. He, X., Chua, T.S.: Neural factorization machines for sparse predictive analytics. In: Proceedings of the 40th International ACM SIGIR Conference on Research and Development in Information Retrieval, pp. 355–364 (2017)
13. Jamali, M., Ester, M.: A matrix factorization technique with trust propagation for recommendation in social networks. In: Proceedings of the 4th ACM Conference on Recommender Systems, pp. 135–142 (2010)
14. Jannach, D., Manzoor, A., Cai, W., Chen, L.: A survey on conversational recommender systems. ACM Comput. Surv. **54**(5), 1–36 (2021)
15. Kipf, T.N., Welling, M.: Semi-supervised classification with graph convolutional networks. In: 5th International Conference on Learning Representations (2017)
16. Koren, Y., Bell, R., Volinsky, C.: Matrix factorization techniques for recommender systems. Computer **42**(8), 30–37 (2009)
17. Lei, W., et al.: Estimation-action-reflection: towards deep interaction between conversational and recommender systems. In: Proceedings of the 13th International Conference on Web Search and Data Mining, pp. 304–312 (2020)

18. Lei, W., He, X., de Rijke, M., Chua, T.S.: Conversational recommendation: formulation, methods, and evaluation. In: Proceedings of the 43rd International ACM SIGIR Conference on Research and Development in Information Retrieval, pp. 2425–2428 (2020)
19. Lei, W., et al.: Interactive path reasoning on graph for conversational recommendation. In: Proceedings of the 26th ACM SIGKDD International Conference on Knowledge Discovery & Data Mining, pp. 2073–2083 (2020)
20. Li, R., Kahou, S., Schulz, H., Michalski, V., Charlin, L., Pal, C.: Towards deep conversational recommendations. In: Proceedings of the 32nd International Conference on Neural Information Processing Systems, pp. 9748–9758 (2018)
21. Luo, K., Yang, H., Wu, G., Sanner, S.: Deep critiquing for VAE-based recommender systems. In: Proceedings of the 43rd International ACM SIGIR Conference on Research and Development in Information Retrieval, pp. 1269–1278 (2020)
22. Ma, H., Yang, H., Lyu, M.R., King, I.: SoRec: social recommendation using probabilistic matrix factorization. In: Proceedings of the 17th ACM Conference on Information and Knowledge Management, pp. 931–940 (2008)
23. Marsden, P.V., Friedkin, N.E.: Network studies of social influence. Sociol. Meth. Res. **22**(1), 127–151 (1993)
24. Mnih, V., et al.: Human-level control through deep reinforcement learning. Nature **518**(7540), 529–533 (2015)
25. Pu, P., Faltings, B.: Decision tradeoff using example-critiquing and constraint programming. Constraints **9**(4), 289–310 (2004)
26. Rendle, S.: Factorization machines. In: 2010 IEEE International Conference on Data Mining, pp. 995–1000 (2010)
27. Rendle, S., Freudenthaler, C., Gantner, Z., Schmidt-Thieme, L.: BPR: bayesian personalized ranking from implicit feedback. In: Proceedings of the 25th Conference on Uncertainty in Artificial Intelligence, pp. 452–461 (2009)
28. Sun, Y., Zhang, Y.: Conversational recommender system. In: The 41st International ACM SIGIR Conference on Research & Development in Information Retrieval, pp. 235–244 (2018)
29. Tang, J., Hu, X., Liu, H.: Social recommendation: a review. Soc. Netw. Anal. Min. **3**(4), 1113–1133 (2013)
30. Wang, H., Wu, Q., Wang, H.: Factorization bandits for interactive recommendation. In: Thirty-First AAAI Conference on Artificial Intelligence (2017)
31. Wu, L., Sun, P., Fu, Y., Hong, R., Wang, X., Wang, M.: A neural influence diffusion model for social recommendation. In: Proceedings of the 42nd International ACM SIGIR Conference on Research and Development in Information Retrieval, pp. 235–244 (2019)
32. Xu, K., Yang, J., Xu, J., Gao, S., Guo, J., Wen, J.R.: Adapting user preference to online feedback in multi-round conversational recommendation. In: Proceedings of the 14th ACM International Conference on Web Search and Data Mining, pp. 364–372 (2021)
33. Zhang, Y., Chen, X., Ai, Q., Yang, L., Croft, W.B.: Towards conversational search and recommendation: system ask, user respond. In: Proceedings of the 27th ACM International Conference on Information and Knowledge Management, pp. 177–186 (2018)
34. Zhou, S., et al.: Interactive recommender system via knowledge graph-enhanced reinforcement learning. In: Proceedings of the 43rd International ACM SIGIR Conference on Research and Development in Information Retrieval, pp. 179–188 (2020)

Forest Image Classification Based on Deep Learning and XGBoost Algorithm

Clopas Kwenda(✉), Mandlenkosi Victor Gwetu, and Jean Vincent Fonou-Dombeu

School of Mathematics, Statistics and Computer Science, University of KwaZulu-Natal, Pietermaritzburg, South Africa
221072651@stu.ukzn.ac.za, {gwetum,fonoudombeuj}@ukzn.ac.za

Abstract. Deep learning and machine learning methods have been recently used in forest classification problems, and have shown significant improvement in terms of efficacy. However, as attributed from the literature, they have the challenge of having insufficient model variance and restricted generalization capabilities. The goal of this study is to improve the accuracy of forest image classification through the development of a hybrid model that incorporates both deep learning and machine learning techniques. This study has proposed an ensemble approach of the Deep Learning technique (ResNet50 in particular), and machine learning model (specifically XGBoost) to increase the prediction capability of classifying satellite forest images. The sole purpose of ResNet50 is to generate a set of features that will in turn be used by the XGBoost algorithm to perform the classification process. The XGBoost algorithm was compared against a fully connected ResNet50 model and other classifiers such as random forest (RF) and light gradient boost machine (LGBM). The best classification results were obtained from XGBoost (0.77), followed by RF (0.74), LGBM (0.73), and ResNet50 (0.59).

Keywords: Machine learning · feature extraction · Convolutional Neural Networks · Image Processing

1 Introduction

Forests remain a key natural resource for both developing and developed countries as their wood and forestry products contribute significantly towards a country's Gross National Product (GDP). Both satellite and aerial images play a pivotal role when it comes to monitoring and evaluation of forests and other vegetation. Such images have made huge significant progress in solving remote sensing science classification problems. Data obtained from features such as spectral, radiometric, and spatial is usually used to perform the forest classification process [1]. Image classification refers to the process of labeling each image into its corresponding category or class [2]. Image segmentation is centered on pixel level classification, whereas image classification involves classifying the entire object into one of the given classes. In general, the majority of classification methods employ the technique of assessing and evaluating the image's content

J. Mikyška et al. (Eds.): ICCS 2023, LNCS 14076, pp. 217–229, 2023.
https://doi.org/10.1007/978-3-031-36027-5_16

and then marshaling pixels into their respective categories. The new instance is classified based on an already trained data set whose classes are known. In general, an image is classified into only one of the predefined classes; however, in some cases, an image can be classified into multiple classes, which are referred to as multi-label classes [2]. In spite of the existence of many algorithms used in the classification of vegetation images, there are limited studies that have employed the ensemble machine learning approach in the classification of satellite forest images. Therefore, the purpose of this study is to report findings obtained from an ensemble approach of XGBoost algorithm and ResNet50 technique for the classification of satellite forest images. The new ensemble classifier approach's performance is evaluated against other classifiers such as Random Forest (RF) and Light Gradient Boost Machine (LGBM) in terms of classification accuracy. Different classes (bare-land, logged forest, shrubs, woodlands, and degraded forest) have been identified, and the ensemble learning approach for satellite forest image classification has been assessed by estimating image classification accuracy for different class labels. The rest of the paper is structured as follows. Section 2 deals with related work. Section 3 describes the flow of the proposed study. Section 4 describes the overview of the model architecture. Results and Discussion are presented in Sect. 5. Section 6 concludes the paper.

2 Related Studies

A study by [3] adopted the Random Forest (RF) algorithm to perform image classification on multi-spectral images obtained from Ikonos and QuickBoard data sets. The algorithm's performance was evaluated against results obtained from Gentle AdaBoost (GAB), Maximum Likelihood Classification (MLC), and Support Vector Machine (SVM) algorithms, and RF gave the best result compared to others. The major issue arising in their study was feature extraction. Features were generated using the Random Feature Selection technique. The main limitation of such a technique is giving equal or similar importance to correlated features. To solve this problem, the proposed study has adopted ResNet50 deep learning technique which excels at producing apt and specific features required to solve image classification problems.

[4] employed a deep learning supervised approach on Unmanned Aerial Vehicle (UAV) satellite images for forest area classification. The deep learning stacked Auto-encoder showed significant potential with regard to forest area classification accuracy. However, the major limitation of the deep learning model is that it requires high computational facilities as compared to machine learning algorithms. As a way of solving this challenge, this study is designed in such a way that the image classification process which is the major task that requires high computational capabilities is performed by the XGBoost machine learning algorithm, while the feature extraction part is performed by the ResNet50.

[5] developed a deep learning model for image classification of VHR (very high resolution) images obtained using UAV. The study was against the backdrop that UAV data sets have been found to be very useful for forest feature identification

attributed to their high spatial resolution. Pre-processed data sets of forests of Nagli area were used for the study. The deep learning model incorporated a stacked Auto-encoder to perform image classification. Results showed that the deep learning technique outperformed other machine learning algorithms in terms of accuracy. Through Cross Validation the deep learning model achieved an accuracy 97%. The study's limitation was that it included all features for classification rather than only appropriate features, resulting in an overhead in terms of the model's time complexity. To address this problem, this study adopted the ResNet50 model to generate a set of features required for the forest image classification problem. The learning process of this model is such that the upper first layers are designed to learn general features and the last lower layers are designed to learn specific features. The final feature vector obtained from the ResNet50 model is specifically related to solving a specific classification problem.

When applied to image classification, traditional artificial neural networks, and machine learning approaches face difficulties in processing massive images for feature extraction, resulting in low efficiency and classification accuracy [6]. [6] proposed a deep learning model for image classification with the goal of providing support for classifying large image datasets. The study discussed various types of convolutional neural networks and their applications in image processing. The model was refined by adjusting parameters for feature extraction and by undergoing a process of noise reduction. This study optimized the proposed deep learning model in order to improve the model's classification efficiency and accuracy. The proposed model outperformed other models such as AleXNet and LeNet in terms of classification accuracy. Classification accuracy was also assessed before and after the optimization of the deep learning model. The results revealed that the optimized model significantly improved image classification accuracy. However, the model had challenges in classifying dynamic targets in a complex environment.

Convolutional neural networks adopted under transfer learning usually compress high-resolution input images [7]. A downsampling operation like this usually results in information loss, which affects image classification accuracy. [7] proposed a CNN model based on wavelets domain inputs to solve this problem. During the image pre-processing stage, the wave packet transform was used to extract information from input images. Some subband image channels were chosen as inputs for conventional CNNs with the first several convolutional layers removed, allowing the networks to learn directly in the wavelet domain. The model achieved a classification improvement of 2.15% and 10.26%, respectively on Caltech-256 dataset and Describable Textures Dataset. However, the model suffered huge problems in terms of training costs due to wavelet transform operations that were applied to each image generated through the augmentation process. To address this issue in the proposed study, output images were obtained from the third batch normalization layer of the ResNet50 architecture, where an image would not have been significantly compressed. [8] used an object-based random forest algorithm to identify eight forest types from freely available remote sensing images in Wuhan, China. The images were obtained using Sentinel-1A,

Sentinel-2A, and Landsat 8 sensors. Results obtained indicated that a single sensor cannot obtain satisfactory results. Phenological and topographic information were used in the hierarchical classification to improve discrimination between different forest types. The final forest-type map was obtained using a hierarchical strategy and had an overall accuracy of 82.78%. However, the model encountered the issue of misclassification on types with similar spectral characteristics. This issue is attributed to the study's use of only the NDVI as the primary feature indicator for image classification. This challenge again is addressed in this study by adopting the ResNet50 model.

3 Proposed Model

The study proposes a hybrid machine learning technique for forest-type image classification that combines convolutional deep learning, specifically ResNet50, and traditional machine learning (XGBoost). Convolutional neural networks are widely used in generating features for solving specific classification problems [9]. Therefore in the same vein, the ResNet50 model was adopted in this study to generate a set of features for the XGBoost to perform the image classification task. The XGBoost algorithm was adopted only to perform the image classification process task. Traditional machine learning algorithms outperform deep learning techniques in terms of classification accuracy for a limited data set. Hence the study adopted the XGBoost (machine learning algorithm) to perform the classification task. Because the study uses limited forest images, the basic idea of the model is that CNN produces a feature vector, and then the XGBoost performs the image classification process. Traditional machine learning algorithms used in image classification include Support Vector Machines (SVM), decision trees (DT), extreme gradient boost (XGBoost), random forest (RF), and k-nearest neighbor (KNN). [10] conducted a study to compare the efficacy and effectiveness of LGBM and XGBoost in remote sensing image classification to RF, KNN, and SVM. Efficacy levels of XGBoost and LGBM were above 90%, while the other algorithms had efficacy levels below 90%. It is against this backdrop that the proposed model has advocated towards XGBoost. The ensemble model was used to perform multi-label image classification on forest images from the categories of logged forest, bare land, degraded forest, woodlands, shrubs, and grassland. The proposed algorithm is shown in Fig. 1. The proposed ensemble learning approach for multi-label image classification has the following key features.

- ResNet50 is adopted under the transfer learning technique
- ResNet50 is used for feature extraction and XGBoost is used to perform the classification task.

3.1 Multi-label Image Classification

Multi-label classification is when a test forest image is assigned to a correct category from a set of categories. Fine-tuning done to the model to enable the

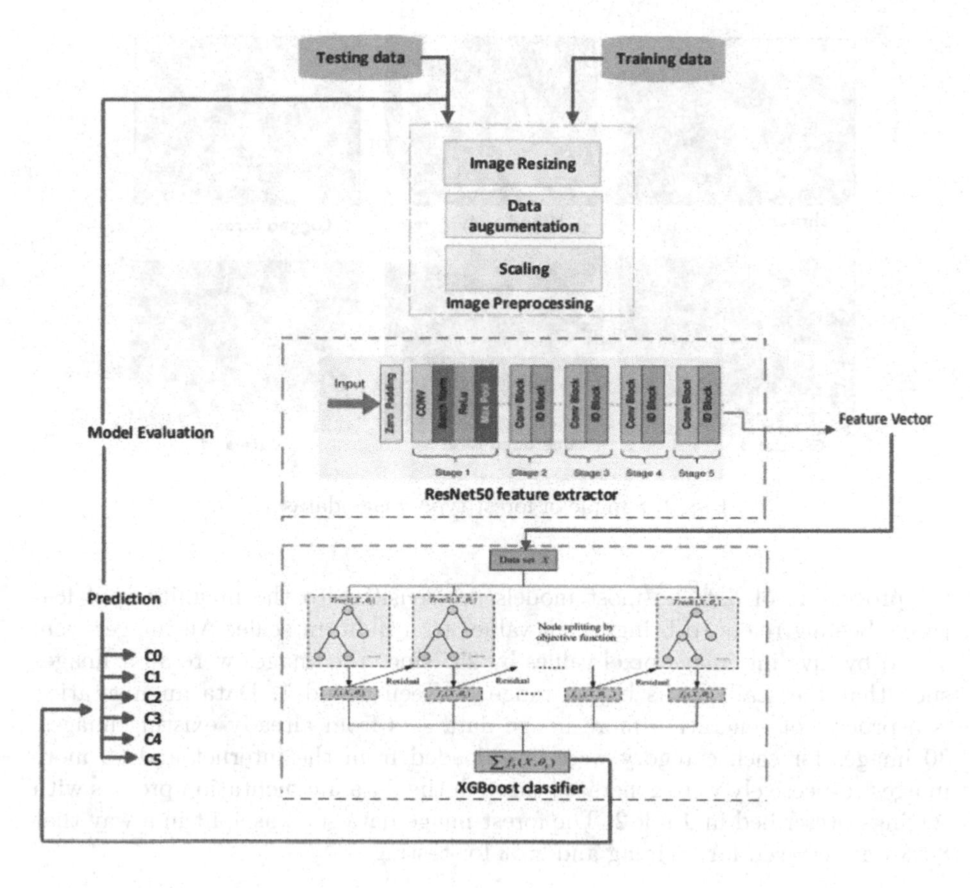

Fig. 1. Proposed ensemble hybrid algorithm for forest image classification.

classification processes involves converting class label strings to integer discrete values. Such conversion is made possible by applying the transform LabelEncoder function adopted from sklearn in Python on the class label vector set. The inverse transform function was invoked in the prediction phase for visualization purpose. Figure 2 shows a sample of forest-type image data set that was used in the study.

3.2 Pre-processing

Since there is no publicly available forest image data set [11], different types of forest images were obtained from the internet. All images were resized to 256 X 256 pixels since the images were of different sizes. Class labels were used as categorical data in this study, and the label encoder technique was used to convert non-numeric categorical data to numeric values. The class labels were transformed into a vector of values 0 through 5. Most machine learning algorithms require labels to numerical integer values. Table 1 represents the labeled classes. Scaling features in machine learning is one of the most critical steps in

Fig. 2. Sample of forest type image dataset.

pre-processing of data as most models are sensitive to the magnitude of features. Scaling refers to bringing all values to a uniform scale. All images were scaled by dividing image pixel values by 255 since the images were 8-bit images such that the scaling was in the range between 0 and 1. Data augmentation is a process of generating more image data sets from already existing images. 30 images for each category were downloaded from the internet and 90 more images respectively were generated through the data augmentation process with settings prescribed in Table 2. The forest image data set was split in a way that 80% was reserved for training and 20% for testing.

Table 1. Labels of forest type images

Value	Class
0	bareland
1	degraded forest
2	grassland
3	woodlands
4	logged forest
5	shrubs

4 Overview of the Model Architecture

This section provides a description of algorithms that were harmonized together to form the proposed hybrid model.

Table 2. Data Augmentation Properties

Property	Value
rotation_ range	45
width_shift_range	0.2
height_shift_range	0.2
zoom_range	0.2
horizontal_flip	True
fill_mode	reflect

4.1 The XGBOOST Algorithm

XGBoost has become predominant in the fraternity of machine learning. It is highly preferred as an alternative to Light Gradient Boost Machines (LGBMs) due to its high execution speed and performance. During the CPU's running time, the XGBoost algorithm employs a parallel computing technique for subsequent tree construction. It uses the 'maxdepth' criteria, instead of the traditional stopping criterion first, and the tree pruning process is initiated from a backward direction. Such a technique significantly improves the computational speed of XGBoost over other LGBM frameworks. Another strength of XGBoost is that it uses the training loss function to automatically learn the best missing values, hence it has the ability to handle different sparsity patterns in the data provided as input efficiently. The XGBoost algorithm uses the following equations for classification:

$$x(t) \approx x(s) + x'(s)(t-a) + \frac{1}{2}x^{n}(s)(t-s)^{2}, \tag{1}$$

$$\zeta \simeq \sum_{i=1}^{n}[l(q_i, q^{t-1}) + r_i x_t(t_i) + \frac{1}{2}s_i x_t^2(m_i)] + \omega(x_t + C), \tag{2}$$

where C is constant, m_i is the input, Ω(x) is the complexity of the tree. r_i and s_i are defined as follows:

$$r_i = \delta \hat{z}_i^{(b-1)} . \int (z_i \hat{z}_i^{n(b-1)}, \tag{3}$$

$$s_i = \delta \hat{z}_i^{(b-1)} . \int (z_i \hat{z}_i^{n(b-1)}, \tag{4}$$

where z_i represents the real value obtained from the training data set. [12] conducted a comparative performance assessment of the XGBoost algorithm, random forest, logistic regression, and standard gradient boosting, and the XGBoost algorithm was found to be most efficient against all other algorithms. It is against this backdrop that the study has settled for the XGBoost technique.

4.2 ResNet50 Network Architecture

A CNN composed of 50 layers is referred to as ResNet-50. Such a deep network with so many layers suffer from network degradation problem. The network is made up of stalked residual blocks. It performs its function with identity short-cut connections that jump one or more layers during the training phase using the residual connections. Intermediate layers have the learning ability to self-adjust their weights to values closer to zero such that the residual block becomes an identity function. The residual skip connections in the ResNet50 architecture helps solves the problem of vanishing gradient experienced in deep neural net-works. It is against this backdrop that the study has adopted the ResNet50 model in the framework. Due to the limited labeled training data set, the study sped up the learning process by adopting the ResNet50 under the transfer learn-ing technique pre-trained on the ImageNet database. ResNet50 architecture is widely known for producing good features for solving classification problems. Features are produced in a way that the upper learns lower-level features and lower layers learn specific features.

5 Metrics for the Study

The multi-class classification evaluation metrics used are accuracy, precision, recall, F1-score, mean average error, root mean square, and confusion matrix. Mean Absolute Error is a measure of the difference between the predicted value and the actual value. It gives an error associated with a predicted image

The root mean square error (RMSE) and the mean absolute error (MAE)are two widely standard metrics used to assess the performance of a model. MEA gives an error associated with a predicted image while RMSE as the name sug-gests gives the mean square of all errors. Considering a set of m observations $x(x_i, i = 1, 2, 3...m)$ and the corresponding model predictions $\hat{x}$ the MAE and RMSE are

$$MAE = \frac{1}{m}\sum_{i=1}^{m}|x_i - \hat{x}_i| \tag{5}$$

$$RMSE = \frac{1}{m}\sqrt{\sum_{i=1}^{m}(x_i - \hat{x_i})^2} \tag{6}$$

Precision being closely related to the measure of quality and recall to the measure of quantity, these two metrics are expressed as follows:

$$precision = \frac{TP}{TP + FP} \tag{7}$$

$$recall = \frac{TP}{TP + FN} \tag{8}$$

where TP is true positives, FP denotes false positives and FP denotes false negatives. F1-Score calculates the harmonic average between recall and precision rates and is expressed as follows:

$$F1 - Score = 2 * \frac{precision * recall}{precision + recall} \quad (9)$$

Accuracy is the overall measure of the model performance and it is expressed as:

$$accuracy = \frac{TP + TN}{TP + TN + FP + FN} \quad (10)$$

Table 3. Metrics for Random Forest

Label	Precision	Recall	F1-Score
0	0.88	0.54	0.67
1	0.50	0.55	0.52
2	0.67	0.80	0.73
3	0.78	0.70	0.74
4	0.71	0.85	0.77
5	0.83	1.00	0.88

6 Results and Discussion

Discussion and results obtained by the study are presented in this section. Table 3 shows that the RF algorithm returns the highest precision for category 0 and subsequently followed by category 5, 3, and 4 respectively, and performed poorly for category 1. Precision is also referred to as the measure of quality. Most of the images in Category 1 were misclassified into Category 0 and this is most likely due to image ambiguity between bare land and degraded forests. However, for recall, category 5 received the most relevant images followed by categories 4, 2, and 3 respectively. Recall is also referred to as the measure of quantity. F1-score provides a balance between precision and recall in relation to positive classes. RF achieved the highest F1-Score for category 5, followed by categories 4, 3, 2, and 1 respectively. Table 4 shows that XGBoost obtained high precision for category 0 with 0.86, i.e. slightly lower than RF. Categories 2, 3, and 4 obtained good quality results in terms of precision as all the scores are above 0.7. Similar to the RF algorithm, the XGBoost algorithm obtained poor results for category 1, and the same reason attributed to poor results in category 1 in RF is also attributed here. The general performance of XGBoost in terms of recall and F1-score is generally the same as with RF. Table 5 shows that LGBM performed poorly for category 1 in terms of precision, recall, and F1- score. That is, the algorithms

Table 4. Metrics for XGBoost

Label	Precision	Recall	F1-Score
0	0.86	0.69	0.77
1	0.50	0.64	0.56
2	0.82	0.70	0.76
3	0.82	0.70	0.76
4	0.75	0.90	0.82
5	0.79	0.95	0.86

Table 5. Metrics for LGBM

Label	Precision	Recall	F1-Score
0	0.75	0.69	0.72
1	0.33	0.18	0.24
2	0.78	0.70	0.74
3	0.82	0.70	0.76
4	0.63	0.85	0.72
5	0.80	1.00	0.89

failed to distinguish clearly between bare land and degraded forests. For the remaining categories, the algorithm obtained good promising results because on average the values obtained were above 70% for all the metrics. The performance of a fully linked ResNet50 was subpar in comparison to that of other classifiers. As presented in Table 6, the model performed the poorest in Category 1 as it registered a zero for all the metrics that were considered.

Table 6. Metrics for ResNet50

Label	Precision	Recall	F1-Score
0	0.37	0.38	0.38
1	0.00	0.00	0.00
2	0.44	0.60	0.51
3	0.75	0.75	0.75
4	0.76	0.65	0.70
5	0.76	0.95	0.84

Accuracy, MAE, and RSME are the most commonly used metrics to evaluate the performance of the model. The hybrid model with XGBoost outperformed the other algorithms in terms of Accuracy, MAE, and RMSE, which obtained values of 0.77, 0.56, and 1.30, respectively as presented in Table 7. Our proposed model's accuracy outperformed the model proposed by [13]. The model

Table 7. Metrics for Classifiers

Classifier	MAE	RMSE	Accuracy
Random Forest	0.63	1.86	0.74
XGBoost	0.56	1.30	0.77
LGBM	0.67	0.40	0.73
ResNet50	0.97	1.68	0.59

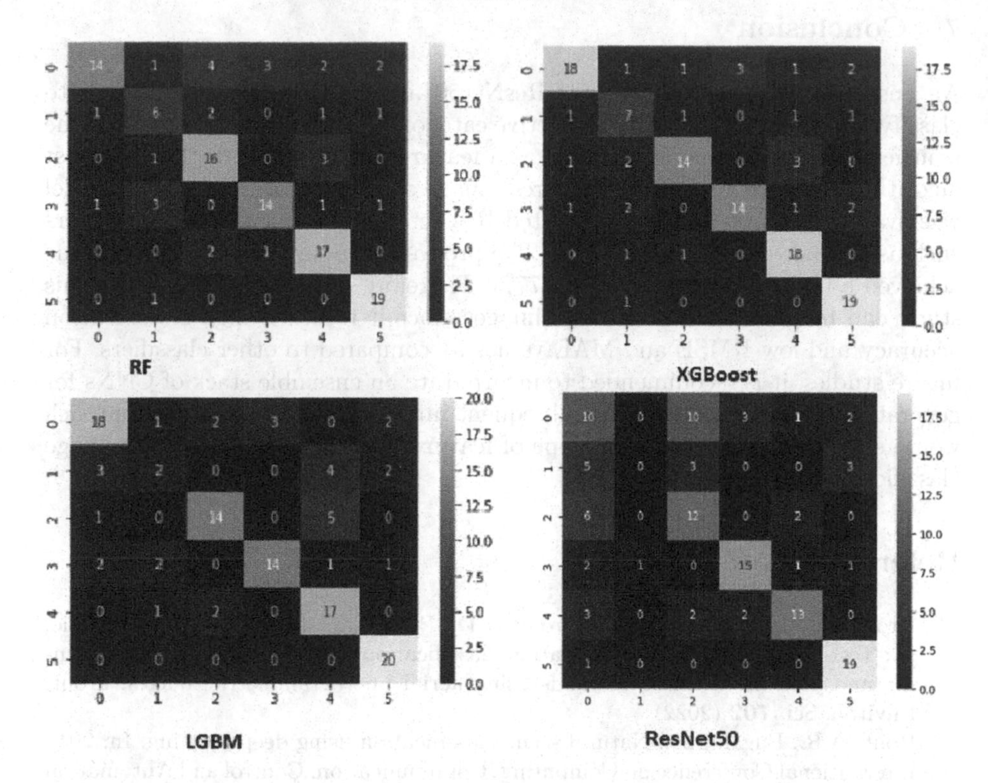

Fig. 3. Confusion Matrix results obtained from RF, XGBoost, LGBM, and ResNet50.

used CNN and Multitemporal High-Resolution Remote Sensing Images to classify individual Tree Species and it obtained an overall accuracy of 75.1% for seven tree species using only the WorldView-3 image data set. The classification accuracy of our proposed model also performed better compared to the results obtained by [3]. Their study achieved a classification accuracy of 68% for classifying multispectral images using Support Vector Machine (SVM). However, the ResNet50 deep learning model proposed by [11] for classifying forest image data set outperformed our model as it achieved an accuracy of 92% for classifying forest images belonging to 3 categories. Such high accuracy could be attributed to the fact that the model was applied on only 3 categories whilst our model was applied to 6 different categories. The performance of a classification algorithm is reflected in a two-dimensional table called the confusion matrix. It is important for summarizing and visualizing a classification algorithm's results. The confusion matrix results as presented in Fig. 3 show that there was high misclassification for category 1 by all the algorithms, with the worst performance by ResNet50. Apart from Category 1, the performance of LGBM and XGBoost in the other categories is generally the same.

7 Conclusion

An ensemble learning approach of ResNet50 and XGBoost was developed to classify forest images into their respective categories. ReNet50 adopted under the transfer learning technique was used as a feature generator, while the XGBoost algorithm was used to perform the forest image classification process. The model was evaluated against a fully connected ResNet50 and other baseline classifiers such as LGBM and Random Forest. The proposed ensemble learning technique achieved a classification accuracy of 77%. Therefore the proposed model in this study can be used to classify forest images since it recorded high classification accuracy and low RMSE and MAE values as compared to other classifiers. For future studies, it is recommended to incorporate an ensemble stack of CNNs for generating plausible features for subsequent image classification. This approach would significantly increase the scope of features required to perform the image classification process.

References

1. Drobnjak, S., Stojanović, M., Djordjević, D., Bakrač, S., Jovanović, J., Djordjević, A.: Testing a new ensemble vegetation classification method based on deep learning and machine learning methods using aerial photogrammetric images. Front. Environ. Sci. **702** (2022)
2. Rout, A.R., Bagal, S.B.: Natural scene classification using deep learning. In: 2017 International Conference on Computing, Communication, Control and Automation (ICCUBEA), pp. 1–5. IEEE (2017)
3. Akar, Ö., Güngör, O.: Classification of multispectral images using random forest algorithm. J. Geodesy Geoinf. Sci. **1**(2), 105–112 (2012)
4. Haq, M.A., Rahaman, G., Baral, P., Ghosh, A.: Deep learning based supervised image classification using UAV images for forest areas classification. J. Indian Soc. Remote Sens. **49**(3), 601–606 (2021)
5. Zhang, X., Chen, G., Wang, W., Wang, Q., Dai, F.: Object-based land-cover supervised classification for very-high-resolution UAV images using stacked denoising autoencoders. IEEE J. Sel. Top. Appl. Earth Obs. Remote Sens. **10**(7), 3373–3385 (2017)
6. Lv, Q., Zhang, S., Wang, Y.: Deep learning model of image classification using machine learning. Adv. Multimed. **2022** (2022)
7. Wang, L., Sun, Y.: Image classification using convolutional neural network with wavelet domain inputs. IET Image Process. **16**(8), 2037–2048 (2022)

8. Liu, Y., Gong, W., Hu, X., Gong, J.: Forest type identification with random forest using Sentinel-1A, Sentinel-2A, multi-temporal Landsat-8 and DEM data. Remote Sens. **10**(6), 946 (2018)
9. Wang, P., Fan, E., Wang, P.: Comparative analysis of image classification algorithms based on traditional machine learning and deep learning. Pattern Recogn. Lett. **141**, 61–67 (2021)
10. Łoś, H., et al.: Evaluation of XGBoost and LGBM performance in tree species classification with sentinel-2 data. In: 2021 IEEE International Geoscience and Remote Sensing Symposium (IGARSS), pp. 5803–5806. IEEE (2021)
11. Tang, Y., Feng, H., Chen, J., Chen, Y.: ForestResNet: a deep learning algorithm for forest image classification. J. Phys: Conf. Ser. **2024**(1), 012053 (2021). IOP Publishing
12. Morde, V.: XGBoost algorithm: long may she reign! (1999)
13. Guo, X., Li, H., Jing, L., Wang, P.: Individual tree species classification based on convolutional neural networks and multitemporal high-resolution remote sensing images. Sensors **22**(9), 3157 (2022)

Radius Estimation in Angiograms Using Multiscale Vesselness Function

Piotr M. Szczypiński(✉)

Institute of Electronics, Lodz University of Technology, Lodz, Poland
piotr.szczypinski@p.lodz.pl

Abstract. This paper presents a new method for estimating the radius of blood vessels using vesselness functions computed at multiple scales. The multiscale vesselness technique is commonly used to enhance blood vessels and reduce noise in angiographic images. The corrected and binarized image resulting from this technique is then used to construct a 3D vector model of the blood vessel tree. However, the accuracy of the model and consequently the accuracy of radii estimated from the model may be limited by the image voxel spacing. To improve the accuracy of the estimated vessel radii, the method proposed in this study makes use of the vesselness functions that are already available as by-products of the preceding enhancement procedure. This approach speeds up the estimation process and maintains sub-voxel accuracy. The proposed method was validated and compared with two other state-of-the-art methods. The quantitative comparison involved artificially generated images of tubes with known geometries, while the qualitative assessment involved analyzing a real magnetic resonance angiogram. The results obtained demonstrate the high accuracy and usefulness of the proposed method. The presented algorithm was implemented, and the source code was made freely available to support further research.

Keywords: Radius estimation · Vesselness · Angiogram analysis

1 Introduction

One of the key challenges in angiographic image analysis is to accurately estimate the radius of veins and arteries [3,6]. These estimated radii form an input data for three-dimensional modeling of vessel structures [4,6,11,27,30,37] which are used for data visualization, surgical planning, and medical diagnosis support [13, 17,18,29,30]. Analyzing the radii along the blood vessels facilitates the detection of lesions such as stenosis and aneurysm.

Angiograms, MRA or CT scans, are raster images of voxels arranged in three-dimensional arrays, which are difficult to quantify directly for medical diagnosis.

Supplementary Information The online version contains supplementary material available at https://doi.org/10.1007/978-3-031-36027-5_17.

J. Mikyška et al. (Eds.): ICCS 2023, LNCS 14076, pp. 230–244, 2023.
https://doi.org/10.1007/978-3-031-36027-5_17

They need to be converted into a more convenient vector representation. One common way to represent image objects in vector form is by surfaces of triangular meshes that define approximate boundaries between anatomical structures [7,8,26]. Alternatively, elongated, tubular structures can be described by their centerlines and radii functions defined along these centerlines [20,37,40]. Both approaches can be used as vector models for the blood vessel system. As blood flows under certain pressure, the vessel walls are inflated and adopt a circular cross-sectional shape. Therefore, the second representation with centerlines and radii is reasonable, and what's more, it directly facilitates quantitative morphological attributes for stenosis or aneurysm detection.

Converting the angiographic raster data to a vector representation involves several processing steps. The first step is filtering the raster image with multiscale vesselness, which enhances blood vessels and reduces image noise [5,10,15,33]. Next, the image is binarized [23,32,35,43] to assign each voxel a label indicating its location inside or outside the blood vessel. Image binarization can be performed using algorithms such as active contours [42,44] , level-set methods [8,39], or deep learning neural networks [12,24,25,31,34], There are two alternative methods for producing centerlines from the binary image. The first approach is to apply marching cubes or a similar algorithm to find the polygonal mesh of the vessel's surface, which is then smoothed and collapsed to form the centerline [14,23,29]. The other method is skeletonization [13,22,41] of the binary image, which thins the structures of blood vessels consisting of white voxels to form a single-voxel-thick line. The resulting centerline may appear rough and requires smoothing in the next step through low-pass filtering of voxel chain coordinates.

This paper presents an original algorithm for radius estimation from multiscale vesselness (REMV), assuming that centerlines and sub-results of multiscale vesselness computation are available. The proposed method is validated and compared with two state-of-the-art methods, one extracting information from the binary image and the other estimating the radius from the original gray-scale image with presumably sub-voxel accuracy. The experiments use artificially generated images of tubes with known geometries for quantitative comparison and a real magnetic resonance angiogram for practical yet qualitative assessment of the results.

2 Methods

2.1 Vesselness-Radius Relationship

The vesselness function algorithm [1,5,10,15,21,33] is used to identify tubular structures in a raster image, and it is particularly useful in the analysis of angiograms. This algorithm enhances the contrast between blood vessels and surrounding tissues, reduces image noise, and conceals anatomical structures that are not cylindrical in shape.

The computation of the vesselness function involves several steps. First, the image is blurred using the Gaussian kernel function with a standard deviation σ. The value of σ should be adjusted according to the radius of the vessel. Second, the Hessian matrices, which are square matrices of second-order partial derivatives of the image intensity function, are computed at every voxel of the image. The matrices are eigendecomposed, and it has been observed [10] that the relations between eigenvalues determine the shape of the structures that are locally present in the image. In tubular structures, one of the eigenvalues is close to zero, while the absolute values of the other two eigenvalues are significantly greater and their values are close to each other.

Several formulas have been proposed to compute the vesselness function from these eigenvalues, including those developed by Erdt [5] , Frangi [9,10] and Sato [33]. In this context, we will discuss the formula developed by Sato. In his approach, the eigenvalues are ordered from the largest to the smallest $\lambda_1 \geq \lambda_2 \geq \lambda_3$. The vesselness formula proposed by Sato has a form similar to Eq. (1). However, the equation presented below includes an additional factor σ^2.

$$F = \sigma^2 \begin{cases} \lambda_c \exp\left(-\frac{\lambda_1^2}{2(\alpha_1 \lambda_c)^2}\right) & : \lambda_1 \leq 0,\ \lambda_c \neq 0 \\ \lambda_c \exp\left(-\frac{\lambda_1^2}{2(\alpha_2 \lambda_c)^2}\right) & : \lambda_1 > 0,\ \lambda_c \neq 0 \\ 0 & : \lambda_c = 0 \end{cases} \tag{1}$$

$$\lambda_c = \min(-\lambda_2, -\lambda_2) \quad \alpha_1 = 0.5 \quad \alpha_2 = 2$$

Setting a proper value of the standard deviation σ is essential for correct enhancement of structures having a particular radius. This parameter should be set according to the actual radius of the blood vessels being enhanced. Figure 1 illustrates this principle using cross-section of an example cylinder for which the vesselness was computed for various values of σ. If this parameter is significantly smaller than the radius of the cylinder, only the elements near the walls are exposed. If it is much larger than the radius, the vesselness appears blurry. The vessel is bright and properly exposed if the σ value is close to the actual radius of the cylinder. Because different values of σ allow for proper enhancement of structures with different radii, σ is sometimes referred to as the scale of the vesselness function. The dependence of the vesselness function on the value of σ is essential for the proposed in here radius estimation algorithm.

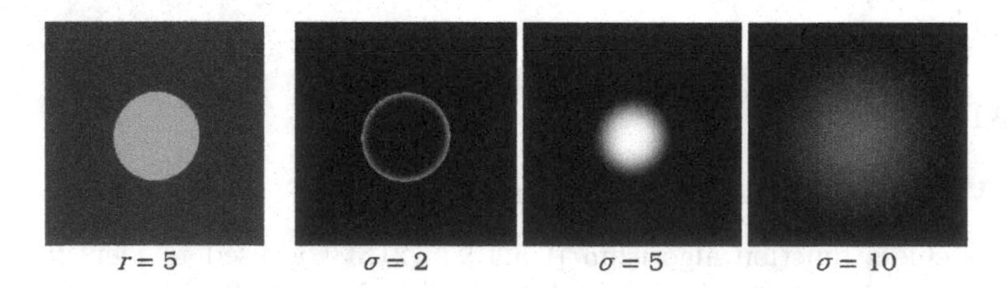

Fig. 1. A cross-section of a cylindrical structure and its vesselness function for various standard deviations of the Gaussian blurring.

If an angiogram consists of a number of vessels having different radii, several vesselness functions can be computed for varying values of the σ parameter.

The range of σ variability should cover the range of radii variability. The resulting images obtained for different values of σ are then combined in a so-called multiscale vesselness function.

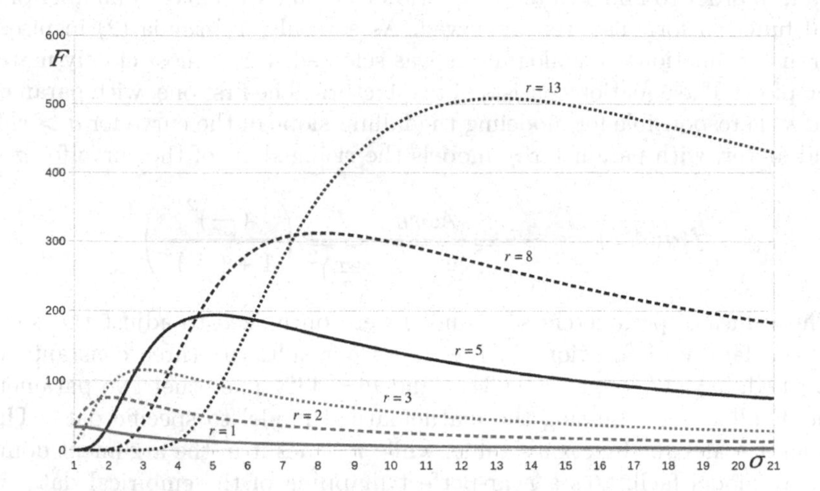

Fig. 2. The vesselness computed at centers of cylinders of various radii as a function of the standard deviation.

An experiment was carried out to compute the vesselness along cylinders of various radii ($r = 1, 2, 3, 5, 8, 13$). Ten cylinder images with random orientation were synthesized for each radius. The vesselness functions were computed for the images using standard deviations σ ranging from 1 to 21. The resulting plots of vesselness obtained at the centers of the cylinders are shown in Fig. 2. It was observed that the plots reached their maxima at the σ argument equal to the radius of the cylinder. This property leads to the conclusion that the radius of a blood vessel can be estimated from such a plot by finding the maximum. Thus, the proposed algorithm consists of computing the vesselness function (1) as a function of the standard deviation of the Gaussian blurring filter. The argument σ at the maximum of the function becomes an estimate of the radius.

2.2 Curve Fitting to Estimate Vessel Radius

To indicate the maximum of the vesselness function with sufficiently high accuracy, the function should be computed for densely distributed values of the σ parameter. Such the approach requires numerous computations of the vesselness functions which is computationally demanding and thus time consuming. What is more, the multiscale vesselness function intended for image enhancement is usually computed for a limited number of scales. To keep the method computationally efficient and accurate, a limited number of vesselness measurements can be approximated with a curve. The σ argument at the maximum of the curve would then estimate the radius of the vessel.

Vesselness computation is a complex procedure, and furthermore, raster images cannot be defined in an analytic form. Therefore, it is infeasible to derive an equation or mathematical model that would accurately define such the curve. Instead, in order to find a suitable equation to define the curve, a number of unimodal function formulas were reviewed. As a result, a formula (2) inspired by the transfer functions of analog filters was selected, which most closely matched all the plots. The equation consists of two factors. The first one, with parameters A and κ, is responsible for modeling the falling slope of the curve for $\sigma > r$. The second factor, with parameter η, models the rising slope of the curve for $\sigma < r$.

$$f(\sigma; A, r) = \frac{A\omega r\kappa}{\sqrt{1+\kappa^2\left(\frac{\sigma}{\omega r}-\frac{\omega r}{\sigma}\right)^2}}\left(\frac{\left(\frac{\sigma}{\omega r}\right)^2}{1+\left(\frac{\sigma}{\omega r}\right)^2}\right)^\eta \tag{2}$$

The values of parameters κ, ω and η were optimized to adjust the formula to fit the family of functions in Fig. 2. As a result, the three constants were established: $\kappa = 17.289$, $\omega = 0.03411$ and $\eta = 432$. The other two parameters, A and r, allow for adjusting the mathematical model to specific data. The A parameter scales the function's value, while r scales it in the argument domain. Such the model facilitates a near-perfect mapping of the empirical data, with the mean square error not exceeding 0.28. It should be noted that the function (2) obtains the maximum when σ equals r. Therefore, after fitting the formula, the parameter r becomes an estimator of the radius.

2.3 Reference Methods

Our goal is to compute the radii of vessels along their centerlines. Typically, radii are estimated for a number of evenly spaced points along the centerline. For each of these points, the local direction of the centerline can be estimated, taking into account the location of neighboring points. Therefore, it is possible to determine the plane perpendicular to this direction, which contains the considered point. This reduces the problem of radius computation in the three-dimensional image space to the two-dimensional space of the vessel's cross-section [3,13,23,38].

In the 2D cross-sectional image, radius estimation consists of determining the distance between the centerline point and the vessel wall. A number of such distances are sampled around the point along concentrically distributed directions, or rays, which are evenly distributed with equal angular intervals. The local radius may be computed by averaging the distances. Typically, the radius is computed as the square root of the averaged squares of the distances. Thus, the resulting radius corresponds to a circle with an area equal to the area of the actual cross-section of the vessel.

Finding the location of the vessel wall along the ray can be accomplished in several ways. One approach involves using a binary image as input, with white voxels representing the inside of the blood vessel and black voxels representing the surrounding tissue. Assuming that the voxels have the shape of small cuboids, the border between the black and white voxels approximates the shape of the wall with accuracy determined by the image raster. The distance sample is defined as the distance between the centerline's point and the point of ray intersection with the nearest boundary between white and black voxels.

The algorithm for locating vessel wall samples in a binary image is computationally efficient and robust. It should be noted that the algorithm operates on binary images, which are visually verified and corrected by human expert to ensure that they accurately reflect the real anatomical structure. While this method accurately reflects the real topology of the blood vessel system, its accuracy is still limited by the resolution of the image. To achieve greater accuracy, information about the brightness distribution in the original image should be utilized. By incorporating this information into the search procedure, it is possible to locate anatomical structures with sub-voxel accuracy.

To find the position of the vessel wall with higher accuracy, a brightness profile along the ray is considered. The profile should be determined to a distance two times longer than the expected radius. This way, the profile equally represents the brightness of the inner and outer regions of the vessel. The profile is characterized by higher brightnesses in areas belonging to the inside of the vessel and significantly lower brightnesses representing the surrounding tissue. Therefore, a distinct slope in the brightness function, where the brightness suddenly drops, should indicate the localization of the vessel's wall.

The brightness profile is affected by various phenomena, including blurring caused by imperfections in the imaging devices. These phenomena can be modeled using the point spread function, which describes the response of the imaging system to a point source signal. Another factor that affects the brightness profile is the partial volume effect. Each voxel occupies a certain volume of three-dimensional space, and if it lies on the edge of two regions, it partially covers a piece of each region. As a result, its brightness is a weighted average of the brightness of both regions, taking into account the partial volumes of these regions covered by the voxel. One acceptable simplification is to model the point spread function of the device with a linear Gaussian blur filter. This allows the shape of the brightness profile near the border to be approximated with a complementary error function (erfc). The erfc function can be fitted to the brightness profile using formula (3), by adjusting the values of its four parameters (Δ_V, Δ_R, V_0 and r) to minimize the mean squared error between the model and the brightness profile [3,27]. After fitting the model to the brightness profile, the parameter r, which determines the shift in the function's domain, indicates the position of the blood vessel wall (Fig. 3).

$$u(d; \Delta_V, \Delta_R, V_0, r) = V_0 + \Delta_V \, erfc\left(\frac{(d-r)}{\Delta_R}\right) \tag{3}$$

In the following experiment, the proposed procedure, which applies the vesselness function for radius estimation, is compared with ray-casting methods. Two alternative approaches for vessel wall localization are used as reference methods: one to locate the wall in a binary image and the other to approximate the brightness profile with the complementary error function.

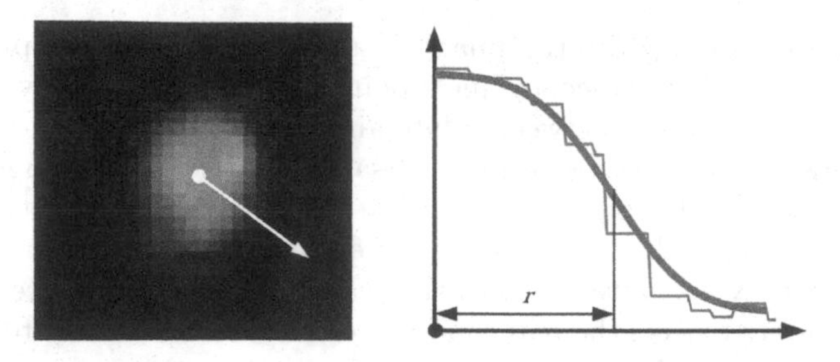

Fig. 3. Matching the complementary error function to the brightness profile.

3 Results

The three algorithms: radius estimation from multiscale vesselness (REMV), cross-sectional ray-casting in binary image (CRB) and cross-sectional ray-casting with erfc matching (CREM) were tested on artificially generated images of tubes and on real magnetic resonance angiogram (MRA). Artificial images of structures of known geometries enabled quantitative comparison and MRA image enabled qualitative, visual assessment. In ray-casting 360 directions were used. In CREM, the erfc was fitted to the brightness profiles consisting of 100 points each, linearly interpolated from the original image. In REMV, the vesselness function was calculated for geometric sequence of 7 scales of σ ranging from 0.5 to 15.

3.1 Radius Estimation Results in Images of Cylinders

In the first experiment, similarly to [3,19,27], synthetic images of cylinders (two examples are in Fig. 4) with dimensions 100×100×100 voxels and 256 gray levels were generated. Each cylinder was positioned at the center of the image. Cylinders had varying radii of 1, 2, 3, 5, 8, and 13. For each radius value, 10 images were produced, each showing the cylinder in a different orientation. It was observed that for blood vessels located on the verge of two volumes with different brightness levels it is difficult to segment the image and accurately determine the radius of the vessel. Therefore, in every image, a sphere with a radius of 150, representing an internal organ, was located in such a way that its surface passed through the center of the image. The cylinders were set to a gray-level of 192, the sphere to 128, and the background to a dark gray value of 64. Partial volume effect was simulated. Finally, Gaussian noise with standard deviation values of 1, 2, 3, 5, 8, and 13 was added to the images. For the CRB procedure, the images were pre-binarized using a brightness threshold of 160, which is a value between the brightness of the sphere and the cylinder.

Figure 4 shows the estimated radii profiles for selected cases of cylinders with real radii of 2 and 5. The radius estimations were computed at 80 points along the centerline. First 40 measurements correspond to the cylinder fragment outside the sphere, the other measurements are from the inside.

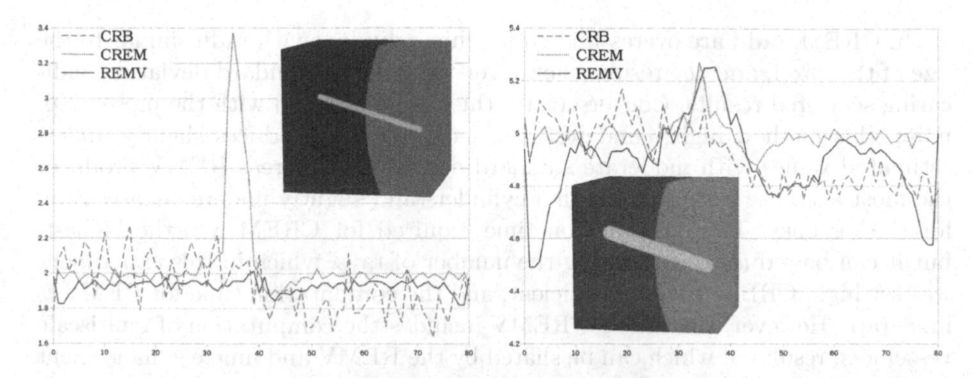

Fig. 4. Example cylinder images with radii of 2 and 5, and their estimated radii profiles.

CREM overestimates the radius in the dark region near the surface of the sphere, which is more pronounced in the thinner cylinder and less so in the thicker one. Both CREM and CRB underestimate the radius inside the bright sphere. The estimated radius plots in CRB are jittery, suggesting vulnerability to noise. In comparison, REMV seems insensitive to changes in the background gray level in the thinner cylinder, but in the thicker one, the radius is overestimated on one side and underestimated on the other side, similar to the results produced by the reference methods.

To make a quantitative comparison of the algorithms, the averages and standard deviations of radius estimates were computed for images with the same noise level and cylinders of the same radius. Table 1 presents these statistics for each algorithm with respect to the real radii and noise levels. The table also shows the total computation time for each algorithm.

Table 1. Average radius estimates and their standard deviations (in brackets) for the selected radii and noise levels.

Radius	Noise	CREM	CRB	REMV
1	1	1.52 (0.90)	0.92 (0.20)	1.02 (0.08)
1	5	1.51 (0.88)	0.93 (0.20)	1.03 (0.09)
1	13	1.44 (0.74)	0.94 (0.21)	1.06 (0.13)
5	1	5.01 (0.08)	4.91 (0.11)	4.85 (0.19)
5	5	5.01 (0.08)	4.91 (0.11)	4.86 (0.19)
5	13	5.01 (0.10)	4.90 (0.12)	4.86 (0.20)
13	1	12.94 (0.10)	12.91 (0.09)	12.63 (0.53)
13	5	12.44 (0.53)	12.91 (0.09)	12.64 (0.53)
13	13	8.21 (2.00)	12.90 (0.09)	12.64 (0.53)
	Time	02:06:41	00:00:39	00:18:36

In CREM, radii are overestimated for thin cylinders with radii similar to the size of the voxel, and the results show a relatively large standard deviation, indicating scattered results. Conversely, for thick cylinders and with the presence of noise, this method highly underestimates radii. CRB produces slightly underestimated values with moderate standard deviations, whereas REMV produces the most accurate estimates for thin cylinders and slightly underestimates radii for thicker ones. The computation time required for CREM was the highest, but it can be reduced by reducing the number of rays, which in this experiment was set high. CRB is the most efficient, and the computation time for REMV is moderate. However, the time for REMV includes the computation of multiscale vesselness, results of which can be shared by the REMV and image enhancement algorithms.

3.2 Radius Estimation Results in Bifurcation Image

In the next experiment, an image of a bifurcation (Fig. 5) was synthesized, which is a structure that often appears in a blood vessel system and is difficult to analyze, estimate radii, and model [3,15,19]. Bifurcations involve three branches of varying diameters, with one usually larger than the other two thinner ones positioned opposite to each other to enable undistorted blood flow. The cross-section of the structure is not circular and becomes increasingly elongated as the cross-section moves towards the two thinner vessels, eventually taking on a shape similar to the digit 8 before splitting into two circular contours. Estimating the radius of a bifurcation using methods that assume a circular cross-sectional shape is difficult due to this non-circularity. Additionally, it is not possible to unambiguously determine the orientation of the vessel at the branching point, as the three vessels that meet there have different orientations.

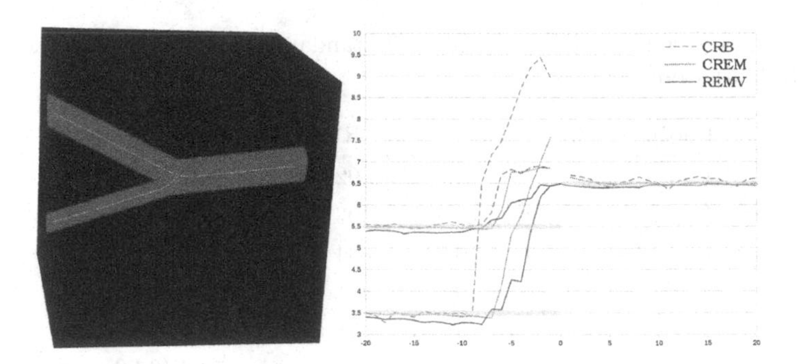

Fig. 5. The synthesized image of bifurcation and its estimated radii profiles.

The image of the bifurcation was synthesized with a background brightness of 64 and a structure brightness of 192, accounting for partial volume effects, and with additive Gaussian noise with a standard deviation of 1. Before applying the radius estimation procedures, the image was binarized using a gray-level threshold of 128 and then skeletonized to extract centerlines [13,22,41]. The binary image was used for radius estimation by CRB, while the other methods estimated radii from the original gray-level image.

When comparing the results (Fig. 5), it can be observed that both CRB and CREM overestimate the radii of the thin blood vessels, and the estimated values near the bifurcation point exceed the radius of the thick vessel. In CRB, the estimated value is over three times higher than the actual one. Moreover, both methods exhibit a discontinuity in the radius estimation function at the branching point. The most accurate mapping of the radius near the bifurcation was obtained by REMV. Although the radii of thinner vessels are overestimated in the vicinity of the bifurcation point, the obtained values do not exceed the radii of the thicker vessel.

3.3 Radius Estimation in MRA

No artificially generated model can capture all the properties and complexities of real angiograms. Therefore, it is necessary to verify if algorithms are capable of correctly estimating radii from genuine medical data. In the case of real images, no ground truth is available on the accurate values of blood vessels' radii. Therefore, the results presented in this section are visually inspected, and the evaluation is qualitative.

In this experiment, courtesy of Prof. Jürgen R. Reichenbach from University Hospital Jena, Germany, a 3D magnetic resonance angiogram (MRA) of a head is used (Fig. 6). The image was previously presented and used for validation of image processing algorithms in other publications [16,20]. It was acquired using the Time of Flight (ToF) Multiple Overlapping Thin Slab Acquisition (MOTSA) [2,28] procedure. The contrast between the flowing blood and the surrounding tissue is achieved without the use of a contrast agent, exclusively due to the dynamics of the flowing blood. In this technique, the stationary tissue is magnetically saturated and does not emit an electromagnetic echo. Fresh blood flowing into the saturated area, agitated by a radio frequency pulse, can emit this signal in contrast to the surrounding tissue. This imaging modality is minimally invasive and relatively comfortable and safe for patients. However, it may expose artifacts such as limited contrast if blood travels a long way in the saturation region, travels parallel to the slabs, or it may expose uneven brightness at the places where the slabs overlap. These artifacts make the image difficult to analyze.

The angiogram is characterized by relatively low spatial resolution. The voxel spacing is $0.42\times0.42\times1.2$ mm. This means that blood vessels less than 0.42 mm in diameter are poorly represented. Moreover, if vessels are located close to each other, at a distance of less than 0.42 mm, they may create false connections or merge into a single trace. The image shows vessels of various diameters, partly straight or curved, running close to each other or close to other structures. In addition, the occipital part of the image is brighter than the fronto-facial part. This data was deliberately selected as it is difficult to interpret and analyze,

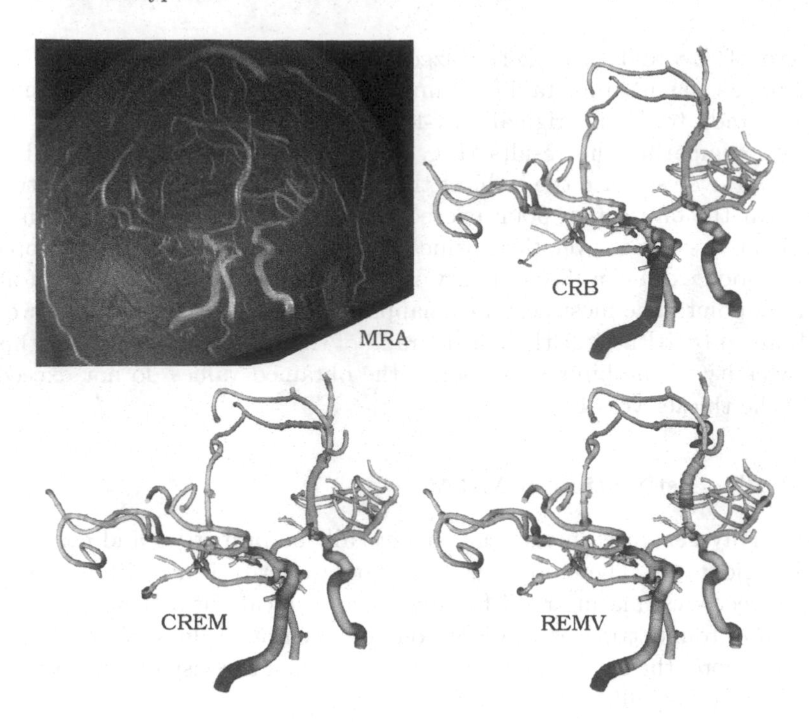

Fig. 6. The input MRA image and the resulting vessel system models.

which should enable the identification of shortcomings and potential limitations of the proposed algorithm.

The image was first enhanced and then binarized using a threshold value that was selected to extract blood vessels from other structures. A flood-fill algorithm was then applied to one of the main arteries to identify all vessels connected to it. The resulting volume was skeletonized to extract centerlines. In REMV, the vesselness function was computed for seven scales of σ ranging from 0.5 to 2.5. Figure 6 displays the resulting models of blood vessel trees, which consist of tubes created around the centerlines. The tube radius varies locally based on the estimations generated by the algorithms under consideration.

There are two types of errors that can be visually identified. The first type of errors consists of moderate discrepancies scattered along the longer fragments of vessels. The second type consists of focal, yet distinct, radius overestimations, which appear as discs transversely oriented with respect to the course of the vessel. In the REMV model, most of the tube fragments appear smooth in comparison to the CRB or CREM models. Therefore, REMV does not have many visible errors of the first type. However, the number of second type of errors is the most pronounced in this method.

One of the unique structures of the blood vessel system in the brain are the two anterior cerebral arteries, which run close to each other. In the angiogram, the two arteries merge in the section where the distance between them is less than the voxel size. The resulting single centerline is therefore incorrect since in reality there are two separate arteries. In CRB and CREM, the radius of the tube covering both arteries is small, and its surface is fairly smooth. In REMV, the radius is larger and uneven, which highlights the incorrectly extracted centerline.

4 Summary and Conclusions

The paper presents an original REMV algorithm for estimating the radii of tubular structures in three-dimensional raster images, particularly medical angiographic images. The standard deviation of the Gaussian window at the maximum of the vesselness function, is a good estimator for vessel radius. A formula to approximate the function from a limited number of scales was found, making REMV computationally effective. Unlike CRB or CREM algorithms, which operate in the cross-sectional plane perpendicular to the vessel, REMV does not require prior information on the vessel's orientation. Moreover, the proposed method estimates the radius from the original grayscale image and does not require the image to be binarized, making it the method with the least requirements for initial input data.

Quantitative results from comparisons with reference methods show that REMV is the most accurate in estimating thin vessels. In comparison with CREM, another method that does not require image binarization, REMV seems to be immune to image noise and it is computationally more efficient. Furthermore, REMV yields the most accurate results in radius estimation in bifurcation, with radius overestimation being significantly smaller than in the other methods. Models derived from the real MRA were qualitatively assessed, and the REMV-based model is smooth; however, it exposes a few focal yet evident radius overestimations.

Table 2. Properties of the radius estimation algorithms.

	CRB	CREM	REMV
Input data requirements:			
Centerline orientation	Yes	Yes	No
Binary image	Yes	No	No
Estimation accuracy	Good	Fair	Good
Immunity to noise	N/A	Low	High
Computation speed	Fast	Slow	Moderate

Table 2 presents an overview comparison of the presented algorithms, taking into account three aspects: input data required by the method, accuracy of radius estimation for artificial images of cylinders, and computation time. It must be noted that immunity to noise of CRB depends on the preceding image segmentation method and is not an intrinsic property of this particular algorithm.

All the compared radius estimation algorithms were implemented in the Vessel knife program [36], which enables the visualization of three-dimensional raster images, their segmentation, extraction of blood vessel centerlines, estimation of radii, and visualization of the resulting models.

References

1. Antiga, L.: Generalizing vesselness with respect to dimensionality and shape. Insight J. **3**, 1–14 (2007)
2. Blatter, D.D., Parker, D.L., Robison, R.O.: Cerebral MR angiography with multiple overlapping thin slab acquisition. Part I. Quantitative analysis of vessel visibility. Radiology **179**(3), 805–811 (1991)
3. Blumenfeld, J., Kocinski, M., Materka, A.: A centerline-based algorithm for estimation of blood vessels radii from 3D raster images. In: 2015 Signal Processing: Algorithms, Architectures, Arrangements, and Applications (SPA), pp. 38–43 (2015)
4. Decroocq, M., Frindel, C., Ohta, M., Lavoué, G.: Modeling and hexahedral meshing of arterial networks from centerlines. arXiv:2201.08279 [cs] (2022)
5. Erdt, M., Raspe, M., Suehling, M.: Automatic hepatic vessel segmentation using graphics hardware. In: International Workshop on Medical Imaging and Virtual Reality, pp. 403–412 (2008)
6. Yuan, F., Yanling Chi, S., Huang, J.L.: Modeling n-Furcated Liver vessels from a 3-D segmented volume using hole-making and subdivision methods. IEEE Trans. Biomed. Eng. **59**(2), 552–561 (2012)
7. Forkert, N.D., Säring, D., Fiehler, J., Illies, T., Möller, D., Handels, H.: Automatic brain segmentation in time-of-flight MRA images. Meth. Inf. Med. **48**(05), 399–407 (2009)
8. Forkert, N.D., et al.: 3D cerebrovascular segmentation combining fuzzy vessel enhancement and level-sets with anisotropic energy weights. Magn. Reson. Imaging **31**(2), 262–271 (2013)
9. Frangi, A.F., Niessen, W.J., Nederkoorn, P.J., Bakker, J., Mali, W.P., Viergever, M.A.: Quantitative analysis of vascular morphology from 3D MR angiograms: in vitro and in vivo results. Magn. Reson. Med. **45**(2), 311–322 (2001)
10. Frangi, A.F., Niessen, W.J., Vincken, K.L., Viergever, M.A.: Multiscale vessel enhancement filtering. In: Wells, W.M., Colchester, A., Delp, S. (eds.) MICCAI 1998. LNCS, vol. 1496, pp. 130–137. Springer, Heidelberg (1998). https://doi.org/10.1007/BFb0056195
11. Han, X., Bibb, R., Harris, R.: Design of bifurcation junctions in artificial vascular vessels additively manufactured for skin tissue engineering. J. Vis. Lang. Comput. **28**, 238–249 (2015)
12. Hilbert, A., et al.: BRAVE-NET: fully automated arterial brain vessel segmentation in patients with cerebrovascular disease. Front. Artif. Intell. **3**, 78 (2020)
13. Hong, Q., et al.: High-quality vascular modeling and modification with implicit extrusion surfaces for blood flow computations. Comput. Meth. Prog. Biomed. **196**, 105598 (2020)
14. Izzo, R., Steinman, D., Manini, S., Antiga, L.: The vascular modeling toolkit: a python library for the analysis of tubular structures in medical images. J. Open Source Softw. **3**(25), 745 (2018)

15. Jerman, T., Pernuš, F., Likar, B., Špiclin, Ž: Enhancement of vascular structures in 3D and 2D angiographic images. IEEE Trans. Med. Imaging **35**(9), 2107–2118 (2016)
16. Klepaczko, A., Szczypiński, P., Deistung, A., Reichenbach, J.R., Materka, A.: Simulation of MR angiography imaging for validation of cerebral arteries segmentation algorithms. Comput. Meth. Programs Biomed. **137**, 293–309 (2016)
17. Klepaczko, A., Szczypiński, P., Dwojakowski, G., Strzelecki, M., Materka, A.: Computer simulation of magnetic resonance angiography imaging: model description and validation. PLoS ONE **9**(4), e93689 (2014)
18. Klepaczko, A., Szczypiński, P., Strzelecki, M., Stefańczyk, L.: Simulation of phase contrast angiography for renal arterial models. Biomed. Eng. Online **17**(1), 41 (2018)
19. Kociński, M., Klepaczko, A., Materka, A., Chekenya, M., Lundervold, A.: 3D image texture analysis of simulated and real-world vascular trees. Comput. Meth. Programs Biomed. **107**(2), 140–154 (2012)
20. Kociński, M., Materka, A., Deistung, A., Reichenbach, J.R.: Centerline-based surface modeling of blood-vessel trees in cerebral 3D MRA. In: 2016 Signal Processing: Algorithms, Architectures, Arrangements, and Applications (SPA), pp. 85–90 (2016)
21. Lamy, J., Merveille, O., Kerautret, B., Passat, N., Vacavant, A.: Vesselness filters: a survey with benchmarks applied to liver imaging. In: 2020 25th International Conference on Pattern Recognition (ICPR), pp. 3528–3535 (2021)
22. Lee, T.C., Kashyap, R.L., Chu, C.N.: Building skeleton models via 3-D medial surface axis thinning algorithms. CVGIP: Graph. Models Image Proc. **56**(6), 462–478 (1994)
23. Lesage, D., Angelini, E.D., Bloch, I., Funka-Lea, G.: A review of 3D vessel lumen segmentation techniques: models, features and extraction schemes. Med. Image Anal. **13**(6), 819–845 (2009)
24. Litjens, G., et al.: State-of-the-art deep learning in cardiovascular image analysis. JACC: Cardiovasc. Imaging **12**(8), 1549–1565 (2019)
25. Livne, M., et al.: A U-Net deep learning framework for high performance vessel segmentation in patients with cerebrovascular disease. Front. Neurosci. **13**, 97 (2019)
26. Lorensen, W.E., Cline, H.E.: Marching cubes: a high resolution 3D surface construction algorithm. ACM Siggraph Comput. Graph. **21**(4), 163–169 (1987)
27. Materka, A., et al.: Automated modeling of tubular blood vessels in 3D MR angiography images. In: 2015 9th International Symposium on Image and Signal Processing and Analysis (ISPA), pp. 54–59 (2015)
28. Parker, D.L., Yuan, C., Blatter, D.D.: MR angiography by multiple thin slab 3D acquisition. Magn. Reson. Med. **17**(2), 434–451 (1991)
29. Piccinelli, M., Veneziani, A., Steinman, D., Remuzzi, A., Antiga, L.: A framework for geometric analysis of vascular structures: application to cerebral aneurysms. IEEE Trans. Med. Imaging **28**(8), 1141–1155 (2009)
30. Quarteroni, A., Manzoni, A., Vergara, C.: The cardiovascular system: mathematical modelling, numerical algorithms and clinical applications*. Acta Numer. **26**, 365–590 (2017)
31. Ronneberger, O., Fischer, P., Brox, T.: U-net: convolutional networks for biomedical image segmentation. In: International Conference on Medical Image Computing and Computer-Assisted Intervention, pp. 234–241 (2015)

32. Rudyanto, R.D., et al.: Comparing algorithms for automated vessel segmentation in computed tomography scans of the lung: the vessel12 study. Med. Image Anal. **18**(7), 1217–1232 (2014)
33. Sato, Y., et al.: 3D multi-scale line filter for segmentation and visualization of curvilinear structures in medical images. In: Troccaz, J., Grimson, E., Mösges, R. (eds.) CVRMed/MRCAS -1997. LNCS, vol. 1205, pp. 213–222. Springer, Heidelberg (1997). https://doi.org/10.1007/BFb0029240
34. da Silva, M.V., Ouellette, J., Lacoste, B., Comin, C.H.: An Analysis of the Influence of Transfer Learning When Measuring the Tortuosity of Blood Vessels. arXiv:2111.10255 [cs, eess] (2022)
35. Smistad, E., Falch, T.L., Bozorgi, M., Elster, A.C., Lindseth, F.: Medical image segmentation on GPUs - a comprehensive review. Med. Image Anal. **20**(1), 1–18 (2015)
36. Szczypinski, P.M.: Vesselknife. https://gitlab.com/vesselknife/vesselknife
37. Vinhais, C., Kociński, M., Materka, A.: Centerline-radius polygonal-mesh modeling of bifurcated blood vessels in 3D images using conformal mapping. In: 2018 Signal Processing: Algorithms, Architectures, Arrangements, and Applications (SPA), pp. 180–185 (2018)
38. Wink, O., Niessen, W., Viergever, M.: Fast delineation and visualization of vessels in 3-D angiographic images. IEEE Trans. Med. Imaging **19**(4), 337–346 (2000)
39. Woźniak, T., Strzelecki, M., Majos, A., Stefańczyk, L.: 3D vascular tree segmentation using a multiscale vesselness function and a level set approach. Biocybern. Biomed. Eng. **37**(1), 66–77 (2017)
40. Yang, G., et al.: Automatic centerline extraction of coronary arteries in coronary computed tomographic angiography. Int. J. Cardiovasc. Imaging **28**(4), 921–933 (2012)
41. Zasiński, P., Kociński, M., Materka, A.: On extracting skeletons from binary 3D images. In: 2017 International Conference on Systems, Signals and Image Processing (IWSSIP), pp. 1–5 (2017)
42. Zeng, Y.Z., et al.: Automatic liver vessel segmentation using 3D region growing and hybrid active contour model. Comput. Biol. Med. **97**, 63–73 (2018)
43. Zhao, J., Zhao, J., Pang, S., Feng, Q.: Segmentation of the true lumen of aorta dissection via morphology-constrained stepwise deep mesh regression. IEEE Trans. Med. Imaging **41**, 1826–1836 (2022)
44. Zhao, Y., Rada, L., Chen, K., Harding, S.P., Zheng, Y.: Automated vessel segmentation using infinite perimeter active contour model with hybrid region information with application to retinal images. IEEE Trans. Med. Imaging **34**(9), 1797–1807 (2015)

3D Tracking of Multiple Drones Based on Particle Swarm Optimization

Tomasz Krzeszowski[1(✉)], Adam Switonski[2], Michal Zielinski[3], Konrad Wojciechowski[3], and Jakub Rosner[3]

[1] Faculty of Electrical and Computer Engineering, Rzeszow University of Technology, al. Powstancow Warszawy 12, 35-959 Rzeszow, Poland
tkrzeszo@prz.edu.pl

[2] Department of Computer Graphics, Vision and Digital Systems, Silesian University of Technology, ul. Akademicka 16, 44-100 Gliwice, Poland
adam.switonski@polsl.pl

[3] Polish-Japanese Academy of Information Technology, ul. Koszykowa 86, 02-008 Warsaw, Poland
{konrad.wojciechowski,jrosner}@pja.edu.pl

Abstract. This paper presents a method for the tracking of multiple drones in three-dimensional space based on data from a multi-camera system. It uses the Particle Swarm Optimization (PSO) algorithm and methods for background/foreground detection. In order to evaluate the developed tracking algorithm, the dataset consisting of three simulation sequences and two real ones was prepared. The sequences contain from one to ten drones moving with different flight patterns. The simulation sequences were created using the Unreal Engine and the AirSim plugin, whereas the real sequences were registered in the Human Motion Lab at the Polish-Japanese Academy of Information Technology. The lab is equipped with the Vicon motion capture system, which was used to acquire ground truth data. The conducted experiments show the high efficiency and accuracy of the proposed method. For the simulation data, tracking errors from 0.086 m to 0.197 m were obtained, while for real data, the error was 0.101–0.124 m. The system was developed for augmented reality applications, especially games. The dataset is available at http://bytom.pja.edu.pl/drones/.

Keywords: unmanned aerial vehicle · drone · particle swarm optimization · multiple drones · multiple cameras · 3D tracking · 3D localization · motion capture · Vicon

M. Zielinski is currently a student of computer science at the Silesian University of Technology.

J. Mikyška et al. (Eds.): ICCS 2023, LNCS 14076, pp. 245–258, 2023.
https://doi.org/10.1007/978-3-031-36027-5_18

1 Introduction

Recently, the use of drones (also referred to as Unmanned Aerial Vehicles - UAVs) has significantly increased, both in civil and military applications. This is also the reason for the increased interest of researchers in various types of drone-related issues, including the problems of tracking these flying vehicles. The development of drone tracking methods is primarily driven by the need to develop effective systems for detecting, identifying, and disabling drones, which are currently extremely decent due to the increasing number of vehicles of this type [4]. These are primarily military and security applications. However, there are also other applications of drone tracking methods, e.g. augmented reality (AR) games [2].

There are many methodologies for detecting and tracking drones, but the most popular are: vision cameras, hyper-spectral images, radars, acoustic sensors, radio frequency techniques (RF), thermal techniques, and hybrid systems [4]. Among these approaches, optical methods stand out, which are considered the most convenient way to deal with this challenge due to their robustness, accuracy, range, and interpretability [17]. For example, Schilling et al. [10] proposed a vision-based detection and tracking algorithm that enables groups of drones to navigate without communication or visual markers. They equipped the drones with multiple cameras to provide omnidirectional visual inputs and utilized the convolutional neural network to detect and localize nearby agents. Another paper [17], presented an interesting concept of using two cameras (static wide-angle and low-angle mounted on a rotating turret) for autonomous drone detection and tracking. The single lightweight YOLO detector was used to detect drones. In [15], Srigrarom et al. described a multiple-camera real-time system for detecting, tracking, and localizing multiple moving drones simultaneously in a 3D space. They utilized a hybrid combination of the blob detection method and the YoloV3 Tiny model to detect drones on images and cross-correlated cameras to obtain global 3D positions of all the tracked drones. Another approach was presented by the authors of the study [8]. They introduce a real-time trinocular system to control rotary wing UAVs based on the 3D information extracted by cameras positioned on the ground. The drone detection and tracking are based on color landmarks and are achieved by using the CamShift algorithm. In [14], a hybrid detecting and tracking system that is made especially for small and fast-moving drones is proposed. In this method, a discrete-time Extended Kalman Filter (EKF) is used to track the positions and velocities of detected moving drones. The Kalman Filter is also used by Son et al. [13]. Sie et al. [12] presented the use of Correlation Filters and an Integrated Multiple Model (IMM) for filtering the position measurement of fast-moving drones. Another approach was proposed by Ganti and Kim [3]. They designed a low-cost system for detecting and tracking small drones. The system used low-cost commercial-off-the-shelf devices and open-source software. Moreover, it utilized image-processing algorithms to detect moving objects and the SURF method to distinguish drones from other objects.

The review of the literature shows that there is a lack of published studies devoted to the problems of tracking multiple drones in three-dimensional space. In addition, no publications were found that measured the accuracy of 3D drone tracking. As far as we know, the Particle Swarm Optimization (PSO) algorithm has also not been used to track drones before. Hence, the development of a method for tracking multiple drones in 3D space and the proposal to use the PSO algorithm in it, are the main motivations for our work. We also prepared simulation sequences for testing the proposed drone tracking method.

2 A Method for Tracking Multiple Drones

The purpose of tracking is to obtain information about the position of the drone in the defined search space. If the drone is tracked in a two-dimensional image space, the complexity of the problem is relatively small and in this case, Kalman Filter [1,14] is often used. However, if the goal is to acquire a 3D position for many drones, the problem becomes much more complicated. For example, if four drones are to be tracked, the search space has 12 dimensions, while in the case of 10 drones, it already has 30 dimensions. In such cases, various optimization algorithms are used.

The proposed drone tracking method is based on the ordinary Particle Swarm Optimization algorithm [5], data from multiple cameras, and image processing methods used to extract the drone from the image. The PSO algorithm and its modifications have many applications, e.g. in signal and image processing, design and modeling, and robotics, as well as in problems related to motion tracking [6,7,9]. In the case of drone applications, flight controllers are the most common [16]. In order to estimate the exact position of drones in 3D space, the system requires at least three cameras, placed around the area in which the drones are moving.

2.1 Dataset

In order to evaluate the developed tracking algorithm, an appropriate dataset was collected. The dataset consists of three simulation sequences (S1, S2, and S3) and two real ones (S4 and S5), which differ in the number of drones and their pattern of moving on scene. The simulation sequences were created in an environment based on the Unreal Engine and the AirSim plugin [11], which is an open-source project created by Microsoft for high-fidelity simulation of autonomous vehicles. The scene for the simulation sequences was prepared using a model of the Human Motion Lab (HML) at the Polish-Japanese Academy of Information Technology in Bytom, in which real sequences were registered. The scene plan is presented in Fig. 1, except that the real lab contains only four cameras (cam_1, cam_2, cam_3, and cam_4) and the simulated lab contains all eight cameras. In addition, for the S3 sequence, the scene dimensions were doubled for the purpose of accommodating a higher amount of drones without an issue of potential collisions between each other. Statistics of sequences are

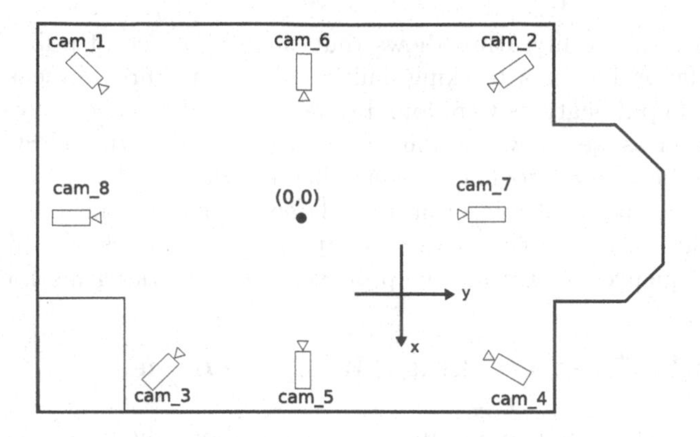

Fig. 1. Top view of the laboratory scene with the position of cameras

Table 1. Sequences metadata

seq. ID	#drones	#frames	length [s]	#cameras	resolution	FPS
S1	2	440	17.6	8	1920 × 1080	25
S2	4	215	8.6	8	1920 × 1080	25
S3	10	230	9.2	8	1920 × 1080	25
S4	1	1415	56.6	4	1924 × 1082	25
S5	1	2885	115.4	4	1924 × 1082	25

summarized in Table 1. Ground truth data for real sequences were acquired using a Vicon motion capture system. Calibration and synchronization of the motion capture system and video cameras were carried out using software and hardware provided by Vicon. The dataset can be downloaded from http://bytom.pja.edu.pl/drones/.

2.2 Particle Swarm Optimization

Particle Swarm Optimization is a metaheuristic method developed by Kennedy and Eberhart [5,6]. The concept of the method was taken from the social behavior of animals living in groups, such as shoals of fish, swarms of bees, or flocks of birds. In the PSO, the solution is found based on a set of particles, each representing a hypothetical solution to the problem. Each of the particles remembers the current position ($\mathbf{x}$), the velocity ($\mathbf{v}$), and the best position it has found so far (**pbest**). In addition, the particles have access to the position of the best particle in the entire swarm (**gbest**). The velocity $\mathbf{v}^k$ and $\mathbf{x}^k$ of the kth particle in the iteration t are updated using the following equation:

$$\mathbf{v}_{t+1}^k = \omega \mathbf{v}_t^k + c_1 \mathbf{r}_1 (\mathbf{pbest}_t^k - \mathbf{x}_t^k) + c_2 \mathbf{r}_2 (\mathbf{gbest}_t - \mathbf{x}_t^k), \tag{1}$$

$$\mathbf{x}_{t+1}^k = \mathbf{x}_t^k + \mathbf{v}_{t+1}^k, \tag{2}$$

where $\omega = 0.75$ is the inertia weight, c_1, c_2 are the cognitive and social coefficients, respectively, equal to 2.05, and $\mathbf{r}_1$, $\mathbf{r}_2$ are vectors of random numbers in the range [0,1]. In subsequent iterations of the algorithm, the particles explore the search space and exchange information in order to find the optimal solution to the problem. The proposals are evaluated based on the fitness function (see Sect. 2.3). The initialization of particles in the first frame of the sequence is based on the known position of the drone. In subsequent frames, the position estimated in the previous frame is used to initialize the algorithm. The structure of the particle proposed for the purpose of tracking n drones is shown in Fig. 2.

2.3 Fitness Function

The fitness function determines the degree of similarity of the solution proposed by the algorithm to the actual position of the tracked drones. For drone silhouette extraction the background/foreground detection algorithm proposed by Zivkovic and van der Heijden [18] and implemented in the OpenCV library was used. In addition, the images obtained in this way are subjected to morphological operations to remove noise and improve the extraction of the silhouettes of drones. In the next step, the hypothetical positions of drones, generated by the PSO algorithm, are projected into 2D image space, and then the rectangles (bounding boxes) approximating the size of the drones in the image from a given camera are determined. Having the bounding boxes representing the drones and the silhouettes of the real drones, their degree of overlap is calculated, and then the average overlap value for all drones in the image is established. Finally, the overlap value is averaged over all cameras. The process of calculating the fitness function for camera c can be described by the following equation:

$$f_c(\mathbf{x}) = \frac{1}{n} \sum_{d \in \mathbf{x}} g(\mathbf{I}_c(box_{c,d}))/size(box_{c,d}), \tag{3}$$

where $\mathbf{I}_c$ is the image with extracted drones for the camera c, n is the number of tracked drones, d is the drone in the particle $\mathbf{x}$ ($\mathbf{x}$ contains the positions of all drones), $box_{c,d}$ is a bounding box defining the drone d on the image of camera c, and $\mathbf{I}_c(box_{c,d})$ is part of the image representing the region of interest of $box_{c,d}$. The function $g(\mathbf{I})$ is defined as

$$g(\mathbf{I}) = \sum_{p} u(\mathbf{I}(p)), \quad u(x) = \begin{cases} 1, & \text{if } x > 0 \\ 0, & \text{otherwise} \end{cases}, \tag{4}$$

where p is the pixel position in the image $\mathbf{I}$.

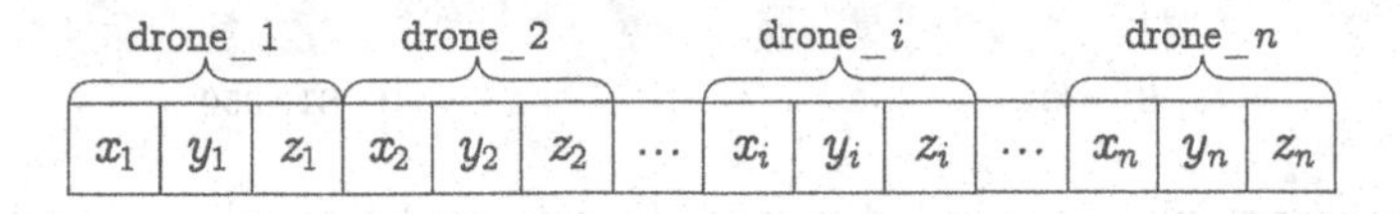

Fig. 2. The structure of a particle

3 Experiment Results

The developed method for tracking multiple drones was tested on 5 video sequences with a different number of drones (see Sect. 2.1). The quality of tracking was assessed by analyzing both qualitative visual evaluations and using ground truth data. The experiments took into account different numbers of particles and iterations of the PSO algorithm. The estimation time of drone position for a single frame ranged from 0.02 to 1 s depending on algorithm configuration, number of drones, and number of cameras. The calculations were performed on a workstation equipped with Intel(R) Core(TM) i7-11800H and 64 GB RAM.

3.1 Simulation Dataset

The example tracking results for the simulation sequences are shown in Fig. 3, 4, and 5, while the obtained errors are presented in Table 2. The mean error value and the standard deviation calculated for individual sequences, averaged over

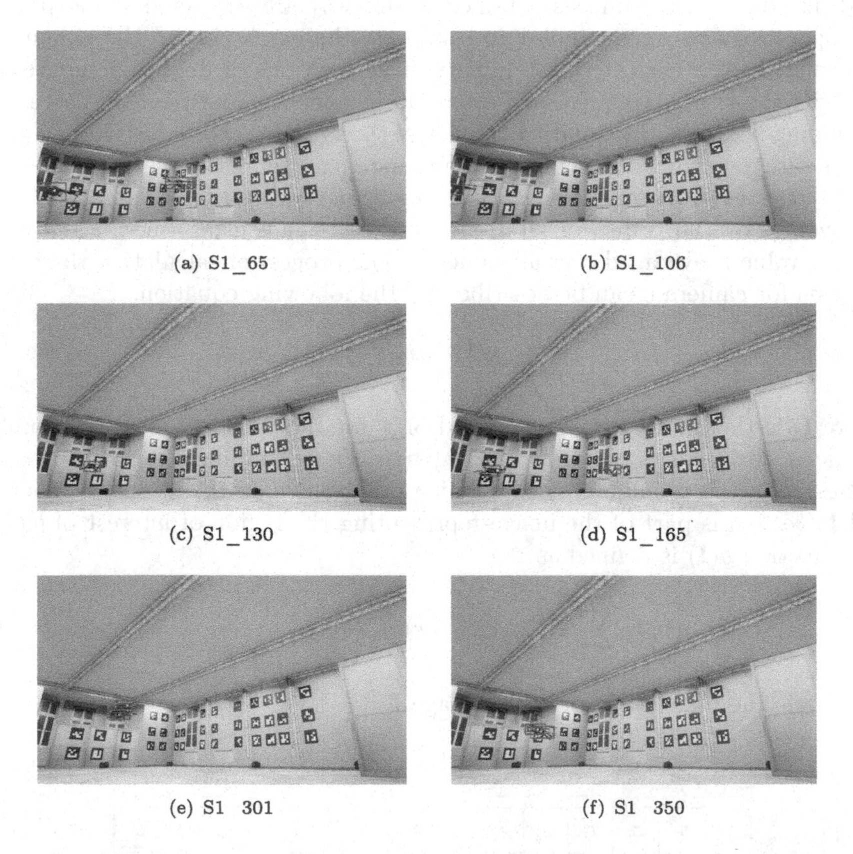

(a) S1_65 (b) S1_106

(c) S1_130 (d) S1_165

(e) S1_301 (f) S1_350

Fig. 3. Tracking results for selected frames of sequences S1, number of particles: 100, number of iterations: 70

Table 2. Tracking errors for sequences S1, S2, and S3

part.	iter.	S1		S2		S3	
		mean [m]	std [m]	mean [m]	std [m]	mean [m]	std [m]
30	30	0.2525	0.3101	0.3567	0.5006	0.4676	0.3745
	50	0.1960	0.2177	0.2207	0.2586	0.3266	0.2463
	70	0.0857	0.0337	0.0965	0.1058	0.2670	0.2118
50	30	0.1419	0.1267	0.1972	0.2804	0.4823	0.3980
	50	0.1408	0.1261	0.1095	0.1293	0.2917	0.2212
	70	0.0859	0.0345	0.1065	0.1359	0.2267	0.1542
100	30	0.1413	0.1263	0.1804	0.2428	0.2970	0.2220
	50	0.1407	0.1261	0.0890	0.0925	0.2202	0.1686
	70	0.0859	0.0343	0.0784	0.0827	0.1972	0.1138

10 runs of the tracking algorithm are depicted. Depending on the configuration of the tracking algorithm and the sequence, the average tracking error obtained varies from 0.078 m to 0.482 m. As expected, the best results are obtained with the configuration with the highest number of particles (100) and iterations (70). For this configuration, the estimation time of the drones' positions for a single frame and sequence with 10 drones was approximately 1 s.

The smallest tracking error of 0.078 m was obtained for sequence S2, in which 4 drones were tracked. The algorithm in this configuration had 100 particles and 70 iterations. Analyzing the results obtained for the sequence with two drones (S1), it can be seen that for the configuration with fewer particles and iterations, worse results are observed than for the sequence with four drones (S2). This is due to the fact that in the S1 sequence, there is a situation in which the drones are very close to each other (Fig. 3(e)) and the algorithm has switched them in some runs. Increasing the number of iterations solved this problem. It is also worth noting that for the S2 sequence, the algorithm with the configuration of 50 particles and 70 iterations achieved an insignificantly worse tracking result than the configuration with fewer particles (30 particles and 70 iterations). This is due to the randomness of the algorithm. When analyzing the detailed tracking results of individual drones, it was observed that drone_1 has a larger tracking error for the configuration with more particles. This drone flies out of the field of view of some cameras, which causes tracking failure. For the configuration of 50 particles and 70 iterations, the problem occurred in four of 10 algorithm runs, while for the configuration of 30 particles and 70 iterations, it was only in three of 10 trials.

The worst results are obtained for the S3 sequence (0.197–0.482 m), which is primarily due to the large number of tracked drones, of which there are 10 in this case. Analyzing the detailed results for the configuration of 100 particles and 70 iterations (see Tables 3), it can be seen that for most drones the results are satisfactory (average error below 0.15 m), and the large error value is caused

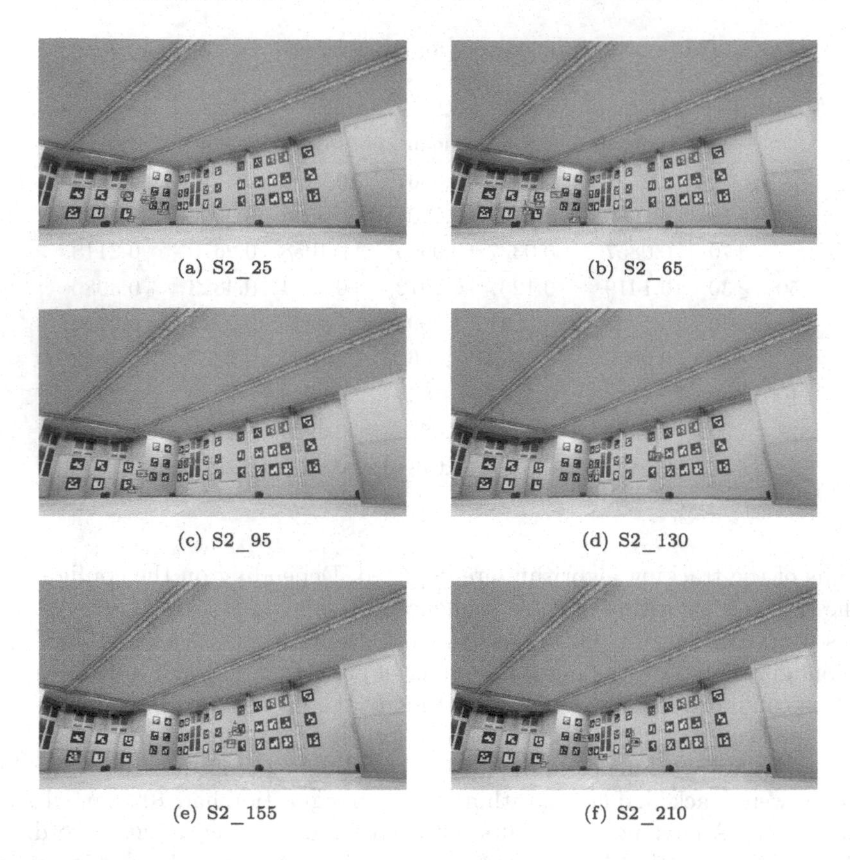

(a) S2_25 (b) S2_65 (c) S2_95 (d) S2_130 (e) S2_155 (f) S2_210

Fig. 4. Tracking results for selected frames of sequence S2, number of particles: 100, number of iterations: 70

by difficulties in the tracking of drones number 1, 2, and 9. These are drones that move on the edge of the scene, so they are not visible in some cameras, and at the same time they are far from the cameras on the opposite side of the scene, so they are smaller and contribute less to the value of the fitness function for a given camera (losing them has less impact on the value of the fitness function than in the case of a drone that is closer and therefore larger). In addition, drone_1 and drone_2 are flying close to each other (see Fig. 5), which resulted in substitutions, or the algorithm tracked one of the drones twice. Similar conclusions can be drawn by analyzing the graphs presented in Fig. 6, which shows the mean error value and standard deviation in subsequent frames of the sequence determined for 10 repetitions of the tracking algorithm. A large value of the standard deviation for some drones (drone_1, drone_2, and drone_9) indicates the occurrence of tracking errors in some of the algorithm runs. It can also be seen that for some runs the algorithm has temporary problems with tracking drone_5 around #40 and #110 frames and drone_10 near frame #200.

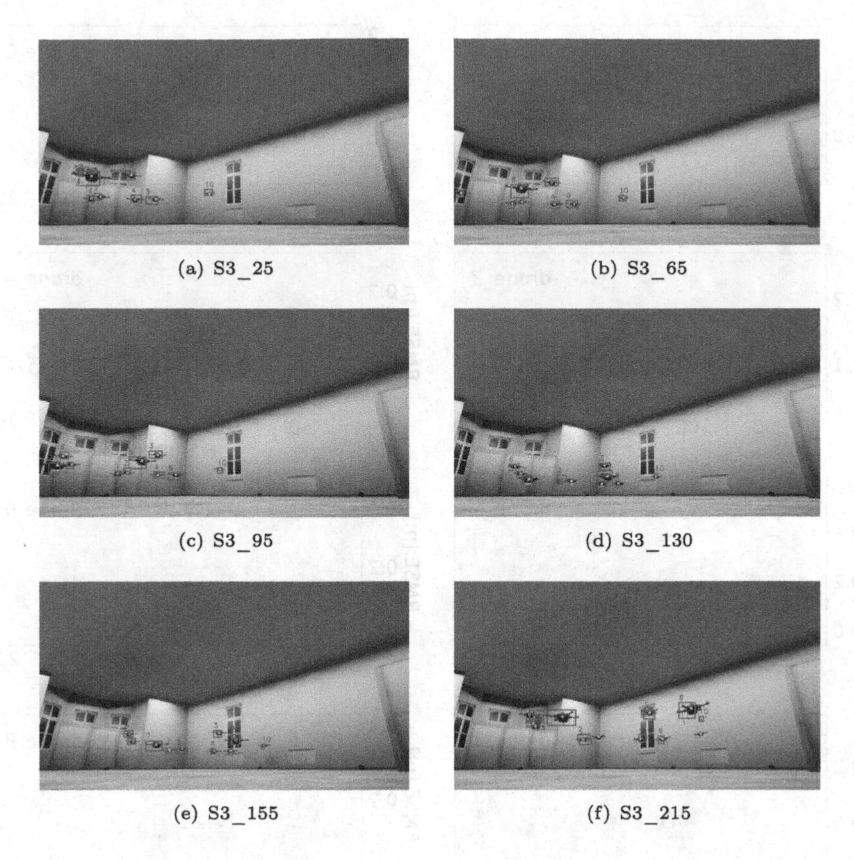

(a) S3_25 (b) S3_65 (c) S3_95 (d) S3_130 (e) S3_155 (f) S3_215

Fig. 5. Tracking results for selected frames of sequence S3, number of particles: 100, number of iterations: 70

Table 3. Detailed error statistics for sequence S3, number of particles: 100, number of iterations: 70

	mean [m]	std [m]	min [m]	25% [m]	50% [m]	75% [m]	max [m]
drone_1	0.3277	0.2116	0.0158	0.1157	0.3848	0.5039	3.7280
drone_2	0.5047	0.3560	0.0265	0.2216	0.4618	0.7104	4.3550
drone_3	0.1258	0.0516	0.0224	0.0904	0.1220	0.1544	0.3648
drone_4	0.1174	0.0507	0.0191	0.0830	0.1100	0.1470	0.3340
drone_5	0.1417	0.0891	0.0171	0.0873	0.1201	0.1680	0.9538
drone_6	0.1358	0.0647	0.0180	0.0896	0.1269	0.1722	0.5953
drone_7	0.1129	0.0471	0.0162	0.0803	0.1092	0.1402	0.4001
drone_8	0.1116	0.0528	0.0197	0.0766	0.1053	0.1353	0.6030
drone_9	0.2599	0.1276	0.0229	0.1776	0.2914	0.3400	1.2981
drone_10	0.1341	0.0866	0.0191	0.0813	0.1138	0.1566	1.2737
Mean	0.1972	0.1138	0.0197	0.1103	0.1945	0.2628	4.3550

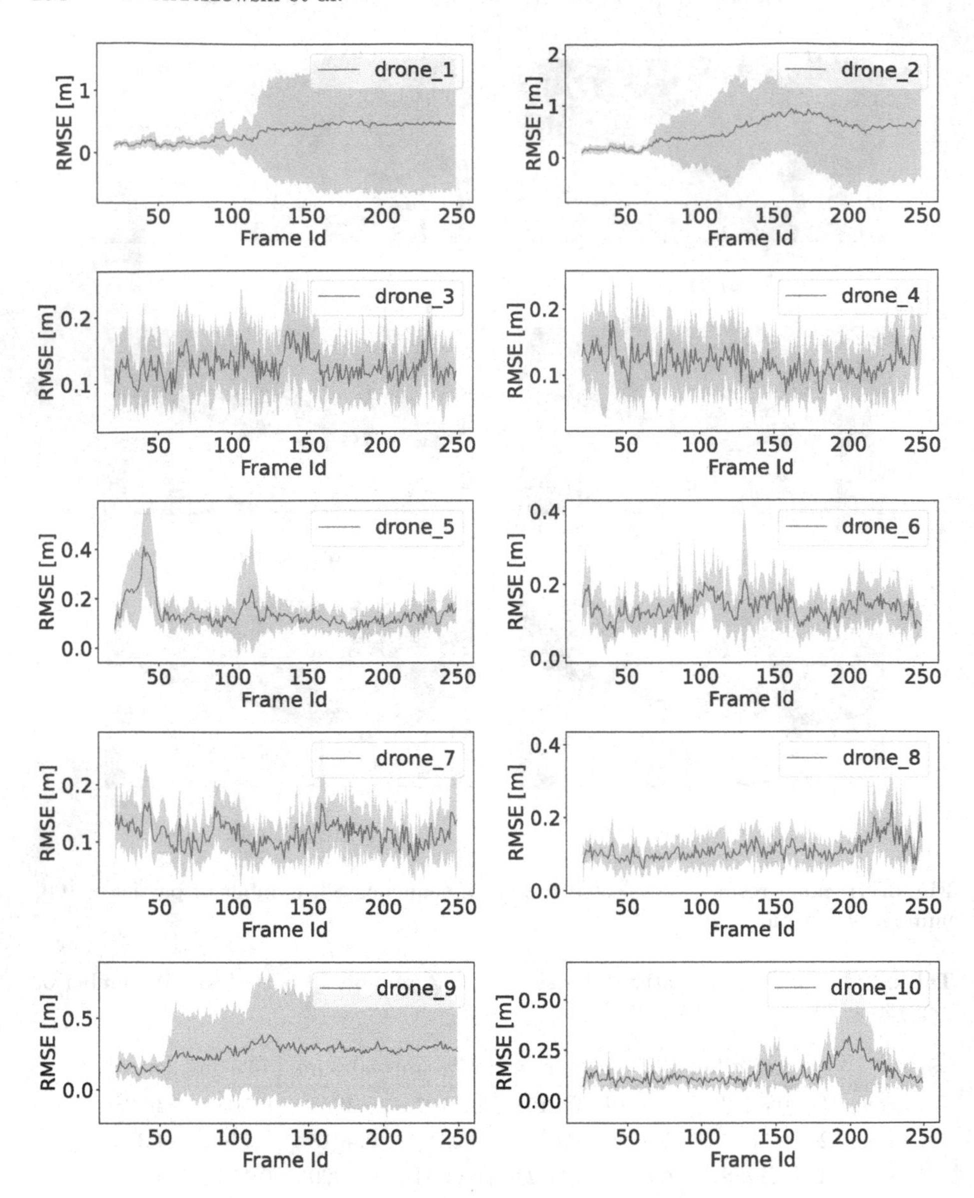

Fig. 6. Average tracking errors in consecutive frames for sequence S3, number of particles: 100, number of iterations: 70

3.2 Real Dataset

The example tracking results for the real sequences are shown in Fig. 7 and 8, while the obtained errors are presented in Table 4. In this case, we focused on the selected configuration of the algorithm (number of particles: 30, number of iterations: 20) and presented detailed results for it. For real sequences, a single drone was tracked, therefore the proposed configuration is sufficient. The number of drones has been limited to one due to space constraints in the laboratory, as well as the risk of collisions or damage to laboratory equipment in the case of a larger number of drones. These restrictions do not occur in simulation sequences. The errors obtained are similar to those obtained for the simulation sequences and range from 0.101 to 0.124m. It would seem that for a single drone, the errors should be smaller since for the simulation sequences with two and four drones errors of 0.08m were obtained. However, it should be noted, that in the case of real sequences, there are additional errors related to the inaccuracy of

Table 4. Detailed error statistics for sequence S4 and S5, number of particles: 30, number of iterations: 20

seq	mean [m]	std [m]	min [m]	25% [m]	50% [m]	75% [m]	max [m]
S4	0.1240	0.0490	0.0091	0.0915	0.1175	0.1488	0.5969
S5	0.1005	0.0362	0.0209	0.0747	0.0945	0.1205	0.4726

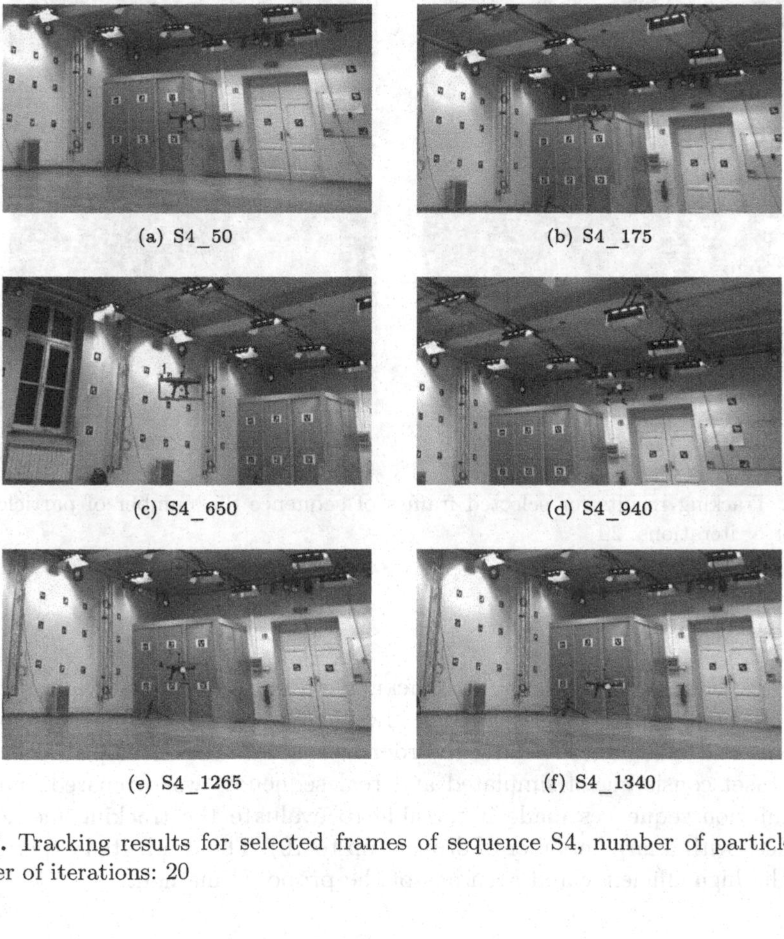

(a) S4_50

(b) S4_175

(c) S4_650

(d) S4_940

(e) S4_1265

(f) S4_1340

Fig. 7. Tracking results for selected frames of sequence S4, number of particles: 30, number of iterations: 20

the motion capture system, synchronization errors, and calibration errors, which can be observed in Fig. 7(a), 7(d), 7(e), 8(b), 8(c), and 8(e). In these cases, it may be noted that the position of the drone determined by the motion capture system (represented by the blue dot) does not coincide with the center of the drone. Another advantage related to the use of simulation sequences can be seen here.

(a) S5_50 (b) S5_522 (c) S5_778 (d) S5_1101 (e) S5_2415 (f) S5_2830

Fig. 8. Tracking results for selected frames of sequence S5, number of particles: 30, number of iterations: 20

4 Conclusions

In the paper, the method for 3D tracking of multiple drones was proposed. The presented approach uses Particle Swarm Optimization, image processing methods, and multi-camera data. In order to evaluate the developed algorithm, the dataset consisting of simulated and real sequences was prepared. The use of simulation sequences made it possible to evaluate the tracking method on sequences with a large number of drones (up to 10). The conducted experiments show the high efficiency and accuracy of the proposed method.

Despite obtaining satisfactory results, the method has some limitations. The main problem is tracking the drones at the edge of the scene where camera coverage is insufficient. This can lead to a decrease in accuracy or even loss of the drone. If the drones are close to each other, they can be switched. Another problem is related to the specificity of the background/foreground detection methods used in the objective function. If the drone stops moving, as a result of updating the background model, it will be treated as a background element over time and will disappear from the extracted images. Problems with drone extraction may also occur if it is similar in color to the background.

Future work will focus on improving the algorithm to remove or eliminate the described limitations by developing hybrid methods, that use other types of sensors than just video cameras.

References

1. Akyon, F.C., Eryuksel, O., Ozfuttu, K.A., Altinuc, S.O.: Track boosting and synthetic data aided drone detection. In: AVSS 2021–17th IEEE International Conference on Advanced Video and Signal-Based Surveillance, pp. 12–16 (2021). https://doi.org/10.1109/AVSS52988.2021.9663759
2. Eichhorn, C., Jadid, A., Plecher, D.A., Weber, S., Klinker, G., Itoh, Y.: Catching the Drone-A tangible augmented reality game in superhuman sports. In: Adjunct Proceedings of the 2020 IEEE International Symposium on Mixed and Augmented Reality, ISMAR-Adjunct 2020, pp. 24–29 (2020). https://doi.org/10.1109/ISMAR-Adjunct51615.2020.00022
3. Ganti, S.R., Kim, Y.: Implementation of detection and tracking mechanism for small UAS. In: 2016 International Conference on Unmanned Aircraft Systems (ICUAS). IEEE (2016)
4. Jurn, Y.N., Mahmood, S.A., Aldhaibani, J.A.: Anti-drone system based different technologies: architecture, threats and challenges. In: Proceedings - 2021 11th IEEE International Conference on Control System, Computing and Engineering, ICCSCE 2021, pp. 114–119, August 2021. https://doi.org/10.1109/ICCSCE52189.2021.9530992
5. Kennedy, J., Eberhart, R.: Particle swarm optimization. In: Proceedings of ICNN'95 - International Conference on Neural Networks, vol. 4, pp. 1942–1948 (1995). https://doi.org/10.1109/ICNN.1995.488968
6. Krzeszowski, T., Przednowek, K., Wiktorowicz, K., Iskra, J.: Estimation of hurdle clearance parameters using a monocular human motion tracking method. Comput. Methods Biomech. Biomed. Eng. **19**(12), 1319–1329 (2016). https://doi.org/10.1080/10255842.2016.1139092
7. Kwolek, B., Krzeszowski, T., Gagalowicz, A., Wojciechowski, K., Josinski, H.: Real-time multi-view human motion tracking using particle swarm optimization with resampling. In: Perales, F.J., Fisher, R.B., Moeslund, T.B. (eds.) AMDO 2012. LNCS, vol. 7378, pp. 92–101. Springer, Heidelberg (2012). https://doi.org/10.1007/978-3-642-31567-1_9
8. Martínez, C., Campoy, P., Mondragón, I., Olivares-Méndez, M.A.: Trinocular ground system to control UAVs. In: 2009 IEEE/RSJ International Conference on Intelligent Robots and Systems, IROS 2009, pp. 3361–3367, December 2009. https://doi.org/10.1109/IROS.2009.5354489

9. Saini, S., Zakaria, N., Rambli, D.R.A., Sulaiman, S.: Markerless human motion tracking using hierarchical multi-swarm cooperative particle swarm optimization. PLOS ONE **10**, 1–22 (2015). https://doi.org/10.1371/journal.pone.0127833
10. Schilling, F., Schiano, F., Floreano, D.: Vision-based drone flocking in outdoor environments. IEEE Robot. Autom. Lett. **6**(2), 2954–2961 (2021). https://doi.org/10.1109/LRA.2021.3062298
11. Shah, S., Dey, D., Lovett, C., Kapoor, A.: AirSim: high-fidelity visual and physical simulation for autonomous vehicles. In: Field and Service Robotics (2017). https://arxiv.org/abs/1705.05065
12. Sie, N.J., et al.: Vision-based drones tracking using correlation filters and linear integrated multiple model. In: 2021 18th International Conference on Electrical Engineering/Electronics, Computer, Telecommunications and Information Technology (ECTI-CON), pp. 1085–1090 (2021). https://doi.org/10.1109/ECTI-CON51831.2021.9454735
13. Son, S., Kwon, J., Kim, H.Y., Choi, H.: Tiny drone tracking framework using multiple trackers and Kalman-based predictor. J. Web Eng. **20**(8), 2391–2412 (2021). https://doi.org/10.13052/jwe1540-9589.2088
14. Srigrarom, S., Hoe Chew, K.: Hybrid motion-based object detection for detecting and tracking of small and fast moving drones. In: 2020 International Conference on Unmanned Aircraft Systems, ICUAS 2020, pp. 615–621 (2020). https://doi.org/10.1109/ICUAS48674.2020.9213912
15. Srigrarom, S., Sie, N.J.L., Cheng, H., Chew, K.H., Lee, M., Ratsamee, P.: Multi-camera multi-drone detection, tracking and localization with trajectory-based re-identification. In: 2021 2nd International Symposium on Instrumentation, Control, Artificial Intelligence, and Robotics, ICA-SYMP 2021 (2021). https://doi.org/10.1109/ICA-SYMP50206.2021.9358454
16. Tran, V.P., Santoso, F., Garratt, M.A.: Adaptive trajectory tracking for quadrotor systems in unknown wind environments using particle swarm optimization-based strictly negative imaginary controllers. IEEE Trans. Aerosp. Electron. Syst. **57**(3), 1742–1752 (2021). https://doi.org/10.1109/TAES.2020.3048778
17. Unlu, E., Zenou, E., Riviere, N., Dupouy, P.-E.: Deep learning-based strategies for the detection and tracking of drones using several cameras. IPSJ Trans. Comput. Vis. Appl. **11**(1), 1–13 (2019). https://doi.org/10.1186/s41074-019-0059-x
18. Zivkovic, Z., van der Heijden, F.: Efficient adaptive density estimation per image pixel for the task of background subtraction. Pattern Recogn. Lett. **27**(7), 773–780 (2006). https://doi.org/10.1016/j.patrec.2005.11.005

Sun Magnetograms Retrieval from Vast Collections Through Small Hash Codes

Rafał Grycuk(✉) and Rafał Scherer

Czestochowa University of Technology, al. Armii Krajowej 36, Czestochowa, Poland
{rafal.grycuk,rafal.scherer}@pcz.pl

Abstract. We propose a method for retrieving solar magnetograms based on their content. We leverage data collected by the SDO Helioseismic and Magnetic Imager using the SunPy and PyTorch libraries to create a vector-based mathematical representation of the Sun's magnetic field regions. This approach enables us to compare short vectors instead of comparing full-disk images. To reduce retrieval time, we use a fully-connected autoencoder to compress the 144-element descriptor to a 36-element semantic hash. Our experimental results demonstrate the efficiency of our approach, which achieved the highest precision value compared to other state-of-the-art methods. Our proposed method is not only applicable for solar image retrieval but also for classification tasks.

Keywords: Fast image hash · Solar activity analysis · Solar image description · CBIR of solar images · Magnetogram Image Descriptor · Magnetogram Image Hash

1 Introduction

Analysing the Sun is crucial for understanding many aspects of our solar system and the universe. By analysing the Sun's internal structure and surface features, we can better understand how the Sun generates and releases energy through nuclear fusion. This information is essential for predicting and understanding the Sun's behaviour, such as the occurrence of solar flares, coronal mass ejections, and other space weather phenomena that can affect Earth and the space environment.

Solar activity can cause disturbances in Earth's magnetic field, leading to geomagnetic storms that can affect power grids, satellites, and communication systems. By monitoring the Sun's activity and predicting space weather, we can take measures to mitigate these effects. By studying the Sun's properties, researchers can better understand the range of conditions that can support life in other star systems.

The project financed under the program of the Polish Minister of Science and Higher Education under the name "Regional Initiative of Excellence" in the years 2019–2023 project number 020/RID/2018/19, the amount of financing 12,000,000.00 PLN.

J. Mikyška et al. (Eds.): ICCS 2023, LNCS 14076, pp. 259–273, 2023.
https://doi.org/10.1007/978-3-031-36027-5_19

NASA's Living With a Star (LWS) Program has a mission to explore how solar activity affects Earth, and the Solar Dynamics Observatory (SDO) is a crucial component of this endeavour. The SDO provides extensive data on the solar atmosphere at different wavelengths and with high spatial and temporal resolution, allowing for thorough analysis of the Sun's impact on our planet. One of the key instruments on the SDO is the Helioseismic and Magnetic Imager (HMI), which specializes in examining oscillations and the magnetic field of the solar surface. By generating dopplergrams, continuum filtergrams, and magnetograms (which are maps of the photospheric magnetic field), the HMI enables researchers to obtain valuable insights into the Sun's behaviour.

The magnetograms created by the HMI are particularly important to researchers. However, with the sheer volume of data generated by the SDO spacecraft, it is impossible to manually annotate and search through the entire collection. Although some image retrieval methods exist, they are primarily designed for real-life images and do not suit the needs of solar research. Instead, researchers have turned to semantic hashing to reduce dimensionality by creating short codes that preserve similarity and reflect the content of the input data. This approach was initially introduced by Salakhutdinov and Hinton [17] and has since been used to describe any short codes that reflect content-similarity.

The goal of semantic hashing is to produce compact vectors that accurately reflect the semantic content of objects. This approach enables the retrieval of similar objects through the search for similar hashes, which is a faster and more memory-efficient process than working directly with the objects themselves. Previous works has used multilayer neural networks to generate hashes. Recently, learned semantic hashes have become popular for image retrieval, as demonstrated by research such as that by [18].

Initially, we found out that generating hashes from full-disk solar images would be impractical due to the large size of the image collections in terms of resolution and quantity. As a result, we developed the hand-crafted intermediate descriptors discussed earlier.

A full-disk content-based image retrieval system is described in [1]. The authors checked eighteen image similarity measures with various image features resulting in one hundred and eighty combinations. The experiments shed light on what metrics are suitable for comparing solar images to retrieve or classify various phenomena.

In [3], a general-purpose retrieval engine called Lucene is utilized to retrieve solar images. Each image is treated as a document with 64 elements (representing the rows of each image), and each image-document is unique. Wild-card characters are used in query strings to search for similar solar events. While the Lucene engine is compared to distance-based image retrieval methods in [4], no clear winner emerged. Each tested method has its advantages and disadvantages in terms of accuracy, speed, and applicability. There is a significant trade-off between accuracy and speed, with retrieval times of several minutes required for accurate results.

In [10], a sparse model representation of solar images is presented. The method utilized the sparse representation technique proposed in [14] and showed superior performance in both accuracy and speed compared to previous solar image retrieval approaches. The authors of [12] focused on tracking multiple solar events across images with 6-minute cadence by selecting specific solar image parameters. Additionally, sparse codes for AIA images were used in [11], where ten texture-based image parameters were computed for regions determined by a 64×64 grid for nine wavelengths. A dictionary of k elements was learned for each wavelength, and a sparse representation was then computed using the learned dictionary. In [13], a new method for image retrieval using fuzzy sets and boosting is proposed. To overcome the curse of dimensionality that affects solar data, the authors use the Minkowski norm and carefully choose the appropriate value for the p parameter. They also employ a 256-dimensional descriptor, which has been shown to be both efficient and accurate compared to previous approaches.

In this paper, we propose a method for automating solar image retrieval and enabling their fast classification using a solar image hash generated from one-dimensional hand-crafted features by a fully-connected autoencoder. The resulting hash is very compact, consisting of only 36 real-valued elements, but experiments have shown that this is sufficient to accurately describe the images. In the dataset used, the images are annotated only by their timestamp, making it difficult to extract any other meaning or interpret the trained system [15]. The timestamp is treated as a measure of similarity, and after training, the algorithm allows retrieval of images by their visual similarity, regardless of the timestamp proximity. The paper is organized into three sections: the proposed method for generating learned solar hashes is introduced in Sect. 2; experiments on the SDO solar image collection are described in Sect. 3; and finally, the paper concludes with Sect. 4.

2 Solar Magnetic Intensity Hash for Solar Image Retrieval

The Solar Dynamics Observatory's (SDO) instruments are not only the Atmospheric Imaging Assembly (AIA) but also Helioseismic and Magnetic Imager (HMI), which allows for creation of magnetograms of the Sun. In active regions, the magnetic field can be significantly stronger than the average magnetic field of the Sun, with some regions having magnetic fields over 1,000 times stronger. By providing information about the magnetic fields of the entire solar disk, magnetograms are useful in many areas of solar analysis. Taking into consideration the noise present in regular active region images due to bright pixels that represent flares extending beyond the solar disk, the utilization of magnetograms in solar image description or solar image hashing seems like a viable solution to enhance the precision of the hash. By providing information about the magnetic fields of the entire solar disk, magnetograms can effectively reduce the unwanted noise in solar images. As such, using magnetograms to analyze the Sun's activity (as shown in Fig. 1) appears to be a more justifiable approach.

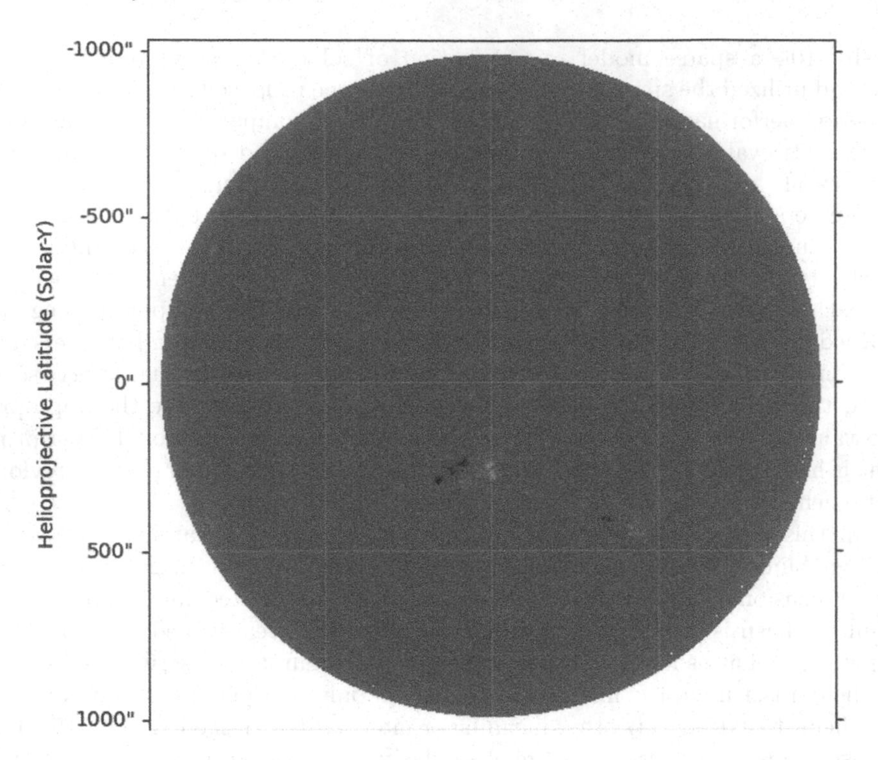

Fig. 1. Example magnetogram image. As can be seen, the image is difficult to analyze without any pre-processing.

Solar images are highly similar to each other, and using general-purpose descriptors for retrieval may not yield accurate results. To address this issue, we propose a novel solar magnetogram hash for solar image retrieval in large datasets. Our experimental setup includes a GeForce RTX 2080 Ti 11 GB GDDR6 graphics card, which allows us to utilize 11 GB of memory. Initially, we attempted to design a full-disc autoencoder, but the higher mini-batch values caused out-of-memory exceptions, and the learning time was several days compared to the minutes required in our proposed approach. Therefore, we developed a preprocessing stage to calculate the solar magnetogram descriptor, which was then used to reduce the hand-crafted vectors to x-element real-valued hashes using the autoencoder (see Sect. 2.3). This approach preserves the significant information about active regions while reducing the dimensionality of the data for efficient retrieval. The presented algorithm is composed of four main stages: active region detection, calculating solar image hand-crafted descriptors, encoding to hash, and retrieval.

2.1 Magnetic Region Detection

In this section, we describe the fundamental steps involved in the hashing process. The first step entails adjusting the magnetogram image to more clearly annotate the magnetic regions, as illustrated in Fig. 2. We refer to this process as magnetic region detection, which we conduct by utilizing magnetogram images obtained via the SunPy library [19,20]. This step enables us to determine the intensity of the magnetic field. Figure 1 and Fig. 2 both depict the increase in magnetic field strength around active regions. To define the magnetic field strength, we utilize color intensities, as illustrated in Fig. 3. Throughout the solar cycle, the magnetic field undergoes twisting, tangling, and reorganization. It is important to note that magnetic regions (MR) have a strong correlation with Coronal Mass Ejections (CMEs) and solar flares, which makes analyzing them critical for understanding the impact on life on Earth. As illustrated in Fig. 3, magnetic region detection (MRD) enables us to determine the north (red) or south (blue) polarities, between which CMEs are most likely to originate. Additionally, tracking and analyzing MRs is valuable for predicting solar flares. By

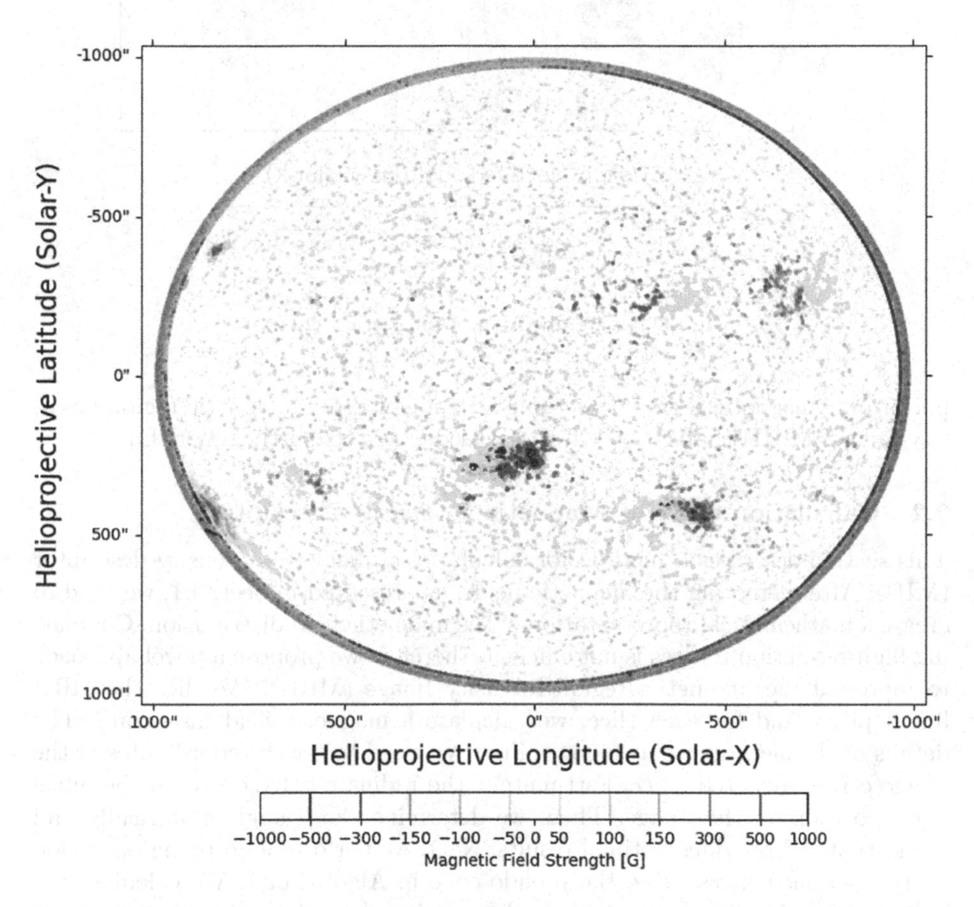

Fig. 2. The magnetic region detection and annotation process. The magnetic regions can be clearly visible. We can observe the polarities (red and blue) and their intensities. (Color figure online)

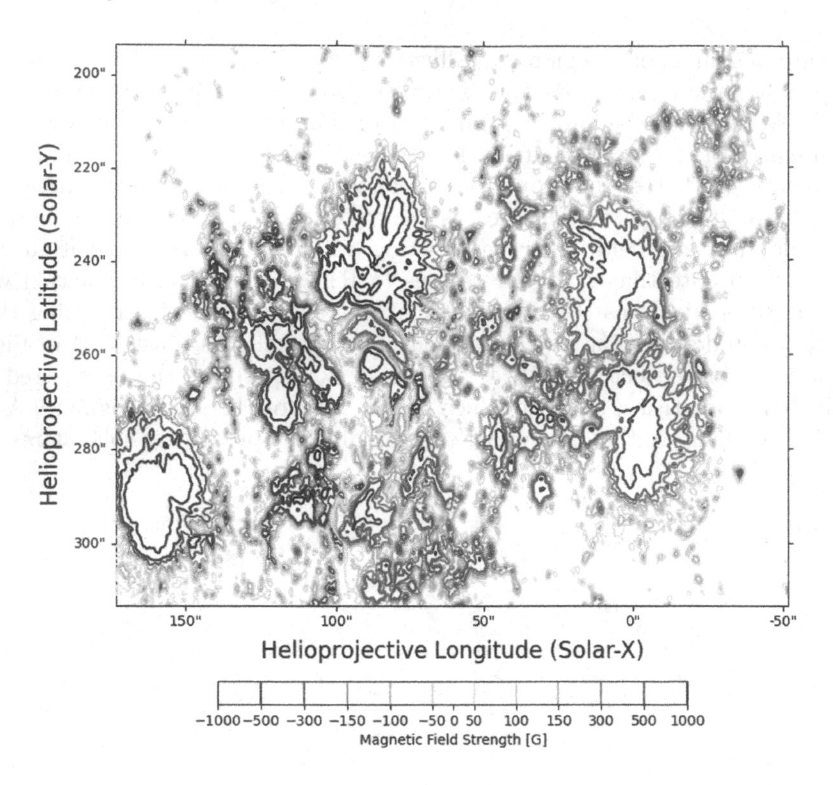

Fig. 3. A magnification of magnetic regions.

performing magnetic region detection, we can generate a magnetic region intensity image (MRII) that is used in the subsequent steps of the algorithm.

2.2 Calculation of Solar Magnetic Intensity Descriptor

This section describes a method for calculating a magnetic intensity descriptor (MID). After detecting the magnetic fields, as presented in Sect. 2.1, we need to create a mathematical representation of the magnetic field distribution. Comparing high-resolution images is not efficient, therefore we propose a novel approach to represent the Magnetic Region Intensity Image (MRII). We slice the MRII like a pizza, and for each slice, we calculate a magnetic field histogram. The details of the method are as follows. First, we need to set the coordinates of the image center, denoted as *cc*. Fortunately, the radius r is fixed due to the Sun's fixed position on the image. Then, we determine the θ angle empirically and found that 30° provides optimal results. Next, we perform a cropping operation on the obtained slices using the pseudo-code in Algorithm 1. We calculate the arc points of the slice (sector) *aps* and *ape* using the following formulas:

$$ape_x = cc_x - 1.5 * r * \sin\theta, \tag{1}$$

$$ape_y = cc_y - 1.5 * r * \cos\theta, \tag{2}$$

The trigonometric functions sin and cos are used in these formulas to calculate the row and column coordinates of two points on the arc. A factor of 1.5 is applied to extend the arc beyond the circle's radius slightly. After dividing the MRII into slices as described in the previous section, the process of cropping is repeated for each subsequent circle segment (slice) until the entire image is covered. This results in a list of MRII slices, each containing the magnetic field intensities for that particular segment. The next step is to create a magnetic field histogram (MFH) for each slice, which allows us to represent the distribution of magnetic fields for each segment of the MRII. The histogram is created with the same scale as the magnetic field intensities, ranging from $[-1000; 1000]$ accordingly to the magnetic field range presented in Fig. 3. Finally, all histograms are combined into a single vector called the Magnetic Intensity Descriptor (MID). This vector represents the overall magnetic field distribution for the entire MRII image. The entire process, including cropping and histogram creation, is illustrated in Fig. 4 and described in Algorithm 2.

INPUT: $MRII$ - magnetic region intensity image
r - radius
cc - center coordinates of MRII
θ - angle of the slice
Local Variables:
MC - mask circle matrix
$MMRII$ - mask MRII matrix
ape - coordinates of starting point on the arc
OUTPUT: $CMRII$ - cropped slice of MRII
$MS := CreateBooleanCircleMatrix(cc, r)$
$MMRII := CreatePolygonMatrix([cc_x, aps_x, ape_x, cc_x],$
$[cc_y, aps_y, ape_y, cc_y])$
$CMRII := CombineMasks(MS, MMRII)$

Algorithm 1: Algorithm for cropping the MRII slice.

The proposed method enables the generation of a hand-crafted hash for a magnetogram input image, called the magnetic intensity descriptor (MID). By setting θ to 30 deg, the resulting MID vector consists of 144 integer-valued elements, which is significantly smaller than the full-disc image.

2.3 Hash Generation

This section outlines the hash generation process, which takes as input a Solar Magnetic Intensity Descriptor (MID) that is later utilized to generate the corresponding hash. The goal of this step is to obtain a representative hash that describes the solar image and, more specifically, the magnetic regions of the Sun at a given timestamp. This step is crucial because it enables the

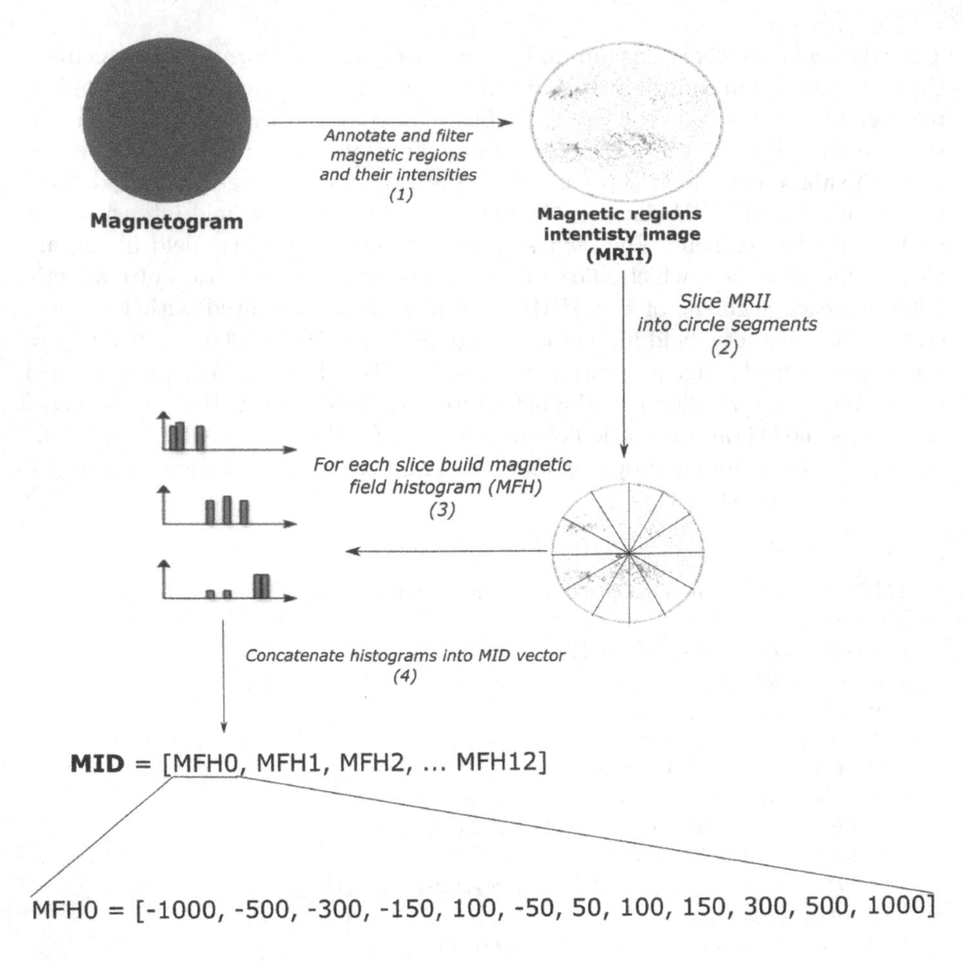

Fig. 4. Steps for calculating the magnetic intensity descriptor (MID).

reduction of data in the retrieval stage (see Sect. 2.4). To perform this operation, we utilized a fully-connected autoencoder (AE) to encode the previously acquired MID. Autoencoders are utilized in various machine learning tasks, such as image compression, dimensionality reduction, feature extraction, and image reconstruction [5,7,16]. Since autoencoders use unsupervised learning, they are ideal for generating semantic hashes. We present the autoencoder model architecture in Table 1. The AE model should be analysed from top to bottom. As shown, the model is relatively simple but enables the reduction of hash length without a significant loss of information about magnetic regions of the magnetogram. We would like to emphasize that only the latent space (encoded) part of the trained AE is used for hash generation, while the decoding part of AE is solely used for training purposes. Through a series of experiments, we determined that 40 epochs are sufficient to achieve a satisfactory level of generalization without the overfitting phenomenon.

INPUT: MI - magnetogram image
Local Variables:
$MRII$ - magnetic region intensity image
$SliceList$ - list of slices
$HistogramList$ - list of histograms
OUTPUT: MID - magnetic Intensity descriptor vector
$MRII := MagneticRegionDetection(MI)$
$SliceList := CroppingTheSlices(MRII)$
foreach $MRIISlice \in SliceList$ **do**
| $MFH := BuildSliceIntanceHist(MRIISlice)$
| $HistogramList.Add(MFH)$
end
$MID = ConcatHist(HistogramList)$

Algorithm 2: Algorithm for calculating a magnetic intensity descriptor.

Table 1. Tabular representation of the fully-connected autoencoder model.

Layer (type)	Output	Filters (in, out)	Params
$Input(InputLayer)$	[1, 144]		0
$Linear\_1(Linear)$	[1, 72]	144, 72	10440
$ReLU\_1$	[1, 72]	0	
$Linear\_2(Linear)$	[1, 72]	72, 36	2628
$ReLU\_2$	[1, 36]	0	
$Encoded(latent - space)$	[1, 36]		
$Linear\_4(Linear)$	[1, 36]	36, 72	2664
$ReLU\_4$	[1, 72]	0	
$Linear\_5(Linear)$	[1, 144]	72, 144	10,512
$ReLU\_5$	[1, 144]	0	
$Decoded(Tanh)$	[1, 144]		

Table 1 illustrates the use of a convolutional autoencoder for generating hashes, with the top layer serving as input. A one-dimensional autoencoder was employed because magnetic intensity descriptors are one-dimensional vectors, which reduces computational complexity. This process effectively shortens the hash length while retaining significant information about the active regions of the solar image. The mean squared error function was utilized as the loss function, and we discovered that training the model for 40 epochs was adequate for achieving the necessary level of generalization and preventing network over-fitting. Following the training process, each image descriptor was fed into the latent space (encoded) layers of the autoencoder, yielding a 36-element hash (known as the Solar Magnetic Intensity Hash). This hash can be used for content-based retrieval applications involving solar images. Moreover, the chosen autoencoder architecture was specifically chosen to achieve optimal generalization.

2.4 Retrieval

In the final phase of the proposed method, we employ the previously generated hashes for solar image retrieval. Following the preceding steps, we assume that each solar image in our database has an assigned hash. The retrieval process entails executing an image query by comparing the distances between the hash of the query image and the hashes created for all images stored in the dataset. To perform this retrieval, we must have a database of solar images that have undergone hash generation. In the subsequent step, we compute the distance (d) between the query image hash and every hash in the database. The cosine distance measure is utilized for this purpose (see [9] for additional information).

$$\cos(QH_j, IH_j) = \sum_{j=0}^{n} \frac{(QH_j \bullet IH_j)}{\|QH_j\| \, \|IH_j\|}, \tag{3}$$

where $\bullet$ is dot product, QH_j is the query image hash, and IH_j a consecutive image hash. Upon computing the cosine distance, the images in the database are sorted in ascending order by their distance to the query (query hash). The final step of the presented technique enables the retrieval of n images closest to the query, which are then returned to the user. To execute the query, the user must provide the value of the parameter n. Algorithm 3 illustrates the complete process as pseudo-code. An alternative method involves retrieving images based on a threshold. To use this approach, a threshold parameter must be supplied instead of n, and images are retrieved if their cosine distance to the query is below the threshold. The proposed method can also accommodate this method; however, the first method is preferable because it is better suited for system users.

INPUT: $ImageHashes$, $QueryImage$, n
OUTPUT: $RetrievedImages$
foreach $ImageHash \in ImageHashes$ **do**
  $QueryImageHash = CalculateHash(QueryImage)$
  $D[i] = Cos(QueryImageHash, ImageHash)$
end
$SortedDistances = SortAscending(D)$
$RetrievedImages = TakeFirst(n)$

Algorithm 3: Image retrieval steps.

3 Experimental Results

This section presents simulation results and a solution for evaluating unlabelled images. Since there was a lack of labelled data, unsupervised learning was employed for encoding descriptors. Consequently, evaluating the proposed method against state-of-the-art approaches was challenging. To overcome this problem, we leveraged the Sun's rotation movement to identify a set of similar images (SI). We hypothesized that consecutive images captured within a small time window would exhibit similar active regions, albeit with slight shifts. The provided solar images at 6-minute intervals, which we could assume were similar due to the nature of the Sun's movement. The only requirement was to adjust the time window difference. Through experimentation, we determined that images captured within a 48-hour window could be regarded as similar. Let us take under consideration an image taken at 2015-10-15, 00:00:00. Based on the above assumptions, we can assume that every image in 24 h before and in 24 h after is similar. Only for evaluation purposes, images are identified by the timestamps. The process of determining similar images is presented in Table 2. We conducted a series of experiments to determine image similarity using the proposed method. A single experiment consisted of the following steps:

Table 2. Defining image similarity. Based on experiments, we determined that images within a 48-hour window can be treated as similar. This allows to evaluate the method.

Timestamp	SI (similar image)/NSI (not similar image)
2015-10-13, 23:54:00	NSI
2015-10-14, 00:00:00	SI
2015-10-14, 00:06:00	SI
2015-10-14, 00:12:00	SI
2015-10-14, 00:18:00	SI
2015-10-14, 00:24:00	SI
2015-10-14, 00:30:00	SI
....	SI
2015-10-15, 00:00:00	QI (query image)
....	SI
2015-10-15, 23:24:00	SI
2015-10-15, 23:30:00	SI
2015-10-15, 23:36:00	SI
2015-10-15, 23:42:00	SI
2015-10-15, 23:48:00	SI
2015-10-15, 23:54:00	SI
2015-10-16, 00:00:00	NSI

1. Executing an image query to retrieve images.
2. Comparing the timestamp of each retrieved image with the query image timestamp.

3. If the timestamp fell within a 48-h window, the image was deemed similar to the query.

After defining similar images (SI), we can define performance measures *precision* and *recall* [6,21] based on following sets:

- SI - set of similar images,
- RI - set of retrieved images for query,
- $PRI(TP)$ - set of positive retrieved images (true positive),
- $FPRI(FP)$ - false positive retrieved images (false positive),
- $PNRI(FN)$ - positive, not retrieved images,
- $FNRI(TN)$ - false, not retrieved images (TN).

Afterwards, we can define *precision* and *recall* for CBIR systems

$$precision = \frac{|PRI|}{|PRI + FPRI|}, \tag{4}$$

$$recall = \frac{|PRI|}{|PRI + PNRI|}. \tag{5}$$

$$F_1 = 2 * \frac{precision * recall}{precision + recall}. \tag{6}$$

We present the results of our experiments in Table 3, which demonstrate the effectiveness of our method. The experimental outcomes presented in Table 3 exhibit promise, as denoted by the mean F_1 score and the high precision values. Our approach yielded an average precision of 0.92581, surpassing the results of Banda [4] (0.848) and Angryk [2] (0.850). Additionally, we outperformed the previous works by Grycuk [8]. Our approach achieves a high value of the precision measure, indicating that most of the images close to the query are correctly retrieved. However, as images move farther from the query, more positive but not retrieved images (PNRI) are retrieved. This phenomenon is due to the Sun's rotation, which results in missing active regions between consecutive images. In the 48-h cadence, significant active regions may change their position, leading to a significant impact on the hash and an increased distance from the query. We implemented the simulation environment in Python using PyTorch and ran it on hardware consisting of an Intel Core i9-9900k 3.6 GHz processor, 32 GB of RAM, and a GeForce RTX 2080 Ti 11 GB graphics card, all running on Windows Server 2016. Hash creation for 525,600 images took approximately 35 min, while the encoding stage took approximately 3 h. On average, the retrieval time was approximately 350 ms.

Table 3. Experiment results for the proposed algorithm. Due to lack of space, we present only a part of all queries.

Timestamp	RI	SI	PRI(TP)	FPRI(FP)	PNRI(FN)	Precision	Recall	F_1
2018-01-01 00:00:00	238	241	200	38	41	0.84	0.83	0.83
2018-01-04 07:00:00	448	481	401	47	80	0.90	0.83	0.86
2018-01-07 16:00:00	433	481	383	50	98	0.88	0.80	0.84
2018-01-13 07:00:00	403	481	394	9	87	0.98	0.82	0.89
2018-01-20 22:06:00	383	481	377	6	104	0.98	0.78	0.87
2018-01-24 06:12:00	431	481	384	47	97	0.89	0.8	0.84
2018-01-31 11:12:00	436	481	400	36	81	0.92	0.83	0.87
2018-02-05 14:18:00	428	481	391	37	90	0.91	0.81	0.86
2018-02-11 21:24:00	432	481	398	34	83	0.92	0.83	0.87
2018-02-18 10:24:00	417	481	381	36	100	0.91	0.79	0.85
2018-02-21 16:30:00	428	481	395	33	86	0.92	0.82	0.87
2018-02-27 12:36:00	392	481	386	6	95	0.98	0.80	0.88
2018-03-07 19:36:00	434	481	391	43	90	0.90	0.81	0.85
2018-03-08 20:42:00	440	481	394	46	87	0.90	0.82	0.86
2018-03-13 09:42:00	398	481	384	14	97	0.96	0.80	0.87
2018-03-19 11:42:00	448	481	405	43	76	0.90	0.84	0.87
2018-03-24 20:42:00	438	481	401	37	80	0.92	0.83	0.87
2018-03-31 04:42:00	407	481	390	17	91	0.96	0.81	0.88
2018-04-01 14:48:00	392	481	386	6	95	0.98	0.80	0.88
2018-04-09 17:48:00	434	481	398	36	83	0.92	0.83	0.87
2018-04-13 18:48:00	439	481	390	49	91	0.89	0.81	0.85
2018-04-16 20:54:00	435	481	399	36	82	0.92	0.83	0.87
2018-04-19 22:00:00	410	481	394	16	87	0.96	0.82	0.88
2018-04-28 02:00:00	422	481	382	40	99	0.91	0.79	0.85
2018-05-04 07:00:00	432	481	394	38	87	0.91	0.82	0.86
2018-05-09 04:00:00	421	481	395	26	86	0.94	0.82	0.88
2018-05-14 00:06:00	428	481	404	24	77	0.94	0.84	0.89
2018-05-18 16:12:00	452	481	405	47	76	0.90	0.84	0.87
2018-05-22 23:12:00	436	481	398	38	83	0.91	0.83	0.87
2018-05-25 14:18:00	407	481	388	19	93	0.95	0.81	0.87
2018-05-26 21:18:00	410	481	393	17	88	0.96	0.82	0.88
2018-06-01 16:24:00	400	481	391	9	90	0.98	0.81	0.89
2018-06-03 08:24:00	425	481	379	46	102	0.89	0.79	0.84
2018-06-11 10:30:00	425	481	379	46	102	0.89	0.79	0.84
2018-06-13 04:30:00	436	481	396	40	85	0.91	0.82	0.86
2018-06-15 10:30:00	425	481	380	45	101	0.89	0.79	0.84
2018-06-18 18:30:00	426	481	395	31	86	0.93	0.82	0.87
2018-06-23 01:30:00	434	481	400	34	81	0.92	0.83	0.87
2018-06-27 01:36:00	441	481	395	46	86	0.90	0.82	0.86
2018-07-04 09:36:00	437	481	400	37	81	0.92	0.83	0.87
2018-07-08 12:42:00	419	481	406	13	75	0.97	0.84	0.9
Avg.						**0.92581**	**0.81674**	**0.86720**

4 Conclusions

We introduced a new method for retrieving solar magnetograms that employs a semantic hash. Our technique uses data from the SDO Helioseismic and Magnetic Imager and is implemented using SunPy and PyTorch libraries. We represent the magnetic regions of the Sun as 144-dimensional vectors, allowing for faster comparison and retrieval. To further expedite the process, we employ a fully-connected autoencoder to convert the 144-element magnetic intensity descriptor (MID) to a 36-element solar magnetic intensity hash. Our experiments, as detailed in Table 3, demonstrate the superior precision of our approach compared to other state-of-the-art methods. Additionally, this method can also be used for classification tasks. By utilizing magnetograms instead of images from the Atmospheric Imaging Assembly, which captures the solar atmosphere in one or multiple wavelengths, the proposed method is more robust against noise.

References

1. Banda, J., Angryk, R., Martens, P.: Steps toward a large-scale solar image data analysis to differentiate solar phenomena. Sol. Phys. **288**(1), 435–462 (2013)
2. Banda, J.M., Angryk, R.A.: Large-scale region-based multimedia retrieval for solar images. In: Rutkowski, L., Korytkowski, M., Scherer, R., Tadeusiewicz, R., Zadeh, L.A., Zurada, J.M. (eds.) ICAISC 2014. LNCS (LNAI), vol. 8467, pp. 649–661. Springer, Cham (2014). https://doi.org/10.1007/978-3-319-07173-2_55
3. Banda, J.M., Angryk, R.A.: Scalable solar image retrieval with lucene. In: 2014 IEEE International Conference on Big Data (Big Data), pp. 11–17. IEEE (2014)
4. Banda, J.M., Angryk, R.A.: Regional content-based image retrieval for solar images: traditional versus modern methods. Astron. Comput. **13**, 108–116 (2015)
5. Brunner, C., Kő, A., Fodor, S.: An autoencoder-enhanced stacking neural network model for increasing the performance of intrusion detection. J. Artif. Intell. Soft Comput. Res. **12**(2), 149–163 (2022). https://doi.org/10.2478/jaiscr-2022-0010
6. Buckland, M., Gey, F.: The relationship between recall and precision. J. Am. Soc. Inf. Sci. **45**(1), 12 (1994)
7. Grycuk, R., Galkowski, T., Scherer, R., Rutkowski, L.: A novel method for solar image retrieval based on the Parzen kernel estimate of the function derivative and convolutional autoencoder. In: 2022 International Joint Conference on Neural Networks (IJCNN), pp. 1–7. IEEE (2022)
8. Grycuk, R., Scherer, R.: Grid-Based concise hash for solar images. In: Paszynski, M., Kranzlmüller, D., Krzhizhanovskaya, V.V., Dongarra, J.J., Sloot, P.M.A. (eds.) ICCS 2021. LNCS, vol. 12744, pp. 242–254. Springer, Cham (2021). https://doi.org/10.1007/978-3-030-77967-2_20
9. Kavitha, K., Rao, B.T.: Evaluation of distance measures for feature based image registration using alexnet. arXiv preprint arXiv:1907.12921 (2019)
10. Kempoton, D., Schuh, M., Angryk, R.: Towards using sparse coding in appearance models for solar event tracking. In: 2016 19th International Conference on Information Fusion (FUSION), pp. 1252–1259 (2016)
11. Kempton, D.J., Schuh, M.A., Angryk, R.A.: Describing solar images with sparse coding for similarity search. In: 2016 IEEE International Conference on Big Data (Big Data), pp. 3168–3176. IEEE (2016)

12. Kempton, D.J., Schuh, M.A., Angryk, R.A.: Tracking solar phenomena from the SDO. Astrophys. J. **869**(1), 54 (2018)
13. Korytkowski, M., Senkerik, R., Scherer, M.M., Angryk, R.A., Kordos, M., Siwocha, A.: Efficient image retrieval by fuzzy rules from boosting and metaheuristic. J. Artif. Intell. Soft Comput. Res. **10**(1), 57–69 (2020). https://doi.org/10.2478/jaiscr-2020-0005
14. Mairal, J., Bach, F., Ponce, J., Sapiro, G.: Online learning for matrix factorization and sparse coding. J. Mach. Learn. Res. **11**(Jan), 19–60 (2010)
15. Mikołajczyk, A., Grochowski, M., Kwasigroch, A.: Towards explainable classifiers using the counterfactual approach - global explanations for discovering bias in data. J. Artif. Intell. Soft Comput. Res. **11**(1), 51–67 (2021). https://doi.org/10.2478/jaiscr-2021-0004
16. Najgebauer, P., Scherer, R., Rutkowski, L.: Fully convolutional network for removing DCT artefacts from images. In: 2020 International Joint Conference on Neural Networks (IJCNN), pp. 1–8. IEEE (2020)
17. Salakhutdinov, R., Hinton, G.: Semantic hashing. Int. J. Approx. Reason. **50**(7), 969–978 (2009). https://doi.org/10.1016/j.ijar.2008.11.006. Special Section on Graphical Models and Information Retrieval
18. de Souza, G.B., da Silva Santos, D.F., Pires, R.G., Marananil, A.N., Papa, J.P.: Deep features extraction for robust fingerprint spoofing attack detection. J. Artif. Intell. Soft Comput. Res. **9**(1), 41–49 (2019). https://doi.org/10.2478/jaiscr-2018-0023
19. Mumford, S., Freij, N., et al.: SunPy: a python package for solar physics. J. Open Sour. Softw. **5**(46), 1832 (2020). https://doi.org/10.21105/joss.01832
20. The SunPy Community, et al.: The SunPy project: open source development and status of the version 1.0 core package. Astrophys. J. **890**, 1–12 (2020). https://doi.org/10.3847/1538-4357/ab4f7a, https://iopscience.iop.org/article/10.3847/1538-4357/ab4f7a
21. Ting, K.M.: Precision and recall. In: Sammut, C., Webb, G.I. (eds.) Encyclopedia of Machine Learning, p. 781. Springer, Boston (2011). https://doi.org/10.1007/978-0-387-30164-8_652

Cerebral Vessel Segmentation in CE-MR Images Using Deep Learning and Synthetic Training Datasets

Artur Klepaczko(✉)

Institute of Electronics, Lodz University of Technology, Łódź, Poland
artur.klepaczko@p.lodz.pl

Abstract. This paper presents a novel architecture of a convolutional neural network designed for the segmentation of intracranial arteries in contrast-enhanced magnetic resonance angiography (CE-MRA). The proposed architecture is based on the V-Net model, however, with substantial modifications in the bottleneck and the decoder part. In order to leverage multiscale characteristics of the input vessel patterns, we postulate to pass the network embeddings generated on the encoder path output through the atrous spatial pyramid pooling block. We motivate that this mechanism allows the decoder part to rebuild the segmentation mask based on local features, however, determined at various ranges of voxels neighborhoods. The ASPP outputs are aggregated using a simple gated recurrent unit which, on the other hand, facilitates the learning of feature maps' relevance with respect to the final output. We also propose to enrich the global context information provided to the decoder by including a vessel-enhancement block responsible for filtering out background tissues. In this study, we also aimed to verify if it is possible to train an effective deep-learning vessel segmentation model based solely on synthetic data. For that purpose, we reconstructed 30 realistic cerebral arterial tree models and used our previously developed MRA simulation framework.

Keywords: Vessel segmentation · Cerebral arterial models · Contrast-Enhanced MR angiography · MR angiography simulation

1 Introduction

Vessel segmentation is a key step in the analysis of angiographic images. It allows the reconstruction and visualization of the topology of the vascular system of body organs and helps to detect anatomical malformations such as stenoses, occlusions, or aneurysms. One of the routinely used techniques in the diagnosis of cerebral vessels is contrast-enhanced magnetic resonance angiography (CE-MRA). This method requires the application of a gadolinium-based contrast agent injected intravenously as a bolus, i.e. a dose of the agent with a volume of several to 20 ml administered over a short period of time. In the ionized state, gadolinium is a paramagnetic medium, which effectively shortens the T_1

J. Mikyška et al. (Eds.): ICCS 2023, LNCS 14076, pp. 274–288, 2023.
https://doi.org/10.1007/978-3-031-36027-5_20

relaxation time of the penetrated body fluids and tissues. Although gadolinium itself is toxic, its clinically used form is *chelated*, which means that a Gd^{3+} ion is bound by a molecular cage giving the agent the required safety profile. An advantage of CE-MRI in comparison with non-contrast techniques, such as Time-of-Flight (ToF) or Phase Contrast Angiography (PCA), is that the former provides excellent contrast of the vessels with respect to other tissues, is free from or less pronounced to flow- and motion-related artifacts and allows good separation of the signal from arteries and veins. The latter is achieved by proper configuration of scanning times so that the signal is acquired at peak concentration of the agent bolus in the arteries before it reaches the capillary bed during its passage through the vascular system. Moreover, CE-MRA does not suffer from other artifactual signal loss mechanisms characteristic of ToF and PCA, such as intra-voxel dephasing and blood spins saturation due to slow, in-plane, recirculating blood flow. On the drawbacks side, one must note that CE-MRI is not recommended in patients with severe renal impairment or after kidney replacement therapy, and in those who may develop a hypersensitive reaction to the gadolinium agent. The risk factors of the latter include multiple allergies, asthma, and a history of hypersensitive reactions to gadolinium and iodinated contrast agents.

This paper focuses on CE-MRA in application to the imaging of cerebral arteries. It must be noted that each imaging modality (such as MRI and computed tomography, e.g.) and a specific method within a given modality (like ToF and CE-MRA) possess characteristic intensity patterns, both with respect to vessels and the surrounding tissues. Hence, machine-learning-based vessel segmentation methods, which rely not only on geometrical features describing tubular elongated shapes of arteries and veins but also on the texture information related to the vessels' lumen and surrounding context, must be developed on the training data corresponding to the target modality. There can be two cases distinguished in regard to the availability of training examples. If there are sufficient images, then the development of semantic segmentation models is conceptually straightforward. On the other hand, when the training images of the target modality are scarce, one could use transfer learning—first, pre-train the model using, e.g. CT data and then fine-tune it with a relatively smaller set of MR images. In any case, the most tedious work is the data annotation so that the model can learn how to differentiate vessel pixels (or voxels in 3D) from background tissues.

In order to avoid the effort of manual data annotation, some researchers use synthetically generated vessel trees. The geometrical tree models are converted into images and ground-truth annotations. Usually, these images are oversimplified, both in terms of specific patterns characteristic of a given imaging modality and the complexity of the vessel system. Hence, the reliability of such models is limited to cases consisting of only a few straight large-scale branches, such as e.g. carotid arteries. Therefore the goal of this work is to show that it is possible to develop a deep-learning multi-scale vessel segmentation model ensuring state-of-the-art performance using the simulated CE-MRA images reconstructed for the realistic arterial tree models. The details of our concept are provided in the subsequent sections as follows. In Sect. 2, we give a summary of the published works concerning vessel segmentation, giving focus on deep-learning technology. This section completes with a list of contributions of our study in light of the presented SOTA methods. Section 3 presents the proposed architecture of a neural network dedicated to the segmentation of multi-scale vessel structures. Then we describe the developed framework for CE-MRA image synthesis which is based on our previous achievements in non-contrast MRA simulations. The results of experiments are presented in Sect. 4, and finally, Sect. 5 concludes.

2 Related Work

2.1 State-of-the-Art Methods

The importance of the vessel segmentation problem in the analysis of angiographic images resulted in numerous contributions over the last two decades and it is still an active research field. Modern approaches are comprehensively reviewed, e.g., in [13,20]. Recently, with the advent of deep learning technology, there is an observed growing trend toward the application of artificial neural networks for semantic segmentation of the vascular system. Since it is also an area of interest in the current study, below some examples most relevant to our research are recalled. For the description of conventional approaches, which utilize vessel enhancement [6], template matching [28], region growing [1], or level-set methods [8,16], the reader is referred to the provided citations.

A large part of published works concerns vessel segmentation in retinal images. For example, in [22] the authors employed a standard architecture of a convolutional neural network (CNN), with 3 convolutional layers, each followed by a max-pool operator, and then a head composed of two consecutive fully connected layers. The network input is a 61×61 pixel patch cropped from the original full-resolution OCT image and the output of the last, 2-neuron classification layer provides the probability estimate that the central patch pixel belongs either to a vessel or a background tissue. This kind of network is an example of architecture, where image pixels are classified based on the context embraced by an image patch. The effectiveness of such an approach is poor due to the need to iterate through multiple overlapping image patches, thus repeating the convolution operations for the same image regions. Nonetheless, such a methodology was utilized in a series of studies, also in the context of other organs and imaging modalities, e.g. liver (CT) [10], esophagus (NBI microscopy) [30], or carotid arteries (US) [24].

An extension to regular CNN architecture with a fully connected output classification layer is the Fully Convolutional Network (FCN) architecture. Instead of the fully connected layer, a tensor of deconvolution filters forms the FCN head, with each filter responsible for activating an individual pixel in the resulting segmentation mask. This mechanism allows for avoiding repeated processing of the overlapping image patches and provides finer delineation of the vessel walls. However, the output mask resolution is usually lower than that of the input image. FCN was reported to ensure satisfactory results in application to the segmentation of e.g. left ventricle vasculature [29], or cardiovascular and pulmonary vessels [17] in MRI and CT images.

Advances in encoder-decoder architectures resulted, on other hand, in numerous semantic segmentation network models. The most common architecture of this kind is the classic U-Net model [23], initially developed for segmenting neuronal structures in electron microscopic 2D images. In regard to vessel segmentation, this model was applied, e.g. to recognize retinal vessels in scanning laser ophthalmoscopy images [18], cerebral arteries in TOF-MR angiograms [14] or renal vasculature in CT data [27]. A 3-dimensional variant of U-Net was utilized in application to cerebral arteries also in TOF-MR angiograms [5].

Authors not only employ standard U-Net configuration with convolutional layers in both encoder and decoder paths with skip connections between the corresponding branch levels but also introduce custom modifications. For example, Wang et al. replace some of the convolution layers in the contracting path with ResNest blocks [9] to increase the saliency of the extracted image embeddings. Moreover, the skip connections which in the regular U-Net perform simple concatenation are enhanced by inserting the frequency channel attention (FCA) units between the encoding layers output and the corresponding decoder inputs. The role of the FCA blocks is to compensate for the potential mismatch between low- and high-level image features. A similar model is proposed in [4] for segmenting 3D coronary CTA images. Here the attention-guided modules fuse embeddings generated by adjacent levels of the encoder-decoder hierarchy to automatically capture the most relevant information inferred by the two paths. Moreover, the feature map on the output of the network bottleneck, before passing it on to the decoder, is processed by the scale-aware feature enhancement layer. This component extracts and selects image embeddings sensitive to various vessel scales. Eventually, the last layer of the decoder is formed by the multi-scale aggregation module, capable of adaptive fusing information from various decoder stages.

The fundamental challenge in any deep-learning task is preparing the representative training data set. The number of image-segmentation mask pairs should be large enough to cover the variability of the input domain. Apparently, according to the review presented in [13], the number of subjects involved in the DL experiments ranges from just a few scans up to several hundred. Hence, in most of the experiments, data augmentation techniques are involved to artificially increase the training set size to thousands of 2D images. In the study described in [7], more than 18 thousand scans were collected, but such big data sets are rather scarce, and usually not available for academic research groups. Therefore, some researchers postulate using synthetic data sets instead of real ones. Such approaches can be found in [26] and [29] where the image simulation procedures mimic CT and MR angiography, respectively. The intrinsic advantage of simulator-generated training examples is that ground-truth segmentation masks are automatically available, as the underlying vessel tree models are needed to synthesize the images. Thus, neither data annotation effort nor its quality check is required. In the published works, however, no study devoted explicitly to realistic contrast-enhance MR angiography is reported. In [29], where the coreMRI simulation platform [3] was utilized, only the 2D image slices were generated using a single-shot balanced steady-state free precession pulse sequence together with the 4D-XCAT whole-body anatomical model, from which the cardiac vasculature and tissue region was extracted.

2.2 Current Contribution

Considering the above-presented state-of-the-art, there are three aspects that can be distinguished as potential fields of improvement. Firstly, for the need of vessel segmentation in contrast-enhanced MR angiography images, a dedicated data set must be collected. Similarly to other researchers, we postulate to create a database of simulated images. For that purpose, we extended our MR angiography simulation framework [12] to the contrast-enhanced acquisition method. Moreover, as we focus on the intracranial arterial system, we constructed a series of cerebral vascularity models both normally appearing as well as with various lesions, such as aneurysms and stenoses. Thus, in contrast to previous works which used arterial models synthesized according to vessel tree-growth simulation algorithms, our approach is based on realistic anatomies reconstructed from true patient data. Secondly, in order to better exploit the multi-scale nature of the vessel system, we introduce a custom modification of the V-Net's encoder path. In our model, the last extracted feature map is passed through the atrous spatial pyramid pooling block, as in the backbone of the DeepLabV3 architecture [2]. Our solution stems from the observation that features must be extracted at multiple scales on the initial steps of the processing path and not only at the decoder output so that all the subsequent decoding stages can reconstruct coarse and fine geometrical vessel details. Then, the multi-scale feature maps representing various levels of geometry are stacked and aggregated with the use of a GRU module. Thirdly, instead of expanding the network architecture with an attention mechanism, criticized for the increased computational cost attributed to the need of learning unimportant features, we propose to enrich the input context information by manually-engineered image embeddings. This task is realized by the vessel-enhancement block which contains a bank of filters sensitive to tubular structures [6].

3 Methods and Materials

3.1 Vessel Segmentation Model

Base Architecture. The objective of the current study was to develop an efficient vessel segmentation model that would incorporate multiscale image feature extraction and a robust mechanism to account for the global context of an imaged organ vasculature. For that purpose, we propose the architecture visualized in Fig. 1. It is essentially based on the concept of V-Net—a volumetric

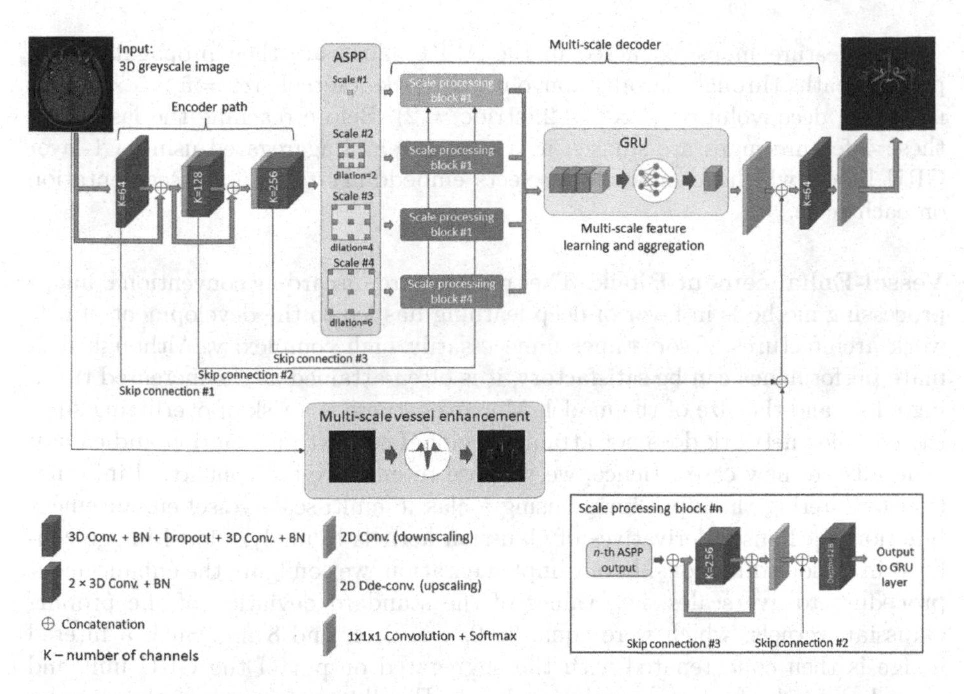

Fig. 1. Overview of the proposed vessel segmentation network architecture.

fully convolutional network [19]. The main computational units in the encoder path realize $5 \times 5 \times 5$ convolution with stride = 1. At each stage, there are 2 convolution layers followed by batch normalization and dropout layer (with probability $p = 0.2$). The activation function for the convolutional layers is the parameterized ReLU. The number of filters in the first stage equals 64 for both layers and doubles in consecutive stages. As proposed in the original paper, the input to each stage is passed over and added to this stage's output, thus enabling the learning of a residual function and decreasing the convergence time. The resolution of image embedding maps is reduced after each stage by a convolution layer with $2 \times 2 \times 2$ kernel with stride 2. Eventually, the outputs of the contractive path stages are concatenated with the inputs of the corresponding blocks in the expansion path.

Multi-scale Feature Extraction and Aggregation. The network embeddings extracted at the third stage of the encoder already comprise rich semantic information. Further reduction of the input resolution poses a risk of hiding varied geometrical patterns characteristic of vessels at different scales. Hence, instead of another contractive step, the proposed model contains the atrous spatial pyramid pooling block, which provides feature encoding at various scales and simultaneously leverages high-resolution context information around the vessels. We adopt the idea of ASPP blocks to 3 dimensions and use 4 kernels of sizes $1 \times 1 \times 1$ (pointwise), and $3 \times 3 \times 3$ with dilation rates r = 2, 4, and 6.

The feature maps extracted in the ASPP block are then propagated in 4 parallel paths through decoder convolution blocks (kernel size = $5 \times 5 \times 5$) and upscaling deconvolution ($2 \times 2 \times 2$, stride = 2). Before reaching the last layer, these 4 feature maps are stacked in a sequence and aggregated using a 1-layer GRU block, which automatically selects embeddings relevant to segmentation on each scale.

Vessel-Enhancement Block. The trend toward discarding conventional image processing methods in favor of deep learning has led to the development of network architectures of sometimes unnecessarily high complexity. Although ultimate performance can be satisfactory, it is often attained at the increased training effort and the size of the model. Moreover, there is a risk of overfitting since the complex network does not attain the global context information and cannot generalize to new cases. Hence, we propose to enhance the contextual information by filtering the input image using a classic multi-scale vessel enhancement function [6]. It uses derivatives of Gaussian kernels to analyze local image contrast in various directions. In our implementation, we configure the enhancement procedure to five scales, i.e., values of the standard deviation of the probing Gaussian kernels, which were equal to 0.5, 1, 2, 4, and 8 mm. Such a filtered image is then concatenated with the aggregated output of the GRU unit and passed through the last convolution block. Finally, the output of the decoder is convolved channel-wise with the filter of size $1 \times 1 \times 1$. The obtained values are consumed by the soft-max function which generates the output segmentation mask. The network training is accomplished by minimizing the standard cross-entropy loss function.

3.2 MR Angiography Simulation

Cerebral Arterial Models. The input to the simulation algorithm, apart from the imaging parameters (such as sequence timing parameters, the field of view, etc.), is the model of the imaged organ. In the case of angiography simulation, the model comprises the geometry of the arterial tree, its functional description, i.e. blood flow through the vessel system, and the stationary tissue mimicking the perfusion volume. The first problem is therefore the reconstruction of a realistic arterial tree geometrical model.

We approach this challenge by manually annotating cerebral vessels in a high-resolution CE-MRA image. For the need of this study, we exploit the IXI dataset [15] - an open database containing 570 MRA images of healthy subjects. Since this study is a preliminary research directed toward the utilization of synthetic images for the training of deep-learning segmentation models, we selected a subset of 30 MRA volumes for the development of training examples. The in-plane resolution of the images was equal to 0.46875×0.46875 mm^2 with the matrix size = 512×512 pixels. The number of axial slices was 100 and the spacing between slices, as well as slice thickness, was equal to 0.8 mm.

Given the vessel system segmentation, its geometric description can be obtained. Firstly, in our algorithm, the segmentation mask is skeletonized to

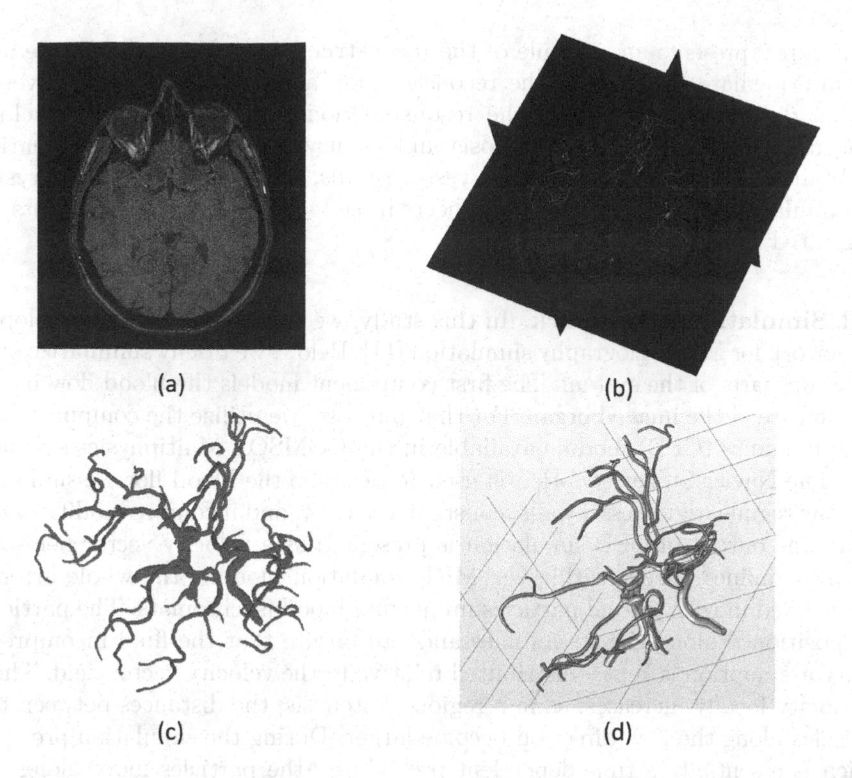

Fig. 2. Reconstruction of a cerebral arterial tree using VesselKnife and Comsol software. Manual annotation of arteries in the real CE-MRA volume (a). Skeletonization of the binary annotation mask and centerline determination (b). Radius estimation along the centerlines (c). Geometry reconstruction in the Comsol software to simulate the blood flow.

roughly identify the courses of the vessel centerlines. These centerlines were defined on the discrete raster, hence they then must be smoothed to obtain a continuous description. The smoothing process involves minimizing the second derivatives calculated at the subsequent nodes of centerlines.

Next, the vessel radii are determined along the smoothed centerlines using the algorithm implemented in the *VesselKnife* software [25]. For a given node point on the centerline, a set of rays passing through that point is defined. The rays are led in multiple directions in order to equally penetrate the space around the node. The number of rays can be adjusted to the user's needs and it is a tradeoff between computational effort and precision of radius estimation. Each ray intersects the vessel wall which is determined as a boundary between white and black regions in the segmentation binary mask. The covariance matrix is then constructed for the distribution of the intersection points. The principal component decomposition of this covariance matrix indicates the local vessel orientation (component with the highest eigenvalue) and its radius, which is approximated as the square root of the smallest eigenvalue multiplied by 2.

Figure 2 presents an example of the reconstructed cerebral arterial tree and the intermediate products of the reconstruction algorithm. Moreover, a vessel system directly obtained from the reconstruction can be modified to include abnormal structures, such as stenosis and aneurysms. The first malformation is obtained by locally decreasing a vessel radius, whereas saccular aneurysms are simulated by adding ellipsoidal objects linked with short vessel segments, as illustrated in Fig. 3.

MR Simulation Framework. In this study, we use our previously developed framework for MR angiography simulation [11]. Below, we briefly summarize only the main parts of the system. The first component models the blood flow in the vasculature of the imaged organ. For that purpose, we utilize the computational fluid dynamics (CFD) module available in the COMSOL Multiphysics software [28]. The Navier-Stokes equation is used to describe the blood flow, assuming a laminar regime, rigid vessel walls, constant viscosity, and incompressibility of the fluid. The output of CFD simulation is presented as a velocity vector field and flow streamlines. Then, within the MRI simulation module, the whole arterial tree is filled in with virtual particles mimicking blood isochromats. The particles are positioned along the streamlines and—to ensure that the fluid incompressibility assumption is kept—distributed relative to the velocity vector field. Thus, if velocity locally increases, as in a region of stenosis, the distances between the particles along the flow direction become larger. During the simulation process–which is essentially a time-dependent procedure—the particles move along the streamlines from the vessel tree inlet to its outlet terminals. Whenever a particle leaves its trajectory, it is replaced by a new one at that trajectory input node. The system keeps track of all particles and can determine their current locations with arbitrary temporal resolution. The number of streamlines and the number of particles positioned thereon depend on the size of the vasculature model.

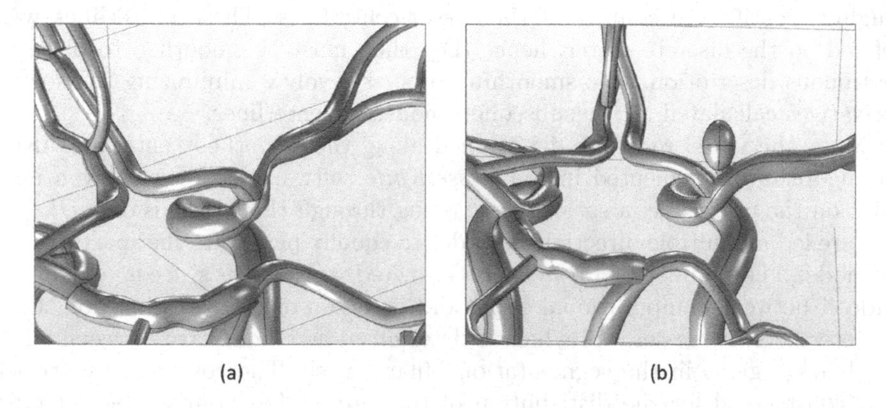

Fig. 3. Malformations introduced into normally-appearing arterial models: stenosis (a) and aneurysm (b) in a middle cerebral artery.

Table 1. MRI parameter values used in experiments.

Tissue type	ρ [%]		T_1 [msec]		T_2 [msec]	
	μ	σ	μ	σ	μ	σ
Blood	100	–	200	10	50	5
Stationary	83	8	920	90	100	10

Based on our past experiments with non-contrast-enhanced sequences [11], we adjust the particle number to obtain the density of 15–20 particles per 1 mm^3.

From the vessel segmentation algorithm perspective, it is also important that the final image comprises also static tissue. Only then, the segmentation model can learn not only the geometry of the vessels but also the background context. Therefore, apart from moving particles located on vasculature streamlines, the simulated object is composed of stationary particles surrounding the arteries. Each particle is assigned magnetic parameters – proton density (ρ), longitudinal (T_1), and transverse (T_2) relaxation times. We set these parameters to values characteristic of the gadolinium-enhanced blood and gray matter in the field strength $B_0 = 1.5$ T. The parameter means and standard deviations used in the experiments under the current study are provided in Table 1. In order to achieve higher variability of the intensity patterns in the synthesized images, the exact values for a given object are randomly sampled from the Gaussian distribution.

The MRI simulation block handles the time passage and manages the imaging events, i.e. RF excitation pulses, phase, and frequency encoding gradients, as well as free precession stages. The specific events are switched on and off according to the prescribed acquisition method and the corresponding sequence parameters, i.e. echo and relaxation times (TE and TR), and the flip angle (FA). Similarly to object attributes, sequence parameters are randomly sampled using mean and standard deviation values, i.e. TR = 20 ($\sigma = 2$) msec, TE = 7 ($\sigma = 1.2$) msec, FA = $15(\sigma = 3)^\circ$. The chosen parameter ranges correspond to the T_1-weighted spoiled gradient echo sequence, with which the CE-MRA acquisitions are routinely accomplished.

The magnetic vectors M_p of all spin isochromats associated with object particles p are initially aligned with the z-axis. At a given time t of the simulation, M_p vectors are flipped toward the transverse plane, and the new state after excitation is calculated as

$$M_p\left(t+\delta t\right) = R_{\mathrm{RF}}\left(\alpha_{\mathrm{eff}}\right) M_p\left(t\right), \tag{1}$$

where δt indicates the temporal resolution of the simulation, R_{RF} rotates M_p about the x-axis. α_{eff} is the effective flip angle that encapsulates off-resonance conditions causing changes in the assumed flip angle α.

The post-excitation evolution of the magnetization vectors is governed by the analytical solution to the Bloch equation as

$$M_p\left(t+\delta t\right) = Rot_z\left(\theta_g\right) Rot_z\left(\theta_i\right) R^{12}_{\mathrm{relax}} M_p\left(t\right) + R^{1}_{\mathrm{relax}} M_0, \tag{2}$$

where Rot_z rotates M_p around the z-axis due to the phase or frequency encoding gradient (θ_g) and field inhomogeneity effect (θ_i), whereas R^{12}_{relax} and R^{1}_{relax} account for the transverse and longitudinal relaxation phenomena. Finally, signal acquisition is accomplished by integrating contributions from every particle p in the k-space:

$$s\left(t\right)=\sum\nolimits_{p=1}^{n_p} M_p\left(t\right)\boldsymbol{x}+j\sum\nolimits_{p=1}^{n_p} M_p\left(t\right)\boldsymbol{y}, \tag{3}$$

with n_p indicating the total number of particles. The readout window is partitioned into nx segments, which correspond to the size of k-space in the frequency encoding direction. During the intervals between these sampling steps, the magnetization vectors of blood particles evolve as they move. Once the readout window is completed, one line of k-space is filled in. After each TR cycle, the remnant transverse magnetization is numerically canceled out before issuing the next RF pulse to achieve the behavior of the spoiled gradient echo method. The final step involves performing an FFT transform on the collected k-space data to reconstruct the image. Before reconstructing the image, it is possible to add thermal noise with a user-defined standard deviation to the measured signal. In this study, we used three different levels of noise, which led to reconstructed images with signal-to-noise ratios (SNRs) of 5:1, 10:1, or 20:1.

4 Experimental Results

4.1 Simulated Training Images

The arterial tree models designed in this study consisted of 40 to 60 branches and comprised, apart from multiple smaller vessels, the main cerebral arteries, such as basilar artery (BA), internal carotid arteries (ICA), posterior cerebral and communicating arteries, anterior cerebral arteries, anterior communicating artery, and middle cerebral arteries (MCA). The simulation of blood flow was driven by setting pressure values in the circulatory stems inlets (BA, left and right ICA) and all the outlets. The pressure difference was adjusted to the values reported in the study [40], where the flow distribution in a real subject was measured with the use of phase-contrast angiography.

Figure 4 presents cross-sections and maximum intensity projection of an example simulated image with added noise at the SNR level = 20:1. All simulated images were formed with the matrix size = 256 × 256 and the number of slices = 100. The number of sequence parameter combinations reached 100. Consequently, 3,000 volumes were simulated which represented 30 reconstructed normally-appearing cerebral arterial trees. For each imaging parameters combination, an arterial tree was randomly rotated about all three spatial axes by an angle drawn from the interval $(-15°; +15°)$, so that virtual acquisitions are effectively made at arbitrary orientations. Each tree was also modified to contain one kind of lesion – stenosis or aneurysm. This intervention resulted in another 3,000 synthesized images. The raw k-space data were then corrupted by the Gaussian white noise, hence, in addition to noise-free images, the simulated dataset comprised 3 noisy variants. Overall, there were 24,000 synthetic CE-MRA volumes available for training and validation.

4.2 Tests of the Segmentation Model

The test images were selected as another subset of the IXI Database, disjunctive from the training subset. The tests of the proposed network were conducted on 50 volumes, for which the ground-truth masks were manually annotated by an observer experienced in image processing and later verified by a radiologist. Then, in order to gain insight into the capability of the model with reference to conventional image processing methods, all test volumes were filtered by the aforementioned vessel enhancement algorithm (with the same set of filter scales) and segmented automatically using the level-set segmentation method. The segmentation results obtained by both level-set and our proposed model were compared against ground-truth masks using the intersection-over-union (IoU) metric, given as

$$IoU = \frac{\sum_{i=1}^{K} y_i \wedge y_i^{pred}}{\sum_{i=1}^{N} y_i \vee y_i^{pred}} \tag{4}$$

where K designates the number of voxels in a processed image and y_i^{pred} is the predicted voxel category. Here, categories are Boolean-valued and a voxel is labeled *True* if it belongs to the artery, *False* otherwise. The averaged results are presented in Table 2. It must also be noted that the performance of various

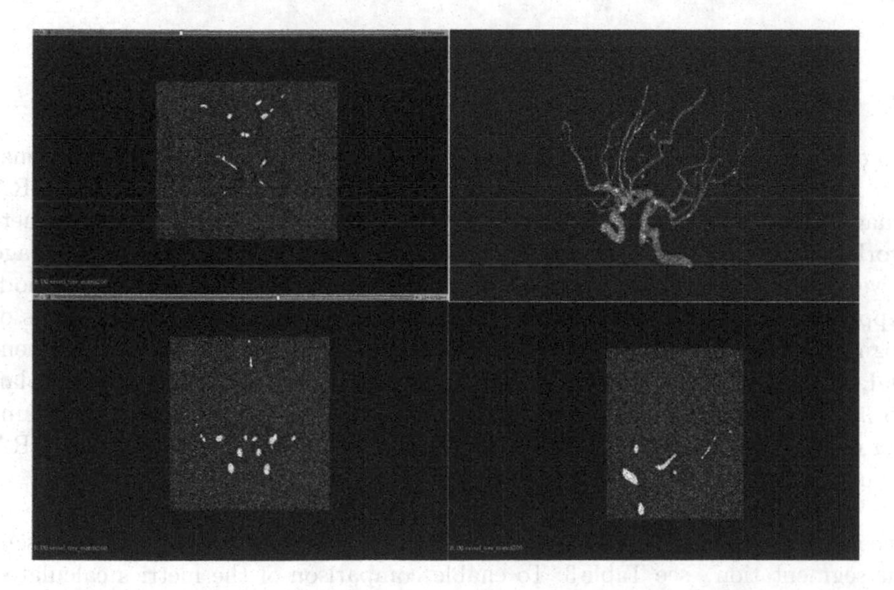

Fig. 4. Anatomical cross-sections and maximum intensity projection (upper-right) of an example simulated CE-MRA image.

segmentation methods may differ with respect to the scale of a vessel. Hence, apart from reporting the overall mean IoU value, we investigated the behavior of both tested methods at 4 distinguished radii ranges of vessel segments.

Table 2. IoU values obtained for the reference method and the proposed model.

Segmentation method	All radii	$r \leq 1$	$1 < r \leq 1.5$	$1.5 < r \leq 2$	$r > 2$
Level-set [16]	0.79	0.62	0.64	0.83	0.89
Proposed	0.75	0.72	0.73	0.76	0.77

r – vessel radius [mm]

Table 3. Comparison of the obtained segmentation results with previous research.

Source	Modality	Architecture	IoU
Fan et al. [5]	ToF-MRA	3D U-Net/HMRF*	0.66
Tetteh et al. [26]	ToF-MRA	3D FCN/cross-hair conv	0.76
Zhang et al. [31]	ToF-MRA	Dilated Dense CNN	0.95
Nazir et al. [21]	CTA	3D FCN/Inception	0.83
Current study	CE-MRA	3D V-Net/Vessel enhancement	0.75

* – hidden Markov random fields

5 Conclusions

To conclude, in this paper, we presented a novel architecture of a convolutional neural network specialized in the segmentation of cerebral arteries in CE-MRA images. We showed that it is feasible to obtain reasonable results using our network trained solely on synthetic data. The calculated IoU values are on average lower than the scores obtained with the use of the conventional level-set method. Apparently, however, the proposed network ensures better stability in terms of segmentation accuracy across various scales of vessel radii. This shows the potential of the designed multi-scale network configuration. In future works, we plan to improve the generalization capability of our model by conducting the training on a dataset reconstructed from a larger and more varied sample of MRA volumes.

When referred to other published studies, the obtained *IoU* ratios are in accordance with the state-of-the-art results in the domain of cerebrovascular segmentation—see Table 3. To enable comparison of the metrics calculated within the current contribution, the Dice coefficients (or F1-scores) reported in other papers were converted to *IoU*. As it can be seen, our segmentation model, although applied to a different imaging modality, performs either better or at a similar level than approaches proposed in [5] and [26]. On the other hand, the solution described in [31] achieved higher accuracy over all others. It must be noted, however, that their experiments involved significantly smaller training and test datasets, composed of only 109 ToF-MRA volumes.

References

1. Cetin, S., Unal, G.: A higher-order tensor vessel tractography for segmentation of vascular structures. IEEE Trans. Med. Imaging **34**(10), 2172–2185 (2015)
2. Chen, L.C., Papandreou, G., Schroff, F., Adam, H.: Rethinking atrous convolution for semantic image segmentation (2017). https://arxiv.org/abs/1706.05587
3. CoreMRI: Advanced MR simulations on the cloud (2023). https://www.coremri.com/. Accessed 3 Mar 2023
4. Dong, C., Xu, S., Dai, D., Zhang, Y., Zhang, C., Li, Z.: A novel multi-attention, multi-scale 3D deep network for coronary artery segmentation. Med. Image Anal. **85**, 102745 (2023)
5. Fan, S., Bian, Y., Chen, H., Kang, Y., Yang, Q., Tan, T.: Unsupervised cerebrovascular segmentation of TOF-MRA images based on deep neural network and hidden Markov random field model. Frontiers in Neuroinformatics **13**, 77 (2020)
6. Frangi, A.F., Niessen, W.J., Vincken, K.L., Viergever, M.A.: Multiscale vessel enhancement filtering. In: Wells, W.M., Colchester, A., Delp, S. (eds.) MICCAI 1998. LNCS, vol. 1496, pp. 130–137. Springer, Heidelberg (1998). https://doi.org/10.1007/BFb0056195
7. Fu, F., et al.: Rapid vessel segmentation and reconstruction of head and neck angiograms using 3D convolutional neural network. Nat. Commun. **11**(1), 4829 (2020)
8. Gao, X., Uchiyama, Y., Zhou, X., Hara, T., Asano, T., Fujita, H.: A fast and fully automatic method for cerebrovascular segmentation on time-of-flight (TOF) MRA image. J. Digit. Imaging **24**(4), 609–625 (2011)
9. He, K., Zhang, X., Ren, S., Sun, J.: Deep residual learning for image recognition (2015). https://arxiv.org/abs/1512.03385
10. Kitrungrotsakul, T., et al.: VesselNet: a deep convolutional neural network with multi pathways for robust hepatic vessel segmentation. Comput. Med. Imaging Graph. **75**, 74–83 (2019)
11. Klepaczko, A., Materka, A., Szczypiński, P., Strzelecki, M.: Numerical modeling of MR angiography for quantitative validation of image-driven assessment of carotid stenosis. IEEE Trans. Nucl. Sci. **62**(3), 619–627 (2015)
12. Klepaczko, A., Szczypiński, P., Deistung, A., Reichenbach, J.R., Materka, A.: Simulation of MR angiography imaging for validation of cerebral arteries segmentation algorithms. Comput. Methods Programs Biomed. **137**, 293–309 (2016)
13. Li, H., Tang, Z., Nan, Y., Yang, G.: Human treelike tubular structure segmentation: a comprehensive review and future perspectives. Comput. Biol. Med. **151**, 106241 (2022)
14. Livne, M., et al.: A U-net deep learning framework for high performance vessel segmentation in patients with cerebrovascular disease. Front. Neurosci. **13**, 97 (2019)
15. London, I.C.: Brain Development (2023). https://brain-development.org/ixi-dataset/. Accessed 3 Mar 2023

16. Manniesing, R., Velthuis, B., van Leeuwen, M., van der Schaaf, I., van Laar, P., Niessen, W.: Level set based cerebral vasculature segmentation and diameter quantification in CT angiography. Med. Image Anal. **10**(2), 200–214 (2006)
17. Merkow, J., Marsden, A., Kriegman, D., Tu, Z.: Dense volume-to-volume vascular boundary detection. In: Ourselin, S., Joskowicz, L., Sabuncu, M.R., Unal, G., Wells, W. (eds.) MICCAI 2016. LNCS, vol. 9902, pp. 371–379. Springer, Cham (2016). https://doi.org/10.1007/978-3-319-46726-9_43
18. Meyer, M.I., Costa, P., Galdran, A., Mendonça, A.M., Campilho, A.: A deep neural network for vessel segmentation of scanning laser ophthalmoscopy images. In: Karray, F., Campilho, A., Cheriet, F. (eds.) ICIAR 2017. LNCS, vol. 10317, pp. 507–515. Springer, Cham (2017). https://doi.org/10.1007/978-3-319-59876-5_56
19. Milletari, F., Navab, N., Ahmadi, S.: V-net: fully convolutional neural networks for volumetric medical image segmentation. In: 2016 Fourth International Conference on 3D Vision (3DV), pp. 565–571. IEEE Computer Society (2016)
20. Moccia, S., De Momi, E., El Hadji, S., Mattos, L.S.: Blood vessel segmentation algorithms - review of methods, datasets and evaluation metrics. Comput. Methods Programs Biomed. **158**, 71–91 (2018)
21. Nazir, A., et al.: OFF-eNET: an optimally fused fully end-to-end network for automatic dense volumetric 3D intracranial blood vessels segmentation. IEEE Trans. Image Process. **29**, 7192–7202 (2020)
22. Prentašić, P., et al.: Segmentation of the foveal microvasculature using deep learning networks. J. Biomed. Opt. **21**(7), 075008 (2016)
23. Ronneberger, O., Fischer, P., Brox, T.: U-net: convolutional networks for biomedical image segmentation. In: Navab, N., Hornegger, J., Wells, W.M., Frangi, A.F. (eds.) MICCAI 2015. LNCS, vol. 9351, pp. 234–241. Springer, Cham (2015). https://doi.org/10.1007/978-3-319-24574-4_28
24. Smistad, E., Løvstakken, L.: Vessel detection in ultrasound images using deep convolutional neural networks. In: Carneiro, G., et al. (eds.) LABELS/DLMIA -2016. LNCS, vol. 10008, pp. 30–38. Springer, Cham (2016). https://doi.org/10.1007/978-3-319-46976-8_4
25. Szczypiński, P.: VesselKnife (2023). http://eletel.p.lodz.pl/pms/SoftwareVesselKnife.html. Accessed 3 Mar 2023
26. Tetteh, G., et al.: DeepVesselNet: vessel segmentation, centerline prediction, and bifurcation detection in 3-D angiographic volumes. Fron. Neurosci. **14**, 592352 (2020). https://doi.org/10.3389/fnins.2020.592352
27. Wang, C., et al.: Precise estimation of renal vascular dominant regions using spatially aware fully convolutional networks, tensor-cut and Voronoi diagrams. Comput. Med. Imaging Graph. **77**, 101642 (2019)
28. Worz, S., Rohr, K.: Segmentation and quantification of human vessels using a 3-D cylindrical intensity model. IEEE Trans. Image Proc. **16**(8), 1994–2004 (2007)
29. Xanthis, C.G., Filos, D., Haris, K., Aletras, A.H.: Simulator-generated training datasets as an alternative to using patient data for machine learning: an example in myocardial segmentation with MRI. Comput. Methods Programs Biomed. **198**, 105817 (2021)
30. Xue, D.X., Zhang, R., Feng, H., Wang, Y.L.: CNN-SVM for microvascular morphological type recognition with data augmentation. J. Med. Biol. Eng. **36**(6), 755–764 (2016)
31. Zhang, B., et al.: Cerebrovascular segmentation from TOF-MRA using model- and data-driven method via sparse labels. Neurocomputing **380**, 162–179 (2020)

Numerical Method for 3D Quantification of Glenoid Bone Loss

Alexander Malyshev[1(✉)] and Algirdas Noreika[2]

[1] Department of Mathematics, University of Bergen, PB 7803, 5020 Bergen, Norway
alexander.malyshev@uib.no
[2] Indeform, K. Petrausko st. 26, 44156 Kaunas, Lithuania
algirdas.noreika@indeform.com

Abstract. Let a three-dimensional ball intersect a three-dimensional polyhedron given by its triangulated boundary with outward unit normals. We propose a numerical method for approximate computation of the intersection volume by using voxelization of the interior of the polyhedron. The approximation error is verified by comparison with the exact volume of the polyhedron provided by the Gauss divergence theorem. Voxelization of the polyhedron interior is achieved by the aid of an indicator function, which is very similar to the signed distance to the boundary of the polyhedron. The proposed numerical method can be used in 3D quantification of glenoid bone loss.

Keywords: Polyhedron · Volume computation · Glenoid bone loss

1 Introduction

Studies show that about 90% of shoulder joints with recurrent anterior shoulder dislocation have an abnormal glenoid shape [13]. The preoperative assessment of anterior glenoid bone loss is a critical step in surgical planning for patients with recurrent anterior glenohumeral instability. The currently accepted gold standard for glenoid structural assessment among most orthopaedic surgeons is the use of 3-dimensional reconstructed computed tomography images with the humeral head digitally subtracted, yielding an en face sagittal oblique view of the glenoid. Several methods have been reported to quantify the amount of glenoid bone loss [4]. One of the most commonly used concepts described in the literature uses the diameter of the "best-fit circle" circumscribed around the inferior glenoid. To quantify the amount of glenoid bone loss, reported as a percentage of either total surface area or diameter, the following measures are used for the diameter-based method and surface area method, respectively: Percent bone loss = (Defect width/Diameter of inferior glenoid circle) × 100% and Percent bone loss = (Defect surface area/Surface area of inferior glenoid

Supported by the Research Executive Agency of the European Commission, grant 778035 - PDE-GIR - H2020-MSCA-RISE-2017.

J. Mikyška et al. (Eds.): ICCS 2023, LNCS 14076, pp. 289–296, 2023.
https://doi.org/10.1007/978-3-031-36027-5_21

circle) × 100%. Comparison of the diameter-based method and surface area method is carried out in Table 1 in [4].

To measure glenoid bone loss, surgeons typically use radiographs and 2D CT scans and apply the surface area method, where certain part of the glenoid bone is positioned within the best-fit circle and the missing area in the circle represents size of the bone defect; see Fig. 1(a).

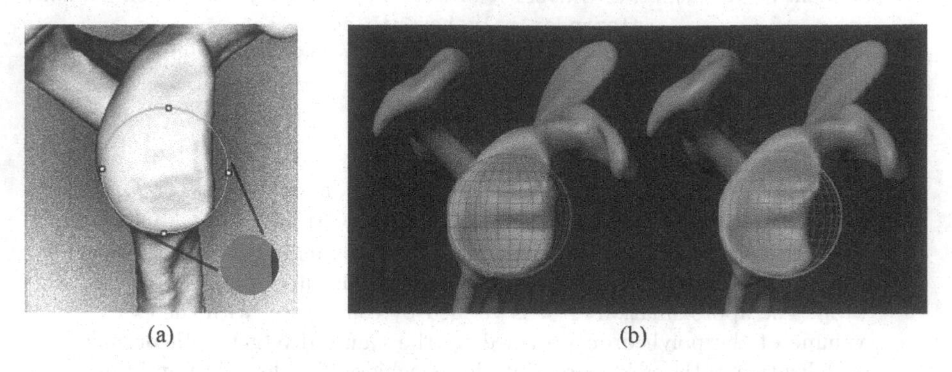

Fig. 1. (a) Best-fit circle method for evaluation of glenoid bone defect. (b) 3D volume measurement of a glenoid bone

Magnetic resonance imaging (MRI) has been the gold standard for visualizing soft tissue lesions, but it has also proven to be an accurate modality for measuring glenoid bone loss in recent clinical studies [3,12,15]. With the development of three-dimensional bone reconstruction technologies, more accurate measurements are being explored, for example, using the best-fit spherical volume instead of the circle area method. By placing the evaluated bone area inside a sphere and measuring the difference between the reconstructed bone and the interior of the sphere, the bone defect is evaluated in terms of 3D volume instead of 2D area; see Fig. 1(b).

In the present paper, we describe a numerical method for measuring the volume portion of a bone inside a given 3D sphere. Our contribution is a careful selection of fast and robust algorithms for this method and a posteriori accuracy test for the computed result. The design of the numerical method was inspired by [6] and can be considered as a simpler alternative to the algorithm in [1] equipped with the accuracy test.

2 Statement of the Problem

We are given a three-dimensional solid object B, such as the glenoid bone, embedded into a three-dimensional parallelepiped Ω. The boundary ∂B of B is assumed to be a two-dimensional triangulated surface without boundary, i.e. it consists of plane triangles and is watertight. The triangulation of ∂B is given in the STL format by means of three matrices P, T and F, which denote respectively the set of vertices, the triangulation connectivity list and the set of outward unit face normals. The i-th row of the real $N \times 3$ matrix P contains the coordinates of vertex i. The k-th row of the integer $K \times 3$ matrix T contains the vertex numbers for the k-th triangle. The k-th row of the real $K \times 3$ matrix F contains the outward unit normal n_k to the k-th triangle. Orientation of the k-th triangle given by the row k of T is consistent with the direction of the normal vector n_k. Recall that a solid object inside a triangulated surface is usually referred to as a polyhedron.

Intersection of a polyhedron with a ball. Consider a three-dimensional open ball $C = \{x \in R^3 \colon \|x - a\|_2 < r\}$, where $a \in R^3$ is a center, $r > 0$ is a radius and $\|v\|_2 = \sqrt{v_1^2 + v_2^2 + v_3^2}$ is the Euclidean vector norm.

Our main problem is how to compute the volume of the intersection $B \cap C$. It is worth emphasizing here that we do not need to evaluate this volume exactly or with high accuracy. For example, a relative error not exceeding one percent may be quite satisfactory.

3 Volume of a Polyhedron via the Gauss Formula

The Gauss-Ostrogradsky divergence theorem for a function $f(v) = (f_1, f_2, f_3)^T$ of $v = (x_1, x_2, x_3)^T$ reads

$$\int_B \left(\frac{\partial f_1}{\partial x_1} + \frac{\partial f_2}{\partial x_2} + \frac{\partial f_3}{\partial x_3} \right) dv = \int_{\partial B} (f_1 n_1 + f_2 n_2 + f_3 n_3)\, ds, \tag{1}$$

where $n = (n_1, n_2, n_3)^T$ is the outward unit normal to the surface ∂B and ds is the surface area element. If $f = (0, 0, x_3)$, then $\int_B dv = \int_{\partial B} x_3 n_3 ds$. Each triangle $T^{(k)}$ in ∂B is given by a sequence of vertex vectors $p_1^{(k)}, p_2^{(k)}, p_3^{(k)}$ ordered consistently with the outward unit normal $n^{(k)}$. Let us denote by $m_3^{(k)}$ the third component of the mean vector $(p_1^{(k)} + p_2^{(k)} + p_3^{(k)})/3$. It follows that $\int_{\partial B} x_3 n_3 ds = \sum_k m_3^{(k)} n_3^{(k)} \mathrm{area}(T^{(k)})$. The vector product $w^{(k)} = (w_1^{(k)}, w_2^{(k)}, w_3^{(k)})^T$ of the vectors $p_2^{(k)} - p_1^{(k)}$ and $p_3^{(k)} - p_2^{(k)}$ satisfies $w^{(k)} = \|w^{(k)}\|_2 n^{(k)}$ and $\|w^{(k)}\|_2 = 2\,\mathrm{area}(T^{(k)})$. Hence the volume of B equals

$$V = \int_B dv = \frac{1}{2} \sum_k m_3^{(k)} w_3^{(k)}. \tag{2}$$

More formulas derived by means of the divergence theorem are found in [10].

4 The Voxelization Approach

The ideal (exact) method could be to describe the boundary of the intersection $B \cap C$ exactly and apply a variant of the Gauss divergence formula similar to

(2). However, theoretical derivation of this method seems to be too involved. Alternatively, one can try to use results from the advanced theory of boolean operations with polyhedra, see e.g. [5,8,14,16].

Since we admit approximate evaluation of the volume of $B \cap C$, we follow a much simpler approach based on the so called voxelization of the solid object. Namely, having defined a uniform 3-D mesh $\mathcal{M}$ in the parallelepiped Ω containing the object B, we find the set $\mathcal{M}_i$ of all interior points of $\mathcal{M}$, i.e. the mesh points belonging to B. Let us denote the number of interior points by $|\mathcal{M}_i|$. Assume that a single voxel in $\mathcal{M}$ is a cube with the side length h so that its volume equals h^3. Then the product $h^3|\mathcal{M}_i|$ approximates the volume of object B. The approximation error converges to zero as $\text{area}(\partial B)O(h)$ at most, where $\text{area}(\partial B)$ is the surface area of ∂B. Note that the voxelization approach is also used for more complicated problems such as boolean operations with polyhedra, see e.g. [11].

The volume of $B \cap C$ is approximated by $h^3|\mathcal{M}_i \cap C|$, where $|\mathcal{M}_i \cap C|$ is the number of interior mesh points inside the ball C.

The main problem in our voxelization approach is the 3-D point location problem that is the determination whether $x \in B$ or not for any point $x \in \Omega$. There exist efficient combinatorial solutions such as the method developed in [7]. We choose the more widespread tools based on the distance maps and the closest neighbor maps combined with the outward surface normals. Namely, we take advantage of the indicator function from Sect. 5. This indicator function has been used, for instance, in [1,6].

5 Description of Our Numerical Method

We use the level set method, in which the surface ∂B is represented by the 0-level set of a function $u(x)$, i.e. $\partial B = \{x \in \Omega \colon u(x) = 0\}$. Moreover, we want to construct $u(x)$ such that $u(x) < 0$ inside B and $u(x) > 0$ outside B. Let us call a function $u(x)$ satisfying these properties an indicator function.

The numerical method consists of the following steps:

- Read the STL data for the surface ∂B.
- Sample (discretize) the surface ∂B into a 3D point cloud $\mathcal{P}$ with outward normals.
- Determine a 3D parallelepiped Ω enclosing the point cloud $\mathcal{P}$ and choose a uniform grid $\mathcal{M}$ in Ω of voxel size h.
- Downsample the point cloud $\mathcal{P}$ onto the grid points of $\mathcal{M}$.
- Compute the closest point map for all grid points of $\mathcal{M}$ with the closest points in the downsampled point cloud.
- Compute the distance map to the downsampled point cloud for all grid points of $\mathcal{M}$.
- Compute the indicator function $u(x)$.
- Remove (possible) impulse noise by the median filter or by the TVG-L1 filter.
- Determine the interior grid points in B by zero thresholding $u(x)$.

- Compute the volume of B as the number of interior points times h^3 and compare it with the volume defined by the divergence theorem.
- Determine the number of interior points lying in C and calculate the volume of the intersection $B \cap C$.

Let us look at these steps in more detail below.

5.1 Randomized Sampling of the Surface

In contrast to [1], where the Euclidean distance map is computed with respect to the triangulated surface, we sample the triangulated surface into a point cloud. Such decision is mainly due to the further use of standard efficient algorithms for computation of the distance map.

We simply choose several points randomly on each triangle $T^{(k)}$ of the triangulation and use the same normal vector $n^{(k)}$ for all of them. The number of chosen points depends on the precomputed area of $T^{(k)}$ in order to get sufficiently dense sampling of the surface.

5.2 Downsampling the Point Cloud

The continuous solid object is discretized on a uniform regular 3D grid $\mathcal{M}$. We specify a grid number for the space direction that has the largest range and the other two will be determined by their related range. The voxels are cubes with the side length h.

To speed up the computation, a simple downsampling preprocessing based on the interpolation to the closest neighbor on the grid $\mathcal{M}$ is applied. By such a preprocessing, the quality of the resulting point cloud does not change when h is sufficiently small.

5.3 Distance Function and Closest Point Map

The distance function, or the distance map, of a point $x \in \Omega$ to a point set $\mathcal{P} = \{p_i\}_{i=1}^N$ is defined as

$$d(x) = \min_i \|x - p_i\|. \tag{3}$$

We use the Euclidean norm in (3). Note that $d(x) \geq 0$ and $d(x)$ is exactly 0 at the point cloud. The function `bwdist` in MATLAB implements the algorithm from [9] for computation of $d(x)$.

A very useful byproduct of the distance function is the closest point map. At a point $x \in \Omega$, we denote by $\mathrm{cp}(x)$ the point $p_i \in \mathcal{P}$ that is closest to x,

$$\mathrm{cp}(x) = \arg\min_i \|x - p_i\|. \tag{4}$$

5.4 Indicator Function

Assume that all points in the cloud $\mathcal{P}$ have a unit outward normal. For any $x \in \Omega$, we define the indicator function

$$f(x) = (x - \mathrm{cp}(x)) \cdot n(\mathrm{cp}(x)), \tag{5}$$

which equals the dot product of the vector $x - \mathrm{cp}(x)$ with the unit outward normal $n(\mathrm{cp}(x))$ at $\mathrm{cp}(x)$. The indicator function $f(x)$ is negative when x is inside B and positive when x is outside B. The function (5) is called the inner product field in [6]. The function $\mathrm{sign}(f(x))d(x)$ is computed exactly for polyhedra in [1] and called the signed distance function there.

5.5 Denoising

The indicator function $f(x)$ in (5) is not stably defined when the vectors $x - \mathrm{cp}(x)$ and $n(\mathrm{cp}(x))$ are orthogonal or almost orthogonal. Therefore, $f(x)$ can be subject to impulse noise in such locations x. The noise is removed by the median filter.

More expensive denoising filter TVG-L1 is proposed in [2]. It produces a smoother 0-level set of $f(x)$ than the median filter. We remark that the median filter provides a quite satisfactory result because our goal is computing the volume of B, not the boundary ∂B. Moreover, the use of $f(x)$ without denoising is often satisfactory. A test of accuracy is given in Sect. 5.6.

The TVG-L1 filter applied to $f(x)$ is a solution $u(x)$ of the variational model

$$\min_{u} \int_{\Omega} g(x)|\nabla u(x)| + \lambda |u - f| dx, \tag{6}$$

where $g(x) > 0$ is a weight function and λ is a suitable smoothing parameter. Solving (6) directly is not easy, therefore, it is often approximated by the easier variational model

$$\min_{u,v} \int_{\Omega} \left(g(x)|\nabla u| + \lambda |v| + \frac{1}{2\theta}|u + v - f|^2 \right) dx, \tag{7}$$

which converges to (6) for sufficiently small $\theta > 0$. We use $g(x) = d(x)$. A numerical algorithm for solving (7) is found in [2,6]. In our implementation, the distance function and indicator function are scaled to be ranged over the interval $[0, 1]$ and then we set $\lambda = 0.01$ and $\theta = 0.05$ as in [6].

5.6 Test of Accuracy

The set of interior grid points of $\mathcal{M}$ in B is $\mathcal{M}_i = \{x \in \mathcal{M} \colon f(x) < 0\}$, or $\mathcal{M}_i = \{x \in \mathcal{M} \colon u(x) < 0\}$ after denoising. An approximate volume of B equals $h^3|\mathcal{M}_i|$, where $|\mathcal{M}_i|$ is the number of points in $\mathcal{M}_i$. We compare $h^3|\mathcal{M}_i|$ with the exact volume computed by (2). If the relative error is not small enough, the computation of $f(x)$ and $u(x)$ is repeated with a smaller voxel size h.

Finally, we can calculate the number of points from $\mathcal{M}_i$ that belong to the ball B and multiply it by h^3 in order to get the volume of the intersection $B \cap C$.

6 Numerical Illustration

Two examples have been computed in MATLAB to examine the proposed numerical method: one for a normal glenoid bone and another for a defective one. The examples are quite similar so we present only results for a defective glenoid bone. The exact volume of the whole bone evaluated from the STL data by the Gauss formula equals 1.99161. The exact area of the bone surface equals 25.6750. The computed approximate volume of the bone amounts to 1.99281 before denoising, to 1.98723 after denoising with the median filter and to 1.99284 after denoising with the TVG-L1 filter. The uniform grid $\mathcal{M}$ is of size $161 \times 225 \times 283$ and has the voxel size $h = 0.0169$. The number of cloud points on the boundary after downsampling is 124190. Arithmetical complexity of the proposed numerical method is $O(|STL|) + O(|\mathcal{M}|)$, where $|STL|$ is the length of all STL data and $|\mathcal{M}|$ is the number of voxels in Ω (Fig. 2).

Fig. 2. 0-level of the denoised indicator function computed by the MATLAB function `isosurface`

7 Conclusion

We have developed a simple method for computing the volume of the intersection of a three-dimensional ball with a polyhedron given by a triangulated closed surface and outward unit normals. The method is based on voxelization of the interior of the polyhedron and uses the closest point map to a sufficiently fine sampling of the triangulated surface. Arithmetic complexity of the method is linear with respect to the number of voxels in a parallelepiped containing the polyhedron. The computed volume is approximate but its accuracy is guaranteed by a posteriori estimate calculated with the help of the Gauss-Ostrogradsky divergence theorem.

References

1. Bærentzen, J.A., Aanæs, H.: Signed distance computation using the angle weighted pseudonormal. IEEE Trans. Vis. Comput. Graph. **11**(3), 243–253 (2005). https://doi.org/10.1109/TVCG.2005.49

2. Bresson, X., Esedoglu, S., Vanderheynst, P., Thiran, J.P., Osher, S.: Fast global minimization of the active contour/snake model. J. Math. Imaging Vis. **28**, 151–167 (2007). https://doi.org/10.1007/s10851-007-0002-0
3. Gyftopoulos, S., et al.: Use of 3D MR reconstructions in the evaluation of glenoid bone loss: a clinical study. Skeletal Radiol. **43**(2), 213–218 (2013). https://doi.org/10.1007/s00256-013-1774-5
4. Hamamoto, J.T., Leroux, T., Chahla, J., et al.: Assessment and evaluation of glenoid bone loss. Arthrosc. Tech. **5**(4), e947–e951 (2016). https://doi.org/10.1016/j.eats.2016.04.027
5. Jiang, X., Peng, Q., Cheng, X., et al.: Efficient Booleans algorithms for triangulated meshes of geometric modeling. Comput. Aided Des. Appl. **13**(4), 419–430 (2016). https://doi.org/10.1080/16864360.2015.1131530
6. Liang, J., Park, F., Zhao, H.: Robust and efficient implicit surface reconstruction for point clouds based on convexified image segmentation. J. Sci. Comput. **54**, 577–602 (2013). https://doi.org/10.1007/s10915-012-9674-8
7. Magalhães, S.V., Andrade, M.V., Franklin, W.R., Li, W.: PinMesh–fast and exact 3D point location queries using a uniform grid. Comput. Graph. **58**, 1–11 (2016). https://doi.org/10.1016/j.cag.2016.05.017
8. Magalhães, S.V., Franklin, W.R., Andrade, M.V.: An efficient and exact parallel algorithm for intersecting large 3-D triangular meshes using arithmetic filters. Comput. Aided Des. **120**(102801), 1–11 (2020). https://doi.org/10.1016/j.cad.2019.102801
9. Maurer, C., Qi, R., Raghavan, V.: A linear time algorithm for computing exact Euclidean distance transforms of binary images in arbitrary dimensions. IEEE Trans. Pattern Anal. Mach. Intell. **25**(2), 265–270 (2003). https://doi.org/10.1109/TPAMI.2003.1177156
10. Mirtich, B.: Fast and accurate computation of polyhedral mass properties. J. Graph. Tools **1**(2), 31–50 (1996). https://doi.org/10.1080/10867651.1996.10487458
11. Pavić, D., Campen, M., Kobbelt, L.: Hybrid Booleans. Comput. Graph. Forum **29**(1), 75–87 (2010). https://doi.org/10.1111/j.1467-8659.2009.01545.x
12. e Souza, P.M., Brandão, B.L., Brown, E., Motta, G., Monteiro, M., Marchiori, E.: Recurrent anterior glenohumeral instability: the quantification of glenoid bone loss using magnetic resonance imaging. Skeletal Radiol. **43**(8), 1085–1092 (2014). https://doi.org/10.1007/s00256-014-1894-6
13. Sugaya, H., Moriishi, J., Dohi, M., Kon, Y., Tsuchiya, A.: Glenoid rim morphology in recurrent anterior glenohumeral instability. J. Bone Joint Surg. **85**(5), 878–884 (2003). https://doi.org/10.2106/00004623-200305000-00016
14. Xiao, Z., Chen, J., Zheng, Y., Zheng, J., Wang, D.: Booleans of triangulated solids by aboundary conforming tetrahedral mesh generation approach. Comput. Graph. **59**, 13–27 (2016). https://doi.org/10.1016/j.cag.2016.04.004
15. Yanke, A.B., Shin, J.J., Pearson, I., et al.: Three-dimensional magnetic resonance imaging quantification of glenoid bone loss is equivalent to 3-dimensional computed tomography quantification: cadaveric study. Arthroscopy: J. Arthroscopic Relat. Surg. **33**(4), 709–715 (2017). https://doi.org/10.1016/j.arthro.2016.08.025
16. Zhou, Q., Grinspun, E., Zorin, D., Jacobson, A.: Mesh arrangements for solid geometry. ACM Trans. Graph. **35**(4), 1–15 (2016). Article No. 39. https://doi.org/10.1145/2897824.2925901

Artificial Immune Systems Approach for Surface Reconstruction of Shapes with Large Smooth Bumps

Akemi Gálvez[1,2], Iztok Fister Jr.[3], Lihua You[4], Iztok Fister[3], and Andrés Iglesias[1,2](✉)

[1] Department of Applied Mathematics and Computational Sciences, University of Cantabria, 39005 Santander, Spain
{galveza,iglesias}@unican.es

[2] Faculty of Pharmaceutical Sciences, Toho University, 2-2-1 Miyama, Funabashi 274-8510, Japan

[3] Faculty of Electrical Engineering and Computer Science, University of Maribor, Maribor, Slovenia
{iztok.fister1,iztok.fister}@um.si

[4] National Center for Computer Animation, Faculty of Media and Communication, Bournemouth University, Poole BH12 5BB, UK
lyou@bournemouth.ac.uk

Abstract. Reverse engineering is one of the classical approaches for quailty assessment in industrial manufacturing. A key technology in reverse engineering is surface reconstruction, which aims at obtaining a digital model of a physical object from a cloud of 3D data points obtained by scanning the object. In this paper we address the surface reconstruction problem for surfaces that can exhibit large smooth bumps. To account for this type of features, our approach is based on using exponentials of polynomial functions in two variables as the approximating functions. In particular, we consider three different models, given by bivariate distributions obtained by combining a normal univariate distribution with a normal, Gamma, and Weibull distribution, respectively. The resulting surfaces depend on some parameters whose values have to be optimized. This yields a difficult nonlinear continuous optimization problem solved through an artificial immune systems approach based on the clonal selection theory. The performance of the method is discussed through its application to a benchmark comprised of three examples of point clouds.

Keywords: Artificial intelligence · reverse engineering · surface reconstruction · artificial immune systems · bivariate distributions · point clouds · data fitting

1 Introduction

1.1 Motivation

Nowadays, there is a renewed and increasing interest in the fields of artificial intelligence (AI) and machine learning (ML). This popularity is due in large

J. Mikyška et al. (Eds.): ICCS 2023, LNCS 14076, pp. 297–310, 2023.
https://doi.org/10.1007/978-3-031-36027-5_22

part to the extraordinary advances of AI and ML in areas such as pattern recognition, computer vision, robotics, healthcare, self-driving cars, natural language processing, automatic machine translation, and many others. The industrial sector is at the core of most of these innovations, with initiatives such as Industry 4.0, and Internet of Things (IoT), paving the way to a new field commonly known as *industrial artificial intelligence*. By this term we refer to the application of AI and ML methods and developments to industrial processes in order to improve the production and manufacturing systems.

One of the most interesting applications of industrial artificial intelligence arises in *quality assessment*, where typically AI methods are applied to analyze the quality of a digital design or a manufactured workpiece to determine whether or not certain aesthetic and/or functional objectives are met. In many industrial settings, quality assessment is carried out through *reverse engineering*, where the goal is to obtain a digital replica of a manufactured good. For instance, reverse engineering is widely used in the design and manufacturing of CAD models for car bodies in the automotive industry, plane fuselages for the aerospace industry, ship hulls in shipbuilding industry, moulds and lasts in footwear industry, components for home appliances, cases for consumer electronics, and in many other fields [32].

A key technology in reverse engineering is *surface reconstruction* [36]. Starting with a cloud of 3D data points obtained by 3D scanning of the physical object, surface reconstruction aims to recover the underlying shape of the real object in terms of mathematical equations of the surfaces fitting these data points, which is a much better way to store and manipulate the geometric information than using the discrete data points directly. These mathematical equations can then be efficiently used for different computer-assisted quality assessment processes, such as shape interrogation, shape analysis, failure detection and diagnosis, and many others. This approach is also used for intellectual property right assessment, industrial property plagiarism control, and other industrial and legal issues.

A central problem in surface reconstruction is the selection of the approximating functions. Classical choices are the free-form parametric polynomial surfaces, such as the Bézier and the B-spline surfaces, which are widely used in computer graphics and geometric design. However, it has been noticed that depending on the geometry of the point cloud, other choices might also be adequate. For instance, exponential functions are particularly suitable to model surfaces with large smooth bumps, as evidenced by the shape of the Gaussian function. Yet, it is difficult to manipulate shapes with a simple exponential function. The exponential of polynomial functions of two variables provides more flexibility, as it introduces extra degrees of freedom that can be efficiently used to modify the global shape of the surface while handling local features as well. Owing to these reasons, this is the approach followed in this paper.

1.2 Aims and Structure of this Paper

In this paper we address the problem of surface reconstruction from data points by using exponentials of polynomial functions in two variables as the approximating functions. In particular, we consider three different models of exponentials of polynomial functions, given by bivariate distributions obtained by combining a normal univariate distribution with a normal, Gamma, and Weibull distribution, respectively. The resulting surfaces depend on some parameters whose values have to be optimized. This yields a difficult nonlinear continuous optimization problem that will be solved through an artificial immune systems approach called ClonalG. The performance of the method will be discussed through its application to a benchmark comprised of three examples of point clouds.

The structure of this paper is as follows: Sect. 2 summarizes the previous work in the field. Section 3 describes the optimization problem addressed in this work. Our method to solve it is described in detail in Sect. 4. The performance of the method is illustrated through three illustrative examples, which are discussed in Sect. 5. The paper closes in Sect. 6 with the main conclusions and some ideas for future work in the field.

2 Previous Work

The issue of surface reconstruction for shape quality assessment has been as topic of research for decades. Early computational algorithms were introduced the 60s and 70s, mostly based on numerical methods [11,33,34]. Subsequent advances during the 80s and 90s applied more sophisticated techniques, although they failed to provided general solutions [3,10]. From a mathematical standpoint, this issue can be formulated as a least-squares optimization problem [26,28,31]. However, classical mathematical optimization techniques had little success in solving it beyond rather simple cases, so the scientific community focused on alternative approaches, such as error bounds [29], dominant points [30] or curvature-based squared distance minimization [37]. These methods provide acceptable results but they need to meet strong conditions such as high differentiability and noiseless data that are not so common in industrial settings.

More recently, methods based on artificial intelligence and soft computing are receiving increasing attention. Some approaches are based on neural networks [18], self-organizing maps [19], or the hybridization of neural networks with partial differential equations [2]. These neural approaches have been extended to functional networks in [20,27] and hybridized with genetic algorithms [16]. Other approaches are based on support vector machines [25] and estimation of distribution algorithms [40]. Other techniques include genetic algorithms [17,38,39], particle swarm optimization [12,13], firefly algorithm [14], cuckoo search algorithm [22,24], artificial immune systems [23], and hybrid techniques [15,21,35]. It is important to remark that none of the previous approaches addressed the problem discussed in this paper.

3 The Optimization Problem

As explained above, our approach to the surface reconstruction problem is to consider exponential of bivariate polynomial functions as the fitting functions.

A suitable way to proceed in this regards is to consider bivariate distributions whose conditionals belong to such families of basis functions [4,5]. In particular, we consider the combination of a normal univariate distribution with the normal, Gamma, and Weibull distributions, respectively. In the first case, the approximating function takes the form:

$$f(x,y) = e^{\frac{C_0}{2} - C_2\,x + C_1\,\frac{x^2}{2} - C_3\,\frac{y^2}{2} - C_4\,\frac{x^2y^2}{2} + C_5xy^2 + C_6y + C_7x^2y - 2C_8xy} \tag{1}$$

which is a model depending on 9 parameters. However, these parameters are not fully free, as they have to fulfill some constraints such as non-negativity and integrability, leading to the following constraints (see [1] for details):

$$C_4 > 0; \quad C_3C_4 > C_5^2 \quad ; \quad -C_1C_4 > C_7^2 \tag{2}$$

In the second case, the approximating function is given by:

$$f(x,y) = e^{F + Ay - Cy^2 + (G + By - Dy^2)x + (H + Jy - Ky^2)log(x)} \tag{3}$$

which depends on 9 parameters, with the constraints:

$$C > 0; \quad D > 0; \quad G < \frac{-B^2}{4D}; \quad H > -1; \quad J = 0; \quad K = 0 \tag{4}$$

The third model is given by:

$$f(x,y) = e^{D + L\,x + F\,x^2 - \left(A + B\,x + G\,x^2\right)(y-K)^C}\,(y-K)^{C-1} \tag{5}$$

which is a model depending on 8 parameters with the following constraints:

$$G \geqslant 0; \quad 4GA \geqslant B^2; \quad C > 0; \quad F < 0 \tag{6}$$

Once the approximating function is selected, the surface reconstruction procedure requires to compute the parameters of the function to obtain an accurate mathematical representation of the function $f(x,y)$ approximating the point cloud accurately. This condition can be formulated as the minimization problem:

$$min\left\{\sum_{p=1}^{P}\left[(x_p - \hat{x_p})^2 + (y_p - \hat{y_p})^2 + (z_p - f(\hat{x_p}, \hat{y_p}))^2\right]\right\} \tag{7}$$

where (x_p, y_p, z_p) and $(\hat{x_p}, \hat{y_p}, \hat{z_p})$ denote the original and reconstructed data points, respectively, and $\hat{x_p}$ and $\hat{y_p}$ can be obtained by projecting the point cloud onto a flat surface $B(x,y)$ determined by principal component analysis. Also, our minimization problem is restricted to the support of the function $f(x,y) - B(x,y)$ and subjected to some parametric constraints given by the pairs of Eqs. (1)–(2), Eqs. (3)–(4), and Eqs. (5)–(6), respectively.

This minimization problem is very difficult to solve, as it becomes a constrained, multivariate, nonlinear, multimodal continuous optimization problem. As a consequence, usual gradient-based mathematical techniques are not suitable to solve it. In this paper, we apply a powerful artificial immune systems algorithm called ClonalG to solve this problem. It is explained in detail in next section.

4 The Proposed Method: ClonalG Algorithm

The *ClonalG algorithm* is a computational method of the family of artificial immune systems (AIS), which are nature-inspired metaheuristic methods based on different aspects and features of the natural immune systems of humans and other mammals [6,7]. In particular, the ClonalG algorithm is based on the widely accepted clonal selection theory, used to explain how the immune system reacts to antigenic stimulus [8,9]. When a new antigen Ag attacks the human body, our immune system elicits an immunological response in the form of antibodies Ab, which are initially only slightly specific to the antigen. A measure of the affinity between the antibodies and the antigen determines which antibodies will be selected for proliferation: those with the highest affinity with the antigen, with the rest being removed from the pool. The selected antibodies undergo an affinity maturation process that enhances their affinity to the antigen over the time. A somatic mutation on the selected antibodies promotes higher diversity of the population of antibodies, so that the affinity improves further during the process. This mutation process is carried out at a much higher (about five or six orders of magnitude) rate than normal mutation, and is therefore called somatic hypermutation.

The ClonalG method was originally envisioned for pattern recognition tasks, using this natural process of immune response as a metaphor. The patterns to be learned (or input patterns) play the role of antigens, which are presented to the computational system (a metaphor of the human body). Whenever a pattern A is to be recognized, it is presented to a population of antibodies B_i, and the affinity between the couples (A, B_i) is computed based on a measure of the pattern similarity (for instance, the Hamming distance between images).

The algorithm is population-based, as it maintains a population of antibodies representing the potential matching patterns, and proceeds iteratively, along generations. It is summarized as follows (see [8,9] for further details):

1. An antigen Ag_j is randomly selected and presented to the collection of antibodies Ab_i, with $i = 1, \ldots, M$, where M is the size of the set of antibodies.
2. A vector affinity $\mathbf{f}$ is computed, as $f_i = Af(Ab_i, Ag_j)$ where Af represents the affinity function.
3. The N highest affinity components of $\mathbf{f}$ are selected for next step.
4. The selected antibodies are cloned adaptively, with the number of clones proportional to the affinity.
5. The clones from the previous step undergo somatic hypermutation, with the maturation rate inversely proportional to the affinity.
6. A new vector affinity $\mathbf{f'}$ on the new matured clones is computed.
7. The highest affinity antibodies from set of matured clones are selected for a memory pool. This mechanism is intended as an elitist strategy to preserve the best individuals for next generations.
8. The antibodies with the lowest affinity are replaced by new random individuals and inserted into the whole population along with the memory antibodies.

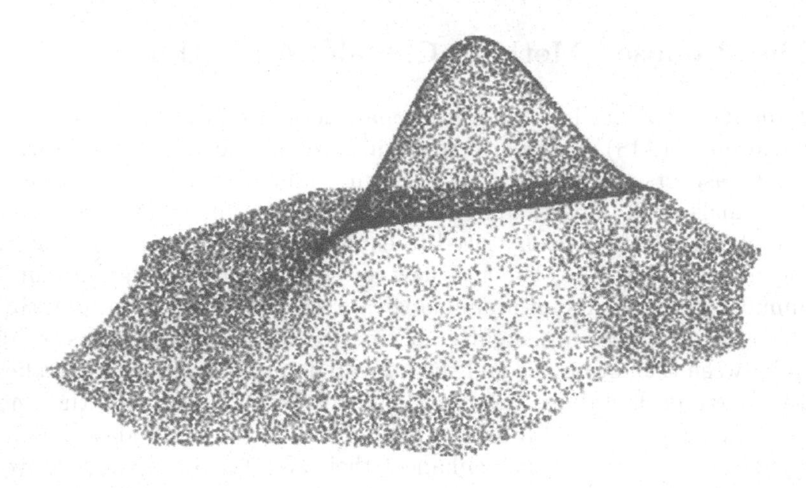

Fig. 1. Original point cloud of *Example I.*

The algorithm is repeated for a given number of generations, N_{gen}, which is a parameter of the method. This algorithm has proved to be efficient for pattern recognition tasks. With some modifications, it can also be applied to solve optimization problems. In short, the main modification is that, for optimization problems, there is no pattern to be learned; instead, a fitness function has to be optimized. In that case, the whole population can be cloned, although it is convenient to preserve an unmuted copy of the best individuals during the maturation step to speed up the method convergence.

Regarding the parameter tuning of the method, the only parameters of the ClonalG algorithm are the population size and the maximum number of iterations. We applied a fully empirical approach for the choice of these values: they have been determined after conducting several computer simulations for different parameter values. After this step, we selected a population of 50 individuals (antibodies) for the method, and a total number of 500 iterations, which have been more than enough to reach convergence in all our simulations. The best solution reached at the final iteration is selected as the optimal solution of the minimization problem in this work.

5 Experimental Results

The method described in the previous section has been applied to several examples of point clouds. For limitations of space, we restrict our discussion to three illustrative examples, fitted according to the models in Eqs. (1), (3), and (5) for *Example I*, *Example II*, and *Example III*, respectively, as discussed in the following paragraphs.

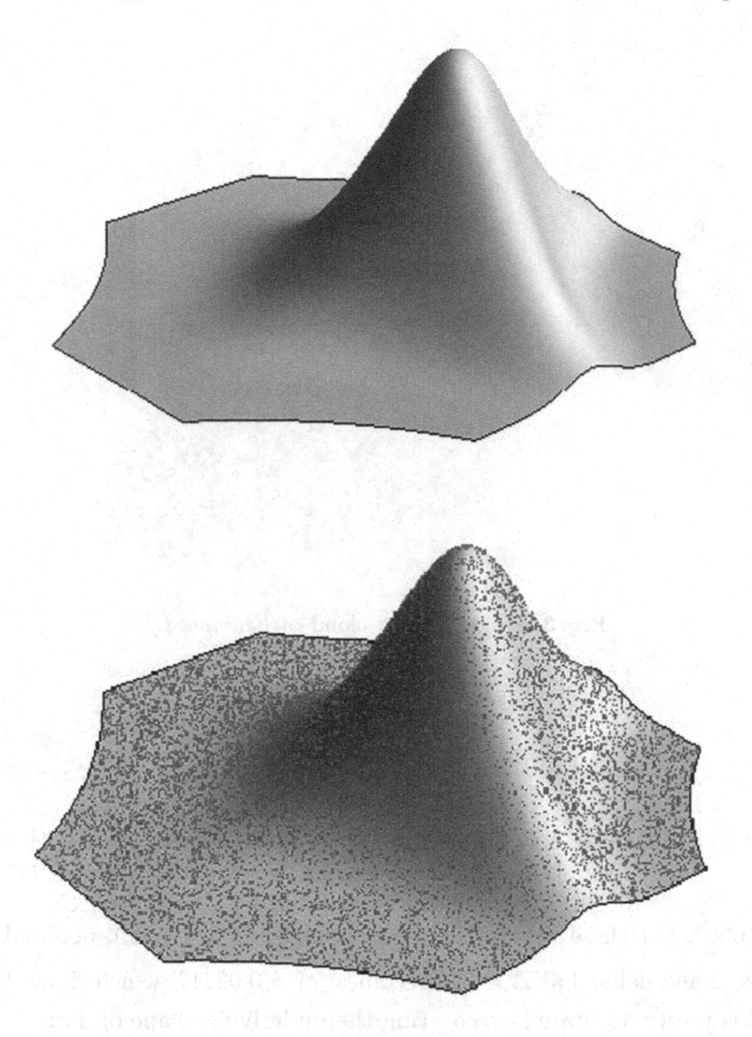

Fig. 2. *Example I*: (top) Reconstructed surface; (bottom) reconstructed surface and data points.

5.1 Example I

This example consists of a cloud of $R = 32,167$ three-dimensional data points displayed in Fig. 1. The data points do not follow a uniform parameterization and are affected by white noise of low intensity (SNR = 10). The point cloud is fitted with model 1 according to Eq. (1) with the constraints in Eq. (2). Therefore, the resulting optimization problem consists of minimizing the functional:

$$\Xi = \sum_{p=1}^{P} \left[(x_p - \hat{x_p})^2 + (y_p - \hat{y_p})^2 + (z_p - f(\hat{x_p}, \hat{y_p}))^2 \right]$$

Fig. 3. Original point cloud of *Example II.*

for

$$f(\hat{x_p}, \hat{y_p}) = e^{\frac{C_0}{2} - C_2 \hat{x_p} + C_1 \frac{\hat{x_p}^2}{2} - C_3 \frac{\hat{y_p}^2}{2} - C_4 \frac{\hat{x_p}^2 \hat{y_p}^2}{2} + C_5 \hat{x_p} \hat{y_p}^2 + C_6 y + C_7 \hat{x_p}^2 \hat{y_p} - 2C_8 \hat{x_p} \hat{y_p}}$$

Applying our method to the minimization of the functional Ξ with the constraints in Eq. (2) we obtained the values: $C_0 = -3.1815; C_1 = -0.9936; C_2 = -0.8977; C_3 = 1.0249; C_4 = 0.9892; C_5 = 0.0073; C_6 = 0.9105; C_7 = 0.0104; C_8 = 0.0027$. For these values, the mean squared error (MSE), denoted in this paper as Δ and defined as $\Delta = \frac{\Xi}{P}$, becomes: $\Delta = 0.02417$, which shows that the method is pretty accurate in recovering the underlying shape of data.

The best reconstructed surface is displayed in Fig. 2(top), where the bottom picture shows the superposition of the fitting surface and the original point cloud for better visualization. From that figure, the good numerical accuracy of the method is visually confirmed, as the fitting surface reproduces the global shape of the point cloud with very good visual fidelity. Note also that the shape of the prominent bump at the center of the surface is faithfully reconstructed.

5.2 Example II

The second example consists of a cloud of $R = 30,753$ three-dimensional data points shown in Fig. 3. The point cloud is fitted with model 2 according to Eq. (3) with the constraints in Eq. (4). Application of the ClonalG method described in Sect. 4 yields the values: $A = 2.9782; B = 2.0273; C = 4.0792; D = 2.9885; K = 0.0; F = 3.9811; G = -1.0337; H = 0.0101; J = 0.0$, for which the mean squared

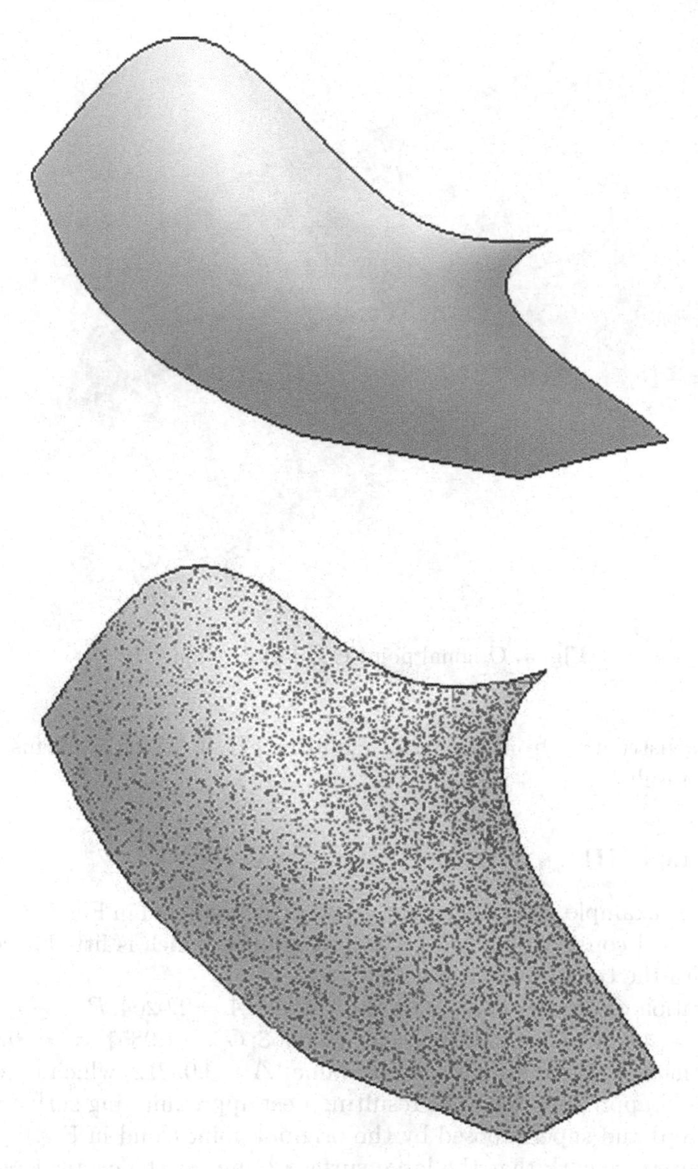

Fig. 4. *Example II*: (top) Reconstructed surface; (bottom) reconstructed surface and data points.

error takes the value: $\Delta = 0.01352$, an excellent indicator of good fitting. Note also that $J = K = 0$ is not directly obtained from the optimization method but given as an input to the method via the constraints in Eq. (4).

Figure 4 shows the optimal reconstructed surface (top) and its superposition with the point cloud (bottom). Note again the excellent visual quality of the

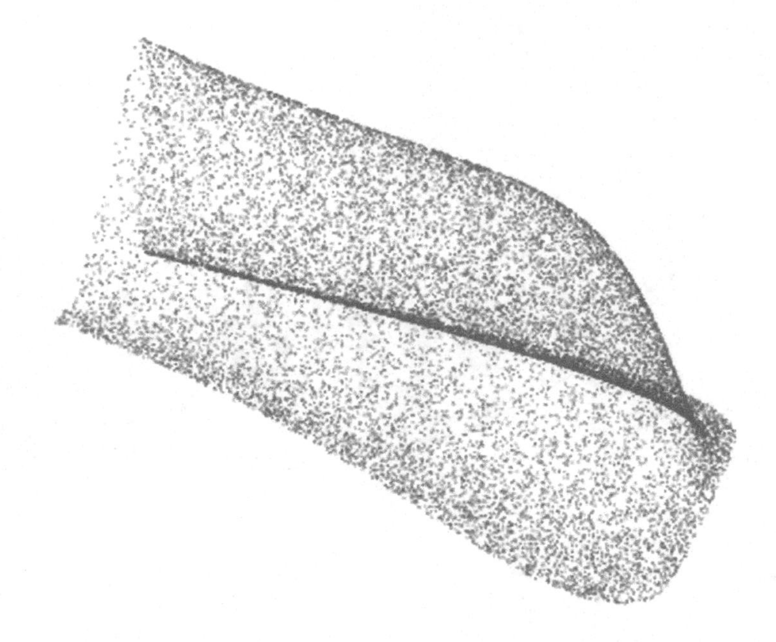

Fig. 5. Original point cloud of *Example III.*

surface reconstruction from the point cloud in Fig. 3, which confirms our good numerical results.

5.3 Example III

For the third example, we consider the cloud point depicted in Fig. 5. In this case, the point cloud consists of $R = 30,679$ data points, which is fitted according to Eq. (5) with the constraints in Eq. (6).

Application of our method yields the values: $A = 2.0204; B = -3.9926; C = 1.9905; D = 3.0168; L = 0.0116; F = -2.0068; G = 1.9851; K = 0.0104$, for which the mean squared error takes the value: $\Delta = 0.09212$, which is considered a satisfactory approximation. The resulting best approximating surface is shown in Fig. 6 (top) and superimposed by the original point cloud in Fig. 6 (bottom). Once again, we remark that the large surface bump at the center is accurately reconstructed.

5.4 Implementation Issues

The computations in this paper have been carried out on a PC desktop with a processor Intel Core i9 running at 3.7 GHz and with 64 GB of RAM. The source code has been implemented by the authors in the programming language of the scientific program *Mathematica* version 12. About the computational times, our method is quite fast. Each execution of the method takes only a few seconds of

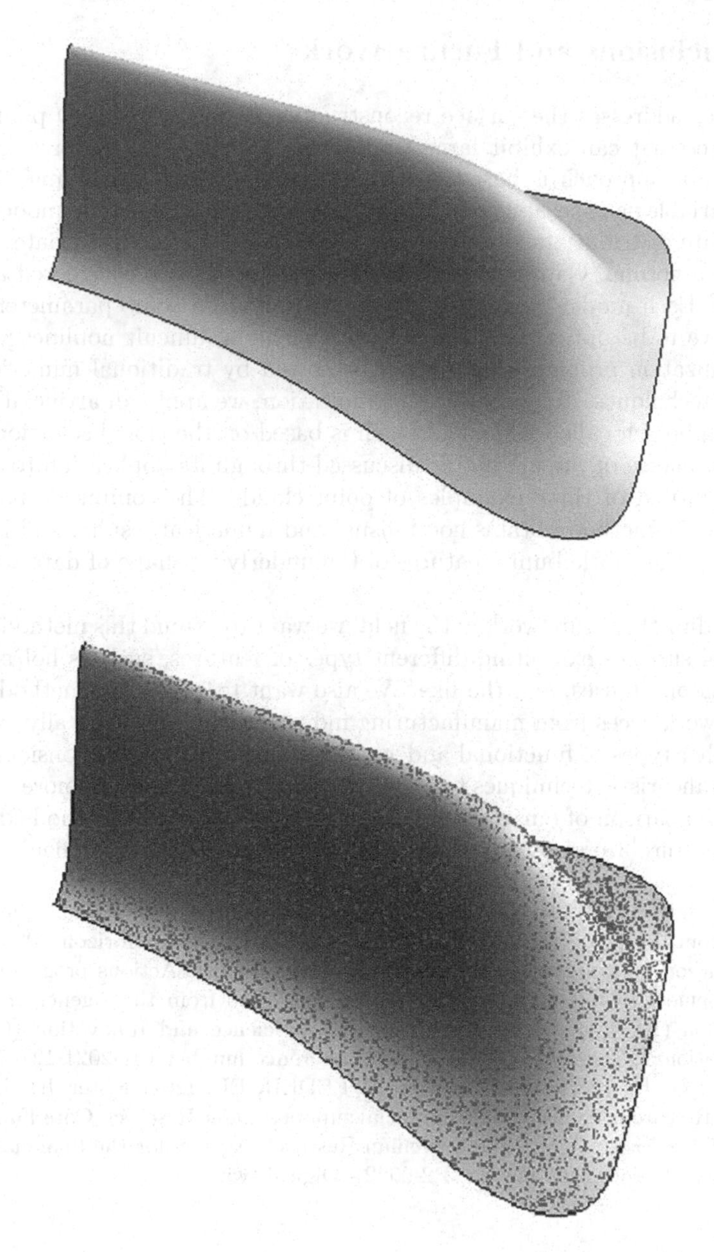

Fig. 6. *Example III*: (top) Reconstructed surface; (bottom) reconstructed surface and data points.

CPU time, depending on the population size, number of iterations, complexity of the problem, and other factors. For reference, the executions of the examples in this paper take about 3–6 s.

6 Conclusions and Future Work

This paper addresses the surface reconstruction problem from 3D point clouds for surfaces that can exhibit large smooth bumps. To account for this type of features, our approach is based on using exponentials of polynomial functions in two variables as the approximating functions. Three different models, given by bivariate distributions obtained by combining a normal univariate distribution with a normal, Gamma, and Weibull distribution, are considered as fitting functions. Each model leads to surfaces depending on some parameters whose values have to be optimized. However, this yields a difficult nonlinear continuous optimization problem that cannot be solved by traditional numerical optimization techniques. To overcome this limitation, we apply an artificial immune systems approach called ClonalG, which is based on the clonal selection theory. The performance of the method is discussed through its application to a benchmark comprised of three examples of point clouds. The computational results show that the method obtains good visual and numerical results, and is able to reconstruct the subtle bump features of the underlying shape of data with good accuracy.

Regarding the future work in the field, we want to extend this method to other families of surfaces exhibiting different types of features, such as holes, critical points, discontinuities, and the like. We also want to apply this methodology to complex workpieces from manufacturing industries that can typically require to satisfy other types of functional and/or design constraints. The consideration of other metaheuristic techniques to solve the optimization problem more efficiently and the comparison of our results with other state-of-the-art methods described in the literature are also part of our plans for future work in the field.

Acknowledgment. Akemi Gálvez, Lihua You and Andrés Iglesias thank the financial support from the project PDE-GIR of the European Union's Horizon 2020 research and innovation programme, in the Marie Sklodowska-Curie Actions programme, with grant agreement of reference number 778035, and also from the Agencia Estatal de Investigación (AEI) of the Spanish Ministry of Science and Innovation (Computer Science National Program), for the grant of reference number PID2021-127073OB-I00 of the MCIN/AEI/10.13039/501100011033/FEDER, EU. Iztok Fister Jr. thanks the Slovenian Research Agency for the financial support under Research Core Funding No. P2-0057. Iztok Fister thanks the Slovenian Research Agency for the financial support under Research Core Funding No. P2-0042 - Digital twin.

References

1. Arnold, B., Castillo, E., Sarabia, J.M.: Conditionally Specified Distributions. Lecture Notes in Statistics, vol. 73. Springer, Berlin (1992). https://doi.org/10.1007/978-1-4612-2912-4
2. Barhak, J., Fischer, A.: Parameterization and reconstruction from 3D scattered points based on neural network and PDE techniques. IEEE Trans. Vis. Comput. Graph. **7**(1), 1–16 (2001)

3. Barnhill, R.E.: Geometric Processing for Design and Manufacturing. SIAM, Philadelphia (1992)
4. Castillo, E., Iglesias, A.: Some characterizations of families of surfaces using functional equations. ACM Trans. Graph. **16**(3), 296–318 (1997)
5. Castillo, E., Iglesias, A., Ruiz, R.: Functional Equations in Applied Sciences. Mathematics in Science and Engineering, vol. 199, Elsevier Science, Amsterdam (2004)
6. Dasgupta, D. (ed.): Artificial Immune Systems and Their Applications. Springer, Berlin (1999). https://doi.org/10.1007/978-3-642-59901-9
7. De Castro, L.N., Timmis, J.: Artificial Immune Systems: A New Computational Intelligence Approach. Springer, London (2002)
8. De Castro, L.N., Von Zuben, F.J.: Artificial immune systems: part I - basic theory and applications. Technical report-RT DCA 01/99 (1999)
9. De Castro, L.N., Von Zuben, F.J.: Learning and optimization using the clonal selection principle. IEEE Trans. Evol. Comput. **6**(3), 239–251 (2002)
10. Dierckx, P.: Curve and Surface Fitting with Splines. Oxford University Press, Oxford (1993)
11. Farin, G.: Curves and Surfaces for CAGD, 5th edn. Morgan Kaufmann, San Francisco (2002)
12. Gálvez, A., Iglesias, A.: Efficient particle swarm optimization approach for data fitting with free knot B-splines. Comput. Aided Des. **43**(12), 1683–1692 (2011)
13. Gálvez, A., Iglesias, A.: Particle swarm optimization for non-uniform rational B-spline surface reconstruction from clouds of 3D data points. Inf. Sci. **192**(1), 174–192 (2012)
14. Gálvez A., Iglesias A.: Firefly algorithm for polynomial Bézier surface parameterization. J. Appl. Math. **2013**, Article ID 237984 (2012)
15. Gálvez, A., Iglesias, A.: A new iterative mutually-coupled hybrid GA-PSO approach for curve fitting in manufacturing. Appl. Soft Comput. **13**(3), 1491–1504 (2013)
16. Gálvez, A., Iglesias, A., Cobo, A., Puig-Pey, J., Espinola, J.: Bézier curve and surface fitting of 3D point clouds through genetic algorithms, functional networks and least-squares approximation. In: Gervasi, O., Gavrilova, M.L. (eds.) ICCSA 2007. LNCS, vol. 4706, pp. 680–693. Springer, Heidelberg (2007). https://doi.org/10.1007/978-3-540-74477-1_62
17. Gálvez, A., Iglesias, A., Puig-Pey, J.: Iterative two-step genetic-algorithm method for efficient polynomial B-spline surface reconstruction. Inf. Sci. **182**(1), 56–76 (2012)
18. Gu, P., Yan, X.: Neural network approach to the reconstruction of free-form surfaces for reverse engineering. Comput. Aided Des. **27**(1), 59–64 (1995)
19. Hoffmann, M.: Numerical control of Kohonen neural network for scattered data approximation. Numer. Algorithms **39**, 175–186 (2005)
20. Iglesias, A., Echevarría, G., Gálvez, A.: Functional networks for B-spline surface reconstruction. Futur. Gener. Comput. Syst. **20**(8), 1337–1353 (2004)
21. Iglesias, A., Gálvez, A.: Hybrid functional-neural approach for surface reconstruction. Math. Probl. Eng. (2014) Article ID 351648, 13 pages
22. Iglesias, A., Gálvez, A., Avila, A.: Hybridizing mesh adaptive search algorithm and artificial immune systems for discrete rational Bézier curve approximation. Vis. Comput. **32**, 393–402 (2016)
23. Iglesias, A., Gálvez, A., Avila, A.: Immunological approach for full NURBS reconstruction of outline curves from noisy data points in medical imaging. IEEE/ACM Trans. Comput. Biol. Bioinf. **15**(6), 929–1942 (2018)

24. Iglesias, A., et al.: Cuckoo search algorithm with Lévy flights for global-support parametric surface approximation in reverse engineering. Symmetry **10**(3), PaperID 58 (2018)
25. Jing, L., Sun, L.: Fitting B-spline curves by least squares support vector machines. In: Proceedings of the 2nd International Conference on Neural Networks & Brain, Beijing, China, pp. 905–909. IEEE Press (2005)
26. Jupp, D.L.B.: Approximation to data by splines with free knots. SIAM J. Numer. Anal. **15**, 328–343 (1978)
27. Knopf, G.K., Kofman, J.: Adaptive reconstruction of free-form surfaces using Bernstein basis function networks. Eng. Appl. Artif. Intell. **14**(5), 577–588 (2001)
28. Ma, W.Y., Kruth, J.P.: Parameterization of randomly measured points for least squares fitting of B-spline curves and surfaces. Comput. Aided Des. **27**(9), 663–675 (1995)
29. Park, H.: An error-bounded approximate method for representing planar curves in B-splines. Comput. Aided Geomet. Design **21**, 479–497 (2004)
30. Park, H., Lee, J.H.: B-spline curve fitting based on adaptive curve refinement using dominant points. Comput. Aided Des. **39**, 439–451 (2007)
31. Patrikalakis, N.M., Maekawa, T.: Shape Interrogation for Computer Aided Design and Manufacturing. Springer, Heidelberg (2002). https://doi.org/10.1007/978-3-642-04074-0
32. Pottmann, H., Leopoldseder, S., Hofer, M., Steiner, T., Wang, W.: Industrial geometry: recent advances and applications in CAD. Comput. Aided Des. **37**, 751–766 (2005)
33. Powell, M.J.D.: Curve fitting by splines in one variable. In: Hayes, J.G. (ed.) Numerical Approximation to Functions and Data. Athlone Press, London (1970)
34. Rice, J.R.: The Approximation of Functions, vol. 2. Addison-Wesley, Reading (1969)
35. Sarfraz, M., Raza, S.A.: Capturing outline of fonts using genetic algorithms and splines. In: Proceedings of Fifth International Conference on Information Visualization, IV 2001, pp. 738–743. IEEE Computer Society Press (2001)
36. Varady, T., Martin, R.: Reverse engineering. In: Farin, G., Hoschek, J., Kim, M. (eds.) Handbook of Computer Aided Geometric Design. Elsevier Science (2002)
37. Wang, W.P., Pottmann, H., Liu, Y.: Fitting B-spline curves to point clouds by curvature-based squared distance minimization. ACM Trans. Graph. **25**(2), 214–238 (2006)
38. Yoshimoto, F., Moriyama, M., Harada, T.: Automatic knot adjustment by a genetic algorithm for data fitting with a spline. In: Proceedings of Shape Modeling International 1999, pp. 162–169. IEEE Computer Society Press (1999)
39. Yoshimoto, F., Harada, T., Yoshimoto, Y.: Data fitting with a spline using a real-coded algorithm. Comput. Aided Des. **35**, 751–760 (2003)
40. Zhao, X., Zhang, C., Yang, B., Li, P.: Adaptive knot adjustment using a GMM-based continuous optimization algorithm in B-spline curve approximation. Comput. Aided Des. **43**, 598–604 (2011)

Machine Learning and Data Assimilation for Dynamical Systems

Clustering-Based Identification of Precursors of Extreme Events in Chaotic Systems

Urszula Golyska and Nguyen Anh Khoa Doan(✉)

Delft University of Technology, 2629HS Delft, The Netherlands
n.a.k.doan@tudelft.nl

Abstract. Abrupt and rapid high-amplitude changes in a dynamical system's states known as extreme events appear in many processes occurring in nature, such as drastic climate patterns, rogue waves, or avalanches. These events often entail catastrophic effects, therefore their description and prediction is of great importance. However, because of their chaotic nature, their modelling represents a great challenge up to this day. The applicability of a data-driven modularity-based clustering technique to identify precursors of rare and extreme events in chaotic systems is here explored. The proposed identification framework based on clustering of system states, probability transition matrices and state space tessellation was developed and tested on two different chaotic systems that exhibit extreme events: the Moehliss-Faisst-Eckhardt model of self-sustained turbulence and the 2D Kolmogorov flow. Both exhibit extreme events in the form of bursts in kinetic energy and dissipation. It is shown that the proposed framework provides a way to identify pathways towards extreme events and predict their occurrence from a probabilistic standpoint. The clustering algorithm correctly identifies the precursor states leading to extreme events and allows for a statistical description of the system's states and its precursors to extreme events.

Keywords: Machine Learning · Extreme events · Clustering · Chaotic Systems · Precursors

1 Introduction

Extreme events, defined as sudden transient high amplitude change in the system states, are present in many nonlinear dynamic processes that can be observed in everyday life, ranging from rogue waves [3], through drastic climate patterns [2,4], to avalanches and even stock market crashes. These kinds of instances are omnipresent and often result in catastrophes with a huge humanitarian and financial impact. As such, the prediction of these events can lead to a timely response, and therefore the prevention of their disastrous consequences.

Unfortunately, many of these problems are so complex that the creation of a descriptive mathematical model is nearly impossible. Over the years, three main approaches have been developed in an attempt to tackle this problem. Statistical methods, such as Large Deviation Theory [16] or Extreme Value Theory [12]

J. Mikyška et al. (Eds.): ICCS 2023, LNCS 14076, pp. 313–327, 2023.
https://doi.org/10.1007/978-3-031-36027-5_23

focus on trying to predict the associated heavy-tail distribution of the probability density function of the observable. Extreme events can be quantified with this method, but are not predicted in a time-accurate manner. Time-accurate forecasting, on the other hand, focuses on predicting the evolution of the system state before, during, and after extreme events. These methods are often based on Reduced Order Modeling and deep learning to try to accuractely reproduce the dynamics of the systems, including extreme events [5,6,13,17]. Theses approaches generally consider the full system states or a partial physical description which may not always be available. The last approach is precursor identification, which aims at identifying an observable of the system which will indicate the occurrence of extreme events within a specific time horizon. The precursors are typically identified based on characteristics of the system closely related to the quantity of interest which determines extreme incidents. Such precursors have traditionally been identified from physical consideration [7,10] which made them case-specific and non-generalizable. Recently, a clustering-based method [14,15] was developed that aims to identify such precursor in a purely data-driven manner by grouping the system's states by their similarity using a parameter called modularity and identifying the pathways to extreme states. This approach showed some potential in providing a purely data-driven approach to this problem of precursor identification which could be generalized to other chaotic systems but was limited to a low-order system.

The purpose of this work is to re-explore this clustering-based approach, apply it to high dimensional systems and assess its predictive capability (statistically and in terms of prediction horizon). The two systems explored here are the Moehlis-Faisst-Eckhardt (MFE) model [9] of the self-sustaining process in wall-bounded shear flows, also investigated in [14,15], and the Kolmogorov flow, a two dimensional flow which exhibits intermittent bursts of energy and dissipation. To the authors' knowledge, it is the first time this latter system is analysed using such clustering-based approach.

The structure of the paper is as follows. First, the methodology behind the developed algorithm is presented, with the specific steps made to reduce the problem at hand. The criteria for the extreme events and the construction of the transition matrix are described. The concept of modularity is then presented, together with the process of clustering the graph form of the data. Then, in Sect. 3, the results for the two discussed systems are shown, followed by a statistical description of the clusters found by the algorithm. The paper is concluded with a summary of the main findings and directions for future work.

2 Methodology

2.1 Preparatory Steps

The initial step consists in generating a long time evolution of the considered systems. These are obtained using ad-hoc numerical methods which will be discussed in Sect. 3 to create appropriate datasets on which to apply the proposed precursor identification technique.

The obtained dataset is then represented in a phase space which is chosen such that the system's dynamics is accurately represented, with the extreme events clearly visible. A combination of the time-dependent states from the data or a modification of them (from physical consideration) is chosen to span the phase space, allowing for a clear representation of the system's trajectory.

The phase space is then tessellated to reduce the system's size as a given system trajectory (in the phase space) only occupies a limited portion of the full phase space. A singular (high-dimensional) volume resulting from the tessellation of the phase space is referred to as a hypercube. This allows for a precise discretization of the system's trajectory, with no overlaps or gaps.

The time series defining the trajectory in phase space is then analyzed, translating the original data into indices of the respective hypercubes. The result is a time series of the hypercube index in which the system lies at a given time. At this point, the indices of the extreme events are also defined. For a certain point to be considered extreme, all of its quantities of interest must surpass a given specified threshold. These quantities of interest are some (or all) of the parameters spanning the phase space, which are used for defining the extreme events. The criteria for the extreme events are calculated based on user input and are here defined as:

$$\bigcap_{\alpha=1}^{N_\alpha} |x_\alpha| \geq \mu_\alpha + d \cdot \sigma_\alpha, \tag{1}$$

where μ_α and σ_α are the mean and standard deviation of the considered state parameter x_α. The total number of variables used to define the extreme events is N_α. Note that N_α could be smaller than the dimension of the phase space as one could use only a subset of the variables of the system to define an extreme event. The constant d is given by the user and is equal for all considered quantities of interest. All the criteria are then calculated (according to Eq. 1) and applied. A certain region in phase space, for which all of the quantities of interest fulfill the extreme criteria, is labeled as extreme. The tessellated hypercubes which fall into that region are also marked as extreme, which is later used for extreme cluster identification.

2.2 Transition Probability Matrix and Graph Interpretation

The tessellated data is then translated into a transition probability matrix. The algorithm calculates elements of the transition probability matrix $\mathbf{P}$ as:

$$P_{ij} = \frac{m(B_i \cap \mathcal{F}^1(B_j))}{m(B_i)} \qquad i, j = 1, ..., N, \tag{2}$$

where P_{ij} describes the probability of transitioning from hypercube B_i to hypercube B_j, N is the total number of hypercubes on the system trajectory and $\mathcal{F}^1$ is the temporal forward operator. The notation $m(B_i)$ represents the number of instances (phase space points) laying in hypercube i.

The result of this process is a sparse transition probability matrix $\mathbf{P}$ of the size M^n, where n is the number of dimensions of the phase space and M is the number of tessellation sections per dimension. The **P** matrix is generally highly diagonal as for the majority of the time, the trajectory will stay within a given hypercube for multiple consecutive time steps. The transitions to other hypercubes are represented by non-zero off-diagonal elements and are therefore much less frequent.

The transition probability matrix can then be interpreted as a weighted and directed graph. The nodes of the graph are the hypercubes of the tessellated trajectory and the graph edges represent the possible transition between hypercubes. The edge weights are the values of the probabilities of transitioning from one hypercube to another. This representation allows to interpret the system's trajectory as a network and further analyze it as one. This approach also preserves the essential dynamics of the system.

2.3 Modularity-Based Clustering

The clustering method adopted for this project is based on modularity maximization, as proposed in [11] and applied to identify clusters (also called communities) in the weighted directed graph, represented by the transition probability matrix **P** obtained in the previous subsection. The Python Modularity Maximization library [1] was re-used here. It implements the methods described in [11] and [8].

The chosen algorithm is based on maximizing the parameter called *modularity* which is a measure of the strength of the division of a network into communities. This parameter was created with the idea that a good division is defined by fewer than expected edges between communities rather than simply fewer edges. It is the deviation (from a random distribution) of the expected number of edges between communities that makes a division interesting. The more statistically surprising the configuration, the higher modularity it will result in [11].

The modularity value for a division of a directed graph into two communities can be written as:

$$Q = \frac{1}{m} \sum_{ij} \left(A_{ij} - \frac{k_i^{in} k_j^{out}}{m} \right) \delta_{s_i, s_j}, \tag{3}$$

where m is the total number of edges and A_{ij} is an element of the adjacency matrix and expresses the number of edges between two nodes. In this equation, s_i determines the specific node affiliation, such that when using the Kronecker delta symbol δ_{s_i,s_j} only nodes belonging to the same community contribute to the modularity. Notations k_i^{in} and k_j^{out} are the in- and out-degrees of the vertices for an edge going from vertex j to vertex i with probability $\frac{k_i^{in} k_j^{out}}{m}$. A community division that maximizes the modularity Q will be searched for.

The method proposed in [11] and used here solves this modularity maximization problem by using an iterative process that continuously divides the network into two communities at each iteration, using the division that increases the modularity the most at that given iteration. The process is finished once no further division increases the overall modularity of the network.

Matrix Deflation. After the new communities are identified by the algorithm, the transition probability matrix is deflated. A community affiliation matrix $\mathbf{D}$ is created, which associates all the nodes of the original graph to a given community (or cluster). The new transition probability matrix $\mathbf{P}_{(1)}$, which described the dynamics amongst the newly found communities, is then created by deflating the original matrix $\mathbf{P}$ according to:

$$\mathbf{P}_{(1)} = \mathbf{D}^{\mathrm{T}}\mathbf{P}\mathbf{D}. \tag{4}$$

$\mathbf{P}_{(1)}$ can then again be interpreted as a graph. The process of clustering, deflating the transition probability matrix and interpreting it as a graph is repeated iteratively until one of the two criteria is reached - the maximum number of iterations is exceeded or the number of communities falls below a specified value (making it humanly tractable). After this step, the matrix will still remain strongly diagonal, indicating that most of the transitions remain inside the clusters. Off-diagonal elements will then indicate the transitions to other clusters.

2.4 Extreme and Precursor Clusters Identification

Once the clustering algorithm is done, the extreme clusters are identified. The clusters containing nodes which were flagged as extreme are considered extreme. This is a very general definition, since even clusters containing only one extreme node will be marked as extreme. The deflated transition probability matrix is then used to identify all the clusters which transition to the extreme clusters. These are considered *precursor clusters*, since they are direct predecessors of the extreme clusters. All the other clusters are considered as *normal clusters* which correspond to a normal state of the system, i.e. far away from the extreme states. Compared to earlier work that also used a similar clustering-based approach [15], we identify the extreme and precursor clusters and perform a further statistical analysis of them, in terms of time spent in each cluster, its probability of transitioning from a precursor cluster to an extreme one and over which time such a transition will take place. Details on this additional analysis will be provided in Sect. 3.

3 Results

The results of applying the proposed clustering method applied to two chaotic systems exhibiting extreme events, the MFE system, also investigated in [15], and the Kolmogorov flow are now discussed.

3.1 MFE System

The MFE system is a model of the self-sustaining process in wall-bounded shear flows. It is governed by the evolution of nine modal coefficients, a_i, whose governing equations can be found in [9]. From those nine modal coefficients, the velocity field in the flow can be reconstructed as $\mathbf{u}(\boldsymbol{x},t) = \sum_{i=1}^{9} a_i(t)\boldsymbol{\phi}_i(\boldsymbol{x})$ where $\boldsymbol{\phi}_i$ are spatial modes whose definitions are provided in [9]. The governing equations of the MFE system are solved using a Runge-Kutta 4 scheme and are simulated for a total duration of 100,000 time units (400,000 time steps).

The turbulent kinetic energy $k = 0.5\sum_{i=1}^{9} a_i^2$ and energy dissipation $D = tr\left(\nabla\mathbf{u}(\nabla\mathbf{u} + \nabla\mathbf{u}^T)\right)$ were calculated and used as observables. This allowed for a reduction of the size of the system to consider from 9 to 2 dimensions. While such a dimension reduction of the system may seem like a simplification of the problem, it is actually a practical challenge for the precursor identification method as trajectories which were distinct in the full space may now be projected onto a similar reduced space making it harder to appropriately identify a precursor region. The evolution of the system is displayed in the k-D phase space in Fig. 1 (left). The phase space was then tessellated into 20 sections per dimension, which is also seen in Fig. 1 (right), where k and D are normalized using their respective minimum and maximum and divided into 20 sections. Both k and D are used for clustering as well as the determination of the extreme events. From the evolution shown in Fig. 1, the probability transition matrix, $\mathbf{P}$, is constructed as described in Sect. 2 and interpreted as a directed and weighted graph.

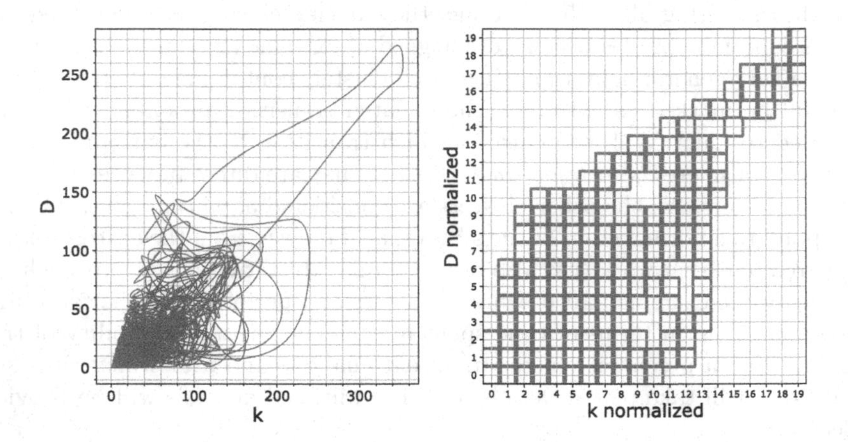

Fig. 1. The trajectory of the Moehlis-Faisst-Eckhart model represented in the phase space (left) and in a 20×20 tessellated phase space (right)

The resulting graph consisted of 165 nodes and 585 edges. This was reduced to 27 nodes and 121 edges after the deflation step, which resulted in a deflation of the transition probability matrix from 400×400 to 27×27, a decrease by a factor of over 200. The clustering algorithm identified 27 clusters, out of which 14 were extreme, 6 were precursor clusters and 7 were considered normal state clusters. The clustered trajectory of the MFE system, in tessellated and non-tessellated phase space, is displayed in Fig. 2, where each cluster is represented by a different color, and the cluster numbers are displayed in their centers. In the tessellated phase space (Fig. 2 (right)), the criteria of the extreme event classification are

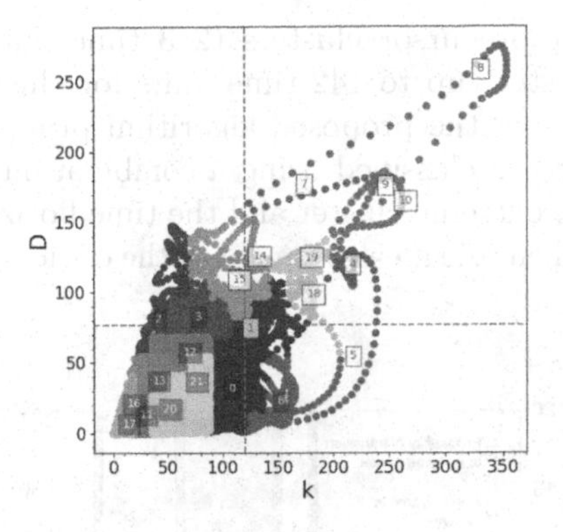

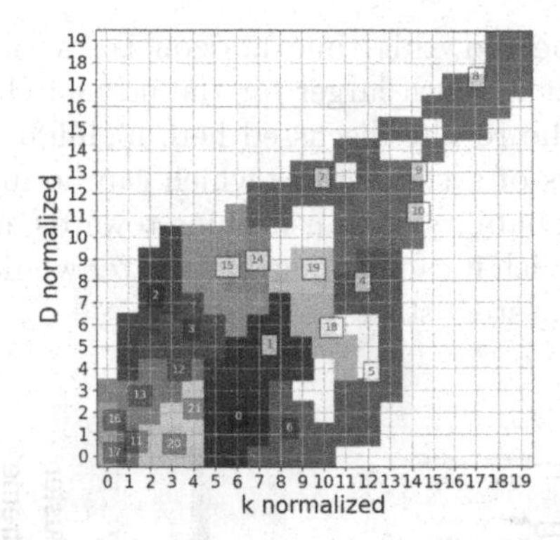

Fig. 2. (left) Trajectory of the MFE system in phase space and (right) in a 20×20 tessellated phase space. Red dashed lines represent the region considered as extreme. The cluster numbers used for their identification are shown in the cluster centers; the extreme cluster numbers are marked in red. (Color figure online)

also shown as red dashed lines. These criteria had to be confirmed visually, ensuring that an adequate part of the trajectory lies within them. The majority of the trajectory is located in the bottom left corner, in clusters 20, 25 and 26, which represent the normal state experiencing low k and D. The closer to the top right extreme section, the less frequent but more severe the events get.

In Fig. 3, the statistics of the most important clusters are provided. Namely, we compute: (i) the average time spent in each cluster; (ii) the percentage of time spent in each cluster (in proportion to the entire duration of the dataset); (iii) the probability of transition from the given cluster to an extreme one; and (iv) the average time for such a transition to occur from the given cluster. In Fig. 3 (left) where the first two quantities are plotted, we can observe that there are a few *normal* clusters (clusters 20, 21, 25 and 26 in blue) where the system spends the largest portion of the time (in average or as a proportion of the total time series duration) while for the precursor (orange) and extreme (red) clusters the proportion is much smaller. This confirms that the proposed approach is efficient in determining clusters which contain the essential dynamics of the system and furthermore, the obtained division helps to contain most of the normal states in one (or few) cluster(s) and isolate the extreme and precursor states.

In Fig. 3 (right), among the precursor clusters (in orange), cluster 23 stands out as having a very large probability of transitioning towards an extreme cluster indicating that this is a very critical cluster as, if the system enters it, it will nearly certainly results in an extreme event. For the other precursor clusters, the probability is not as high indicating that, while they are precursor clusters, entering in those does not necessarily results in the occurrence of an extreme event as it is possible that the system will transition again towards a normal cluster. An additional metric shown in Fig. 3 (right) is the average time before the transition to an extreme event which is a proxy for the time horizon which would be available to prevent the occurrence of the extreme events. As could be expected, this time horizon is low for precursor clusters (2–3 time units) while it is much larger for the normal cluster (up to 142 time units for cluster 19). The results discussed here highlight how the proposed algorithm provides clusters of system states which can be further classified using a combination of the probability of transitioning towards an extreme cluster and the time horizon over which it can happen. This latter would enable an assessment of the criticality of the system state and cluster.

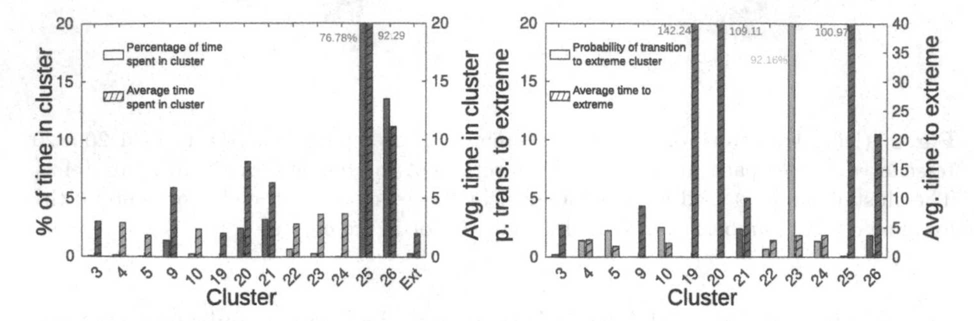

Fig. 3. Statistics of the MFE clusters showing (left) the percentage of the time series and average time spent in cluster and (right) the maximum probability of transitioning to an extreme event and the minimum average time to an extreme event. The extreme clusters are colored red and their direct precursors orange. (Color figure online)

To help illustrate the observation made above, the time series of the kinetic energy and dissipation of the MFE system for two extreme events is plotted in Fig. 4 where the plots are colored by the types of cluster the system is in at a given time (normal, precursor or extreme). One can observe that, according to the precursor cluster definition, each extreme cluster is preceded by a precursor cluster. This time during which the system lies in a precursor state varies in the two cases but illustrates the horizon over which actions could be taken to prevent the occurrence of the extreme event.

The false positive rate was also calculated as the ratio of the identified precursors without a following extreme cluster transition to all identified precursor clusters. It showed to be very high, at over 75%. It was noticed that often not only is there a precursor state preceding, but also following the extreme events, when the trajectory is getting back to the normal states, as can be seen in Fig. 4 (left) around $t = 40$. These transitions were accounted for and the corrected rate of false positives was calculated and reduced down to 50%. Other observations included the fact that extreme events come in batches, often with a false positive slightly preceding it (see Fig. 4 for example where in both cases shown, there is an identified precursor state approximately 10 to 20 time units before the occurrence of the extreme event). This might be leveraged as an additional indication of the upcoming extreme events.

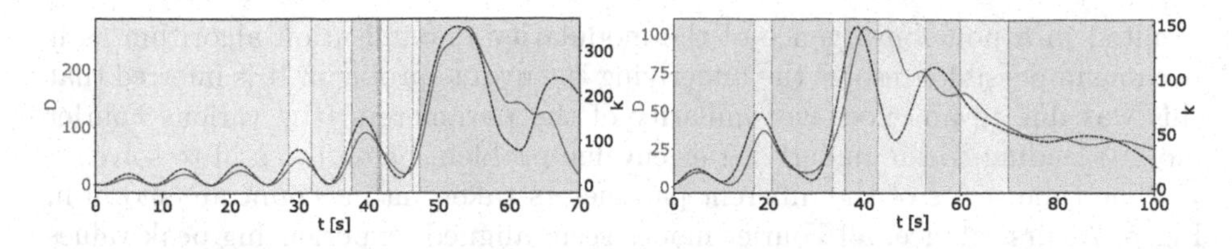

Fig. 4. Time series sections of the kinetic energy (dotted black line, right axis) and energy dissipation (red line, left axis) for the MFE system. Background color indicates the type of clusters the system is in (blue: normal; orange: precursor; red: extreme). (Color figure online)

3.2 Kolmogorov Flow

The Kolmogorov flow is a two-dimensional flow governed by the incompressible Navier-Stokes equations with the addition of a harmonic volume force in the momentum equation [5]. These are here solved on a domain $\Omega \equiv [0, 2\pi] \times [0, 2\pi]$, (using a grid of 24×24 with periodic boundary conditions (same conditions as in [5]), with a forcing term of the shape $f = (\sin(k_f y), 0)$, with $k_f = 4$ and $Re = 40$. The dataset of the Kolmogorov flow is obtained using a pseudo-spectral code and it is generated for a duration of 5,000 time units (500,000 time steps).

The intermittent bursts of energy dissipation, D, are of interest for the Kolmogorov flow, with the kinetic energy, k, as an observable. In [5], additional observables related to the extreme events were identified to be the modulus of the three Fourier modes of the velocity field: $a(1,0)$, $a(0,4)$ and $a(1,4)$, which form a triad (where $k_f = 4$ is the forcing wavenumber and where $a(i,j)$ indicates the i-th,j-th coefficient of the 2D Fourier mode). This approach was applied in the current work, and the algorithm was tested by including one of these three Fourier modes, together with k and D, for the clustering analysis, reducing the system dimensionality to 3. This resulted in five different cases, depending on which/how the Fourier mode is considered, which are detailed below:

- **Case 1** - k, D, and modulus of Fourier mode $|a(1,0)|$
- **Case 2** - k, D, and absolute value of the real part of Fourier mode $|Re(a(1,0))|$
- **Case 3** - k, D, and absolute value of the imaginary part of Fourier mode $|Im(a(1,0))|$
- **Case 4** - k, D, and modulus of Fourier mode $|a(0,4)|$
- **Case 5** - k, D, and modulus of Fourier mode $|a(1,4)|$

A comparative analysis between the different cases was also performed which showed which Fourier mode is the most effective in helping identify precursors of extreme events. Cases 2 and 3, which include the real and imaginary part of the Fourier mode rather than its modulus, were chosen because of an error encountered for case 1 in the clustering process, which will be detailed hereunder. In all cases, the extreme events were defined using only the values of k and D.

It should be noted that it would be possible to consider all Fourier modes simultaneously, in addition to k and D which was attempted. However, this resulted in a non-convergence of the modularity maximisation algorithm as it became impossible to solve the underlying eigenvalue problem. It is inferred that this was due to an excessive similarity of the parameters (the various Fourier modes) leading to the underlying eigenvalue problem being too stiff to solve.

The time series of the different parameters taken into account are shown in Fig. 5. At first glance, all Fourier modes seem aligned, experiencing peak values correlated with the peaks in k and D. This could lead to the conclusion that all the cases considered above would lead to similar performance in precursor identification. This was not the case, as will be discussed in more detail below.

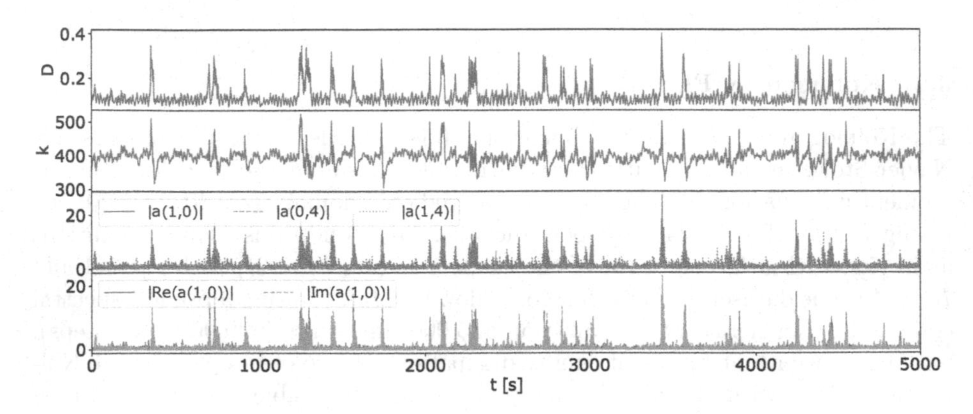

Fig. 5. Time series of the energy, energy dissipation and different Fourier modes taken into account for the analysis of the Kolmogorov flow.

The trajectory of the Kolmogorov flow is shown in a phase space spanned by k and D, which makes for an easier comparison of the results obtained from different cases, since it is common for all of them. Once again, the tessellation was performed using 20 sections per dimension. The phase space, together with its tessellated form (where k and D are normalized using their respective minimum and maximum and divided into 20 sections), is presented in Fig. 6.

The case including the modulus of $|a(1,0)|$ was tested first (case 1), as this is the one used in [5]. The clustering converged successfully, but resulted in abnormally large values of time spent in both the extreme and precursor clusters. Over a quarter of the total time is spent in precursor clusters. This makes the clustering unusable, since little information can be gained about the probability and time of transitioning to an extreme event. This abnormal result was attributed to a cluster located in the middle of the trajectory, which includes a single node positioned in the extreme section of the phase space (see Fig. 7).

The other cases yielded much better results, as will be discussed next, although slightly different from each other, varying in the number of clusters and percentage of time spent in extreme and precursor clusters. This highlights the necessity of appropriately choosing the variables used to define the phase space trajectory of the system.

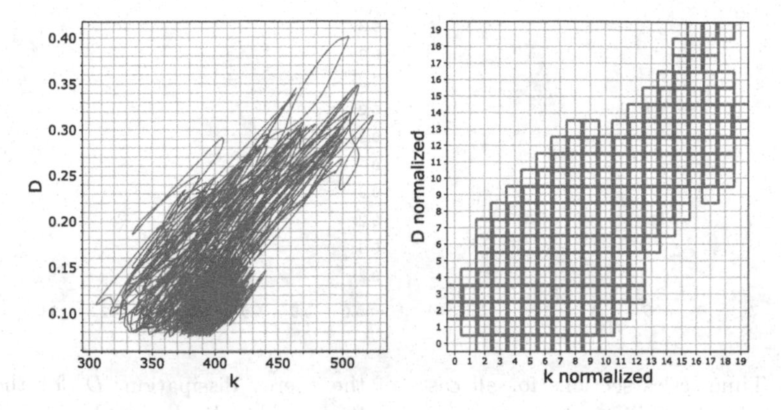

Fig. 6. The trajectory of the Kolmogorov flow represented in the phase space (left) and in a 20×20 tessellated phase space (right).

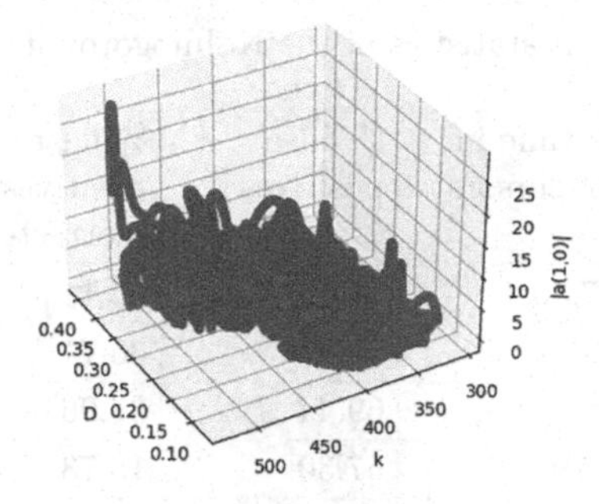

Fig. 7. The clustered phase space of the Kolmogorov flow for case 1, with the faulty extreme cluster colored in red. (Color figure online)

Comparison Between All Cases: To understand the differences in the results for the discussed cases of the Kolmogorov flow, segments of the time series of the energy dissipation are plotted with the transitions to different types of clusters marked (Fig. 8). This plot shows a significant difference between the first and other considered cases. At a macroscopic level, cases 2 to 5 seem to identify the peaks of D at the same instances, while for case 1 the transitions are occurring constantly throughout the considered time. This shows the results of incorrectly clustering the system's trajectory making it impossible to deduct any reasonable conclusions and recommendations.

Case 1 shows a very chaotic behavior identifying extreme events where no peaks of D are present. The other cases identify the extreme clusters in a narrower scope, with the smoothest transitions appearing in case 4. For the analysis of the extreme events and their precursors, the percentage of time spent in the precursor and extreme clusters, as well as the false positive and false negative rates for all of the discussed cases are presented in Table 1.

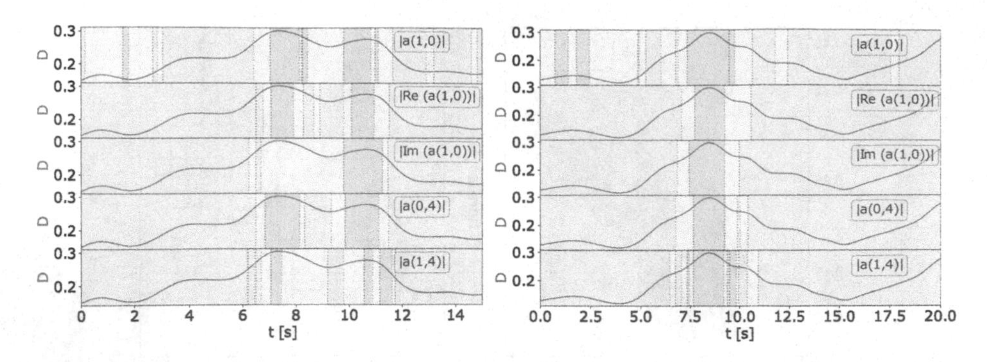

Fig. 8. Time series sections for all cases of the energy dissipation, D, for the Kolmogorov flow for 2 different extreme events (left and right). Background color indicates the type of clusters the system is in (blue: normal; orange: precursor; red: extreme). (Color figure online)

Table 1. Extreme event statistics of the Kolmogorov flow for different cases.

Case	% time in extreme	% time in precursors	% false positives	% false positives (corrected)	% false negatives
1	2.78	27.85	77.15	55.11	0.0
2	0.71	0.87	66.67	35.71	0.0
3	0.76	1	69.44	37.96	0.0
4	0.85	0.79	57.30	15.73	0.0
5	0.81	1	59.70	22.39	0.0

As mentioned already, case 1 was disregarded, due to faulty clustering resulting in a high percentage of time spent in the precursor clusters. Table 1 also shows that the rate of false positives is the highest for this case, which is explained by the positioning of the faulty extreme cluster (located in the center of the phase space trajectory). All of the discussed cases have a false negative rate of 0 % which is due to the used definition of precursor cluster as being always followed by an extreme cluster. Case number 4, which shows the lowest false positive rates both before and after applying the correction (57.3 % and 15.73 % respectively) was chosen for further analysis hereunder.

It should be noted that compared to earlier work [5], we observe here that using $|a(1,0)|$ did not enable the identification of a precursor of extreme events, while using $|a(0,4)|$ did. This contrasting view may be related to the fact that we use directly the magnitude of the Fourier mode compared to the energy flux between modes used in [5]. Our analysis suggests that when considering the magnitude directly, $|a(0,4)|$ becomes more adapted as is detailed next.

Case 4: The original trajectory of the flow expressed in graph form contained 1,799 nodes and 6,284 edges. After deflation, this decreased to 105 nodes and 719 edges. The transition probability matrix was reduced from $8{,}000 \times 8{,}000$ to 105×105. The clustering algorithm identified 105 clusters, out of which 11 were extreme and 8 were precursor clusters. The clustered trajectory of the Kolmogorov flow is shown in Fig. 9, with each cluster represented by a different color and the cluster numbers displayed in their centers. As for the MFE system, the statistics of the main clusters are shown in Fig. 10.

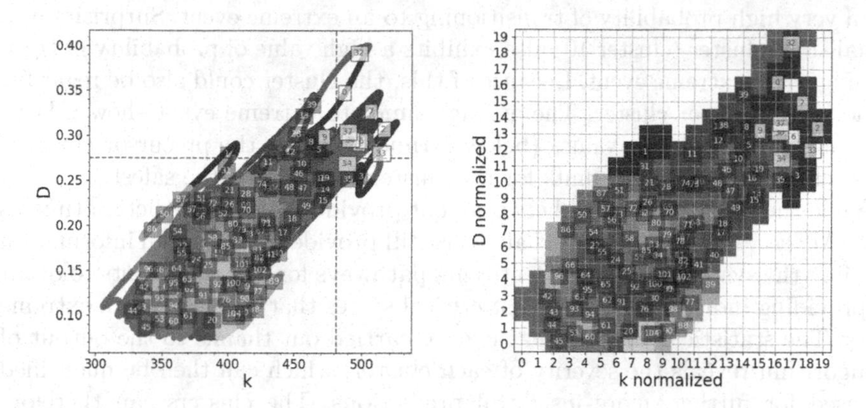

Fig. 9. (left) Trajectory of the Kolmogorov flow (case 4) system in phase space and (right) in a 20×20 tessellated phase space. Red dashed lines represent the region considered as extreme. The cluster numbers used for their identification are shown in the cluster centers; the extreme cluster numbers are marked in red. (Color figure online)

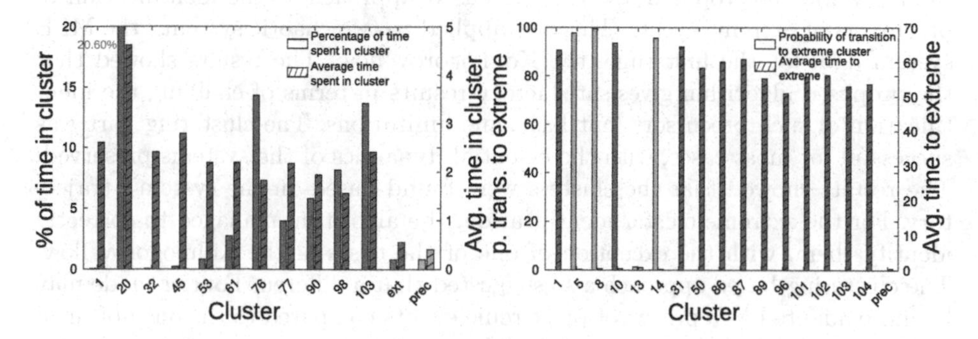

Fig. 10. Statistics of the Kolmogorov flow clusters (case 4) showing (left) the percentage of the time series and average time spent in cluster and (right) the maximum probability of transitioning to an extreme event and the minimum average time to an extreme event. Red: extreme clusters; orange: precursors cluster. (Color figure online)

The resulting statistics showed that most of the time the trajectory is within the normal states, with a few dominating clusters (Fig. 10 (left)). There is one extreme cluster (cluster 32) where the system spends on average a large time which is related to a very critical extreme event where the flow exhibit a long dissipation rate (related to a long quasi-laminar state of the Kolmogorov flow). Compared to the MFE system, there is no single (or few) clusters in which the Kolmogorov flow spends the majority of the time which is related to the higher complexity of its dynamics. Regarding the precursor, their combined contribution is shown and, on average, the system spends less than 1 time unit within these precursors states indicating a low predictability horizon. A finer analysis is provided in Fig. 10 (right) which shows the presence of three precursor clusters with a very high probability of transitioning to an extreme event. Surprisingly, a normal state cluster (cluster 10) also exhibits a high value of probability of transitioning to an extreme event. Because of this, the cluster could also be regarded as another precursor cluster. The average times to extreme event show a large range of values, with low values (below 1 time unit) for the precursor clusters, but reaching up to 64 time units for the clusters considered the safest.

While the cluster identified here may not provide as much prediction time as for the MFE system, the proposed analysis still provided some useful information regarding the system dynamics identifying pathways for extreme events to occur and providing indications of the most critical states that will lead to an extreme events. The statistical analysis that can be carried out thanks to the output of the algorithm reveals the severity of each cluster, which can then be quantified and used for further, more insightful predictions. The clusters can therefore be classified based on any of the calculated parameters to track the real-time progression of the system.

4 Conclusions

In this work, we proposed a clustering-based approach to the identification of precursors of extreme events. This was applied to two chaotic systems, the MFE system and, for the first time, the Kolmogorov flow. The results showed that the proposed algorithm gives satisfactory results in terms of enabling the identification of such precursors, but has some limitations. The clustering part was successful for most cases, with the essential dynamics of the systems preserved. The results showed that the clusters were found based on the system's trajectory. For the extreme cluster identification, the algorithm managed to correctly identify them, with the exception of one of the cases of the Kolmogorov flow. The clustering-based approach also suggested that a different Fourier mode may be more adapted as a precursor of extreme events compared to the one obtained from a variational-based approach in [5]. The clustering-based precursors were identified correctly and, due to their adopted definition, the false negative rate was equal to zero for all cases. Compared to earlier work [15], a further statistical analysis of the identified precursor cluster allowed to obtain an understanding of their critical nature, depending on their actual probability of transition to an extreme state and the time horizon over which such transition could occur.

Further work will be dedicated to investigating how to improve the identification and assessment of precursor clusters and applying the proposed method to experimental measurements obtained from systems exhibiting extreme events. Application to real-time data should also be explored, to verify current cluster identification and prediction of its upcoming transitions in such a setting.

References

1. Brown, K., Zuo, Z.: Python implementation of Newman's spectral methods to maximize modularity. https://github.com/thelahunginjeet/python-modularity-maximization. Accessed 15 Feb 2022
2. Dakos, V., Scheffer, M., van Nes, E.H., Brovkin, V., Petoukhov, V., Held, H.: Slowing down as an early warning signal for abrupt climate change. Proc. Natl. Acad. Sci. **105**, 14308–14312 (2008)
3. Dysthe, K., Krogstad, H.E., Müller, P.: Oceanic rogue waves. Annu. Rev. Fluid Mech. **40**, 287–310 (2008)
4. Easterling, D.R., et al.: Observed variability and trends in extreme climate events: a brief review. Bull. Am. Meteor. Soc. **81**, 417–425 (2000)
5. Farazmand, M., Sapsis, T.P.: A variational approach to probing extreme events in turbulent dynamical systems. Sci. Adv. **3**, 1–8 (2017)
6. Farazmand, M., Sapsis, T.P.: Extreme events: Mechanisms and prediction. Appl. Mech. Rev. **71**, 050801 (2019)
7. Kobayashi, T., Murayama, S., Hachijo, T., Gotoda, H.: Early detection of thermoacoustic combustion instability using a methodology combining complex networks and machine learning. Phys. Rev. Appl. **11**, 1 (2019)
8. Leicht, E.A., Newman, M.E.: Community structure in directed networks. Phys. Rev. Lett. **100**, 1–4 (2008)
9. Moehlis, J., Faisst, H., Eckhardt, B.: A low-dimensional model for turbulent shear flows. New J. Phys. **6**, 56 (2004)
10. Murugesan, M., Sujith, R.I.: Detecting the onset of an impending thermoacoustic instability using complex networks. J. Propul. Power **32**, 707–712 (2016)
11. Newman, M.E.J.: Modularity and community structure in networks. Proc. Natl. Acad. Sci. **103**, 8577–8582 (2006)
12. Nicodemi, M.: Extreme value statistics. In: Encyclopedia of Complexity and Systems Science, pp. 1066–1072 (2015)
13. Racca, A., Magri, L.: Data-driven prediction and control of extreme events in a chaotic flow. Phys. Rev. Fluids **7**, 1–24 (2022)
14. Schmid, P.J., Garciá-Gutierrez, A., Jiménez, J.: Description and detection of burst events in turbulent flows. J. Phys.: Conf. Ser. **1001**, 012015 (2018)
15. Schmid, P.J., Schmidt, O., Towne, A., Hack, P.: Analysis and prediction of rare events in turbulent flows. In: Proceedings of the Summer Program. Center for Turbulence Research (2018)
16. Varadhan, S.R.S.: Large deviations. Ann. Probab. **36**, 397–419 (2008)
17. Wan, Z.Y., Vlachas, P., Koumoutsakos, P., Sapsis, T.P.: Data-assisted reduced-order modeling of extreme events in complex dynamical systems. PLoS One **13**, 1–22 (2018)

Convolutional Autoencoder for the Spatiotemporal Latent Representation of Turbulence

Nguyen Anh Khoa Doan[1(✉)], Alberto Racca[2,3], and Luca Magri[3,4]

[1] Delft University of Technology, 2629HS Delft, The Netherlands
n.a.k.doan@tudelft.nl
[2] University of Cambridge, Cambridge CB2 1PZ, UK
[3] Imperial College London, London SW7 2AZ, UK
[4] The Alan Turing Institute, London NW1 2DB, UK

Abstract. Turbulence is characterised by chaotic dynamics and a high-dimensional state space, which make this phenomenon challenging to predict. However, turbulent flows are often characterised by coherent spatiotemporal structures, such as vortices or large-scale modes, which can help obtain a latent description of turbulent flows. However, current approaches are often limited by either the need to use some form of thresholding on quantities defining the isosurfaces to which the flow structures are associated or the linearity of traditional modal flow decomposition approaches, such as those based on proper orthogonal decomposition. This problem is exacerbated in flows that exhibit extreme events, which are rare and sudden changes in a turbulent state. The goal of this paper is to obtain an efficient and accurate reduced-order latent representation of a turbulent flow that exhibits extreme events. Specifically, we employ a three-dimensional multiscale convolutional autoencoder (CAE) to obtain such latent representation. We apply it to a three-dimensional turbulent flow. We show that the Multiscale CAE is efficient, requiring less than 10% degrees of freedom than proper orthogonal decomposition for compressing the data and is able to accurately reconstruct flow states related to extreme events. The proposed deep learning architecture opens opportunities for nonlinear reduced-order modeling of turbulent flows from data.

Keywords: Chaotic System · Reduced Order Modelling · Convolutional Autoencoder

1 Introduction

Turbulence is a chaotic phenomenon that arises from the nonlinear interactions between spatiotemporal structures over a wide range of scales. Turbulent flows are typically high-dimensional systems, which may exhibit sudden and unpredictable bursts of energy/dissipation [2]. The combination of these dynamical properties makes the study of turbulent flows particularly challenging. Despite these complexities, advances have been made in the analysis of turbulent flows

J. Mikyška et al. (Eds.): ICCS 2023, LNCS 14076, pp. 328–335, 2023.
https://doi.org/10.1007/978-3-031-36027-5_24

through the identification of coherent (spatial) structures, such as vortices [15], and modal decomposition techniques [13]. These achievements showed that there exist energetic patterns within turbulence, which allow for the development of reduced-order models.

To efficiently identify these patterns, recent works have used machine learning [3]. Specifically, Convolutional Neural Networks (CNNs) have been used to identify spatial features in flows [10] and perform nonlinear modal decomposition [7,11]. These works showed the advantages of using CNNs over traditional methods based on Principal Component Analysis (PCA) (also called Proper Orthogonal Decomposition in the fluid mechanics community), providing lower reconstruction errors in two-dimensional flows. More recently, the use of such CNN-based architecture has also been extended to a small 3D turbulent channel flow at a moderate Reynolds number [12]. The works highlight the potential of deep learning for the analysis and reduced-order modelling of turbulent flows, but they were restricted to two-dimensional flows or weakly turbulent with non-extreme dynamics. Therefore, the applicability of deep-learning-based techniques to obtain an accurate reduced-order representation of 3D flows with extreme events remains unknown. Specifically, the presence of extreme events is particularly challenging because they appear rarely in the datasets. In this paper, we propose a 3D Multiscale Convolutional Autoencoder (CAE) to obtain such a reduced representation of the 3D Minimal Flow Unit (MFU), which is a flow that exhibit such extreme events in the form of a sudden and rare intermittent quasi-laminar flow state. We explore whether the Multiscale CAE is able to represent the flow in a latent space with a reduced number of degrees of freedom with higher accuracy than the traditional PCA, and whether it can also accurately reconstruct the flow state during the extreme events.

Section 2 describes the MFU and its extreme events. Section 3 presents in detail the Multiscale CAE framework used to obtain a reduced-order representation of the MFU. The accuracy of the Multiscale CAE in reconstructing the MFU state is discussed in Sect. 4. A summary of the main results and directions for future work are provided in Sect. 5.

2 Minimal Flow Unit

The flow under consideration is the 3D MFU [9]. The MFU is an example of prototypical near-wall turbulence, which consists of a turbulent channel flow whose dimensions are smaller than conventional channel flow simulations. The system is governed by the incompressible Navier-Stokes equations

$$\begin{aligned} \nabla \cdot \boldsymbol{u} &= 0, \\ \partial_t \boldsymbol{u} + \boldsymbol{u} \cdot \nabla \boldsymbol{u} &= \frac{1}{\rho}\boldsymbol{f_0} - \frac{1}{\rho}\nabla p + \nu \Delta \boldsymbol{u}, \end{aligned} \tag{1}$$

where $\boldsymbol{u} = (u, v, w)$ is the 3D velocity field and $\boldsymbol{f_0} = (f_0, 0, 0)$ is the constant forcing in the streamwise direction, x; ρ, p, and ν are the density, pressure, and kinematic viscosity, respectively. In the wall-normal direction, y, we impose a no-slip boundary condition, $\boldsymbol{u}(x, \pm\delta, z, t) = 0$, where δ is half the channel width. In the streamwise, x, and spanwise, z, directions we have periodic boundary conditions. For this study, a channel with dimension $\Omega \equiv \pi\delta \times 2\delta \times 0.34\pi\delta$ is considered, as in [2] with $\delta = 1.0$. The Reynolds number of the flow, which is based on the bulk velocity and the half-channel width, is set to $Re = 3000$, which corresponds to a friction Reynolds number $Re_\tau \approx 140$. An in-house code similar to that of [1] is used to simulate the MFU, and generate the dataset on which the Multiscale CAE is trained and assessed.

The extreme events in the MFU are quasi-relaminarization events, which take place close to either wall. A typical evolution of the flow during a quasi-relaminarization event is shown in Fig. 1(a–g). Time is normalized by the eddy turnover time. During an extreme event, (i) the flow at either wall (the upper wall in Fig. 1) becomes laminar (Fig. 1(a–c)); (ii) the flow remains laminar for some time (Fig. 1(c–f)), which results in a larger axial velocity close to the centerline (and therefore an increase in kinetic energy); (iii) the greater velocity close to the centerline makes the effective Reynolds number of the flow larger, which in turn makes the flow prone to a turbulence burst on the quasi-laminar wall; (iv) the turbulence burst occurs on the quasi-laminar wall, which results in a large increase in the energy dissipation rate; and (v) the flow close to that quasi-laminar wall becomes turbulent again, which leads to a decrease in the kinetic energy (Fig. 1(g)).

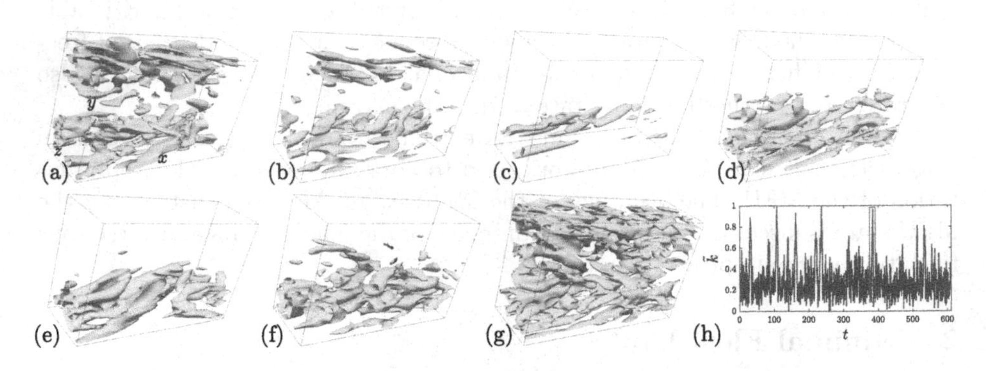

Fig. 1. (a–g) Snapshots of the Q-criterion isosurface (with value $Q = 0.1$) during an extreme event, where $Q = 0.5(||\boldsymbol{\omega}||^2 - ||\boldsymbol{S}||^2)$, $\boldsymbol{\omega}$ is the vorticity vector, and $\boldsymbol{S}$ is the strain-rate tensor. (h) Evolution of kinetic energy, k, of the MFU. The red box indicates the event whose evolution is shown in (a–g). (Color figure online)

These quasi-relaminarisation are accompanied by bursts in the total kinetic energy, $k(t) = \int\int\int_\Omega \frac{1}{2}\boldsymbol{u} \cdot \boldsymbol{u} dxdydz$, where Ω is the computational domain. This can be seen in Fig. 1h, where the normalized kinetic energy, $\tilde{k} = (k - \min(k))/(\max(k) - \min(k))$ is shown.

The dataset of the MFU contains 2000 eddy turnover times (i.e., 20000 snapshots) on a grid of $32 \times 256 \times 16$, which contains 50 extreme events. The first

200 eddy turnover times of the dataset (2000 snapshots) are employed for the training of the Multiscale CAE, which contains only 4 extreme events.

3 Multiscale Convolutional Autoencoder

We implement a 3D convolutional autoencoder. It should be noted that we only consider CNN-based autoencoder here and not other approaches such as transformer-based ones (like in [4]) as the latter have not yet shown to be widely applicable to dataset with a strong spatial based information (such as images or flow dataset). A schematic of the proposed architecture is shown in Fig. 2.

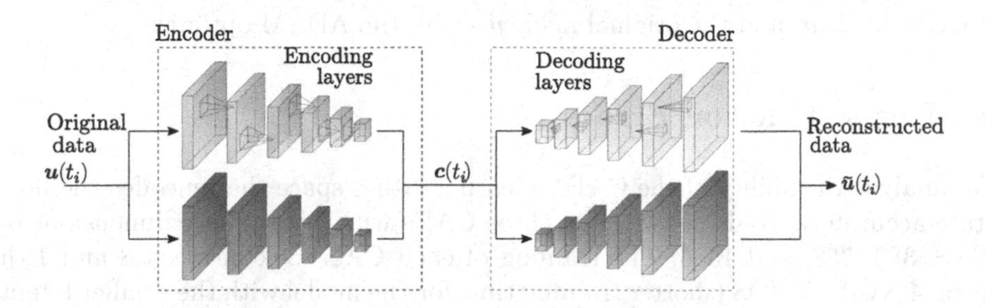

Fig. 2. Schematic of the 3D multiscale convolutional autoencoder.

The three dimensional convolution autoencoder (CAE) learns an efficient reduced-order representation of the original data, which consists of the flow state $\boldsymbol{u} \in \mathbb{R}^{N_x \times N_y \times N_z \times N_u}$, where N_x, N_y and N_z are the number of grid points, and $N_u = 3$ is the number of velocity components. On one hand, the encoder (blue box in Fig. 2) reduces the dimension of the data down to a latent state, $\boldsymbol{c}$, with small dimension $N_c \ll N_x N_y N_z N_u$. This operation can be symbolically expressed as $\boldsymbol{c} = \mathcal{E}(\boldsymbol{u}; \boldsymbol{\phi}_E)$, where $\boldsymbol{\phi}_E$ represents the weights of the encoder. On the other hand, the decoder (green box in Fig. 2) reconstructs the data from the latent state back to the original full flow state. This operation is expressed as $\widetilde{\boldsymbol{u}} = \mathcal{D}(\boldsymbol{c}; \boldsymbol{\phi}_D)$, where $\boldsymbol{\phi}_D$ are the trainable weights of the decoder. We employ a multiscale autoencoder, which was originally developed for image-based super-resolution analysis [5]. We use here a multiscale autoencoder and not a standard one as previous works [6,8] have demonstrated the ability of the multiscale version in leveraging the multiscale information in turbulent flows to better reconstruct the flow from the latent space. It relies on the use of convolutional kernels of different sizes to analyse the input and improve reconstruction in fluids [8,14]. In this work, two kernels, $(3 \times 5 \times 3)$ and $(5 \times 7 \times 5)$, are employed (represented schematically by the two parallel streams of encoder/decoder in the blue and green boxes in Fig. 2). This choice ensures a trade-off between the size of the 3D multiscale autoencoder and the reconstruction accuracy (see Sect. 4). To reduce the dimension of the input, the convolution operation in each layer of the encoder is applied in a strided manner, which means that the convolutional neural network (CNN) kernel is progressively applied to the input by moving the CNN kernel by $(s_x, s_y, s_z) = (2, 4, 2)$ grid points. This results in an output of a smaller dimension than the input. After each convolution layer, to fulfill the boundary conditions of the MFU, periodic padding is applied in the x and z directions, while zero padding is applied in the y direction. Three successive layers of CNN/padding operations are applied to decrease the dimension of the original field from $(32, 256, 16, 3)$ to $(2, 4, 2, N_f)$, where N_f is the specified number of filters in the last encoding layer. As a result, the dimension of the latent space is $N_c = 16 \times N_f$. The decoder mirrors the architecture of the encoder, where transpose CNN layers [16] are used, which increase the dimension of the latent space up to the original flow dimension. The end-to-end autoencoder is trained by minimizing the mean squared error (MSE) between the reconstructed velocity field, $\widetilde{\boldsymbol{u}}$, and the original field, $\boldsymbol{u}$ using the ADAM optimizer.

4 Reconstruction Error

We analyze the ability of the CAE to learn a latent space that encodes the flow state accurately. To do so, we train three CAEs with latent space dimensions of N_c = 384, 768, and 1536. The training of each CAE took between 8 and 16 h using 4 Nvidia V100S (shorter training time for the model with the smaller latent space). After training, the CAE can process 100 samples in 1.8 s to 5 s depending on the dimension of the latent space (the faster processing time corresponding to the CAE with the smaller latent space). We compute their reconstruction errors on the test set based on the MSE. A typical comparison between a reconstructed velocity field obtained from the CAE with N_c = 1536 is shown in Fig. 3. The CAE is able to reconstruct accurately the features of the velocity field.

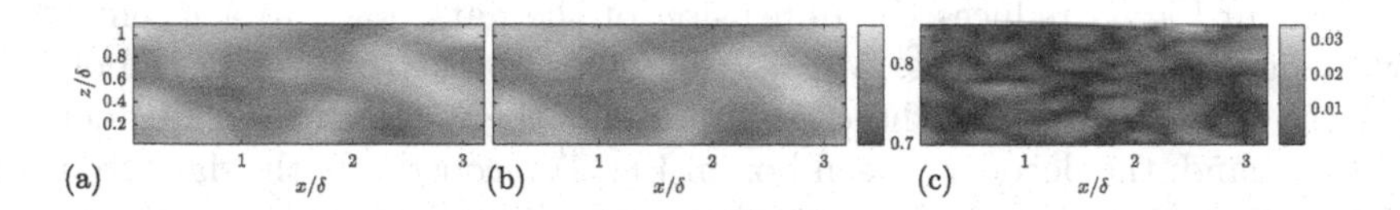

Fig. 3. Comparison of (a) the actual velocity magnitude (ground truth), (b) the CAE-reconstructed velocity magnitude, (c) the root-squared difference between (a) and (b) in the mid-y plane for a typical snapshot in the test set.

To provide a comparison with the CAE, we also compute the reconstruction error obtained from PCA, whose principal directions are obtained with the method of snapshots [13] on the same dataset. Figure 4 shows the reconstruction error. PCA decomposition requires more than 15000 PCA components to reach the same level of accuracy as the CAE with a latent space of dimension 1536. This highlights the advantage of learning a latent representation with nonlinear operations, as in the autoencoder, compared with relying on a linear combination of components, as in PCA.

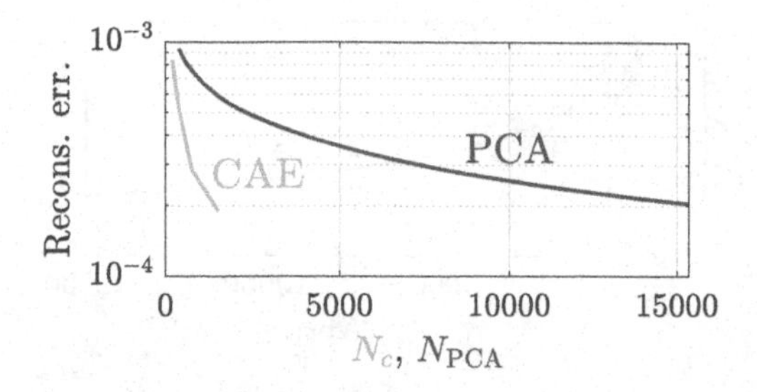

Fig. 4. Reconstruction error with a PCA-based method (blue) and the autoencoder (yellow) for different dimensions of the latent space, N_c, or number of retained PCA components, N_{PCA}. The reconstruction error is computed as the mean squared error between the reconstructed velocity field and the exact field, averaged over the test set. (Color figure online)

The better performance of the CAE with respect to PCA is evident when we consider the reconstruction accuracy for flow states that correspond to extreme events, which we extract from the test set. Here, we define an extreme event as a flow state with normalized kinetic energy above a user-selected threshold value of 0.7. Hence, for the selected snapshots in the test set, the mean squared error (MSE) between the reconstructed velocity and the truth is computed using the CAE and PCA as a function of the latent space size. The resulting MSE is shown in Fig. 5, where the CAE exhibits an accuracy for the extreme dynamics similar to the reference case of the entire test set (see Fig. 4). On the other hand, the accuracy of the PCA is lower than the reference case of the entire dataset. This lack of accuracy is further analysed in Fig. 6, where typical velocity magnitudes, i.e. the norm of the velocity field $\boldsymbol{u}$, are shown for the mid-z plane during a representative extreme event. The velocity reconstructed with the CAE (Fig. 6b) is almost identical to the true velocity field (Fig. 6a). The CAE captures the smooth variation at the lower wall indicating a quasi-laminar flow state in that region. In contrast, the velocity field reconstructed using PCA (Fig. 6c) completely fails at reproducing those features. This is because extreme states are rare in the training set used to construct the PCA components (and the CAE). Because of this, the extreme states only have a small contribution to the PCA components and are accounted for only in higher order PCA components, which are neglected in the 1536-dimensional latent space.

This result indicates that only the CAE and its latent space can be used for further applications that requires a reduced-order representation of the flow, such as when trying to develop a reduced-order model. This is also supported by findings in [14] where it was shown that a similar CAE could be used in combination with a reservoir computer to accurately forecast the evolution of a two dimensional flow.

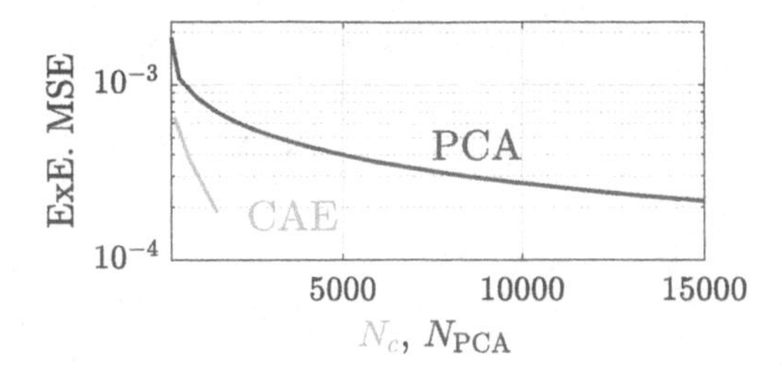

Fig. 5. Reconstruction error on the extreme events flow states with PCA-based method (blue) and autoencoder (yellow) for different dimensions of the latent space, N_c, or number of retained PCA components, N_{PCA}. The reconstruction error is computed as in Fig. 4 but only on the flow state corresponding to extreme events. (Color figure online)

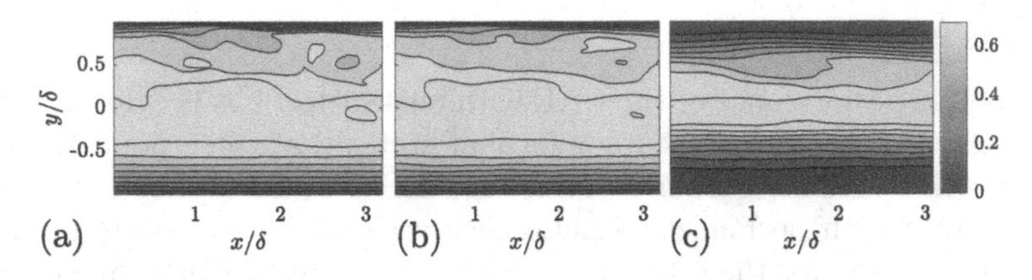

Fig. 6. Comparison of (a) the velocity magnitude (ground truth), (b) the CAE-reconstructed velocity magnitude, (c) the PCA-reconstructed velocity magnitude in the mid-z plane for a typical extreme event snapshot in the test set.

5 Conclusion

In this work, we develop a nonlinear autoencoder to obtain an accurate latent representation of a turbulent flow that exhibits extreme events. We propose the 3D Multiscale CAE to learn the spatial features of the MFU, which exhibits extreme events in the form of near-wall quasi-relaminarization events. The model consists of a convolutional autoencoder with multiple channels, which learn an efficient reduced latent representation of the flow state. We apply the framework to a three-dimensional turbulent flow with extreme events (MFU). We show that the Multiscale CAE is able to compress the flow state to a lower-dimensional latent space by three orders of magnitude to accurately reconstruct the flow state from this latent space. This constitutes a key improvement over principal component analysis (PCA), which requires at least one order of magnitude more PCA components to achieve an accuracy similar to the CAE. This improvement in reconstruction accuracy is crucial for the reconstruction of the flow state during the extreme events of the MFU. This is because extreme states are rare and, thus, require a large number of PCA components to be accurately reconstructed.

The proposed method and results open up possibilities for using deep learning to obtain an accurate latent reduced representation of 3D turbulent flows. Future work will be devoted to physically interpreting the latent space discovered by the Multiscale CAE, and learning the dynamics in this latent space.

Acknowledgements. The authors thank Dr. Modesti for providing the flow solver. N.A.K.D and L.M acknowledge that part of this work was performed during the 2022 Stanford University CTR Summer Program. L.M. acknowledges the financial support from the ERC Starting Grant PhyCo 949388. A.R. is supported by the Eric and Wendy Schmidt AI in Science Postdoctoral Fellowship, a Schmidt Futures program.

References

1. Bernardini, M., Pirozzoli, S., Orlandi, P.: Velocity statistics in turbulent channel flow up to Re_{τ} =4000. J. Fluid Mech. **742**, 171–191 (2014)
2. Blonigan, P.J., Farazmand, M., Sapsis, T.P.: Are extreme dissipation events predictable in turbulent fluid flows? Phys. Rev. Fluids **4**, 044606 (2019)
3. Brunton, S.L., Noack, B.R., Koumoutsakos, P.: Machine learning for fluid mechanics. Annu. Rev. Fluid Mech. **52**(1), 477–508 (2020)
4. Dosovitskiy, A., et al.: An image is worth 16x16 words: transformers for image recognition at scale. In: ICLR2021 (2021)
5. Du, X., Qu, X., He, Y., Guo, D.: Single image super-resolution based on multi-scale competitive convolutional neural network. Sensors **18**(3), 1–17 (2018)
6. Fukami, K., Nabae, Y., Kawai, K., Fukagata, K.: Synthetic turbulent inflow generator using machine learning. Phys. Rev. Fluids **4**(6), 1–18 (2019)
7. Fukami, K., Nakamura, T., Fukagata, K.: Convolutional neural network based hierarchical autoencoder for nonlinear mode decomposition of fluid field data. Phys. Fluids **32**(9), 1–12 (2020)
8. Hasegawa, K., Fukami, K., Murata, T., Fukagata, K.: Machine-learning-based reduced-order modeling for unsteady flows around bluff bodies of various shapes. Theor. Comput. Fluid Dyn. **34**(4), 367–383 (2020). https://doi.org/10.1007/s00162-020-00528-w
9. Jiménez, J., Moin, P.: The minimal flow unit in near-wall turbulence. J. Fluid Mech. **225**, 213–240 (1991)
10. Morimoto, M., Fukami, K., Zhang, K., Nair, A.G., Fukagata, K.: Convolutional neural networks for fluid flow analysis: toward effective metamodeling and low-dimensionalization. Theor. Comput. Fluid Dyn. **35**, 633–658 (2021)
11. Murata, T., Fukami, K., Fukagata, K.: Nonlinear mode decomposition with machine learning for fluid dynamics. J. Fluid Mech. **882**, A13 (2020)
12. Nakamura, T., Fukami, K., Hasegawa, K., Nabae, Y., Fukagata, K.: Convolutional neural network and long short-term memory based reduced order surrogate for minimal turbulent channel flow. Phys. Fluids **33**, 025116 (2021)
13. Berkooz, G., Holmes, P., Lumley, J.L.: The proper orthogonal, decomposition in the analysis of turbulent flows. Annu. Rev. Fluid Mech. **25**, 539–575 (1993)
14. Racca, A., Doan, N.A.K., Magri, L.: Modelling spatiotemporal turbulent dynamics with the convolutional autoencoder echo state network. arXiv (2022)
15. Yao, J., Hussain, F.: A physical model of turbulence cascade via vortex reconnection sequence and avalanche. J. Fluid Mech. **883**, A51 (2020)
16. Zeiler, M.D., Krishnan, D., Taylor, G.W., Fergus, R.: Deconvolutional networks. In: Proceedings of 2010 IEEE Computer Society Conference on Computer Vision and Pattern Recognition, pp. 2528–2535 (2010)

Graph Neural Network Potentials for Molecular Dynamics Simulations of Water Cluster Anions

Alfonso Gijón[1(✉)], Miguel Molina-Solana[1,2], and Juan Gómez-Romero[1]

[1] Department of Computer Science and AI, University of Granada, Granada, Spain
alfonso.gijon@ugr.es

[2] Department of Computing, Imperial College London, London, UK

Abstract. Regression of potential energy functions is one of the most popular applications of machine learning within the field of materials simulation since it would allow accelerating molecular dynamics simulations. Recently, graph-based architectures have been proven to be especially suitable for molecular systems. However, the construction of robust and transferable potentials, resulting in stable dynamical trajectories, still needs to be researched. In this work, we design and compare several neural architectures with different graph convolutional layers to predict the energy of water cluster anions, a system of fundamental interest in chemistry and biology. After identifying the best aggregation procedures for this problem, we have obtained accurate, fast-evaluated and easy-to-implement graph neural network models which could be employed in dynamical simulations in the future.

Keywords: Graph Neural Networks · Atomic Potentials · Molecular Dynamics · Message Passing · Convolutional layers

1 Introduction

Calculating macroscopic properties (mechanical, thermodynamic, electronic...) of materials through simulation of their microscopic components is a very active field with high impact in science and engineering. This process is usually done employing statistical physics by taking averages of the instantaneous properties on the phase space (in which all possible states of a system are represented), sampled with a suitable distribution. One of the most powerful and widely used technique to sample the phase space and obtain dynamical trajectories is molecular dynamics (MD). In a few words, MD is an algorithm that receives an initial state as input and produces an output state at the following time-step by solving the motion equations. Another popular approach is Monte Carlo (MC), where the phase space is sampled by generating random moves which are accepted or rejected following the Metropolis algorithm. In both MC and MD methods, the potential energy function describes the underlying interactions of the system and is a key ingredient for the quality of a simulation. The most time-consuming

J. Mikyška et al. (Eds.): ICCS 2023, LNCS 14076, pp. 336–343, 2023.
https://doi.org/10.1007/978-3-031-36027-5_25

part of a simulation is the evaluation of the energy for MC and both energy and forces for MD. Forces acting on each atom are computed as derivatives of the potential energy with respect to the atomic positions and are necessary to update the state. Accurate first principles potentials require significant computational resources and it is challenging for simulations to converge when the system size increases.

Different machine learning (ML) architectures have been proposed in recent years as a powerful tool for predicting material properties and molecular interactions [2,9]. In particular, graph neural networks (GNNs) have shown to be especially helpful for evaluating energy and forces of molecular systems, which can reduce the computational cost of molecular dynamics by multiple orders of magnitude [6,7]. Graphs are non-Euclidean data structures able to model systems with complex interactions, such as molecules, as well as respect symmetries like permutational, rotational and translational invariance.

Although the accuracy of complex architectures for predicting static properties has been extensively proven on benchmark datasets [6,11], the robustness of GNN potentials when employed in real MD simulations has still to be improved [10]. Therefore, further investigations are necessary to obtain stable and transferable potentials based on GNNs. Flexible and complex architectures equipped with physics constraints could be a good direction to explore in future work.

In this work, we analyze the performance of different GNN architectures on a dataset of water cluster anions, with chemical formula $(H_2O)_N^-$. This system has been intensively investigated as model for the hydrated electron in bulk water, a species of fundamental interest, which is involved in important chemical and biological electron transfer processes [1,3]. However, despite the huge experimental and theoretical effort made to understand how the excess electron or other negatively-charge species are bound to water clusters [8], there are still open questions.

This paper is organized as follows. The system under study is presented in Sect. 2, together with the basis of graph neural networks. Some methodological details are explained in Sect. 3, while the main results are exposed in Sect. 4. Lastly, some conclusions and possible future work are pointed out in Sect. 5.

2 Background

The properties of a material are determined by its underlying interactions, which can be codified into the potential energy function $V(\mathbf{r}_1, \ldots, \mathbf{r}_N)$, a quantity dependent on the atomic position coordinates $\mathbf{r} = (x, y, z)$. The knowledge of the forces acting on each atom (computed as derivatives of the potential energy function with respect to the atomic positions) of the system would allow to numerically integrate the motion equations and obtain the updated positions. If this process is repeated iteratively, one generates a dynamical trajectory of the system which can be employed to compute macroscopic properties.

In water cluster anions, the excess electron is not attached to a specific water molecule, but is bound globally to several water molecules or to the whole cluster. Depending on the position of the excess electron we can distinguish surface

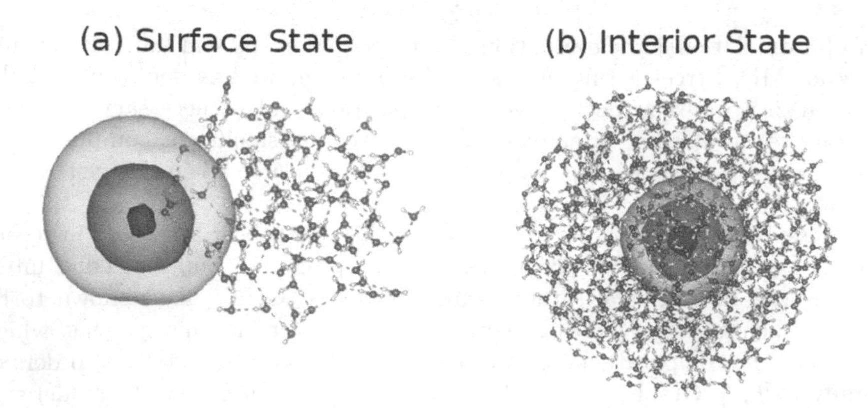

Fig. 1. Schemes of surface (a) and interior (b) states of the excess electron of two water clusters anions.

and interior states, see Fig. 1. The exact description of the interactions is unaffordable, but is not necessary for our purposes. The neutral water molecules are approximated as classical entities interacting through a SPC/F effective potential, while the excess electron is represented by its wave function as a quantum particle living in the pseudopotential generated by the atoms [1,3]. This way, the total potential energy can be considered as the sum of two terms, the first one describing the atomic interactions and the second one taking account the electronic contribution of the excess electron:

$$E(\mathbf{r}_1, \ldots, \mathbf{r}_N) = V_{\mathrm{SPC/F}}(\mathbf{r}_1, \ldots, \mathbf{r}_N) + E_0(\mathbf{r}_1, \ldots, \mathbf{r}_N). \tag{1}$$

While the atomic term is evaluated very fast, the electronic contribution involves the numerical resolution of a 3D single-particle Schrödinger equation[1], which is time-expensive and frequently becomes a bottleneck to converge MD simulations and therefore to compute macroscopic properties.

Graphs serve to model a set of objects (nodes) and their interactions (edges). When applied to molecules, each atom is considered a node and edges are usually assigned to chemical bonds. Both nodes and edges are defined by their features, $\mathbf{x}_i$ and $\mathbf{e}_{i,j}$, respectively. Deep learning (DL) methods can be applied to graph-structured data to perform classification and regression tasks, among others, to give rise to graph neural networks, able to learn about local relations among nodes through the exchange of messages carrying information on the environment of each node [5]. A convolutional (Conv) or message passing (MP) layer updates the node features taking into account information from neighbor nodes

[1] The 3D one-particle Schrödinger equation yields:

$$H\psi_e(\boldsymbol{r}_1, \ldots, \boldsymbol{r}_N; \boldsymbol{r}_e) = E_0\psi_e(\boldsymbol{r}_1, \ldots, \boldsymbol{r}_N; \boldsymbol{r}_e),$$

where the hamiltonian operator is $H = -\nabla^2/2m_e + V_{\mathrm{W}-e}(\boldsymbol{r}_1, \ldots, \boldsymbol{r}_N; \boldsymbol{r}_e)$.

and edges:

$$\mathbf{x}'_i = \texttt{UPDATE}\left[\mathbf{x}_i, \texttt{AGGREGATE}_{j\in\mathcal{N}(i)}\left(\mathbf{x}_i, \mathbf{x}_j, \mathbf{e}_{ij}\right)\right] \quad (2)$$

$$= \gamma\left(\mathbf{x}_i, \square_{j\in\mathcal{N}(i)}\,\phi\left(\mathbf{x}_i, \mathbf{x}_j, \mathbf{e}_{ij}\right)\right) = \mathrm{Conv}(\mathbf{x}_i, \mathbf{x}_j, \mathbf{e}_{ij}). \quad (3)$$

Symbol $\square$ denotes a differentiable, permutation invariant function (sum, mean or max), γ and ϕ represent differentiable functions such as MLP (multilayer perceptron) or a different operation involving node and edge features, and $\mathcal{N}(i)$ represents the set of nodes connected to node i. There are plenty of different convolutional layers which could be used to construct a GNN model, so we perform a systematic study of six of them in next section. Apart from the architecture of a GNN, the embedding process is crucial for the goodness of a model. That is, the encoding of the original data into a graph structure should capture the important features and relations among nodes.

3 Methods

We employ a database of water cluster structures with size ranging from 20 to 237 water molecules. Each element of the database consists of the Cartesian position coordinates of all atoms composing the cluster, along with the atomic and electronic contributions of the total energy associated to that geometry. The geometries are extracted from previous MD equilibrium simulations at 50, 100, 150 and 200 K of temperature, carried out with a software available at https://github.com/alfonsogijon/WaterClusters_PIMD. Finally, the database is formed by 1280 cluster geometries, splitted in 80 % for training and 20 % for testing purposes. Throughout this work, we employ atomic units for energy and distance, unless the opposite is explicitly said.

Each cluster geometry is encoded into a graph structure, where nodes represent atoms and edges represent chemical bonds. Each node contains 4 features, describing the chemical properties of each atom: the first one identifies the atom type through its atomic number (8 for oxygen and 1 for hydrogen) and the 3 remaining features correspond to the atom's position, that is, $\mathbf{x}_i = (Z_i, x_i, y_i, z_i)$. Only one edge feature is considered, as the distance between two connected atoms:

$$e_{ij} = \begin{cases} r_{ij} & i \,\text{and}\, j \,\text{are connected,} \\ 0 & \text{elsewhere.} \end{cases} \quad (4)$$

Each atom is connected to all atoms inside a cutoff sphere (if $r_{ij} < r_c$). The cutoff radius is a tunable parameter, which is set to 6, which implies an average number of 5 neighbor or connected atoms. The cutoff r_c was chosen as the minimum value so that a bigger value does not affect significantly the accuracy of the model. Together with the edge features, for some convolutional layers is convenient to define a binary adjacency matrix, non-zero when two atoms are connected and zero elsewhere:

$$A_{ij} = \begin{cases} 1 & i \text{ and } j \text{ are connected,} \\ 0 & \text{elsewhere.} \end{cases} \quad (5)$$

Our GNN models receive a graph as input and provide a value of the energy as output. The model architecture is summarized in 3 steps:

(i) 2 convolutional layers:

$$\mathbf{x}_i^{k+1} = \text{Conv}\left(\mathbf{x}_i^k, \mathbf{x}_j^k, \mathbf{e}_{ij}^k\right)_{j \in \mathcal{N}(i)}.$$

(ii) Global sum pooling:

$$\mathbf{x}^F = \text{GlobalSumPool}\left(\mathbf{x}_1^2, \ldots, \mathbf{x}_N^2\right) = \sum_{i=1}^{N} \mathbf{x}_i^2.$$

(iii) 2 fully connected layers:

$$E = \text{NN}\left(\mathbf{x}^{\text{F}}\right).$$

First, two convolution layers are applied to the node features, which are updated to account for each node environment. Then, the node features are pooled into a final vector of node features. Finally, a fully connected neural network with 2 hidden layers predicts the energy from the final feature vector. We fixed the number of convolutional and hidden layers to 2 because more layers do not produce an important improvement in the final performance. For each hidden layer of the dense neural network, we used 128 units.

In the results section, the performance of several available types of convolutional layers is tested, while keeping constant the rest of the architecture. To define and train the graph-models the Spektral software [4] was employed, an open-source project freely available on https://graphneural.network/.

4 Results and Discussion

To analyze the performance of different convolutional layers for the energy regression task, up to 6 types of Conv layers were used, namely CrystalConv, GATConv, GCSConv, ECCConv, AGNNCong and GeneralConv. Specific details on each architecture can be consulted on the official Spektral website. For each architecture, two different networks were used to predict each contribution to the total energy, remind Eq. 1. Each model was trained until the loss function (set as MAE) converged to a constant value, which typically occurred between 100 and 150 epochs.

The accuracy of the models on the test dataset is shown in Table 1. As can be seen observing the MAPE columns, the regression of the atomic energy is better than the electronic contribution. This was expected because the atomic term comes from a simple potential (the SPC/F model is sum of two-body terms) while

Table 1. Accuracy of different models with respect to different metrics.

	Atomic Energy			Electronic Energy		
	MAE (Ha)	MSE (Ha^2)	MAPE (%)	MAE (Ha)	MSE (Ha^2)	MAPE (%)
CrystalConv	0.10	0.0232	8.0	0.110	0.06571	101.8
GATConv	0.02	0.0008	1.6	0.005	0.00005	7.6
GCSConv	0.04	0.0027	2.9	0.007	0.00007	10.2
ECCConv	0.25	0.1569	16.1	0.018	0.00053	36.5
AGNNConv	0.06	0.0063	4.2	0.017	0.00059	23.9
GeneralConv	0.03	0.0015	2.2	0.014	0.00028	26.9

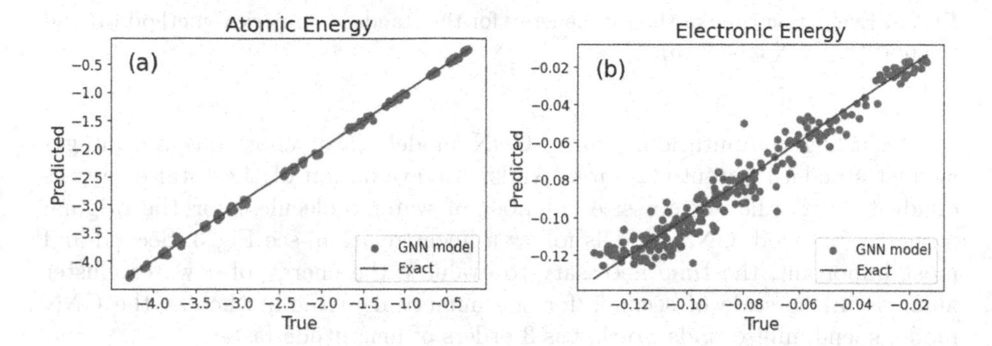

Fig. 2. Comparison between predicted and true energies for (a) atomic and (b) electronic contributions on the test dataset, for the GATConv model.

the electronic term has a more complex nature, such as a Schrödinger equation. Comparing the convolutional layers, GATConv is the best architecture, obtainig a MAPE of 1.6% for the atomic contribution and 7.6% for the electronic one. The energy predictions for that architecture can be visualized in Fig. 2, where predicted energies are compared to the true values.

GATConv layer uses an attention mechanism to weight the adjacency matrix that seems to learn very well different connections of each node (O-O, O-H, H-H bonds). It is remarkable that CrystalConv layer, specially designed for material properties [11], does yield poor results, but this can be explained by the fact that our system is finite as we are working with clusters, and CrystalConv layer was designed to represent crystal, that is, infinite periodical systems. As a matter of fact, only GATConv and GCSConv produces acceptable results for both the atomic and the electronic energies. The former weights the adjacency matrix and the latter has a trainable skip connection layer.

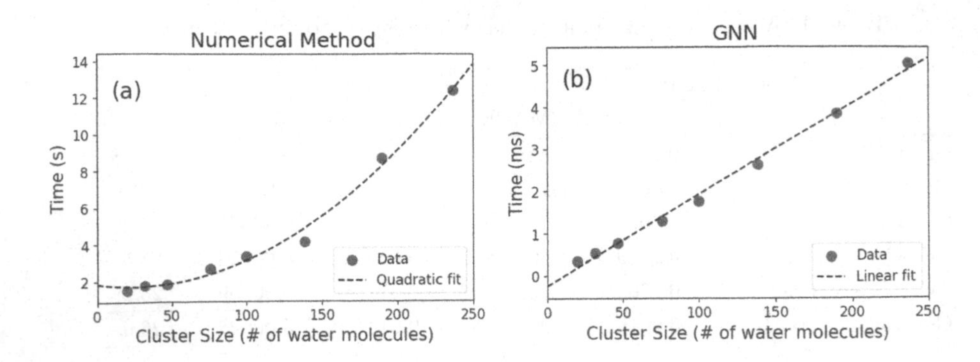

Fig. 3. Evaluation time of the total energy for the standard numerical method (a) and the optimal GNN model (b).

Regarding computational time, GNN models have enormous advantages against standard potentials. First, whilst the evaluation of the total energy is quadratic with the system size (number of water molecules) for the original numerical method, GNN models follow a linear relation, see Fig. 3. Second and most important, the time necessary to evaluate the energy of a water cluster anion is in the scale of seconds for the numerical method, whereas the GNN model spend milliseconds, so that is 3 orders of magnitude faster.

5 Conclusions

As a first step, we have identified an optimum architecture to construct a GNN model able to predict accurate energies for water cluster anions. Besides, the model is easily implemented, in contrast to the numerical method needed to solve the one-electron Schrödinger equation to obtain the energy via standard numerical methods. Our GNN model is also much more faster to evaluate and could make possible to converge long simulations involving many atoms, which are necessary to compute some macroscopic properties.

Our results prove the importance of choosing an appropriate architecture, specially a suitable convolutional scheme, when constructing a GNN model for finite molecular systems as water cluster anions. GATConv and GCSConv layers yield good results and could be improved including more node features (atomic charge, atomic mass...) and edge features (type of bond, angular and dihedrical information...).

As future work, we plan to implement the calculation of the forces in a consistent way, as derivatives of the energy with respect to the atomic coordinates. This would allow to carry out molecular dynamics simulations, as long as improving the learning process and obtain more robust energy potentials.

References

1. Gijón, A.G.: Classical and quantum molecular dynamics simulations of condensed aqueous systems. PhD thesis, Digital CSIC (2021). https://hdl.handle.net/10261/251865
2. Behler, J.: Four generations of high-dimensional neural network potentials. Chem. Rev. **121**(16), 10037–10072 (2021). https://doi.org/10.1021/acs.chemrev.0c00868. PMID: 33779150
3. Gijón, A., Hernandez, E.R.: Quantum simulations of neutral water clusters and singly-charged water cluster anions. Phys. Chem. Chem. Phys. **24**, 14440–14451 (2022). https://doi.org/10.1039/D2CP01088G
4. Grattarola, D., Alippi, C.: Graph neural networks in tensorflow and keras with spektral [application notes]. Comp. Intell. Mag. **16**(1), 99–106 (2021). https://doi.org/10.1109/MCI.2020.3039072
5. Hamilton, W.L.: Graph representation learning. Syn. Lect. Artif. Intell. Mach. Learn. **14**(3), 1–159
6. Klicpera, J., Becker, F., Günnemann, S.: Gemnet: universal directional graph neural networks for molecules. In: Beygelzimer, A., Dauphin, Y., Liang, P., Vaughan, J.W. (eds.) Advances in Neural Information Processing Systems (2021). https://openreview.net/forum?id=HS_sOaxS9K-
7. Li, Z., Meidani, K., Yadav, P., Barati Farimani, A.: Graph neural networks accelerated molecular dynamics. J. Chem. Phys. **156**(14), 144103 (2022). https://doi.org/10.1063/5.0083060
8. Rodríguez-Segundo, R., Gijón, A., Prosmiti, R.: Quantum molecular simulations of micro-hydrated halogen anions. Phys. Chem. Chem. Phys. **24**, 14964–14974 (2022). https://doi.org/10.1039/D2CP01396G
9. Schmidt, J., Marques, M.R.G., Botti, S., Marques, M.A.L.: Recent advances and applications of machine learning in solid-state materials science. npj Comput. Mater. **5**(1), 83 (2019). https://doi.org/10.1038/s41524-019-0221-0
10. Stocker, S., Gasteiger, J., Becker, F., Günnemann, S., Margraf, J.T.: How robust are modern graph neural network potentials in long and hot molecular dynamics simulations? Mach. Learn.: Sci. Technol. **3**(4), 045010 (2022). https://doi.org/10.1088/2632-2153/ac9955
11. Xie, T., Grossman, J.C.: Crystal graph convolutional neural networks for an accurate and interpretable prediction of material properties. Phys. Rev. Lett. **120**, 145301 (2018). https://doi.org/10.1103/PhysRevLett.120.145301

Bayesian Optimization of the Layout of Wind Farms with a High-Fidelity Surrogate Model

Nikolaos Bempedelis[1(✉)] and Luca Magri[1,2]

[1] Department of Aeronautics, Imperial College London, London SW7 2AZ, UK
n.bempedelis20@imperial.ac.uk
[2] The Alan Turing Institute, London NW1 2DB, UK

Abstract. We introduce a gradient-free data-driven framework for optimizing the power output of a wind farm based on a Bayesian approach and large-eddy simulations. In contrast with conventional wind farm layout optimization strategies, which make use of simple wake models, the proposed framework accounts for complex flow phenomena such as wake meandering, local speed-ups and the interaction of the wind turbines with the atmospheric flow. The capabilities of the framework are demonstrated for the case of a small wind farm consisting of five wind turbines. It is shown that it can find optimal designs within a few iterations, while leveraging the above phenomena to deliver increased wind farm performance.

Keywords: Wind farm layout optimization · Large-eddy simulations · Bayesian optimization

1 Introduction

Today, the need for renewable energy sources is more urgent than ever. In the UK, wind power is the largest source of renewable electricity, and the UK government has committed to a further major expansion in capacity by 2030. However, currently installed wind farms do not produce as much power as expected because the majority of the turbines operate within the wake field of other turbines in the farm. A wind turbine operating within a wake field is an issue for two reasons. First, the reduction of its power output due to wind speed deceleration, and, second, the increase of fatigue loads due to increased wind fluctuations.

Wake effects can be minimised by optimally arranging the wind turbines over the available land. Typically, wind farm layout optimization (WFLO) is carried out with low-fidelity flow solvers (wake models) as objective function (farm power output) evaluators [9,10]. Wake models are based on simplified assumptions for the wakes of porous disks, and do not account for several mechanisms including unsteadiness, non-linear interactions, or blockage, to name a few. As a result, optimization based on wake models misses out on a number of opportunities for performance gains through manipulation and, possibly, exploitation of these phenomena. Furthermore, wake models typically provide discontinuous solutions, which renders their use within gradient-based optimization algorithms problematic. Nevertheless, wake models are almost invariably used in layout optimization

J. Mikyška et al. (Eds.): ICCS 2023, LNCS 14076, pp. 344–352, 2023.
https://doi.org/10.1007/978-3-031-36027-5_26

studies due to their low computational cost (a single evaluation typically runs in under a second).

More accurate approximations of the wind farm flow have been considered in only a limited number of works [1,6]. In these works, the wind farm layout was optimized by an adjoint approach and steady-state Reynolds-averaged Navier-Stokes (RANS) simulations. The use of steady RANS allowed capturing a number of the aforementioned phenomena. Nevertheless, the assumption of a steady flow means that the wake dynamics and the wake-to-wake and atmosphere-to-wake interactions, which are critical for the wind farm layout optimization problem, were not appropriately accounted for.

In this work, we present a gradient-free framework for optimizing the output of a wind farm based on a Bayesian approach and high-fidelity large-eddy simulations of the flow around the wind farm. Bayesian optimization is a suitable optimization strategy due to the multi-modality of the WFLO problem (which makes gradient-based methods prone to getting stuck in local extrema), and the high cost of evaluating the objective function (at least when an accurate model of the flow field is desired, as in our study). The structure of the paper is as follows. The data-driven optimization framework is described in Sect. 2. Section 3 discusses its application to a wind farm layout optimization problem. Finally, Sect. 4 summarises the present study.

2 Methodology

2.1 The Optimization Problem

We aim to maximise the overall power output P from N different wind turbines experiencing K different wind states by controlling their position $\boldsymbol{c} = [\boldsymbol{x}, \boldsymbol{y}]^T$, with $\boldsymbol{x} = (x_1, \ldots, x_N)$ and $\boldsymbol{y} = (y_1, \ldots, y_N)$, within a given space $\boldsymbol{X}$. The space $\boldsymbol{X}$ corresponds to the available land where the wind turbines may be installed. To avoid overlap between the different wind turbines, we enforce a constraint that ensures that their centers (i.e. their position) are spaced at least one turbine diameter D apart. The optimization problem can be expressed as

$$\begin{aligned} &\arg\max_{\boldsymbol{c}} && \sum_{n=1}^{N}\sum_{k=1}^{K} a_k P_{n,k} \\ &\text{s.t.} && \boldsymbol{c} \in \boldsymbol{X} \\ & && ||\boldsymbol{c}_i - \boldsymbol{c}_j|| > D, \text{ for } i,j = 1,\ldots,N \text{ and } i \neq j \end{aligned} \tag{1}$$

with a_k being the weight (i.e. probability) of each wind state, which are obtained from the local meteorological data (obtained via a measurement mast).

2.2 Flow Solver

The power output of the wind turbines is computed with the open-source finite-difference framework `Xcompact3D` [2], which solves the incompressible filtered

Navier-Stokes equations on a Cartesian mesh using sixth-order compact schemes and a third-order Adams-Bashforth method for time advancement [7]. Parallelisation is achieved with the `2Decomp & FFT` library, which implements a 2D pencil decomposition of the computational domain [8]. The wind turbines are modelled with the actuator disk method. The Smagorinsky model is used to model the effects of the unresolved fluid motions. Finally, to realistically model the interaction of the wind farm with the atmospheric flow, a precursor simulation of a fully-developed neutral atmospheric boundary layer is performed to generate the inlet conditions. For more details on the numerical solver and a related validation study, the reader is referred to [3].

2.3 Bayesian Optimization Algorithm

Bayesian optimization is a gradient-free optimization technique that consists of two main steps. First, a surrogate model (here a Gaussian Process) of the objective function is computed given knowledge of its value for a set of parameters. This also quantifies the uncertainty of the approximation. Second, it proposes points in the search space where sampling is likely to yield an improvement. This is achieved by minimising an acquisition function. In this work, we make use of the `GPyOpt` library [11]. In particular, we use the Matérn 5/2 kernel and the Expected Improvement acquisition function, which is minimised using the L-BFGS algorithm. Bayesian optimization requires a number of initial samples to start. In this work, these are obtained using both the Latin hypercube sampling technique and a custom function that targets a uniform distribution of the wind turbines over the available land $\boldsymbol{X}$ whilst favouring placement on the domain boundaries (the initial layouts used in this study are shown in Fig. 4). For a more thorough description of Bayesian optimization, along with an example of it being used to optimize a chaotic fluid-mechanical system, the reader is referred to [5].

3 Results

The data-driven optimization framework is deployed on the following problem. The available land is a square of size $6D \times 6D$, where $D = 100$ m is the turbine diameter. The wind blows from a single direction (westerly) and at constant speed. The atmospheric boundary layer is characterised by friction velocity $u^* = 0.442$ m/s, height $\delta = 501$ m and roughness length $z_0 = 0.05$ m, which correspond to conditions in the North Sea [12]. The velocity at the hub height of the turbines, $h = 100$ m, is $U_h = 8.2$ m/s and the turbulence intensity at the same level is $TI_h = 7.4\%$ (see Fig. 1). The wind turbines operate at a thrust coefficient $C_T = 0.75$. The size of the computational domain is $2004 \times 1336 \times 501$ m in the streamwise, spanwise, and vertical directions. It is discretised with $193 \times 128 \times 49$ points, respectively (amounting to $\approx 5 \times 10^6$ degrees of freedom at each time step). The land where the turbines can be placed starts at $x = 3.68D$ from the upstream boundary. Periodic conditions are enforced on the lateral

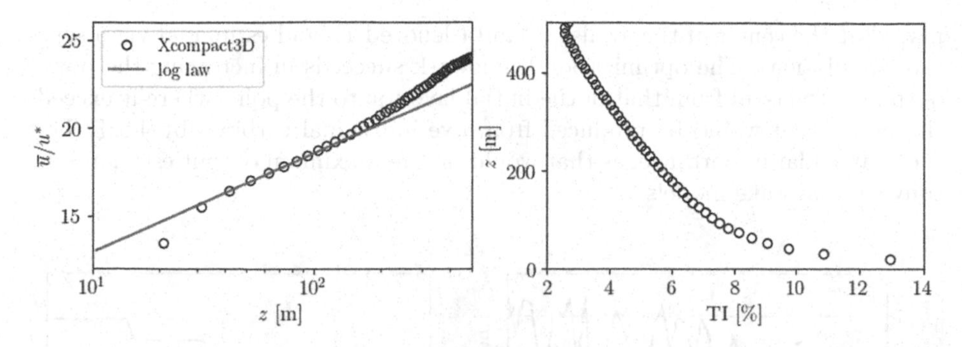

Fig. 1. Mean velocity (left) and turbulence intensity (right) of the simulated atmospheric boundary layer.

domain boundaries. A time step $\Delta t = 0.15$ s is used, with the maximum CFL number remaining under 0.18. Statistics are averaged over a one-hour time period following one hour of initialisation (each period corresponds to $\approx$ 15 flow-through times based on the hub-height velocity).

We consider a wind farm consisting of five wind turbines. Five layouts are used to initialise the Bayesian optimization. Four are generated with the Latin hypercube sampling technique and one with the custom function (see Sect. 2). A snapshot of the streamwise velocity in the latter case is presented in Fig. 2, which shows the turbulent nature of the flow field.

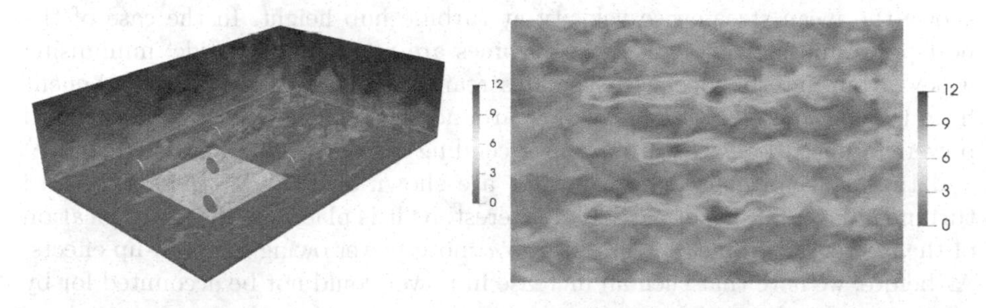

Fig. 2. Instantaneous streamwise velocity for layout #5. Three-dimensional view including a square indicating the available land and five disks indicating the turbine rotors (left). Horizontal cut at the turbine hub height (right).

The optimization runs in batch mode [4], with three sets of parameter values to be explored proposed at every step. This means that convergence is potentially sub-optimal; however, it allows for three numerical experiments to be run in parallel. The optimization is stopped after 50 iterations. Each simulation required $\approx$ 70 CPU hours on ARCHER2. Figure 3 shows the optimization history. The average farm output is normalised by the power produced by a single turbine placed at the center of the available land (denoted P_0 and estimated via a separate simulation). The optimization framework succeeds in increasing the power output of the farm from that of the initial layouts, to the point where it exceeds the power that would be produced from five individual turbines by 4.5%. This is of particular importance, as that would be the maximum output estimated by conventional wake models.

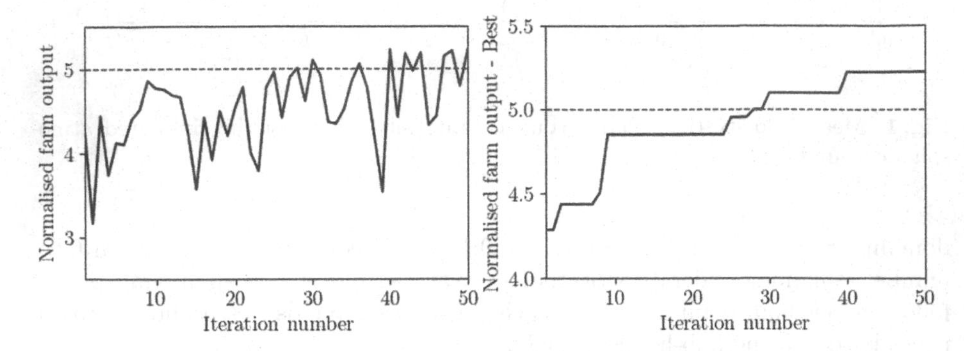

Fig. 3. Normalised average wind farm power output (left) and best performance history (right).

The initial layouts together with two explored during the optimization and the best-performing one are presented in Fig. 4. For each layout, the figure also shows the mean streamwise velocity at turbine hub height. In the case of the best-performing layout (#40), the turbines are placed side-by-side, minimising wake-turbine interference, with a small streamwise offset so that they can benefit from the local acceleration of the flow around their neighbours. The normalised power produced by each turbine (in ascending order with streamwise distance) and the total normalised farm output are shown in Table 1. Here, the third turbine of layout #5 is of particular interest, as it is placed at the exact location of the reference turbine, but produces 5% more power owing to speed-up effects. As before, we note that such an increase in power could not be accounted for by conventional wake models.

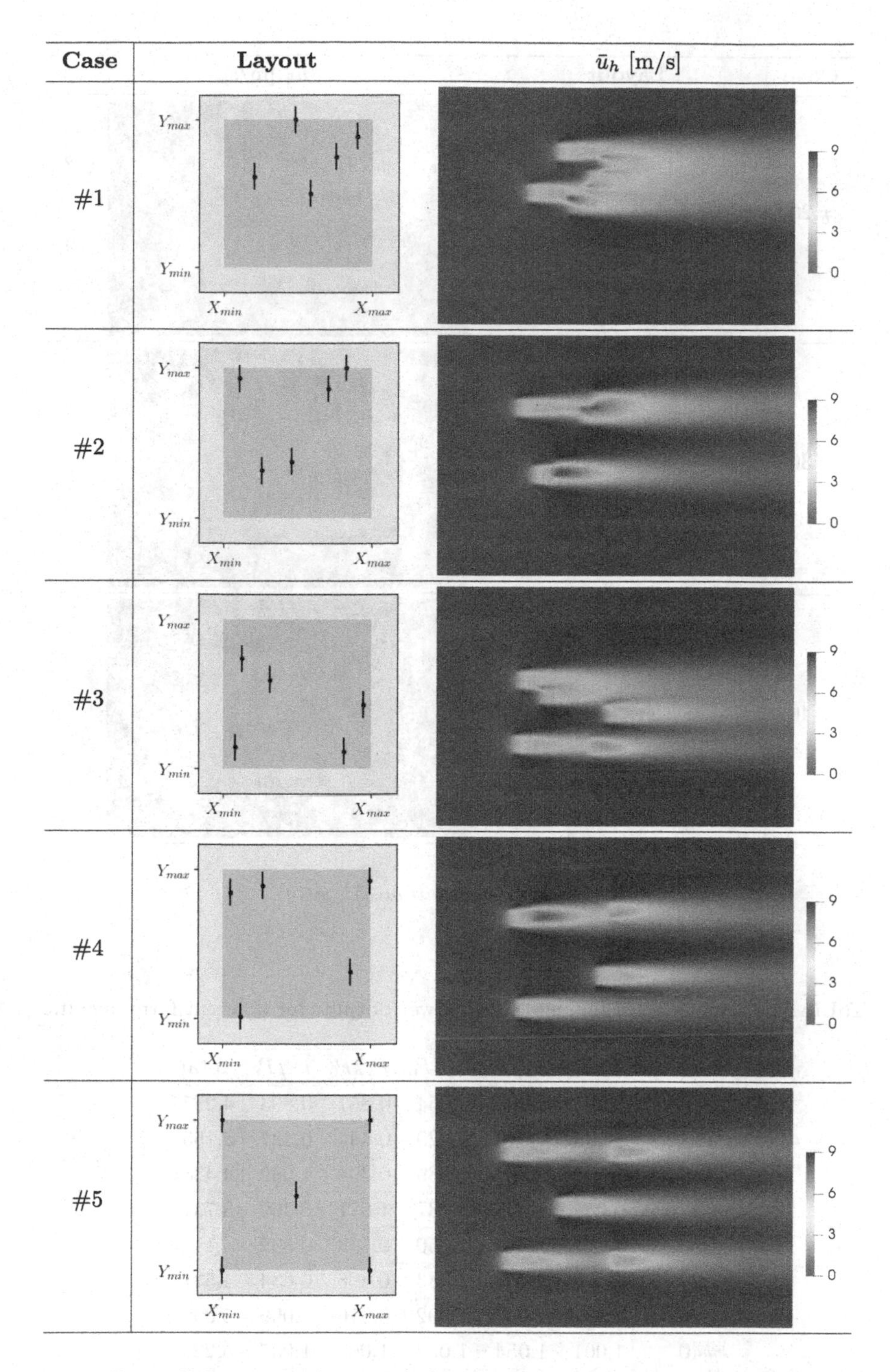

Fig. 4. Example layouts and associated mean streamwise velocity at hub height.

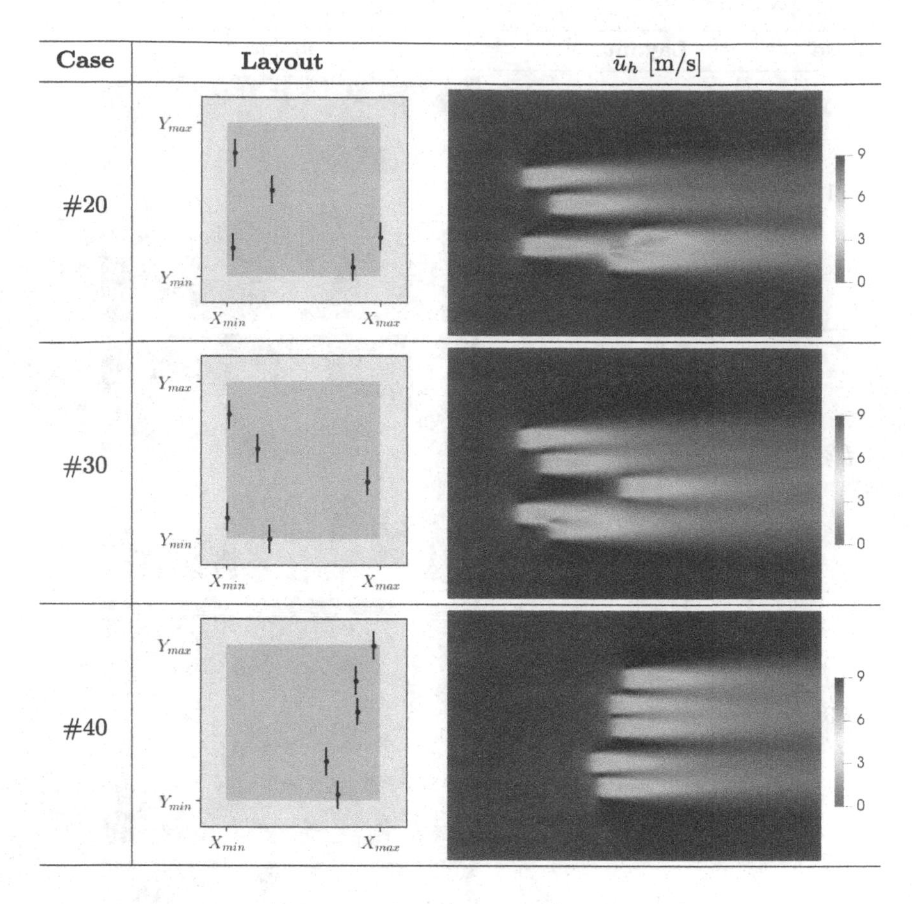

Fig. 4. (*continued*)

Table 1. Normalised turbine and farm power outputs for different farm layouts.

Layout	P_1/P_0	P_2/P_0	P_3/P_0	P_4/P_0	P_5/P_0	Total
#1	0.975	1.000	0.654	0.857	0.803	4.287
#2	0.986	0.979	0.329	0.444	0.427	3.165
#3	1.005	0.987	0.989	0.398	1.059	4.438
#4	0.951	1.003	0.237	1.071	0.469	3.731
#5	1.001	1.016	1.050	0.513	0.532	4.113
#20	1.006	1.003	1.083	0.806	0.634	4.531
#30	0.984	1.007	1.092	0.916	1.099	5.099
#40	1.001	1.054	1.033	1.068	1.067	5.224

4 Conclusions

This study proposes a gradient-free data-driven framework that optimizes the power output of a wind farm using a Bayesian approach and large-eddy simulations of the flow around the farm. Unlike traditional wind farm optimization strategies, which use simple wake models, this framework considers turbulent flow dynamics including wake meandering, wake-to-wake and atmosphere-to-wake interactions. The effectiveness of the framework is demonstrated through a case study of a small wind farm with five turbines. It is shown that the optimization can quickly find optimal designs whilst improving wind farm performance by taking into account these complex flow phenomena. In the future, the framework will be applied to configurations with complex wind roses (multiple wind directions and velocities) and large wind farms with more turbines in order to assess its scalability and computational efficiency and compare it with other optimisation strategies.

Acknowledgement. NB and LM are supported by EPSRC, Grant No. EP/W026686/1. LM also acknowledges financial support from the ERC Starting Grant No. PhyCo 949388. The authors would like to thank the UK Turbulence Consortium (EP/R029326/1) for providing access to ARCHER2.

References

1. Antonini, E.G.A., Romero, D.A., Amon, C.H.: Continuous adjoint formulation for wind farm layout optimization: a 2D implementation. Appl. Energy **228**, 2333–2345 (2018)
2. Bartholomew, P., Deskos, G., Frantz, R.A.S., Schuch, F.N., Lamballais, E., Laizet, S.: Xcompact3D: an open-source framework for solving turbulence problems on a Cartesian mesh. SoftwareX **12**, 100550 (2020)
3. Bempedelis, N., Laizet, S., Deskos, G.: Turbulent entrainment in finite-length wind farms. J. Fluid Mech. **955**, A12 (2023)
4. González, J., Dai, Z., Hennig, P., Lawrence, N.: Batch Bayesian optimization via local penalization. In: Artificial Intelligence and Statistics, pp. 648–657. PMLR (2016)
5. Huhn, F., Magri, L.: Gradient-free optimization of chaotic acoustics with reservoir computing. Phys. Rev. Fluids **7**(1), 014402 (2022)
6. King, R.N., Dykes, K., Graf, P., Hamlington, P.E.: Optimization of wind plant layouts using an adjoint approach. Wind Energy. Sci. **2**(1), 115–131 (2017)
7. Laizet, S., Lamballais, E.: High-order compact schemes for incompressible flows: a simple and efficient method with quasi-spectral accuracy. J. Comput. Phys. **228**(16), 5989–6015 (2009)
8. Laizet, S., Li, N.: Incompact3d: a powerful tool to tackle turbulence problems with up to $\mathcal{O}(10^5)$ computational cores. Int. J. Numer. Meth. Fluids **67**(11), 1735–1757 (2011)
9. Porté-Agel, F., Bastankhah, M., Shamsoddin, S.: Wind-turbine and wind-farm flows: a review. Bound.-Layer Meteorol. **174**(1), 1–59 (2020)
10. Shakoor, R., Hassan, M.Y., Raheem, A., Wu, Y.K.: Wake effect modeling: a review of wind farm layout optimization using Jensen's model. Renew. Sustain. Energy Rev. **58**, 1048–1059 (2016)

11. The GPyOpt authors: GPyOpt: A Bayesian Optimization framework in Python (2016). http://github.com/SheffieldML/GPyOpt
12. Wu, Y.T., Porté-Agel, F.: Modeling turbine wakes and power losses within a wind farm using LES: an application to the Horns Rev offshore wind farm. Renew. Energy **75**, 945–955 (2015)

An Analysis of Universal Differential Equations for Data-Driven Discovery of Ordinary Differential Equations

Mattia Silvestri(✉), Federico Baldo, Eleonora Misino, and Michele Lombardi

University of Bologna, Bologna, Italy
{mattia.silvestri4,federico.baldo2,eleonora.misino2, michele.lombardi2}@unibo.it

Abstract. In the last decade, the scientific community has devolved its attention to the deployment of data-driven approaches in scientific research to provide accurate and reliable analysis of a plethora of phenomena. Most notably, Physics-informed Neural Networks and, more recently, Universal Differential Equations (UDEs) proved to be effective both in system integration and identification. However, there is a lack of an in-depth analysis of the proposed techniques. In this work, we make a contribution by testing the UDE framework in the context of Ordinary Differential Equations (ODEs) discovery. In our analysis, performed on two case studies, we highlight some of the issues arising when combining data-driven approaches and numerical solvers, and we investigate the importance of the data collection process. We believe that our analysis represents a significant contribution in investigating the capabilities and limitations of Physics-informed Machine Learning frameworks.

1 Introduction

Physics-informed Machine Learning has gained high attention in the last few years [3,5,14,16–18,23], enabling the integration of physics knowledge into machine learning models. Purely data-driven methods, like Deep Neural Networks (DNNs), have huge representational power and can deal with noisy high dimensional raw data; however, they may learn observational biases, leading to physically inconsistent predictions and poor generalization performance. On the other hand, despite the relentless progress in the field, solving real-world partial differential equations (PDEs) using traditional analytical or computational approaches requires complex formulations and prohibitive costs. A lot of effort has been devoted to bridging DNNs with differential equations in end-to-end trainable frameworks. However, less attention has been paid to analyze the advantages and limitations of the proposed approaches.

We view the lack of an in-depth analysis of physics-informed techniques as a major issue. We make a contribution in this area by performing an analysis on

M. Silvestri, F. Baldo and E. Misino—These authors have contributed equally.

J. Mikyška et al. (Eds.): ICCS 2023, LNCS 14076, pp. 353–366, 2023.
https://doi.org/10.1007/978-3-031-36027-5_27

the Universal Differential Equation (UDE) [17] framework in the context of data-driven discovery of ODEs. We focus on UDE since its general formulation allows to express other existing frameworks. In particular, we focus on: 1) *evaluating two training approaches in terms of accuracy and efficiency*; 2) *testing the effect of the numerical solver accuracy in the parameters approximation*, 3) *analyzing the impact of the data collection process regarding the approximation accuracy*, and in 4) *exploring the effectiveness of UDE in reconstructing a functional dependence between a set of observables and the unknown parameters.*

The paper is structured as follows. In Sect. 2, we provide an overview of the existing work in physics-informed machine learning and system identification. We briefly introduce the UDE framework in Sect. 3, and we describe our research questions in Sect. 4. In Sect. 5, we present the experiments and report the results. Finally, in Sect. 6, we draw some conclusions and discuss future directions.

2 Related Work

In this section, we briefly present some of the most promising trends in Physics-informed Machine Learning. For an exaustive literature overview, we refer the reader to [10].

Physics-Informed Loss Function. The most straightforward way to enforce constraints in Neural Networks is via an additional term in the loss function. In [11] the authors propose Physics-guided Neural Network, a framework that exploits physics-based loss functions to increase deep learning models' performance and ensure physical consistency of their predictions. Similarly, the work of Chen et al. [9] generalizes Recurrent Neural Networks adding a regularization loss term that captures the variation of energy balance over time in the context of lake temperature simulation. Work of [1] proposes to enforce physics constraints in Neural Networks by introducing a penalization term in the loss function defined as the mean squared residuals of the constraints.

Physics-Informed Neural Architectures. Recent works focus on designing deep learning frameworks that integrate physics knowledge into the architecture of deep learning models [3,5,14,16–18,23]. Neural Ordinary Differential Equations (Neural ODEs) [5] bridge neural networks with differential equations by defining an end-to-end trainable framework. In a Neural ODE, the derivative of the hidden state is parameterized by a neural network, and the resulting differential equation is numerically solved through an ODE solver, treated as a black-box. Neural ODEs have proven their capacity in time-series modeling, supervised learning, and density estimation. Moreover, recent works adopt Neural ODEs for system identification by learning the discrepancy between the prior knowledge of the physical system and the actual dynamics [14] or by relying on a two-stage approach to identify unknown parameters of differential equations [3]. Recently, O'Leary et al. [16] propose a framework that learns hidden physics and dynamical models of stochastic systems. Their approach is based on Neural ODEs, moment-matching, and mini-batch gradient descent to approximate the unknown hidden physics. Another approach is represented by the Physics-informed Neural Network (PINN) framework [18] which approximates the hidden state of a physical system through a neural network. The authors show how to use PINNs both to solve a PDE given the model parameters and to discover the model parameters from data. Zhang et al. [23] further extend PINNs by accounting for the uncertainty quantification of the solution. In particular, the authors focus on the *parametric* and *approximation uncertainty*. Universal Differential Equations (UDEs) [17] represent a generalization of Neural ODE where part of a differential equation is described by a universal approximator, such as a neural network. The formulation is general enough to allow the modeling of time-delayed, stochastic, and partial differential equations. Compared to PINNs, this formalism is more suitable to integrate recent and advanced numerical solvers, providing the basis for a library that supports a wide range of scientific applications.

System Identification. Research towards the automated dynamical system discovery from data is not new [6]. The seminal works on system identification through genetic algorithms [2,22] introduce symbolic regression as a method to discover nonlinear differential equations. However, symbolic regression is limited in its scalability. Brunton and Lipson [4] propose a sparse regression-based method for identifying ordinary differential equations, while Rudy et al. [21] and Schaeffer [8] apply sparse regression to PDEs discovering. Recent works [3,14,15] focus on applying physics-informed neural architectures to tackle the system discovery problem. Lu et al. [15] propose a physics-informed variational autoencoder to learn unknown parameters of dynamical systems governed by partial differential equations. The work of Lai et al. [14] relies on Neural ODE for structural-system identification by learning the discrepancy with respect to the true dynamics, while Bradley at al. [3] propose a two-stage approach to identify unknown parameters of differential equations employing Neural ODE.

3 Universal Differential Equations

The Universal Differential Equation (UDE) [17] formulation relies on embedded universal approximators to model forced stochastic delay PDEs in the form:

$$\mathcal{N}\left[u(t), u(\alpha(t)), \mathrm{W}(t), U_\theta(u, \beta(t))\right] = 0 \tag{1}$$

where $u(t)$ is the system state at time t, $\alpha(t)$ is a delay function, and $\mathrm{W}(t)$ is the Wiener process. $\mathcal{N}[\cdot]$ is a nonlinear operator and $U_\theta(\cdot)$ is a universal approximator parameterized by θ. The UDE framework is general enough to express other frameworks that combine physics knowledge and deep learning models. For example, by considering a one-dimensional UDE defined by a neural network, namely $u' = U_\theta(u(t), t)$, we retrieve the Neural Ordinary Differential Equation framework [5,12,20].

UDEs are trained by minimizing a cost function C_θ defined on the current solution $u_\theta(t)$ with respect to the parameters θ. The cost function is usually computed on discrete data points (t_i, y_i) which represent a set of measurements of the system state, and the optimization can be achieved via gradient-based methods like ADAM [13] or Stochastic Gradient Descent (SGD) [19].

4 UDE for Data-Driven Discovery of ODEs

In this section, we present the UDE formulation we adopt, and we describe four research questions aimed at performing an in-depth analysis of the UDEs framework in solving data-driven discovery of ODEs.

Formulation. We restrict our analysis to dynamical systems described by ODEs with no stochasticity or time delay. The corresponding UDE formulation is:

$$u' = f(u(t), t, U_\theta(u(t), t)). \tag{2}$$

where $f(\cdot)$ is the known dynamics of the system, and $U_\theta(\cdot, \cdot)$ is the universal approximator for the unknown parameters. As cost function, we adopt the Mean Squared Error (MSE) between the current approximate solution $u_\theta(t)$ and the true measurement $y(t)$, formally:

$$C_\theta = \sum_i \|u_\theta(t_i) - y(t_i)\|_2^2 . \tag{3}$$

We consider discrete time models, where the differential equation in (2) can be solved via numerical techniques. Among the available solvers, we rely on the Euler method, which is fully differentiable and allows for gradient-based optimization. Moreover, the limited accuracy of this first-order method enlightens the effects of the integration technique on the unknown parameter approximation. Our analysis starts from a simplified setting, in which we assume that the unknown parameters are fixed. Therefore, the universal approximator in Eq. 2 reduces to a set of learnable variables, leading to:

$$u' = f(u(t), t, U_\theta). \tag{4}$$

Training Procedure. Given a set of state measurements y in the discrete interval $[t_0, t_n]$, we consider two approaches to learn Eq. (4), which we analyze in terms of *accuracy* and *efficiency*. The first approach, mentioned by [18] and named here FULL-BATCH, involves 1) applying the Euler method on the whole temporal series with $y(t_0)$ as the initial condition, 2) computing the cost function C_θ, and 3) optimizing the parameters θ via full-batch gradient-based methods. An alternative approach, named MINI-BATCH, consists of splitting the dataset into pairs of consecutive measurements $(y(t_i), y(t_{i+1}))$, and considering each pair as a single initial value problem. Then, by applying the Euler method on the single pair, we can perform a mini-batch training procedure, which helps in mitigating the gradient vanishing problem [7]. Conversely to the FULL-BATCH approach, which requires data to be ordered and uniform in observations, the MINI-BATCH method has less strict requirements and can be applied also to partially ordered datasets.

Solver Accuracy. In the UDE framework, the model is trained to correctly predict the system evolution by learning an approximation of the unknown parameters that minimizes the cost function C_θ. The formulation relies on the integration method to approximate the system state $u(t)$. However, the numerical solver may introduce approximation errors that affect the whole learning procedure. Here, we want to investigate the *impact of the solver accuracy on the unknown parameters approximation.* Since the Euler method is a first-order method, its error depends on the number of iterations per time step used to estimate the value of the integral, and, thus, we can perform our analysis with direct control on the trade-off between execution time and solver accuracy.

Functional Dependence. By relying on the universal approximator in Eq. 2, the UDE framework is able to learn not only fixed values for the unknown parameters, but also functional relationships between them and the observable variables. Thus, we add a level of complexity to our analysis by considering the system parameters as functions of observable variables, and we *evaluate the UDE accuracy in approximating the unknown functional dependence.*

Data Sampling. Since UDE framework is a data-driven approach, it is important to investigate the effectiveness of the UDE framework under different data samplings. In particular, *can we use the known dynamics of the system under analysis to design the data collection process in order to increase the approximation accuracy?*

5 Empirical Analysis

Here, we report the results of our analysis performed on two case studies: 1) *RC circuit*, i.e., estimating the final voltage in a first-order resistor-capacitor circuit; 2) *Predictive Epidemiology*, i.e., predicting the number of infected people during a pandemic. We start by describing the two case studies; then, we illustrate the evaluation procedure and the experimental setup. Finally, we present the experiments focused on the research questions highlighted in Sect. 4, and we report the corresponding results.

RC Circuit. We consider a first-order RC circuit with a constant voltage generator. The state evolution of the system is described by

$$\frac{dV_C(t)}{dt} = \frac{1}{\tau}(V_s - V_C(t)) \tag{5}$$

where $V_C(t)$ is the capacitor voltage at time t, V_s is the voltage provided by the generator, and τ is the time constant which defines the circuit response.

We use the UDE formulation to approximate τ and V_s by writing Eq. (5) as

$$u' = \frac{1}{U_{\theta_1}(t)}(U_{\theta_2}(t) - u(t)) \tag{6}$$

where u_t is a short notation for $V_C(t)$, $U_{\theta_1}(t)$ and $U_{\theta_2}(t)$ are the neural networks approximating τ and V_s respectively. The cost function is defined as

$$C_{\theta_1,\theta_2} = \sum_i (u_{\theta_1,\theta_2}(t_i) - y_i)^2 \tag{7}$$

where $u_{\theta_1,\theta_2}(t_i)$ and y_i are the current solution and the discrete-time measurements of the capacitor voltage at time t_i, respectively.

Predictive Epidemiology. Among the different compartmental models used to describe epidemics, we consider the well-known Susceptible-Infected-Recovered (SIR) model, where the disease spreads through the interaction between susceptible and infected populations. The dynamics of a SIR model is described by the following set of differential equations:

$$\begin{aligned} \frac{dS}{dt} &= -\beta\, \frac{S \cdot I}{N}, \\ \frac{dI}{dt} &= \beta\, \frac{S \cdot I}{N} - \gamma\, I, \\ \frac{dR}{dt} &= \gamma\, I, \end{aligned} \tag{8}$$

where S, I, and R refer to the number of susceptible, infected, and recovered individuals in the population. The population is fixed, so $N = S + I + R$. The parameter $\gamma \in [0, 1]$ depends on the average recovery time of an infected subject, while $\beta \in [0, 1]$ is the number of contacts needed per time steps to have a new infected in the susceptible population. β determines the spreading coefficient of the epidemic and is strongly affected by different environmental factors (e.g., temperature, population density, contact rate, etc.). The introduction of public health measures that directly intervene on these environmental factors allows to contain the epidemic spreading.

We rely on the UDE framework to i) perform system identification on a simulated SIR model, and ii) estimate the impact of *Non-Pharmaceutical Interventions* (NPIs) on the epidemic spreading. We define the state of the system at time t as $\mathbf{u}_t = (S_t, I_t, R_t)$ and we formulate the Equations in (8) as

$$\mathbf{u}' = f(\mathbf{u}_t, t, U_\theta(\mathbf{u}_t, t, X_t)) \tag{9}$$

where X_t is the set of NPIs applied at time t. We assume γ to be fixed and known, and we approximate the SIR model parameter β with a neural network $U_\theta(\mathbf{u}_t, t, X_t)$. The cost function for this case study is defined as

$$C_\theta = \sum_i (u_\theta(t_i) - \hat{y}_i)^2 \tag{10}$$

where $u_\theta(t_i)$ and y_i are the current solution and the discrete-time measurements of the system state at time t_i, respectively.

Evaluation and Experimental Setup. We evaluate the model accuracy by relying on two metrics: the *Absolute Error* (AE), to evaluate the estimation of the parameters, and the *Root Mean Squared Error* (RMSE), to study the approximation of the state of the dynamic systems. For each experiment, we perform 100 trials, normalize the results, and report mean and standard deviation. All the experiments are run on a Ubuntu virtual environment equipped with 2 T V100S, both with a VRAM of 32 GB. We work in a `Python 3.8.10` environment, and the neural models are implemented in `TensorFlow 2.9.0`. The source code is available at https://github.com/ai-research-disi/ode-discovery-with-ude.

5.1 Training Procedure

We compare FULL-BATCH and MINI-BATCH methods to assess which is the most accurate and efficient. We rely on high-precision simulation to generate data for both case studies. For the RC circuit, we set $V_c(0) = 0$, and we sample 100 values of V_s and τ in the range $[5, 10]$ and $[2, 6]$, respectively. Then, we generate data by relying on the analytical solution of Eq. 5. From each of the resulting curves, we sample 10 data points $(V_c(t), t)$ equally spaced in the temporal interval $[0, 5\tau]$. Concerning the epidemic case study, the data generation process relies on a highly accurate Euler integration with 10.000 iterations per time step. We use the same initial condition across all instances, namely 99% of susceptible and 1% of infected on the entire population, and we assume γ to be equal to 0.1, meaning that the recovery time of infected individuals is on average 10 days. We create 100 epidemic curves, each of them determined by the sampled value of β in the interval $[0.2, 0.4]$. The resulting curves contain daily data points in the temporal interval from day 0 to day 100 of the outbreak evolution.

We evaluate the accuracy of UDE in approximating the unknown parameters and the system state, and we keep track of the total computation time required to reach convergence. We believe it is relevant to specify that the MINI-BATCH has an advantage compared to the FULL-BATCH. The evaluation of the latter involves predicting the whole state evolution given only the initial one u_0; whereas, the first approach reconstructs the state evolution if provided with intermediate values. Thus, to have a fair comparison, the predictions of the MINI-BATCH method are fed back to the model to forecast the entire temporal series given only u_0. As shown in Table 1, for the RC circuit case study, both FULL-BATCH and MINI-BATCH approximate quite accurately V_s and $V_c(t)$, whereas the approximation of

τ has a non-negligible error. However, FULL-BATCH requires almost 3 times the computational time to converge. In the SIR use case (Table 2), the two training procedures achieve very similar accuracies, but FULL-BATCH is more than 8 times computationally expensive. Since both the methods have very similar estimation accuracy, we can conclude that MINI-BATCH *is a more efficient method to train the UDE compared to* FULL-BATCH. Thus, we rely on the MINI-BATCH method in the remaining experiments.

Table 1. Comparison between MINI-BATCH and FULL-BATCH methods in RC circuit use case. We report the AE of V_s and τ approximation, RMSE for $V_c(t)$ prediction, and computational time in seconds.

	V_s	τ	$V_c(t)$	Time
MINI-BATCH	0.027 ± 0.013	0.163 ± 0.101	0.021 ± 0.010	9.21 ± 39.49
FULL-BATCH	0.018 ± 0.021	0.200 ± 0.081	0.014 ± 0.020	26.19 ± 5.69

Table 2. Comparison between MINI-BATCH and FULL-BATCH methods in epidemic use case. We report the AE of β approximation, RMSE for $SIR(t)$ prediction, and computational time in seconds.

	β	$SIR(t)$	Time
MINI-BATCH	0.0030 ± 0.0019	0.017 ± 0.0046	1.28 ± 0.23
FULL-BATCH	0.0065 ± 0.0053	0.019 ± 0.0079	10.23 ± 2.50

5.2 Solver Accuracy

In the context of ODE discovery, we are interested in approximating the unknown system parameters. Despite an overall accurate estimation of the system state, the results of the previous analysis show that UDE framework does not reach high accuracy in approximating the system parameters. The model inaccuracy might be caused by the approximation error introduced by the integration method. Thus, to investigate the *impact of the solver accuracy on the unknown parameters approximation*, we test different levels of solver accuracy by increasing the number of iterations between time steps in the integration process. A higher number of iterations per time step of the Euler method should lead to more accurate solutions of the ODE; however, this comes also at the cost of a higher computational time as shown in Fig. 1.

In this experiment, we use the same data generated for *Training procedure* experiment. In Fig. 2, we report the approximation error of the UDE framework when applying the Euler method with an increasing number of steps. As expected, in both use cases, by increasing the precision of the Euler method, the ODE parameters estimation becomes more accurate, until reaching a plateau after 10 iterations per time step.

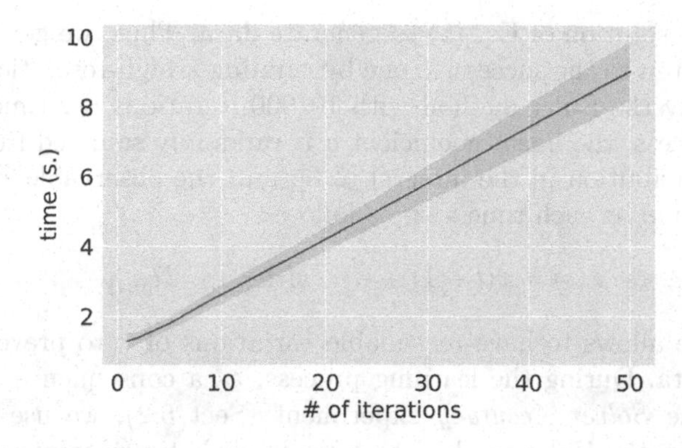

Fig. 1. UDE training time as a function of the number of iterations per time step of the Euler method.

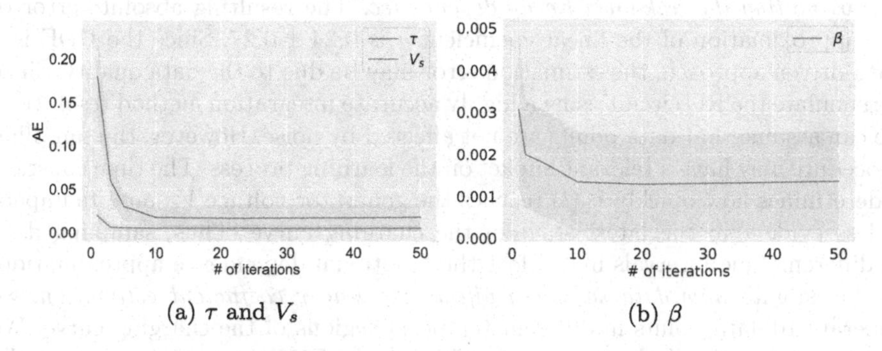

Fig. 2. Average and standard deviation of the AE as a function of the number iterations per time step of the Euler method.

5.3 Functional Dependence and Data Sampling

In a real-world scenario, the dynamical systems that we are analyzing often depend on a set of external variables, or *observables*, that influence the behaviour of the system. These elements can be environmental conditions or control variables which affect the evolution of the system state. We study the UDE framework in presence of observables, assuming two kinds of relationship between the independent and dependent variable, namely, a *linear* and a *non-linear* dependence.

Linear Dependence. For the RC circuit, we consider a simple and controlled setup where τ is a linear function of a continuous input variable x changing over time, namely $\tau(x) = ax$, where a and x are scalar values. Conversely to the previous experiments, we assume V_s to be known and equal to 1; we perform this design choice to focus our analysis on the approximation accuracy of the linear relationship solely. Since the value of τ changes over time, we can not rely on the analytic solution of Eq. (5) to generate data. Thus, we generate samples from one timestep to the successive one by running a high-resolution integration method, namely the Euler method with 10,000 iterations per time step. In the generation process, the linear coefficient a is randomly sampled from a uniform probability distribution in the interval [2, 6], and the observable x is initialized to 1, and updated at each time step as follows:

$$x(t) = x(t-1) + \epsilon, \quad \text{with} \quad \epsilon \sim \mathcal{U}_{[0,1]}.$$

This procedure allows to have reasonable variations of τ to prevent physically implausible data. During the learning process, as a consequence of the results obtained in the *Solver Accuracy* experiment (Sect. 5.2), we use 10 iterations per time step in the Euler method as a trade-off between numerical error and computational efficiency.

In this experiment, we are interested in *evaluating the UDE accuracy in approximating the unknown linear dependence.* The resulting absolute error of the approximation of the linear coefficient a is 0.24 ± 0.27. Since the UDE is a data-driven approach, the estimation error may be due to the data quality. Since we simulate the RC circuit using a highly accurate integration method resolution, we can assume that data points are not affected by noise. However, the sampling procedure may have a relevant impact on the learning process. The time constant τ determines how quickly $V_c(t)$ reaches the generator voltage V_s, and its impact is less evident in the latest stage of the charging curve. Thus, sampling data in different time intervals may affect the functional dependence approximation. To investigate *how data sampling affects the linear coefficient estimation*, we generate 10 data points in different temporal regions of the charging curve. We consider intervals of the form $[0, EOH]$, where $EOH \in (0, 5\tau]$ refers to the end-of-horizon of the measurements; since τ changes over time, we consider the maximum as a reference value to compute the EOH.

As shown in Fig. 3, the linear model approximation is more accurate if the data points are sampled in an interval with $EOH \in [1.5\tau, 3\tau]$, where $V_c(t)$ approximately reaches respectively the 77% and 95% of V_s. With higher values of EOH, the sampled data points are closer to the regime value V_s, and the impact of τ is less relevant in the system state evolution. Thus, the learning model can achieve high prediction accuracy of $V_c(t)$ without correctly learning the functional dependence.

Non-linear Dependence. Here, we test the UDE framework under the assumption of a non-linear dependence between the observable and the β parameter of the epidemic model. The observable is a set of Non-Pharmaceutical Interventions (NPIs), which affects the virus spreading at each time step. To generate the epidemic data, we define the following time series representing the variation at time t of the inherent infection rate of the disease, $\hat{\beta}$, under the effect of two different NPIs per time instance:

$$\beta(t, \mathbf{x}^t, \mathbf{e}, \hat{\beta}) = \hat{\beta} \cdot e_1^{x_1^t} \cdot e_2^{x_2^t} \tag{11}$$

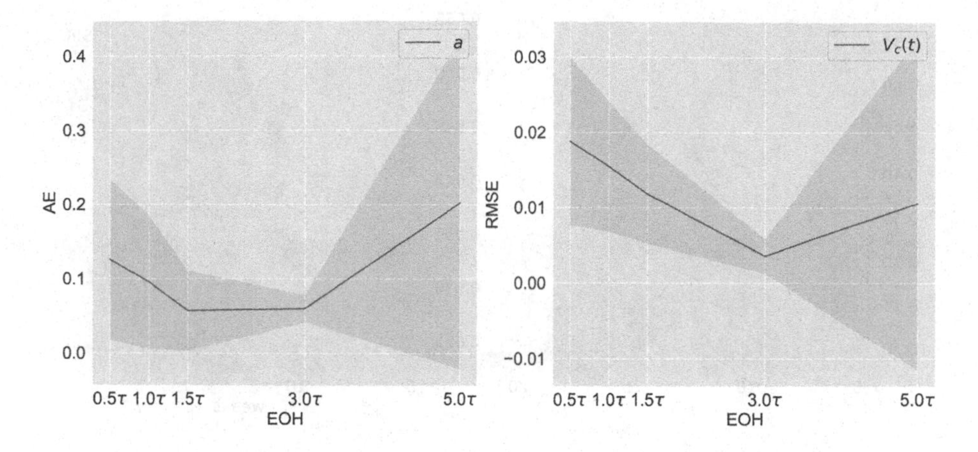

Fig. 3. Linear coefficients and predictions error and a function of the EOH.

where $\mathbf{x}^t \in \{0,1\}^2$ is the binary vector indicating whether the corresponding NPIs are active at time t. The vector $\mathbf{e} \in [0,1]^2$ represents the effects of the two NPIs in reducing the infection rate. We compute 100 different time-series for β by assuming that the vector of NPIs, $\mathbf{x}^t$, randomly changes each week. For each of the resulting time series, we generate 20 data points equally spaced from day 0 to day 140 of the outbreak evolution. The generation process relies on a highly accurate Euler integration with 10.000 iterations per time step and uses the same initial condition and γ value described in the *Training Procedure* experiment (Sect. 5.1). To approximate the non-linear dependence in Eq. 11, we rely on a DNN which forecasts the value of β based on $\mathbf{x}^t$ and the state of the system at time $t-1$. Thus, the resulting universal approximator of the UDE framework is a black-box model able to capture complex functional dependencies, but lacking interpretability.

The experimental results show that the UDE framework is able to estimate the dynamic system state with high accuracy (the RMSE of the state prediction is 0.037 ± 0.013); however, the model is unable to provide an accurate estimation of the β time-series, which RMSE is equal to 0.37 ± 0.17. Similarly to the RC circuit, we investigate the effect of the data sampling frequency to the parameter approximation accuracy of the UDE. We consider 4 different time horizons, namely $5, 10, 15$, and 20 weeks of the outbreak evolution, in which we sample 20 equally spaced data points. We train the model on the resulting data, and we compute the reconstruction error (RMSE) on the complete epidemic curve of 20 weeks. We report both the parameter approximation error and the curve reconstruction error in Fig. 4.

Conversely to RC circuit, the sampling process does not seem to a have a significant impact on the model accuracy. The reason for this result may be found in the complexity of the function to be approximated, and in the impact of β parameter to the epidemic curve. In the RC-circuit, the system state evolution

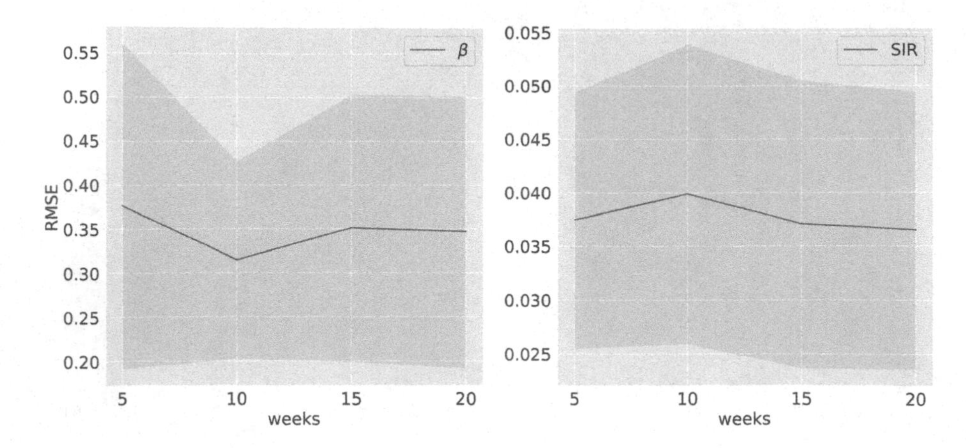

Fig. 4. Non-Linear dependence: β and prediction errors with different sampling frequencies.

is an exponential function of the unknown parameter τ, and we can design the collection process to cover the temporal interval where the impact of τ is more relevant. In the SIR model, we do not have a closed-form of the epidemic curve, and thus it is harder to select the most relevant temporal horizon.

6 Conclusions

In this paper, we perform an in-depth analysis of the UDEs framework in solving data-driven discovery of ODEs. We experimentally probe that MINI-BATCH gradient descent is faster than the FULL-BATCH version without compromising the final performances. We highlight some issues arising when combining data-driven approaches and numerical integration methods, like the discrepancy in accuracy between state evolution prediction and system parameter approximations. We investigated the integration method precision as a possible source of error, and we discuss the trade-off between approximation accuracy and computational time. Moreover, we study the importance of the data collection process in reaching higher parameter approximation accuracy.

We believe that our analysis can foster the scientific community to further investigate the capabilities and limitations of Physics-informed machine learning frameworks in the context of differential equation discovery.

In the future we plan to extend our analysis by i) testing different numerical integration solvers (e.g., higher-order Runge-Kutta or adjoint-state methods), ii) considering the unknown parameters to be stochastic, rather than deterministic, iii) extending the analysis to PDEs.

Acknowledgement. Research partly supported by European ICT-48-2020 Project TAILOR - g.a. 952215.

References

1. Beucler, T., Pritchard, M., Rasp, S., Ott, J., Baldi, P., Gentine, P.: Enforcing analytic constraints in neural networks emulating physical systems. Phys. Rev. Lett. **126**(9), 098302 (2021)
2. Bongard, J.C., Lipson, H.: Automated reverse engineering of nonlinear dynamical systems. Proc. Natl. Acad. Sci. U.S.A. **104**(24), 9943–9948 (2007)
3. Bradley, W., Boukouvala, F.: Two-stage approach to parameter estimation of differential equations using neural odes. Industr. Eng. Chem. Res. **60**(45), 16330–16344 (2021)
4. Champion, K.P., Zheng, P., Aravkin, A.Y., Brunton, S.L., Kutz, J.N.: A unified sparse optimization framework to learn parsimonious physics-informed models from data. IEEE Access **8**, 169259–169271 (2020)
5. Chen, R.T., Rubanova, Y., Bettencourt, J., Duvenaud, D.K.: Neural ordinary differential equations. Adv. Neural Inf. Process. Syst. **31** (2018)
6. Crutchfield, J.P., McNamara, B.S.: Equations of motion from a data series. Complex Syst. **1**(3) (1987). https://www.complex-systems.com/abstracts/v01_i03_a03.html
7. Glorot, X., Bengio, Y.: Understanding the difficulty of training deep feedforward neural networks. In: Proceedings of the Thirteenth International Conference on Artificial Intelligence and Statistics, pp. 249–256. JMLR Workshop and Conference Proceedings (2010)
8. Hayden, S.: Learning partial differential equations via data discovery and sparse optimization. Proc. R. Soc. A. 4732016044620160446 (2017). https://doi.org/10.1098/rspa.2016.0446
9. Jia, X., et al.: Physics guided RNNs for modeling dynamical systems: A case study in simulating lake temperature profiles. In: Proceedings of the 2019 SIAM International Conference on Data Mining, pp. 558–566. SIAM (2019)
10. Karniadakis, G.E., Kevrekidis, I.G., Lu, L., Perdikaris, P., Wang, S., Yang, L.: Physics-informed machine learning. Nat. Rev. Phys. **3**(6), 422–440 (2021)
11. Karpatne, A., Watkins, W., Read, J.S., Kumar, V.: Physics-guided neural networks (PGNN): an application in lake temperature modeling. CoRR abs/1710.11431 (2017). https://arxiv.org/abs/1710.11431
12. Kidger, P., Morrill, J., Foster, J., Lyons, T.: Neural controlled differential equations for irregular time series. In: Larochelle, H., Ranzato, M., Hadsell, R., Balcan, M., Lin, H. (eds.) Advances in Neural Information Processing Systems, vol. 33, pp. 6696–6707. Curran Associates, Inc. (2020). https://proceedings.neurips.cc/paper/2020/file/4a5876b450b45371f6cfe5047ac8cd45-Paper.pdf
13. Kingma, D.P., Ba, J.: Adam: a method for stochastic optimization. In: ICLR (2015)
14. Lai, Z., Mylonas, C., Nagarajaiah, S., Chatzi, E.: Structural identification with physics-informed neural ordinary differential equations. J. Sound Vib. **508**, 116196 (2021)
15. Lu, P.Y., Kim, S., Soljačić, M.: Extracting interpretable physical parameters from spatiotemporal systems using unsupervised learning. Phys. Rev. X **10**(3), 031056 (2020)
16. O'Leary, J., Paulson, J.A., Mesbah, A.: Stochastic physics-informed neural ordinary differential equations. J. Comput. Phys. **468**, 111466 (2022). https://doi.org/10.1016/j.jcp.2022.111466
17. Rackauckas, C., et al.: Universal differential equations for scientific machine learning. arXiv preprint arXiv:2001.04385 (2020)

18. Raissi, M., Perdikaris, P., Karniadakis, G.E.: Physics-informed neural networks: a deep learning framework for solving forward and inverse problems involving nonlinear partial differential equations. J. Comput. Phys. **378**, 686–707 (2019)
19. Robbins, H., Monro, S.: A stochastic approximation method. Ann. Math. Stat. **22**, 400–407 (1951)
20. Rubanova, Y., Chen, R.T.Q., Duvenaud, D.K.: Latent ordinary differential equations for irregularly-sampled time series. In: Wallach, H., Larochelle, H., Beygelzimer, A., d' Alché-Buc, F., Fox, E., Garnett, R. (eds.) Advances in Neural Information Processing Systems, vol. 32. Curran Associates, Inc. (2019). https://proceedings.neurips.cc/paper/2019/file/42a6845a557bef704ad8ac9cb4461d43-Paper.pdf
21. Rudy, S.H., Brunton, S.L., Proctor, J.L., Kutz, J.N.: Data-driven discovery of partial differential equations. Sci. Adv. **3**(4), e1602614 (2017). https://doi.org/10.1126/sciadv.1602614, https://www.science.org/doi/abs/10.1126/sciadv.1602614
22. Schmidt, M., Lipson, H.: Distilling free-form natural laws from experimental data. Science **324**(5923), 81–85 (2009). https://doi.org/10.1126/science.1165893, https://www.science.org/doi/abs/10.1126/science.1165893
23. Zhang, D., Lu, L., Guo, L., Karniadakis, G.E.: Quantifying total uncertainty in physics-informed neural networks for solving forward and inverse stochastic problems. J. Comput. Phys. **397**, 108850 (2019)

Learning Neural Optimal Interpolation Models and Solvers

Maxime Beauchamp(✉), Quentin Febvre, Joseph Thompson, Hugo Georgenthum, and Ronan Fablet

IMT Atlantique Bretagne Pays de la Loire, Brest, France
maxime.beauchamp@imt-atlantique.fr

Abstract. The reconstruction of gap-free signals from observation data is a critical challenge for numerous application domains, such as geoscience and space-based earth observation, when the available sensors or the data collection processes lead to irregularly-sampled and noisy observations. Optimal interpolation (OI), also referred to as kriging, provides a theoretical framework to solve interpolation problems for Gaussian processes (GP). The associated computational complexity being rapidly intractable for n-dimensional tensors and increasing numbers of observations, a rich literature has emerged to address this issue using ensemble methods, sparse schemes or iterative approaches. Here, we introduce a neural OI scheme. It exploits a variational formulation with convolutional auto-encoders and a trainable iterative gradient-based solver. Theoretically equivalent to the OI formulation, the trainable solver asymptotically converges to the OI solution when dealing with both stationary and non-stationary linear spatio-temporal GPs. Through a bi-level optimization formulation, we relate the learning step and the selection of the training loss to the theoretical properties of the OI, which is an unbiased estimator with minimal error variance. Numerical experiments for 2D+t synthetic GP datasets demonstrate the relevance of the proposed scheme to learn computationally-efficient and scalable OI models and solvers from data. As illustrated for a real-world interpolation problems for satellite-derived geophysical dynamics, the proposed framework also extends to non-linear and multimodal interpolation problems and significantly outperforms state-of-the-art interpolation methods, when dealing with very high missing data rates.

Keywords: optimal interpolation · differentiable framework · variational model · optimizer learning

1 Introduction

Interpolation problems are critical challenges when dealing with irregularly-sampled observations. Among others, Space earth observation, geoscience, ecology, fisheries generally monitor a process of interest through partial observations due to the characteristics of the sensors and/or the data collection process. As illustrated in Fig. 3 for satellite-based earth observation, missing

J. Mikyška et al. (Eds.): ICCS 2023, LNCS 14076, pp. 367–381, 2023.
https://doi.org/10.1007/978-3-031-36027-5_28

data rates may be greater than 90%, which makes the interpolation problem highly challenging.

Optimal Interpolation (OI) also referred to as kriging [6], provides a theoretical framework to address such interpolation problems for Gaussian processes. Given the covariance structure of the process of interest along with the covariance of the observation noise, one can derive the analytical OI solution. For high-dimensional states, such as space-time processes, the computation of this analytical solution rapidly becomes intractable as it involves the inversion of a $N \times N$ matrix with N the number of observation points. When dealing with space-time processes, OI also relates to data assimilation [2]. In this context, Kalman methods, including ensemble-based extensions, exploit the sequential nature of the problem to solve the OI problem allowing for dynamical flow propagation of the uncertainties.

Data-driven and learning-based approaches have also received a growing interest to address interpolation problems [3,15], while Image and video inpainting are popular interpolation problems in computer vision [22]. As they typically relate to object removal applications or restoration problems, they usually involve much lower missing data rates than the ones to be delt with in natural images, which are likely not representative of space-time dynamics addressed in geoscience, meteorology, ecology... A recent literature has also emerged to exploit deep learning methods to solve inverse problems classically stated as the minimization of a variational cost. This includes neural architectures based on the unrolling of minimization algorithms [17,21].

Here, we introduce a neural OI framework. Inspired by the neural method introduced in [8] for data assimilation, we develop a variational formulation based on convolutional auto-encoders and introduce an associated trainable iterative gradient-based solver. Our key contributions are four-fold:

- We show that our variational formulation is equivalent to OI when dealing with Gaussian processes driven by linear dynamics. Under these assumptions, our trainable iterative gradient-based solver converges asymptotically towards the OI solution;
- Regarding the definition of the training losses, we relate the learning step of the proposed neural architecture to the properties of the OI solution as an unbiased estimator with minimal error variance;
- Our framework extends to learning optimal interpolation models and solvers for non-linear/non-Gaussian processes and multimodal observation data;
- Numerical experiments for a 2D+t Gaussian process support the theoretical equivalence between OI and our neural scheme for linear Gaussian case-studies. They also illustrate the targeted scalable acceleration of the interpolation.
- We report a real-world application to the interpolation of sea surface dynamics from satellite-derived observations. Our neural OI scheme significantly outperforms the state-of-the-art methods and can benefit from multimodal observation data to further improve the reconstruction performance.

To make easier the reproduction of our results, an open-source version of our code is available[1].

This paper is organized as follows. Section 2 formally introduces optimal interpolation and related work. We present the proposed neural OI framework in Sect. 3. Section 4 reports numerical experiments for both synthetic GP datasets and real-world altimetric sea surface observations. We discuss further our main contributions in Sect. 5.

2 Problem Statement and Related Work

For a n-dimensional Gaussian process $\mathbf{x}$ with mean μ and covariance $\mathbf{P}$, the optimal interpolation states the reconstruction of state $\mathbf{x}$ from noisy and partial observations $\mathbf{y}$ as the minimization of a variational cost:

$$\widehat{\mathbf{x}} = \arg\min_{\mathbf{x}} \|\mathbf{y} - H_\Omega \cdot \mathbf{x}\|^2_{\mathbf{R}} + \|\mathbf{x} - \mu\|^2_{\mathbf{P}} \tag{1}$$

with H_Ω denotes the observation matrix to map state $\mathbf{x}$ over domain $\mathcal{D}$ to the observed domain Ω. $\|\cdot\|^2_{\mathbf{R}}$ is the Mahanalobis norm w.r.t the covariance of the observation noise $\mathbf{R}$ and $\|\cdot\|^2_{\mathbf{P}}$ the Mahanalobis distance with *a priori* covariance $\mathbf{P}$. The latter decomposes as a 2-by-2 block matrix $[\mathbf{P}_{\Omega,\Omega}\mathbf{P}_{\Omega,\overline{\Omega}}; \mathbf{P}_{\overline{\Omega},\overline{\Omega}}\mathbf{P}^T_{\Omega,\overline{\Omega}}]$ with $\mathbf{P}_{\mathcal{A},\mathcal{A}'}$ the covariance between subdomains $\mathcal{A}$ and $\mathcal{A}$ of domain $\mathcal{D}$.

The OI variational cost (1) being linear quadratic, the solution of the optimal interpolation problem is given by:

$$\widehat{\mathbf{x}} = \mu + \mathbf{K} \cdot \mathbf{y} \tag{2}$$

with $\mathbf{K}$ referred to as the Kalman gain $\mathbf{P}H_\Omega^\mathrm{T}(H_\Omega\mathbf{P}H_\Omega^\mathrm{T} + \mathbf{R})^{-1}$, where $\mathbf{P}H_\Omega^\mathrm{T}$ is the (grid,obs) prior covariance matrix, $H_\Omega\mathbf{P}H_\Omega^\mathrm{T}$ is the (obs,obs) prior covariance matrix. For high-dimensional states, such as nD and nD+t states, and large observation domains, the computation of the Kalman gain becomes rapidly intractable due to the inversion of a $|\Omega| \times |\Omega|$ covariance matrix. This has led to a rich literature to solve minimization (1) without requiring the above-mentioned $|\Omega| \times |\Omega|$ matrix inversion, among others gradient-based solvers using matrix-vector multiplication (MVMs) reformulation [18], methods based on sparse matrix decomposition with tapering [19] or precision-based matrix parameterizations [16].

Variational formulations have also been widely explored to solve inverse problems. Similarly to (1), the general formulation involves the sum of a data fidelity term and of a prior term [2]. In a model-driven approach, the latter derives from the governing equations of the considered processes. For instance, data assimilation in geoscience generally exploits PDE-based terms to state the prior on some hidden dynamics from observations. In signal processing and

[1] To be made available in a final version.

computational imaging, similar formulations cover a wide range of inverse problems, including inpainting issues [4]:

$$\widehat{\mathbf{x}} = \arg\min_{\mathbf{x}} \mathcal{J}_{\Phi}(\mathbf{x},\mathbf{y},\mathbf{\Omega}) = \arg\min_{\mathbf{x}} \mathcal{J}^{o}(\mathbf{x},\mathbf{y}) + \lambda \left\|\mathbf{x} - \Phi(\mathbf{x})\right\|^2$$

$\mathcal{J}^{o}(\mathbf{x},\mathbf{y})$ is the data fidelity term which is problem-dependent. The prior regularization term $\|\mathbf{x} - \Phi(\mathbf{x})\|^2$ can be regarded as a projection operator. This parameterization of the prior comprises both gradient-based priors using finite-difference approximations, proximal operators as well as plug-and-play priors [17]. As mentioned above, these formualtions have also gained interest in the deep learning literature for the definition of deep learning schemes based on the unrolling of minimization algorithms [1] for (??). Here, we further explore the latter category of approaches to solve optimal problems stated as (1), including when covariance **P** is not known a priori.

3 Neural OI Framework

This section presents the proposed trainable OI framework. We first introduce the proposed neural OI solver (Sect. 3.1) and the associated learning setting (Sect. 3.2). We then describe extensions to non-linear and multimodal interpolation problems.

3.1 Neural OI Model and Solver

Let us introduce the following variational formulation to reconstruct state $\mathbf{x}$ from partial observations $\mathbf{y}$:

$$\widehat{\mathbf{x}} = \arg\min_{\mathbf{x}} \mathcal{J}_{\Phi}(\mathbf{x},\mathbf{y},\Omega) = \arg\min_{\mathbf{x}} \|\mathbf{y} - H_{\Omega}\cdot\mathbf{x}\|^2 + \lambda \left\|\mathbf{x} - \Phi(\mathbf{x})\right\|^2 \qquad (3)$$

where λ is a positive scalar to balance the data fidelity term and the prior regularization. $\Phi(\cdot)$ is a linear neural auto-encoder which states the prior onto the solution.

Variational formulation (3) is equivalent to optimal interpolation problem (1) when considering a matrix parameterization of the prior $\Phi(x) = (\mathbf{I} - \mathbf{L})\mathbf{x}$ with $\mathbf{L}$ the square-root (as a Cholesky decomposition) of $\mathbf{P}$ and a spherical observation covariance, *i.e.* $\mathbf{R} = \sigma^2\mathbb{1}$. The proof comes immediately when noting that the regularization term of the variational cost also writes: $\mathbf{x}^{\mathsf{T}}\mathbf{P}^{-1}\mathbf{x} = \mathbf{x}^{\mathsf{T}}\mathbf{L}^{\mathsf{T}}\mathbf{L}\mathbf{x} = ||\mathbf{L}\mathbf{x}||^2 = ||\mathbf{x} - \Phi\mathbf{x}||^2$.

Lemma 1. *For a stationary Gaussian process and a Gaussian observation noise with* $\mathbf{R} = \sigma^2\mathbf{I}$, *we can restate the associated optimal interpolation problem (1) as minimization problem (3) with neural operator* $\Phi(\cdot)$ *being a linear convolutional network.*

The proof results from the translation-invariant property of the covariance of stationary Gaussian processes. Computationally, we can derive Φ as the inverse Fourier transform of the square-root of the Fourier transform of covariance $\mathbf{P}$ in (1) as exploited in Gaussian texture synthesis [10].

Lemma 1 provides the basis to learn a solver of variational formulation (3) and address optimal interpolation problem (1). We benefit from automatic differentiation tools associated with neural operators to investigate iterative gradient-based solvers as introduced in meta-learning [12], see Algorithm 1. The latter relies on an iterative gradient-based update where neural operator $\mathcal{G}$ combines an LSTM cell [20] and a linear layer to map the hidden state of the LSTM cell to the space spanned by state $\mathbf{x}$. Through the LSTM may capture long-term dependencies, operator $\mathcal{G}$ defines a momentum-based gradient descent. Overall, this class of generic learning-based methods was explored and referred to as 4DVarNet schemes in [8] for data assimilation problems. Here, as stated in Lemma 2, we parameterize weighting factors $a(\cdot)$ and $\omega(\cdot)$ such that the LSTM-based contribution dominates for the first iterations while for a greater number of iterations the iterative update reduces to a simple gradient descent. Hereafter, we refer to the proposed neural OI framework as 4DVarNet-OI.

Algorithm 1. Iterative gradient-based solver for (3) given initial condition $\mathbf{x}^{(0)}$, observation $\mathbf{y}$ and sampled domain Ω. Let $a(\cdot)$ and $\omega(\cdot)$ be positive scalar functions and $\mathcal{G}$ a LSTM-based neural operator.

$\mathbf{x} \leftarrow \mathbf{x}^{(0)}$
$i \leftarrow 0$
while $i \leq K$ **do**
  $i \leftarrow i + 1$
  $\mathbf{g} \leftarrow \nabla_{\mathbf{x}} \mathcal{J}_{\Phi}\left(\mathbf{x}^{(i)}, \mathbf{y}, \Omega\right)$
  $\mathbf{x} \leftarrow \mathbf{x} - a(i) \cdot [\omega(i) \cdot \mathbf{g} + (1 - \omega(i)) \cdot \mathcal{G}(\mathbf{g})]$
end while

Lemma 2. *Let us consider the following parameterizations for functions $a(\cdot)$ and $\omega(\cdot)$*

$$a(i) = \frac{\nu \cdot K_0}{K_0 + i} \; ; \omega(i) = \tanh\left(\alpha \cdot (i - K_1)\right) \tag{4}$$

where ν and α are positive scalars, and $K_{0,1}$ positive integers. If $\mathcal{G}(\cdot)$ is a bounded operator and $\Phi(\cdot)$ is a linear operator given by $\mathbf{I} - \mathbf{P}^{1/2}$, then Algorithm 1 converges towards the solution (2) of the minimization of optimal interpolation cost (3).

The proof of this lemma derives as follows. As $\mathcal{G}(\cdot)$ is bounded, the considered parameterization of the gradient step in Algorithm 1 is asymptotically equivalent to a simple gradient descent with a decreasing step size. Therefore, it satisfies the convergence conditions towards the global minimum for a linear-quadratic variational cost [5]. We may highlight that the same applies with a stochastic version of Algorithm 1 and a convex variational cost [5].

The boundedness of operator $\mathcal{G}$ derives from that of the LSTM cell. Therefore, Lemma 2 guarantees that Algorithm 1 with a LSTM-based parameterization for operator $\mathcal{G}$ converges to the minimum of optimal interpolation cost (3) whatever the parameters of $\mathcal{G}$.

In this setting, operator $\mathcal{G}$ aims at accelerating the convergence rate of the gradient descent towards analytical solution (2). Overall, we define $\Theta = \{\Phi, \mathcal{G}\}$ and $\Psi^K_\Theta(\mathbf{x}^{(0)}, \mathbf{y}, \Omega)$ the interpolated state resulting from the application of Algorithm 1 with K iterations from initial condition $\mathbf{x}^{(0)}$ given observation data $\{\mathbf{y}, \Omega\}$.

3.2 Learning Setting

Formally, we state the training of the considered neural OI scheme (3) according to a bi-level optimization problem

$$\widehat{\Theta} = \arg\min_{\Theta} \mathcal{L}\left(\mathbf{x}_k, \mathbf{y}_k, \Omega_k, \widehat{\mathbf{x}}_k\right) \text{ s.t. } \widehat{\mathbf{x}}_k = \arg\min_{\mathbf{x}_k} \mathcal{J}_\Phi\left(\mathbf{x}_k, \mathbf{y}_k, \Omega_k\right) \tag{5}$$

where $\mathcal{L}(\{\mathbf{x}_k, \mathbf{y}_k, \widehat{\mathbf{x}}_k\})$ defines a training loss and k denotes the time index along the data assimilation window (DAW) $[t - k\Delta t; t + k\Delta t]$.

Let us consider Optimal Interpolation problem (1) with a spherical observation covariance $\mathbf{R} = \sigma^2 \cdot \mathbf{I}$ and prior covariance $\mathbf{P}$. Let us parameterize trainable operator Φ in (3) as a linear convolution operator. Optimal interpolation (2) is then solution of a bi-level optimization problem (5) with $\Phi = \mathbf{I} - \mathbf{P}^{1/2}$ for each of the following training losses:

$$\begin{aligned}
\mathcal{L}_1\left(\mathbf{x}_k, , \widehat{\mathbf{x}}_k\right) &= \sum_{k=1}^{N} \|\mathbf{x}_k - \widehat{\mathbf{x}}_k\|^2 \\
\mathcal{L}_2\left(\mathbf{x}_k, \mathbf{y}_k, \widehat{\mathbf{x}}_k\right) &= \sum_{k=1}^{N} \|\mathbf{y}_k - H_\Omega \cdot \mathbf{x}_k\|^2_{\mathbf{R}} + \|\mathbf{x}_k\|^2_{\mathbf{P}}
\end{aligned} \tag{6}$$

where $\mathcal{L}_1$ denotes the mean squared error (MSE) w.r.t true states and $\mathcal{L}_2$ stands for the OI variational cost. This results from the equivalence between variational formulations (1) and (3) under parameterization $\Phi = \mathbf{I} - \mathbf{P}^{1/2}$ and the property that OI solution (2) is a minimum-variance unbiased estimator.

Such a formulation motivates the following training setting for the proposed scheme:

$$\widehat{\Theta} = \arg\min_{\Theta} \mathcal{L}\left(\mathbf{x}_k, \mathbf{y}_k, \widehat{\mathbf{x}}_k\right) \text{ s.t. } \widehat{\mathbf{x}}_k = \Psi^K_\Theta\left(\mathbf{x}_k^{(0)}, \mathbf{y}_k, \Omega_k\right) \tag{7}$$

where $\mathcal{L}$ is either $\mathcal{L}_1$ or $\mathcal{L}_2$ and $\mathbf{x}_k^{(0)}$ is an initial condition. Let stress that $\mathcal{L}_2$ relates to unsupervised learning but requires the explicit definition and parameterization of prior covariance $\mathbf{P}$. In such situations, the proposed training framework aims at delivering a fast and scalable computation of (2). If using training loss $\mathcal{L}_1$, it only relies on the true states with no additional hypothesis on the underlying covariance, which makes it more appealing for most supervised learning-based applications, see the experimental conclusions of Sect. 4. To train jointly the solver component $\mathcal{G}$ and operator Φ, we vary initial conditions between some initialization $\mathbf{x}_k^{(0)}$ of the state and detached outputs of Algorithm 1 for a predefined number of total iteration steps. This strategy also provides a practical solution to the memory requirement, which rapidly increases with the number of iterations during the training phase due to the resulting depth of the computational graph. In all the reported experiments, we use Adam optimizer over 200 epochs.

3.3 Extension to Non-linear and Multimodal Optimal Interpolation

While the analytical derivation of solution (2) requires to consider a linear-quadratic formulation in both (1) and (3), Algorithm 1 applies to any differentiable parameterization of operator Φ. This provides the basis to investigate optimal interpolation models and solvers for non-linear and/or non-Gaussian processes through a non-linear parameterization for operator Φ. Here, we benefit from the variety of neural auto-encoder architectures introduced in the deep learning literature, such as simple convolutional auto-encoders, U-Nets [7], ResNets [11]... For such parameterization, the existence of a unique global minimum for minimization (3) may not be guaranteed and Algorithm 1 will converge to a local minimum depending on the considered initial condition.

Multimodal interpolation represents another appealing extension of the proposed framework. Let us assume that some additional gap-free observation data $\mathbf{z}$ is available such that $\mathbf{z}$ is expected to partially inform state $\mathbf{x}$. We then introduce the following multimodal variational cost:

$$\widehat{\mathbf{x}} = \arg\min_{\mathbf{x}} \lambda_1 \left\|\mathbf{y} - H_\Omega \cdot \mathbf{x}\right\|^2 + \lambda_2 \left\|g(\mathbf{z}) - h(\mathbf{x})\right\|^2 + \lambda_3 \left\|\mathbf{x} - \Phi(\mathbf{x})\right\|^2 \quad (8)$$

where $g(\cdot)$ and $h(\cdot)$ are trainable neural operators which respectively extract features from state $\mathbf{x}$ and observation $\mathbf{z}$. In this multimodal setting, Θ in (7) comprises the trainable parameters of operators Φ, $\mathcal{G}$, g and h. Given this reparameterization of the variational cost, we can exploit the exact same architecture for the neural solver defined by Algorithm 1 and the same learning setting.

4 Experiments

We report numerical experiments for the interpolation of 2D+t Gaussian process (GP) for which we can compute the analytical OI solution (2), as well as a real-world case-study for the reconstruction of sea surface dynamics from irregularly-sampled satellite-derived observations.

4.1 2D+t GP Case-Study

We use as synthetic dataset the stochastic partial differential equation (SPDE) approach introduced by [16] to generate a spatio-temporal Gaussian Process (GP). Let $\mathbf{x}$ denote the SPDE solution, we draw from the classic isotropic formulation to introduce some diffusion in the general fractional operator:

$$\left\{\frac{\partial}{\partial t} + \left\{\kappa^2(\mathbf{s},t) - \nabla \cdot \mathbf{H}(\mathbf{s},t)\nabla\right\}^{\alpha/2}\right\}\mathbf{x}(\mathbf{s},t) = \tau\mathbf{z}(\mathbf{s},t) \tag{9}$$

with parameters $\kappa = 0.33$ and regularization variance $\tau = 1$. To ensure the GP to be smooth enough, we use a value of $\alpha = 4$. Such a formulation enables to generate GPs driven by local anisotropies in space leading to non stationary spatio-temporal fields with eddy patterns. Let denote this experiment GP-DIFF2 where $\mathbf{H}$ is a 2-dimensional diffusion tensor. We introduce a generic decomposition of $\mathbf{H}(\mathbf{s},t)$, see e.g. [9], through the equation $\mathbf{H} = \gamma\mathbf{I}_2 + \beta\mathbf{v}(\mathbf{s})^{\mathsf{T}}\mathbf{v}(\mathbf{s})$ with $\gamma = 1$, $\beta = 25$ and $\mathbf{v}(\mathbf{s}) = (v_1(\mathbf{s}), v_2(\mathbf{s}))^{\mathsf{T}}$ using a periodic formulation of its two vector fields components: it decomposes the diffusion tensor as the sum of an isotropic and anisotropic effects, the latter being described by its amplitude and magnitude. This is a valid decomposition for any symmetric positive-definite 2×2 matrix.

We use the Finite Difference Method (FDM) in space coupled with an Implicit Euler scheme (IES) in time to solve for the equation. Let $\mathcal{D} = [0, 100] \times [0, 100]$ be the square spatial domain of simulation and $\mathcal{T} = [0, 500]$ the temporal domain. Both spatial and temporal domains are discretized so that the simulation is made on a uniform Cartesian grid consisting of points (x_i, y_j, t_k) where $x_i{=}i\Delta x$, $y_j{=}j\Delta j$, $t_k{=}k\Delta t$ with Δx, Δy and Δt all set to one.

To be consistent with the second dataset produced in Sect. 4.2, we sample pseudo-observations similar to along-track patterns produced by satellite datasets, with a periodic sampling leading to spatial observational rate similar to the along-track case-study. Observational noise is negligible and taken as $\mathbf{R} = \sigma^2\mathbf{I}$ with $\sigma^2 = 1\text{E}-3$ to compute the observational term of the variational cost.

For the dataset GP-DIFF2, we involve spatio-temporal sequences of length 5 as data assimilation window (DAW) to apply our framework and benchmark the following methods: analytical OI, as a solution of the linear system, the gradient descent OI solution, a direct CNN/UNet interpolation using a zero-filling initialization and different flavors of 4DVarnet using either a UNet trainable prior or a known precision-based prior coupled with LSTM-based solvers. As already stated in Sect. 3.2, we use two training losses: the mean squared error (MSE) w.r.t to the groundtruth and the OI variational cost of Eq. 1. Regarding the performance metrics, we assess the quality of a model based on both OI cost value for the known SPDE precision matrix and the MSE score w.r.t to the groundtruth. We also provide the computational GPU time of all the benchmarked models on the test period. For training-free models (analytical and gradient-based OI), there is no training time. The training period goes from timestep 100 to 400 and the optimization is made on 20 epochs with

Adam optimizer, with no significant improvements if trained longer. During the training procedure, we select the best model according to metrics computed over the validation period from timestep 30 to 80. Overall, the set of metrics is computed on a test period going from timestep 450 to 470. Let note that by construction, the analytical OI solution is optimal regarding the OI variational cost.

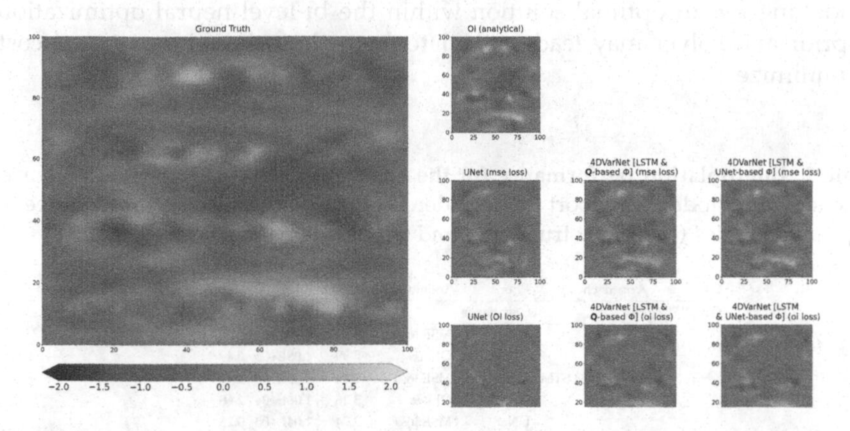

Fig. 1. SPDE-based GP spatio-temporal field (Ground Truth) and its reconstructions based on a 5 time lag assimilation window

Table 1 displays the performance evaluation for the experiment GP-DIFF2. Figure 1 shows the interpolation obtained at the middle of the test period for all the benchmarked models. Because we use spatio-temporal assimilation windows of length 5, we only display the reconstructions at the center of the DAW. We also provide in Fig. 2a) the scatterplot of the global MSE w.r.t OI variational cost throughout the iteration process, and in Fig. 2b) the OI variational cost vs the number of iterations of the algorithm. Both figures represent how the methods behaves once the training phase is finished: for learning-based gradient-descent approaches, their corresponding line plot then illustrates how the trained recurrent schemes is efficient to mimic and speed up the traditional gradient descent. The mapping clearly indicates that direct inversion by CNN schemes is not efficient for this reconstruction task. On the opposite, LSTM-based iterative solvers are all consistent with the optimal solution, with potential variations that can be explained by the training loss used. While using MSE w.r.t true states quickly converges in a very few number of iterations, 20 typically, involving the OI variational cost as a training loss implies to increase the number of gradient steps, up to about a hundred, to reach satisfactory performance. This makes sense because using global MSE relates to supervised learning while the variational cost-based training is not.

In addition, when a similar LSTM-based solver is used, the trainable prior considerably speeds up the convergence towards the optimal solution compared to the known precision matrix-based prior. This leads to three key conclusions:

- using MSE as training loss with trainable neural priors is enough for a reconstruction task and can even speed up the iterative convergence compared to the known statistical prior parametrization;
- the GP-based experiments demonstrates the relevance of an LSTM-based solver to speed up and accurate iterative solutions of minimization problems;
- looking for an optimal solution within the bi-level neural optimization of prior and solver may lead to deviate from the original variational cost to minimize.

Table 1. Interpolation performance for the synthetic 2D+T GP case-study: For each benchmarked model, we report the considered performance metrics for the three training loss strategies (MSE w.r.t true states and OI variational cost)

GP	Approach	Prior	Training loss	MSE_x	OI-score	Comp. time (mins)
GP-DIFF2	OI	Covariance		**2.72**	**9.8E+03**	2.41
	UNet	N/A	MSE loss	3.50	6.10E+05	0.08
			OI loss	5.99	1.08E+05	0.49
	4DVarNet-LSTM	Covariance	MSE loss	2.84	4.52E+04	2.4
			OI loss	3.26	1.06E+04	2.68
		UNet	MSE loss	2.74	1.04E+05	0.25
			OI loss	3.17	1.27E+04	0.48

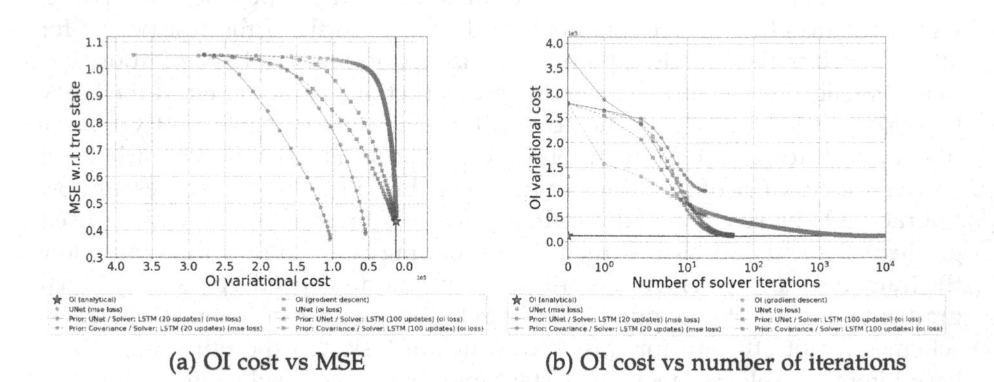

(a) OI cost vs MSE

(b) OI cost vs number of iterations

Fig. 2. Optimal Interpolation derived variational cost vs Mean Squared Error (MSE) loss (a) and OI variational cost vs number of iterations (b) for all the benchmarked methods at timestep 16 of the test period throughout their iterations process. For the analytical Optimal Interpolation and direct UNet neural formulations, there is no iteration, so a single point is displayed. Direct UNet trained with MSE loss does not appear in the Figures because it does not scale properly compared to the other methods

4.2 Satellite Altimetry Dataset

We also apply our neural OI scheme to a real-world dataset, namely the interpolaton of sea surface height (SSH) fields from irregularly-sampled satellite altimetry observations. The SSH relates to sea surface dynamics and satellite altimetry data which are characterized by an average missing data rate above 90%. We exploit the experimental setting defined in [14][2]. It relies on a groundtruthed dataset given by the simulation of realistic satellite altimetry observations from numerical ocean simulations. Overall, this dataset refers to 2D+t states for a $10° \times 10°$ domain with $1/20°$ resolution corresponding to a small area in the Western part of the Gulf Stream. Regarding the evaluation framework, we refer the reader to SSH mapping data challenge above mentioned for a detailed presentation of the datasets and evaluation metrics. The latter comprises the average RMSE-scores μ(RMSE) (the higher the better), the minimal spatial scales resolved λ_x (degree) (the lower the better) and the minimal temporal scales resolved λ_t (days) (the lower the better). We also look for the relative gains τ_{SSH} (%) and $\tau_{\nabla SSH}$ (%) w.r.t DUACS OI for SSH and its gradient. For learning-based approaches, the training dataset spans from mid-February 2013 to October 2013, while the validation period refers to January 2013. All methods are tested on the test period from October 22, 2012 to December 2, 2012.

For benchmarking purposes, we consider the operational baseline (DUACS) based on an optimal interpolation, multi-scale OI scheme MIOST, model-driven interpolation schemes BFN and DYMOST. We also include a state-of-the-art UNet architecture to train a direct inversion scheme. For all neural schemes, we consider 29-day space-time sequences to account for time scales considered in state-of-the-art OI schemes. Regarding the parameterization of our framework, we consider a bilinear residual architecture for prior Φ, a classic UNet flavor as well as a simple linear convolutional prior. Similarly to the GP case-study, we use a 2D convolutional LSTM cell with 150-dimensional hidden states. Besides the interpolation scheme using only altimetry data, we also implement a multimodal version of our interpolation framework. It uses sea surface temperature (SST) field as complementary gap-free observations. SST fields are widely acknowledge to convey information on sea surface dynamics though the derivation of an explicit relationship between SSH and SST fields remain a challenge, except for specific dynamical regimes [13]. Our multimodal extension exploits simple ConvnNets for the parameterization of operators $g(\cdot)$ and $h(\cdot)$ in Eq. 8.

Figure 3 displays the reconstructions of the SSH field and the corresponding gradients on 2012-11-05 for all the benchmarked models. It clearly stresses how our scheme improves the reconstruction when considering a non-linear prior. Especially, we greatly sharpen the gradient along the main meander of the Gulf Stream compared with other interpolation schemes. Oceanic eddies

[2] SSH Mapping Data Challenge 2020a: https://github.com/ocean-data-challenges/2020a_SSH_mapping_NATL60.

are also better retrieved. Table 2 further highlights the performance gain of the proposed scheme. The relative gain is greater than 50% compared to the operational satellite altimetry processing. We outperform by more than 20% in terms of relative gain to the baseline MIOST and UNet schemes, which are the second best interpolation schemes. Interestingly, our scheme is the only one to retrieve time scales below 10 days when considering only altimetry data.

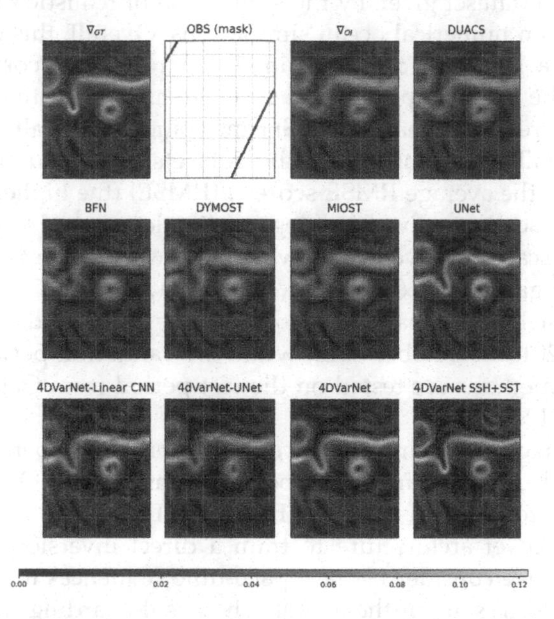

Fig. 3. Gradient SSH reconstructions (2012-11-05) for all benchmarked models based on 4 along-track nadirs pseudo-observations in the Gulf Stream domain. Trom top left to bottom right: Ground Truth, Observations, naive OI, DUACS, BFN, DYMOST, MIOST, UNet, 4DVarNet with a linear CNN-based prior, 4DVarNet with UNet prior, 4DVarNet with BiLin-Res prior and multimodal 4DVarNet with BiLin-Res prior embedding SSH-SST synergies in the variational cost

As stressed by last line and map of Table 2 and Fig. 3, the multimodal version of the proposed interpolation scheme further improves the interpolation performance. Our trainable OI solver learns how to extract fine-scale features from SST fields to best reconstruct the fine-scale structure of SSH fields and brings a significant improvement on all performance metrics.

Table 2. Interpolation performance for the satellite altimetry case-study: For each benchmarked models, we report the considered performance metrics averaged on the test period when learning-based methods are trained on the MSE loss (true states and its gradient). Metrics obtained from SOTA DA methods (top lines in the Table) can be found in the BOOST-SWOT 2020a SSH Mapping Data Challenge: https://github.com/ocean-data-challenges/2020a_SSH_mapping_NATL60

Approach	Prior	MSE	λ_x (degree)	λ_t (days)	τ_{SSH} (%)	$\tau_{\nabla SSH}$ (%)
DUACS	–	0.92	1.42	12.13	–	–
BFN	–	0.92	1.23	10.82	7.93	23.69
DYMOST	–	0.91	1.36	11.91	−10.88	0.38
MIOST	–	0.93	1.35	10.41	25.63	11.16
UNet	–	0.924	1.25	11.33	20.13	26.16
4DVarNet-LSTM	Linear CNN	0.89	1.46	12.63	−84.14	−10.24
	UNet	0.89	1.4	12.45	0.24	0.01
	BiLin-Res	**0.94**	**1.17**	**6.86**	**54.79**	**55.14**
Multimodal interpolation models (SSH+SST)						
UNet	–	0.55	2.36	35.72	−2741.29	−355.24
4DVarNet-LSTM	BiLin-Res	**0.96**	**0.66**	**2.97**	**79.80**	**75.71**

5 Conclusion

This paper addresses the end-to-end learning of neural schemes for optimal interpolation. We extend the neural scheme introduced in [8] for data assimilation to optimal interpolation with theoretical guarantees so that the considered trainable solvers asymptotically converge towards the analytical OI solution. The computation of the analytical OI solution is challenging when dealing with high-dimensional states. Whereas classic gradient-based iterative methods may suffer from a relatively low convergence rate, our experiments support the relevance of the proposed trainable solvers to speed up the convergence and reach good interpolation performance with only 10 to 100 gradient steps. Importantly, the convolutional architecture of the trainable solver also guarantees their scalability and a linear complexity with respect to the size of the spatial domain as well as the number of observations. Our GP experiment highlight the relevance of the bi-level formulation of the OI problem.We greatly speed up the interpolation time, when considering a UNet-based parameterization of the inner cost and the interpolation error as the outer performance metrics. The latter strategy greatly simplifies the application of the proposed framework to real datasets, where the underlying covariance model is not known and/or a Gaussian process approximation does not apply. As illustrated for our application to ocean remote sensing data, the proposed framework greatly outperforms all SOTA techniques, especially when benefitting from additional multimodal observations. Whereas in the GP case, we know the variational OI cost to be the optimal variational formulation to solve the interpolation, no such theoretical result exists in most non-Gaussian/non-linear cases. The proposed end-to-end learning framework provides new means to explore the reduction of estimation biases in Bayesian setting. Especially, our experiments on ocean remote sensing data suggest that the prior term in the inner variational formulation shall be adapted to the observation configuration rather than considering generic plug-and-play priors. This works also supports new avenues thanks to the connection made between neural Optimal Interpolation and trainable solvers. Indeed, while the GP experiment used in the paper is entirely controlled, in the sense that the parameters of the stochastic PDE driving the GP are known, future works may consider to also train the SPDE parameters so that the prior operator Φ would be linear, though stochastic. It would open the gate to uncertainty quantification and fast huge ensemble-based formulations. In addition, it would pave the way to a full stochastic neural formulation of the framework, when making the explicit link between diffusion-based generative models and SPDE that are in fact linear diffusion models.

References

1. Andrychowicz, M., et al.: Learning to learn by gradient descent by gradient descent. In: Advances in Neural Information Processing Systems, pp. 3981–3989 (2016)
2. Asch, M., Bocquet, M., Nodet, M.: Data assimilation. In: Fundamentals of Algorithms, Society for Industrial and Applied Mathematics (2016). https://doi.org/10.1137/1.9781611974546
3. Barth, A., Alvera-Azcárate, A., Troupin, C., Beckers, J.M.: DINCAE 2.0: multivariate convolutional neural network with error estimates to reconstruct sea surface temperature satellite and altimetry observations. Geosci. Model Dev. **15**(5), 2183–2196 (2022). https://doi.org/10.5194/gmd-15-2183-2022
4. Bertalmio, M., Bertozzi, A.L., Sapiro, G.: Navier-Stokes, fluid dynamics, and image and video inpainting. In: IEEE CVPR, pp. 355–362 (2001)
5. Borkar, V.S.: Stochastic Approximation. TRM, vol. 48. Hindustan Book Agency, Gurgaon (2008). https://doi.org/10.1007/978-93-86279-38-5
6. Chilès, J., Delfiner, P.: Geostatistics: Modeling Spatial Uncertainty, 2nd edn. Wiley, New York (2012)
7. Cicek, O., Abdulkadir, A., Lienkamp, S., Brox, T., Ronneberger, O.: 3D U-Net: learning dense volumetric segmentation from sparse annotation. In: Proc. MICCAI, pp. 424–432 (2016)
8. Fablet, R., Beauchamp, M., Drumetz, L., Rousseau, F.: Joint interpolation and representation learning for irregularly sampled satellite-derived geophysical fields. Front. Appl. Math. Stat. **7**, 25 (2021). https://doi.org/10.3389/fams.2021.655224
9. Fuglstad, G.A., Lindgren, F., Simpson, D., Rue, H.: Exploring a new class of non-stationary spatial gaussian random fields with varying local anisotropy. Stat. Sin. **25**(1), 115–133 (2015). https://doi.org/10.5705/ss.2013.106w
10. Galerne, B., Gousseau, Y., Morel, J.: Random phase textures: theory and synthesis. IEEE Trans. Image Process. **20**(1), 257–267 (2011). https://doi.org/10.1109/TIP.2010.2052822
11. He, K., Zhang, X., Ren, S., Sun, J.: Deep residual learning for image recognition. In: 2016 IEEE Conference on Computer Vision and Pattern Recognition (CVPR), pp. 770–778 (2016). https://doi.org/10.1109/CVPR.2016.90
12. Hospedales, T., Antoniou, A., Micaelli, P., Storkey, A.: Meta-learning in neural networks: a survey. arXiv:2004.05439 (2020)
13. Isern-Fontanet, J., Chapron, B., Lapeyre, G., Klein, P.: Potential use of microwave sea surface temperatures for the estimation of ocean currents. Geophys. Res. Lett. **33**, l24608 (2006). https://doi.org/10.1029/2006GL027801

14. Le Guillou, F., et al.: Mapping altimetry in the forthcoming SWOT era by back-and-forth nudging a one-layer quasi-geostrophic model. Earth and Space Science Open Archive, p. 15 (2020). https://doi.org/10.1002/essoar.10504575.1
15. Lguensat, R., et al.: Data-driven interpolation of sea level anomalies using analog data assimilation (2017). https://hal.archives-ouvertes.fr/hal-01609851
16. Lindgren, F., Rue, H., Lindström, J.: An explicit link between gaussian fields and Gaussian Markov random fields: the stochastic partial differential equation approach. J. R. Stat. Soc. B Stat. Methodol. **73**(4), 423–498 (2011). https://doi.org/10.1111/j.1467-9868.2011.00777.x
17. McCann, M., Jin, K., Unser, M.: Convolutional neural networks for inverse problems in imaging: a review. IEEE SPM **34**(6), 85–95 (2017). https://doi.org/10.1109/MSP.2017.2739299
18. Pleiss, G., Jankowiak, M., Eriksson, D., Damle, A., Gardner, J.R.: Fast matrix square roots with applications to Gaussian processes and Bayesian optimization (2020). https://doi.org/10.48550/ARXIV.2006.11267
19. Romary, T., Desassis, N.: Combining covariance tapering and lasso driven low rank decomposition for the kriging of large spatial datasets (2018). https://doi.org/10.48550/ARXIV.1806.01558
20. Shi, X., Chen, Z., Wang, H., Yeung, D.Y., Wong, W.K., Woo, W.C.: Convolutional LSTM network: a machine learning approach for precipitation nowcasting. In: Cortes, C., Lawrence, N., Lee, D., Sugiyama, M., Garnett, R. (eds.) Advances in Neural Information Processing Systems, vol. 28. Curran Associates, Inc. (2015). https://proceedings.neurips.cc/paper/2015/file/07563a3fe3bbe7e3ba84431ad9d055af-Paper.pdf
21. Wei, K., Aviles-Rivero, A., Liang, J., Fu, Y., Schönlieb, C.R., Huang, H.: Tuning-free plug-and-play proximal algorithm for inverse imaging problems. In: ICML, pp. 10158–10169 (2020). https://proceedings.mlr.press/v119/wei20b.html. ISSN: 2640-3498
22. Xu, R., Li, X., Zhou, B., Loy, C.C.: Deep flow-guided video inpainting. In: The IEEE Conference on Computer Vision and Pattern Recognition (CVPR) (2019)

Physics-Informed Long Short-Term Memory for Forecasting and Reconstruction of Chaos

Elise Özalp[1(✉)], Georgios Margazoglou[1], and Luca Magri[1,2]

[1] Department of Aeronautics, Imperial College London, London SW7 2AZ, UK
elise.ozalp@imperial.ac.uk
[2] The Alan Turing Institute, London NW1 2DB, UK

Abstract. We present the physics-informed long short-term memory (PI-LSTM) network to reconstruct and predict the evolution of unmeasured variables in a chaotic system. The training is constrained by a regularization term, which penalizes solutions that violate the system's governing equations. The network is showcased on the Lorenz-96 model, a prototypical chaotic dynamical system, for a varying number of variables to reconstruct. First, we show the PI-LSTM architecture and explain how to constrain the differential equations, which is a non-trivial task in LSTMs. Second, the PI-LSTM is numerically evaluated in the long-term autonomous evolution to study its ergodic properties. We show that it correctly predicts the statistics of the unmeasured variables, which cannot be achieved without the physical constraint. Third, we compute the Lyapunov exponents of the network to infer the key stability properties of the chaotic system. For reconstruction purposes, adding the physics-informed loss qualitatively enhances the dynamical behaviour of the network, compared to a data-driven only training. This is quantified by the agreement of the Lyapunov exponents. This work opens up new opportunities for state reconstruction and learning of the dynamics of nonlinear systems.

Keywords: Long short-term memory · Chaos · State reconstruction

1 Introduction

Chaotic dynamics arise in a variety of disciplines such as meteorology, chemistry, economics, and engineering. Their manifestation emerges because of the exponential sensitivity to initial conditions, which makes long-term time-accurate prediction difficult. In many engineering cases, only partial information about the system's state is available, e.g., because of the computational cost or a limited number of sensors in a laboratory experiment and hence, no data is available on the unmeasured variables. Making predictions of the available observations and reconstructing the unmeasured variables is key to understanding and predicting the behaviour of dynamical systems.

This research is supported by the ERC Starting Grant No. PhyCo 949388.

J. Mikyška et al. (Eds.): ICCS 2023, LNCS 14076, pp. 382–389, 2023.
https://doi.org/10.1007/978-3-031-36027-5_29

Neural networks are powerful expressive tools to extract patterns from data and, once trained, they are fast and efficient at making predictions. Suitable for time series and dynamical evolutions are recurrent neural networks (RNNs) and long short-term memory networks (LSTMs), which have shown promising performance in the inference of dynamical systems with multi-scale, chaotic, or turbulent behaviour [16]. The integration of neural networks with knowledge of governing physical equations has given rise to the field of physics-informed neural networks [7,12]. The use of LSTM approaches that embed domain-specific knowledge was applied to nonlinear structures [15,17] and degradation modelling [14], but their generalization to chaotic systems remains unexplored. Other architectures, such as physics-informed RNNs with reservoir computers have been applied successfully in the short-term reconstruction of unmeasured variables [2,11] of chaotic systems.

In this paper, we propose a physics-informed LSTM (PI-LSTM) to make predictions of observed variables and, simultaneously, infer the unmeasured variables of a chaotic system. The quality of the prediction is evaluated by analysing the autonomous long-term evolution of the LSTM and collecting the statistics of the reconstructed variables. Crucial quantities for characterizing chaos are the Lyapunov exponents (LEs), which provide insight into an attractor's dimension and tangent space. In this paper, we extract the LEs from an LSTM, trained on a prototypical chaotic dynamical system. For state reconstruction, we show that it is necessary to embed prior knowledge in the LSTM, which is in the form of differential equations in this study.

The paper is structured as follows. Section 2 provides a brief introduction to the LEs of chaotic systems and the problem setup of unmeasured variables. The PI-LSTM is proposed in Sect. 3. In Sect. 4, we discuss the results for the Lorenz-96 system. Finally, we summarize the work and propose future direction in Sect. 5.

2 Chaotic Dynamical Systems

We consider a nonlinear autonomous dynamical system

$$\frac{d}{dt}\boldsymbol{y}(t) = f(\boldsymbol{y}(t)), \tag{1}$$

where $\boldsymbol{y}(t) \in \mathbb{R}^N$ is the state vector of the physical system and $f : \mathbb{R}^N \to \mathbb{R}^N$ is a smooth nonlinear function. The dynamical system (1) is chaotic if infinitesimally nearby trajectories diverge at an exponential rate. This behaviour is quantified by the LEs, which measure the average rate of stretching of the trajectories in the phase space. The LEs, $\lambda_1 \geq \cdots \geq \lambda_N$, provide fundamental insight into the chaoticity and geometry of an attractor. Chaotic dynamical systems have at least one positive Lyapunov exponent. In chaotic systems, the Lyapunov time $\tau_\lambda = \frac{1}{\lambda_1}$ defines a characteristic timescale for two nearby trajectories to separate, which gives an estimate of the system's predictability horizon. LEs can be computed numerically by linearizing the equations of the dynamical system around a reference point, defining the tangent space, and by extracting the LEs based on the Gram-Schmidt orthogonalization procedure from the corresponding Jacobian [3].

2.1 State Reconstruction

Let $\boldsymbol{y}(t) = [\boldsymbol{x}(t); \boldsymbol{\xi}(t)]$ be the state of a chaotic dynamical system, where $\boldsymbol{x}(t) \in \mathbb{R}^{N_x}$ are the observed variables and $\boldsymbol{\xi}(t) \in \mathbb{R}^{N_\xi}$ are the unmeasured variables with $N = N_x + N_\xi$. Specifically, let us assume that $\boldsymbol{x}(t_i)$ is measured at times $t_i = i\Delta t$ with $i = 0, \ldots N_t$ and constant time step Δt. Based on these observations, we wish to predict the full state $\boldsymbol{y}(t_i) = [\boldsymbol{x}(t_i), \boldsymbol{\xi}(t_i)]$, whilst respecting the governing Eq. (1), using the PI-LSTM.

3 Physics-Informed Long Short-Term Memory

LSTMs have been successfully applied to time forecasting of dynamical systems when full observations are available [13,16]. They are characterized by a cell state $\boldsymbol{c}_{i+1} \in \mathbb{R}^{N_h}$ and a hidden state $\boldsymbol{h}_{i+1} \in \mathbb{R}^{N_h}$ that are updated at each step. In the case of partial observations, the states are updated using the observed variables $\boldsymbol{x}(t_i)$ as follows

$$
\begin{aligned}
\boldsymbol{i}_{i+1} &= \sigma\left(\boldsymbol{W}^i[\boldsymbol{x}(t_i); \boldsymbol{h}_i] + \boldsymbol{b}^i\right), & \tilde{\boldsymbol{c}}_{i+1} &= \tanh\left(\boldsymbol{W}^g[\boldsymbol{x}(t_i); \boldsymbol{h}_i] + \boldsymbol{b}^g\right),\\
\boldsymbol{f}_{i+1} &= \sigma\left(\boldsymbol{W}^f[\boldsymbol{x}(t_i); \boldsymbol{h}_i] + \boldsymbol{b}^f\right), & \boldsymbol{c}_{i+1} &= \sigma\left(\boldsymbol{f}_{i+1} * \boldsymbol{c}_i + \boldsymbol{i}_{i+1} * \tilde{\boldsymbol{c}}_{i+1}\right),\\
\boldsymbol{o}_{i+1} &= \sigma\left(\boldsymbol{W}^o[\boldsymbol{x}(t_i); \boldsymbol{h}_i] + \boldsymbol{b}^o\right), & \boldsymbol{h}_{i+1} &= \tanh\left(\boldsymbol{c}_{i+1}\right) * \boldsymbol{o}_{i+1},
\end{aligned}
$$

where $\boldsymbol{i}_{i+1}, \boldsymbol{f}_{i+1}, \boldsymbol{o}_{i+1} \in \mathbb{R}^{N_h}$ are the input, forget and output gates. The matrices $\boldsymbol{W}^i, \boldsymbol{W}^f, \boldsymbol{W}^o, \boldsymbol{W}^g \in \mathbb{R}^{N_h \times (N_x + N_h)}$ are the corresponding weight matrices, and $\boldsymbol{b}^i, \boldsymbol{b}^f, \boldsymbol{b}^o, \boldsymbol{b}^g \in \mathbb{R}^{N_h}$ are the biases. The full prediction $\tilde{\boldsymbol{y}}(t_{i+1}) = [\tilde{\boldsymbol{x}}(t_{i+1}), \tilde{\boldsymbol{\xi}}(t_{i+1})]$ is obtained by concatenating the hidden state $\boldsymbol{h}_{i+1}$ with a dense layer

$$
\begin{bmatrix} \tilde{\boldsymbol{x}}(t_{i+1}) \\ \tilde{\boldsymbol{\xi}}(t_{i+1}) \end{bmatrix} = \boldsymbol{W}^{dense} \boldsymbol{h}_{i+1} + \boldsymbol{b}^{dense},
$$

where $\boldsymbol{W}^{dense} \in \mathbb{R}^{(N_x + N_\xi) \times N_h}$ and $\boldsymbol{b}^{dense} \in \mathbb{R}^{N_x + N_\xi}$.

LSTMs are universal approximators for an arbitrary continuous objective function [4,5]; however, practically, the network's performance is dependent on the parameters, such as weights and biases, which are computed during the training phase. To train the weights and biases, a data-driven loss is defined on the observed data via the mean-squared error $\mathcal{L}_{dd} = \frac{1}{N_t}\sum_{i=1}^{N_t}\left(\boldsymbol{x}(t_i) - \tilde{\boldsymbol{x}}(t_i)\right)^2$. To constrain the network for the unmeasured dynamics, we add a penalization term

$$
\mathcal{L}_{pi} = \frac{1}{N_t}\sum_{i=1}^{N_t}\left(\frac{d}{dt}\tilde{\boldsymbol{y}}(t_i) - f(\tilde{\boldsymbol{y}}(t_i))\right)^2.
$$

This loss regularizes the network's training to provide predictions that fulfil the governing equation from (1) (up to a numerical tolerance).

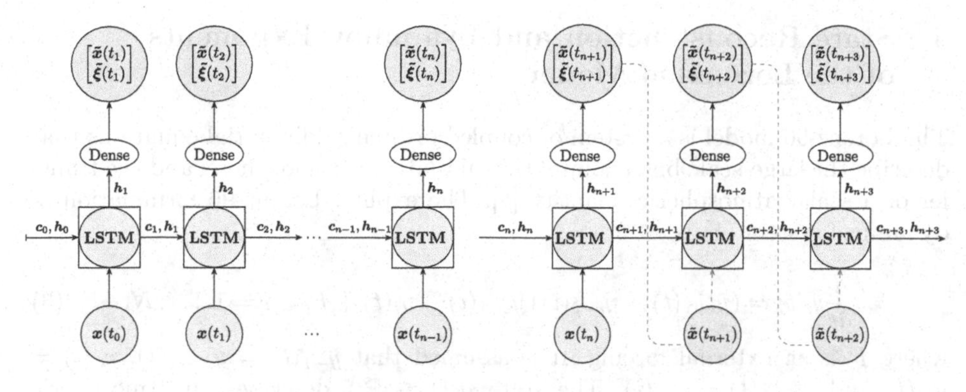

Fig. 1. Open-loop configuration for training and validation.

Fig. 2. Closed-loop configuration for autonomous evolution and testing.

For simplicity, the time derivative $\frac{d}{dt}\tilde{\boldsymbol{y}}$ is computed using a forward difference scheme

$$\frac{d}{dt}\tilde{\boldsymbol{y}}(t_i) \approx \frac{\tilde{\boldsymbol{y}}(t_{i+1}) - \tilde{\boldsymbol{y}}(t_i)}{\Delta t}.$$

(Alternatively, the time derivative can be computed with a higher-order scheme.) Combining the data-driven loss and weighing the physics-informed loss leads to the total loss

$$\mathcal{L} = \mathcal{L}_{dd} + \alpha_{pi}\mathcal{L}_{pi}, \quad \alpha_{pi} \in \mathbb{R}^+, \tag{2}$$

where α_{pi} is a penalty hyperparameter. If $\alpha_{pi} = 0$, the network is not constrained by the governing equations, which is referred to as the 'data-driven LSTM'. The selection of an appropriate weighting for the penalty term is a nontrivial task, and a grid search is employed in this paper. The weights and biases are optimized by minimizing the loss $\mathcal{L}$ with the Adam optimizer [6]. Early stopping is employed to avoid overfitting. During the training and validation, the network is in an open-loop configuration, as shown in Fig. 1. After training, the network is evaluated on test data with fixed weights and biases, operating in a closed-loop configuration, as shown in Fig. 2. In this mode, the network predicts the observed variables and the unmeasured variables, while the observed variables are used as input for the next time step, allowing for an autonomous evolution of the LSTM. This effectively defines a dynamical system and allows for stability analysis to be performed on the LSTM.

Previous work showed that when echo state networks, gated recurrent units or convolutional neural networks are employed to learn and predict full states from chaotic systems, the LEs of the network align with those of the dynamical system [1,9,10,16] (to a numerical tolerance), allowing to gain valuable insight into the behaviour of the network. Thus, we analyse the LEs of the proposed PI-LSTM.

4 State Reconstruction and Lyapunov Exponents of the Lorenz-96 Model

The Lorenz-96 model is a system of coupled ordinary differential equations that describe the large-scale behaviour of the mid-latitude atmosphere, and the transfer of a scalar atmospheric quantity [8]. The explicit Lorenz-96 formulation of dynamical system (1) is

$$\frac{d}{dt}y_i(t) = (y_{i+1}(t) - y_{i-2}(t))\, y_{i-1}(t) - y_i(t) + F, \quad i = 1, \ldots, N, \tag{3}$$

where F is an external forcing. It is assumed that $y_{-1}(t) = y_{N-1}(t)$, $y_0(t) = y_N(t)$ and $y_{N+1}(t) = y_1(t)$. The state $\boldsymbol{y}(t) \in \mathbb{R}^N$ describes an atmospheric quantity in N sectors of one latitude circle. In this study, we set $F = 8$ and $N = 10$, for which the system exhibits chaos with three positive LEs. Both the numerical solution and the reference LEs are computed using a first-order Euler scheme with a time step of $\Delta t = 0.01$. The largest Lyapunov exponent is $\lambda_1 \approx 1.59$. The training set consists of $N_t = 20000$ points, which is equivalent to $125\tau_\lambda$. Hyperparameter tuning was utilized to optimise key network parameters. In particular, the dimension of the hidden and cell state N_h was selected from $\{20, 50, 100\}$ and the weighing of the physics-informed loss α_{pi} varied from 10^{-9} to 1 in powers of 10.

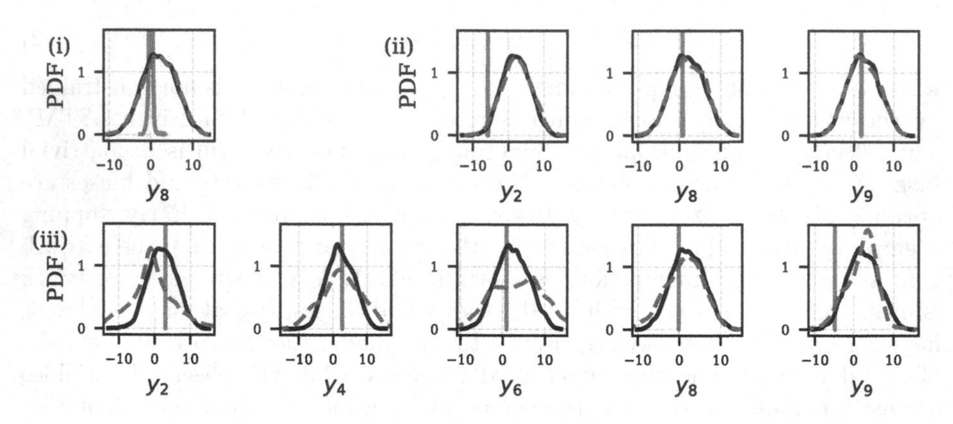

Fig. 3. Statistics reconstruction of unmeasured variables. Comparison of the target (black line), PI-LSTM (red dashed line) and data-driven LSTM (blue line) probability density functions (PDF) of (i) $N_\xi = 1$, (ii) $N_\xi = 3$, (iii) $N_\xi = 5$ unmeasured variables over a $1000\tau_\lambda$ trajectory in closed-loop configuration. (Color figure online)

We deploy the PI-LSTM to reconstruct the unmeasured variables in three test cases: reconstruction of (i) $N_\xi = 1$, (ii) $N_\xi = 3$, (iii) $N_\xi = 5$ unmeasured variables, which are removed from the full state $y(t)$ prior to training. We display networks with parameters (i) $N_h = 100, \alpha_{pi} = 0.01$, (ii) $N_h = 100, \alpha_{pi} = 0.01$, and (iii) $N_h = 50, \alpha_{pi} = 0.001$. We choose test case (i) and (ii) to highlight the capabilities of the PI-LSTM and select (iii) to demonstrate how the network behaves with further limited information. (We remark that, when the full state is available, i.e. $N_\xi = 0$, both data-driven LSTM and PI-LSTM perform equally well in learning the long-term statistics and the LEs.) The reconstruction is based on an autonomous $1000\tau_\lambda$ long trajectory in closed-loop mode.

In Fig. 3 we show the statistics of the reconstructed variables corresponding to $\xi(t)$, which are computed based on the autonomous evolution of the network. To illustrate, in case(i) with $N_\xi = 1$, the reconstructed variable $\xi(t)$ is $y_8(t)$, and accordingly for cases (ii) and (iii). The data-driven LSTM fails to reproduce the solution, in particular, the corresponding delta-like distribution (in blue) indicates a fixed-point. In test cases (i) and (ii), the PI-LSTM accurately reproduces the long-term behaviour of the dynamical system. At each time step, it successfully extrapolates from the partial input to the full state. Case (iii) shows that, by increasing the number of unmeasured observations, the complexity of the reconstruction is increased, and the accuracy of the reconstruction is decreased. However, the PI-LSTM provides a markedly more accurate statistical reconstruction of the target compared to the data-driven LSTM. The PI-LSTM predicts the observed variables well (not shown here), which indicates that incorporating knowledge of the underlying physics enables accurate long-term prediction of the full state.

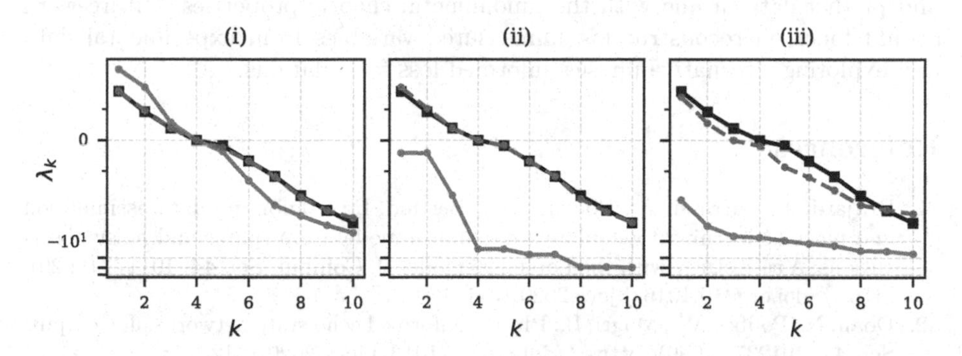

Fig. 4. Comparison of the target (black squares), PI-LSTM (red dots) and data-driven LSTM (blue dots) LEs for (i) $N_\xi = 1$, (ii) $N_\xi = 3$, (iii) $N_\xi = 5$ unmeasured variables. All the vertical axes are in logarithmic scale. (Color figure online)

In Fig. 4, we compare the reference LEs (in black squares) with LEs extracted from the data-driven LSTM (in blue circles) and PI-LSTM (in red circles) in the three test cases. By reconstructing the unmeasured variables the networks effectively reconstruct the tangent space, the properties of which are encapsulated in the LEs. In all cases, the LEs of the data-driven LSTM deviate significantly from the reference LEs, differing more from the target when fewer observations are available. For test cases (i) and (ii), the PI-LSTM reproduces the target LEs with high accuracy, with an error of 0.28% and 8% in λ_1, respectively. When reducing the number of observations further, as in (iii), the accuracy of the PI-LSTM LEs is limited by the lack of information, with an error of 11.6% in λ_1. Figure 4 also shows that in cases (ii) and (iii) of the data-driven LSTM, the leading LE is negative ($\lambda_1 < 0$), resulting in a completely incorrect solution (fixed point solution). This means that in (ii) and (iii) the data-driven LSTM displays no chaotic dynamics, whereas the PI-LSTM reproduces the chaotic behaviour.

5 Conclusions and Future Directions

We propose the physics-informed long short-term memory (PI-LSTM) network to embed the knowledge of the governing equations into an LSTM, by using the network's prediction to compute a first-order derivative. In contrast to physics-informed neural networks, which have no internal recurrent connections, the PI-LSTMs capture temporal dynamics whilst penalizing predictions that violate the system's governing equations. We deploy the PI-LSTM to reconstruct the unmeasured variables of the Lorenz-96 system, which is a chaotic system with three positive Lyapunov exponents. The long-term prediction of the PI-LSTM in closed-loop accurately reconstructs the statistics of multiple unmeasured variables. By computing the Lyapunov exponents of the PI-LSTM, we show the key role of the physics-informed loss in learning the dynamics. This exemplifies how leveraging knowledge of the physical system can be advantageous to reconstruct and predict data in line with the fundamental chaotic properties. Future work should focus on reconstructing unmeasured variables from experimental data and exploring alternative physics-informed loss formulations.

References

1. Brajard, J., Carrassi, A., Bocquet, M., Bertino, L.: Combining data assimilation and machine learning to emulate a dynamical model from sparse and noisy observations: A case study with the Lorenz 96 model. J. Comput. Sci. **44**, 101171 (2020). https://doi.org/10.1016/j.jocs.2020.101171
2. Doan, N., Polifke, W., Magri, L.: Physics-informed echo state networks. J. Comput. Sci. **47**, 101237 (2020). https://doi.org/10.1016/j.jocs.2020.101237
3. Eckmann, J.P., Ruelle, D.: Ergodic theory of chaos and strange attractors. Rev. Mod. Phys. **57**, 617–656 (1985). https://doi.org/10.1103/RevModPhys.57.617
4. Funahashi, K.I.: On the approximate realization of continuous mappings by neural networks. Neural Netw. **2**(3), 183–192 (1989). https://doi.org/10.1016/0893-6080(89)90003-8
5. Hornik, K., Stinchcombe, M., White, H.: Multilayer feedforward networks are universal approximators. Neural Netw. **2**(5), 359–366 (1989). https://doi.org/10.1016/0893-6080(89)90020-8
6. Kingma, D.P., Ba, J.: Adam: a method for stochastic optimization. In: 3rd International Conference on Learning Representations, ICLR 2015, San Diego 7–9 May 2015, Conference Track Proceedings (2015). https://doi.org/10.48550/ARXIV.1412.6980
7. Lagaris, I., Likas, A., Fotiadis, D.: Artificial neural networks for solving ordinary and partial differential equations. IEEE Trans. Neural Netw. **9**(5), 987–1000 (1998). https://doi.org/10.1109/72.712178

8. Lorenz, E.N.: Predictability: a problem partly solved. In: Seminar on Predictability, 4–8 September 1995, vol. 1, pp. 1–18. ECMWF, Shinfield Park, Reading (1996). https://doi.org/10.1017/CBO9780511617652.004
9. Margazoglou, G., Magri, L.: Stability analysis of chaotic systems from data. Nonlinear Dyn. **111**, 8799–8819 (2023). https://doi.org/10.1007/s11071-023-08285-1
10. Pathak, J., Lu, Z., Hunt, B.R., Girvan, M., Ott, E.: Using machine learning to replicate chaotic attractors and calculate Lyapunov exponents from data. Chaos: Interdisc. J. Nonlinear Sci. **27**(12), 121102 (2017). https://doi.org/10.1063/1.5010300
11. Racca, A., Magri, L.: Automatic-differentiated physics-informed echo state network (API-ESN). In: Paszynski, M., Kranzlmüller, D., Krzhizhanovskaya, V.V., Dongarra, J.J., Sloot, P.M.A. (eds.) ICCS 2021. LNCS, vol. 12746, pp. 323–329. Springer, Cham (2021). https://doi.org/10.1007/978-3-030-77977-1_25
12. Raissi, M., Perdikaris, P., Karniadakis, G.E.: Physics-informed neural networks: a deep learning framework for solving forward and inverse problems involving nonlinear partial differential equations. J. Comput. Phys. **378**, 686–707 (2019). https://doi.org/10.1016/j.jcp.2018.10.045
13. Sangiorgio, M., Dercole, F., Guariso, G.: Forecasting of noisy chaotic systems with deep neural networks. Chaos, Solitons Fractals **153**, 111570 (2021). https://doi.org/10.1016/j.chaos.2021.111570
14. Shi, J., Rivera, A., Wu, D.: Battery health management using physics-informed machine learning: online degradation modeling and remaining useful life prediction. Mech. Syst. Signal Process. **179**, 109347 (2022)
15. Tsai, L.W., Alipour, A.: Physics-informed long short-term memory networks for response prediction of a wind-excited flexible structure. Eng. Struct. **275**, 114968 (2023). https://doi.org/10.1016/j.engstruct.2022.114968
16. Vlachas, P., et al.: Backpropagation algorithms and reservoir computing in recurrent neural networks for the forecasting of complex spatiotemporal dynamics. Neural Netw. **126**, 191–217 (2020). https://doi.org/10.1016/j.neunet.2020.02.016
17. Zhang, R., Liu, Y., Sun, H.: Physics-informed multi-LSTM networks for metamodeling of nonlinear structures. Comput. Methods Appl. Mech. Eng. **369**, 113226 (2020). https://doi.org/10.1016/j.cma.2020.113226

Rules' Quality Generated by the Classification Method for Independent Data Sources Using Pawlak Conflict Analysis Model

Małgorzata Przybyła-Kasperek(✉) and Katarzyna Kusztal

Institute of Computer Science, University of Silesia in Katowice, Będzińska 39, 41-200 Sosnowiec, Poland
{malgorzata.przybyla-kasperek,kkusztal}@us.edu.pl

Abstract. The study concerns classification based on dispersed data, more specifically data collected independently in many local decision tables. The paper proposes an approach in which coalitions of local tables are generated using Pawlak's conflict analysis model. Decision trees are built based on tables aggregated within the coalitions The paper examines the impact of the stop criterion (determined by the number of objects in a node) on the quality classification and on the rules' quality generated by the model. The results are compared with the baseline approach, in which decision trees are built independently based on each of the decision tables. The paper shows that using the proposed model, the generated decision rules have much greater confidence than the rules generated by the baseline method. In addition, the proposed model gives better classification quality regardless of the stop criterion compared to the non-coalitions approach. Moreover, the use of higher values of the stop criterion for the proposed model significantly reduces the length of the rules while maintaining the classification accuracy and the rules' confidence at a high level.

Keywords: Pawlak conflict analysis model · independent data sources · coalitions · decision trees · stop criterion · dispersed data

1 Introduction

Nowadays, huge volumes of data are often dealt with, most dispersed and available from various independent sources. Processing such fragmented data is a big challenge both in terms of achieving satisfactory classification quality and understanding the basis for generated predictions. In particular, clear presentation of the extracted knowledge is difficult when we deal with dispersed data. It is assumed that fragmented data is available as a set of local decision tables collected by independent units. For real dispersed data the simple aggregation is not possible – it leads to inconsistencies and conflicts. In this study, we deal with an approach that generates coalitions and makes it possible to aggregate local tables for coalitions. Another important issue addressed in this paper is the

J. Mikyška et al. (Eds.): ICCS 2023, LNCS 14076, pp. 390–405, 2023.
https://doi.org/10.1007/978-3-031-36027-5_30

analysis of the quality of the patterns generated based on dispersed data. Studies are presented on how valuable the knowledge generated by the proposed model is. Among the many different machine learning models, those that are characterized by high interpretability and transparency of the generated knowledge can be distinguished. Approaches such as rule–based models [18,29] or decision trees [7] can be mentioned here. There are also models that give a very good quality of classification, but without the simple interpretation and understanding the basis of the decisions made. Neural networks and deep learning can be mentioned here [6,8]. In many real–world applications, the ability to justify the prediction is very important. Sometimes the knowledge generated by the model itself is more of a value than the prediction of new unknown objects. In the case when data is available in the form of multiple decision tables, the situation is even more difficult. For example, let's consider a medical application. Each clinic has a separate decision table. Models can be generated separately based on the local tables, but the knowledge generated in this way will be fragmented and will not allow to make simple interpretations. Aggregating all local tables into one table is not possible due to inconsistencies that occur. The proposed model can be applied in such situation.

This paper proposes an approach that gives a more condensed representation of knowledge, as it allows for partial aggregation of local tables. The proposed approach is designed for set of local tables in which the conditional attributes are identical. Coalitions of decision tables that store compatible values are determined. Identification of such coalitions is performed using Pawlak's conflict model. Aggregation of tables from one coalition is performed for a coalition. This allows, on the one hand, to combine data and reduce fragmentation and on the other hand to provide for the model a larger training set based on witch certain patterns/knowledge are generated. A decision tree is built based on partially aggregated data. Decision tree model was chosen because of the transparency and easy interpretability of the generated result. By using the proposed approach and doing data aggregation still, a few decision trees for dispersed data is received. However, as research shows, the number of obtained trees is smaller than if we generate trees separately based on individual local tables. Coalitions are not disjoint, which means that one local table can be included in several coalitions at the same time. Generated coalitions of local tables represent consistent data on common concepts. Decision trees describing a single concept are generated based on coalition – complete information about coherent knowledge, which also contributes to trees' quality. The paper [22] proposed the first study on this subject, in which for the basic approach (without the model's parameter optimizing) the classification quality was analyzed. The paper focused not only on the classification accuracy but also on the knowledge that the proposed model generates. Based on decision trees, decision rules are determined. Various parameters evaluating the quality of generated rules are analyzed, i.e. support, confidence and rule length. The main contribution and the novelty of this paper is a comprehensive analysis of the quality of rules generated by the model as well as the analysis of the impact of the stop criterion used during the construction of the tree on the accuracy of classification and the quality of decision rules generated based on decision trees. The size of the constructed trees has a very significant impact on the time and memory complexity of the method. The presented study shows that using the proposed model, the generated decision rules have much greater confidence than rules generated by using decision trees for each decision table separately. In addition, the proposed model gives better classification accuracy regardless of the stop criterion compared to the non-coalitions approach. The use of higher values of the stop criterion for the proposed model significantly reduces the length of the rules while maintaining the classification accuracy and the rules' confidence at a high level. The study presented in the paper proves that the proposed classification model not only provides high classification accuracy but also generates more interesting knowledge – rules with large confidence – than is the case when coalitions are not used.

The paper is organized as follows. Section 2 contains a literature review. In Sect. 3, the proposed classification dispersed system and Pawlak's conflict model are described. Section 4 addresses the data sets that were used, presents the conducted experiments and discussion on obtained results. Section 5 is on conclusions and future research plans.

2 Related Works

Distributed data is used in the concept of ensemble of classifiers [3,9] and federated learning [19]. In the classifier ensembles approach, the interpretation and the justification of the results obtained is not obvious and clear. There is no aggregation of data, the models are built locally in a hierarchical [34] or parallel manner [10], and the final decision is generated by the chosen fusion method [13]. There are many different fusion methods, some are based on fuzzy set theory [26] or mathematical measures [11]. Others fusion methods are based on issues related to evidence theory [27] and voting power [21]. However, a simple and concise justification with using some pattern like decision rule or decision tree is not possible. In the classifier ensembles approach, there is no aggregation of data and no effort is given to generate a concise representation of knowledge. On the other hand, in federated learning a common model is built, but without local data exchange, as data protection is one of the primary issues in this approach [12]. In this approach, only models' parameters are exchanged (models are generated locally) to a central space/server. The models are then aggregated and sent back to the local destinations. This iterative algorithms leads to convergence and agreement on a common central model [15]. The approach proposed in this paper can be seen as an intermediate solution. No single common model is generated as in federated learning. Instead, several common models are generated, one for each coalition. A coalition is a set of local tables containing compatible data – data on a single concept. Another difference between the proposed approach and the federated learning approach is the exchange of data within the coalition. Here, the model for a coalition is generated based on aggregated table.

The proposed approach also differs from the ensemble of classifiers approach, as it is based on coalitions formation and data aggregation. The second important difference is the form of local tables. In the classifier ensembles, it is assumed that initially the data is accumulated in a single decision table. The process of fragmentation/dispersion is realized in some controlled and planned way so as to increase the accuracy of the ensemble classification [5]. In contrast, in the proposed approach, we have no control over the form of local tables. They are provided by independent units. They do not satisfy the constraints of separability or equality of sets of objects. Nor is it possible to guarantee diversity or focus on the most difficult objects as it is in the case with many approaches known from classifier ensembles [14].

The method of conflict analysis that is used in the study was proposed in [17] by Pawlak. This approach was chosen for its simplicity and very good capabilities in identifying natural coalitions [24,25]. In the approach one of only three values is assigned to conflicting issues by the conflicting sides. These three values correspond to three possible views: support for the issue, neutrality towards the issue and being against the issue. This relatively simple approach provides tools for efficient conflict analysis and to determine the sets of coalitions and the strength of the conflict intensity. Pawlak's conflict analysis approach is very popular and widely used and developed, for example, in the three-way decisions [33] or in approach proposed by Skowron and Deja in [28]. In papers [24,25] Pawlak's conflict analysis was also used for dispersed data. However, the approach proposed there is quite different from the one considered in this paper. The main differences are the form of the local tables that are being considered and the basis for recognizing the conflict situation. In papers [24,25] it was assumed that the sets of conditional attributes appearing in the local tables are of any form, no restrictions are imposed. In contrast, in the present study, we assume that the conditional attributes are the same in all local tables. In papers [24,25] conflicts were considered in terms of decisions made by local classifiers defined based on local tables – the k-nearest neighbor classifiers were used there. In the present study, in contrast, no pre-determined classifiers were used. The basis for calculating the strength of conflicts is the values that are present in the local tables. We form coalitions in terms of a common concept – compatible values that are present in these tables.

In this study, rule generation and knowledge representation are very important concepts. In the literature, there are two main approaches to generate rules: directly from data in tables or from decision trees. A decision rule consists of two parts: a premise and a conclusion. We assume that rules are in the form of Horn clauses, i.e., the premise is a conjunction of conditions on conditional attributes, while the conclusion is a single value of a decision attribute. There are many algorithms for generating decision rules: based on the rough set theory [30], based on covering [4] or associative approaches [32]. We distinguish between exhaustive [16] and approximate [31] methods of rule generation. Decision rules may also be generated based on decision trees. The division criterion is crucial when building decision trees. The most popular are the Gini index, entropy and statistical measures approach [20]. A method of limiting tree's growth and thus overfitting of the tree is using the stop criterion [23]. In this way we can generate approximate rules. The best situation would be to generate high quality and short rules (short means with a minimum number of conditions in the premise). Generating minimal rules is an NP-hard problem, so it is possible to apply the algorithm only to small decision tables. In the literature, we can find various measures to determine the quality of rules, among others, we distinguish confidence and support, gain, variance and chi-squared value and others [2]. In this study, confidence, support and rule length are used to determine rules quality.

3 Model and Methods

In this section, we discuss the proposed hierarchical classification model for dispersed data. This model was first considered in the paper [22], where a detailed description, discussion of computational complexity and an illustrative example can be found. In the model, we assume that a set of local decision tables (collected independently, containing inconsistencies) with the same conditional attributes are given $D_i = (U_i, A, d), i \in \{1, \dots, n\}$, where U_i is the universe, a set of objects; A is a set of conditional attributes; d is a decision attribute. Based on the values of conditional attributes stored in local tables, coalitions of tables containing compatible data are defined. For this purpose, Pawlak's conflict model is used [17,22]. In this model, each agent (in our case a local table) determines its view of a conflict issue by assigning one of three values $\{-1, 0, 1\}$. The conflict issues will be conditional attributes, and the views of local tables will be assigned with respect to the values stored in the tables. For each quantitative attribute $a_{quan} \in A$, we determine the average of all attribute's values occurring in local table D_i, for each $i \in \{1, \dots, n\}$. Let us denote this value as $\overline{Val}^i_{a_{quan}}$. We also calculate the global average and the global standard deviation. Let us denote them as $\overline{Val}_{a_{quan}}$ and $SD_{a_{quan}}$. For each qualitative attribute $a_{qual} \in A$, we determine a vector over the values of that attribute. Suppose attribute a_{qual} has c values $val_1, \dots, val_c$. The vector $Val^i_{a_{qual}} = (n^i_1, \dots, n^i_c)$ represents the number of occurrences of each of these values in the decision table D_i.

Then an information system is defined $S = (U, A)$, where U is a set of local decision tables and A is a set of conditional attributes (qualitative and quantitative) occurring in local tables. For the quantitative attribute $a_{quan} \in A$ a function $a_{quan} : U \to \{-1, 0, 1\}$ is defined

$$a_{quan}(D_i) = \begin{cases} 1 & \text{if } \overline{Val}_{a_{quan}} + SD_{a_{quan}} < \overline{Val}^i_{a_{quan}} \\ 0 & \text{if } \overline{Val}_{a_{quan}} - SD_{a_{quan}} \leq \overline{Val}^i_{a_{quan}} \leq \overline{Val}_{a_{quan}} + SD_{a_{quan}} \\ -1 & \text{if } \overline{Val}^i_{a_{quan}} < \overline{Val}_{a_{quan}} - SD_{a_{quan}} \end{cases} \tag{1}$$

For the quantitative attribute a_{quan} and tables, which have lower average values than typical, we assign -1. For tables with higher average values than typical, we assign 1; and for tables with typical values, we assign 0. Whereas, for the qualitative attribute a_{qual} we use the 3−means clustering algorithm for vectors $Val^i_{a_{qual}}, i \in \{1,\ldots,n\}$. This is done in order to define three groups of tables with similar distribution of the attribute's a_{qual} values. Then for the attribute a_{qual} and the tables in the first group are assigned 1, in the second group 0, in the third group −1.

After defining the information system that determines the conflict situation, the conflict intensity between pairs of tables are calculated using the function $\rho(D_i, D_j) = \frac{card\{a \in A: a(D_i) \neq a(D_j)\}}{card\{A\}}$. Then coalitions are designated, a coalition is a set of tables that for every two tables D_i, D_j, $\rho(D_i, D_j) < 0.5$ is satisfied. An aggregated decision table is defined for each coalition. This is done by summing objects from local tables in the coalition. Based on the aggregated table the classification and regression tree algorithm is used with Gini index. In this way we obtain k models $M_1, \ldots, M_k$, where k is the number of coalitions. The final result $\hat{d}(x)$ is the set of decisions that were most frequently indicated by models $M_1, \ldots, M_k$. This means that there may be a tie, we do not resolve it in any way. In the experimental part the relevant measures for evaluating the quality of classification, which takes into account the possibility of occurring draws, were used. The results obtained using the proposed method are compared with the results generated by an approach without any conflict analysis. In the baseline approach, based on each local table the classification and regression tree algorithm is used. The final result is the set of decisions that were most frequently indicated by trees. Ties can arise, but analogously as before, we do not resolve them in any way.

4 Data Sets, Results and Discussion

The experiments were carried out on the data available in the UC Irvine Machine Learning Repository [1]. Three data sets were selected for the analysis – the Vehicle Silhouettes, the Landsat Satellite and the Soybean (Large) data sets. In the case of Landsat Satellite and Soybean data sets, training and test sets are in the repository. The Vehicle data set was randomly split into two disjoint subsets, the training set (70% of objects) and the test set (30% of objects). The Vehicle Silhouettes data set has eighteen quantitative conditional attributes, four decision classes and 846 objects – 592 training, 254 test set. The Landsat Satellite data set has thirty-six quantitative conditional attributes, six decision classes and 6435 objects – 4435 training, 1000 test set. The Soybean data set has thirty-five quantitative conditional attributes, nineteen decision classes and 683 objects – 307 training, 376 test set. The training sets of the above data sets were dispersed. Only objects are dispersed, whereas the full set of conditional attributes is included in each local table. We use a stratified mode for dispersion. Five different dispersed versions with 3, 5, 7, 9 and 11 local tables were prepared to check for different degrees of dispersion for each data set.

The quality of classification was evaluated based on the test set. Three measures were used. The classification accuracy is the ratio of correctly classified objects from the test set to their total number in this set. When the correct decision class of an object is contained in the generated decision set, the object is considered to be correctly classified. The classification ambiguity accuracy is also the ratio of correctly classified objects from the test set to their total number in this set. With the difference that this time when only one correct decision class was generated, the object is considered to be correctly classified. The third measure allows to assess the frequency and number of draws generated by the classification model. A very important aspect discussed in the paper is the rules' quality. Rules are generated based on decision trees that are built by the model. The quality of decision rules was evaluated based on the test set using the following measures. The rule confidence is the ratio of objects from the test set that matching the rule's conditions and its decision to the number of objects that satisfy the rule's conditions. The confidence is a measure of the strength of the relation between conditions and decision of rule. This is a very important measure in the context of classification because it indicates the quality of the knowledge represented by the rule – how strongly is it justified to make a certain decision based on given conditions. The rule support is the ratio of objects from the test set that matching the rule's conditions and its decision to their total number in this set. The support is a measure of the frequency of a rule. Support proves the popularity of rules. But common rules do not always constitute new and relevant knowledge. That is why confidence becomes so important. Another rules' related measure analyzed was the length of the rule indicated by the number of conditions occurring in the rule and the total number of rules generated by the model. The experiments were carried out according to the following scheme. For both the proposed and the baseline methods for five degrees of dispersion (3, 5, 7, 9, 11 local tables) different stop criterion were analyzed. The initial stop value was 2, and the step was 5. For smaller step values, non-significant differences in results were noted. The following stop values were tested: $2, 7, 12$. The classification quality was evaluated using decision trees. Then, rules were generated based on decision trees and the rules' quality was estimated. For the Landsat Satellite and the Soybean data sets with 3 local tables, the proposed model did not generate coalitions, so the results obtained using the model are the same as for the baseline model. These results were omitted in the rest of the paper.

4.1 Classification Quality

Table 1 shows the values of classification quality measures obtained for the proposed model and the baseline model. The higher value of classification accuracy is shown in bold. As can be seen, in the vast majority of cases, the proposed model generates better results. Improvements in classification accuracy were obtained regardless of the degree of dispersion, the used stop criterion or the analyzed data set. Statistical test was performed in order to confirm the importance in the differences in the obtained results *acc*. The received classification accuracy values were divided into two dependent data samples, each consisting of 39 observations. It was confirmed by the Wilcoxon test that the difference between the classification accuracy for both groups is significant with the level $p = 0.0001$.

Table 1. Classification accuracy *acc*, classification ambiguity accuracy acc_{ONE} and the average number of generated decisions set $\bar{d}$ for the proposed method with coalitions and the baseline approach without coalitions. SC – Stop criteria, T – No. of tables

T	SC	Proposed method $acc/acc_{ONE}/\bar{d}$	Baseline method $acc/acc_{ONE}/\bar{d}$	Proposed method $acc/acc_{ONE}/\bar{d}$	Baseline method $acc/acc_{ONE}/\bar{d}$
		Landsat Satellite		Soybean	
5	2	**0.890**/0.816/1.104	0.875/0.838/1.049	**0.889**/0.780/1.223	0.865/0.777/1.152
	7	**0.891**/0.821/1.103	0.877/0.842/1.046	**0.814**/0.682/1.274	0.764/0.699/1.125
	12	**0.885**/0.812/1.094	0.862/0.831/1.045	**0.733**/0.615/1.291	0.672/0.601/1.145
7	2	**0.896**/0.824/0.824	0.877/0.844/1.038	**0.902**/0.814/1.111	0.814/0.726/1.139
	7	**0.880**/0.813/1.090	0.867/0.833/1.043	**0.858**/0.807/1.105	0.659/0.601/1.122
	12	**0.885**/0.808/1.097	0.865/0.834/1.043	**0.804**/0.699/1.115	0.601/0.574/1.115
9	2	**0.885**/0.847/1.055	0.875/0.853/1.032	**0.905**/0.851/1.064	0.807/0.713/1.145
	7	0.868/0.832/1.052	**0.872**/0.842/1.041	**0.861**/0.828/1.044	0.635/0.584/1.135
	12	**0.876**/0.831/1.064	0.867/0.840/1.039	**0.774**/0.747/1.034	0.601/0.557/1.118
11	2	**0.895**/0.861/1.040	0.872/0.851/1.031	**0.868**/0.841/1.057	0.791/0.740/1.074
	7	**0.887**/0.856/1.036	0.870/0.848/1.029	**0.872**/0.845/1.044	0.615/0.571/1.122
	12	**0.884**/0.853/1.037	0.867/0.847/1.025	**0.801**/0.794/1.010	0.341/0.304/1.071
		Vehicle Silhouettes			
3	2	**0.819**/0.516/1.382	0.780/0.665/1.236		
	7	**0.839**/0.488/1.417	0.795/0.673/1.252		
	12	**0.827**/0.480/1.413	0.776/0.657/1.236		
5	2	**0.768**/0.701/1.142	0.756/0.669/1.098		
	7	**0.760**/0.673/1.197	0.736/0.634/1.110		
	12	**0.732**/0.642/1.189	**0.732**/0.622/1.110		
7	2	0.776/0.697/1.181	**0.783**/0.705/1.114		
	7	**0.772**/0.638/1.268	0.748/0.677/1.118		
	12	**0.780**/0.657/1.252	0.736/0.661/1.114		
9	2	0.740/0.689/1.067	**0.752**/0.669/1.118		
	7	0.720/0.701/1.047	**0.732**/0.685/1.063		
	12	0.717/0.677/1.055	**0.732**/0.665/1.114		
11	2	**0.791**/0.744/1.087	0.717/0.665/1.083		
	7	**0.783**/0.732/1.063	0.736/0.677/1.071		
	12	**0.756**/0.720/1.047	0.713/0.650/1.075		

4.2 Rules' Quality

Tables 2, 3 and 4 show the values of measures determining the rules' quality obtained for the data for the proposed and baseline approach. For confidence, support and length, the minimum and the maximum values obtained for the generated rules, as well as the average of values obtained for all rules with the standard deviation, are given. Results are given for different degrees of dispersion (number of tables) and different values of the stop criteria. For each degree of dispersion, the best average value (highest for confidence and support, lowest for length) obtained and the smallest number of rules generated are shown in bold. As can be seen, for lower values of the stop criterion, we get rules with higher confidence but smaller support. This can be concluded that rules generated based on more expanded trees, better justify the connection between conditions and decision. The next conclusion is quite natural, for larger values of the stop criterion (by limiting the trees' growth) we get shorter rules and their number is smaller.

Table 2. Confidence, support and length of the rules generated by the proposed and the baseline method (Vehicle Silhouettes). SC – Stop criteria, T – No. of tables

T	SC	Rules' confidence Min/Max/AVG/SD	Rules' support Min/Max/AVG/SD	Rules' length Min/Max/AVG/SD	No. rules
Proposed method					
3	2	0.200/1/**0.660**/0.273	0.004/0.173/0.021/0.033	3/12/6.938/2.358	64
	7	0.111/1/0.625/0.258	0.004/0.177/0.023/0.035	3/11/6.525/2.227	59
	12	0.200/1/0.639/0.241	0.004/0.173/**0.025**/0.035	3/11/**6.173**/2.101	**52**
5	2	0.143/1/**0.655**/0.281	0.004/0.185/0.020/0.033	3/14/6.907/2.529	97
	7	0.111/1/0.605/0.268	0.004/0.185/0.024/0.035	3/14/6.578/2.475	83
	12	0.111/1/0.593/0.270	0.004/0.185/**0.027**/0.037	3/12/**5.986**/2.223	**71**
7	2	0.125/1/**0.644**/0.270	0.004/0.161/0.026/0.037	1/13/6.781/2.587	73
	7	0.091/1/0.584/0.276	0.004/0.161/0.028/0.038	1/12/6.212/2.390	66
	12	0.091/1/0.596/0.263	0.004/0.154/**0.034**/0.040	1/11/**5.804**/2.271	**56**
9	2	0.111/1/**0.649**/0.276	0.004/0.177/0.022/0.034	2/11/6.333/1.868	150
	7	0.100/1/0.605/0.288	0.004/0.177/0.024/0.036	2/10/5.852/1.774	135
	12	0.091/1/0.589/0.285	0.004/0.177/**0.027**/0.036	2/10/**5.593**/1.714	**118**
11	2	0.071/1/**0.630**/0.273	0.004/0.193/0.027/0.038	2/9/5.878/1.756	164
	7	0.091/1/0.611/0.263	0.004/0.193/0.033/0.040	2/9/5.348/1.636	138
	12	0.091/1/0.592/0.260	0.004/0.193/**0.037**/0.043	1/8/**4.992**/1.497	**120**
Baseline method					
3	2	0.111/1/**0.673**/0.273	0.004/0.173/0.027/0.037	2/12/6.167/1.979	72
	7	0.167/1/0.641/0.269	0.004/0.177/0.03/0.038	2/12/5.848/2.091	66
	12	0.167/1/0.605/0.261	0.004/0.177/**0.032**/0.039	2/10/**5.441**/1.844	**59**
5	2	0.111/1/**0.633**/0.633	0.004/0.185/0.030/0.038	2/9/5.388/1.470	103
	7	0.067/1/0.58/0.268	0.004/0.185/0.035/0.042	2/8/4.874/1.413	87
	12	0.067/1/0.581/0.271	0.004/0.185/**0.044**/0.044	2/6/**4.29**/1.105	**69**
7	2	0.071/1/**0.582**/0.271	0.004/0.181/0.033/0.040	1/10/5.165/1.697	127
	7	0.091/1/0.574/0.259	0.004/0.181/0.04/0.043	1/8/4.608/1.509	102
	12	0.111/1/0.566/0.252	0.004/0.181/**0.052**/0.045	1/7/**4**/1.271	**78**
9	2	0.050/1/**0.553**/0.263	0.004/0.217/0.038/0.044	2/9/4.674/1.449	132
	7	0.05/1/0.537/0.273	0.004/0.232/0.048/0.047	2/8/4.151/1.257	106
	12	0.048/1/0.551/0.264	0.004/0.217/**0.061**/0.050	1/7/**3.675**/1.163	**83**
11	2	0.036/1/0.523/0.252	0.004/0.181/0.041/0.046	2/8/4.437/1.369	151
	7	0.036/1/0.518/0.243	0.004/0.185/0.052/0.049	2/8/3.866/1.173	119
	12	0.095/1/**0.525**/0.242	0.004/0.185/**0.067**/0.055	1/6/**3.44**/1.061	**91**

In order to confirm the importance in the differences of the obtained measures (confidence, support and length) in relation to different stop criterion, statistical

Table 3. Confidence, support and length of the rules generated by the proposed and the baseline method (Landsat Satellite). SC – Stop criteria, T – No. of tables

T	SC	Rules' confidence Min/Max/AVG/SD	Rules' support Min/Max/AVG/SD	Rules' length Min/Max/AVG/SD	No. rules
Proposed method					
5	2	0.043/1/**0.646**/0.288	0.001/0.187/0.010/0.028	3/18/8.645/2.763	313
	7	0.043/1/0.637/0.283	0.001/0.188/0.012/0.030	3/17/8.209/2.838	268
	12	0.043/1/0.623/0.281	0.001/0.187/**0.014**/0.032	3/17/**7.927**/2.899	**233**
7	2	0.050/1/**0.665**/0.280	0.001/0.186/0.010/0.026	3/19/8.907/2.891	332
	7	0.050/1/0.634/0.276	0.001/0.186/0.011/0.027	3/19/8.524/2.971	290
	12	0.037/1/0.626/0.276	0.001/0.186/**0.013**/0.029	3/19/**8.313**/3.060	**252**
9	2	0.067/1/**0.663**/0.272	0.001/0.185/0.011/0.029	3/19/8.959/3.361	365
	7	0.053/1/0.654/0.270	0.001/0.184/0.013/0.031	2/19/8.477/3.314	302
	12	0.053/1/0.634/0.266	0.001/0.184/**0.015**/0.032	2/19/**8.348**/3.410	**273**
11	2	0.045/1/**0.656**/0.276	0.001/0.196/0.010/0.027	3/17/8.728/2.752	643
	7	0.048/1/0.630/0.277	0.001/0.195/0.012/0.029	3/17/8.336/2.761	560
	12	0.048/1/0.632/0.270	0.001/0.195/**0.013**/0.031	2/17/**8.109**/2.814	**494**
Baseline method					
5	2	0.083/1/**0.633**/0.270	0.001/0.187/0.012/0.030	3/15/8.161/2.410	348
	7	0.056/1/0.597/0.277	0.001/0.188/0.013/0.032	2/14/7.679/2.312	296
	12	0.056/1/0.603/0.277	0.001/0.187/**0.016**/0.034	2/14/**7.325**/2.411	**252**
7	2	0.063/1/**0.608**/0.279	0.001/0.187/0.014/0.033	3/15/7.731/2.383	401
	7	0.063/1/0.604/0.266	0.001/0.187/0.018/0.036	2/14/7.201/2.396	318
	12	0.080/1/0.606/0.261	0.001/0.187/**0.020**/0.038	2/14/**7.080**/2.441	**275**
9	2	0.045/1/**0.599**/0.286	0.001/0.196/0.017/0.038	3/14/7.005/2.144	419
	7	0.040/1/0.583/0.280	0.001/0.195/0.022/0.042	2/14/6.571/2.188	324
	12	0.045/1/0.575/0.281	0.001/0.194/**0.025**/0.045	2/13/**6.279**/2.177	**276**
11	2	0.053/1/**0.594**/0.280	0.001/0.199/0.020/0.041	3/14/6.668/1.985	440
	7	0.043/1/0.573/0.286	0.001/0.199/0.024/0.044	2/13/6.184/1.926	358
	12	0.043/1/0.581/0.276	0.001/0.199/**0.029**/0.047	2/13/**5.950**/2.004	**300**

tests were performed. The results were grouped depending on the stop criterion, three dependent samples, were created. The Friedman test was used. There were statistically significant differences in the results of confidence, support and length obtained for different stop criterion being considered. The following results were obtained: for confidence $\chi^2(26,2) = 36.538, p = 0.000001$; for support $\chi^2(26,2) = 46.231, p = 0.000001$; for length $\chi^2(26,2) = 52, p = 0.000001$. Additionally, we confirmed (by using the Wilcoxon test) that the differences in confidence, support and length are significant between each-pair of stop criterion values. Comparative box-whiskers charts for the results with three values of stop criterion were created (Fig. 1). Based on the charts, we can see that the greatest differences in results divided into stop criterion were obtained for support and length (here the boxes are located farthest from each other). Thus, using a larger stop criterion reduces the length of the rules and increases support to a greater extent than it reduces confidence.

Table 5 presents a comparison of the average values of rules' confidence, support and length generated by the proposed and baseline method. The higher

Table 4. Confidence, support and length of the rules generated by the proposed and the baseline method (Soybean). SC – Stop criteria, T – No. of tables

T	SC	Rules' confidence Min/Max/AVG/SD	Rules' support Min/Max/AVG/SD	Rules' length Min/Max/AVG/SD	No. rules
Proposed method					
5	2	0.091/1/**0.734**/0.282	0.003/0.142/0.032/0.034	3/12/5.337/1.872	89
	7	0.100/1/0.668/0.286	0.003/0.142/0.041/0.037	2/9/4.603/1.453	63
	12	0.100/1/0.621/0.286	0.003/0.169/**0.053**/0.043	2/7/**3.886**/1.112	**44**
7	2	0.074/1/**0.798**/0.284	0.003/0.142/0.033/0.035	2/12/5.521/1.785	94
	7	0.125/1/0.749/0.287	0.003/0.142/0.041/0.038	2/11/4.944/1.786	72
	12	0.093/1/0.692/0.277	0.003/0.145/**0.050**/0.044	2/9/**4.364**/1.577	**55**
9	2	0.125/1/**0.758**/0.302	0.003/0.142/0.033/0.036	3/10/5.457/1.724	138
	7	0.075/1/0.733/0.285	0.003/0.142/0.040/0.036	2/9/4.850/1.440	113
	12	0.083/1/0.627/0.277	0.003/0.149/**0.049**/0.040	2/8/**4.300**/1.364	**80**
11	2	0.077/1/**0.770**/0.286	0.003/0.132/0.031/0.036	2/12/5.917/1.950	132
	7	0.071/1/0.739/0.291	0.003/0.155/0.038/0.040	2/11/5.272/1.855	103
	12	0.083/1/0.747/0.278	0.003/0.172/**0.047**/0.043	1/9/**4.910**/1.834	**78**
Baseline method					
5	2	0.071/1/**0.723**/0.297	0.003/0.142/0.034/0.038	2/10/5.225/1.650	102
	7	0.083/1/0.640/0.281	0.003/0.142/0.045/0.041	2/7/4.362/1.285	69
	12	0.100/1/0.557/0.297	0.003/0.169/**0.062**/0.049	2/6/**3.444**/0.858	**45**
7	2	0.056/1/**0.685**/0.281	0.003/0.132/0.044/0.031	2/9/5.282/1.657	103
	7	0.016/1/0.576/0.286	0.003/0.135/0.046/0.038	1/8/4.067/1.352	60
	12	0.014/1/0.499/0.305	0.007/0.169/**0.070**/0.046	1/5/**2.824**/0.785	**34**
9	2	0.016/1/**0.584**/0.330	0.003/0.169/0.033/0.035	1/10/4.817/1.798	109
	7	0.056/1/0.514/0.279	0.003/0.169/0.057/0.045	2/6/3.525/1.168	61
	12	0.016/1/0.416/0.241	0.010/0.172/**0.078**/0.055	1/5/**2.611**/0.980	**36**
11	2	0.006/1/**0.566**/0.308	0.003/0.149/0.034/0.036	1/8/4.552/1.445	125
	7	0.014/1/0.492/0.292	0.007/0.169/0.063/0.051	1/6/3.034/0.920	59
	12	0.015/0.9/0.352/0.240	0.007/0.172/**0.083**/0.061	1/3/**1.970**/0.577	**33**

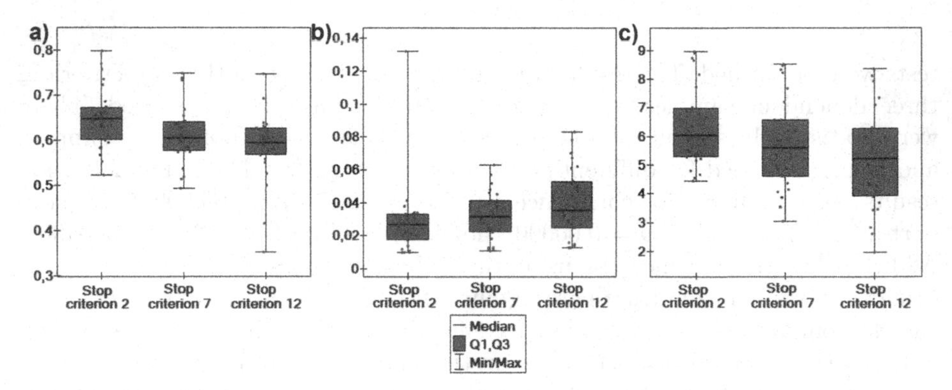

Fig. 1. Comparison of (a) the average rules' confidence (b) the average rules' support (c) the average rules' length obtained for different stop criterion.

values of confidence, support and lower values of rules' length are shown in bold. As can be seen in the vast majority of cases better confidence was obtained for the proposed method. The creation of coalitions and aggregated tables for the coalitions made it possible to generate rules that are better representation of knowledge hidden in data. The confidence is the most important measure, it shows how much the rule's conditions actually indicates the decision. Higher values of support were noted for the baseline approach. However, this measure only shows the fraction of objects supporting the rule's antecedent and decision – it does not indicate the actual connection between conditions and decision. The baseline method produces shorter rules than the proposed method. In most cases, the average number of conditions in rules is greater by one condition for the proposed method than for the baseline method. However, this measure is also less important than confidence when evaluating the quality of generated rules. Statistical tests were performed in order to confirm the importance in the differences in the obtained results of rules' confidence, support and length. At first, the average values of rules' confidence in two dependent groups were analysed – the proposed and the baseline methods. Both groups contained 39 observations each – all results for dispersed data sets. It was confirmed by the Wilcoxon test that the difference between the averages of rules' confidence for both groups is significant with the level $p = 0.0001$. In an analogous way – using the Wilcoxon test – the statistical significance of the differences between the averages of rules' support and rules' length were confirmed, with the level $p = 0.0001$ in both cases. Additionally, comparative box-whiskers chart for the values of rules' confidence, support and length was created (Fig. 2). The biggest difference can be noticed in the case of rules' confidence – the boxes are located in different places, do not overlap in any part. The difference in rules' confidence is the most significant for us, it shows that the knowledge generated when using the proposed method is of much better quality than for the non-coalitions approach.

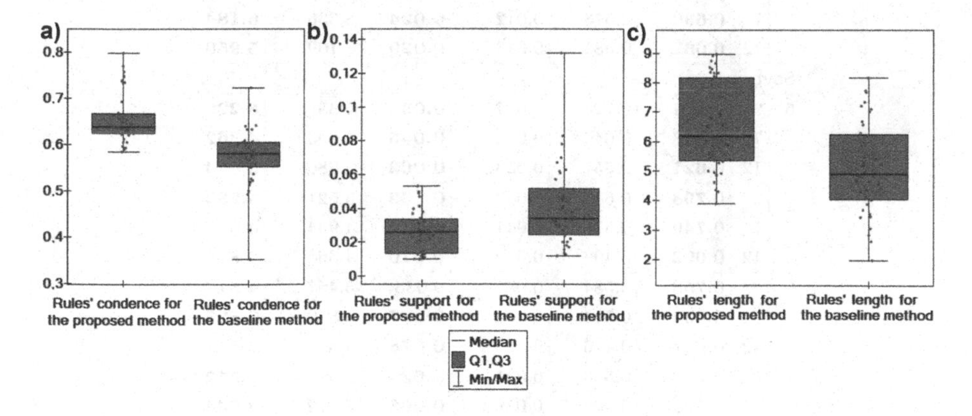

Fig. 2. Comparison of (a) the average rules' confidence (b) the average rules' support (c) the average rules' length obtained for the proposed method and the baseline method.

Table 5. Comparison of the average confidence, support and length of rules obtained for the proposed and the baseline methods. SC – Stop criteria, T – No. of tables

T	SC	Rules' confidence		Rules' support		Rules' length	
		Proposed	Baseline	Proposed	Baseline	Proposed	Baseline
Vehicle Silhouettes							
3	2	0.660	**0.673**	0.021	**0.027**	6.938	**6.167**
	7	0.625	**0.641**	0.023	**0.03**	6.525	**5.848**
	12	**0.639**	0.605	0.025	**0.032**	6.173	**5.441**
5	2	**0.655**	0.633	0.020	**0.030**	6.907	**5.388**
	7	**0.605**	0.58	0.024	**0.035**	6.578	**4.874**
	12	**0.593**	0.581	0.027	**0.044**	5.986	**4.290**
7	2	**0.644**	0.582	0.026	**0.033**	6.781	**5.165**
	7	**0.584**	0.574	0.028	**0.04**	6.212	**4.608**
	12	**0.596**	0.566	0.034	**0.052**	5.804	**4**
9	2	**0.649**	0.553	0.022	**0.038**	6.333	**4.674**
	7	**0.605**	0.537	0.024	**0.048**	5.852	**4.151**
	12	**0.589**	0.551	0.027	**0.061**	5.593	**3.675**
11	2	**0.630**	0.523	0.027	**0.041**	5.878	**4.437**
	7	**0.611**	0.518	0.033	**0.052**	5.348	**3.866**
	12	**0.592**	0.525	0.037	**0.067**	4.992	**3.44**
Landsat Satellite							
5	2	**0.646**	0.633	0.010	**0.012**	8.645	**8.161**
	7	**0.637**	0.597	0.012	**0.013**	8.209	**7.679**
	12	**0.623**	0.603	0.014	**0.016**	7.927	**7.325**
7	2	**0.665**	0.608	0.010	**0.014**	8.907	**7.731**
	7	**0.634**	0.604	0.011	**0.018**	8.524	**7.201**
	12	**0.626**	0.606	0.013	**0.020**	8.313	**7.080**
9	2	**0.663**	0.599	0.011	**0.017**	8.959	**7.005**
	7	**0.654**	0.583	0.013	**0.022**	8.477	**6.571**
	12	**0.634**	0.575	0.015	**0.025**	8.348	**6.279**
11	2	**0.656**	0.594	0.010	**0.020**	8.728	**6.668**
	7	**0.630**	0.573	0.012	**0.024**	8.336	**6.184**
	12	**0.632**	0.581	0.013	**0.029**	8.109	**5.950**
Soybean							
5	2	**0.734**	0.723	0.032	**0.034**	5.337	**5.225**
	7	**0.668**	0.640	0.041	**0.045**	4.603	**4.362**
	12	**0.621**	0.557	0.053	**0.062**	3.886	**3.444**
7	2	**0.798**	0.685	0.033	**0.132**	5.521	**5.282**
	7	**0.749**	0.576	0.041	**0.046**	4.944	**4.067**
	12	**0.692**	0.499	0.050	**0.070**	4.364	**2.824**
9	2	**0.758**	0.584	0.033	0.033	5.457	**4.817**
	7	**0.733**	0.514	0.040	**0.057**	4.850	**3.525**
	12	**0.627**	0.416	0.049	**0.078**	4.300	**2.611**
11	2	**0.770**	0.566	0.031	**0.034**	5.917	**4.552**
	7	**0.739**	0.492	0.038	**0.063**	5.272	**3.034**
	12	**0.747**	0.352	0.047	**0.083**	4.910	**1.970**

5 Conclusion

The paper presents a classification model for data stored independently in several decision tables. We assume that the sets of conditional attributes in all tables are equal. The proposed model creates coalitions of tables containing similar data – more precisely, similar attributes' values. For the coalitions aggregated tables are created. Decision trees are generated based on these tables. The study compared the proposed model with a model in which coalitions are not used. An analysis of the quality of rules generated by the model and the effect of the stop criterion on the results was also made. It was shown that the proposed model generates significantly better classification accuracy than the model without coalitions. Also, the rules generated by the proposed model have significantly higher confidence than the rules generated by the baseline model. The use of larger values of stop criterion has less effect on reducing rules' confidence, while it has greater effect on increasing support and reducing rules' length. In the future work, it is planned to consider the variation of conditional attributes' values within each decision class for generating coalitions of local tables.

References

1. Asuncion, A., Newman, D.J.: UCI machine learning repository. University of Massachusetts Amherst, USA (2007)
2. Bayardo Jr., R.J., Agrawal, R.: Mining the most interesting rules. In: Proceedings of the 5th ACM SIGKDD, pp. 145–154 (1999)
3. Czarnowski, I., Jędrzejowicz, P.: Ensemble online classifier based on the one-class base classifiers for mining data streams. Cybern. Syst. **46**(1–2), 51–68 (2015)
4. Dembczyński, K., Kotłowski, W., Słowiński, R.: ENDER: a statistical framework for boosting decision rules. Data Min. Knowl. Disc. **21**(1), 52–90 (2010)
5. Freund, Y., Schapire, R.E.: A decision-theoretic generalization of on-line learning and an application to boosting. J. Comput. Syst. Sci. **55**(1), 119–139 (1997)
6. Gao, J., Lanchantin, J., Soffa, M.L., Qi, Y.: Black-box generation of adversarial text sequences to evade deep learning classifiers. In: 2018 IEEE Security and Privacy Workshops (SPW), pp. 50–56. IEEE (2018)
7. Gilpin, L.H., Bau, D., Yuan, B.Z., Bajwa, A., Specter, M., Kagal, L.: Explaining explanations: an overview of interpretability of machine learning. In: IEEE 5th Int. Conf. Data Sci. Adv. Anal. (DSAA), pp. 80–89. IEEE (2018)
8. Gut, D., Tabor, Z., Szymkowski, M., Rozynek, M., Kucybała, I., Wojciechowski, W.: Benchmarking of deep architectures for segmentation of medical images. IEEE Trans. Med. Imaging **41**(11), 3231–3241 (2022)
9. Kozak, J.: Decision Tree and Ensemble Learning Based on Ant Colony Optimization. SCI, vol. 781. Springer, Cham (2019). https://doi.org/10.1007/978-3-319-93752-6
10. Krawczyk, B., Woźniak, M., Cyganek, B.: Clustering-based ensembles for one-class classification. Inf. Sci. **264**, 182–195 (2014)
11. Kuncheva, L.I.: Combining Pattern Classifiers: Methods and Algorithms. Wiley, Hoboken (2014)
12. Li, Z., Sharma, V., Mohanty, S.P.: Preserving data privacy via federated learning: challenges and solutions. IEEE Consum. Electron. Mag. **9**(3), 8–16 (2020)

13. Meng, T., Jing, X., Yan, Z., Pedrycz, W.: A survey on machine learning for data fusion. Inf. Fusion **57**, 115–129 (2020)
14. Nam, G., Yoon, J., Lee, Y., Lee, J.: Diversity matters when learning from ensembles. Adv. Neural. Inf. Process. Syst. **34**, 8367–8377 (2021)
15. Nguyen, H.T., Sehwag, V., Hosseinalipour, S., Brinton, C.G., Chiang, M., Poor, H.V.: Fast-convergent federated learning. IEEE J. Sel. Areas Commun. **39**(1), 201–218 (2020)
16. Pawlak, Z., Skowron, A.: Rough sets and Boolean reasoning. Inf. Sci. **177**(1), 41–73 (2007)
17. Pawlak, Z.: Conflict analysis. In: Proceedings of the Fifth European Congress on Intelligent Techniques and Soft Computing (EUFIT 1997), pp. 1589–1591 (1997)
18. Pięta, P., Szmuc, T.: Applications of rough sets in big data analysis: an overview. Int. J. Appl. Math. Comput. Sci. **31**(4), 659–683 (2021)
19. Połap, D., Woźniak, M.: Meta-heuristic as manager in federated learning approaches for image processing purposes. Appl. Soft Comput. **113**, 107872 (2021)
20. Priyanka, Kumar, D.: Decision tree classifier: a detailed survey. Int. J. Inf. Decis. Sci. **12**(3), 246–269 (2020)
21. Przybyła-Kasperek, M., Smyczek, F.: Comparison of Shapley-Shubik and Banzhaf-Coleman power indices applied to aggregation of predictions obtained based on dispersed data by k-nearest neighbors classifiers. Procedia Comput. Sci. **207**, 2134–2143 (2022)
22. Przybyła-Kasperek, M., Kusztal, K.: New classification method for independent data sources using Pawlak conflict model and decision trees. Entropy **24**(11), 1604 (2022). https://doi.org/10.3390/e24111604
23. Przybyła-Kasperek, M., Aning, S.: Stop criterion in building decision trees with bagging method for dispersed data. Procedia Comput. Sci. **192**, 3560–3569 (2021)
24. Przybyła-Kasperek, M.: Coalitions' weights in a dispersed system with Pawlak conflict model. Group Decis. Negot. **29**(3), 549–591 (2020)
25. Przybyła-Kasperek, M.: Three conflict methods in multiple classifiers that use dispersed knowledge. Int. J. Inf. Technol. Decis. Mak. **18**(02), 555–599 (2019)
26. Ren, P., Xu, Z., Kacprzyk, J.: Group Decisions with Intuitionistic Fuzzy Sets. In: Kilgour, D.M., Eden, C. (eds.) Handbook of Group Decision and Negotiation, pp. 977–995. Springer, Cham (2021). https://doi.org/10.1007/978-3-030-49629-6_43
27. Skokowski, P., Łopatka, J., Malon, K.: Evidence theory based data fusion for centralized cooperative spectrum sensing in mobile ad-hoc networks. In: 2020 Baltic URSI Symposium (URSI), pp. 24–27. IEEE (2020)
28. Skowron, A., Deja, R.: On some conflict models and conflict resolutions. Rom. J. Inf. Sci. Technol. **3**(1–2), 69–82 (2002)
29. Słowiński, R., Greco, S., Matarazzo, B.: Rough set analysis of preference-ordered data. In: Alpigini, J.J., Peters, J.F., Skowron, A., Zhong, N. (eds.) RSCTC 2002. LNCS (LNAI), vol. 2475, pp. 44–59. Springer, Heidelberg (2002). https://doi.org/10.1007/3-540-45813-1_6
30. Stefanowski, J.: On rough set based approaches to induction of decision rules. Rough Sets Knowl. Disc. **1**(1), 500–529 (1998)
31. Ślęzak, D., Wróblewski, J.: Order based genetic algorithms for the search of approximate entropy reducts. In: Wang, G., Liu, Q., Yao, Y., Skowron, A. (eds.) RSFDGrC 2003. LNCS (LNAI), vol. 2639, pp. 308–311. Springer, Heidelberg (2003). https://doi.org/10.1007/3-540-39205-X_45
32. Wieczorek, A., Słowiński, R.: Generating a set of association and decision rules with statistically representative support and anti-support. Inf. Sci. **277**, 56–70 (2014)

33. Yao, Y.: Rough sets and three-way decisions. International Conference, RSKT, 62–73, Springer, Cham, 2015
34. Zou, X., Zhong, S., Yan, L., Zhao, X., Zhou, J., Wu, Y.: Learning robust facial landmark detection via hierarchical structured ensemble. In: Proc. IEEE Int. Conf. Comput. Vis., pp. 141–150 (2019)

Data-Driven Stability Analysis of a Chaotic Time-Delayed System

Georgios Margazoglou[1(✉)] and Luca Magri[1,2,3]

[1] Department of Aeronautics, Imperial College London, London SW7 2AZ, UK
{g.margazoglou,l.magri}@imperial.ac.uk
[2] The Alan Turing Institute, London NW1 2DB, UK
[3] Isaac Newton Institute for Mathematical Sciences, Cambridge CB3 0EH, UK

Abstract. Systems with time-delayed chaotic dynamics are common in nature, from control theory to aeronautical propulsion. The overarching objective of this paper is to compute the stability properties of a chaotic dynamical system, which is time-delayed. The stability analysis is based only on data. We employ the echo state network (ESN), a type of recurrent neural network, and train it on timeseries of a prototypical time-delayed nonlinear thermoacoustic system. By running the trained ESN autonomously, we show that it can reproduce (i) the long-term statistics of the thermoacoustic system's variables, (ii) the physical portion of the Lyapunov spectrum, and (iii) the statistics of the finite-time Lyapunov exponents. This work opens up the possibility to infer stability properties of time-delayed systems from experimental observations.

Keywords: Echo State Networks · Chaos · Time-delayed systems

1 Introduction

Chaotic systems with time-delayed dynamics appear in a range of scientific fields [8]. Because of their dependence on both the present and past states, these systems have rich and intricate dynamics. Their chaotic behaviour can be assessed with stability analysis, which is a mathematical tool that quantifies the system's response to infinitesimal perturbations. Stability analysis relies on the linearization of the time-delayed dynamical equations, which spawns the Jacobian of the system. From the Jacobian, we compute the Lyapunov Exponents (LEs), which are the key quantities to quantifying chaos [2].

A data-driven method, which considers the sequential nature of the dataset (e.g. timeseries) to infer chaotic dynamics, is the recurrent neural network (RNN). Such networks have been successfully applied to learn chaotic dynamics (with no time delay) for different applications [12,13,17]. The majority of RNNs require backpropagation through time for training, which can lead to vanishing or exploding gradients, as well as long training times [17]. Instead, the echo state network (ESN), which we employ here, is trained via ridge regression, which eliminates backpropagation and provides a faster training [7,9,14]. The objective of this paper is to train an ESN with data from a prototypical chaotic thermoacoustic system, which is a nonlinear time-delayed wave equation. We further assess the capabilities of the ESN to accurately learn the ergodic and

This research has received financial support from the ERC Starting Grant No. PhyCo 949388.

J. Mikyška et al. (Eds.): ICCS 2023, LNCS 14076, pp. 406–413, 2023.
https://doi.org/10.1007/978-3-031-36027-5_31

stability properties of the thermoacoustic system, by calculating fundamental quantities, such as the Lyapunov exponents. Other RNN architectures for time-series forecasting are the Long-Short term memory and the Gated Recurrent units [3], which require back-propagation. In this paper, we take advantage of the simple training of ESNs, which do not require backpropagation. We briefly review stability analysis for time-delayed systems in Sect. 2 and present the considered thermoacoustic system in Sect. 2.1. In Sect. 3, we discuss the ESN architecture and properties. We present the results in Sect. 4 and conclude in Sect. 5.

2 Stability Analysis for Time-Delayed Systems

We consider a physical state $\boldsymbol{x}(t) \in \mathbb{R}^D$, which is the solution of a nonlinear time-delayed dynamical system

$$\frac{d\boldsymbol{x}}{dt} = f(\boldsymbol{x}(t), \boldsymbol{x}(t-\tau)), \quad \boldsymbol{x}(t) = \boldsymbol{x}_0, \ \forall\, t \leq 0, \tag{1}$$

where τ is a constant time-delay. We analyse the system's stability by perturbing the state with infinitesimal perturbations $\boldsymbol{u} \sim \mathcal{O}(\varepsilon)$, $\varepsilon \to 0$, as $\boldsymbol{x}+\boldsymbol{u}$, with $\boldsymbol{x} \sim \mathcal{O}(1)$. Hence, we obtain the tangent linear equation

$$\frac{d\boldsymbol{U}}{dt} = \boldsymbol{J}(\boldsymbol{x}(t), \boldsymbol{x}(t-\tau))\,\boldsymbol{U}, \tag{2}$$

which involves the time-marching of $K \leq D$ tangent vectors, $\boldsymbol{u}_i \in \mathbb{R}^D$, as columns of the matrix $\boldsymbol{U} \in \mathbb{R}^{D\times K}$, $\boldsymbol{U} = [\boldsymbol{u}_1, \boldsymbol{u}_2, \ldots, \boldsymbol{u}_K]$. This is a linear basis of the tangent space. The linear operator $\boldsymbol{J} \in \mathbb{R}^{D\times D}$ is the Jacobian of the system, which is time-dependent in chaotic attractors. As shown in [12], we can extract the Jacobian of a reservoir computer by linearizing Eqs. (7), which are the evolution equations of the ESN.

We periodically orthonormalize the tangent space basis during time evolution by using a QR-decomposition of $\boldsymbol{U}$, as $\boldsymbol{U}(t) = \boldsymbol{Q}(t)\boldsymbol{R}(t,\Delta t)$ and by updating the columns of $\boldsymbol{U}$ with the columns of $\boldsymbol{Q}$, i.e. $\boldsymbol{U} \leftarrow \boldsymbol{Q}$ [2]. The matrix $\boldsymbol{R}(t,\Delta t) \in \mathbb{R}^{K\times K}$ is upper-triangular and its diagonal elements $[\boldsymbol{R}]_{i,i}$ are the local growth rates over a time span Δt of $\boldsymbol{U}$. The LEs are the time averages of the logarithms of the diagonal of $[\boldsymbol{R}]_{i,i}$, i.e.,

$$\lambda_i = \lim_{T\to\infty} \frac{1}{T}\int_{t_0}^{T} \ln[\boldsymbol{R}(t,\Delta t)]_{i,i}dt. \tag{3}$$

The FTLEs are defined as $\Lambda_i = \frac{1}{\Delta t}\ln[\boldsymbol{R}]_{i,i}$, which quantify the expansion and contraction rates of the tangent space on finite-time intervals, $\Delta t = t_2 - t_1$.

2.1 Time-Delayed Thermoacoustic System

As a practical application of time-delayed systems, we consider a thermoacoustic system, which is composed of three interacting subsystems, the acoustics, the flame and the hydrodynamics (see, e.g., [10]). The interaction of these sub systems can result in a positive feedback loop, which manifests itself as a thermoacoustic instability. If uncontrolled, this instability can lead to structural failure. Typical paradigms from engineering include gas-turbine and rocket-motor engines, and their configuration requires design optimization to prevent instabilities [10]. We consider a prototypical time-delayed thermoacoustic system with a longitudinal acoustic cavity and a heat source modelled with a time-delayed model, following the same setup as in [5,6,11]. The system is governed by the conservation of momentum, mass, and energy. Upon re-arrangement [10], thermoacoustic dynamics are governed by the nondimensional partial differential equations

$$\frac{\partial u}{\partial t}+\frac{\partial p}{\partial x}=0, \qquad \frac{\partial p}{\partial t}+\frac{\partial u}{\partial x}+\zeta p-\dot{q}\delta(x-x_f)=0, \tag{4}$$

where u is the non-dimensional velocity, p the pressure, $\dot{q}$ the heat-release rate, $x\in[0,1]$ the axial coordinate and t the time; ζ is the damping coefficient, which takes into account all the acoustic dissipation. The heat source is assumed to be small compared to the acoustic wavelength, and it is modelled as a point in the grid, via the Dirac delta distribution $\delta(x-x_f)$, located at $x_f=0.2$. The heat-release rate is provided by a modified King's law, $\dot{q}(t)=\beta\left(\sqrt{|1+u_f(t-\tau)|}-1\right)$, which is a nonlinear time-delayed model. For the numerical studies of this paper, we set $\beta=7.0$ for the heat parameter, and $\tau=0.2$ for the time delay. Those values ensure chaotic evolution, and encapsulate all information about the heat source, base velocity and ambient conditions.

As in [5], we transform the time-delayed problem into an initial value problem. This is mathematically achieved by modelling the advection of a perturbation v with velocity τ^{-1} as

$$\frac{\partial v}{\partial t}+\frac{1}{\tau}\frac{\partial v}{\partial X}=0, \quad 0\le X\le 1, \quad v(X=0,t)=u_f(t). \tag{5}$$

We discretise Eqs. (4) by a Galerkin method. First, we separate the acoustic variables in time and space as $u(x,t)=\sum_{j=1}^{N_g}\eta_j(t)\cos(j\pi x)$, and $p(x,t)=-\sum_{j=1}^{N_g}\mu_j(t)\sin(j\pi x)$, in which the spatial functions are the acoustic eigenfunctions of the configuration under investigation. Then, we project Eqs. (4) onto the Galerkin spatial basis $\{\cos(\pi x),\cos(2\pi x),\ldots,\cos(N_g\pi x)\}$ to obtain

$$\dot{\eta}_j-j\pi\mu_j=0, \qquad \dot{\mu}_j+j\pi\eta_j+\zeta_j\mu_j+2\dot{q}\sin(j\pi x_f)=0. \tag{6}$$

The system has $2N_g$ degrees of freedom. The time-delayed velocity becomes $u_f(t-\tau)=\sum_{k=1}^{N_g}\eta_k(t-\tau)\cos(k\pi x_f)$, and the damping, ζ_j, is modelled by $\zeta_j=c_1 j^2+c_2 j^{1/2}$, where $c_1=0.1$ and $c_2=0.06$. The equation for linear advection, Eq. (5), is discretised using N_c+1 points with a Chebyshev spectral method. This discretisation adds N_c degrees of freedom, thus a total of $D=2N_g+N_c=30$ in our case, as $N_g=N_c=10$. We integrate Eqs. (6) with a fourth order Runge-Kutta scheme and timestep $dt=0.01$.

3 Echo State Network

By applying the method of [12] to time-delayed problems, we linearize the Echo State Network (ESN) [7] to calculate the stability properties of chaotic systems. ESNs are proven effective for accurate learning of chaotic dynamics (see e.g. [1,4,12–14,17]).

The ESN is a reservoir computer [7]. It has a sparsely-connected single-layer hidden state, which is termed "reservoir". The reservoir weights, $\mathbf{W}$, as well as the input-to-reservoir weights, $\mathbf{W}_{\text{in}}$, are randomly assigned and remain fixed through training and testing. The reservoir-to-output weights, $\mathbf{W}_{\text{out}}$, are trained via ridge regression. The evolution equations of the reservoir and output are, respectively

$$\boldsymbol{r}(\mathrm{t}_{i+1}) = \tanh\left([\hat{\mathbf{y}}_{\text{in}}(\mathrm{t}_i); b_{\text{in}}]^T \mathbf{W}_{\text{in}} + \boldsymbol{r}(\mathrm{t}_i)^T \mathbf{W}\right), \qquad \mathbf{y}_{\text{p}}(\mathrm{t}_{i+1}) = [\boldsymbol{r}(\mathrm{t}_{i+1}); 1]^T \mathbf{W}_{\text{out}}, \quad (7)$$

where at any discrete time t_i the input vector, $\mathbf{y}_{\text{in}}(\mathrm{t}_i) \in \mathbb{R}^{N_y}$, is mapped into the reservoir state $\boldsymbol{r} \in \mathbb{R}^{N_r}$, by the input matrix, $\mathbf{W}_{\text{in}}$, where $N_r \gg N_y$ [7,14]. Here, $(\hat{\ })$ indicates normalization by the component-wise maximum-minus-minimum range of the target in training set, ^{T} indicates matrix transposition, and the semicolon indicates array concatenation. The dimensions of the weight matrices are $\mathbf{W}_{\text{in}} \in \mathbb{R}^{(N_y+1)\times N_r}$, $\mathbf{W} \in \mathbb{R}^{N_r \times N_r}$ and $\mathbf{W}_{\text{out}} \in \mathbb{R}^{(N_r+1)\times N_y}$. The hyperparameter input bias, $b_{\text{in}} = 1$, is selected to have the same order of magnitude as the normalized inputs, $\hat{\mathbf{y}}_{\text{in}}$. The dimensions of the input and output vectors are equal to the dimension of the dynamical system; here described by Eqs. (6), i.e. $N_y \equiv D$. Furthermore, $\mathbf{W}_{\text{out}}$ is trained via the minimization of the mean square error $\text{MSE} = \frac{1}{N_{\text{tr}} N_y} \sum_{i=0}^{N_{\text{tr}}} ||\mathbf{y}_{\text{p}}(\mathrm{t}_i) - \mathbf{y}_{\text{in}}(\mathrm{t}_i)||^2$ between the outputs and the data over the training set, where $||\cdot||$ is the L_2 norm, $N_{\text{tr}} + 1$ is the total number of data in the training set, and $\mathbf{y}_{\text{in}}$ the input data on which the ESN is trained.

Training the ESN is performed by solving with respect to $\mathbf{W}_{\text{out}}$ via ridge regression of the equation $(\mathbf{R}\mathbf{R}^T + \beta\mathbb{I})\mathbf{W}_{\text{out}} = \mathbf{R}\mathbf{Y}_{\text{d}}^T$. In the previous expression $\mathbf{R} \in \mathbb{R}^{(N_r+1)\times N_{\text{tr}}}$ and $\mathbf{Y}_{\text{d}} \in \mathbb{R}^{N_y \times N_{\text{tr}}}$ are the horizontal concatenation of the reservoir states with bias, $[\boldsymbol{r}(\mathrm{t}_i); 1]$, $\mathrm{t}_i \in [0, T_{\text{train}}]$, and of the output data, respectively; $\mathbb{I}$ is the identity matrix and β is the Tikhonov regularization parameter [16]. Therefore the ESN does not require backpropagation. The ESN can run in two configurations, either open-loop or closed-loop. In open-loop, which is necessary for the training stage, the input data is given at each step, allowing for the calculation of the reservoir timeseries $\boldsymbol{r}(\mathrm{t}_i)$, $\mathrm{t}_i \in [0, T_{\text{train}}]$. In closed-loop the output $\mathbf{y}_{\text{p}}$ at time step t_i, is recurrently used as an input at time step t_{i+1}, allowing for the autonomous temporal evolution of the network. The closed-loop configuration is used for validation (i.e. hyperparameter tuning) and testing, but not for training.

Regarding validation, we use the chaotic recycle validation (RVC), as introduced in [14]. It has proven to be a robust strategy, providing enhanced performance of the ESN, compared to standard strategies, as recently successfully applied in [12,15]. Briefly, in RVC the network is trained only once on the entire training dataset (in open-loop), and validation is performed on multiple intervals already used for training (but now in closed-loop). The validation interval simply shifts as a small multiple of the first Lyapunov exponent, $N_{\text{val}} = 3\lambda_1$ here. The key hyperparameters that we tune are the input scaling σ_{in} of the input matrix $\mathbf{W}_{\text{in}}$, the spectral radius ρ of the matrix $\mathbf{W}$, and the Tikhonov parameter β. Furthermore, σ_{in} and ρ are tuned via Bayesian Optimization in the hyperparameter space $[\sigma_{\text{in}}, \rho] = [-1, 1] \times [0.1, 1.4]$, while for β we perform a grid search $\{10^{-6}, 10^{-8}, 10^{-10}, 10^{-12}\}$ within the optimal $[\sigma_{\text{in}}, \rho]$. The reservoir size is $N_r = 300$. The connectivity of matrix $\mathbf{W}$ is set to $d = 80$. We further add to the training and validation data a Gaussian noise with zero mean and standard deviation, $\sigma_n = 0.0006\sigma_y$, where σ_y is the standard deviation of the data component-wise (noise regularizes the problem, see [12,14,17] for more details). The ESN is trained on a training set (open-loop) of size $500\tau_\lambda$, and is tested on a test set (closed-loop) of size $4000\tau_\lambda$, where $\tau_\lambda = 1/\lambda_1$ is the Lyapunov time, which is the inverse of the maximal Lyapunov exponent $\lambda_1 \approx 0.13$ in our case.

4 Results

We analyse the statistics produced by the autonomous temporal evolution of the ESN and the target time-delayed system. The selected observables are the statistics of the system's chaotic variables, the Lyapunov exponents (LEs), and the statistics of the finite-time Lyapunov exponents (FTLEs).

First, we test the capabilities of the ESN to learn the long-term statistical properties of the thermoacoustic system by measuring the probability density function (PDF) of the learned variables, $\mathbf{y}$. In Fig. 1, we show the PDF of the first three components of the Galerkin modes (see Eq. (6)) η_i, μ_i for $i = 1,2,3$, in which the black line corresponds to the target and the dashed red to the ESN. The ESN predictions are in agreement with the target, including the variables that are not shown here.

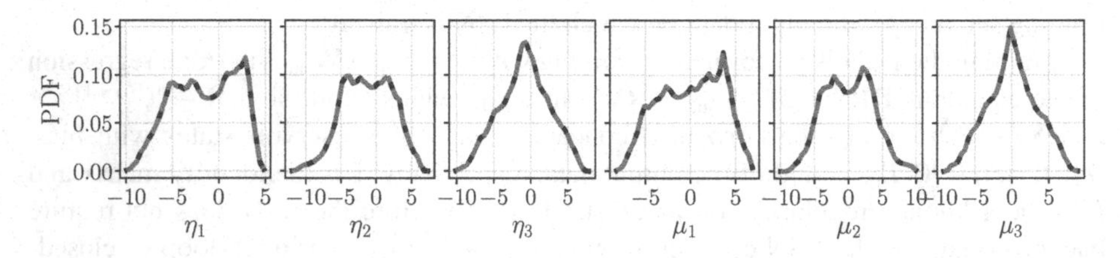

Fig. 1. Probability density functions (via histograms) of the three first Galerkin modes, η_i, μ_i, $i = 1,2,3$. Black is for target and dashed red for ESN. The statistics for ESN are collected in autonomous evolution on unseen data, after training and validation.

Testing the accuracy of the calculated LEs in autonomous evolution is a harder consistency check for the ESN. Indeed, the ESN has been trained only on timeseries of the variables η_i, μ_i, and v. Therefore, a good agreement of the LEs means that the ESN is capable to accurately reproduce intrinsic chaotic properties of the system's attractor. The LEs of the ESN are calculated following [12], and in Fig. 2 we compare the first $K = 14$ LEs. We also add an inner plot showing, in additional detail, the first 6 LEs. Each LE is the average of the measured LEs from seven selected independent ESNs used for the analysis. We train the ESNs independently on different chaotic target sets. The shaded region corresponds to the standard deviation per λ_i from those seven ESNs. There is close agreement for the first 8 exponents. In particular, we measure the leading, and only positive, exponent $\lambda_1^{\text{targ}} \approx 0.130$ and $\lambda_1^{\text{ESN}} \approx 0.127$ for ESN, which gives a 2.3% absolute error. The ESN also provides an accurate estimate of the neutral exponent ($\lambda_2 = 0$) with $\lambda_2^{\text{ESN}} \approx 0.008$. The rest of the exponents, $\lambda_i, i \geq 3$, are negative and the ESN achieves a small 5.4% mean absolute percentage error for all. Note that a gradual disagreement of the negative exponents, which are sensitive due to the numerical method, between ESN and target has also been reported in [13,17] for the one-dimensional Kuramoto-Sivashinsky equation.

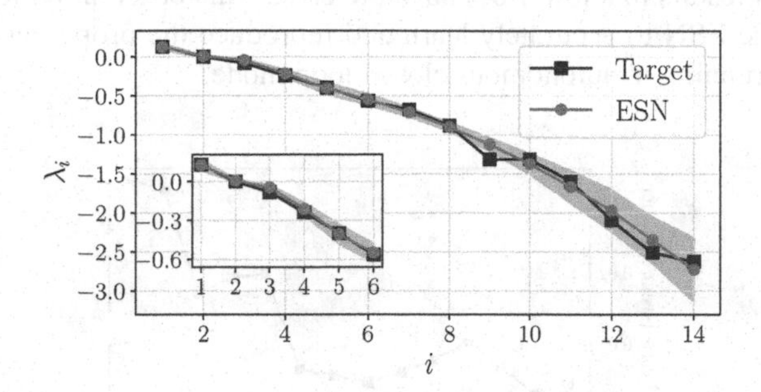

Fig. 2. The first 14 (outer plot) and the first 6 (inner plot) Lyapunov exponents. The shaded red region indicates the error based on the ensemble of 5 ESNs. The statistics for ESN are collected in closed-loop mode. (Color figure online)

Fig. 3 shows the PDF of the first 14 FTLEs of the target (black line) and ESN (dashed red line). We collect the statistics from 7 independently trained ESNs, thus creating 7 histograms of FTLEs. We then average those histograms bin-wise, and the standard deviation of each averaged bin is given by the shaded regions. Note that the mean of each PDF should correspond to each Lyapunov exponent λ_i (i.e. of Fig. 2), which is indeed the case. In Fig. 4, we quantify the difference of the estimated FTLE statistics by computing the Wasserstein distance between the two distributions for each Λ_i. The difference is small. We observe a gradual increase in Λ_i, $i \geq 10$, which is in agreement with the trend of the Lyapunov exponents in Fig. 2.

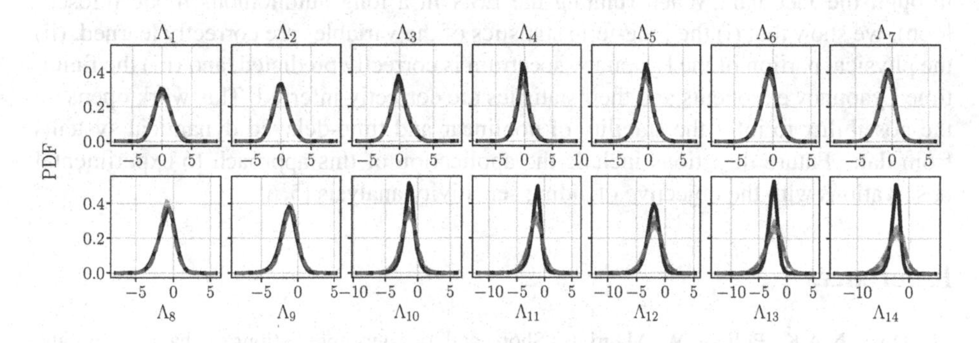

Fig. 3. Probability density functions of the first 14 finite-time Lyapunov exponents. Black is for target and dashed red for ESN. (Color figure online)

We also report the values for the Kaplan-Yorke dimension for both ESN and target. This dimension is an upper bound of the attractor's fractal dimension [2]. It is given by

$D_{\mathrm{KY}} = k + \frac{\sum_{i=1}^{k} \lambda_i}{|\lambda_{i+1}|}$, where k is such that the sum of the first k LEs is positive and the sum of the first $k+1$ LEs is negative. We obtain $D_{\mathrm{KY}}^{\mathrm{targ}} \approx 3.37$ for target and $D_{\mathrm{KY}}^{\mathrm{ESN}} \approx 3.43$ for ESN, which results in a low 1.8% absolute error. This observation further confirms the ability of the ESN to accurately learn and reproduce the properties of the chaotic attractor when running in autonomous closed loop mode.

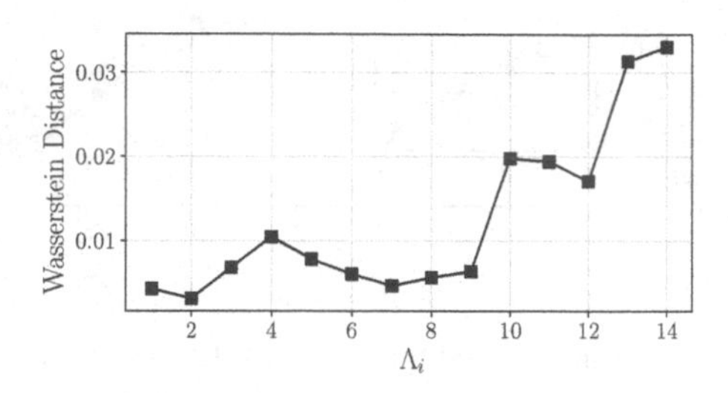

Fig. 4. Wasserstein distance between the PDFs of target versus ESN of the first 14 finite-time Lyapunov exponents (see Fig. 3).

5 Conclusion

We propose a method to compute the stability properties of chaotic time-delayed systems using only data. We use the echo state network (ESN) as a surrogate model for learning the chaotic dynamics from time series observations of the system and calculate its long-term statistical and stability properties. Viewing the ESN as a discrete dynamical system, we linearize the map (7) to derive the tangent evolution of the attractor through the Jacobian. When running the ESN in a long autonomous mode (closed-loop), we show that (i) the long-term statistics of the variables are correctly learned, (ii) the physical portion of the Lyapunov spectrum is correctly predicted, and (iii) the finite-time Lyapunov exponents and their statistics are correctly inferred. This work opens up the possibility to infer the stability of nonlinear and time-delayed dynamical systems from data. Future directions include the application of this approach to experimental observations with the objective of aiding sensitivity analysis [5,6].

References

1. Doan, N.A.K., Polifke, W., Magri, L.: Short- and long-term predictions of chaotic flows and extreme events: a physics-constrained reservoir computing approach. Proc. R. Soc. A Math. Phys. Eng. Sci. **477**(2253), 20210135 (2021). https://doi.org/10.1098/rspa.2021.0135
2. Eckmann, J.P., Ruelle, D.: Ergodic theory of chaos and strange attractors. Rev. Mod. Phys. **57**, 617–656 (1985). https://doi.org/10.1103/RevModPhys.57.617
3. Goodfellow, I., Bengio, Y., Courville, A.: Deep Learning. MIT Press, Cambridge (2016). http://www.deeplearningbook.org

4. Huhn, F., Magri, L.: Learning ergodic averages in chaotic systems. In: Krzhizhanovskaya, V.V., Závodszky, G., Lees, M.H., Dongarra, J.J., Sloot, P.M.A., Brissos, S., Teixeira, J. (eds.) ICCS 2020. LNCS, vol. 12142, pp. 124–132. Springer, Cham (2020). https://doi.org/10.1007/978-3-030-50433-5_10
5. Huhn, F., Magri, L.: Stability, sensitivity and optimisation of chaotic acoustic oscillations. J. Fluid Mech. **882**, A24 (2020). https://doi.org/10.1017/jfm.2019.828
6. Huhn, F., Magri, L.: Gradient-free optimization of chaotic acoustics with reservoir computing. Phys. Rev. Fluids **7**, 014402 (J2022). https://doi.org/10.1103/PhysRevFluids.7.014402
7. Jaeger, H., Haas, H.: Harnessing nonlinearity: predicting chaotic systems and saving energy in wireless communication. Science **304**(5667), 78–80 (2004). https://doi.org/10.1126/science.1091277
8. Lakshmanan, M., Senthilkumar, D.V.: Dynamics of Nonlinear Time-Delay Systems. Springer Science & Business Media, Berlin, Heidelberg (2011). https://doi.org/10.1007/978-3-642-14938-2
9. Lukoševičius, M.: A Practical Guide to Applying Echo State Networks, pp. 659–686. Springer, Berlin, Heidelberg (2012). https://doi.org/10.1007/978-3-642-35289-8_36
10. Magri, L.: Adjoint methods as design tools in thermoacoustics. Appl. Mech. Rev. **71**(2), 020801 (2019). https://doi.org/10.1115/1.4042821
11. Magri, L., Juniper, M.P.: Sensitivity analysis of a time-delayed thermo-acoustic system via an adjoint-based approach. J. Fluid Mech. **719**, 183–202 (2013). https://doi.org/10.1017/jfm.2012.639
12. Margazoglou, G., Magri, L.: Stability analysis of chaotic systems from data. Nonlinear Dyn. **111**, 8799–8819 (2023). https://doi.org/10.1007/s11071-023-08285-1
13. Pathak, J., Lu, Z., Hunt, B.R., Girvan, M., Ott, E.: Using machine learning to replicate chaotic attractors and calculate lyapunov exponents from data. Chaos Interdiscip. J. Nonlinear Sci. **27**(12), 121102 (2017). https://doi.org/10.1063/1.5010300
14. Racca, A., Magri, L.: Robust optimization and validation of echo state networks for learning chaotic dynamics. Neural Netw. **142**, 252–268 (2021). https://doi.org/10.1016/j.neunet.2021.05.004
15. Racca, A., Magri, L.: Data-driven prediction and control of extreme events in a chaotic flow. Phys. Rev. Fluids **7**, 104402 (2022). https://doi.org/10.1103/PhysRevFluids.7.104402
16. Tikhonov, A.N., Goncharsky, A., Stepanov, V., Yagola, A.G.: Numerical Methods for the Solution of Ill-posed Problems, vol. 328. Springer Science & Business Media, Berlin, Heidelberg (1995). https://doi.org/10.1007/978-94-015-8480-7
17. Vlachas, P., et al.: Backpropagation algorithms and reservoir computing in recurrent neural networks for the forecasting of complex spatiotemporal dynamics. Neural Netw. **126**, 191–217 (2020). https://doi.org/10.1016/j.neunet.2020.02.016

Learning 4DVAR Inversion Directly from Observations

Arthur Filoche[1], Julien Brajard[2], Anastase Charantonis[3], and Dominique Béréziat[1](✉)

[1] Sorbonne Université, CNRS, LIP6, Paris, France
dominique.bereziat@lip6.fr
[2] NERSC, Bergen, Norway
[3] ENSIIE, CNRS, LAMME, Evry, France

Abstract. Variational data assimilation and deep learning share many algorithmic aspects. While the former focuses on system state estimation, the latter provides great inductive biases to learn complex relationships. We here design a hybrid architecture learning the assimilation task directly from partial and noisy observations, using the mechanistic constraint of the 4DVAR algorithm. Finally, we show in an experiment that the proposed method was able to learn the desired inversion with interesting regularizing properties and that it also has computational interests.

Keywords: Data Assimilation · Unsupervised Inversion · Differentiable Physics

1 Introduction

Data Assimilation [5] is a set of statistical methods solving particular inverse problems, involving a dynamical model and imperfect data obtained through an observation process, with the objective to estimate a considered system state. It produces state-of-the-art results in various numerical weather prediction tasks and is mostly used in operational meteorological centers.

Although they are not initially designed for the same purpose, variational data assimilation [11] and deep learning share many algorithmic aspects [1]. It has already been argued that both methods can benefit from each other [10,17]. Data assimilation provides a proper Bayesian framework to combine sparse and noisy data with physics-based knowledge while deep learning can leverage a collection of data extracting complex relationships from it. Hybrid methods have already been developed either to correct model error [6,8], to jointly estimate parameters and system state [4,14] or to fasten the assimilation process [20]. Most of these algorithms rely on iterative optimization schemes alternating data assimilation and machine learning steps.

In this work we design a hybrid architecture bridging a neural network and a mechanistic model to directly learn system state estimation from a collection

J. Mikyška et al. (Eds.): ICCS 2023, LNCS 14076, pp. 414–421, 2023.
https://doi.org/10.1007/978-3-031-36027-5_32

of partial and noisy observations. We optimize it in only one step still using the variational assimilation loss function. Finally, We show in an experiment using the chaotic Lorenz96 dynamical system, that the proposed method is able to learn the variational data assimilation with desirable regularizing properties, then providing a computationally efficient inversion operator.

2 Related Work

Hybridizing Data Assimilation with Machine Learning. While deep learning has proven to be extremely useful for a variety of inverse problems where the ground truth is available, unsupervised inversion is still being investigated [15]. For instance, when data are highly-sparse, neural architectures may be hard to train. On the other hand, data assimilation can provide dense data. From this statement, approaches have naturally emerged in the data assimilation community, iterating data assimilation steps and machine learning steps for simultaneous state and parameters estimation [4,14]. But end-to-end learning approaches are also investigated, in [7] the architecture is constrained to internally behave like a 4DVAR pushing the hybridization further.

Mechanistically Constrained Neural Networks. Variational data assimilation has a pioneering expertise in PDE-constrained optimization [11], making use of automatic differentiation to retro-propagate gradients through the dynamical system. In [3,13] the output of a neural network is used as input in a dynamical model, and architectures are trained with such gradients, in a supervised and adversarial manner, respectively. Similar methods have been used to learn accurate numerical simulations still using differentiable mechanistic models [18,19]. Also, Physically-consistent architectures are developed to enforce the conservation of desired quantity by neural architectures [2].

3 Data Assimilation and Learning Framework

3.1 State-Space System

A system state $\mathbf{X}_t$ evolves over time according to a considered perfectly known dynamics $\mathbb{M}_t$ and observations $\mathbf{Y}_t$ are obtained through an observation operator $\mathbb{H}$ up to an additive noise ε_{R_t}, as described in Eqs. 1 and 2,

$$\text{Dynamics:} \qquad \mathbf{X}_{t+1} = \mathbb{M}_t(\mathbf{X}_t) \tag{1}$$

$$\text{Observation:} \qquad \mathbf{Y}_t = \mathbb{H}_t(\mathbf{X}_t) + \varepsilon_{R_t} \tag{2}$$

We denote the trajectory $\mathbf{X} = [\mathbf{X}_0, \ldots, \mathbf{X}_T]$, a sequence of state vectors over a temporal window, and $\mathbf{Y}$ the associated observations. The objective of data assimilation is to provide an estimation of the posterior probability $p(\mathbf{X} \mid \mathbf{Y})$ leveraging the information about the mechanistic model $\mathbb{M}$. The estimation can later be used to produce a forecast.

3.2 The Initial Value Inverse Problem

When considering the dynamics perfect, the whole trajectory only depends on the initial state $\mathbf{X}_0$, the assimilation is then said with strong-constraint. The whole process to be inverted is summed up in the simple Eq. 3, where $\mathcal{F}$ is the forward model, combining $\mathbb{M}_t$ and $\mathbb{H}_t$. More precisely, by denoting multiple model integrations between two times $\mathbb{M}_{t_1 \to t_2}$, we can rewrite the observation equation as in Eq. 4.

$$\mathbf{Y} = \mathcal{F}(\mathbf{X}_0) + \varepsilon_R \tag{3}$$

$$\mathbf{Y}_t = \mathbb{H}_t \circ \mathbb{M}_{0 \to t}(\mathbf{X}_0) + \varepsilon_{R_t} \tag{4}$$

The desired Bayesian estimation now requires a likelihood model $p(\mathbf{X} \mid \mathbf{Y})$ and a prior model $p(\mathbf{X}) = p(\mathbf{X}_0)$. We assume the observation errors uncorrelated in time so that $p(\mathbf{X} \mid \mathbf{Y}) = \prod_t p(\varepsilon_{R_t})$ and we here make no particular assumption on $\mathbf{X}_0$ corresponding to a uniform prior.

3.3 Variational Assimilation with 4DVAR

The solve this problem in a variational manner, it is convenient to also assume white and Gaussian observational errors ε_{R_t}, of known covariance matrices $\mathbf{R}_t$, leading to the least-squares formulation given in Eqs. 5, where $\|\varepsilon_{R_t}\|^2_{R_t}$ stands for the Mahalanobis distance associated with the matrix $\mathbf{R}_t$. The associated loss function is denoted $\mathcal{J}_{4DV}$, Eq. 6, and minimizing it corresponds to a maximum a posteriori estimation, here equivalent to a maximum likelihood estimation.

$$-\log p(\mathbf{X} \mid \mathbf{Y}) = \frac{1}{2} \sum_{t=0}^{T} \|\varepsilon_{R_t}\|^2_{\mathbf{R}_t} - \log K \quad \text{s.t.} \quad \mathbb{M}(\mathbf{X}_t) = \mathbf{X}_{t+1} \tag{5}$$

$$\mathcal{J}_{4DV}(\mathbf{X}_0) = \frac{1}{2} \sum_{t=0}^{T} \|\mathbb{H}_t \circ \mathbb{M}_{0 \to t}(\mathbf{X}_0) - \mathbf{Y}_t\|^2_{\mathbf{R}_t} \tag{6}$$

This optimization is an optimal control problem where the initial state $\mathbf{X}_0$ plays the role of control parameters. Using the adjoint state method, we can derive an analytical expression of $\nabla_{\mathbf{X}_0} \mathcal{J}_{4DV}$ as in Eq. 7. It is worth noting that the mechanism at stake here is equivalent to the back-propagation algorithm used to train neural networks. The algorithm associated with this optimization is named 4DVAR.

$$\nabla_{\mathbf{X}_0} \mathcal{J}_{4DV}(\mathbf{X}_0) = \sum_{t=0}^{T} \left[\frac{\partial(\mathbb{H}_t \circ \mathbb{M}_{0 \to t})}{\partial \mathbf{X}_0} \right]^{\top} \mathbf{R}_t^{-1} \varepsilon_{R_t} \tag{7}$$

3.4 Learning Inversion Directly from Observations

We now consider independent and identically distributed trajectories denoted and the dataset of observations $\mathcal{D} = \{\mathbf{Y}^{(i)}, \mathbf{R}^{-1(i)}\}_{i=1}^{N}$. The associated ground truth $\mathcal{T} = \{\mathbf{X}^{(i)}\}_{i=1}^{N}$ is not available so the supervised setting is not an option. The posteriors for each trajectory are then also independent as developed in this equation: $\log p(\mathcal{T} \mid \mathcal{D}) = \sum_{i=0}^{N} \log p(\mathbf{X}^{(i)} \mid \mathbf{Y}^{(i)})$.

Our objective is to learn a parameterized pseudo-inverse $\mathcal{F}_{\theta}^{\star} : (\mathbf{Y}, \mathbf{R}^{-1}) \mapsto \mathbf{X}_0$ that should output initial condition from observations and associated errors covariance, which is exactly the task solved by 4DVAR. Such modeling choice corresponds to the prior $p(\mathbf{X}_0) = \delta(\mathbf{X}_0 - \mathcal{F}_{\theta}^{\star}(\mathbf{Y}, \mathbf{R}^{-1}))$, as we do not use additional regularization, where δ is the Dirac measure.

To learn the new control parameters $\boldsymbol{\theta}$, we leverage the knowledge of the dynamical model $\mathbb{M}$ as in 4DVAR. After outputting the initial condition $\mathbf{X}_0$ we forward it with the dynamical model and then calculate the observational loss. A schematic view of the performed integration is drawn in Fig. 1.

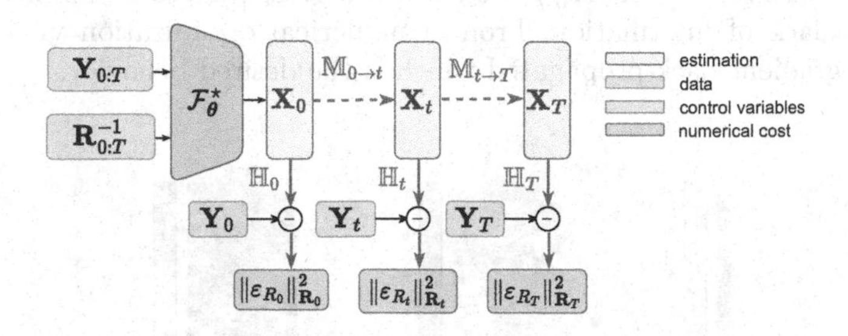

Fig. 1. Schematic view of the hybrid architecture learning the 4DVAR inversion

Then the cost function associated with the MAP estimation can be developed as in Eq. 8. A simple way of thinking it is to run multiple 4DVAR in parallel to optimize a common set of control parameters $\boldsymbol{\theta}$.

$$\mathcal{J}(\boldsymbol{\theta}) = \sum_{\mathcal{D}} \mathcal{J}_{4DV}(\mathbf{X}_0^{(i)}) \quad \text{s.t. } \mathcal{F}_{\boldsymbol{\theta}}(\mathbf{Y}^{(i)}, \mathbf{R}^{-1(i)}) = \mathbf{X}_0^{(i)} \tag{8}$$

To calculate $\nabla_{\boldsymbol{\theta}} \mathcal{J}$ we simply use the linearity of the gradient then the chain rule (Eq. 9) and finally we can re-use $\nabla_{\mathbf{X}_0} \mathcal{J}_{4DV}$ calculated before (Eq. 7). Gradients are back-propagated through the dynamical model first and then through the parameterized pseudo-inverse. Calculating the gradient on the whole dataset at each iteration may be computationally too expensive so one could instead use mini-batch gradient descent.

$$\nabla_{\boldsymbol{\theta}} \mathcal{J} = \sum_{\mathcal{D}} \nabla_{\boldsymbol{\theta}} \mathcal{J}_{4DV} = \sum_{\mathcal{D}} \nabla_{\mathbf{X}_0} \mathcal{J}_{4DV} \nabla_{\boldsymbol{\theta}} \mathbf{X}_0 = \sum_{\mathcal{D}} \nabla_{\mathbf{X}_0} \mathcal{J}_{4DV} \nabla_{\boldsymbol{\theta}} \mathcal{F}_{\boldsymbol{\theta}}^{\star} \tag{9}$$

4 Experiments and Results

4.1 Lorenz96 Dynamics and Observations

We use the Lorenz96 dynamics [12] as an evolution model Lorenz96, $\frac{d\mathbf{X}_{t,n}}{dt} = (\mathbf{X}_{t,n+1} - \mathbf{X}_{t,n-2})\mathbf{X}_{t,n-1} - \mathbf{X}_{t,n} + F$, numerically integrated with a fourth-order

Runge Kutta scheme. Here n indexes a one-dimensional space. On the right-hand side, the first term corresponds to an advection, the second term represents damping and F is an external forcing. We use the parameters $dt = 0.1$ and $F = 8$ corresponding to a chaotic regime [9]. Starting from white noise and after integrating during a spin-up period to reach a stationary state, we generate ground truth trajectories.

To create associated observations, we use a randomized linear projector as an observation operator, making the observation sparse to finally add a white noise. Noises at each point in time and space can have different variances, $\varepsilon_{R_{n,t}} \sim \mathcal{N}(0, \sigma_{n,t})$, and we use the associated diagonal variance matrix defined by $\mathbf{R}_{n,t}^{-1} = \frac{1}{\sigma_{n,t}^2}$. Figure 2 displays an example of simulated observations. Variances are sampled uniformly such that $\sigma_{n,t} \sim \mathcal{U}(0.25, 1)$. When a point in the grid is not observed we fix "$\mathbf{R}_{n,t}^{-1} = 0$", which corresponds to an infinite variance meaning a lack of information. From a numerical optimization view, no cost means no gradient back-propagated which is the desired behavior.

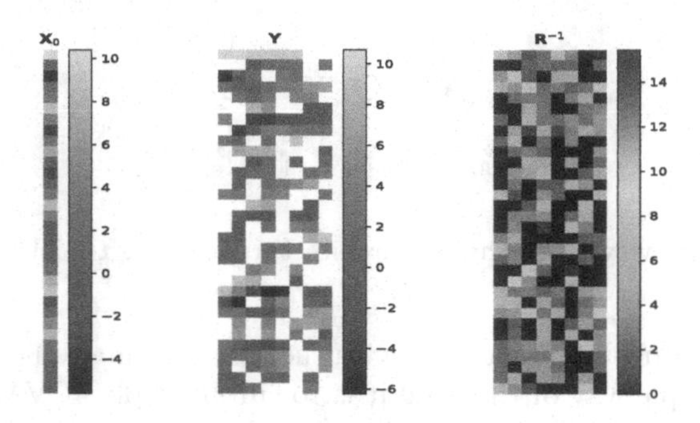

Fig. 2. Observation generated with the Lorenz96 model, a randomized linear projector as observation operator, and a white noise.

4.2 Algorithm Benchmarks

We evaluate our method (NN-4DVAR-e2e) on the assimilation task which is estimating $\mathbf{X}_0$. We compare it with a 4DVAR, a 4DVAR with additional $\mathcal{L}_2$ regularization (4DVAR-B), a neural network trained on the output of both 4DVAR estimations (NN-4DVAR-iter and NN-4DVAR-B-iter), and a neural network trained with the ground truth (NN-perfect). The latter should represent the best-case scenario for the chosen architecture while NN-4DVAR-iter plays the role of the iterative method. The same neural architecture is used for all the methods involving learning. Its design is fairly simple, being composed of 5 convolutional layers using 3×3 kernels, ReLu activation, no down-scaling, and a last layer flattening the two-dimensional maps into the shape of $\mathbf{X}_0$. We use 250, 50, and 250 samples for training, validation, and testing, respectively. When learning is involved, the Adam optimizer is used while 4DVAR is optimized with the L-BFGS solver. We notice here that once learned, both NN-4DVAR-iter and NN-4DVAR-e2e provide

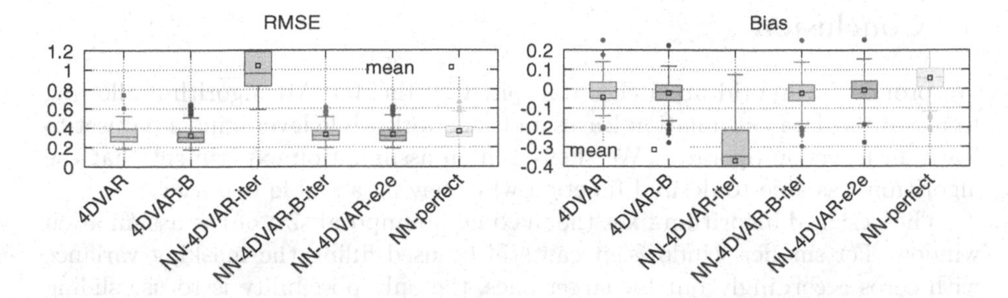

Fig. 3. Boxplot of assimilation accuracy of each algorithm, RMSE and Bias scores, on the 250 samples of the test set

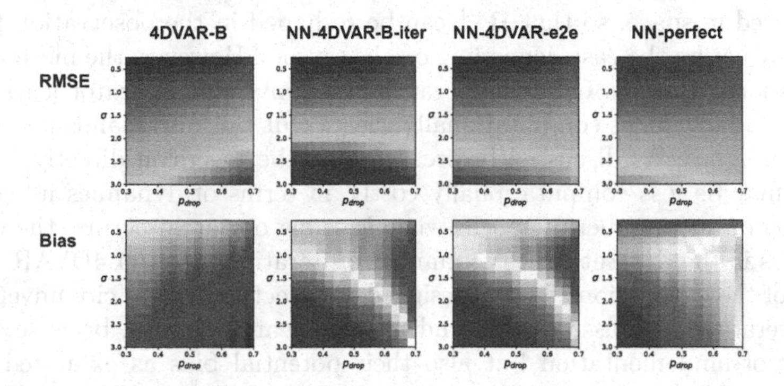

Fig. 4. Sensitivity of the assimilation regarding noise and sparsity levels (σ, p_{drop}). At each pixel, levels are constant and scores are averaged on 25 samples. $\sigma > 1$ not seen in training.

a computationally cheap inversion operator. For their learning, the computationally intensive step was the forward integration of the dynamical model. Denoting n_iter the number of iterations done in 4DVAR and n_epoch the number of epochs in our learning process, NN-4DVAR-iter, and 4DVAR-e2e cost $N \times n\_iter$ and $N \times n\_epoch$ dynamics integration, respectively. Depending on these parameters, one approach or the other will be less computationally intensive. In our case, we used $n\_iter < 150$ and $n\_epoch = 50$.

4.3 Results

The accuracy of the $\mathbf{X}_0$ estimation on the test set is quantified using the RMSE and the average bias (see Fig. 3). We notice first that when 4DVAR is not regularized, some samples induce bad estimations which disturb 4DVAR-NN-iter learning over them. The others methods involving produce RMSE scores on par with 4DVAR-B, the best estimator. However, our 4DVAR-NN-e2e is the less biased algorithm. It is to be noted that 4DVAR-NN-e2e has no additional regularization and still stays robust regarding difficult samples, highlighting desirable properties from the neural architecture.

5 Conclusion

We proposed a hybrid architecture inspired by the 4DVAR algorithm allowing to use of the data assimilation Bayesian framework while leveraging a dataset to learn an inversion operator. We showed in an assimilation experiment that the algorithm was able to desired function while having a stable behavior.

The designed algorithm fixes the maximum temporal size of the assimilation window. For smaller windows, it can still be used filling the masking variance with zeros accordingly but for larger ones, the only possibility is to use sliding windows, then raising to question of the coherence in time. Typically, the method in that form can not fit quasi-static strategies [16] employed in variational assimilation. Also, We made the convenient hypothesis that observational errors are uncorrelated in space, so that $\mathbf{R}^{-1}$ can be reshaped in the observation format, which may not be the case depending on the sensors. However, the method has a computational interest. Once the parameterized inversion operator learned, the inversion task becomes computationally cheap. But this also stands for the iterative approaches. As discussed before, learning the inversion directly with our method may be less computationally costly, in terms of dynamics integration, depending on the number of epochs when learning our architecture, the number of samples in the dataset, and the number of iterations used in 4DVAR.

One of the motivations for the designed architecture was to circumvent algorithms iterating data assimilation and machine learning steps, because of their difficulty of implementation but also their potential bias as exhibited in the experiment. However, we made the debatable, simplifying, perfect model hypothesis. Usually, the forward operator is only partially known and we ambition to develop the proposed framework further to relax such a hypothesis. In Fig. 4, we performed an accuracy sensitivity experiment regarding noise and sparsity levels. Particularly, we tested noise levels out of the dataset distribution. We see that learning-based approaches are more sensitive to noise increases while 4DVAR is more concerned by sparsity. Also, we notice that our NN-4DVAR-e2e methods generalize better than NN-4DVAR-B-iter to unseen levels of noise.

References

1. Abarbanel, H., Rozdeba, P., Shirman, S.: Machine learning: deepest learning as statistical data assimilation problems. Neural Comput. **30**(8), 2025–2055 (2018)
2. Beucler, T., Pritchard, M., Rasp, S., Ott, J., Baldi, P., Gentine, P.: Enforcing analytic constraints in neural networks emulating physical systems. Phys. Rev. Lett. **126**(9), 1079–7114 (2021)
3. de Bézenac, E., Pajot, A., Gallinari, P.: Deep learning for physical processes: incorporating prior scientific knowledge. J. Stat. Mech: Theory Exp. **2019**(12), 124009 (2019)
4. Bocquet, M., Brajard, J., Carrassi, A., Bertino, L.: Bayesian inference of chaotic dynamics by merging data assimilation, machine learning and expectation-maximization. Found. Data Sci. **2**(1), 55–80 (2020)

5. Carrassi, A., Bocquet, M., Bertino, L., Evensen, G.: Data assimilation in the geosciences: an overview of methods, issues, and perspectives. Wiley Interdisc. Rev.: Clim. Change **9**(5), e535 (2018)
6. Düben, P., et al.: Machine learning at ECMWF: A roadmap for the next 10 years. ECMWF Tech. Memoranda **878** (2021)
7. Fablet, R., Amar, M., Febvre, Q., Beauchamp, M., Chapron, B.: End-to-end physics-informed representation learning for satellite ocean remote sensing data: applications to satellite altimetry and sea surface currents. ISPRS **3**, 295–302 (2021)
8. Farchi, A., Laloyaux, P., Bonavita, M., Bocquet, M.: Using machine learning to correct model error in data assimilation and forecast applications. Q. J. R. Meteorol. Soc. **147**(739), 3067–3084 (2021)
9. Fertig, E., Harlim, J., Ramon, H.: A comparative study of 4D-VAR and a 4D ensemble Kalman filter: perfect model simulations with Lorenz-96. Tellus (2007)
10. Geer, A.: Learning earth system models from observations: machine learning or data assimilation? Philos. Trans. the Roy. Soc. A **379** (2021)
11. Le Dimet, F.X., Talagrand, O.: Variational algorithms for analysis and assimilation of meteorological observations: theoretical aspects. Tellus A **38**(10), 97 (1986)
12. Lorenz, E.: Predictability: a problem partly solved. In: Seminar on Predictability, vol. 1, pp. 1–18. ECMWF (1995)
13. Mosser, L., Dubrule, O., Blunt, M.: Stochastic seismic waveform inversion using generative adversarial networks as a geological prior. Math. Geosci. **52**, 53–79 (2018)
14. Nguyen, D., Ouala, S., Drumetz, L., Fablet, R.: Assimilation-based learning of chaotic dynamical systems from noisy and partial data. In: ICASSP (2020)
15. Ongie, G., Jalal, A., Metzler, C., Baraniuk, R., Dimakis, A., Willett, R.: Deep learning techniques for inverse problems in imaging. IEEE J. Sel. Areas Inf. Theory **1**(1), 39–56 (2020)
16. Pires, C., Vautard, R., Talagrand, O.: On extending the limits of variational assimilation in nonlinear chaotic systems. Tellus A **48**, 96–121 (1996)
17. Reichstein, M., Camps-Valls, G., Stevens, B., Jung, M., Denzler, J., Carvalhais, N.: Deep learning and process understanding for data-driven earth system science. Nature **566**(7743), 195–204 (2019)
18. Tompson, J., Schlachter, K., Sprechmann, P., Perlin, K.: Accelerating Eulerian fluid simulation with convolutional networks. In: ICML, pp. 5258–5267 (2017)
19. Um, K., Brand, R., Fei, Y., Holl, P., Thuerey, N.: Solver-in-the-loop: learning from differentiable physics to interact with iterative PDE-solvers. In: Advances in Neural Information Processing Systems (2020)
20. Wu, P., et al.: Fast data assimilation (FDA): data assimilation by machine learning for faster optimize model state. J. Comput. Sci. **51**, 101323 (2021)

Human-Sensors & Physics Aware Machine Learning for Wildfire Detection and Nowcasting

Jake Lever[1,2,3], Sibo Cheng[1,2], and Rossella Arcucci[1,2,3](✉)

[1] Data Science Institute, Imperial College London, London, UK
r.arcucci@imperial.ac.uk
[2] Leverhulme Centre for Wildfires, Environment and Society, London, UK
[3] Department of Earth Science and Engineering, Imperial College London, London, UK

Abstract. This paper proposes a wildfire prediction model, using machine learning, social media and geophysical data sources to predict wildfire instances and characteristics with high accuracy. We use social media as a predictor of wildfire ignition, and a machine learning based reduced order model as a fire spead predictor. We incorporate social media data into wildfire instance prediction and modelling, as well as leveraging reduced order modelling methods to accelerate wildfire prediction and subsequent disaster response effectiveness.

Keywords: Wildfires · Data Science · Machine Learning

1 Introduction

Real-time forecasting of wildfire dynamics has attracted increasing attention recently in fire safety science. Twitter and other social media platforms are increasingly being used as real-time human-sensor networks during natural disasters, detecting, tracking and documenting events [13]. Current wildfire models currently largely omit social media data, representing a shortcoming in these models, as valuable and timely information is transmitted via this channel [14]. Rather, these models use other data sources, mainly satellites, to track the ignition and subsequently model the progression of these events. This data can often be incomplete or take an infeasible amount of preprocessing time or computation. Subsequently, the computation of current wildfire models is extremely challenging due to the complexities of the physical models and the geographical features [8]. Running physics-based simulations for large-scale wildfires can be computationally difficult. The combination of these factors often makes wildfire modelling computationally infeasible, due to both the unavailability or delay in the data, and the computational complexity of the subsequent modelling.

We show that by including social data as a real-time data source, setting up a Twitter based human sensor network for just the first days of a massive wildfire

J. Mikyška et al. (Eds.): ICCS 2023, LNCS 14076, pp. 422–429, 2023.
https://doi.org/10.1007/978-3-031-36027-5_33

event can predict the ignition point to a high degree of accuracy. We also propose a novel algorithm scheme, which combines reduced-order modelling (ROM) and recurrent neural networks (RNN) for real-time forecasting/monitoring of the burned area. An operating cellular automata (CA) simulator is used to compute a data-driven surrogate model for forecasting fire diffusions. A long-short-term-memory (LSTM) neural network is used to build sequence-to-sequence predictions following the simulation results projected/encoded in a reduced-order latent space. This represents a fully coupled wildfire predictive & modelling framework. The performance of the proposed algorithm has been tested in a recent massive wildfire event in California - The Chimney Fire.

In using social data coupled with a fast and efficient model, we aim to help disaster managers make more informed, socially driven decisions. We implement machine learning in a wildfire prediction model, using social media and geophysical data sources to predict wildfire instances and characteristics with high accuracy. We demonstrate that social media is a predictor of wildfire ignition, and present aforementioned novel modelling methods which accurately simulate these attributes. This work contributes to the development of more socially conscious wildfire models, by incorporating social media data into wildfire instance prediction and modelling, as well as leveraging a ROM to accelerate wildfire prediction and subsequent disaster response effectiveness.

2 Background and Literature Review

Sentiment analysis has been used in several studies [3,4,16] to predict natural disasters. These studies analyse social media content to detect sentiments related to natural disasters and identify potential warnings or updates. [15] used Twitter data and sentiment analysis to identify tweets related to wildfires and classify them based on the level of urgency. The study utilized natural language processing techniques to extract features from the tweets, such as location, hashtags, and keywords, and trained a machine learning model to classify the tweets into categories such as warnings, updates, or irrelevant. In addition to sentiment analysis, some studies have also used other techniques such as image analysis and environmental data integration to improve the accuracy of wildfire detection [12,18]. For example, [23] used a combination of machine learning models and environmental data such as temperature and humidity to predict the occurrence and spread of wildfires accurately. In summary, sentiment analysis has shown promise to predict and detect wildfires. By analyzing social media content and identifying relevant sentiments, researchers can improve the efficiency and accuracy of real-time detection systems which is crucial for real-time wildfire nowcasting. Machine learning (ML)-based reduced order surrogate models [5,17] have become a popular tool in geoscience for efficiently simulating and predicting the behavior of complex physical systems. Advantages of these models include their ability to effectively capture the nonlinear relationships and high-dimensional input-output mappings in geoscience problems and their ability to operate in real-time [6]. These models can be easily combined with data assimilation techniques to perform real-time corrections [9]. Additionally, machine learning algorithms can be trained on large amounts of data, making it possible to effectively incorporate a wide range of observations and simulations into the modeling process [8]. ML-based ROMs have shown promise in the application of wildfire forecasting [8,24,25]. These models can effectively capture the complex relationships between inputs such as weather patterns, topography, and vegetation, and outputs such as fire spread and intensity. Advantages of these models include the ability to operate in real-time, the ability to effectively incorporate a wide range of observations and simulations, and the ability to incorporate the effects of time-varying variables such as wind speed and direction.

In this paper, we apply a surrogate model based on offline cellular automata (CA) simulations [2] in a given ecoregion. The physics-based CA model takes into account the impact of geophysical variables such as vegetation density and slope elevation on fire spread. The surrogate model consists of POD (Proper Orthogonal Decomposition) [19] (also known as Principle Component Analysis (PCA)) and LSTM (Long Short-Term Memory) neural network [20]. Wildfire spread dynamics are often chaotic and non-linear. In such cases, the use of advanced forecasting models, such as LSTM networks, can be highly beneficial in providing accurate and timely predictions of fire spread.

3 Methods

The pipeline of the proposed Human-Sensors & Physics aware machine learning framework is shown in Fig. 1. The main two component consist of the ignition point prediction using the Human Sensor network and the fire spread prediction using a ML based ROM.

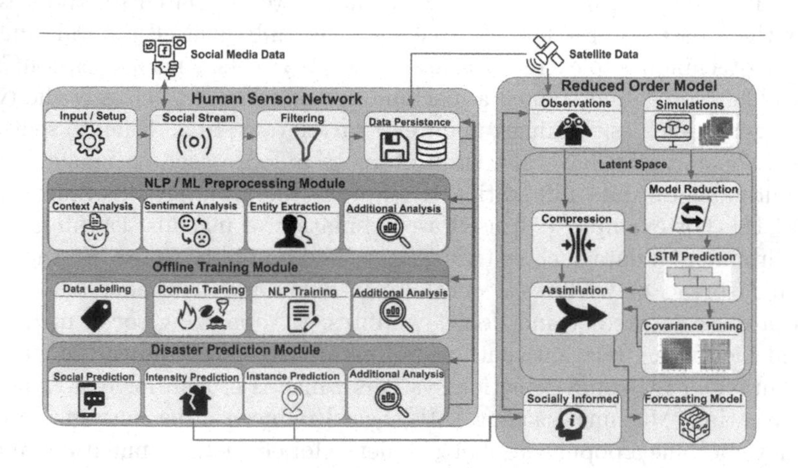

Fig. 1. Workflow of the proposed Human-Sensors & Physics aware Machine Learning framework

Ignition Point Prediction Using Social Media: The data used for the instance prediction stage of the coupled model is taken from Twitter, using an academic licence for the V2 API, allowing for a full archive search of all tweets from the period of the event. The twitter data queried was geotagged with a place name, which was subsequently geocoded to generate coordinates for the place of posting. The historical tweets were queried using the following query: '$(ChimneyORFireORWildfireOR\#ChimneyFire)place : Californiahas : geo'$, meaning the query was for the first keywords, posted in the region of California, and containing geolocation data. The query was run on a full historical search from public accounts from 12th-13th August 2016 - the first day of the wildfire event. For this period, in this location, 154 Tweets were downloaded and analysed. The resulting tweets then formed the dataset for the prediction. These tweets were analysed using a BERT transformer model for disaster classification, using the same methodology as used in [1]. This model analyses the text content of the message, and allocates disaster related labels based on the content of the post. Only posts with labels, i.e. classified by the model to be disaster related, were considered. Following this, the place name was extracted from the filtered Tweets via the Tweet metadata. The place name is given by a $place_id$, a hash code allocated by the API. Some of these also contained coordinates, but those that didn't were geocoded using the API to generate *lon* and *lat* coordinates for the tweet. Finally, all of the filtered, disaster related Tweets for the first day of the event had been allocated coordinates. Following this, named entity extraction was performed using google Entity Analysis API. This API uses NLP to extract named entities from a string of text. For this stage of the analysis, only entities with the type 'LOCATION' were extracted. These entities extracted were then geocoded using the Google Maps Geocoding API. Finally, the coordinate list for the tweet location and named entity location were averaged and used to make the final prediction for the wildfire ignition point. The averages were performed with a higher weighting on the locations extracted from the Entity Analysis, as these had been shown to be more indicative of accurate wildfire reporting.

Physics Aware Wildfire Nowcasting. The ignition point computed in the first part, is here used as input to predict the fire spread. The physics simulation is implemented using a CA model, a mathematical model used for simulating the behavior of complex systems [2,21]. In this study, the basic idea behind CA is to divide the landscape into a grid of cells, each of which can be in a certain state (e.g., unburned, burning, burned) [2,8]. At each time step, the state of a cell is updated based on the states of its neighboring cells and a set of rules that determine how fire spreads. To generate the training dataset of machine learning algorithms, we perform offline CA simulations with random ignition points in a given ecoregion. To do so, an ecoregion of dimension 10×10 km^2 (split equally to 128×128 cells for CA) is chosen. 40 fire events with random ignition points are generated using the state-of-the-art CA model [2]. We then train a ML model using the simulations as training data. The ML model can then be used for unseen fire events in the ecoregion. A reduced space is then computed implementing a POD, a mathematical method used in dynamical systems to reduce the complexity and dimensionality of large-scale systems. It represents the system's behavior using a reduced set of orthogonal modes. These modes are obtained through a singular value decomposition of the time-series data as shown in many applications [7,10,26]. In this study, POD is applied to perform dimension reduction for wildfire burned area, where 100 principle components are used to describe the whole system. More precisely, the temporal-spatial snapshots, obtained from offline CA simulations [2] are collected to compute the principle components of the burned area. These offline CA simulations are carried out with randomly selected ignition point in the ecoregion. This ensures the good generalizability of the proposed approach when dealing with unseen data. Finally a ML model is trained in the reduced space using a LSTM [20], a type of recurrent neural network commonly used in machine learning and deep learning applications. In a reduced latent space, LSTMs are used to model complex sequential data in a compact and computationally efficient manner [5,11,22]. After the dimension reduction, by working in the reduced latent space, the LSTM network is able to handle the high-dimensional input data. In this work,we train a sequence-to-sequence LSTM model for burned area forecasting. The model take 4 snapshots (equivalent of 24 h of fire spread) to predict the next 4 time steps of fire propagation. To form the training datasets for both POD and LSTM, 40 simulations with random ignition points are created. Each simulation is performed with 4 d of fire spread after ignition. Once the prediction in the lo.w-dimensional latent space is performed, the predicted latent vector can be decompressed to the full physical space. Forecasting the spread of a wildfire at the beginning of the fire is crucial because it helps decision-makers allocate resources effectively and prioritize evacuation plans.

4 Results: Wildfire Ignition and Spread Prediction

We tested the proposed approach on the Chimney wildfire event. The system has analysed twitted data as detailed in Sect. 3 and the ignition point result in (−119.72232658516701, 37.53677532750952) as (latitude, longtitude) coordinates. This point has been used as starting point for the fire spread prediction detailed later. The predicted fire spread of the first four days is illustrated in Fig. 2. The image background represent the associated vegetation distribution in the ecoregion where the green color refers to a high vegetation density. The dimension of the ecoregion is about $10 \times 10\,\text{km}^2$. As explained in Sect. 3, CA simulations are carried out offline *a priori* to generate training data for the ML surrogate model. The developed ML model then manages to handle unseen ignition scenarios with the closest ecoregion. It can be clearly observed from Fig. 2 that the wildfire prediction model exhibits a high level of performance, with a simulated fire spread that demonstrates a strong alignment with relevant geological features, namely the vegetation distribution. In fact, as described in Sect. 3, the ML model learns the fire-vegetation relationship from the CA simulation. It is clearly shown in Fig. 2 that the area with higher vegetation density will have a higher probability to be burned out. In this study, we focus on the initial phase of

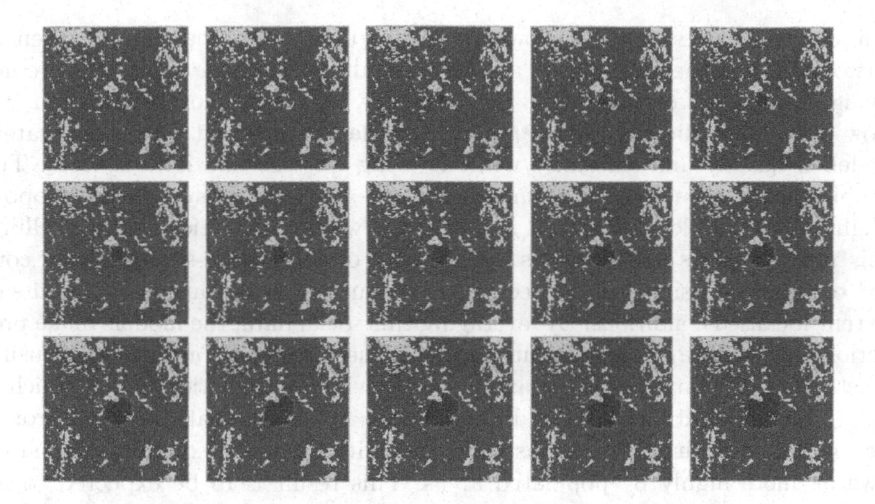

Fig. 2. Predicted sequences of burned area with the ignition point predicted by human sensors. The time interval is 6 h and the first four time steps are provided by the CA simulation as the initial input of the ML model.

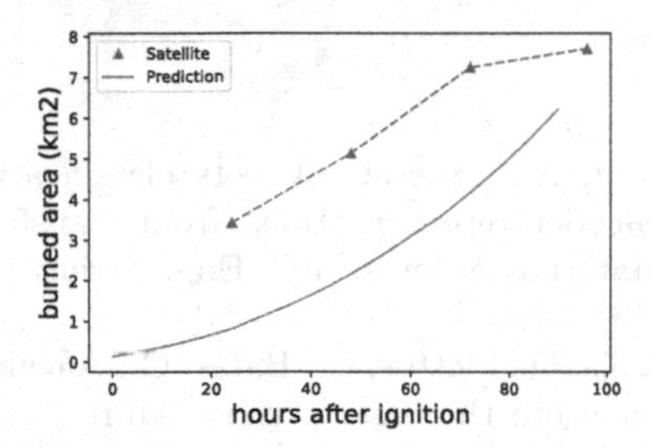

Fig. 3. The burned area after ignition: Satellite vs. ML prediction

fire spread (i.e., the first four days) since providing in-time fire spread nowcasting at the beginning of the fire event is crucial for fire fighting decision making. We also compare the predicted evolution of the burned area in km^2 to the satellite observations in Fig. 3. It can be clearly seen that the predicted burned area in km^2 exhibits a growth trajectory similar to that of the observed values, which demonstrates the robustness of the proposed approach. The comparison between the proposed POD + LSTM approach and the original CA simulations has been performed in [8] for recent massive wildfires in California.

5 Conclusion and Future Work

The human-sensors & physics aware machine learning proposed in this paper provide an end-to-end framework that provides reliable fire ignition detection/localization and early-stage fire spread nowcasting. By including social data as a real-time data source, we show that setting up a Twitter based human sensor network for just the first days of a massive wildfire event can accurately predict the ignition point of the event. Subsequently, using our novel algorithm, we show that the predicted burned area for each day of the event can be accurately modelled quickly and efficiently without using conventional data sources. The combination of using this real-time data source and a ROM system, we propose a lightweight coupled framework for real time wildfire detection and modelling. This work employs and develops the concept of the human-sensor in the context of wildfires, using users' Tweets as noisy subjective sentimental accounts of current localised conditions. By leveraging this social data, the models make predictions on wildfire instances. Subsequently, these instances are modelled using a fast, computationally efficient and accurate wildfire prediction model which is able to predict ignition points and burned area. We found that the main error in the prediction of fire ignition was that the ignition point prediction was biased towards more highly by populated areas. This result is to be expected to an extent, as there would naturally be more viewers and therefore sensors of an event in these locations. To combat this, a future work can aim to improve the methodology by taking into account the population density.

References

1. Ningsih, A.K., Hadiana, A.I.: Disaster tweets classification in disaster response using bidirectional encoder representations from transformer (BERT). In: IOP Conference Series: Materials Science and Engineering, vol. 1115(1), p. 012032 (2021)
2. Alexandridis, A., Vakalis, D., Siettos, C., Bafas, G.: A cellular automata model for forest fire spread prediction: the case of the wildfire that swept through spetses island in 1990. Appl. Math. Comput. **204**(1), 191–201 (2008)
3. Bai, H., Yu, G.: A Weibo-based approach to disaster informatics: incidents monitor in post-disaster situation via Weibo text negative sentiment analysis. Nat. Hazards **83**(2), 1177–1196 (2016)
4. Beigi, G., Hu, X., Maciejewski, R., Liu, H.: An overview of sentiment analysis in social media and its applications in disaster relief. In: Pedrycz, W., Chen, S.-M. (eds.) Sentiment Analysis and Ontology Engineering. SCI, vol. 639, pp. 313–340. Springer, Cham (2016). https://doi.org/10.1007/978-3-319-30319-2_13
5. Cheng, S., et al.: Generalised latent assimilation in heterogeneous reduced spaces with machine learning surrogate models. J. Sci. Comput. **94**(1), 1–37 (2023)
6. Cheng, S., et al.: Parameter flexible wildfire prediction using machine learning techniques: forward and inverse modelling. Remote Sens. **14**(13), 3228 (2022)
7. Cheng, S., Lucor, D., Argaud, J.P.: Observation data compression for variational assimilation of dynamical systems. J. Comput. Sci. **53**, 101405 (2021)
8. Cheng, S., Prentice, I.C., Huang, Y., Jin, Y., Guo, Y.K., Arcucci, R.: Data-driven surrogate model with latent data assimilation: application to wildfire forecasting. J. Comput. Phys. **464**, 111302 (2022)
9. Cheng, S., et al.: Machine learning with data assimilation and uncertainty quantification for dynamical systems: a review. arXiv preprint arXiv:2303.10462 (2023)

10. Gong, H., et al.: An efficient digital twin based on machine learning SVD autoencoder and generalised latent assimilation for nuclear reactor physics. Ann. Nucl. Energy **179**, 109431 (2022)
11. Hasegawa, K., Fukami, K., Murata, T., Fukagata, K.: CNN-LSTM based reduced order modeling of two-dimensional unsteady flows around a circular cylinder at different Reynolds numbers. Fluid Dyn. Res. **52**(6), 065501 (2020)
12. Ko, A., Lee, N., Sham, R., So, C., Kwok, S.: Intelligent wireless sensor network for wildfire detection. WIT Trans. Ecol. Environ. **158**, 137–148 (2012)
13. Lever, J., Arcucci, R.: Sentimental wildfire: a social-physics machine learning model for wildfire nowcasting. J. Comput. Soc. Sci. **5**(2), 1427–1465 (2022)
14. Lever, J., Arcucci, R., Cai, J.: Social data assimilation of human sensor networks for wildfires. In: Proceedings of the 15th International Conference on PErvasive Technologies Related to Assistive Environments, pp. 455–462 (2022)
15. Loureiro, M.L., Alló, M., Coello, P.: Hot in Twitter: assessing the emotional impacts of wildfires with sentiment analysis. Ecol. Econ. **200**, 107502 (2022)
16. Mendon, S., Dutta, P., Behl, A., Lessmann, S.: A hybrid approach of machine learning and lexicons to sentiment analysis: enhanced insights from twitter data of natural disasters. Inf. Syst. Front. **23**, 1145–1168 (2021)
17. Pandey, A., Pokharel, R.: Machine learning based surrogate modeling approach for mapping crystal deformation in three dimensions. Scripta Mater. **193**, 1–5 (2021)
18. Qian, J., Lin, H.: A forest fire identification system based on weighted fusion algorithm. Forests **13**(8), 1301 (2022)
19. Smith, T.R., Moehlis, J., Holmes, P.: Low-dimensional modelling of turbulence using the proper orthogonal decomposition: a tutorial. Nonlinear Dyn. **41**, 275–307 (2005)
20. Staudemeyer, R.C., Morris, E.R.: Understanding LSTM-a tutorial into long short-term memory recurrent neural networks. arXiv preprint arXiv:1909.09586 (2019)
21. Trunfio, G.A.: Predicting wildfire spreading through a hexagonal cellular automata model. In: Sloot, P.M.A., Chopard, B., Hoekstra, A.G. (eds.) ACRI 2004. LNCS, vol. 3305, pp. 385–394. Springer, Heidelberg (2004). https://doi.org/10.1007/978-3-540-30479-1_40
22. Wiewel, S., Becher, M., Thuerey, N.: Latent space physics: towards learning the temporal evolution of fluid flow. In: Computer graphics forum, vol. 38, pp. 71–82. Wiley Online Library (2019)
23. Xu, R., Lin, H., Lu, K., Cao, L., Liu, Y.: A forest fire detection system based on ensemble learning. Forests **12**(2), 217 (2021)
24. Zhang, C., Cheng, S., Kasoar, M., Arcucci, R.: Reduced order digital twin and latent data assimilation for global wildfire prediction. EGUsphere, 1–24 (2022)
25. Zhu, Q., et al.: Building a machine learning surrogate model for wildfire activities within a global earth system model. Geosci. Model Dev. **15**(5), 1899–1911 (2022)
26. Zhuang, Y., et al.: Ensemble latent assimilation with deep learning surrogate model: application to drop interaction in a microfluidics device. Lab Chip **22**(17), 3187–3202 (2022)

An Efficient ViT-Based Spatial Interpolation Learner for Field Reconstruction

Hongwei Fan[1,2], Sibo Cheng[1], Audrey J. de Nazelle[2], and Rossella Arcucci[1,3(✉)]

[1] Data Science Institute, Imperial College London, London, UK
r.arcucci@imperial.ac.uk

[2] Centre for Environmental Policy, Imperial College London, London, UK

[3] Department of Earth Science and Engineering, Imperial College London, London, UK

Abstract. In the field of large-scale field reconstruction, Kriging has been a commonly used technique for spatial interpolation at unobserved locations. However, Kriging's effectiveness is often restricted when dealing with non-Gaussian or non-stationary real-world fields, and it can be computationally expensive. On the other hand, supervised deep learning models can potentially address these limitations by capturing underlying patterns between observations and corresponding fields. In this study, we introduce a novel deep learning model that utilizes vision transformers and autoencoders for large-scale field reconstruction. The new model is named ViTAE. The proposed model is designed specifically for large-scale and complex field reconstruction. Experimental results demonstrate the superiority of ViTAE over Kriging. Additionally, the proposed ViTAE model runs more than 1000 times faster than Kriging, enabling real-time field reconstructions.

Keywords: Field Reconstruction · Vision Transformer · Deep Learning

1 Introduction

Spatial interpolation, which is predicting values of a spatial process in unmonitored areas from local observations, is a major challenge in spatio-temporal statistics. As a reference method of spatial interpolation, Kriging [5,18] provides the best linear unbiased prediction from observations. As a Gaussian process [21] governed by covariance, Kriging interpolates unmonitored areas as a weighted average of observed data. Kriging [5,18] is a geostatistical method that provides the optimal linear unbiased prediction based on observed data. It assumes that the underlying data follows a Gaussian process and is governed by covariance. However, authors [19,26] noted that in many cases, the spatial covariance function of physical fields is non-Gaussian and non-stationary. As a consequence, the

J. Mikyška et al. (Eds.): ICCS 2023, LNCS 14076, pp. 430–437, 2023.
https://doi.org/10.1007/978-3-031-36027-5_34

optimality of Kriging can not be guaranteed in real-world scenarios. Another limitation of Kriging is that its computational complexity can render it impractical for large spatial datasets. In fact, the online implementation of Kriging involves computing the inversion of a $N \times N$ covariance matrix, where N is the number of observed locations [13]. The aforementioned limitations of Kriging pose significant challenges in utilizing this method for generating credible large-field reconstructions in real-time.

Recently, deep learning (DL) [15] or neural network (NN) [14] have become increasingly utilized and powerful prediction tools for a wide range of applications [11], especially in computer vision and natural language processing [15]. DL has witnessed an explosion of architectures that continuously grow in capacity [22]. The utilization of convolutional neural network (CNN)s has become increasingly prevalent due to the rapid progress in hardware. CNNs are well-suited for prediction tasks that involve complex features, such as non-linearity and non-stationarity, and offer computational efficiency when analyzing massive datasets with GPU acceleration. Previous research efforts, such as [23], have explored the use of DL techniques for field reconstructions from observations. However, the traditional CNN-based approaches are inadequate in dealing with the problem of time-varying sensor placement. Re-training is often required when the number or the locations of sensors change, resulting in difficulties of real-time field construction. Therefore, this paper proposes the use of DL for large field reconstruction from random observations in real time.

2 Related Works and Contribution of the Present Work

NNs [15] have become a promising approach for effectively reconstructing fields from sparse measurements. [1,8,25]. While graph neural network (GCN) [24] and multi-layer perception (MLP) [15] could handle sparse data, they are known to scale poorly because of the high computational cost. Moreover, these methodologies require predetermined measurements as input data, which renders them unfeasible for real-world situations where sensor quantities and positions frequently vary over time, ultimately making them impractical [7]. To tackle these two bottlenecks, Fukami et al. [9] utilized Voronoi tessellation to transfer observations to a structured grid representation, which is available for CNN. Despite their effectiveness in capturing features, CNN typically overlook the spatial relationships between different features, thus making it difficult for them to accurately model the spatial dependencies required for reconstructing large-scale fields [16]. This difficulty can be addressed by the introduction of Vision Transformers (ViT) [6]. Transformers were proposed by Vaswani et al. (2017) [22] and have since become the state-of-the-art method in machine translation [20]. Apart from the complex architecture of transformers, the effectiveness of natural language processing (NLP) models heavily relies on the training strategy employed. One of the critical techniques utilized in training is auto-encoding with masks [12], where a subset of data is removed, and the model learns to predict the missing content. This technique has also demonstrated promising results in the field of computer vision, further highlighting its potential for enhancing model performance. By using masked autoencoder (AE)s to drop random patches of the input image and reconstruct missing patches, He et al. [12] demonstrates that it is possible to reconstruct images that appear realistic even when more than 90% of the original pixels are masked. The underlying principle of reconstructing an image from randomly selected patches bears resemblance to the process of reconstructing a field from observations.

Although the ViT model and the AE method [12] have succeeded in image reconstruction, to the best of our knowledge, no previous studies have applied them to field reconstruction task. Inspired by the success of ViT and AE methods, we propose a simple, effective, and scalable form of a Vision Transformer-based autoencoder (ViTAE) for field construction. To address the challenges mentioned earlier, we present a technique ViTAE that incorporates sparse sensor data into a Transformer model by mapping the observed values onto the field grid and masking unobserved areas. The masked observations field is divided into patches and fed into the transformer encoder to obtain representations. These representations are reshaped into patches and concatenated before being fed into the decoder to predict the grid values. Our proposed model, ViTAE, is capable of efficiently and accurately reconstructing fields from unstructured and time-varying observations. We compare the performance of our ViTAE with Kriging-based field reconstruction methods in this study.

The rest of the paper is organized as follow. Section 3 introduces the construction and properties of our ViTAE method. Section 4 presents some studies to show the performance of ViTAE. Section 5 summarizes our main results and suggests directions for future work.

3 Methodology

Our objective is to reconstruct a two-dimensional global field variable Q ($\dim(Q) = n$) from some local observations $\{\tilde{Q}_i\}, i \in O$ where O is a subset of $[1, ..., n]$. The proposed ViTAE is an autoencoder that aims to reconstruct the complete field from its partial observations. To deal with the sparsity of the data and extract meaningful representations, we employ a ViT-based autoencoder, which enables us to process the observations. Figure 1 illustrates the flowchart of the proposed approach.

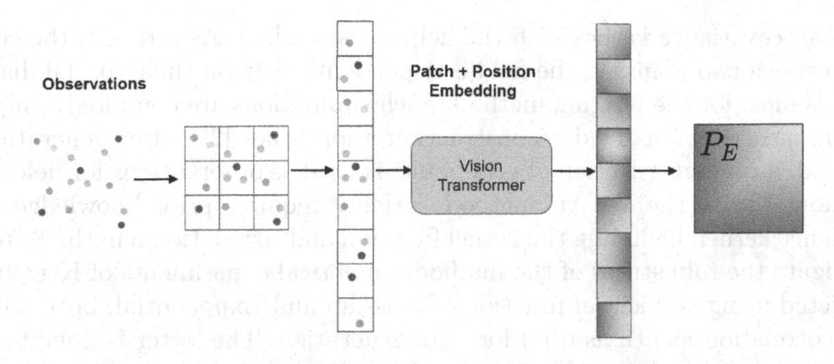

Fig. 1. Model overview. We split an image into fixed-size patches, linearly embed each of them, add position embeddings, and feed the resulting sequence of vectors to a standard Transformer encoder.

ViT-Based Autoencoder. First, to allow computationally tractable use of ViT, we project local sensor measurements into a grid field $\boldsymbol{I}$ according to their location in the field, defined as:

$$\boldsymbol{I}_i\,(i=1,...,n)=\begin{cases}\tilde{Q}_i & \text{if } \tilde{Q}_i \text{ is observable (i.e., } i\in O)\\ 0 & \text{otherwise.}\end{cases} \quad (1)$$

Just as in a standard ViT, we reshape the field $\boldsymbol{I}$ into a sequence of N flattened 2D patches, N is the resulting number of patches, which also serves as the effective input sequence length for the Transformer. The Transformer uses a constant latent vector size D through all of its layers, so we flatten the patches and map to the embedding of dimensions D by a linear projection with added positional embeddings, and then process the resulting set via a series of Transformer blocks to obtain latent representations and map the latent representation back to the predicted field P_E.

Reconstruction Target. Our ViTAE reconstructs the field by predicting the values for entire field grids. Unlike [12] which calculates the loss only on the masked patch, we compute the loss in the entire field. The loss is defined as:

$$L=MSE\,(Q,P_E) \quad (2)$$

it computes the mean squared error (MSE) between the reconstructed P_E and original fields Q in pixel space.

4 Test Cases and Results

This section presents an evaluation of the performance of ViTAE compared to Kriging in stationary simulation data with a structured grid. To create a spatially isotropic physical field, we generated the simulation field data using the

Gaussian covariance kernel with the help of gstool [17]. As a result, the correlation between two points in the field is dependent solely on their spatial distance, which is ideal for the Kriging method. Such simulations are commonly employed for comparing various field reconstruction approaches [2]. After generating the grid field, a random selection of grid points is used as observations for field reconstruction. Unlike the ViTAE method, Kriging requires prior knowledge of the covariance kernel, including the kernel function and correlation length. To further investigate the robustness of the method, numerical experiments of Krigging are conducted using two kernel functions: Gaussian and Exponential, both with the same correlation length as used for data generation. The latter is done to simulate cases where the kernel function is misspecified, as in real-world applications, the exact covariance kernel is often unknown [10].

Our initial focus is on the computational efficiency of the proposed method. Table 1 displays the accuracy of the reconstructing fields using the Gaussian kernel of different sizes with varying numbers of observations. The results indicate that when the field size is larger than 256 and the number of observations exceeds 0.1%, Kriging's computational time grows exponentially, taking thousands of seconds to fit and predict. It should be noted that Kriging must be performed online, i.e., after the observations are available, which poses computational challenges for high-dimensional systems. To compare the reconstruction accuracy of our proposed ViTAE against Kriging, we conducted experiments on a field size of 512×512. We used 0.5%, 1%, 2%, and 5% of the total number of grid points in the field as the number of observations for training. For each observation ratio, we generated 10,000 field snapshots and randomly selected observations from each snapshot. This allowed us to use time-varying observations as input data for Kriging and ViTAE to learn the entire physical field. We randomly partitioned our dataset into training, validation, and testing sets using an 80/10/10 split.

Table 1. Gaussian field reconstruction result of the Gaussian field reconstruction for ViTAE, and Kriging.

Model	ϵ			
Kriging/RBF	0.2243	0.2221	0.2218	0.2215
Kriging/Exp	0.2553	0.2552	0.2550	0.2379
ViTAE-lite/16	0.2431	0.2346	0.2290	0.2242
ViTAE-base/16	0.2280	0.2369	0.2250	0.2234
ViTAE-large/16	**0.2255**	**0.2228**	**0.2213**	0.2202
Sampling Percent	0.5%	1%	2%	5%

Model Variation. For the ViT-based encoder design, we follow the original ViT set up [6] and use "Lite", "Base", and "Large" models, such that ViT-Lite has 8 layers, 32 as hidden size, 8 as Heads, 16 hannels and 16 as Patch size, ViT-Base has 8 layers, 64 as hidden size, 32 as Heards, 16 hannels and 16 as Patch size,

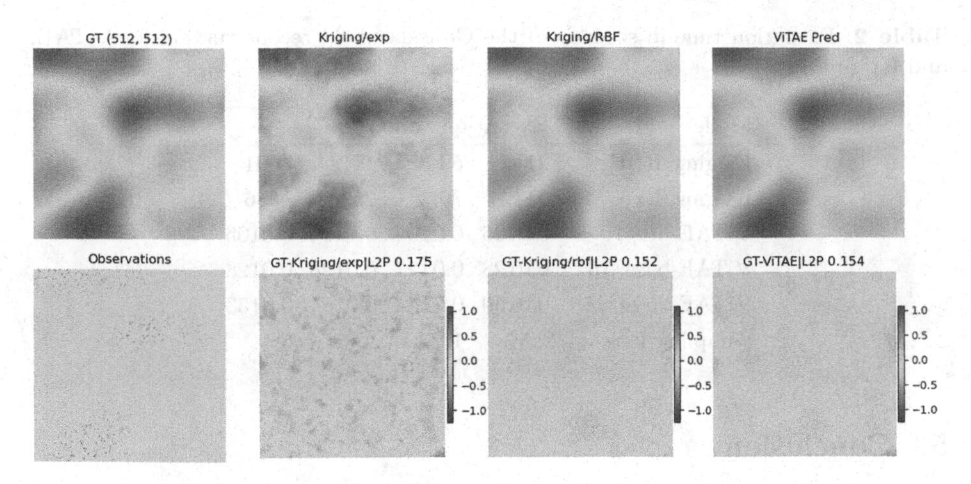

Fig. 2. 512 × 512 Gaussian field reconstruction results of ViTAE and Kriging , 0.5% sampling rate comparet to the GT.

and ViT-Large has 8 layers, 128 as hidden size, 64 as Heads, 16 hannels and 16 as Patch size. In what follows we use brief notation to indicate the model size and the input patch size: for instance, ViT-L/16 denotes the "Large" variant with 16 × 16 input patch size.

The field reconstruction results are shown in Fig. 2. The field reconstructed by ViTAE shows great similarity against the ground truth (GT) without knowing the spatial correlation function *a priori*.

Figure 2 also reports the relative error defined as:

$$\epsilon = \frac{\|Q_{\text{ref}} - Q_{\text{reconstruct}}\|_2}{\|Q_{\text{ref}}\|_2}, \tag{3}$$

where $\|\cdot\|$ denotes the L_2 norm, and Q_{ref} and $Q_{\text{reconstruct}}$ are the reference and reconstructed simulation fields, respectively. This metric of relative error has been widely used in field reconstruction and prediction tasks [3,4]. In this section, we compare the performance of our ViTAE model with two Kriging models using Gaussian and exponential covariance kernels, denoted as Kriging/RBF and Kriging/exp, respectively. As shown in Fig. 2, Kriging/exp is significantly outperformed by Kriging/RBF, which demonstrates the vulnerability of Kriging when the covariance kernel is not perfectly known. On the other hand, our ViTAE model, which does not require prior knowledge of the covariance kernel, achieves reconstruction results that are almost as accurate as Kriging/RBF. Additionally, the online computation of ViTAE is much more efficient than Kriging, as shown in Table 2. For example, when 5% of the field is observable, ViTAE-lite/16 runs 10^6 faster than Kriging.

Table 2. Execution time in seconds of the Gaussian field reconstruction for ViTAE, and Kriging.

Model	Execution time (s)			
Kriging/RBF	21	59	191	1491
Kriging/Exp	31	76	253	1586
ViTAE-lite/16	0.0105	0.0104	0.0105	0.0106
ViTAE-base/16	0.0128	0.0127	0.0128	0.0128
ViTAE-large/16	0.0150	0.0154	0.0151	0.0153
Sampling Percent	0.5%	1%	2%	5%

5 Conclusion

A long-standing challenge in engineering and sciences has been spatial interpolation for large field reconstruction. To tackle this issue, the paper introduces a novel autoencoder based on the ViT architecture, which serves as an efficient learner for spatial interpolation. The results presented in this paper indicate that the proposed ViTAE approach outperforms the Kriging method in spatial interpolation tasks. The method does not require prior knowledge of the spatial distribution and is computationally efficient. This work opens up new possibilities for applying DL to spatial prediction and has potential applications in complex data structures. In addition, the proposed method can be extended to real-world physical systems to investigate relationships and correlations between observations, supporting studies in spatio-temporal statistics and geostatistics.

References

1. Bolton, T., Zanna, L.: Applications of deep learning to ocean data inference and subgrid parameterization. J. Adv. Mod. Earth Syst. **11**(1), 376–399 (2019)
2. Chen, W., Li, Y., Reich, B.J., Sun, Y.: DeepKriging: spatially dependent deep neural networks for spatial prediction. arXiv preprint arXiv:2007.11972 (2020)
3. Cheng, S., et al.: Generalised latent assimilation in heterogeneous reduced spaces with machine learning surrogate models. J. Sci. Comput. **94**(1), 1–37 (2023)
4. Cheng, S., Prentice, I.C., Huang, Y., Jin, Y., Guo, Y.K., Arcucci, R.: Data-driven surrogate model with latent data assimilation: application to wildfire forecasting. J. Comput. Phys. **464**, 111302 (2022)
5. Cressie, N.: The origins of kriging. Math. Geol. **22**, 239–252 (1990)
6. Dosovitskiy, A., et al.: An image is worth 16×16 words: Transformers for image recognition at scale. arXiv preprint arXiv:2010.11929 (2020)
7. Environmental Research Group, I.C.L.: London air quality network (2022). https://www.londonair.org.uk/LondonAir/Default.aspx. Accessed 8 Nov 2022
8. Erichson, N.B., Mathelin, L., Yao, Z., Brunton, S.L., Mahoney, M.W., Kutz, J.N.: Shallow neural networks for fluid flow reconstruction with limited sensors. Proc. Roy. Soc. A **476**(2238), 20200097 (2020)

9. Fukami, K., Maulik, R., Ramachandra, N., Fukagata, K., Taira, K.: Global field reconstruction from sparse sensors with voronoi tessellation-assisted deep learning. Nat. Mach. Intell. **3**(11), 945–951 (2021)
10. Ginsbourger, D., Dupuy, D., Badea, A., Carraro, L., Roustant, O.: A note on the choice and the estimation of kriging models for the analysis of deterministic computer experiments. Appl. Stoch. Model. Bus. Ind. **25**(2), 115–131 (2009)
11. Hadash, G., Kermany, E., Carmeli, B., Lavi, O., Kour, G., Jacovi, A.: Estimate and replace: A novel approach to integrating deep neural networks with existing applications. arXiv preprint arXiv:1804.09028 (2018)
12. He, K., Chen, X., Xie, S., Li, Y., Dollár, P., Girshick, R.: Masked autoencoders are scalable vision learners. arXiv preprint arXiv:2111.06377 (2021)
13. Heaton, M.J., et al.: A case study competition among methods for analyzing large spatial data. J. Agric. Biol. Environ. Stat. **24**(3), 398–425 (2019)
14. Krizhevsky, A., Sutskever, I., Hinton, G.E.: ImageNet classification with deep convolutional neural networks. Commun. ACM **60**(6), 84–90 (2017)
15. LeCun, Y., Bengio, Y., Hinton, G.: Deep learning. Nature **521**(7553), 436–444 (2015)
16. Linsley, D., Kim, J., Veerabadran, V., Windolf, C., Serre, T.: Learning long-range spatial dependencies with horizontal gated recurrent units. In: Advances in Neural Information Processing Systems, vol. 31 (2018)
17. Müller, S.: GeoStat framework (2022). https://geostat-framework.org/. Accessed Nov 8 2022
18. Oliver, M.A., Webster, R.: Kriging: a method of interpolation for geographical information systems. Int. J. Geogr. Inf. Sci. **4**, 313–332 (1990)
19. Paciorek, C.J., Schervish, M.J.: Spatial modelling using a new class of nonstationary covariance functions. Environ. Official J. Int. Environmetrics Soc. **17**(5), 483–506 (2006)
20. Radford, A., Narasimhan, K., Salimans, T., Sutskever, I.: Improving language understanding by generative pre-training (2018)
21. Rasmussen, C.E.: Gaussian processes in machine learning. In: Bousquet, O., von Luxburg, U., Rätsch, G. (eds.) ML -2003. LNCS (LNAI), vol. 3176, pp. 63–71. Springer, Heidelberg (2004). https://doi.org/10.1007/978-3-540-28650-9_4
22. Vaswani, A., et al.: Attention is all you need. In: Advances in Neural Information Processing Systems, vol. 30 (2017)
23. Wu, G., Zhao, M., Wang, L., Dai, Q., Chai, T., Liu, Y.: Light field reconstruction using deep convolutional network on epi. In: Proceedings of the IEEE Conference on Computer Vision and Pattern Recognition, pp. 6319–6327 (2017)
24. Wu, Z., Pan, S., Chen, F., Long, G., Zhang, C., Philip, S.Y.: A comprehensive survey on graph neural networks. IEEE Trans. Neural Netw. Learn. Syst. **32**(1), 4–24 (2020)
25. Yu, J., Hesthaven, J.S.: Flowfield reconstruction method using artificial neural network. AIAA J. **57**(2), 482–498 (2019)
26. Zareifard, H., Khaledi, M.J.: Non-gaussian modeling of spatial data using scale mixing of a unified skew gaussian process. J. Multivar. Anal. **114**, 16–28 (2013)

A Kernel Extension of the Ensemble Transform Kalman Filter

Sophie Mauran[1(✉)], Sandrine Mouysset[2], Ehouarn Simon[1], and Laurent Bertino[3]

[1] Université de Toulouse, INP, IRIT, 2 Rue Charles Camichel, Toulouse, France
sophie.mauran@etu.toulouse-inp.fr
[2] Université de Toulouse, UT3, IRIT, Cr Rose Dieng-Kuntz, Toulouse, France
[3] Nansen Environmental and Remote Sensing Center, Jahnebakken 3, 5007 Bergen, Norway

Abstract. Data assimilation methods are mainly based on the Bayesian formulation of the estimation problem. For cost and feasibility reasons, this formulation is usually approximated by Gaussian assumptions on the distribution of model variables, observations and errors. However, when these assumptions are not valid, this can lead to non-convergence or instability of data assimilation methods. The work presented here introduces the use of kernel methods in data assimilation to model uncertainties in the data in a more flexible way than with Gaussian assumptions. The use of kernel functions allows to describe non-linear relationships between variables. The aim is to extend the assimilation methods to problems where they are currently unefficient. The Ensemble Transform Kalman Filter (ETKF) formulation of the assimilation problem is reformulated using kernels and show the equivalence of the two formulations for the linear kernel. Numerical results on the toy model *Lorenz 63* are provided for the linear and hyperbolic tangent kernels and compared to the results obtained by the ETKF showing the potential for improvement.

Keywords: data assimilation · kernel methods · ensemble Kalman filters

1 Introduction

Data assimilation is a process that consists of estimating the state of a system from observations and a numerical model of this system. This field of research has many applications, for example in meteorology [1–3] or oceanography [2,4,5]. The resolution on this type of large-scale systems involves a large number of state and observation variables, sometimes of the order of a billion [4]. In view of the complexity of these systems, several simplifications in the resolution are made and in particular assumptions of Gaussianity of the uncertainties on the observations and on the predictions [2].

J. Mikyška et al. (Eds.): ICCS 2023, LNCS 14076, pp. 438–452, 2023.
https://doi.org/10.1007/978-3-031-36027-5_35

However, these assumptions are often not satisfied in practice and it happens that the non-respect of these assumptions leads to problems in the assimilation process, either prediction errors or instabilities of the method [6]. To overcome these different problems, several approaches have been put forward, such as Gaussian anamorphoses [7,8] or 2-step Bayesian updates [9]. In this paper, we investigate kernel methods. Recent works [10] proposed a first formulation of an ensemble-based data assimilation algorithm based on "RBF" (radial-basis-function) kernels; [11]). It is then proposed to approximate the innovation term by a linear combination of RBF kernels, which parameters are estimated during the assimilation process. The introduction of RKHS is also proposed to model the temporal evolution of the dynamical system in order to apply ensemble methods in data assimilation [12,13]. Furthermore, the analysis step of the Random Feature Map Data Assimilation algorithm [13] can be viewed as the application of an ensemble Kalman filter in a particular RKHS. In this work, we are reformulating a data assimilation ensemble algorithm, the Ensemble Transform Kalman Filter, as an optimization problem on any RKHS, in order to extend the approach to a nonlinear framework, with limited loss of optimality. First, we reformulate the ETKF as presented in [14] using kernel methods and present a way to reconstruct the ensemble based on this formulation. We then give numerical results for the Lorenz 63 toy model for the linear kernel and the hyperbolic tangent kernel.

2 ETKF Reformulation with Kernel Methods

In this section, we first present the classical ETKF formulation and its resolution and then we present the corresponding kernel version.

2.1 ETKF Formulation and Classical Resolution

The ETKF (Ensemble Transform Kalman Filter) problem is a classical formulation of the data assimilation problem, proposed by [14]. Its formulation derives from:

$$\arg\min_{\mathbf{w}\in\mathbb{R}^N} \mathcal{J}(\mathbf{w}) = \frac{N-1}{2}||\mathbf{w}||_2^2 + \frac{1}{2}||\mathbf{y} - \mathcal{H}(\bar{\mathbf{x}}^{\mathbf{f}} + \mathbf{X}^{\mathbf{f}}\mathbf{w})||^2_{\mathbf{R}^{-1}} \tag{1}$$

where $\bar{\mathbf{x}}^{\mathbf{f}} \in \mathbb{R}^n$ the ensemble mean obtained after the prediction step,

$\mathbf{X}^{\mathbf{f}} = [(\mathbf{x}^{\mathbf{f}}_{\mathbf{1}} - \bar{\mathbf{x}}^{\mathbf{f}}), ..., (\mathbf{x}^{\mathbf{f}}_{\mathbf{N}} - \bar{\mathbf{x}}^{\mathbf{f}})] \in \mathbb{R}^{n\times N}$ the ensemble anomaly matrix obtained after the prediction step, $\mathcal{H} : \mathbb{R}^n \to \mathbb{R}^p$ the transition operator between the model space and the observation space, $\mathbf{w}$ parameterizing $\mathbf{x}^{\mathbf{a}}$: $\mathbf{x}^{\mathbf{a}} = \bar{\mathbf{x}}^{\mathbf{f}} + \mathbf{X}^{\mathbf{f}}\mathbf{w}$, under the assumption that $\mathbf{x}^{\mathbf{a}}$ lies in the affine subspace of $\bar{\mathbf{x}}^{\mathbf{f}}$ generated by $\mathbf{X}^{\mathbf{f}}$, N the ensemble size, $\mathbf{y} \in \mathbb{R}^p$ the observation vector, $\mathbf{R} \in \mathbb{R}^{p\times p}$ the error covariance matrix of the observations and $||.||_{\mathbf{R}^{-1}}$ defined by $||\mathbf{x}||^2_{\mathbf{R}^{-1}} = \mathbf{x}^\top\mathbf{R}^{-1}\mathbf{x}$

For the sake of simplicity, we assume that the observation operator $\mathcal{H}$ is linear. We can then rewrite (1) as:

$$\underset{\mathbf{w}\in\mathbb{R}^N}{\arg\min}\,\mathcal{J}(\mathbf{w}) = \frac{N-1}{2}||\mathbf{w}||_2^2 + \frac{1}{2}||\mathbf{y} - \mathbf{H}\bar{\mathbf{x}}^\mathbf{f} - \mathbf{H}\mathbf{X}^\mathbf{f}\mathbf{w}||_{\mathbf{R}^{-1}}^2 \tag{2}$$

with $\mathbf{H}$ the observation operator.

Here, for the sake of later clarity, we introduce some additional notations:

$$\widetilde{\mathbf{d}} = \mathbf{R}^{-1/2}(\mathbf{y} - \mathbf{H}\bar{\mathbf{x}}^\mathbf{f}) \in \mathbb{R}^p \quad \text{and } \widetilde{\mathbf{H}} = \mathbf{R}^{-1/2}\mathbf{H}\mathbf{X}^\mathbf{f} = \begin{bmatrix} \widetilde{\mathbf{h}_1}^\top \\ \vdots \\ \widetilde{\mathbf{h}_\mathbf{p}}^\top \end{bmatrix} \in \mathbb{R}^{p\times N}.$$

The first order optimality condition yields $\mathbf{w}^*$ solution of (2):

$$\mathbf{w}^* = [(N-1)\mathbf{I} + \widetilde{\mathbf{H}}^\top\widetilde{\mathbf{H}}]^{-1}\widetilde{\mathbf{H}}^\top\widetilde{\mathbf{d}} \tag{3}$$

We then generate the ensemble members noted $\mathbf{x}_\mathbf{i}^\mathbf{a} \in \mathbb{R}^n$, $\forall\, i \in \{1, ..., N\}$:

$$\mathbf{x}_\mathbf{i}^\mathbf{a} = \bar{\mathbf{x}}^\mathbf{f} + \mathbf{X}^\mathbf{f}\mathbf{w}^* + \sqrt{N-1}(\mathbf{X}^\mathbf{f}\nabla^2\mathcal{J})_\mathbf{i} \tag{4}$$

and

$$\nabla^2\mathcal{J} = (N-1)\mathbf{I}_\mathbf{N} + \widetilde{\mathbf{H}}^\top\widetilde{\mathbf{H}} \tag{5}$$

In practice, the anomaly matrix from the analysis $\mathbf{X}^\mathbf{a}$ is obtained by the following formula:

$$\mathbf{X}^\mathbf{a} = \mathbf{X}^\mathbf{f}\mathbf{T} \tag{6}$$

with $\mathbf{T} = (\mathbf{I} + \mathbf{S}\mathbf{S}^\top)^{-1/2} - \mathbf{I}$ and $\mathbf{S} = \frac{1}{\sqrt{N-1}}\mathbf{R}^{-1/2}\mathbf{H}\mathbf{X}^\mathbf{f}$

2.2 Reformulation and Resolution of ETKF with Kernels

Let the previous ETKF problem be embedded in a RKHS (Reproducing Kernel Hilbert Space):

$$(2) \Leftrightarrow \underset{\mathbf{w}\in\mathbb{R}^N}{\arg\min}\,\mathcal{J}(\mathbf{w}) = \frac{N-1}{2}\kappa(\mathbf{w},\mathbf{w}) + \frac{1}{2}\sum_{i=1}^{p}(\kappa(\widetilde{\mathbf{h}_\mathbf{i}},\mathbf{w}) - \widetilde{d}_i)^2 \tag{7}$$

with κ the linear kernel defined by: $\kappa(\mathbf{x},\mathbf{y}) = \mathbf{x}^\top\mathbf{y}$, $\forall\,(\mathbf{x},\mathbf{y}) \in \mathbb{R}^N \times \mathbb{R}^N$.

Considering the RKHS $\mathcal{H}_\kappa$ with reproducing kernel κ, where $\mathcal{H}_\kappa$ is a functional space, (7) is equivalent to:

$$\underset{f\in\mathcal{H}_\kappa}{\arg\min}\,\widetilde{\mathcal{J}}(f) = \frac{N-1}{2}||f||_{\mathcal{H}_\kappa}^2 + \frac{1}{2}\sum_{i=1}^{p}(f(\widetilde{\mathbf{h}_\mathbf{i}}) - \widetilde{d}_i)^2 \tag{8}$$

with $f \in \mathcal{H}_\kappa$ such as: $f : \begin{cases} \mathbb{R}^p \to \mathbb{R} \\ \mathbf{x} \;\mapsto \kappa(\mathbf{x},\mathbf{w}) \end{cases}$

We thus have extended the previous problem, since we may use any kernel, not only the linear one.

We now can apply the representer theorem which leads to an optimisation problem in finite dimension, given by the amount of data $\{(\mathbf{x_j})_{1\leq j\leq n}; (\widetilde{\mathbf{h_l}})_{1\leq l\leq p}\}$. The Eq. (8) can then be rewritten:

$$\underset{\alpha\in\mathbb{R}^{p+n}}{\arg\min}\,\widetilde{\widetilde{\mathcal{J}}}(\alpha) = \frac{N-1}{2}\alpha^\top \mathbf{K}\alpha + \frac{1}{2}\|\widetilde{\mathbf{d}} - \mathbf{\Pi_H K}\alpha\|_2^2 \tag{9}$$

with:

- $\mathbf{\Pi_H} = \begin{bmatrix}\mathbf{0_{nn}} & \mathbf{0_{np}}\\ \mathbf{0_{pn}} & \mathbf{I_p}\end{bmatrix} \in \mathbb{R}^{(n+p)\times(n+p)}$ the projection matrix on the observation space
- $\mathbf{K} = \begin{bmatrix}\mathbf{K_X} & \mathbf{K_{XH}}\\ \mathbf{K_{XH}}^\top & \mathbf{K_H}\end{bmatrix} \in \mathbb{R}^{(n+p)\times(n+p)}$ with
  - $\mathbf{K_X} = (\kappa(\mathbf{a_i^f}, \mathbf{a_j^f}))_{1\leq i,j\leq n} \in \mathbb{R}^{n\times n}$, the kernel applied to state variables
  - $\mathbf{K_{HX}} = (\kappa(\mathbf{a_i^f}, \widetilde{\mathbf{h_j}}))_{1\leq i\leq n, 1\leq j\leq p} \in \mathbb{R}^{n\times p}$, the crossed kernel application
  - $\mathbf{K_H} = (\kappa(\widetilde{\mathbf{h_i}}, \widetilde{\mathbf{h_j}}))_{1\leq i,j\leq p} \in \mathbb{R}^{p\times p}$, the kernel applied to observations

The solution α^* of (9) is then determined with the first order optimality condition:

$$\alpha^* = [(N-1)\mathbf{I_{n+p}} + \mathbf{\Pi_H K}]^{-1}\begin{bmatrix}\mathbf{0_{n1}}\\ \widetilde{\mathbf{d}}\end{bmatrix} \tag{10}$$

which can be simplified as:

$$\begin{cases}\alpha^*_{\mathbf{X}} = \mathbf{0_{n1}}\\ \alpha^*_{\mathbf{H}} = [(N-1)\mathbf{I_p} + \mathbf{K_H}]^{-1}\widetilde{\mathbf{d}}\end{cases} \tag{11}$$

with $\alpha^* = \begin{bmatrix}\alpha^*_{\mathbf{X}}\\ \alpha^*_{\mathbf{H}}\end{bmatrix} \in \mathbb{R}^{p+n}$.

Thus, the mean of the ensemble after the analysis will be:

$$\bar{\mathbf{x}}^{\mathbf{a}} = \bar{\mathbf{x}}^{\mathbf{f}} + \mathbf{K_{XH}}\alpha^*_{\mathbf{H}} \tag{12}$$

$$\Leftrightarrow \bar{\mathbf{x}}^{\mathbf{a}} = \bar{\mathbf{x}}^{\mathbf{f}} + \mathbf{K_{XH}}[(N-1)\mathbf{I_p} + \mathbf{K_H}]^{-1}\widetilde{\mathbf{d}} \tag{13}$$

It requires to solve a linear system of a symmetric positive definite (SPD) matrix.

Ensemble's Reconstruction: Now that we have the solution α^*, we must integrate it into the construction of the ensemble resulting from the analysis. The idea is to extend the ensemble to construct by considering the set of variables to be determined not only as the set of observed and unobserved variables $\mathbf{E^a}$ but also the set of transformed observed variables $\mathbf{HE^a}$, as it has already been done in the determination of α^*.

The principle of this generation is based on the deterministic version of the Ensemble Kalman Filter (EnKF) algorithm [17]. However, following the strategy formalized in [17], the state vector is extended to include both observed and unobserved variables. Formally, it reads:

$$\begin{bmatrix} \mathbf{E^a} \\ \mathbf{HE^a} \end{bmatrix} = \begin{bmatrix} \mathbf{\bar{x}^f} \\ \mathbf{H\bar{x}^f} \end{bmatrix} + \begin{bmatrix} \mathbf{X^f} \\ \mathbf{HX^f} \end{bmatrix} \mathbf{w} + \sqrt{N-1}\mathbf{P^a}^{1/2} \tag{14}$$

with $\mathbf{P^a} = \begin{bmatrix} \mathbf{P^a_X} & \mathbf{P^a_{XH}} \\ \mathbf{P^a_{XH}}^\top & \mathbf{P^a_H} \end{bmatrix} \in \mathbb{R}^{(n+p)\times(n+p)}$ the analysis error covariance matrix.

However, in order to update the ensemble $\mathbf{E^a}$, the block $\mathbf{P^a_X}$ of $\mathbf{P^a}$ is sufficient. Setting $\mathbf{P^a_X} = \mathbf{U\Sigma V}^\top$ the SVD of $\mathbf{P^a_X}$, we truncate $\mathbf{\Sigma}$ to its rank r_Σ ($\widetilde{\mathbf{\Sigma}}$) and the columns of $\mathbf{U}$ are also truncated to r_Σ ($\widetilde{\mathbf{U}}$) to compute the square root of $\mathbf{P^a_X}$. Using (13), this can be written member by member as follows:

$$\forall i \in \{1, ..., N\}, \quad \mathbf{x^a_i} = \mathbf{\bar{x}^f} + \mathbf{K_{XH}}[(N-1)\mathbf{I_p} + \mathbf{K_H}]^{-1}\tilde{\mathbf{d}} + [\widetilde{\mathbf{U}}\widetilde{\mathbf{\Sigma}}^{1/2}\mathbf{W}]_i \tag{15}$$

with $\mathbf{W} \in \mathbb{R}^{r_\Sigma \times N}$ a rotation matrix for column augmentation as implemented by [16] and detailed in Algorithm 3.

However, we still have to determine the expression of $\mathbf{P^a_X}$ in order to perform the SVD.

From the perspective of random variables, we can write $\alpha \sim \mathcal{N}(\mu_\alpha, \mathbf{P}^\alpha)$ and we can interpret (12) as:

$$\mathbf{x^a} = \mathbf{\bar{x}^f} + \mathbf{\Pi_X K}\alpha \tag{16}$$

where $\mathbf{\Pi_X} = \begin{bmatrix} \mathbf{I_n} & \mathbf{0_{np}} \end{bmatrix}$ and $\mathbf{x^a} \sim (\mu_\mathbf{a}, \mathbf{P^a_X})$. Thus,

$$\mathbf{P^a_X} = Cov(\mathbf{x^a_i}, \mathbf{x^a_j}) = \mathbf{\Pi_X K P^\alpha K \Pi_X} \tag{17}$$

where $(\mathbf{x^a_i}, \mathbf{x^a_j})_{1 \leq (i,j) \leq n}$ are randoms draws of $\mathbf{x^a}$.

According to [15], $\mathbf{P}^\alpha$ can be approximated by:

$$\mathbf{P}^\alpha \approx [\nabla^2 \tilde{\tilde{\mathcal{J}}}]^{-1} \tag{18}$$

and

$$\nabla^2 \tilde{\tilde{\mathcal{J}}} = [(N-1)\mathbf{K} + \mathbf{K\Pi_H\Pi_H K}] \tag{19}$$

Case where K is not Invertible: Since $\mathbf{K}$ may be singular, we thus calculate an approximation of $\nabla^2 \tilde{\tilde{\mathcal{J}}}$ from de generalized inverse by truncating the SVD of the matrix at rank of $\mathbf{K}$:

We set $\nabla^2 \tilde{\tilde{\mathcal{J}}} = \mathbf{V_\mathcal{J} \Sigma_\mathcal{J} U_\mathcal{J}^\top} = \widetilde{\mathbf{V_\mathcal{J}}}\widetilde{\mathbf{\Sigma_\mathcal{J}}}\widetilde{\mathbf{U_\mathcal{J}}}^\top$ with $\widetilde{\mathbf{V_\mathcal{J}}}$, $\widetilde{\mathbf{\Sigma_\mathcal{J}}}$ and $\widetilde{\mathbf{U_\mathcal{J}}}^\top$ the respective matrix of $\mathbf{V_\mathcal{J}}, \mathbf{\Sigma_\mathcal{J}}, \mathbf{U_\mathcal{J}}^\top$ truncated at the rank of $\mathbf{K}$. Thus,

$$\mathbf{P^a_X} = \mathbf{\Pi_X K}\widetilde{\mathbf{V_\mathcal{J}}}\widetilde{\mathbf{\Sigma_\mathcal{J}}}\widetilde{\mathbf{U_\mathcal{J}}}^\top\mathbf{K\Pi_X} \tag{20}$$

Depending on the chosen kernel, the rank of $\mathbf{K}$ can be less than N. We then have to augment the columns of $\mathbf{P^a_X}^{1/2}$ in order to keep the same number of ensemble members. Appendix A of [16] gives an algorithm to augment an ensemble covariance matrix $\mathbf{P}^{\mathbf{a}}_{\mathbf{X}}$ while keeping the same empirical covariance $\left(\widetilde{\mathbf{P}^{\mathbf{a}}_{\mathbf{X}}}^{1/2}\right)^T \left(\widetilde{\mathbf{P}^{\mathbf{a}}_{\mathbf{X}}}^{1/2}\right)$ and $\widetilde{\mathbf{P}^{\mathbf{a}}_{\mathbf{X}}}^{1/2}$ centered. We include this algorithm in the ensemble reconstruction.

The case where K is invertible is provided in Annex A, which gives an analytical expression for $\mathbf{P}^{\mathbf{a}}_{\mathbf{X}}$.

The method presented above is summarized in the form of Algorithms 1, 2 and 3.

Algorithm 1. ETKF kernel analysis

$\widetilde{\mathbf{H}} \leftarrow \mathbf{R}^{-1/2}\mathbf{H}\mathbf{X}^{\mathbf{f}}$
$\widetilde{\mathbf{d}} \leftarrow \mathbf{R}^{-1/2}(\mathbf{y} - \mathbf{H}\bar{\mathbf{x}}^{\mathbf{f}})$
Compute $\mathbf{K}$ ▷ depends on the chosen kernel
$\alpha^*_{\mathbf{H}} = [(N-1)\mathbf{I}_{\mathbf{p}} + \mathbf{K}_{\mathbf{H}}]^{-1}\widetilde{\mathbf{d}}$ ▷ Solve a linear system of a SPD matrix
$\bar{\mathbf{x}}^{\mathbf{a}} = \bar{\mathbf{x}}^{\mathbf{f}} + \mathbf{K}_{\mathbf{XH}}{}^{\top}\alpha^*_{\mathbf{H}}$
Compute $\mathbf{\Sigma}$, $\mathbf{U}$ the singular values and vectors of $\mathbf{P}^{\mathbf{a}}_{\mathbf{X}}$ ▷ refer to Algorithm 2
Truncate $\mathbf{\Sigma}$ to its rank r_{Σ} and compute its square root: $\widetilde{\mathbf{\Sigma}}^{1/2}$
$\mathbf{P}^{\mathbf{a}}_{\mathbf{X}}{}^{1/2} \leftarrow \widetilde{\mathbf{U}}\widetilde{\mathbf{\Sigma}}^{1/2}$ ▷ with $\widetilde{\mathbf{U}}$, the first r_{Σ} columns of $\mathbf{U}$
for i = 1... N-r_{Σ} **do**
  rotate $\mathbf{P}^{\mathbf{a}}_{\mathbf{X}}{}^{1/2}$ following the rotation step of Algorithm 3
end for
$\mathbf{E} = \bar{\mathbf{x}}^{\mathbf{a}} + \sqrt{N-1}\mathbf{P}^{\mathbf{a}}_{\mathbf{X}}{}^{1/2}$

Algorithm 2. Computation of $\mathbf{P}^{\mathbf{a}}_{\mathbf{X}}$

---

if K is not invertible **then**
  Compute $\nabla^2\widetilde{\widetilde{\mathcal{J}}} \leftarrow [(N-1)\mathbf{K} + \mathbf{K}\mathbf{\Pi}_{\mathbf{H}}\mathbf{\Pi}_{\mathbf{H}}\mathbf{K}]$
  Compute the SVD of $\nabla^2\widetilde{\widetilde{\mathcal{J}}} = \mathbf{U}_{\mathcal{J}}\mathbf{\Sigma}_{\mathcal{J}}\mathbf{V}_{\mathcal{J}}^{\top}$
  Compute $\widetilde{\mathbf{U}_{\mathcal{J}}}$, $\widetilde{\mathbf{\Sigma}_{\mathcal{J}}}$ and $\widetilde{\mathbf{V}_{\mathcal{J}}}^{\top}$ the respective matrix of $\mathbf{U}_{\mathcal{J}}, \mathbf{\Sigma}_{\mathcal{J}}, \mathbf{V}_{\mathcal{J}}{}^{\top}$ truncated at the rank of $\mathbf{K}$
  $\mathbf{\Sigma} \leftarrow \mathbf{\Sigma}_{\mathcal{J}}^{-1}$
  Compute $\mathbf{U} \leftarrow \mathbf{\Pi}_{\mathbf{X}}\mathbf{K}\widetilde{\mathbf{U}_{\mathcal{J}}}$ ▷ $\mathbf{P}^{\mathbf{a}}_{\mathbf{X}}{}^{1/2} = \mathbf{U}\mathbf{\Sigma}^{1/2}$
else
  Compute $\mathbf{P}^{\mathbf{a}}_{\mathbf{X}} \leftarrow \mathbf{K}_{\mathbf{X}} - \mathbf{K}_{\mathbf{HX}}\mathbf{U}_{\mathbf{H}}diag(\frac{1}{(N-1)+\lambda_i})\mathbf{U}_{\mathbf{H}}{}^{\top}\mathbf{K}_{\mathbf{HX}}{}^{\top}$ ▷ with $[\lambda_i]_{1\leq i\leq n}$ the eigenvalues of $\mathbf{K}_{\mathbf{H}}$
  Compute the SVD of $\mathbf{P}^{\mathbf{a}}_{\mathbf{X}} = \mathbf{U}\mathbf{\Sigma}\mathbf{V}^{\top}$
end if

Algorithm 3. Rotation step of $\mathbf{P}_{\mathbf{X}}^{\mathbf{a}^{1/2}}$, directly derived from Annex A of [16]

Require: $1 \leq i \leq N - r_\Sigma$

$\epsilon \leftarrow 1.0$

Compute $q \leftarrow r_\Sigma + i$

Compute $\theta \leftarrow \frac{\sqrt{q}}{\sqrt{q}-\epsilon}$

Compute $\mathbf{Q}_\epsilon \leftarrow \frac{-\theta}{q} \times \begin{bmatrix} \frac{\epsilon}{\sqrt{q}} & \cdots & \cdots & \cdots & \cdots & \frac{\epsilon}{\sqrt{q}} \\ \vdots & 1-\frac{\theta}{q} & \frac{-\theta}{q} & \cdots & \cdots & \frac{-\theta}{q} \\ \vdots & \frac{-\theta}{q} & 1-\frac{\theta}{q} & \frac{-\theta}{q} & \cdots & \frac{-\theta}{q} \\ \vdots & \vdots & \ddots & \ddots & \ddots & \vdots \\ \vdots & \vdots & & \ddots & 1-\frac{\theta}{q} & \frac{-\theta}{q} \\ \frac{\epsilon}{\sqrt{q}} & \frac{-\theta}{q} & \cdots & \cdots & \frac{-\theta}{q} & 1-\frac{\theta}{q} \end{bmatrix} \in \mathbb{R}^{q \times q}$

$\mathbf{W} \leftarrow \left[\mathbf{0_n}\ \mathbf{P}_{\mathbf{X}}^{\mathbf{a}^{1/2}}\right] \in \mathbb{R}^{n \times q}$

Compute $\mathbf{P}_{\mathbf{X}}^{\mathbf{a}^{1/2}} \leftarrow \mathbf{W}\mathbf{Q}_\epsilon$

3 Numerical Experiments

In the following, we perform numerical experiments in order to compare the Root Mean Square Errors (RMSE) of the ETKF and the proposed Kernel ETKF.

3.1 Experimental Setup

These experiments are performed using the Lorenz 63 model, which is a simplified chaotic and nonlinear model for atmospheric convection, widely used in data assimilation to benchmark. It is defined by the following differential equations:

$$\begin{cases} \frac{dx}{dt} = \sigma(y - x) \\ \frac{dy}{dt} = \rho x - y - xz \\ \frac{dz}{dt} = xy - \beta z. \end{cases} \tag{21}$$

These equations are integrated through a fourth-order Runge-Kutta scheme with a time-step of $\delta t = 0.01$, $\sigma = 10$, $\rho = 28$, $\beta = 8/3$, as set in [18,19]. The initial condition is distributed according to a Gaussian distribution of mean $[1.509, -1.531, 25.46]$ and a covariance matrix set to $\mathbf{C} = 2 * \mathbf{I_3}$.

A graphic representation of the Lorenz 63 model is given in Fig. 1(a). For all experiments, we computed the average RMSE over 10 different seeds (generating observations, initial state...).

In a first set of experiments, we compare the initial ETKF and Kernel ETKF applied to the linear kernel:

$$\forall(\mathbf{x}, \mathbf{y}) \in \mathbb{R}^N \times \mathbb{R}^N, \quad \kappa(\mathbf{x}, \mathbf{y}) = \mathbf{x}^\top \mathbf{y} \tag{22}$$

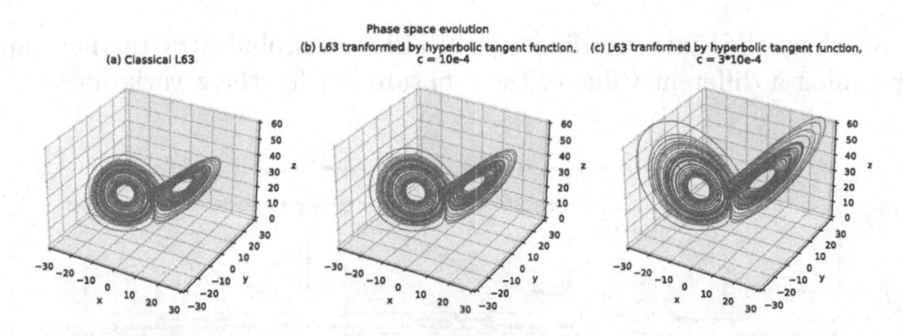

Fig. 1. Phase space evolution of the Lorenz 63 model: (a) classical L63; (b) L63 tranformed by hyperbolic tangent fonction (25) with $c = 10^{-4}$; (c) L63 tranformed by hyperbolic tangent fonction (25) with $c = 3 \times 10^{-4}$ to accentuate the visual effect.

The observations are generated every $\delta to = 0.02$, as set in [19], with $\mathbf{R} = 2 * \mathbf{I_3}$ and we set $\mathbf{H} = \begin{bmatrix} 1\,0\,0 \\ 0\,1\,0 \end{bmatrix}$, observing only the two first variables. A burn-in period of $5 \times 10^3 \times \delta to$ is enforced, as set in [19]. For these experiments, we use 5×10^5 observations vectors which determines the number of cycles. We compare different inflation factors: $infl \in \{1.0, 1.04, 1.1\}$. For each inflation factor, the evaluated ensemble sizes were $N \in \{3, 6, 9, 10, 12, 15\}$.

In a second set of experiments, we compare the performances of the initial ETKF and the Kernel ETKF applied to a non linear kernel: the hyperbolic tangent one. The hyperbolic tangent kernel is defined as in [20] by:

$$\forall (\mathbf{x}, \mathbf{y}) \in \mathbb{D}_c^N \times \mathbb{D}_c^N, \quad \kappa(\mathbf{x}, \mathbf{y}) = \phi(\mathbf{x})^\top \phi(\mathbf{y}) \tag{23}$$

where $\mathbb{D}_c^N$ is the Poincaré ball:

$$\mathbb{D}_c^N = \{z \in \mathbb{R}^N : c||z|| < 1\} \tag{24}$$

and

$$\forall c > 0, \ \forall \mathbf{z} \in \mathbb{D}_c^N, \quad \phi(\mathbf{z}) = tanh^{-1}(\sqrt{c}||\mathbf{z}||)\frac{\mathbf{z}}{\sqrt{c}||\mathbf{z}||} \tag{25}$$

A representation of the Lorenz 63 transformation induced by the hyperbolic tangent kernel with $c = 10^{-4}$ and $c = 3 \times 10^{-4}$ are given in Fig. 1(b) and (c) respectively. One can notice that L63 is sensitive to small variations of the c parameter. Quantile-quantile plots are given in Fig. 2 in order to compare the initial distribution of each variable of the L63 and the distribution after applying the hyperbolic tangent function with $c = 10^{-4}$ and $c = 3 \times 10^{-4}$ to the normal distribution. This last case is presented only to exacerbate the visual effect of the transformation induced by the kernel but will not be used for the Kernel ETKF experiments, because the transformation is too important to allow a regular convergence of the method. The QQ-plots confirm that the L63 variables have shorter tails than the Gaussian distribution but the hyperbolic tangent stretches the tails of the x and y variables closer to the Gaussian. The third variable z still shows short tails after the transformation, probably because the two L63 attractors share the same coordinate z=0. This can probably be further improved by fine tuning a different value of the c parameter for the z variable.

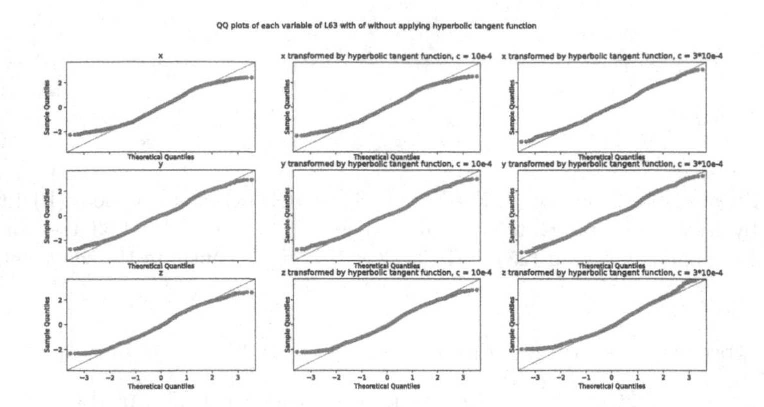

Fig. 2. QQ plots of each variable of the L63 relative to the normal distribution, Left: classical L63; Middle: L63 tranformed by hyperbolic tangent fonction (25) with $c = 10^{-4}$; Right: L63 tranformed by hyperbolic tangent fonction (25) with $c = 3 \times 10^{-4}$ to accentuate the visual effect.

First, we set $\mathbf{H} = \mathbf{I_3}$ and we change the time between the generation of observations to $\delta to = 0.50$ to reinforce the nonlinearities. We use 2×10^4 observation vectors and a burn-in period of $2 \times 10^2 \times \delta to$.

We then compute a last set the experiments with $\mathbf{H} = \begin{bmatrix} 1 & 0 & 0 \\ 0 & 1 & 0 \end{bmatrix}$, observing only the first two variables. For this set, we set $\delta to = 0.25$ and we use 5×10^4 observation vectors. A burn-in period of $5 \times 10^2 \times \delta to$ is enforced.

For these experiments, we computed the average RMSE over 10 different seeds for the two methods, using five ensembles sizes: $N = \{3, 6, 10, 12, 15\}$ for which different inflation factors have been evaluated: $infl \in \{1.0, 1.04, 1.1\}$.

For all the experiments presented here, we used the python package DAPPER [18], which allows benchmarking of performances of data assimilation methods. Several methods of data assimilation, both ensemble-based and variational, are already implemented in the package. We have also implemented our Kernel ETKF method.

3.2 Discussion

Comparison of the Classical ETKF and the Kernel ETKF Applied to the Linear Kernel: In Fig. 3, we represent the average RMSE of both methods with different ensemble sizes and different inflation factors when only two variables observed. We expect similar performances, as the two formulations are theoretically equivalent for the linear kernel. Any differences would be due to the method of ensemble reconstruction.

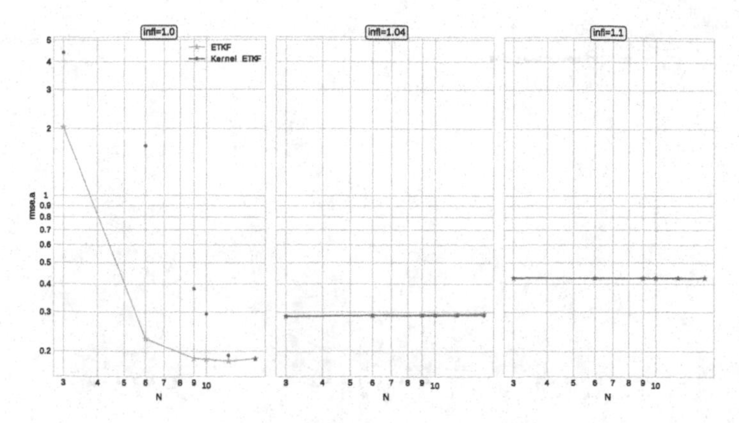

Fig. 3. Average RMSE obtained by ETKF (in green) and linear Kernel ETKF (in blue) assimilation methods when applied to the Lorenz 63 model and observing only the first two variables. The average is computed upon 10 different seeds generating observations, initial state... On each subfigure, a different inflation factor is applied to each method: Left: no inflation (infl = 1.0), Middle: infl = 1.04, Right: infl = 1.1. In each subfigure, different ensemble sizes N were tested, in each case $N \in \{3, 6, 9, 10, 12, 15\}$. (Color figure online)

When $infl \neq 1.0$, in Fig. 3, the results of the two methods are similar, as expected. However, the Kernel ETKF seems more sensitive to the absence of inflation than the classical ETKF. This could be due to the implementation of ensemble reconstruction for the Kernel ETKF method, and in particular to sampling issues.

We obtain similar results when all variables of the L63 are observed: the Kernel ETKF has equivalent results with the classical ETKF, except without inflation where it shows a slower convergence. The presented results show the equivalent performance (when using some inflation) of the classical ETKF and the linear kernel-based approach proposed here as expected.

Experiments with Non Linearity Reinforcement and Non Linear Kernels: In this perspective, we have carried out another set of experiments, this time increasing the time between the generation of observations to reinforce the nonlinearities in the data assimilation experiments. We tested the performance of a non linear kernel. We present here the results for the hyperbolic tangent kernel with $c = 10^{-4}$ which presented the best performances among the tested kernels. In Figs. 4 and 5, we give the average RMSE of both methods with different ensemble sizes and different inflation factors when all variables are observed in Fig. 4 and only two variables observed in Fig. 5.

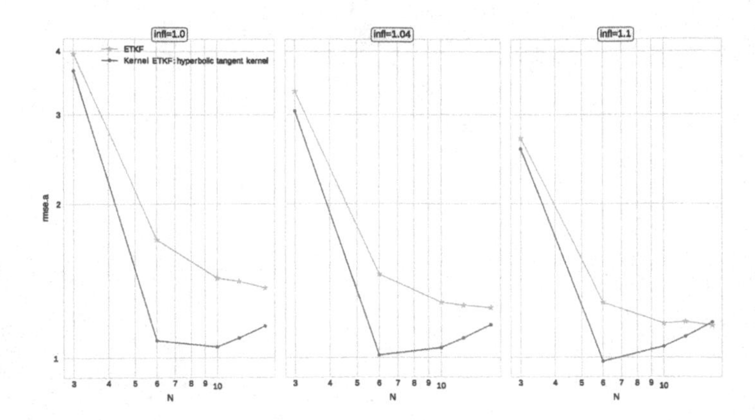

Fig. 4. Average RMSE obtained by ETKF (in green) and Kernel ETKF applied to hyperbolic tangent kernel with $c = 10^{-4}$ (in blue) when applied to the Lorenz 63 model and observing all variables. The average is computed upon 10 different seeds generating observations, initial state... Left: $infl = 1.0$. Middle: $infl = 1.04$. Right: $infl = 1.1$. For each subfigure, different ensemble sizes are tested: $N \in \{3, 6, 10, 12, 15\}$. (Color figure online)

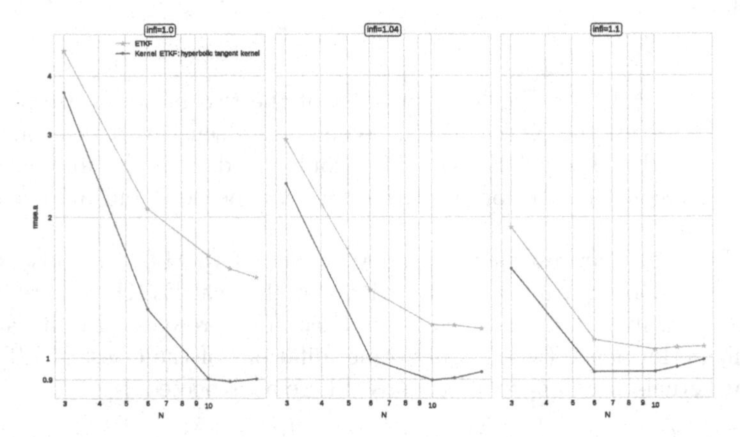

Fig. 5. Average RMSE obtained by ETKF (in green) and Kernel ETKF applied to hyperbolic tangent kernel with $c = 10^{-4}$ (in blue) when applied to the Lorenz 63 model and observing the first two variables. The average is computed upon 10 different seeds generating observations, initial state... Left: $infl = 1.0$. Middle: $infl = 1.04$. Right: $infl = 1.1$. In each one, different ensemble sizes are tested: $N \in \{3, 6, 10, 12, 15\}$. (Color figure online)

In both cases, the hyperbolic tangent Kernel ETKF has a clear advantage over the classical ETKF, especially for small sets, which seems particularly promising for large problems, where set size is a key issue. However, we can note an increase in RMSE for large ensemble sizes for the hyperbolic tangent Kernel ETKF. This increase was already observed in [21], and seems intrinsically linked to the ETKF. The hyperbolic tangent Kernel ETKF being more efficient for small ensemble sizes, it is logical that the above mentioned problem occurs earlier. This problem of RMSE increase can be corrected by introducing a rotation after the analysis, as proposed in [22].

In Fig. 6, we take a closer look at the similarities between the ensemble spread (experimental measure of forecast error) and the RMSE (exact measure of forecast error) for each variable of the system in the case where $N = 10$, $infl = 1.04$ and the two first variables are observed (same as in Fig. 5). We note that the RMSE of the classical ETKF is globally further from the spread than the one of the Kernel ETKF, which explains in this case the overall difference in RMSE between the two methods and the fact that the Kernel ETKF gives an overall better result than the classical ETKF. Some statistics for this experiment are given in Table 1. The advantage of the Kernel ETKF over the classical ETKF is highlighted: the RMSE of each model variable is lower for the Kernel ETKF than for the classical ETKF.

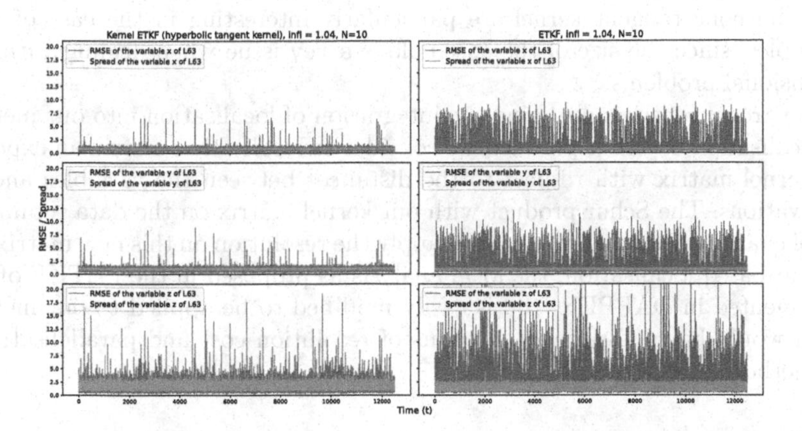

Fig. 6. Ensemble spread (in orange) and RMSE (in blue) obtained by hyperbolic tangent Kernel ETKF (left panel) and classical ETKF (right panel) assimilation methods when applied to the Lorenz 63 model and observing the first two variables in the case where $N = 10$, $infl = 1.04$. The ensemble spread and RMSE are displayed for each three variables of the L63 model individually. (Color figure online)

Table 1. RMSE and Spread of each variables for the hyperbolic tangent Kernel ETKF and the classical ETKF when applied to the Lorenz 63 model and observing the first two variables in the case where $N = 10$, $infl = 1.04$.

	ETKF		Kernel ETKF	
L63 variable	RMSE	Spread	RMSE	Spread
x	1.1 ± 1.0	0.68 ± 0.062	$\mathbf{0.69 \pm 0.28}$	0.75 ± 0.057
y	1.4 ± 1.1	0.93 ± 0.082	$\mathbf{0.93 \pm 0.32}$	1.0 ± 0.069
z	2.2 ± 1.7	1.4 ± 0.37	$\mathbf{1.5 \pm 0.82}$	1.7 ± 0.43

4 Conclusion and Perspectives

In this paper, we have proposed a generalisation of the ETKF problem by introducing kernels into its formulation. We also proposed a reconstruction of the set from the solution of the optimisation problem induced by the formulation. An explicit algorithm for this method is given. In a second part, we compared the performance of the proposed method with the linear kernel and the hyperbolic tangent kernel with that of the ETKF in the Lorenz 63 framework. We obtained similar performances for the linear kernel which was theoretically expected. We also could envisage the interest of using other kernels with the results of the hyperbolic tangent kernel when the model presents important nonlinearities, which is encouraging for the continuation of our research: the interest of the kernel methods will intervene when we will address systems where the use of non-linear kernels (such as the Gaussian or hyperbolic tangent kernels) will be more appropriate for the given data. Moreover, the good results obtained with the hyperbolic tangent kernel are particularly interesting in the case of small ensembles, since the size of the ensemble is a key issue when dealing with high dimensional problems.

We are currently considering the integration of localisation into our method: for localisation by Schur product, we consider as localisation matrix an exponential kernel matrix with respect to the distances between the variables and the observations. The Schur product with our kernel matrix on the data retains the kernel characteristic and we can then apply the resolution on this new matrix. For domain-based localisation, the local analysis as proposed in the LETKF of [14], implemented in DAPPER [18], is easily modified to be applied to our method, which would have advantages in terms of resolution cost and parallelisation of the method.

A Construction of P^a When K is Invertible

We consider the expression of $\mathbf{P^a}$ given by (17):

$$\mathbf{P^a} = \mathbf{K}\mathbf{P}^{\alpha}\mathbf{K}$$

and substitute for $\mathbf{P}^{\alpha}$ its approximation by the hessian (18):

$$\mathbf{P^a} = \mathbf{K}[(N-1)\mathbf{K} + \mathbf{K}\mathbf{\Pi_H}\mathbf{\Pi_H}\mathbf{K}]^{-1}\mathbf{K} \tag{26}$$

Since $\mathbf{K}$ is invertible,

$$\Leftrightarrow \mathbf{P^a} = \frac{1}{N-1}(\mathbf{K}^{-1} + \frac{1}{N-1}\mathbf{\Pi_H}\mathbf{\Pi_H})^{-1} \tag{27}$$

We the apply Woodbury identity:

$$\Leftrightarrow \mathbf{P^a} = \frac{1}{N-1}(\mathbf{K} - \mathbf{K}\mathbf{\Pi_H}((N-1)\mathbf{I_{n+p}} + \mathbf{\Pi_H}\mathbf{K}\mathbf{\Pi_H})^{-1}\mathbf{\Pi_H}\mathbf{K}) \tag{28}$$

$$\Leftrightarrow \mathbf{P^a} = \frac{1}{N-1}(\mathbf{K} - \begin{bmatrix} \mathbf{0_n} & \mathbf{K_{HX}} \\ \mathbf{0_{pn}} & \mathbf{K_H} \end{bmatrix} \begin{bmatrix} (N-1)\mathbf{I_n} & \mathbf{0_{np}} \\ \mathbf{0_{pn}} & (N-1)\mathbf{I_p} + \mathbf{K_H} \end{bmatrix}^{-1} \begin{bmatrix} \mathbf{0_n} & \mathbf{0_{np}} \\ \mathbf{K_H X^\top} & \mathbf{K_H} \end{bmatrix}) \tag{29}$$

Let $\mathbf{K_H}$ decompose as $\mathbf{K_H} = \mathbf{U_H \Sigma_H U_H}^\top$ with $\mathbf{\Sigma_H} = diag([\lambda_i]_{1\leq i\leq p})$ and $[\lambda_i]_{1\leq i\leq p}$ the eigenvalues of $\mathbf{K_H}$. We finally obtain:

$$\mathbf{P^a} = \frac{1}{N-1} \begin{bmatrix} \mathbf{K_X} - \mathbf{K_{HX} U_H} diag(\frac{1}{(N-1)+\lambda_i})\mathbf{U_H}^\top \mathbf{K_{HX}}^\top & \mathbf{K_{HX}} - \mathbf{K_{HX} U_H} diag(\frac{\lambda_i}{(N-1)+\lambda_i})\mathbf{U_H}^\top \\ \mathbf{K_{HX}}^\top - \mathbf{U_H} diag(\frac{\lambda_i}{(N-1)+\lambda_i})\mathbf{U_H}^\top \mathbf{K_{HX}}^\top & \mathbf{U_H} diag(\frac{\lambda_i (N-1)}{(N-1)+\lambda_i})\mathbf{U_H}^\top \end{bmatrix} \tag{30}$$

This gives an explicit expression for $\mathbf{P^a}$ which does not require numerical inversion.

References

1. Buehner, M., McTaggart-Cowan, R., Heilliette, S.: An ensemble Kalman filter for numerical weather prediction based on variational data assimilation: VarEnKF. Mon. Weather Rev. **145**(2), 617–635 (2017). https://doi.org/10.1175/MWR-D-16-0106.1
2. Carrassi, A., Bocquet, M., Bertino, L., Evensen, G.: Data assimilation in the geosciences: an overview of methods, issues, and perspectives. Wiley Interdiscip. Rev. Clim. Change **9**(5), e535 (2018). https://doi.org/10.1002/wcc.535
3. Tsuyuki, T., Miyoshi, T.: Recent progress of data assimilation methods in meteorology. J. Meteorol. Soc. Japan **85**, 331–361 (2007). https://doi.org/10.2151/jmsj.85B.331
4. Sakov, P., Counillon, F., Bertino, L., Lisæter, K.A., Oke, P.R., Korablev, A.: TOPAZ4: an ocean-sea ice data assimilation system for the North Atlantic and Arctic. Ocean Sci. **8**, 633–656 (2012). https://doi.org/10.5194/os-8-633-2012
5. Barth, A., et al.: Assimilation of sea surface temperature, ice concentration and ice drift in a model of the Southern Ocean. Ocean Model. **93**, 22–39 (2015)
6. Lei, J., Bickel, P., Snyder, C.: Comparison of ensemble Kalman filters under non-Gaussianity. Mon. Weather Rev. **138**(4), 1293–1306 (2010). https://doi.org/10.1175/2009MWR3133.1
7. Bertino, L., Evensen, G., Wackernagel, H.: Sequential data assimilation techniques in oceanography. Int. Stat. Rev. **71**(2), 223–241 (2003). https://doi.org/10.1111/j.1751-5823.2003.tb00194.x
8. Simon, E., Bertino, L.: Application of the Gaussian anamorphosis to assimilation in a 3-D coupled physical-ecosystem model of the North Atlantic with the EnKF: a twin experiment. Ocean Sci. **5**(4), 495–510 (2009). https://doi.org/10.5194/os-5-495-2009
9. Grooms, I.: A comparison of nonlinear extensions to the ensemble Kalman filter. Comput. Geosci. **26**(3), 633–650 (2022). https://doi.org/10.1007/s10596-022-10141-x
10. Luo, X.: Ensemble-based kernel learning for a class of data assimilation problems with imperfect forward simulators. PLoS ONE **14**(7), 1–40 (2019). https://doi.org/10.1371/journal.pone.0219247
11. Broomhead, D.S., Lowe, D.: Radial basis functions, multivariable functional interpolation and adaptive networks. In: Royal Signals and Radar Establishment Malvern, UK (1988)

12. Gottwald, G.A., Reich, S.: Supervised learning from noisy observations: combining machine-learning techniques with data assimilation. Phys. D Nonlinear Phenom. **423**, 132911 (2021). https://doi.org/10.1016/j.physd.2021.132911
13. Hug, B., Mémin, E., Tissot, G.: Ensemble forecasts in reproducing kernel Hilbert space family: dynamical systems in Wonderland. arXiv preprint arXiv:2207.14653 (2022)
14. Hunt, B.R., Kostelich, E.J., Szunyogh, I.: Efficient data assimilation for spatio-temporal chaos: a local ensemble transform Kalman filter. Phys. D Nonlinear Phenom. **230**(1), 112–126 (2007). https://doi.org/10.1016/j.physd.2006.11.008
15. Didier Auroux Homepage. https://math.unice.fr/~auroux/Work/These/html/node40.html. Accessed 29 Jan 2023
16. Farchi, A., Bocquet, M.: On the efficiency of covariance localisation of the ensemble Kalman filter using augmented ensembles. Front. Appl. Math. Stat. **5**, 3 (2019). https://doi.org/10.3389/fams.2019.00003
17. Evensen, G.: Data Assimilation: The Ensemble Kalman Filter, 2nd edn., pp. 273–274. Springer, Heidelberg (2009). https://doi.org/10.1007/978-3-642-03711-5
18. Raanes, P.N., Chen, Y., Grudzien, C., Tondeur, M., Dubois, R.: v. 1.2.1. https://doi.org/10.5281/zenodo.2029296
19. Fillion, A., Bocquet, M., Gratton, S.: Quasi-static ensemble variational data assimilation: a theoretical and numerical study with the iterative ensemble Kalman smoother. Nonlinear Process. Geophys. **25**(2), 315–334 (2018). https://doi.org/10.5194/npg-25-315-2018
20. Fang, P., Harandi, M., Petersson, L.: Kernel methods in hyperbolic spaces. In: Proceedings of the IEEE/CVF International Conference on Computer Vision (ICCV), pp. 10665–10674 (2021). https://doi.org/10.1109/ICCV48922.2021.01049
21. Sakov, P., Oke, P.R.: A deterministic formulation of the ensemble Kalman filter: an alternative to ensemble square root filters. Tellus A Dyn. Meteorol. Oceanogr. **60**(2), 361–371 (2008). https://doi.org/10.1111/j.1600-0870.2007.00299.x
22. Sakov, P., Oke, P.R.: Implications of the form of the ensemble transformation in the ensemble square root filters. Mon. Weather Rev. **136**, 1042–1053 (2008). https://doi.org/10.1175/2007MWR2021.1

Fixed-Budget Online Adaptive Learning for Physics-Informed Neural Networks. Towards Parameterized Problem Inference

Thi Nguyen Khoa Nguyen[1,2,3](✉), Thibault Dairay[2], Raphaël Meunier[2], Christophe Millet[1,3], and Mathilde Mougeot[1,4]

[1] Universite Paris-Saclay, ENS Paris-Saclay, CNRS, Centre Borelli, 91190 Gif-sur-yvette, France
thi.nguyen2@ens-paris-saclay.fr
[2] Michelin, Centre de Recherche de Ladoux, 63118 Cébazat, France
[3] CEA, DAM, DIF, 91297 Arpajon, France
[4] ENSIIE, 91000 Évry-Courcouronnes, France

Abstract. Physics-Informed Neural Networks (PINNs) have gained much attention in various fields of engineering thanks to their capability of incorporating physical laws into the models. The partial differential equations (PDEs) residuals are minimized on a set of collocation points which distribution appears to have a huge impact on the performance of PINNs and the assessment of the sampling methods for these points is still an active topic. In this paper, we propose a Fixed-Budget Online Adaptive Learning (FBOAL) method, which decomposes the domain into sub-domains, for training collocation points based on local maxima and local minima of the PDEs residuals. The numerical results obtained with FBOAL demonstrate important gains in terms of the accuracy and computational cost of PINNs with FBOAL for non-parameterized and parameterized problems. We also apply FBOAL in a complex industrial application involving coupling between mechanical and thermal fields.

Keywords: Physics-informed neural networks · Adaptive learning · Rubber calendering process

1 Introduction

In the last few years, Physics-Informed Neural Networks (PINNs) [8] have become an attractive and remarkable scheme of solving inverse and ill-posed partial differential equations (PDEs) problems. The applicability of PINNs has been demonstrated in various fields of research and industrial applications [3]. However, PINNs suffer from significant limitations. The training of PINNs takes high computational cost, that is, a standard PINN must be retrained for each PDE problem, which is expensive and the numerical physics-based methods can

J. Mikyška et al. (Eds.): ICCS 2023, LNCS 14076, pp. 453–468, 2023.
https://doi.org/10.1007/978-3-031-36027-5_36

strongly outperform PINNs for forward modeling tasks. There are continuing efforts to overcome this limitation by proposing to combine with reduced order methods so that the model has a strong generalization capacity [2]. Furthermore, the theoretical convergence properties of PINNs are still poorly understood and need further investigations [9]. As PINNs integrate the PDEs constraints by minimizing the PDE residuals on a set of collocation points during the training process, it has been shown that the location of these collocation points has a great impact on the performance of PINNs [1]. To the best of the authors' knowledge, the first work that showed the improvement of PINNs performance by modifying the set of collocation points is introduced by Lu et al. (2021) [5]. This work proposed the Residual-based Adaptive Refinement (RAR) that adds new training collocation points to the location where the PDE residual errors are large. RAR has been proven to be very efficient to increase the accuracy of the prediction but however leads to an uncontrollable amount of collocation points and computational cost at the end of the training process. In this work, we propose a Fixed-Budget Online Adaptive Learning (FBOAL) that fixes the number of collocation points during the training. The method adds and removes the collocation points based on the PDEs residuals on sub-domains during the training. By dividing the domain into smaller sub-domains it is expected that local maxima and minima of the PDEs residuals will be quickly captured by the method. Furthermore, the stopping criterion is chosen based on a set of reference solutions, which leads to an adaptive number of iterations for each specific problem and thus avoids unnecessary training iterations. The numerical results demonstrate that the use of FBOAL help to reduce remarkably the computational cost and gain significant accuracy compared to the conventional method of non-adaptive training points. In the very last months, several works have also introduced a similar idea of adaptive re-sampling of the PDE residual points during the training [1,6,10]. Wu et al. (2023) [10] gave an excellent general review of these methods and proposed two adaptive re-sampling methods named Residual-based Adaptive Distribution (RAD) and Residual-based adaptive refinement with distribution (RAR-D), which are the generalization of all existing methods of adaptive re-sampling for the collocation points. These approaches aim to minimize the PDEs residuals at their global maxima on the entire domain. Besides that, the existing studies did not investigate the parameterized PDE problems (where the parameter of interest is varied). In this study, we first compare the performance of RAD, RAR-D, and FBOAL in an academic test case (Burgers equation). We illustrate a novel utilization of these adaptive sampling methods in the context of parameterized problems. The following of this paper is organized as follows. In Sect. 2, we briefly review the framework of PINNs and introduce the adaptive learning strategy (FBOAL) for the collocation points. We then provide the numerical results of the performance of the studied methods and comparison to the classical PINNs and other adaptive re-sampling methods such as RAD and RAR-D in a test case of Burgers equation. The application to an industrial use case is also represented in this section. Finally, we summarize the conclusions in Sect. 4.

2 Methodology

In this section, the framework of Physics-Informed Neural Networks (PINNs) [8] is briefly presented. Later, Fixed-Budget Online Adaptive Learning (FBOAL) for PDE residual points is introduced.

2.1 Physics-Informed Neural Networks

To illustrate the methodology of PINNs, let us consider the following parameterized PDE defined on the domain $\Omega \subset \mathbb{R}^d$: $\boldsymbol{u}_t + \mathcal{N}_{\mathbf{x}}(\boldsymbol{u}, \boldsymbol{\lambda}) = 0$ for $\mathbf{x} \in \Omega, t \in [0, T]$ where $\boldsymbol{\lambda} \in \mathbb{R}^p$ is the PDE parameter vector with the boundary condition $\mathcal{B}(\boldsymbol{u}, \mathbf{x}, t) = 0$ for $\mathbf{x} \in \partial\Omega$ and the initial condition $\boldsymbol{u}(\mathbf{x}, 0) = g(\mathbf{x})$ for $\mathbf{x} \in \Omega$. In the conventional framework of PINNs, the solution $\boldsymbol{u}$ of the PDE is approximated by a fully-connected feed-forward neural network $\mathcal{NN}$ and the prediction for the solution can be represented as $\hat{\boldsymbol{u}} = \mathcal{NN}(\mathbf{x}, t, \boldsymbol{\theta})$ where $\boldsymbol{\theta}$ denotes the trainable parameters of the neural network. The parameters of the neural network are trained by minimizing the cost function $L = L_{pde} + w_{ic}L_{ic} + w_{bc}L_{bc}$, where the terms L_{pde}, L_{ic}, L_{bc} penalize the loss in the residual of the PDE, the initial condition, the boundary condition, respectively, and w_{ic}, w_{bc} are the positive weight coefficients to adjust the contribution of each term to the cost function: $L_{pde} = \frac{1}{N_{pde}} \sum_{i=1}^{N_{pde}} |\hat{\boldsymbol{u}}_{t_i} + \mathcal{N}_{\mathbf{x_i}}(\hat{\boldsymbol{u}}, \boldsymbol{\lambda})|^2$, $L_{ic} = \frac{w_{ic}}{N_{ic}} \sum_{i=1}^{N_{ic}} |\hat{\boldsymbol{u}}(\mathbf{x}_i, 0) - g(\mathbf{x}_i)|^2$, and $L_{bc} = \frac{w_{bc}}{N_{bc}} \sum_{i=1}^{N_{bc}} |\mathcal{B}(\hat{\boldsymbol{u}}, \mathbf{x}_i, t_i)|^2$ where N_{ic}, N_{bc}, N_{data} denote the numbers of learning points for the initial condition, boundary condition, and measurements (if available), respectively, and N_{pde} denotes the number of residual points (or collocation points or unsupervised points) of the PDE.

We note that PINNs may provide different performances with different network initialization. In this work when comparing the results of different configurations of PINNs, we train PINNs five times and choose the mean and one standard deviation values of the performance criteria of five models for visualization and numerical comparison.

2.2 Adaptive Learning Strategy for PDEs Residuals

Fixed-Budget Online Adaptive Learning. Motivated by the work of Lu et al. (2021) [5] which proposed the Residual-based Adaptive Refinement (RAR) that adds progressively during the training more collocation points at the locations that yield the largest PDE residuals, our primary idea is to control the number of potentially added training points by removing the collocation points that yield the smallest PDE residuals so that the number of collocation points remains the same during the training. With this approach, the added points tend to be placed at nearly the same location corresponding to the global maximum value of the PDE residual. We suggest considering a set of sub-domains in order to capture not only the global extrema but also the local extrema of the PDEs residuals. More precisely, we propose a Fixed-Budget Online Adaptive Learning

Algorithm 1. Fixed-Budget Online Adaptive Learning (FBOAL)

Require: The number of sub-domains d, the number of added and removed points m, the period of resampling k, a testing data set, a threshold s.
1: Generate the set $\mathcal{C}$ of collocation points on the studied domain Ω.
2: Divide the domain into d sub-domains $\Omega_1 \cup \Omega_2 ... \cup \Omega_d = \Omega$.
3: **for** lr_i in lr **do**
4: **repeat**
5: Train PINNs for k iterations with the learning rate lr_i.
6: Generate a new set $\mathcal{C}'$ of random points inside the domain Ω.
7: Compute the PDE residual at all points in the set $\mathcal{C}'$ and the set $\mathcal{C}$.
8: On each subdomain Ω_i, take 1 point of the set $\mathcal{C}'$ which yield the largest PDE residuals on the subdomain. Gather these points into a set $\mathcal{A}$.
9: On each subdomain Ω_i, take 1 point of the set $\mathcal{C}$ which yield the smallest PDE residuals on the subdomain. Gather these points into a set $\mathcal{R}$.
10: Add m points of the set $\mathcal{A}$ which yield the largest errors for the residuals to the set $\mathcal{C}$, and remove m points of the set $\mathcal{R}$ which yield the smallest errors for the residuals.
11: **until** The maximum number of iterations K is reached or the error of the prediction on the testing data set of reference is smaller than some threshold s.
12: **end for**

(FBOAL) that adds and removes collocation points that yield the largest and the smallest PDE residuals on the sub-domains during the training (see Algorithm 1). With the domain decomposition step, the algorithm is capable of detecting the local extrema inside the domain. Another remarkable advancement of this method is that in the parameterized problem, the collocation points can be relocated to the values of the parameter for which the solution is more complex (see Sect. 3.1 for an illustration on Burgers equation). To minimize the cost function, we adopt Adam optimizer with a learning rate decay strategy, which is proven to be very efficient in training deep learning models. In this work, we choose a set of learning rate values $lr = \{10^{-4}, 10^{-5}, 10^{-6}\}$. The way to divide the domain and the number of sub-domains can play important roles in the algorithm. If we dispose of expert knowledge on the PDEs problem, we can divide the domain as a finite-element mesh such that it is very fine at high-gradient locations and coarse elsewhere. In this primary work, we dispose of no knowledge *a priori* and use square partitioning for the domain decomposition. We note this partitioning is not optimal when dealing with multidimensional space and different scales of spatial and/or temporal dimensions. The stopping criterion is chosen based on a number of maximum iterations for each value of the learning rate and an error criterion computed on a testing data set of reference. The detail of this stopping criterion is specified in each use case. With this stopping criterion, the number of training iterations is also an adaptive number in each specific case, which helps to avoid unnecessary training iterations. We note that when dealing with multi-physics problems which involve systems of PDEs, we separate the set of collocation points into different subsets for each equation and then effectuate the process independently for each subset. Separating the set of collocation points helps to avoid the case when added points for one equation are removed for another. The code in this study is available from the GitHub repository https://github.com/nguyenkhoa0209/PINNs_FBOAL.

Residual-Based Adaptive Distribution and Residual-Based Adaptive Refinement with Distribution. Wu et al. (2023) [10] proposed two residual-based adaptive sampling approaches named Residual-based Adaptive Distribution (RAD) and Residual-based Adaptive Refinement with Distribution (RAR-D). In these approaches, the training points are randomly sampled according to a probability density function which is based on the PDE residuals. In RAD, all the training collocation points are re-sampled. While in RAR-D, only a few new points are sampled and then added to the training data set. The main differences between RAD, RAR-D, and our proposed method FBOAL lie in the domain decomposition step in FBOAL and the percentage of modified collocation points after every time we effectuate the re-sampling step. The decomposition step helps FBOAL to be able to detect local maxima and local minima of the residuals inside the domain (which, however, depends on the way we divide the domain), and thus FBOAL gives equal concentration for all local maxima of the PDE residuals.

3 Numerical Results

In this section, we first compare and demonstrate the use of adaptive sampling methods (FBOAL, RAD, and RAR-D) to solve the viscous Burgers equation in both non-parameterized context and parameterized context (i.e. the viscosity is fixed or not). We then illustrate the performance of FBOAL in a realistic industrial case: a system of PDEs that is used in the rubber calendering process. In the following, unless specifying otherwise, to compare the numerical performance of each methodology, we use the relative $\mathcal{L}^2$ error defined as $\epsilon_w = \dfrac{||w - \hat{w}||_2}{||w||_2}$ where w denotes the reference simulated field of interest and $\hat{w}$ is the corresponding PINNs prediction.

3.1 Burgers Equation

We consider the following Burgers equation:

$$\begin{cases} u_t + uu_x - \nu u_{xx} = 0 \quad \text{for } x \in [-1, 1], t \in [0, 1] \\ u(x, 0) = -\sin(\pi x) \\ u(-1, t) = u(1, t) = 0 \end{cases}$$

where ν is the viscosity. For a small value of ν, the solution is very steep close to $x = 0$. For higher values, the solution becomes smoother. Thus when ν is fixed, it is expected that the collocation points are located close to $x = 0$ during the training to better capture the discontinuity. When ν is varied, it is expected that the number of collocation points is more important for smaller values of ν while being located close to $x = 0$. In the following, we assess whether FBOAL can relocate the collocation points to improve the performance of PINNs.

In our experiment, to simplify the cost function and guarantee the boundary and initial conditions, these conditions are forced to be automatically satisfied by using the following representation for the prediction $\hat{\boldsymbol{u}} = t(x-1)(x+1)\mathcal{NN}(.) - \sin(\pi x)$. Interested readers may refer to the work in [4] for the general formulations. With this strategy, we do not need to adjust different terms in the loss function as there is only the loss for PDE residuals which is left. For the architecture of PINNs, we use a feedforward network with 4 hidden layers with 50 neurons per layer with *tanh* activation function. The results are obtained with 50,000 epochs with the learning rate $lr = 10^{-3}$, 200,000 epochs with the learning rate $lr = 10^{-4}$ and 200,000 epochs with $lr = 10^{-5}$. For a fair comparison, the initialization of the training collocation points is the same for all methodologies. The learning data set and the testing data set are independent in all cases.

Non-parameterized Problem. We first illustrate the performance of adaptive sampling approaches in a context where ν is fixed. We take 10 equidistant values of $\nu \in [0.0025, 0.0124]$. For each ν, we compare the performance of classical PINNs, PINNs with RAD, RAR-D, and PINNs with FBOAL. For the training of PINNs, we take $N_{pde} = 1024$ collocation points that are initialized equidistantly inside the domain. We take a testing set of reference solutions on a 10×10 equidistant spatio-temporal mesh and stop the training when either the number of iterations surpasses $K = 500,000$ or the relative $\mathcal{L}^2$ error between PINNs prediction and the testing reference solution is smaller than the threshold $s = 0.02$. The following protocol allows us to compare fairly all adaptive sampling methods. For the training of FBOAL, we divide the domain into $d = 200$ subdomains as squares of size 0.1. After every $k = 2,000$ iterations, we add and remove $m = 2\% \times N_{pde} \approx 20$ collocation points based on the PDE residuals. For the training of RAD, we take $k = 1$ and $c = 1$ and effectuate the process after $2,000$ iterations. For the training of RAR-D, we take $k = 2$ and $c = 0$ and after every $2,000$ iterations, 5 new points are added to the set of training collocation points. At the end of the training, the number of collocation points for FBOAL and RAD remains the same as at the beginning ($N_{pde} = 1024$), while for RAR-D, this number increases gradually until the stopping criterion is satisfied.

Figure 1 shows the PDE residuals for $\nu = 0.0025$ after the training process on the line $x = 0$ and at the instant $t = 1$ where the solution is very steep. The curves and shaded regions represent the geometric mean and one standard deviation of five runs. On the line $x = 0$ (Fig. 1a), with classical PINNs where the collocation points are fixed during the training, the PDE residuals are very high and obtain different peaks (local maxima) at different instants. It should be underlined that all the considered adaptive sampling methods are able to decrease the values of local maxima of the PDEs residuals. Among the approaches where the number of collocation points is fixed, FBOAL is able to obtain the smallest values for the PDE residuals, and for its local maxima, among all the adaptive resampling methodologies. At the instant $t = 1$ (Fig. 1b), the same conclusion can be drawn. However, we note that at this instant, the classical PINNs outperform other methods at the zone where there are no discontinuities (the zones that are not around $x = 0$). This is because the collocation points are fixed during the training process for classical PINNs, while with other methods, the collocation points are either re-assigned (with FBOAL and RAD) or some points are added (with RAR-D) to the high gradient regions. Thus, for adaptive resampling methods, the training networks have to additionally balance the errors for high-gradient locations and low-gradient locations and thus diminish the accuracy at the zones where there are low gradients as the cost function is a sum of the PDE errors at all points.

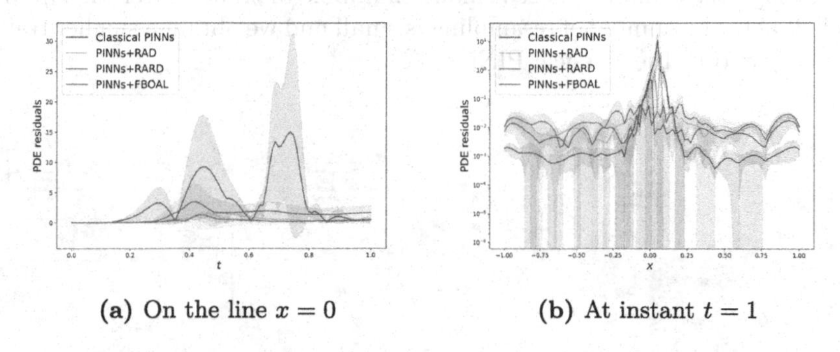

(a) On the line $x = 0$ **(b)** At instant $t = 1$

Fig. 1. Burgers equation: Absolute value of PDE residuals for $\nu = 0.0025$.

To assess the overall performance of PINNs, we evaluate the errors of the prediction on a 256×100 spatio-temporal mesh (validation mesh). Figure 2a shows the relative $\mathcal{L}^2$ error between PINNs predictions and reference solution. Figure 2b shows the number of training iterations for PINNs to meet the stopping criterion. As expected, when ν increases, which means the solution becomes smoother, the accuracy of PINNs in all methodologies increases and the models need less number of iterations to meet the stopping criterion. We note that when ν is large and the solution is very smooth, the classical PINNs are able to give comparable performance to PINNs with adaptative methodologies. However, when ν becomes smaller, it is clear that PINNs with adaptive sampling methods outperform classical PINNs in terms of accuracy and robustness. Among these strategies, FBOAL provides the best accuracy in terms of errors and also needs the least iterations to stop the algorithm. Table 1 illustrates the training time of each methodology for $\nu = 0.0025$. We observe that by using FBOAL a huge amount of training time is gained compared to other approaches. Figure 2c illustrates the cost function during the training process for $\nu = 0.0025$ and the errors of the prediction on the testing mesh. For clarity, only the best cost function (which yields the smallest values after the training process in five runs) of each methodology is plotted. After every $k = 2,000$ epochs, as the adaptive sampling methods relocate the collocation points, there are jumps in the cost function. The classical PINNs minimize the cost function better than other methodologies because the position of collocation points is fixed during the training. This leads to the over-fitting of classical PINNs on the training collocation points and does not help to increase the accuracy of the prediction on the testing mesh. While with adaptive sampling methods, the algorithm achieves better performance on the generalization to different meshes (the testing and validation meshes). Table 1 provides the training time and the number of resampling of each methodology with the hyperparameter values mentioned previously, which are optimal for all adaptive resampling methods. We see that the resampling does affect the training time. More precisely, with RAR-D and RAD, a huge number of resampling is effectuated, which leads to a long training time compared to the classical PINNs (even though these methods need smaller numbers of training iterations). While with FBOAL, the number of resampling is small and we obtain a smaller training time compared to the classical PINNs.

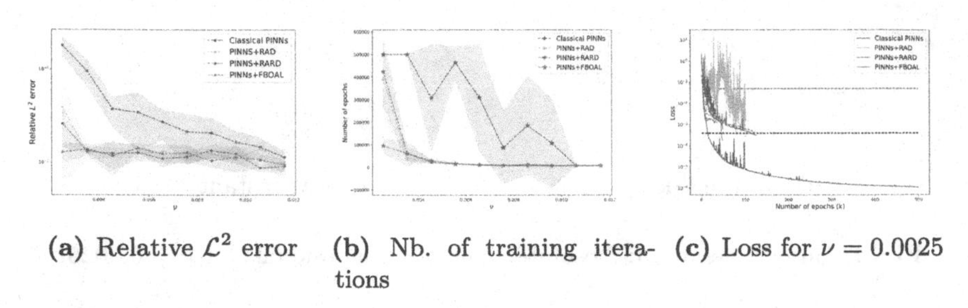

(a) Relative $\mathcal{L}^2$ error **(b)** Nb. of training iterations **(c)** Loss for $\nu = 0.0025$

Fig. 2. Burgers equation: Comparison of classical PINNs and PINNs with adaptive sampling approaches. In (c) the solid lines show the cost function during the training, the dashed lines show the errors on the testing mesh, and the black line shows the threshold to stop the training.

Table 1. Burgers equation: Training time and the number of resampling for $\nu = 0.0025$. The training is effectuated on an NVIDIA V100 GPU card.

	Classical	RAR-D	RAD	FBOAL
Training time (minutes)	33.7 ± 1.5	41.0 ± 5.8	38.8 ± 2.3	$\mathbf{21.5 \pm 2.7}$
Number of resampling	0 ± 0	201 ± 75	210 ± 77	$\mathbf{48 \pm 2}$

In the following, we analyze in detail the performance of FBOAL. Figure 3 illustrates the density of collocation points after the training with FBOAL for different values of ν. We see that, for the smallest value $\nu = 0.0025$, FBOAL relocates the collocation points close to x = 0 where the solution is highly steep. For $\nu = 0.0076$, as there is only one iteration of FBOAL that adds and removes points (see Fig. 2b), there is not much difference with the initial collocation points but we still see few points are added to the center of the domain, where the solution becomes harder to learn. For the biggest value $\nu = 0.0116$, there is no difference with the initial collocation points as PINNs already satisfy the stopping criterion after a few iterations. For the values of the hyperparameters, empirical tests (not shown here for concision) suggest starting with small values of m (number of added and removed points) and k (period of resampling) (for example $k = 1,000$, $m = 0.5\%N_{pde}$) and then increase these values to see whether the predictions can become more accurate or not.

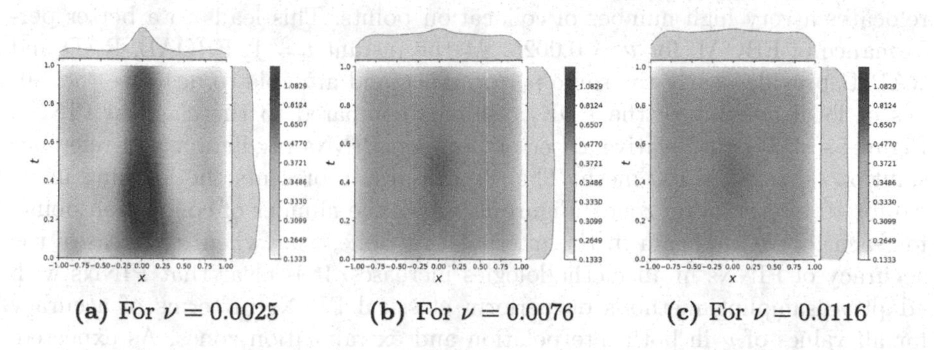

(a) For $\nu = 0.0025$ **(b)** For $\nu = 0.0076$ **(c)** For $\nu = 0.0116$

Fig. 3. Burgers equation: Density of collocation points after FBOAL training.

Parameterized Problem. We now illustrate the performance of PINNs in a parameterized problem where ν can be varied. In this case, ν is also considered as an input of PINNs, i.e. the network in PINNs is represented as $\mathcal{NN}(\mathbf{x}, t, \nu, \boldsymbol{\theta})$. With this configuration, PINNs can predict the solution for different values of ν in one model. For the training, we take 40 values of $\nu \in [0.0025, 0.0124]$. For each ν, we initialize 1024 equidistant collocation points, which leads to $N_{pde} = 1024 \times 40 = 40,960$ collocation points in total. During the training with FBOAL, the number of total collocation points remains the same, however, the number of collocation points can be varied for each ν. We take a testing set of reference solutions on a 10×10 equidistant spatio-temporal mesh and stop the training when either the number of iterations surpasses $K = 2 \times 10^6$ or the sum of relative $\mathcal{L}^2$ error between PINNs prediction and the testing reference solution of all learning values of ν is smaller than the threshold $s = 0.02 \times 40 = 0.8$. For the training of FBOAL, we divide the domain into $d = 200$ sub-domains as squares of size 0.1. After every $k = 2,000$ iterations, we add and remove $m = 0.5\% \times N_{pde} \approx 200$ collocation points based on the PDE residuals. For the training of RAD, we take $k = 1$ and $c = 1$ and effectuate the process after 2,000 iterations. For the training of RAR-D, we take $k = 2$ and $c = 0$ and after every 2,000 iterations, $5 \times 40 = 200$ new points are added to the set of training collocation points. At the end of the training, the number of collocation points for FBOAL and RAD remains the same as the beginning ($N_{pde} = 40,960$), while for RAR-D, this number increases gradually until the stopping criterion is satisfied.

Figure 4 illustrates the absolute value of the PDE residuals for $\nu = 0.0025$ on the line $x = 0$ and at the instant $t = 1$ where the solution is very steep. On the line $x = 0$ (Fig. 4a), with classical PINNs, the PDE residuals are very high and obtain different local maxima at different instants. Again, all the considered adaptive sampling methods are able to decrease the values of local maxima of the PDEs residuals. Among these approaches, FBOAL is able to obtain the smallest values for the PDE residuals. We note that the number of collocation points of each methodology is different (see Fig. 5b). For $\nu = 0.0025$, FBOAL relocates a very high number of collocation points. This leads to a better performance of FBOAL for $\nu = 0.0025$. At the instant $t = 1$, FBOAL, RAD and RAR-D provide nearly the same performance and are able to decrease the values of local maxima of the PDEs residuals compared to the classical PINNs. Figure 5a shows the relative $\mathcal{L}^2$ error between PINNs prediction and reference solution on a 256×100 mesh. The zone in gray represents the learning interval for ν (interpolation zone). Figure 5b shows the number of collocation points for each ν. We see again in the interpolation zone, that when ν increases, the accuracy of PINNs in all methodologies increases. It is clear that PINNs with adaptive sampling methods outperform classical PINNs in terms of accuracy for all values of ν in both interpolation and extrapolation zones. As expected, with the approaches FBOAL and RAD (where N_{pde} is fixed), the number of collocation points for smaller values of ν is more important than the number for higher ones. This demonstrates the capability of FBOAL and RAD of removing unnecessary collocation points for higher values of ν (whose solution is easier to learn) and adding them for smaller values of ν (as the solution becomes harder to learn). When ν tends to the higher extreme, the number of training points increases as there are fewer training values for ν in this interval. While with RAR-D (where N_{pde} increases gradually), we do not see the variation of the number of collocation points between different values of ν as RAR-D tries to add more points at the discontinuity for all ν. Table 2 provides the training time of and the number of resampling of each methodology. It is clear that RAR-D

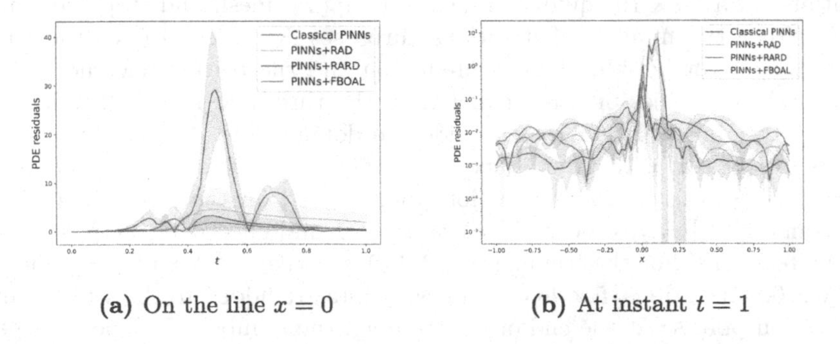

(a) On the line $x = 0$ **(b)** At instant $t = 1$

Fig. 4. Burgers equation: Absolute value of PDE residuals for $\nu = 0.0025$.

outperforms the other approaches in terms of computational time as this approach needs very few numbers of resampling to meet the stopping criterion. However, to achieve this gain, RAR-D added for about 40,000 new collocation points after the training, which is not profitable in terms of memory. RAD and FBOAL provide comparable computational training time in this case and they both outperform the classical PINNs while the number of collocation points in total is fixed. Figure 5c illustrates the cost function during the training process and the errors of the prediction on the testing mesh. For clarity, only the best cost function (which yields the smallest number of iterations after the training process in five runs) of each methodology is plotted. Again, since the position of the collocation points is fixed during the training, classical PINNs minimize the cost function better than PINNs with adaptive sampling methods, which leads to over-fitting on the training data set and does not help to improve the accuracy of the prediction on the testing mesh. We see that FBOAL and RAR-D need a much smaller number of iterations to meet the stopping criterion.

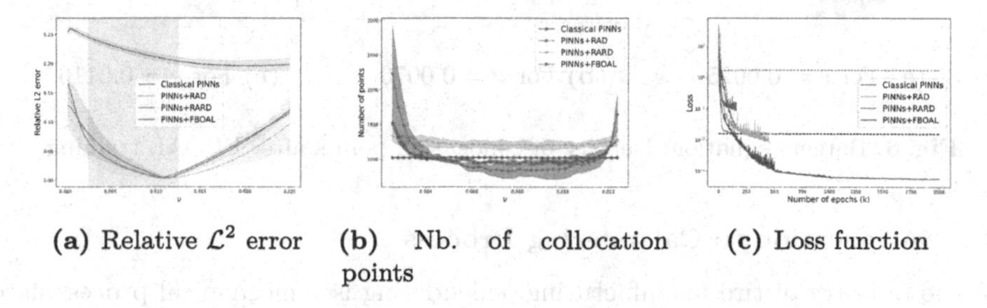

(a) Relative $\mathcal{L}^2$ error **(b)** Nb. of collocation points **(c)** Loss function

Fig. 5. Burgers equation: comparison of classical PINNs and PINNs with adaptive sampling approaches. The zone in gray is the learning interval. In (c) the solid lines show the cost function during the training, the dashed lines show the errors on the testing data set, and the black line shows the threshold to stop the training. (Color figure online)

In the following, we analyze in detail the performance of FBOAL. Figure 6 illustrates the density of collocation points after the training with FBOAL with different values of ν. We see that for all values of ν, FBOAL relocates the collocation points close to the discontinuity. For the values of the hyperparameters, empirical tests (not shown here for concision) suggest starting with small values of m (number of added and removed points) and k (period of resampling) (for example $k = 1,000, m = 0.5\% N_{pde}$) and then increase these values to see whether the predictions can become more accurate or not.

Table 2. Burgers equation: Training time and the number of resampling of each methodology. The training is effectuated on an NVIDIA A100 GPU card.

	Classical	RAR-D	RAD	FBOAL
Training time (hours)	13.2 ± 0.0	**1.4 ± 0.3**	9.1 ± 3.5	7.8 ± 1.9
Number of resampling	0 ± 0	**34 ± 8**	204 ± 71	173 ± 25

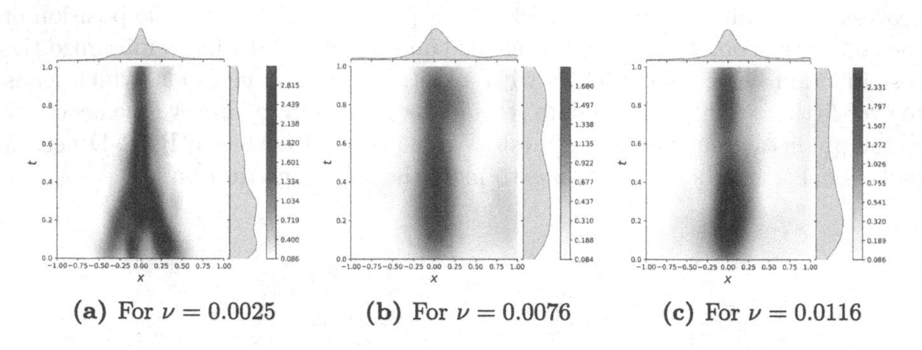

(a) For $\nu = 0.0025$ **(b)** For $\nu = 0.0076$ **(c)** For $\nu = 0.0116$

Fig. 6. Burgers equation: Density of collocation points after FBOAL training.

3.2 Application to Calendering Process

In the industry of tire manufacturing, calendering is a mechanical process used to create and assemble thin tissues of rubber. From the physical point of view, the rubber is assimilated as an incompressible non-Newtonian fluid flow in the present study (in particular, the elastic part of the material is not considered here). Moreover, only the steady-state regime is considered. The goal of the present study is only to model the 2D temperature, velocity, and pressure fields inside rubber materials going through two contra-rotating rolls of the calender as depicted in Fig. 7a. A detailed description of the calendering process and its mathematical formulation can be found in [7]. Here we briefly introduce the dimensionless PDEs system that is used in PINNs training process:

$$\tilde{\nabla}.\left(2\tilde{\eta}(\vec{\tilde{\boldsymbol{u}}},\tilde{T})\bar{\bar{\tilde{\epsilon}}}(\vec{\tilde{\boldsymbol{u}}})\right) - \vec{\tilde{\nabla}}\tilde{p} = \vec{0}$$

$$\tilde{\nabla}.\vec{\tilde{\boldsymbol{u}}} = 0$$

$$\vec{\tilde{\boldsymbol{u}}}\vec{\tilde{\nabla}}\tilde{T} = \frac{1}{Pe}\tilde{\nabla}^2\tilde{T} + \frac{Br}{Pe}\tilde{\eta}(\vec{\tilde{\boldsymbol{u}}},\tilde{T})|\tilde{\gamma}(\vec{\tilde{\boldsymbol{u}}})|^2$$

where $\vec{\tilde{\boldsymbol{u}}} = (\tilde{u}_x, \tilde{u}_y)^T$ is the velocity vector, $\tilde{p}$ is the pressure, $\tilde{T}$ is the temperature, $\bar{\bar{\tilde{\epsilon}}}$ is the strain-rate tensor, $\tilde{\eta}$ is the dynamic viscosity. Pe and Br are the dimensionless Péclet number and Brickman number, respectively.

We tackle an ill-posed configuration problem, i.e. the boundary conditions of the problem are not completely defined, and we dispose of some measurements of the temperature ($N_T = 500$ measurements, which are randomly distributed inside the domain). The goal is to infer the pressure, velocity, and temperature fields at all points in the domain. We note that here, no information on the pressure field is given, only its gradient in the PDE residual is. Thus the pressure is only identifiable up to a constant. As shown in [7], only the information of the temperature is not sufficient to guarantee a unique solution for the velocity and the pressure fields, we take in addition the knowledge of velocity boundary conditions. The authors also showed that taking the collocation points on a finite-element (FE) mesh, which is built thanks to domain expertise and provides *a priori* knowledge of high gradient location, improves significantly the accuracy of the prediction instead of taking the points randomly in the domain. However, to produce the FE mesh, expert physical knowledge is required. In this work, we show that FBOAL and other adaptive sampling methods RAD and RAD-D are able to infer automatically position for collocation points, and thus improve the performance of PINNs without the need for any FE mesh.

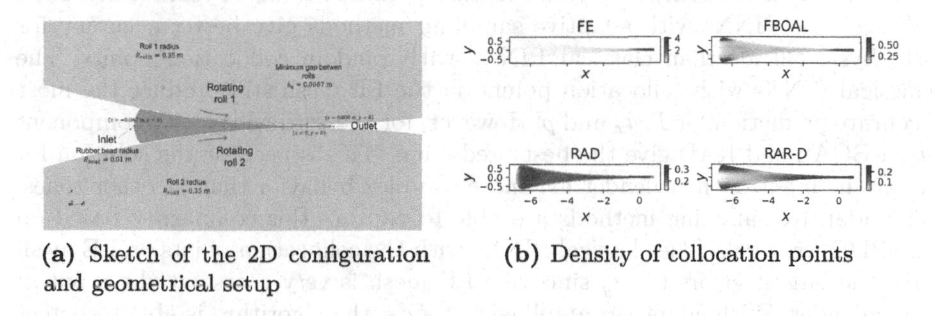

(a) Sketch of the 2D configuration and geometrical setup

(b) Density of collocation points

Fig. 7. Calendering process: Configuration and density of training points.

For the architecture of PINNs, we use a feedforward network with 5 hidden layers and 100 neurons per layer. To minimize the cost function, we adopt Adam optimizer with the learning rate decay strategy. The results are obtained with 50,000 epochs with the learning rate $lr = 10^{-3}$, 100,000 epochs with the learning rate $lr = 10^{-4}$ and 150,000 epochs with $lr = 10^{-5}$. For the training of PINNs, we suppose to dispose of the boundary condition for the velocity and $N_T = 500$ points of measurements for the temperature that are randomly distributed inside the domain. The number of collocation points is fixed as $N_{pde} = 10,000$ points. For the training of FBOAL, in this primary investigation, we do not divide the domain. After every $k = 25,000$ iterations, we add and remove $m = 2.5\% \times N_{pde} = 250$ collocation points based on the PDE residuals. Again, we note that in this case, since there are four PDEs residuals, we divide the set of collocation points into four separated subsets whose cardinal equal to $N_{pde}/4 = 2,500$, and then add and remove $m/4$ points independently for each equation. For the training of RAD, we choose $k = 1$, $c = 1$ and effectuate the re-sampling after every 25,000 iterations. For the training of RAR-D, we choose $k = 2$, $c = 0$ after every 25,000 iterations we add 250 new collocation points to the training set. Figure 7b shows the density of collocation points produced by different methods. The FE mesh is much finer at the output of the calender compared to other methods, where there are high-gradient locations. However, with FBOAL, we see that the collocation points are not only added to the output of the calender, but also to the input where there is contact with the solid rolls. The same conclusion can be drawn with RAR-D. With RAD, as this method redistributes all the collocation points at the same time, the observation for important zones is not as clear as in other methods.

To assess the performance of PINNs, we evaluate the errors of the prediction on a random mesh that contains $N = 162,690$ points. To avoid any artificial high values of the error for fields very close to zero, we use a relative $\mathcal{L}^2$ error that divides the absolute error by the reference field amplitude, which is defined as $\epsilon_w = \frac{||w - \hat{w}||_2}{w_{max} - w_{min}}$ where w denotes the reference simulated field of interest and $\hat{w}$ is the corresponding PINNs prediction. Table 3 shows the performance of PINNs with different cases of collocation points in terms of relative $\mathcal{L}^2$ error. We see that PINNs with adaptive sampling methods give better accuracy for all physical fields than classical PINNs with random collocation points. The classical PINNs with collocation points on the FE mesh still produce the most accurate prediction for T, u_x and p. However, for the vertical velocity component u_y, FBOAL and RAD give the best prediction. This is because the solution for u_y at the input of the calender has more complex behavior than at other zones. The adaptive sampling methods are able to capture this complexity based on the PDEs residuals. The classical PINNs with the collocation points on FE mesh give the largest errors for u_y since the FE mesh is very coarse at the input of the calender. With adaptive sampling methods, the algorithm is able to detect new zones which produce high errors for the PDEs residuals. Figure 8 illustrates the absolute error between the reference solution and the prediction produced by different methodologies for u_y. Again, we see that with adaptive sampling methods, the errors are significantly reduced compared to the classical approach with the collocation points taken on the FE mesh or random points.

Table 3. Calendering process: Relative $\mathcal{L}^2$ errors between reference solution and PINNs prediction with different cases of collocation points.

	ϵ_T	ϵ_{u_x}	ϵ_{u_y}	ϵ_P
Random points	13.6 ± 1.17	24.1 ± 3.77	15.6 ± 2.73	11.9 ± 1.33
FE mesh	**8.54 ± 1.39**	**14.7 ± 2.49**	18.0 ± 2.61	**1.41 ± 0.17**
FBOAL	10.5 ± 1.82	19.3 ± 1.85	**9.63 ± 3.24**	5.13 ± 0.72
RAD	10.2 ± 1.41	20.5 ± 1.96	**9.71 ± 2.55**	5.08 ± 0.76
RAR-D	12.3 ± 1.74	18.7 ± 1.65	11.1 ± 2.90	4.87 ± 0.69

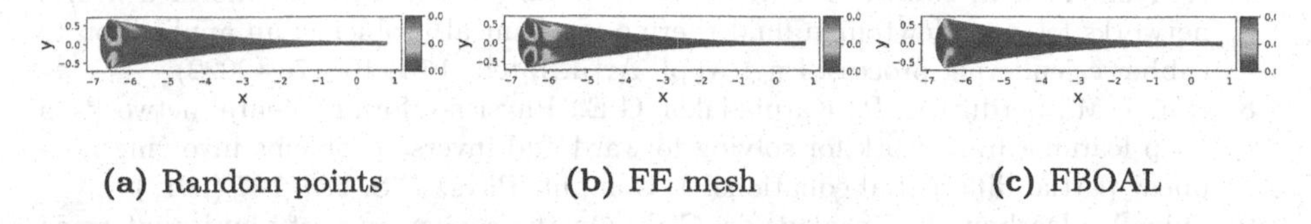

Fig. 8. Calendering process: Absolute error between the reference solution and the prediction produced by different methodologies for the vertical velocity component u_y.

4 Conclusion

In this paper, we introduced a Fixed-Budget Online Adaptive Learning (FBOAL) for the collocation points in PINNs that adds and removes points based on PDEs residuals on sub-domains while fixing the number of training points. The numerical results in academic test cases showed that FBOAL provides better accuracy with fewer iterations than classical PINNs. Besides that, we also compared the performance of FBOAL with other existing methods of adaptive sampling for the collocation points such as Residual-based Adaptive Distribution (RAD) and Residual-based Adaptive Refinement with Distribution (RAR-D). It is shown that in most cases, FBOAL is able to provide a comparable or even better performance than other approaches in terms of accuracy and computation cost. We then apply the methods FBOAL, RAD, and RAR-D to the rubber calendering process simulation. PINNs with these adaptive sampling methods give remarkably better predictions for the vertical velocity component than classical PINNs with the collocation points taken on an expert-designed FE mesh. This promising result demonstrates that the adaptive sampling methods can help to provide *a prior* knowledge of high-gradient locations and improve the conventional numerical solver in the construction of the mesh.

References

1. Daw, A., Bu, J., Wang, S., Perdikaris, P., Karpatne, A.: Rethinking the importance of sampling in physics-informed neural networks. arXiv:2207.02338 (2022)
2. Fu, J., et al.: Physics-data combined machine learning for parametric reduced-order modelling of nonlinear dynamical systems in small-data regimes. Comput. Methods Appl. Mech. Eng. **404**, 115771 (2023)
3. Karniadakis, G.E., Kevrekidis, I.G., Lu, L., Perdikaris, P., Wang, S., Yang, L.: Physics-informed machine learning. Nat. Rev. Phys. **3**(6), 422–440 (2021)
4. Liu, Z., Yang, Y., Cai, Q.D.: Solving differential equation with constrained multilayer feedforward network. arXiv:1904.06619 (2019)
5. Lu, L., Meng, X., Mao, Z., Karniadakis, G.E.: DeepXDE: a deep learning library for solving differential equations. SIAM Rev. **63**(1), 208–228 (2021)
6. Nabian, M.A., Gladstone, R.J., Meidani, H.: Efficient training of physics-informed neural networks via importance sampling. Comput.-Aided Civ. Infrastruct. Eng. **36**(8), 962–977 (2021)

7. Nguyen, T.N.K., Dairay, T., Meunier, R., Mougeot, M.: Physics-informed neural networks for non-Newtonian fluid thermo-mechanical problems: an application to rubber calendering process. Eng. Appl. Artif. Intell. **114**, 105176 (2022)
8. Raissi, M., Perdikaris, P., Karniadakis, G.E.: Physics-informed neural networks: a deep learning framework for solving forward and inverse problems involving non-linear partial differential equations. J. Comput. Phys. **378**, 686–707 (2019)
9. Shin, Y., Darbon, J., Karniadakis, G.E.: On the convergence of physics informed neural networks for linear second-order elliptic and parabolic type PDEs. arXiv:2004.01806 (2020)
10. Wu, C., Zhu, M., Tan, Q., Kartha, Y., Lu, L.: A comprehensive study of non-adaptive and residual-based adaptive sampling for physics-informed neural networks. Comput. Methods Appl. Mech. Eng. **403**, 115671 (2023)

Towards Online Anomaly Detection in Steel Manufacturing Process

Jakub Jakubowski[1,2](✉), Przemysław Stanisz[2], Szymon Bobek[3], and Grzegorz J. Nalepa[3]

[1] Department of Applied Computer Science, AGH University of Science and Technology, 30-059 Krakow, Poland
jjakubow@agh.edu.pl
[2] ArcelorMittal Poland, 31-752 Krakow, Poland
[3] Faculty of Physics, Astronomy and Applied Computer Science, Institute of Applied Computer Science, and Jagiellonian Human-Centered AI Lab (JAHCAI), and Mark Kac Center for Complex Systems Research, Jagiellonian University, ul. prof. Stanisława Łojasiewicza 11, 30-348 Kraków, Poland

Abstract. Data generated by manufacturing processes can often be represented as a data stream. The main characteristics of these data are that it is not possible to store all the data in memory, the data are generated continuously at high speeds, and it may evolve over time. These characteristics of the data make it impossible to use ordinary machine learning techniques. Specially crafted methods are necessary to deal with these problems, which are capable of assimilation of new data and dynamic adjustment of the model. In this work, we consider a cold rolling mill, which is one of the steps in steel strip manufacturing, and apply data stream methods to predict distribution of rolling forces based on the input process parameters. The model is then used for the purpose of anomaly detection during online production. Three different machine learning scenarios are tested to determine an optimal solution that fits the characteristics of cold rolling. The results have shown that for our use case the performance of the model trained offline deteriorates over time, and additional learning is required after deployment. The best performance was achieved when the batch learning model was re-trained using a data buffer upon concept drift detection. We plan to use the results of this investigation as a starting point for future research, which will involve more advanced learning methods and a broader scope in relation to the cold rolling process.

Keywords: data streams · anomaly detection · cold rolling

1 Introduction

Progressing digitalization of the industry has led to the production of enormous amounts of data, which possesses many characteristics of data streams. These data are generated in form of a continuous flow of information from sensor readings, which is produced with high speed and volume and is infinite in nature, as new data will be generated as long as the manufacturing process is in operation.

J. Mikyška et al. (Eds.): ICCS 2023, LNCS 14076, pp. 469–482, 2023.
https://doi.org/10.1007/978-3-031-36027-5_37

In many cases, it is not possible to store all sensor readings and process them in batch mode. This approach might be infeasible due to hardware or software constraints. The manufacturing process may evolve over time (concept drift) due to factors like wear of the asset, changes in production mix, or modifications in the production process. Still, sensor data may give valuable insight into the nature of the process and improve its performance by e.g., adapting the process to new conditions or detecting anomalies in near real time. Therefore, it can be very beneficial for companies to use these data; however, its processing pipeline should be well adjusted to the specific problem and take into account important characteristics of the data.

Learning from data streams is a challenging task that involves dealing with many data problems, i.e., inability to process all data at once, variations in data distribution over time, class imbalance, delayed or inaccessible labels [13]. It often requires the assimilation of most recent data to adapt the model to dynamically changing conditions. To build a robust machine learning model to control the manufacturing process, all the mentioned factors must be carefully analyzed and addressed.

In this work, we present our preliminary results on the machine learning model that controls the steel cold rolling process and detects potential anomalous measurements. Cold rolling is an important step in the steel manufacturing process, where the thickness of the steel is reduced to reach the dimensions requested by the client. One of the critical parameters of this process is the rolling force, which should be carefully monitored to ensure the quality of the final product [24]. To control the rolling forces, we propose a machine learning approach, which learns the proper distribution of forces based on the given steel properties and rolling parameters. The main goal of the proposed solution is to detect anomalies in the rolling force in real time. We exploit three different scenarios of the ML pipeline to handle the problem and evaluate them in terms of their learning capabilities. Our baseline scenario assumes processing the small part of the data in batch mode and evaluating its results on the rest of the data. In the second scenario, we retrain the model in batch mode using a sliding window every time a drift is detected. Finally, we use online ML models to learn from the data continuously as each data point arrives. We also discuss and tackle the problem of data imbalance, which is present in our dataset and is one of the important concerns when learning from streaming data. To the best of our knowledge, there is no paper dealing with the problem of anomaly detection in cold rolling by using data stream learning techniques. One of the issues that we discuss in this paper is the problem of distinction between anomalies and concept drift that might occur in the data stream.

The rest of the paper is organized as follows. In Sect. 2 we briefly discuss state-of-the-art techniques for learning from data streams and provide details on the cold rolling mill. In Sect. 3 we present the details of the proposed anomaly detection model and the learning scenarios. In Sect. 4 we present our preliminary results from the experiments carried out and discuss them. In Sect. 5 we conclude our findings and discuss the potential directions of future research in this topic.

2 Related Works

2.1 Data Streams

Learning from data streams is a well-established problem in the field of machine learning and has been a topic of many studies [10,11]. The main requirements for an ML model in streaming data setting that need to be faced include: (1) the computation time and memory usage should be irrespective of the total number of observed measurements, (2) the model should be able to learn by using each observation only once, (3) the model should perform similarly to equivalent model trained in batch setting, and (4) the model should be able to adapt to any changes in the data distribution, but should not forget relevant information from the past (catastrophic forgetting). In practice, many of these requirements are difficult to meet as some trade-offs might occur. For example, increasing the complexity of the model may allow to increase its accuracy but may deteriorate its ability to learn in real-time. Models such as neural networks benefit greatly from making many passes (epochs) over the same chunk of data, so limiting learning to a single pass over data will usually decrease their performance.

Many state-of-the-art machine learning models, e.g., XGBoost, Random Forest, require access to the entire dataset for training. After the training has ended, it is not possible to adjust the model to the new data while preserving some part of the information learned earlier. To solve this problem, specific algorithms for online learning, which can learn incrementally as new data arrive, were proposed [12,15]. On the other hand, ordinary ML algorithms can also be useful in streaming data applications if we are able to store some part of data in memory in the form of, e.g. sliding window.

Concept dirft is one of the biggest problems faced when dealing with streaming data. Changes in data can be *actual*, when there occurs a change in the decision boundaries, or *virtual*, when the decision boundary remains stable, but the distribution of the data within each class changes [21]. Many concept drift detection methods have been proposed, such as EDDM [2], ADWIN [3] to list a few. They focus on observing the performance of the model and detecting the point in time in which the model starts to deteriorate. This is an indication that the current model no longer preserves its original accuracy and that some adaptations are needed to recover the model.

The problem of data imbalance is well known and has been studied, especially in batch learning [14]. However, in streaming data, it is particularly difficult to deal with class imbalance due to factors like dynamically changing distribution of the data. The class imbalance may also affect the drift detection algorithms [6] where, for example, the classifier has a lower accuracy on rare examples. There are several strategies to deal with imbalanced data. Oversampling strategies aim to generate new samples from minority classes, while undersampling selects only some portion of the data from majority classes to achieve better data balance. There exist also hybrid approaches, which combine the above mentioned strategies.

2.2 Anomaly Detection

Anomaly detection methods aim to discover the outliers in a dataset, that do not fit to the observed distribution and usually constitute a small fraction of the data. In supervised learning scenarios, where each observation has an anomaly label assigned, the problem can be described as a specific type of imbalanced learning with binary classification task. However, in practical applications, especially in manufacturing processes, the data are usually generated without labels. In such cases, it is not possible to use supervised learning techniques.

On the other hand, unsupervised learning methods are very robust in anomaly detection tasks. There exist several well-studied techniques for anomaly detection based on tree or clustering algorithms, which also have their implementation for online learning problems [25,26]. Another approach to detect anomalies is to train a regular ML model to learn the relationship between independent and dependent variables and assume the error of the model as the anomaly level.

2.3 Cold Rolling Process

ArcelorMittal Poland is the largest steel producer in Poland, and its production chain includes all relevant steps from the production of pig iron to the final product in the form of a steel strip. In the steel plant, the thickness of the steel is about 220 mm. One of the important steps in the production of steel strips is to reduce this thickness to obtain the product ordered by customers. First, the steel is hot rolled, where the thickness of the steel is reduced at high temperature to a range of about 2 to 5 mm. If there is a demand for a lower thickness, a cold rolling is necessary. During cold rolling, a metal strip is passed through a rolling mill, consisting of many subsequent pairs of work rolls without prior preheating. The minimum strip thickness produced in the analyzed plant is below 0.4 mm. The simplified scheme of the considered rolling mill is presented in Fig. 1.

During contact of the strip with the work rolls, a force is applied to the rolls, which is transferred to the strip and causes its reduction in thickness and equivalent elongation. In steel manufacturing mills, the cold rolling is performed at very high speeds (sometimes exceeding 1000 m m/min), which requires a very fast adaptation of process parameters to achieve the desired thickness and flatness of the final product. These adaptations are usually made by a PLC, that controls the whole process. The superior goal of the control system is to achieve desired thickness after the last stand, while preserving the safe limits of the speed, forces, etc.

There are several process parameters that can be used to evaluate the condition of the rolling mill. One of the most important parameters is the rolling force applied in each stand. In general, these forces should be within certain limits based on the product characteristics (thickness, width, steelgrade) and process parameters (speed, state of the work rolls). High deviations of the forces from the normal working condition may be a symptom of malfunctions and can lead to dangerous situations, e.g., strip breaks. However, taking into account the

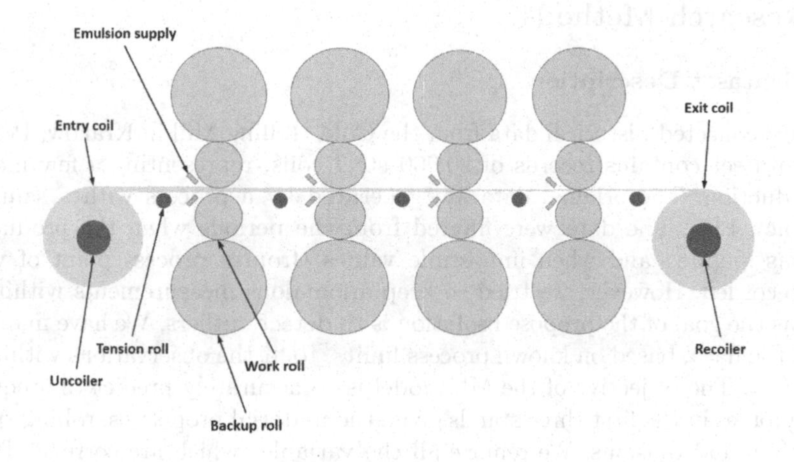

Fig. 1. Schematic diagram of tandem cold mill with four rolling stands.

number of variables that influence the force value, it is infeasible to build simple rules to control their values in a kind of manual mode.

Rolling mechanics has been studied for decades, and there are several state-of-the-art physical models of this process [1,4]. The main problem with the physical models is that they usually rely on some assumptions, which are hard to precisely determine during online production. For example, the Bland and Ford model [4] is highly dependent on friction between the strip and the work roll, which is a kind of stochastic parameter and is almost unmeasurable in the production environment [20]. This implies that more robust machine learning approaches could be used to model the cold rolling process. Several researchers proposed using ML methods in the analysis of the cold rolling process [8,16,19], but we have not found work that used the approach of learning from streaming data. The aforementioned characteristics of the data streams are consistent with the characteristics of the steel rolling process. Important process parameters, for example, steel yield strength, friction coefficient, are prone to systematic variations due to changes in roll roughness, lubrication, or unknown variations in previous production steps. Therefore, it is important that the model developed for the cold rolling process is robust and has the ability to evolve under changing conditions. On the other hand, the model should remember relevant patterns from the past – this is important especially for the rarely produced steels, which can appear in the schedule only a few times in a year.

3 Research Methods

3.1 Dataset Description

We have collected historical data from the Cold Rolling Mill in Kraków, Poland. The data set contains records of 20,000 steel coils, representing a few months of production. The original data are generated by a process with a sampling frequency 1 Hz. The data were filtered from the periods when the production line was stopped and when impossible values (from a process point of view) were recorded. However, we tried to keep anomalous measurements within the data, as the goal of the proposed solution is to detect outliers. We have manually scaled the data, based on known process limits, to fit the observations within the range 0–1. The objective of the ML model is to accurately predict the required rolling forces in the first three stands given the material properties, rolling speed, reduction, and tensions. We remove all the variables which are correlated with force but are not independent of it, e.g., rolling gap, motor torque, and motor load. The final dataset consists of over 250,000 observations and 23 features. Next, on the basis of the mean squared error (MSE) between the predicted and measured rolling forces, the anomaly score is calculated.

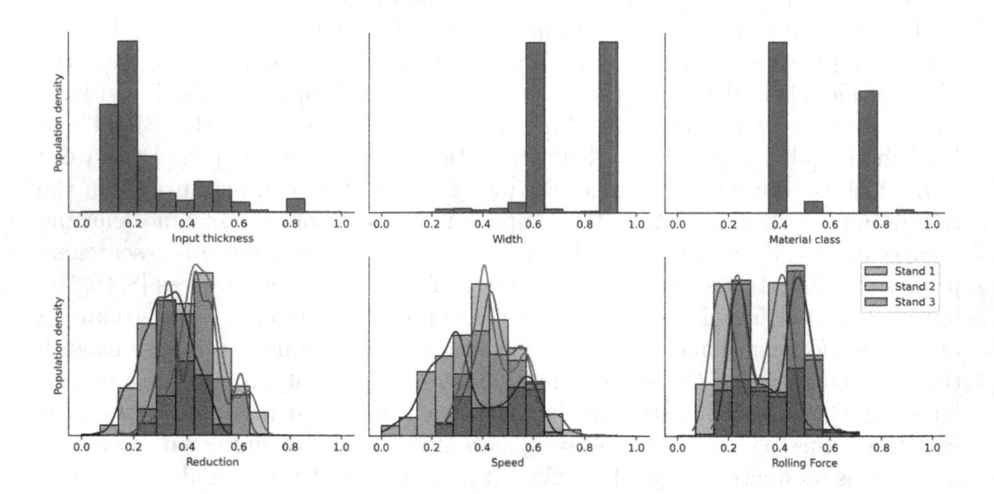

Fig. 2. Normalized data distribution of the most relevant features

Figure 2 presents the distribution of selected features. These distributions show that we have a potential issue with inbalanced data, since most of the production is a thin material with a very specific width and material class. We address this issue and describe our solution in Sect. 3.2.

3.2 Dealing with Imbalanced Data

In Sect. 3.1 we discuss the problem of data imbalance, which is caused by the fact that the majority of production consists of few types of products and the other products are not well represented in the dataset. This could easily lead to overfitting the model to a majority group of products.

To solve this problem, we propose to use a clustering algorithm to group the data based on intrinsic product characteristics, that is, thickness, width, and material class. We use the KMeans model, which is built on the training dataset to assign each observation to a cluster. The number of clusters is determined based on the Silhouette Score [23].

In the next step, we have used a sampling algorithm, which combines the under-sampling and over-sampling methods. We have iterated over each point and computed the repetitions of that sampled point on the basis of the cluster size and the Poisson distribution. In practice, some of the samples from overrepresented clusters were omitted, and the samples from underrepresented clusters were duplicated. To avoid generating too many of the very same points, we limited the number of possible sample repetitions to three.

3.3 Learning Scenarios

To simulate the data stream environment, we have divided the dataset into the following proportions: 15% for training, 5% for validation (hyperparameter tuning) and 80% for testing. It is important to mention that although we divide the data in an ordinary way, in some scenarios the test data is used to learn the model, e.g. in online learning where the model never stops learning. In this work, we analyze three scenarios regarding the model learning process in order to find the best solution to the problem.

Batch Model. This is the baseline scenario, which assumes that we follow an ordinary machine learning pipeline, where the model is trained only once on a training dataset. This scenario assumes that we are able to collect enough data during the model development and that no concept drift will occur in the future.

Batch Model with Retraining. In this scenario, we train the model as described in the above scenario. The trained model is put into production (run on a test dataset) along with a concept drift detection algorithm, which controls if the current model is still valid. We also keep a fixed-size buffer in the form of a sliding window (FIFO). If at any point a drift is detected, the current model is dropped and a new model is built, which is trained on the data stored in the buffer. Therefore, the assimilation of the new data is made upon detection of concept drift.

Online Learning. In this setting, we apply an online learning paradigm, where the model is learned iteratively as new samples arrive. This scenario does not require separation of the data into training, validation, and test sets, but we keep such a structure to be more coherent with previous scenarios to allow for a trustworthy evaluation of the results. In this setting, the assimilation of the recent data is made as soon as it is accessible. Since we assume that we have access to the data stored in the training and validation sets, we can calculate the performance metrics on the validation set to perform hyperparameter tuning.

3.4 Learning Algorithms and Validation

In this work we have used three different types of machine learning algorithms, that is:

- Linear Model (with L1 regularization)
- Multi-Layer Perceptron (MLP), which is one of the basic types of Artificial Neural Networks (ANN)
- Random Forest (RF).

In the latter case, more specifically, we have used a forest of CART [5] in a batch setting and Adaptive Random Forest [12] (ARF) built with Hoeffding Trees [9] in an online setting. We have utilized *sckit-learn* [7] library for training batch models and *river* [22] library for online learning. The choice of models was mainly motivated by the fact that all three types of models have their implementation for batch and online learning. The focus of this work is on selecting the best learning scenario, rather than achieving the best possible performance of the model.

The hyperparameters of each model were determined using a grid search method by fitting the model to the training dataset and evaluating its mean absolute error (MAE) in the validation data set. We have chosen MAE as the main metric for model evaluation to minimize the effect of anomalies, which we want to find in the next step. However, we also evaluate the models based on the root mean squared error ($RMSE$) and the coefficient of determination (R^2).

An important aspect of model evaluation is the proper selection of observations for the calculation of performance metrics. In our use-case we deal with situations where we observe very similar points for a longer period of time due to e.g. production of the same steel strip for a longer period of time. Such a setting results in situation where dummy predictor, which uses previous measurement as the prediction, may have very high performance metrics. To address this issue, we have limited the data used for evaluation purposes. The condition to include the observation in the evaluation data was that there is a significant difference between the current and previous force measurements (set at 3% of the total range).

3.5 Anomaly Detection

The anomaly detection is performed by comparing the measured rolling forces with the values predicted by the model. This assumes that the model has learned the relationship between input and output features precisely. The difference between those values can be treated as the anomaly level. The magnitude of the anomaly, which we will refer to as the *anomaly score*, is computed by calculating MSE between the observed and predicted forces.

An important aspect, which we consider in our work, is that we must be able to distinguish between anomalies and concept drifts. In some cases the rapid changes in the value of the force may not be a clear symptom of anomaly, but a sudden drift which occurred in the process. We note that sometimes it might not be possible to distinguish between them, as a concept drift might be a result of some persistent failure (which from the process point of view should be treated as an anomaly). Moreover, we assume that the model should not update itself on anomalous samples; therefore, once detected, they should not be included in the model learning in the online phase. To resolve these issues, we propose the following methodology. First, we determine the running statistics of the mean anomaly score (μ) and its standard deviation (σ). Next, we compare the anomaly score of a single observation with the distribution of all calculated anomaly scores.

1. If the anomaly score is less than $\mu + 2\sigma$, there is no anomaly, and any ongoing deviations can be treated as concept drift.
2. If the anomaly score is between $\mu + 2\sigma$ and $\mu + 3\sigma$, there is a high risk that we observe an anomaly, but we consider this as a transient state and only raise a warning. The model is not updated with this data.
3. If the anomaly score is above $\mu + 3\sigma$, we raise an alert, which means that the anomaly is detected.

Nevertheless, we use the observations marked as anomalies in the calculation of model performance metrics to avoid the situation when the incorrectly learned relations are discarded from the validation.

4 Results and Discussion

In this section, we present the results of our research. The first step of the machine learning pipeline was the resampling of the original data. To determine which observations to undersample and which to oversample, we have clustered the observations, as discussed in Sect. 3.2, with the KMeans algorithm. For our application, the highest Silhouette score was achieved when the number of clusters was $n = 6$. Figure 3 presents the number of observations in the training set before and after resampling.

Below, we present the achieved performance metrics of each model on the test dataset – each model was verified in terms of its mean absolute error, root mean squared error, and coefficient of determination. We also evaluated each model in terms of the share of anomalies, which is the number of anomalies (determined as described in Sect. 3.5) as a fraction of all observations. The results are listed in Table 1.

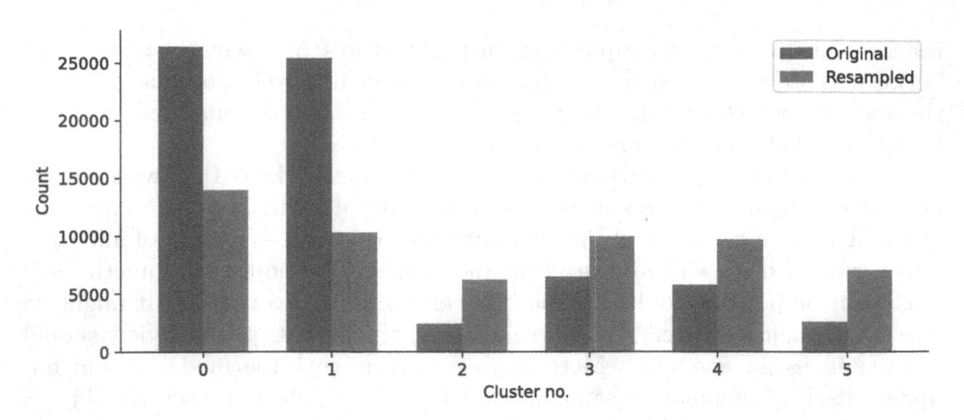

Fig. 3. Count of original and resampled data with respect to the assigned cluster.

The results show that the best performance of the model has been achieved in a setting with a batch model and a concept drift detector. Surprisingly, the best metrics were achieved for Linear Models, which are characterized by the least complexity. This induces the correlations between the variables to be mostly linear, and increasing the complexity of the model might harm the ability of the model to generalize. However, the metrics achieved for the MLP network were comparable, indicating that some more fine-tuning of this algorithm could help outperform the Linear Model. The most important observation is that there is a significant increase in performance of the models, if the model learns continuously. This implies that a model without the ability to adapt as new data come is not able to precisely predict the rolling force and thus potential anomalies. Such a model in production could result in too many false alarms, which is an undesirable scenario.

The proposed methodology for anomaly detection results in having between 1.0% and 2.5% of alarms. These values seem to be in acceptable range; however, this problem should be further addressed to determine optimal anomaly threshold for the analyzed process. We observe that online models tend to give less alarms than other methods, which is most probably due to its ability to quickly adapt to changing conditions.

Figure 4 presents how the anomaly score varies with the number of observations. In the initial stage, we note that all three models have a similar anomaly score. However, after approximately 12,000 observations, a sudden drift occurs, leading to an increase in the anomaly score for batch models. The online model maintains a similar anomaly score as before. Subsequently, the batch model equipped with drift detection is capable of adapting to new data and reducing the anomaly score, which is the situation we expected.

Figure 5 presents the sample of data, where the rolling force in the first stand is plotted over time. Comparison of the measured force with the calculated ones clearly shows that at a given point in time the accuracy of the unadapted batch model is much worse than that of the other two models. Furthermore,

Table 1. Achieved model performance metrics and corresponding percentage of observations marked as anomalies. The best results are highlighted in bold.

Method	ML model	Metrics			Anomaly share
		MAE	RMSE	R^2	
Batch Learning	Linear	0.052	0.066	0.74	1.6%
	ANN	0.050	0.063	0.76	2.5%
	RF	0.047	0.059	0.79	2.2%
Batch Learning with Retraining	Linear	**0.033**	**0.044**	**0.89**	1.8%
	ANN	0.035	0.046	0.88	1.9%
	RF	0.038	0.051	0.85	2.2%
Online Learning	Linear	0.035	0.046	0.88	1.4%
	ANN	0.038	0.051	0.85	1.4%
	ARF	0.036	0.047	0.87	1.0%

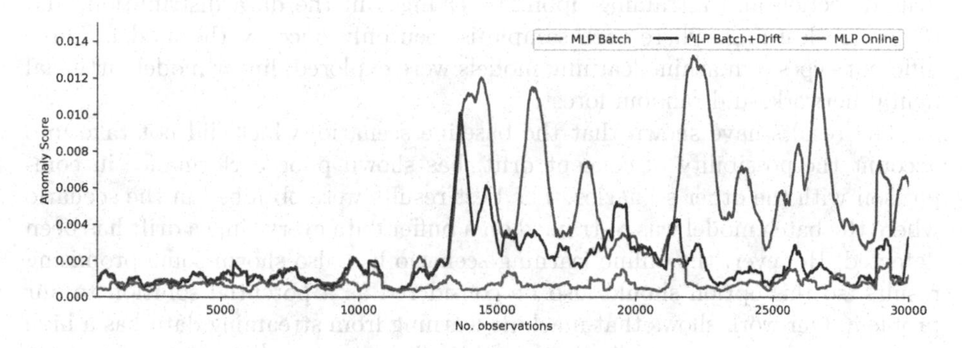

Fig. 4. Examplary anomaly score variations (moving average) for the MLP models.

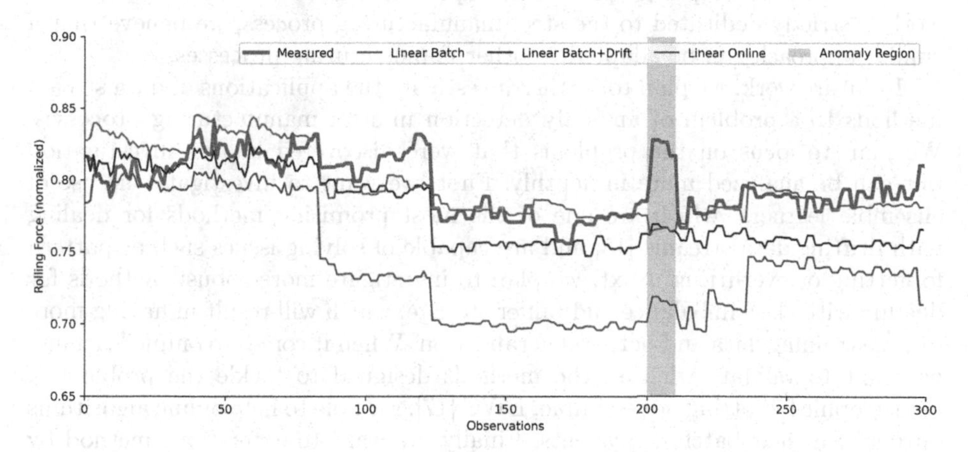

Fig. 5. Examplary rolling forces in the first stand measured and calculated by the linear models.

a discovered anomaly region has been highlighted, where the force has largely increased for a certain period of time, but the models did not predict it to happen, showing that they are able to indicate the anomaly in the process.

5 Conclusion and Future Works

In this paper, we have applied machine learning models to predict rolling forces in a cold rolling process. The goal of this model is to control the manufacturing process online and detect anomalies. The characteristics of the problem imply the use of methods designed for data streams. The main data stream features, which we considered were the inability to store a whole data set in memory, the possibility of concept drift occurring, and imbalances in the data. We have considered three learning scenarios: (1) batch learning without any changes in the model once it is deployed in production, (2) batch learning with concept drift detection and retraining upon the changes in the data distribution, and (3) online learning, where each sample is seen only once by the model. Three different types of machine learning models were explored; linear model, artificial neural network, and random forest.

The results have shown that the baseline scenario, which did not take into account the possibility of concept drift, has shown poor performance in comparison with the other scenarios. The best results were obtained in the scenario where the batch model was re-trained on a buffer data every time a drift has been detected. However, the online learning scenario has also shown some promising results, so this option should also be considered as a potential solution to our problem. Our work shows that machine learning from streaming data has a high potential to discover anomalies in the cold rolling process. Assimilation of recent data is crucial to keep high prediction capabilities of the model. Although our work is strictly dedicated to the steel manufacturing process, we believe that a similar approach can be adapted to other manufacturing processes.

In future work, we plan to further investigate the applications of data stream methods to a problem of anomaly detection in steel manufacturing processes. We plan to focus on the problems that were discovered in this investigation but can be analyzed more in-depthly. First, we want to investigate the use of ensemble learners, which are one of the most promising methods for dealing with drifting data streams [18] and are capable of solving issues such as pattern forgetting or overfitting. Next, we plan to investigate more robust methods for dealing with class imbalance and buffer storage, which will result in having more diverse training data and better generalization. When it comes to online learning, we want to validate whether the methods designed to tackle the problem of catastrophic forgetting, for example, EWC [17], are able to help online algorithms outperform their batch equivalents. Finally, we want to extend our method by detecting anomalies not only in the rolling force measurements but also in other dependent variables such as gap position or motor load.

Acknowledgements. Project XPM is supported by the National Science Centre, Poland (2020/02/Y/ST6/00070), under CHIST-ERA IV programme, which has received funding from the EU Horizon 2020 Research and Innovation Programme, under Grant Agreement no 857925.

References

1. Alexander, J.M.: On the theory of rolling. Proc. R. Soc. Lond. Ser. A Math. Phys. Sci. **326**(1567), 535–563 (1972). http://www.jstor.org/stable/77929
2. Baena-García, M., Campo-Avila, J.D., Fidalgo, R., Bifet, A., Gavalda, R., Morales-Bueno, R.: Early drift detection method (2005)
3. Bifet, A., Gavaldà, R.: Learning from time-changing data with adaptive windowing (2007)
4. Bland, D.R., Ford, H.: The calculation of roll force and torque in cold strip rolling with tensions. Proc. Inst. Mech. Eng. **159**(1) (1948)
5. Breiman, L., Friedman, J.H., Olshen, R.A., Stone, C.J.: Classification and regression trees (1984)
6. Brzezinski, D., Stefanowski, J., Susmaga, R., Szczech, I.: On the dynamics of classification measures for imbalanced and streaming data. IEEE Trans. Neural Netw. Learn. Syst. **31**, 2868–2878 (2020). https://doi.org/10.1109/TNNLS.2019.2899061
7. Buitinck, L., et al.: API design for machine learning software: experiences from the scikit-learn project. In: ECML PKDD Workshop: Languages for Data Mining and Machine Learning, pp. 108–122 (2013)
8. Chen, Z., Liu, Y., Valera-Medina, A., Robinson, F.: Strip snap analytics in cold rolling process using machine learning. In: 2019 IEEE 15th International Conference on Automation Science and Engineering (CASE), pp. 368–373 (2019). https://doi.org/10.1109/COASE.2019.8842967
9. Domingos, P., Hulten, G.: Mining high-speed data streams. In: Proceedings of the Sixth ACM SIGKDD International Conference on Knowledge Discovery and Data Mining, KDD 2000, pp. 71–80. Association for Computing Machinery, New York (2000). https://doi.org/10.1145/347090.347107
10. Domingos, P., Hulten, G.: Catching up with the data: research issues in mining data streams (2001)
11. Gama, J.: Knowledge Discovery from Data Streams, 1st edn (2010)
12. Gomes, H.M., et al.: Adaptive random forests for evolving data stream classification. Mach. Learn. **106**, 1469–1495 (2017). https://doi.org/10.1007/s10994-017-5642-8
13. Gomes, H.M., Read, J., Bifet, A., Barddal, J.P., Gama, J.: Machine learning for streaming data. ACM SIGKDD Explor. Newsl. **21**, 6–22 (2019). https://doi.org/10.1145/3373464.3373470
14. Haixiang, G., Yijing, L., Shang, J., Mingyun, G., Yuanyue, H., Bing, G.: Learning from class-imbalanced data: review of methods and applications. Expert Syst. Appl. **73**, 220–239 (2017). https://doi.org/10.1016/J.ESWA.2016.12.035
15. Hulten, G., Spencer, L., Domingos, P.: Mining time-changing data streams. In: Proceedings of the Seventh ACM SIGKDD International Conference on Knowledge Discovery and Data Mining, KDD 2001, pp. 97–106. Association for Computing Machinery, New York (2001). https://doi.org/10.1145/502512.502529
16. Jakubowski, J., Stanisz, P., Bobek, S., Nalepa, G.J.: Roll wear prediction in strip cold rolling with physics-informed autoencoder and counterfactual explanations. In: Proceedings of the 9th IEEE International Conference on Data Science and Advanced Analytics (DSAA) (2022)

17. Kirkpatrick, J., et al.: Overcoming catastrophic forgetting in neural networks. Proc. Natl. Acad. Sci. **114**(13), 3521–3526 (2017). https://doi.org/10.1073/pnas.1611835114. https://www.pnas.org/doi/abs/10.1073/pnas.1611835114
18. Krawczyk, B., Minku, L.L., Gama, J., Stefanowski, J., Woźniak, M.: Ensemble learning for data stream analysis: a survey. Inf. Fusion **37**, 132–156 (2017). https://doi.org/10.1016/j.inffus.2017.02.004
19. Lee, S., Son, Y.: Motor load balancing with roll force prediction for a cold-rolling setup with neural networks. Mathematics **9**(12), 1367 (2021). https://doi.org/10.3390/math9121367
20. Lenard, J.G.: 9 - tribology. In: Lenard, J.G. (ed.) Primer on Flat Rolling, 2nd edn, pp. 193–266. Elsevier, Oxford (2014). https://doi.org/10.1016/B978-0-08-099418-5.00009-3
21. Lu, J., Liu, A., Dong, F., Gu, F., Gama, J., Zhang, G.: Learning under concept drift: a review. IEEE Trans. Knowl. Data Eng. 1 (2018). https://doi.org/10.1109/TKDE.2018.2876857
22. Montiel, J., et al.: River: machine learning for streaming data in python (2021). https://github.com/online-ml/river
23. Rousseeuw, P.J.: Silhouettes: a graphical aid to the interpretation and validation of cluster analysis. J. Comput. Appl. Math. **20**, 53–65 (1987). https://doi.org/10.1016/0377-0427(87)90125-7. https://www.sciencedirect.com/science/article/pii/0377042787901257
24. Rusnák, J., Malega, P., Svetlík, J., Rudy, V., Šmajda, N.: The research of the rolling speed influence on the mechanism of strip breaks in the steel rolling process. Materials **13**(16), 3509 (2020). https://doi.org/10.3390/ma13163509
25. Tan, S.C., Ting, K.M., Liu, T.F.: Fast anomaly detection for streaming data. In: Proceedings of the Twenty-Second International Joint Conference on Artificial Intelligence, IJCAI 2011, vol. 2, pp. 1511–1516. AAAI Press (2011)
26. Yin, C., Zhang, S., Yin, Z., Wang, J.: Anomaly detection model based on data stream clustering. Clust. Comput. **22**(1), 1729–1738 (2017). https://doi.org/10.1007/s10586-017-1066-2

MeshFree Methods and Radial Basis Functions in Computational Sciences

Biharmonic Scattered Data Interpolation Based on the Method of Fundamental Solutions

Csaba Gáspár(✉)

Széchenyi István University, Egyetem tér 1, Györ 9026, Hungary
gasparcs@math.sze.hu

Abstract. The two-dimensional scattered data interpolation problem is investigated. In contrast to the traditional Method of Radial Basis Functions, the interpolation problem is converted to a higher order (biharmonic or modified bi-Helmholtz) partial differential equation supplied with usual boundary conditions as well as pointwise interpolation conditions. To solve this fourth-order problem, the Method of Fundamental Solutions is used. The source points, which are needed in the method, are located partly in the exterior of the domain of the corresponding partial differential equation and partly in the interpolation points. This results in a linear system with possibly large and fully populated matrix. To make the computations more efficient, a localization technique is applied, which splits the original problem into a sequence of local problems. The system of local equations is solved in an iterative way, which mimics the classical overlapping Schwarz method. Thus, the problem of large and ill-conditioned matrices is completely avoided. The method is illustrated via a numerical example.

Keywords: Scattered data interpolation · Method of Fundamental Solutions · Localization

1 Introduction

The scattered data interpolation problem is a relatively new mathematical problem which goes back to the pioneering work of Shepard [11]. His method was based on weighted averages, the weights of which are inversely proportional to some powers of the distances between the interpolation points and the point in which the interpolation function is to be evaluated. Later, a much more powerful family of methods was developed, the method of Radial Basis Functions (RBFs), see e.g. [2,6]. Here the interpolation function is sought in the following form:

$$u(x) := \sum_{j=1}^{N} \alpha_j \cdot \Phi(x - x_j), \tag{1}$$

where Φ is a predefined radial (i.e., circularly symmetric) function, $x_1, x_2, ..., x_N$ are predefined interpolation points scattered in the plane $\mathbf{R}^2$ without having

J. Mikyška et al. (Eds.): ICCS 2023, LNCS 14076, pp. 485–499, 2023.
https://doi.org/10.1007/978-3-031-36027-5_38

any grid or mesh structure, $x \in \mathbf{R}^2$ is an evaluation point. The method can be defined for higher dimensional interpolation problems in a similar way: here we restrict ourselves to 2D problems.

The a priori unknown coefficients $\alpha_1, \alpha_2, ..., \alpha_N$ can be determined by enforcing the *interpolation conditions*:

$$\sum_{j=1}^{N} \alpha_j \cdot \Phi(x_k - x_j) = u_k, \qquad k = 1, 2, ..., N, \tag{2}$$

where the predefined values $u_1, u_2, ..., u_N$ are associated to the interpolation points $x_1, x_2, ..., x_N$.

For the RBF Φ, several choices have been proposed. Some popular techniques are as follows (written in polar coordinates, for the sake of simplicity):

- Multiquadrics: $\Phi(r) := \sqrt{c^2 + r^2}$;
- Inverse multiquadrics: $\Phi(r) := \frac{1}{\sqrt{c^2+r^2}}$;
- Thin plate splines: $\Phi(r) := r^2 \cdot \log(r)$;
- Polyharmonic splines: $\Phi(r) := r^{2k} \cdot \log(r)$ (where k is a predefined positive integer);
- Gauss functions: $\Phi(r) := e^{-c^2 \cdot r^2}$;

and so forth (in the above formulations, c denotes a predefined scaling constant). The above radial basis functions are *globally supported*, therefore the matrix of the system (2) is fully populated and sometimes severely ill-conditioned, which may cause computational difficulties, especially when the number of the interpolation points is large.

To overcome this difficulty, several methods have been developed. One of them is the use of *compactly supported* radial basis functions (Wendland functions, see e.g. [13]). Thus, the matrix of the system (2) becomes sparse, which is advantageous from computational point of view.

Another technique is a generalization of the concept of the thin plate splines. Utilizing the fact that the radial basis function of the thin plate spline $\Phi(r) = r^2 \cdot \log(r)$ is biharmonic (except for the origin), the interpolation problem can be converted to a problem defined for the biharmonic equation supplied with some usual boundary conditions along the boundary and also with the interpolation conditions at the interpolation points. Thus, instead of a scattered data interpolation problem, a fourth-order partial differential equation is to be solved (supplied with some unusual conditions, i.e. pointwise interpolation conditions).

In its original form, the resulting biharmonic problem is solved by a finite volume method. The cell system which the finite volume method is performed on, is preferably defined by a *quadtree subdivision* algorithm, which automatically generates local refinements in the vicinity of the interpolation points. The solution procedure can be embedded in a natural multi-level context. This makes the method quite economic from computational point of view. Note that the above biharmonic equation can be replaced with more general fourth-order partial differential equations e.g. the modified Helmholtz equation. See [3] for details. However, the method can be considered a 'quasi-meshfree' method only, though the construction of cell system as well as the solution process are completely controlled by the interpolation points.

In this paper, the above outlined strategy based on the solution of a biharmonic equation is connected with a truly *meshless* method, namely, the Method of Fundamental Solutions (MFS), see [5]. This method results in a linear system of equations with a fully populated (but often ill-conditioned) matrix. The approach has been applied to biharmonic equations as well, see [10]. The computational difficulties can be reduced by introducing *localization* techniques which convert the original problem to a set of smaller problems. See e.g. [1,12].

In the following, the MFS-based solution technique is generalized to the biharmonic interpolation problem using a special localization method which is based on the traditional Schwarz alternating method. This results in a special iterative method and splits the original problem into several local (and much less) subproblems. The method is illustrated through a simple example.

2 Biharmonic Interpolation

The main idea of this type of interpolation is to convert original interpolation problem to a higher order partial differential equation supplied with the interpolation conditions as special pointwise boundary conditions. For second-order partial differential equations, this results in an ill-posed problem, but for fourth-order equations, this does not remain the case.

Let $\Omega \subset \mathbf{R}^2$ be a two-dimensional, bounded and sufficiently smooth domain, and let $x_1, x_2, ..., x_N \in \Omega$ be predefined interpolation points. Denote by $u_1, u_2, ..., u_N \in \mathbf{R}$ the values associated to the interpolation points. The biharmonic interpolation function u is expected to satisfy the biharmonic equation in the domain Ω except for the interpolation points:

$$\Delta\Delta u = 0, \qquad \text{in } \Omega \setminus \{x_1, x_2, ..., x_N\}, \tag{3}$$

where Δ denotes the Laplace operator. Along the boundary $\Gamma := \partial\Omega$, some usual boundary condition can be prescribed, e.g. Dirichlet boundary condition:

$$u|_\Gamma = u_0, \qquad \frac{\partial u}{\partial n}|_\Gamma = v_0, \tag{4}$$

or Navier boundary condition:

$$u|_\Gamma = u_0, \qquad \Delta u|_\Gamma = w_0, \tag{5}$$

where u_0, v_0, w_0 are predefined, sufficiently regular boundary functions. At the interpolation points, the interpolation conditions

$$u(x_k) = u_k, \qquad k = 1, 2, ..., N, \tag{6}$$

are prescribed.

It is known that in spite of the pointwise defined interpolation conditions, the problem (3)–(6) has a unique solution in a closed subspace of the Sobolev space $H^2(\Omega)$. See [3] for details. Note that the interpolation conditions defined in discrete points do not make the problem ill-posed due to the fact that the Dirac functionals $u \to u(x_k)$ are continuous in the Sobolev space $H^2(\Omega)$ (but not in $H^1(\Omega)$, which is the usual basis of second-order elliptic boundary value problems). It should be pointed out that the biharmonic Eq. (3) can be replaced with other fourth-order differential equations e.g. the modified bi-Helmholtz equation:

$$(\Delta - c^2 I)^2 u = 0, \qquad \text{in } \Omega \setminus \{x_1, x_2, ..., x_N\}, \tag{7}$$

where I denotes the identity operator and c is a predefined constant which plays some scaling role.

The above idea converts the original *interpolation problem* into the solution of a (fourth-order) *partial differential equation* which seems to be much more difficult from computational point of view. However, if this partial differential equation is solved by a computationally efficient method, e.g. on a non-uniform, non-equidistant cell system using finite volume schemes and multi-level techniques, the necessary computational cost can significantly be reduced. Such a non-equidistant, non-uniform cell system can be created by the help of the well-known *quadtree algorithm* controlled by the interpolation points. The algorithm results in local refinements in the vicinity of the interpolation points and makes it possible to build up a multi-level solution technique in a natural way. For details, see [3,4]. Nevertheless, the accuracy of the above finite volume schemes is moderate. Moreover, the evaluation points of the interpolation function are fixed to be the cell centers. In this paper, another solution technique is presented, which is based on the Method of Fundamental Solutions.

3 The Method of Fundamental Solutions Applied to the Biharmonic Interpolation Problem

First, let us briefly recall the main concepts and ideas of the Method of Fundamental Solutions. For details, see e.g. [5].

The Method of Fundamental Solutions (MFS) is now a quite popular method for solving partial differential equations. It is truly meshless, i.e. it requires neither domain nor boundary grid or mesh structure. If the differential equation has the form

$$Lu = 0, \qquad \text{in } \Omega,$$

where L is a linear partial differential operator, and Φ denotes a fundamental solution of the operator L, then the MFS produces the approximate solution in the following form:

$$u(x) = \sum_{j=1}^{M} \alpha_j \cdot \Phi(x - s_j),$$

where $s_1, s_2, ..., s_M$ are predefined *source points* in the exterior of Ω. The a priori unknown coefficients $\alpha_1, \alpha_2, ..., \alpha_M$ can be computed by enforcing the boundary conditions in some predefined $x_1, x_2, ..., x_N$ *boundary collocation points*. In the simplest case, when L is of second order and Dirichlet boundary condition is prescribed, this results in the following linear system of algebraic equations:

$$\sum_{j=1}^{M} \alpha_j \cdot \Phi(x_k - s_j) = u(x_k), \qquad k = 1, 2, ..., N.$$

Recently, the approach has been generalized also to inhomogeneous problems, see e.g. [14], and also for more general equations, see [8].

In the case of the biharmonic equation

$$\Delta\Delta u = 0, \tag{8}$$

the MFS defines the approximate solution in the following form [10]:

$$u(x) := \sum_{j=1}^{M} \alpha_j \cdot \Phi(x - s_j) + \sum_{j=1}^{M} \beta_j \cdot \Psi(x - s_j). \tag{9}$$

Here $s_1, s_2, ..., s_M$ are again exterior source points. Φ denotes the following harmonic fundamental solution:

$$\Phi(x) := \frac{1}{2\pi} \log ||x||,$$

and Ψ denotes the following biharmonic fundamental solution:

$$\Psi(x) := \frac{1}{8\pi} ||x||^2 \log ||x|| - \frac{1}{8\pi} ||x||^2$$

(the symbol $|| \cdot ||$ denotes the usual Euclidean norm in $\mathbf{R^2}$).

The above definitions imply that $\Delta\Psi = \Phi$. This will simplify the later calculations.

Suppose, for simplicity, that the biharmonic Eq. (8) is supplied with Navier boundary condition:

$$u|_\Gamma = u_0, \qquad \Delta u|_\Gamma = w_0. \tag{10}$$

Then the coefficients $\alpha_1, ..., \alpha_M, \beta_1, ..., \beta_M$ can be calculated by enforcing the boundary conditions. Utilizing the equality $\Delta\Psi = \Phi$, this results in the following linear system of equations:

$$\begin{aligned} \sum_{j=1}^{M} \alpha_j \cdot \Phi(x_k - s_j) + \sum_{j=1}^{M} \beta_j \cdot \Psi(x_k - s_j) &= u_0(x_k), \quad k = 1, 2, ..., N, \\ \sum_{j=1}^{M} \beta_j \cdot \Phi(x_k - s_j) &= w_0(x_k), \quad k = 1, 2, ..., N, \end{aligned} \tag{11}$$

where $x_1, x_2, ..., x_N \in \Gamma$ are predefined boundary collocation points. In general, the numbers N and M need not be equal. If they differ, then the system (11) should be solved e.g. in the sense of least squares. (11) is often overdetermined, i.e., the number of collocation points is (much) greater than the number of sources. Even if $N = M$, the above system of equations may be severely ill-conditioned, especially when the sources are located far from the boundary of the domain Ω. Moreover, the matrix of the system is fully populated. Therefore, the practical implementation of the method may often be difficult from computational point of view.

3.1 Solution of the Biharmonic Problem by Overlapping Schwarz Method

To circumvent the above mentioned computational problems, the traditional Schwarz overlapping method [7] is applied. Such a method splits the original problem into several smaller ones. It was originally defined for second-order partial differential equations, however, the idea can easily be generalized for our biharmonic problem. For the sake of simplicity, assume that Ω is a rectangle, $\Omega = \Omega_1 \cup \Omega_2$, where $\Omega_1 = (-\pi + h, h) \times (0, \pi)$, $\Omega_2 = (-h, \pi - h) \times (0, \pi)$. The predefined value $0 < h < \pi/2$ characterizes the overlap, see Fig. 1. This situation is a *strong overlap*, i.e. the distance of Γ_1 and Γ_2 is positive ($= 2h$), where $\Gamma_1 := \partial\Omega_1 \cap \Omega_2$, and $\Gamma_2 := \partial\Omega_2 \cap \Omega_1$.

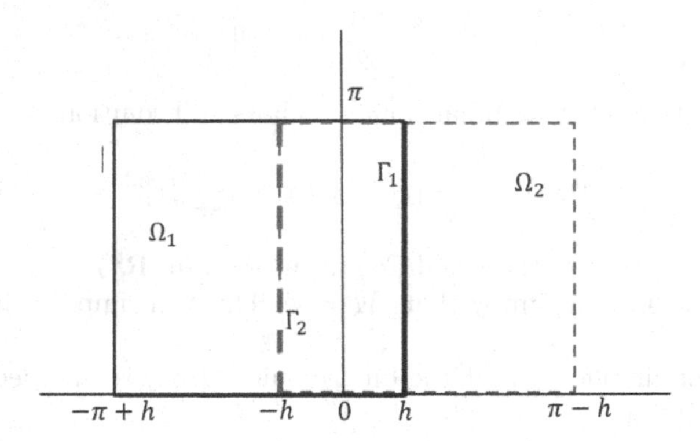

Fig. 1. Strongly overlapping subdomains. The distance of Γ_1 and Γ_2 is positive.

First, consider the pure biharmonic problem supplied with Navier boundary conditions but without interpolation conditions:

$$\begin{aligned} \Delta\Delta u &= 0, \qquad \text{in } \Omega, \\ u|_\Gamma = u_0, &\qquad \Delta u|_\Gamma = w_0. \end{aligned} \tag{12}$$

The Schwarz overlapping method approximates the solution with the following sequence of functions. Starting from a function u_1, for which $u_1 \in H^2(\Omega)$, $\Delta u_1 \in H^1(\Omega)$, define the sequence of the following subproblems for $n = 1, 2, ...$:

$$\begin{aligned} \Delta\Delta u_{n+1/2} &= 0, \qquad \text{in } \Omega_1, \\ u_{n+1/2}|_{\partial\Omega_1\setminus\Gamma_1} = u_0|_{\partial\Omega_1\setminus\Gamma_1}, &\qquad \Delta u_{n+1/2}|_{\partial\Omega_1\setminus\Gamma_1} = w_0|_{\partial\Omega_1\setminus\Gamma_1}, \\ u_{n+1/2}|_{\Gamma_1} = u_n|_{\Gamma_1}, &\qquad \Delta u_{n+1/2}|_{\Gamma_1} = \Delta u_n|_{\Gamma_1}, \end{aligned} \tag{13}$$

$$\begin{aligned} \Delta\Delta u_{n+1} &= 0, \qquad \text{in } \Omega_2, \\ u_{n+1}|_{\partial\Omega_2\setminus\Gamma_2} &= u_0|_{\partial\Omega_2\setminus\Gamma_2}, \qquad \Delta u_{n+1}|_{\partial\Omega_2\setminus\Gamma_2} = w_0|_{\partial\Omega_2\setminus\Gamma_2}, \\ u_{n+1}|_{\Gamma_2} &= u_{n+1/2}|_{\Gamma_2}, \qquad \Delta u_{n+1}|_{\Gamma_2} = \Delta u_{n+1/2}|_{\Gamma_2}. \end{aligned} \tag{14}$$

Now we will prove that the above defined Schwarz sequence converges to the exact solution u^*. Denote by e_n, $e_{n+1/2}$ the corresponding errors: $e_n := u_n - u^*$ (in Ω_2) and $e_{n+1/2} := u_{n+1/2} - u^*$ (in Ω_1). Then, obviously, the error functions satisfy the following biharmonic problems:

$$\begin{aligned} \Delta\Delta e_{n+1/2} &= 0, \qquad \text{in } \Omega_1, \\ e_{n+1/2}|_{\partial\Omega_1\setminus\Gamma_1} &= 0, \qquad \Delta e_{n+1/2}|_{\partial\Omega_1\setminus\Gamma_1} = 0, \\ e_{n+1/2}|_{\Gamma_1} &= e_n|_{\Gamma_1}, \qquad \Delta e_{n+1/2}|_{\Gamma_1} = \Delta e_n|_{\Gamma_1}, \end{aligned} \tag{15}$$

$$\begin{aligned} \Delta\Delta e_{n+1} &= 0, \qquad \text{in } \Omega_2, \\ e_{n+1}|_{\partial\Omega_2\setminus\Gamma_2} &= 0, \qquad \Delta e_{n+1}|_{\partial\Omega_2\setminus\Gamma_2} = 0, \\ e_{n+1}|_{\Gamma_2} &= e_{n+1/2}|_{\Gamma_2}, \qquad \Delta e_{n+1}|_{\Gamma_2} = \Delta e_{n+1/2}|_{\Gamma_2}. \end{aligned} \tag{16}$$

Now express $e_n|_{\Gamma_1}$ and $\Delta e_n|_{\Gamma_1}$ in terms of trigonometric Fourier series (with respect to the variable y). Due to the boundary conditions along the horizontal sides, it is sufficient to use *sinusoidal* Fourier series. Using the traditional notations x, y for the spatial variables, we have:

$$e_n|_{\Gamma_1} = e_n(h,y) = \sum_{k=1}^{\infty} \alpha_k^{(n)} \sin ky, \quad \Delta e_n|_{\Gamma_1} = \Delta e_n(h,y) = \sum_{k=1}^{\infty} \beta_k^{(n)} \sin ky. \tag{17}$$

Straightforward calculations show that, in Ω_1, the solution $e_{n+1/2}$ has the form:

$$\begin{aligned} e_{n+1/2}(x,y) = \sum_{k=1}^{\infty} \Big(& A_k^{(n+1/2)} \sinh k(x+\pi-h) \\ & + B_k^{(n+1/2)} (x+\pi-h) \cosh k(x+\pi-h) \Big) \sin ky \end{aligned} \tag{18}$$

with some coefficients $A_k^{(n+1/2)}$, $B_k^{(n+1/2)}$. Indeed, one can easily check that each term in the Fourier series (as a function of x and y) is biharmonic. Moreover:

$$\Delta e_{n+1/2}(x,y) = \sum_{k=1}^{\infty} 2kB_k^{(n+1/2)} \sinh k(x+\pi-h) \sin ky, \tag{19}$$

therefore both $e_{n+1/2}$ and $\Delta e_{n+1/2}$ vanish along $\partial\Omega_1 \setminus \Gamma_1$.

The equalities

$$e_{n+1/2}|_{\Gamma_1} = e_n|_{\Gamma_1}, \qquad \Delta e_{n+1/2}|_{\Gamma_1} = \Delta e_n|_{\Gamma_1}$$

imply that the vectors of coefficients $\begin{pmatrix} A_k^{(n+1/2)} \\ B_k^{(n+1/2)} \end{pmatrix}$ satisfy the systems of equations:

$$\begin{pmatrix} \sinh k\pi & \pi \cosh k\pi \\ 0 & 2k \sinh k\pi \end{pmatrix} \begin{pmatrix} A_k^{(n+1/2)} \\ B_k^{(n+1/2)} \end{pmatrix} = \begin{pmatrix} \alpha_k^{(n)} \\ \beta_k^{(n)} \end{pmatrix}.$$

Now the traces of $e_{n+1/2}$ and $\Delta e_{n+1/2}$ along Γ_2 can be calculated without difficulty:

$$\begin{aligned} e_{n+1/2}|_{\Gamma_2} = e_{n+1/2}(-h,y) = \sum_{k=1}^{\infty} \Big(A_k^{(n+1/2)} \sinh k(\pi - 2h) \\ + B_k^{(n+1/2)} (\pi - 2h) \cosh k(\pi - 2h) \Big) \sin ky \\ =: \sum_{k=1}^{\infty} \alpha_k^{(n+1/2)} \sin ky, \end{aligned} \tag{20}$$

$$\begin{aligned} \Delta e_{n+1/2}|_{\Gamma_2} = \Delta e_{n+1/2}(-h,y) = \sum_{k=1}^{\infty} 2kB_k^{(n+1/2)} \sinh k(\pi - 2h) \sin ky \\ =: \sum_{k=1}^{\infty} \beta_k^{(n+1/2)} \sin ky. \end{aligned} \tag{21}$$

That is, the vectors of the Fourier coefficients $\begin{pmatrix} \alpha_k^{(n+1/2)} \\ \beta_k^{(n+1/2)} \end{pmatrix}$ can be expressed with the help of $\begin{pmatrix} \alpha_k^{(n)} \\ \beta_k^{(n)} \end{pmatrix}$ as follows:

$$\begin{pmatrix} \alpha_k^{(n+1/2)} \\ \beta_k^{(n+1/2)} \end{pmatrix} = \begin{pmatrix} \sinh k(\pi - 2h) & (\pi - 2h) \cosh k(\pi - 2h) \\ 0 & 2k \sinh k(\pi - 2h) \end{pmatrix} \begin{pmatrix} A_k^{(n+1/2)} \\ B_k^{(n+1/2)} \end{pmatrix}$$

$$= \begin{pmatrix} \sinh k(\pi - 2h) & (\pi - 2h) \cosh k(\pi - 2h) \\ 0 & 2k \sinh k(\pi - 2h) \end{pmatrix} \begin{pmatrix} \sinh k\pi & \pi \cosh k\pi \\ 0 & 2k \sinh k\pi \end{pmatrix}^{-1} \begin{pmatrix} \alpha_k^{(n)} \\ \beta_k^{(n)} \end{pmatrix}.$$

In a quite similar way, the error function e_{n+1} can be expressed in Ω_2 as:

$$\begin{aligned} e_{n+1}(x,y) = \sum_{k=1}^{\infty} \Big(A_k^{(n+1)} \sinh k(x - \pi + h) \\ + B_k^{(n+1)} (x - \pi + h) \cosh k(x - \pi + h) \Big) \sin ky \end{aligned} \tag{22}$$

with some coefficients $A_k^{(n+1)}$, $B_k^{(n+1)}$. Moreover:

$$\Delta e_{n+1}(x,y) = \sum_{k=1}^{\infty} 2kB_k^{(n+1)} \sinh k(x - \pi + h) \sin ky, \tag{23}$$

where the vectors of coefficients $\begin{pmatrix} A_k^{(n+1)} \\ B_k^{(n+1)} \end{pmatrix}$ satisfy the systems of equations:

$$\begin{pmatrix} \sinh k(-\pi) & -\pi \cosh k(-\pi) \\ 0 & 2k \sinh k(-\pi) \end{pmatrix} \begin{pmatrix} A_k^{(n+1)} \\ B_k^{(n+1)} \end{pmatrix} = \begin{pmatrix} \alpha_k^{(n+1/2)} \\ \beta_k^{(n+1/2)} \end{pmatrix}.$$

Calculating the traces of e_{n+1} and Δe_{n+1} along Γ_1, we obtain:

$$\begin{aligned} e_{n+1}|_{\Gamma_1} = e_{n+1}(h,y) = \sum_{k=1}^{\infty} \Big(A_k^{(n+1)} \sinh k(2h-\pi) \\ + B_k^{(n+1)} (2h-\pi) \cosh k(2h-\pi) \Big) \sin ky \\ =: \sum_{k=1}^{\infty} \alpha_k^{(n+1)} \sin ky, \end{aligned} \tag{24}$$

$$\begin{aligned} \Delta e_{n+1}|_{\Gamma_1} = \Delta e_{n+1}(h,y) = \sum_{k=1}^{\infty} 2kB_k^{(n+1)} \sinh k(2h-\pi) \sin ky \\ =: \sum_{k=1}^{\infty} \beta_k^{(n+1)} \sin ky. \end{aligned} \tag{25}$$

Consequently, the vectors of the Fourier coefficients $\begin{pmatrix} \alpha_k^{(n+1)} \\ \beta_k^{(n+1)} \end{pmatrix}$ can be expressed with the help of $\begin{pmatrix} \alpha_k^{(n+1/2)} \\ \beta_k^{(n+1/2)} \end{pmatrix}$ exactly in the same way than $\begin{pmatrix} \alpha_k^{(n+1/2)} \\ \beta_k^{(n+1/2)} \end{pmatrix}$ with the help of $\begin{pmatrix} \alpha_k^{(n)} \\ \beta_k^{(n)} \end{pmatrix}$. This implies that after a complete Schwarz iteration:

$$\begin{pmatrix} \alpha_k^{(n+1)} \\ \beta_k^{(n+1)} \end{pmatrix} = M_k^2 \begin{pmatrix} \alpha_k^{(n)} \\ \beta_k^{(n)} \end{pmatrix},$$

where

$$M_k = \begin{pmatrix} \sinh k(\pi-2h) & (\pi-2h)\cosh k(\pi-2h) \\ 0 & 2k\sinh k(\pi-2h) \end{pmatrix} \begin{pmatrix} \sinh k\pi & \pi\cosh k\pi \\ 0 & 2k\sinh k\pi \end{pmatrix}^{-1}.$$

Standard calculations show that

$$\begin{pmatrix} \sinh k\pi & \pi\cosh k\pi \\ 0 & 2k\sinh k\pi \end{pmatrix}^{-1} = \begin{pmatrix} \frac{1}{\sinh k\pi} & -\frac{\pi\cosh k\pi}{2k(\sinh k\pi)^2} \\ 0 & \frac{1}{2k\sinh k\pi} \end{pmatrix}.$$

Consequently, M_k is an upper triangular matrix, and both diagonal entries are equal to $\frac{\sinh k(\pi-2h)}{\sinh k\pi}$. This implies that both eigenvalues of M_k^2 are equal to $\left(\frac{\sinh k(\pi-2h)}{\sinh k\pi}\right)^2$. And since $0 < h < \frac{\pi}{2}$, these eigenvalues are less than 1, i.e., the Schwarz iteration is convergent. Theorem is proven.

The above theorem can be generalized to the case when the number of subdomains is greater that 2. Details are omitted.

3.2 Localization of the MFS for the Biharmonic Equation Based on Overlapping Schwarz Method

In practice, the idea of the overlapping Schwarz method can be used for creating a special localization technique. Here we briefly outline this method applied to the biharmonic equation. For the localization of the biharmonic equation, without Schwarz method, see e.g. [1,9,12].

Consider a point set $S_0 := \{x_1, x_2, ..., x_N\}$ scattered in the domain Ω. Denote by S_b the set of predefined boundary collocation points: $S_b := \{\hat{x}_1, \hat{x}_2, ..., \hat{x}_M\}$, and define the set $S := S_0 \cup S_b$.

For a given central point $x_0^{(i)} \in S_0$, define the circle centered at $x_0^{(i)}$ with the prescribed radius $R^{(i)}$ (the *local subdomains*):

$$\Omega^{(i)} := \{x \in \mathbf{R}^2 : \ ||x - x_0^{(i)}|| < R^{(i)}\}$$

Define the sets:

$$S_0^{(i)} := S_0 \cap \Omega^{(i)} = \{x_1^{(i)}, ..., x_{N_p^{(i)}}^{(i)}\}, \quad S_b^{(i)} := S_b \cap \Omega^{(i)} = \{\hat{x}_1^{(i)}, ..., \hat{x}_{N_b^{(i)}}^{(i)}\}.$$

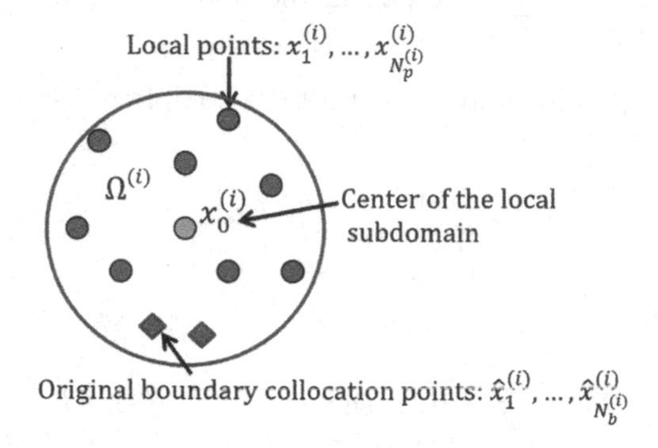

Fig. 2. Local subdomain and local points

Note that the set $S_b^{(i)}$ may be empty (when no boundary collocation points is included in $\Omega^{(i)}$, i.e. $N_b^{(i)} = 0$). See Fig. 2 for illustration. The approximate solutions of the local problems are computed by utilizing the MFS in $\Omega^{(i)}$. To do this, define some local source points $s_1^{(i)}, s_2^{(i)}, ..., s_{N_s^{(i)}}^{(i)}$ (e.g. along the perimeter of a circle centered at $x_0^{(i)}$ with radius which is greater than $R^{(i)}$ (the definition $2R^{(i)}$ is an acceptable value). For illustration, see Fig. 3. The numbers of local sources can be kept at a common value, but $N_s^{(i)} < N_p^{(i)}$ should be satisfied. The approximate local solution is expressed in the following form:

$$u(x) := \sum_{j=1}^{N_s^{(i)}} \alpha_j^{(i)} \Phi(x - s_j^{(i)}) + \sum_{j=1}^{N_s^{(i)}} \beta_j^{(i)} \Psi(x - s_j^{(i)}), \tag{26}$$

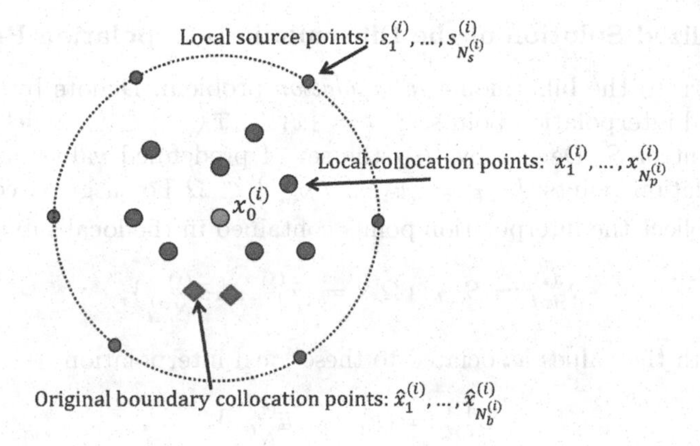

Fig. 3. Local sources along a circle

with some a priori unknown coefficients $\alpha_j^{(i)}$, $\beta_j^{(i)}$, $j = 1, 2, ..., N_s^{(i)}$. The local Navier boundary conditions are approximated by enforcing the following equalities for all local points $x_k^{(i)}$, $k = 1, 2, ..., N_p^{(i)}$:

$$\sum_{j=1}^{N_s^{(i)}} \alpha_j^{(i)} \Phi(x_k^{(i)} - s_j^{(i)}) + \sum_{j=1}^{N_s^{(i)}} \beta_j^{(i)} \Psi(x_k^{(i)} - s_j^{(i)}) = u_k^{(i)} := u(x_k^{(i)}). \tag{27}$$

If the local subdomain $\Omega^{(i)}$ contains boundary collocation points, i.e. $N_b^{(i)} > 0$, then the above equations are completed by the original boundary conditions:

$$\begin{aligned} \sum_{j=1}^{N_s^{(i)}} \alpha_j^{(i)} \Phi(\hat{x}_k^{(i)} - s_j^{(i)}) + \sum_{j=1}^{N_s^{(i)}} \beta_j^{(i)} \Psi(\hat{x}_k^{(i)} - s_j^{(i)}) &= u_0(\hat{x}_k^{(i)}), \\ \sum_{j=1}^{N_s^{(i)}} \beta_j^{(i)} \Phi(\hat{x}_k^{(i)} - s_j^{(i)}) &= w_0(\hat{x}_k^{(i)}), \end{aligned} \tag{28}$$

where $k = 1, 2, ..., N_b^{(i)}$ (since $\Delta\Phi = 0$, except for the origin, and $\Delta\Psi = \Phi$). The system (27)–(28) is overdetermined, if $2N_s^{(i)} < N_p^{(i)} + 2N_b^{(i)}$, and should be solved in the sense of least squares, i.e. by solving the corresponding Gaussian normal equations. Since the local systems are typically small, they can be expected to be only moderately ill-conditioned.

Having solved the ith local system, the approximate solution at the point $x_0^{(i)}$ is updated by:

$$u(x_0^{(i)}) := \sum_{j=1}^{N_s^{(i)}} \alpha_j^{(i)} \Phi(x_0^{(i)} - s_j^{(i)}) + \sum_{j=1}^{N_s^{(i)}} \beta_j^{(i)} \Psi(x_0^{(i)} - s_j^{(i)}), \tag{29}$$

and the iteration should be continued for all local subdomains.

3.3 Localized Solution of the Biharmonic Interpolation Problem

Let us return to the biharmonic *interpolation* problem. Denote by S_{int} the set of predefined interpolation points: $S_{int} := \{\overline{x}_1, ..., \overline{x}_{N_{int}}\} \subset \Omega$, which is assumed to be disjoint of S. Denote by U_{int} the set of predefined values associated to the interpolation points: $U_{int} := \{\overline{u}_1, ..., \overline{u}_{N_{int}}\} \subset \Omega$ For a given central point $x_0^{(i)} \in S_0$, collect the interpolation points contained in the local subdomain $\Omega^{(i)}$:

$$S_{int}^{(i)} := S_{int} \cap \Omega^{(i)} = \{\overline{x}_1^{(i)}, ..., \overline{x}_{N_{int}^{(i)}}^{(i)}\},$$

together with the values associated to these local interpolation

$$U_{int}^{(i)} := \{\overline{u}_1^{(i)}, ..., \overline{u}_{N_{int}^{(i)}}^{(i)}\}.$$

Now the approximate local solution is expressed in the following form:

$$u(x) := \sum_{j=1}^{N_s^{(i)}} \alpha_j^{(i)} \Phi(x - s_j^{(i)}) + \sum_{j=1}^{N_s^{(i)}} \beta_j^{(i)} \Psi(x - s_j^{(i)}) + \sum_{j=1}^{N_{int}^{(i)}} \gamma_j^{(i)} \Psi(x - \overline{x}_j^{(i)}) \tag{30}$$

(the last sum plays the role of the thin plate splines).

Thus, the equalities to be enforced will be expanded as follows.

Collocation at the inner points $x_k^{(i)}$, $k = 1, 2, ..., N_p^{(i)}$:

$$\sum_{j=1}^{N_s^{(i)}} \alpha_j^{(i)} \Phi(x_k^{(i)} - s_j^{(i)}) + \sum_{j=1}^{N_s^{(i)}} \beta_j^{(i)} \Psi(x_k^{(i)} - s_j^{(i)}) + \sum_{j=1}^{N_{int}^{(i)}} \gamma_j^{(i)} \Psi(x_k^{(i)} - \overline{x}_j^{(i)}) = u(x_k^{(i)}). \tag{31}$$

Collocation at the boundary collocation points $\hat{x}_k^{(i)}$, $k = 1, 2, ..., N_b^{(i)}$:

$$\begin{aligned} &\sum_{j=1}^{N_s^{(i)}} \alpha_j^{(i)} \Phi(\hat{x}_k^{(i)} - s_j^{(i)}) + \sum_{j=1}^{N_s^{(i)}} \beta_j^{(i)} \Psi(\hat{x}_k^{(i)} - s_j^{(i)}) \\ &\qquad + \sum_{j=1}^{N_{int}^{(i)}} \gamma_j^{(i)} \Psi(\hat{x}_k^{(i)} - \hat{x}_j^{(i)}) = u_0(\hat{x}_k^{(i)}) \\ &\sum_{j=1}^{N_s^{(i)}} \beta_j^{(i)} \Phi(\hat{x}_k^{(i)} - s_j^{(i)}) + \sum_{j=1}^{N_{int}^{(i)}} \gamma_j^{(i)} \Phi(\hat{x}_k^{(i)} - \overline{x}_j^{(i)}) = w_0(\hat{x}_k^{(i)}). \end{aligned} \tag{32}$$

Collocation at the interpolation points $\overline{x}_k^{(i)}$, $k = 1, 2, ..., N_{int}^{(i)}$:

$$\begin{aligned} \sum_{j=1}^{N_s^{(i)}} \alpha_j^{(i)} \Phi(\overline{x}_k^{(i)} - s_j^{(i)}) + \sum_{j=1}^{N_s^{(i)}} \beta_j^{(i)} \Psi(\overline{x}_k^{(i)} - s_j^{(i)}) + \sum_{j=1}^{N_{int}^{(i)}} \gamma_j^{(i)} \Psi(\overline{x}_k^{(i)} - \overline{x}_j^{(i)}) \\ = \overline{u}_k^{(i)}. \end{aligned} \tag{33}$$

The system (31)–(33) is overdetermined, if $2N_s^{(i)} < N_p^{(i)} + 2N_b^{(i)}$, as earlier. It should be solved in the sense of least squares, i.e. by solving the corresponding Gaussian normal equations.

After solving the ith local system, the approximate solution at the point $x_0^{(i)}$ is updated by:

$$u(x_0^{(i)}) := \sum_{j=1}^{N_s^{(i)}} \alpha_j^{(i)} \Phi(x_0^{(i)} - s_j^{(i)}) + \sum_{j=1}^{N_s^{(i)}} \beta_j^{(i)} \Psi(x_0^{(i)} - s_j^{(i)}) + \sum_{j=1}^{N_{int}^{(i)}} \gamma_j^{(i)} \Psi(x_0^{(i)} - \overline{x}_j^{(i)}), \tag{34}$$

and the iteration should be continued for all local subdomains.

4 A Numerical Example

To illustrate the method, suppose that the original domain Ω is the unit circle. Let $\overline{x}_1, ..., \overline{x}_{N_{int}}$ some interpolation points defined in Ω in a quasi-random way. Let us associate the values $\overline{u}_k$ $(k = 1, 2, ..., N_{int})$ to the interpolation points by $\overline{u}_k := u^*(\overline{x}_k)$, where the test function u^* is defined by

$$u^*(x) := 1 - ||x||^2. \tag{35}$$

Note that the test function (35) itself is biharmonic, and satisfies the following boundary conditions along Γ:

$$u^*|_\Gamma = u_0^* = 0, \qquad \Delta u^*|_\Gamma = w_0^* = -4. \tag{36}$$

Therefore, it is expected that the test function can be reconstructed from its values at the interpolation points more or less exactly, provided that the number of scattered points as well as of interpolation points is large enough.

$N := N_p + N_{int} + N_b$ points were defined in $\overline{\Omega}$, where $N_p := 4000$ is the number of the inner points, while N_{int} denotes the number of interpolation points. The number of boundary collocation points N_b was set to $N_b := 500$. The inner and the interpolation points were defined also in a quasi-random way. To define the local subdomains, the radii of the subdomains were set to 0.1. In each subdomain, the number of local sources was set to the same constant denoted by N_s. Table 1 shows the discrete relative L_2-errors of the approximation

$$\textit{relative } L_2\textit{-error} := \frac{\sqrt{\sum_{j=1}^{N} (u(x_j) - u^*(x_j))^2}}{\sqrt{\sum_{j=1}^{N} (u^*(x_j))^2}}$$

with respect to different numbers of N_s and N_{int}.

Table 1. The relative L_2-errors with different values of the numbers of local sources (N_s) and the numbers of interpolation points (N_{int}).

$N_s \setminus N_{int}$	250	500	1000	2000	4000
6	2.742E−3	1.964E−3	4.018E−4	1.902E−4	7.817E−5
8	1.861E−4	1.543E−4	7.558E−5	4.778E−5	1.159E−5
10	7.512E−5	1.858E−5	1.212E−5	7.193E−6	2.395E−6
12	5.649E−6	3.092E−6	1.899E−6	1.140E−6	5.217E−7

As it was expected, the relative errors decrease when the number of interpolation points increases. The errors also decrease when the number of local sources increases; note, however, that in this case, the computational cost also increases, since the local systems become larger. It is anticipated that the computational cost can be reduced by applying a simple multi-level technique.

5 Conclusions

A computational method for solving the 2D scattered data interpolation problem has been proposed. The method converts the interpolation problem to a biharmonic equation supplied with some boundary conditions along the boundary and pointwise interpolation conditions at the interpolation points. The Method of Fundamental Solutions has been generalized to this special fourth-order problem. A localization technique based on the overlapping Schwarz iteration has been also introduced. This results in an iterative algorithm; in each step, only a local subproblem is to be solved, which makes the computation simpler. In addition to it, the problem of solving large linear systems with dense and possibly ill-conditioned matrices is completely avoided.

References

1. Fan, C.M., Huang, Y.K., Chen, C.S., Kuo, S.R.: Localized method of fundamental solutions for solving two-dimensional Laplace and biharmonic equations. Eng. Anal. Boundary Elem. **101**, 188–197 (2019). https://doi.org/10.1016/j.enganabound.2018.11.008
2. Franke, R.: Scattered data interpolation: tests of some method. Math. Comput. **38**(157), 181–200 (1982). https://doi.org/10.1090/S0025-5718-1982-0637296-4
3. Gáspár, C.: Multi-level biharmonic and bi-Helmholz interpolation with application to the boundary element method. Eng. Anal. Boundary Elem. **24**(7–8), 559–573 (2000). https://doi.org/10.1016/S0955-7997(00)00036-9
4. Gáspár, C.: A meshless polyharmonic-type boundary interpolation method for solving boundary integral equations. Eng. Anal. Boundary Elem. **28**, 1207–1216 (2004). https://doi.org/10.1016/j.enganabound.2003.04.001
5. Golberg, M.A.: The method of fundamental solutions for Poisson's equation. Eng. Anal. Boundary Elem. **16**, 205–213 (1995). https://doi.org/10.1016/0955-7997(95)00062-3

6. Hardy, R.L.: Theory and applications of the multiquadric-biharmonic method 20 years of discovery 1968–1988. Comput. Math. Appl. **19**(8–9), 163–208 (1990). https://doi.org/10.1016/0898-1221(90)90272-L
7. Lions, P.L.: On the Schwarz alternating method I. In: Glowinski, R., Golub, J.H., Meurant, G.A., Periaux, J. (eds.) Domain Decomposition Methods for Partial Differential Equations 1988, pp. 1–42. SIAM, Philadelphia (1988)
8. Liu, S., Li, P.W., Fan, C.M., Gu, Y.: Localized method of fundamental solutions for two- and three-dimensional transient convection-diffusion-reaction equations. Eng. Anal. Boundary Elem. **124**, 237–244 (2021). https://doi.org/10.1016/j.enganabound.2020.12.023
9. Liu, Y.C., Fan, C.M., Yeih, W., Ku, C.Y., Chu, C.L.: Numerical solutions of two-dimensional Laplace and biharmonic equations by the localized Trefftz method. Comput. Math. Appl. **88**, 120–134 (2021). https://doi.org/10.1016/j.camwa.2020.09.023
10. Pei, X., Chen, C.S., Dou, F.: The MFS and MAFS for solving Laplace and biharmonic equations. Eng. Anal. Boundary Elem. **80**, 87–93 (2017). https://doi.org/10.1016/j.enganabound.2017.02.011
11. Shepard, D.: A two-dimensional interpolation function for irregularly-spaced data. In: Proceedings of the 23rd ACM National Conference, pp. 517–524. Association for Computing Machinery Press, New York (1968). https://doi.org/10.1145/800186.810616
12. Wang, F., Fan, C.M., Hua, Q., Gu, Y.: Localized MFS for the inverse Cauchy problems of two-dimensional Laplace and biharmonic equations. Appl. Math. Comput. **364**, 1–14, Article no. 124658 (2020). https://doi.org/10.1016/j.amc.2019.124658
13. Wendland, H.: Piecewise polynomial, positive definite and compactly supported radial functions of minimal degree. Adv. Comput. Math. **4**, 389–396 (1995). https://doi.org/10.1007/BF02123482
14. Zhang, J., Yang, C., Zheng, H., Fan, C.M., Fu, M.F.: The localized method of fundamental solutions for 2D and 3D inhomogeneous problems. Math. Comput. Simul. **200**, 504–524 (2022). https://doi.org/10.1016/j.matcom.2022.04.024

Spatially-Varying Meshless Approximation Method for Enhanced Computational Efficiency

Mitja Jančič[1,2(✉)], Miha Rot[1,2], and Gregor Kosec[1]

[1] Parallel and Distributed Systems Laboratory, Jožef Stefan Institute, Ljubljana, Slovenia
{mitja.jancic,miha.rot,gregor.kosec}@ijs.si
[2] Jožef Stefan International Postgraduate School, Ljubljana, Slovenia

Abstract. In this paper, we address a way to reduce the total computational cost of meshless approximation by reducing the required stencil size through spatially varying computational node regularity. Rather than covering the entire domain with scattered nodes, only regions with geometric details are covered with scattered nodes, while the rest of the domain is discretized with regular nodes. A simpler approximation using solely monomial basis can be used in regions covered by regular nodes, effectively reducing the required stencil size and computational cost compared to the approximation on scattered nodes where a set of polyharmonic splines is added to ensure convergent behaviour.

The performance of the proposed hybrid scattered-regular approximation approach, in terms of computational efficiency and accuracy of the numerical solution, is studied on natural convection driven fluid flow problems. We start with the solution of the de Vahl Davis benchmark case, defined on a square domain, and continue with two- and three-dimensional irregularly shaped domains. We show that the spatial variation of the two approximation methods can significantly reduce the computational demands, with only a minor impact on the accuracy.

Keywords: Collocation · RBF-FD · RBF · Meshless · Hybrid method · Fluid-flow · Natural convection · Numerical simulation

1 Introduction

Although the meshless methods are formulated without any restrictions regarding the node layouts, it is generally accepted that quasi-uniformly-spaced node sets improve the stability of meshless methods [12,23]. Nevertheless, even with quasi-uniform nodes generated with recently proposed node positioning algorithms [14,15,18], a sufficiently large stencil size is required for stable approximation. A stencil with $n = 2\binom{m+d}{m}$ nodes is recommended [1] for the local Radial Basis Function-generated Finite differences (RBF-FD) [20] method in a d-dimensional domain for approximation order m. The performance of RBF-FD method—with approximation basis consisting of Polyharmonic splines (PHS)

J. Mikyška et al. (Eds.): ICCS 2023, LNCS 14076, pp. 500–514, 2023.
https://doi.org/10.1007/978-3-031-36027-5_39

and monomial augmentation with up to and including monomials of degree m—has been demonstrated with scattered nodes on several applications [8,17,24]. On the other hand, approximation on regular nodes can be performed with considerably smaller stencil [10] ($n = 5$ in two-dimensional domain) using only monomial basis.

Therefore, a possible way to enhance the overall computational efficiency and consider the discretization-related error is to use regular nodes far away from any geometric irregularities in the domain and scattered nodes in their vicinity. A similar approach, where the approximation method is spatially varied, has already been introduced, e.g., a hybrid FEM-meshless method [6] has been proposed to overcome the issues regarding the unstable Neumann boundary conditions in the context of meshless approximation. Moreover, the authors of [2,5] proposed a hybrid of Finite Difference (FD) method employed on conventional cartesian grid combined with meshless approximation on scattered nodes. These hybrid approaches are computationally very efficient, however, additional implementation-related burden is required on the transition from cartesian to scattered nodes [9], contrary to the objective of this paper relying solely on the framework of meshless methods.

In this paper we experiment with such *hybrid scattered-regular* method with spatially variable stencil size on solution of natural convection driven fluid flow cases. The solution procedure is first verified on the reference de Vahl Davis case, followed by a demonstration on two- and three-dimensional irregular domains. We show that spatially varying the approximation method can have positive effects on the computational efficiency while maintaining the accuracy of the numerical solution.

2 Numerical Treatment of Partial Differential Equations

To obtain the hybrid scattered-regular domain discretization, we first fill the entire domain with regular nodes. A portion of this regular nodes is then removed in areas where a scattered node placement is desired, i.e., close to the irregular boundaries. Finally, the voids are filled with a dedicated node positioning algorithm [18] that supports variable nodal density and allows us to refine the solution near irregularities in the domain. This approach is rather naive but sufficient for demonstration purposes. A special hybrid fill algorithm is left for future work.

An example of an h-refined domain discretization is shown in Fig. 1. It is worth noting that the width of the scattered node layer δ_h is non-trivial and affects both the stability of the solution procedure and the accuracy of the numerical solution. Although we provide a superficial analysis of δ_h variation in Sects. 4.1 and 4.2, further work is warranted.

After the computational nodes $\boldsymbol{x}_i \in \Omega$ are obtained, the differential operators $\mathcal{L}$ can be locally approximated in point $\boldsymbol{x}_c$ over a set of n neighbouring nodes (stencil) $\{\boldsymbol{x}_i\}_{i=1}^n = \mathcal{N}$, using the following expression

$$(\mathcal{L}u)(\boldsymbol{x}_c) \approx \sum_{i=1}^{n} w_i u(\boldsymbol{x}_i). \tag{1}$$

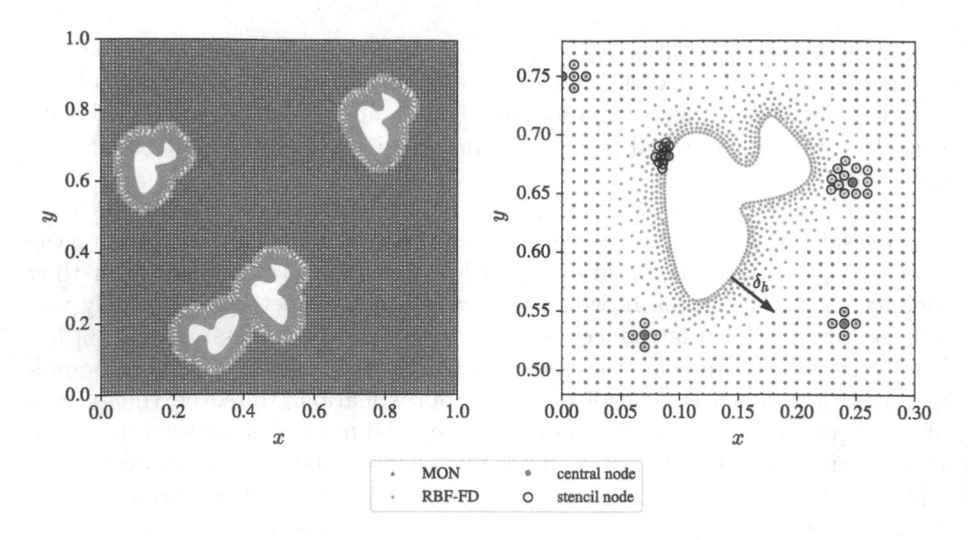

Fig. 1. Irregular domain discretization example *(left)* and spatial distribution of approximation methods along with corresponding example stencils *(right)*.

The approximation (1) holds for an arbitrary function u and yet to be determined weights $\boldsymbol{w}$. To determine the weights, the equality of approximation (1) is enforced for a chosen set of basis functions. Here we will use two variants

(i) a set of Polyharmonic splines (PHS) augmented with monomials to ensure convergent behaviour [1,7], effectively resulting in a popular *radial basis function-generated finite differences* (RBF-FD) approximation method [20].
(ii) a set of monomials centred at the stencil nodes that we will refer to as (MON) [10].

We use the least expensive MON with $2d + 1$ monomial basis functions[1] and the same number of support nodes in each approximation stencil. This setup is fast, but only stable on regular nodes [10,16]. For the RBF-FD part, we also resort to the minimal configuration required for 2nd-order operators, i.e., 3rd-order PHS augmented with all monomials up to the 2nd-order ($m = 2$). According to the standard recommendations [1], this requires a stencil size of $n = 2\binom{m+d}{m}$.

Note the significant difference between stencil sizes—5 vs. 12 nodes in 2D—that only increases in higher dimensions (7 vs. 30 in 3D). This results both in faster computation of the weights $\boldsymbol{w}$—an $\mathcal{O}(N^3)$[2] operation performed only once for each stencil—and in faster evaluation for the $\mathcal{O}(n)$ explicit operator approx-

[1] In 2D, the 5 basis functions are $\{1, x, y, x^2, y^2\}$. The xy term is not required for regularly placed nodes and its omission allows us to use the smaller and completely symmetric 5-node stencil.

[2] $N_{\text{RBF-FD}} \sim 3N_{\text{MON}}$ due to the larger stencil size and the extra PHS in the approximation basis.

imation (1) performed many times during the explicit time stepping. Therefore, a spatially varying node regularity can have desirable consequences on the discretization-related errors and computational efficiency of the solution procedure.

2.1 Computational Stability

By enforcing the equality of approximation (1), we obtain a linear system $\mathbf{M}\boldsymbol{w} = \boldsymbol{\ell}$. Solving the system provides us with the approximation weights $\boldsymbol{w}$, but the stability of such procedure can be uncertain and is usually estimated via the condition number $\kappa(\mathbf{M}) = \|\mathbf{M}\| \, \|\mathbf{M}^{-1}\|$ of matrix $\mathbf{M}$, where $\|\cdot\|$ denotes the L^2 norm.

A spatial distribution of condition numbers is shown in Fig. 2. It can be observed that the RBF-FD approximation method generally results in higher condition numbers than the MON approach. This could be due to the fact that the matrices $\mathbf{M}$ for the RBF-FD part are significantly larger and based on scattered nodes. Nevertheless, it is important to note that the transition from regular to scattered nodes does not appear to affect the conditionality of the matrices.

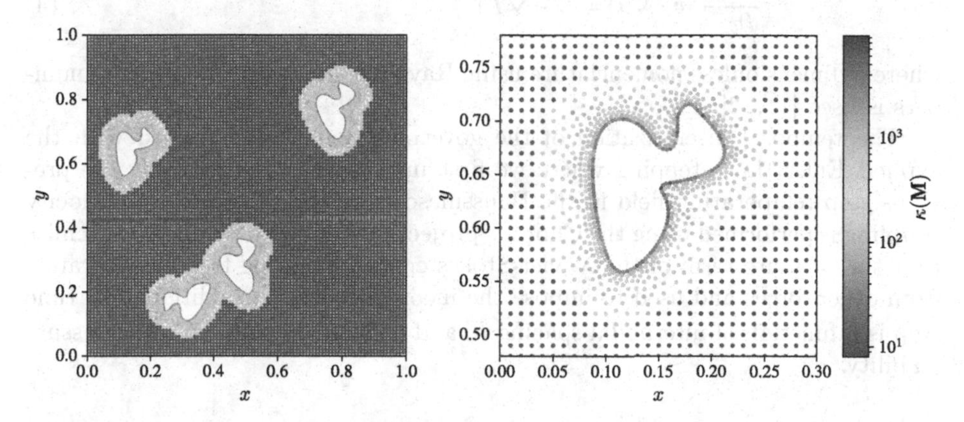

Fig. 2. Condition numbers $\kappa(\mathbf{M})$ for the Laplacian operator: entire computational domain *(left)* and a zoomed-in section around the irregularly shaped obstacle *(right)*.

2.2 Implementation Details

The entire solution procedure employing the hybrid scattered-regular method is implemented in C++. The projects implementation[3] is strongly dependent on our in-house developed meshless C++ framework *Medusa library* [19] supporting all building blocks of the solution procedure, i.e., differential operator approximations, node positioning algorithms, etc.

[3] Source code is available at http://gitlab.com/e62Lab/public/2023_cp_iccs_hybrid_nodes under tag *v1.1*.

We used `g++ 11.3.0 for Linux` to compile the code with `-O3 -DNDEBUG` flags on `Intel(R) Xeon(R) CPU E5520` computer. To improve the timing accuracy we run the otherwise parallel code in a single thread with the CPU frequency fixed at 2.27 GHz, disabled boost functionality and assured CPU affinity using the `taskset` command. Post-processing was done using Python 3.10.6 and Jupyter notebooks, also available in the provided git repository.

3 Governing Problem

To objectively assess the advantages of the hybrid method, we focus on the natural convection problem that is governed by a system of three PDEs that describe the continuity of mass, the conservation of momentum and the transfer of heat

$$\nabla \cdot \boldsymbol{v} = 0, \tag{2}$$

$$\frac{\partial \boldsymbol{v}}{\partial t} + \boldsymbol{v} \cdot \nabla \boldsymbol{v} = -\nabla p + \nabla \cdot (\mathrm{Pr} \nabla \boldsymbol{v}) - \mathrm{RaPr} \boldsymbol{g} T_{\Delta}, \tag{3}$$

$$\frac{\partial T}{\partial t} + \boldsymbol{v} \cdot \nabla T = \nabla \cdot (\nabla T), \tag{4}$$

where a dimensionless nomenclature using Rayleigh (Ra) and Prandtl (Pr) numbers is used [11,22].

The temporal discretization of the governing equations is solved with the explicit Euler time stepping where we first update the velocity using the previous step temperature field in the Boussinesq term [21]. The pressure-velocity coupling is performed using the Chorin's projection method [3] under the premise that the pressure term of the Navier-Stokes equation can be treated separately from other forces and used to impose the incompressibility condition. The time step is a function of internodal spacing h, and is defined as $\mathrm{d}t = 0.1\frac{h^2}{2}$ to assure stability.

4 Numerical Results

The governing problem presented in Sect. 3 is solved on different geometries employing (i) MON, (ii) RBF-FD and (iii) their spatially-varying combination. The performance of each approach is evaluated in terms of accuracy of the numerical solution and execution times. Unless otherwise specified, the MON method is employed using the monomial approximation basis omitting the mixed terms, while the RBF-FD approximation basis consists of Polyharmonic splines or order $k = 3$ augmented with monomials up to and including order $m = 2$.

4.1 The de Vahl Davis Problem

First, we solve the standard de Vahl Davis benchmark problem [22]. The main purpose of solving this problem is to establish confidence in the presented solution procedure and to shed some light on the behaviour of considered approximation methods, the stability of the solution procedure and finally on the computational efficiency. Furthermore, the de Vahl Davis problem was chosen as the basic test case, because the regularity of the domain shape allows us to efficiently discretize it using exclusively scattered or regular nodes and compare the solutions to that obtained with the hybrid scattered-regular discretization.

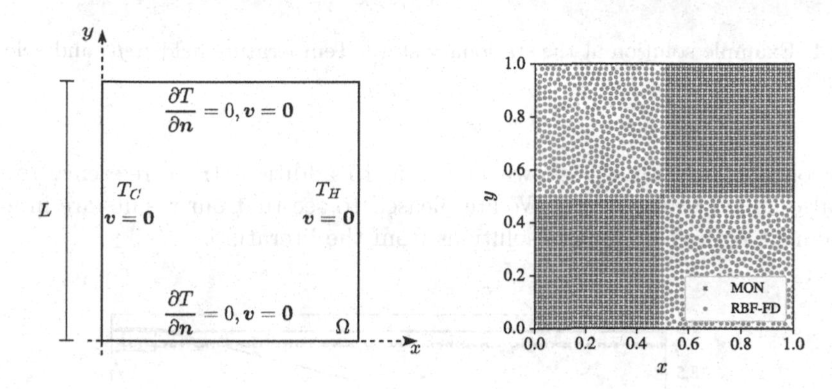

Fig. 3. The de Vahl Davis sketch *(left)* and example hybrid scattered-regular domain discretization *(right)*.

For a schematic representation of the problem, see Fig. 3 *(left)*. The domain is a unit box $\Omega = [0, 1] \times [0, 1]$, where the left wall is kept at a constant temperature $T_C = -0.5$, while the right wall is kept at a higher constant temperature $T_H = 0.5$. The upper and lower boundaries are insulated, and no-slip boundary condition for velocity is imposed on all walls. Both the velocity and temperature fields are initially set to zero.

To test the performance of the proposed hybrid scattered-regular approximation method, we divide the domain Ω into quarters, where each quarter is discretized using either scattered or regular nodes – see Fig. 3 *(right)* for clarity.

An example solution for $\mathrm{Ra} = 10^6$ and $\mathrm{Pr} = 0.71$ at a dimensionless time $t = 0.15$ with approximately $N = 15\,800$ discretization nodes is shown in Fig. 4.

We use the Nusselt number—the ratio between convective and conductive heat transfer—to determine when a steady state has been reached and as a convenient scalar value for comparison with reference solutions. In the following analyses, the average Nusselt number ($\overline{\mathrm{Nu}}$) is calculated as the average of the Nusselt values at the cold wall nodes

$$\mathrm{Nu} = \frac{L}{T_H - T_C} \left| \frac{\partial T}{\partial n} \right|_{x=0}. \tag{5}$$

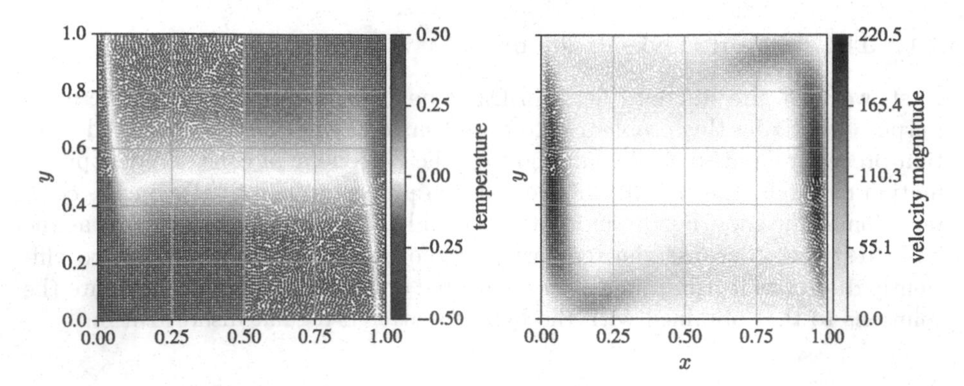

Fig. 4. Example solution at the stationary state. Temperature field *(left)* and velocity magnitude *(right)*.

Its evolution over time is shown in Fig. 5. In addition, three reference results are also added to the figure. We are pleased to see that our results are in good agreement with the reference solutions from the literature.

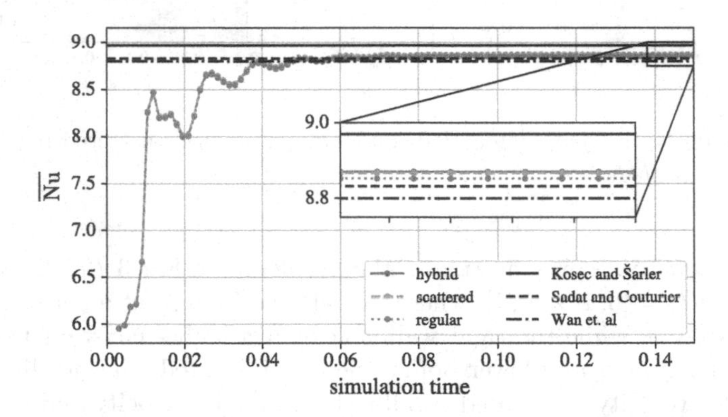

Fig. 5. Time evolution of the average Nusselt number along the cold edge calculated with the densest considered discretization. Three reference results Kosec and Šarler [11], Sadat and Couturier [13] and Wan et. al. [4] are also added.

Moreover, Fig. 5 also shows the time evolution of the average Nusselt number value for cases where the entire domain is discretized using either scattered or regular nodes. We find that all—hybrid, purely scattered and purely regular domain discretizations—yield results in good agreement with the references. More importantly, the hybrid method shows significantly shorter computational time (about 50%) than that required by the scattered discretization employing RBF-FD, as can be seen in Table 1 for the densest considered discretization with $h = 0.00398$.

Table 1. Average Nusselt along the cold edge along with execution times and number of discretization nodes.

Approximation	$\overline{\mathrm{Nu}}$	execution time [h]	N
scattered	8.867	6.23	55 477
regular	8.852	2.42	64 005
hybrid	8.870	3.11	59 694
Kosec and Šarler (2007) [11]	8.97	/	10201
Sadat and Couturier (2000) [13]	8.828	/	22801
Wan et. al. (2001) [4]	8.8	/	10201

To further validate the hybrid method, we show in Fig. 6 the vertical component of the velocity field across the section $y = 0.5$. It is important to observe that the results for the hybrid, scattered and regular approaches overlap, which means that the resulting velocity fields for the three approaches are indeed comparable.

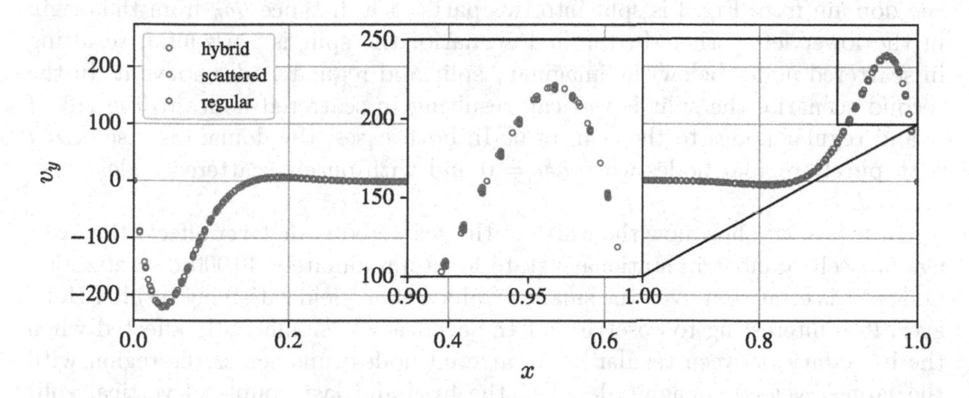

Fig. 6. Vertical velocity component values at nodes close to the vertical midpoint of the domain, i.e., $|y-0.5| \leq h$ for purely scattered, purely regular and hybrid discretizations.

As a final remark, we also study the convergence of the average Nusselt number with respect to the number of discretization nodes in Fig. 7 *(left)*, where we confirm that all our discretization strategies converge to a similar value that is consistent with the reference values. Moreover, to evaluate the computational efficiency of the hybrid approach, the execution times are shown on the right. Note that the same values for h were used for all discretization strategies and the difference in the total number of nodes is caused by the lower density of scattered nodes at the same internodal distance.

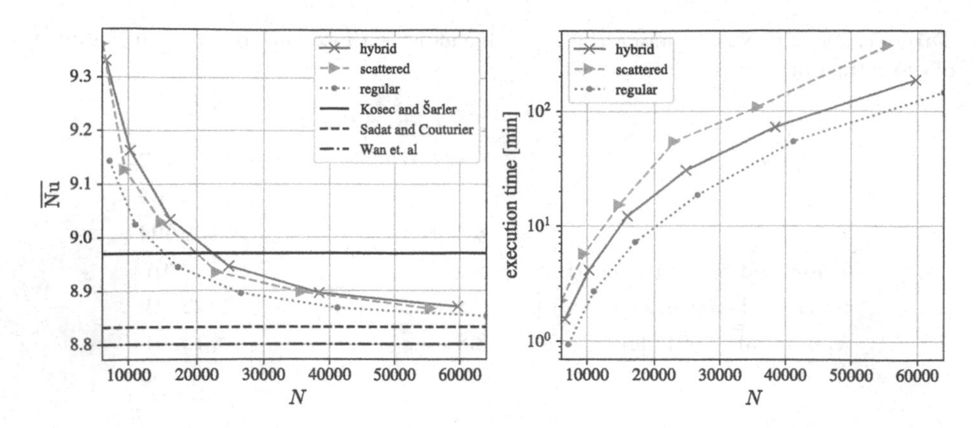

Fig. 7. Convergence of average Nusselt number with respect to discretization quality *(left)* and corresponding execution times *(right)*.

The Effect of the Scattered Nodes Layer Width δ_h. To study the effect of the width of the scattered node layer δ_h, we consider two cases. In both cases, the domain from Fig. 3 is split into two parts at a distance $h\delta_h$ from the origin in the lower left corner. In the first scenario, the split is horizontal, resulting in scattered nodes below the imaginary split and regular nodes above it. In the second scenario, the split is vertical, resulting in scattered nodes to the left of it and regular nodes to the right of it. In both cases, the domain is discretized with purely regular nodes when $h\delta_h = 0$ and with purely scattered nodes when $h\delta_h = L$.

In Fig. 8, we show how the width of the scattered node layer affects the average Nusselt number in stationary state for approximately 40 000 discretization nodes. It is clear that even the smallest values of δ_h yield satisfying results. However, it is interesting to observe that the accuracy is significantly affected when the boundary between regular and scattered nodes runs across the region with the largest velocity magnitudes, i.e., the first and last couple of vertical split data points in Fig. 8.

4.2 Natural Convection on Irregularly Shaped Domains

In the previous section we demonstrated that the hybrid scattered-regular approximation method is computationally more efficient than the pure RBF-FD approximation with only minor differences in the resulting fields. However, to truly exploit the advantages of the hybrid method, irregular domains must be studied. Therefore, in this section, the hybrid scattered-regular approach is

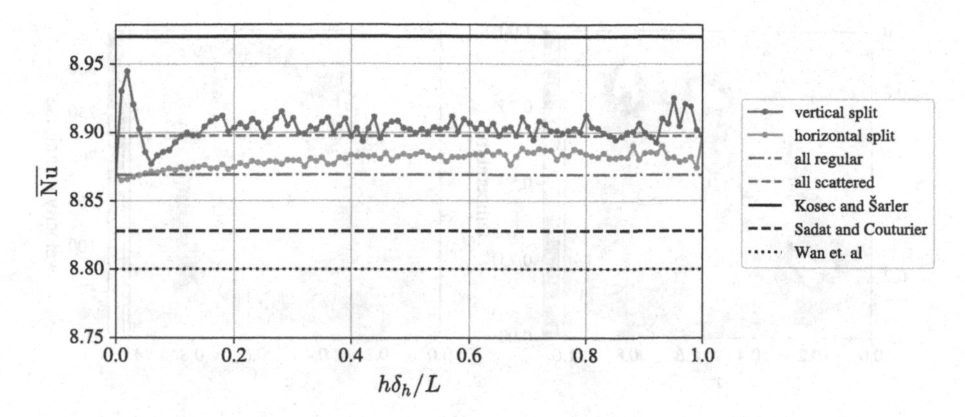

Fig. 8. Demonstration of the scattered node layer width (δ_h) effect on the accuracy of the numerical solution.

employed on an irregularly shaped domain. Let the computational domain Ω be a difference between the two-dimensional unit box $\Omega = [0, 1] \times [0, 1]$ and 4 randomly positioned duck-shaped obstacles introducing the domain irregularity.

The dynamics of the problem are governed by the same set of Eqs. (2–4) as in the previous section. This time, however, all the boundaries of the box are insulated. The obstacles, on the other hand, are subject to Dirichlet boundary conditions, with half of them at $T_C = 0$ and the other half at $T_H = 1$. The initial temperature is set to $T_{\text{init}} = 0$.

We have chosen such a problem because it allows us to further explore the advantages of the proposed hybrid scattered-regular discretization. Generally speaking, the duck-shaped obstacles within the computational domain represent an arbitrarily complex shape that requires scattered nodes for accurate description, i.e., reduced discretization-related error. Moreover, by using scattered nodes near the irregularly shaped domain boundaries, we can further improve the local field description in their vicinity by employing a h-refined discretization. Specifically, we employ h-refinement towards the obstacles with linearly decreasing internodal distance from $h_r = 0.01$ (regular nodes) towards $h_s = h_r/3$ (irregular boundary) over a distance of $h_r\delta_h$. The refinement distance and the width of the scattered node layer are the same, except in the case of fully scattered discretization. Such setup effectively resulted in approximately $N = 11\,600$ computational nodes ($N_s = 3149$ scattered nodes and $N_r = 8507$ regular nodes), as shown in Fig. 1 for a scattered node layer width $\delta_h = 4$. Note that the time step is based on the smallest h, i.e., $\mathrm{d}t = 0.1\frac{h_s^2}{2}$.

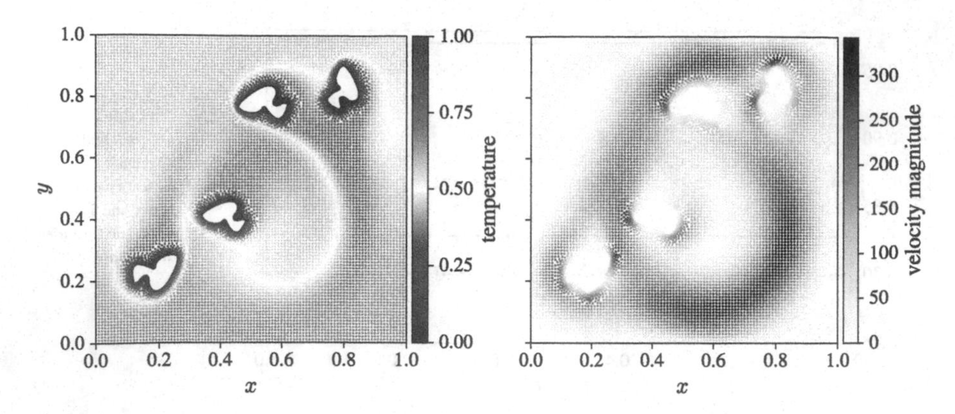

Fig. 9. Example solution on irregular domain. Temperature field *(left)* and velocity magnitude *(right)*.

Figure 9 shows an example solution for an irregularly shaped domain. The hybrid scattered-regular solution procedure was again able to obtain a reasonable numerical solution.

Furthermore, Fig. 10 *(left)* shows the average Nusselt number along the cold duck edges where we can observe that a stationary state has been reached. The steady state values for all considered discretizations match closely but it is interesting to note that in the early stage of flow formation, the fully scattered solutions with different refinement distance δ_h differ significantly more than the hybrid and the fully scattered solutions with the same refinement strategy.

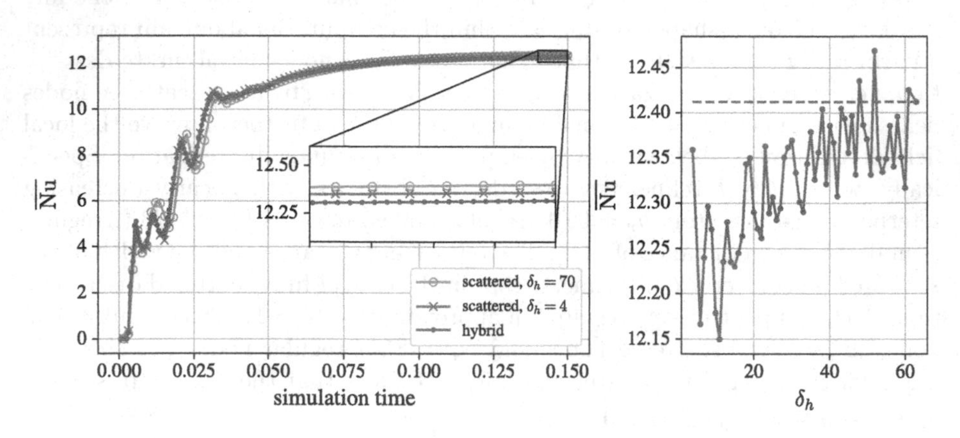

Fig. 10. Time evolution of the average Nusselt number calculated on the cold duck-shaped obstacles of an irregularly shaped domain. *(left)* Changes in stationary state average Nusselt number as the scattered node layer width δ_h increases. *(right)*

It is perhaps more important to note that the execution times gathered in Table 2 show that the hybrid method effectively reduces the execution time for approximately 35 %. The pure regular discretization with MON approximation is omitted from the table because a stable numerical solution could not be obtained.

Table 2. Average Nusselt along the cold duck edges along with execution times. Note that all values in the table were obtained for $\delta_h = 4$.

Approximation	$\overline{\mathrm{Nu}}$	execution time [min]	N
scattered	12.32	46.31	10 534
hybrid	12.36	29.11	11 535

The Effect of the Scattered Nodes Layer Width δ_h. To justify the use of $\delta_h = 4$, we show in Fig. 10 *(right)* that the average value of the Nusselt number at steady state for different values of δ_h. In the worst case, the difference is <2%, justifying the use of the computationally cheaper smaller δ_h. Note that in this particular domain setup, $\delta_h > 64$ already yields a purely scattered domain discretization, while the minimum stable value is $\delta_h = 4$. Note also that the general increase of the Nusselt number with respect to the width of the scattered node layers δ_h may also exhibit other confounding factors. An increase in δ_h leads to a finer domain discretization due to a more gradual refinement, i.e., a fully scattered discretization using $\delta_h = 70$ results in about 35 000 discretization points compared to 11 600 at $\delta_h = 4$, while a decrease in δ_h leads to a more aggressive refinement that could also have a negative effect. This can be supported by observing the difference between the results for the two fully scattered discretizations in Fig. 10.

4.3 Application to Three-Dimensional Irregular Domains

As a final demonstrative example, we employ the proposed hybrid scattered-regular approximation method on a three-dimensional irregular domain. The computational domain Ω is a difference between the three-dimensional unit box $\Omega = [0, 1] \times [0, 1] \times [0, 1]$ and 4 randomly positioned and sized spheres introducing the domain irregularity.

The dynamics are governed by the same set of Eqs. (2–4) as in the two-dimensional case from Sect. 4.2. To improve the quality of the local field description near the irregularly shaped domain boundaries, h-refinement is employed with a linearly decreasing internodal distance from $h_r = 0.025$ (regular nodes) towards $h_s = h_r/2$ (spherical boundaries). Two spheres were set to a constant temperature $T_C = 0$ and the remaining two to $T_H = 1$. The Rayleigh number was set to 10^4.

Although difficult to visualize, an example solution is shown in Fig. 11. Using the hybrid scattered-regular domain discretization, the solution procedure was again able to obtain a reasonable numerical solution.

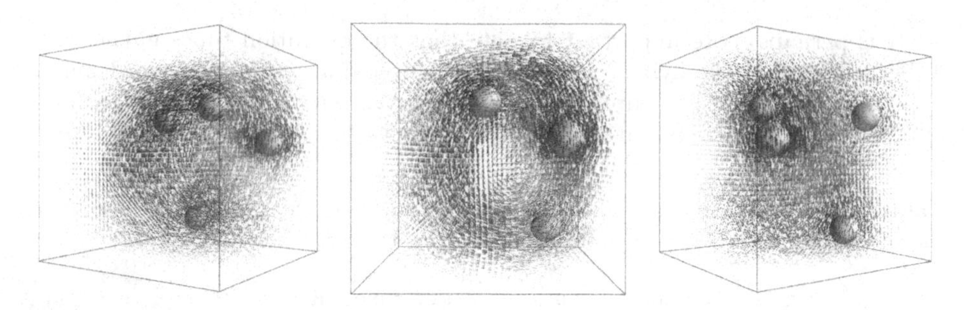

Fig. 11. Example solution viewed from three different angles. The arrows show the velocity in computational nodes and are coloured according to the temperature in that node. The values range from dark blue for T_C to dark red for T_H. For clarity, only a third of the nodes is visualized. (Color figure online)

Note that the scattered method took about 48 h and the hybrid scattered-regular approximation method took 20 h to simulate 1 dimensionless time unit with the dimensionless time step $dt = 7.8125 \cdot 10^{-6}$ and about 75 000 computational nodes with $\delta_h = 4$. For clarity, the data is also gathered in Table 3. Note that the pure regular discretization with MON approximation is again omitted from the table because a stable numerical solution could not be obtained.

Table 3. Average Nusselt along the cold spheres, execution time, and number of computational nodes.

Approximation	$\overline{\mathrm{Nu}}$	execution time [h]	N
scattered	7.36	48.12	65 526
hybrid	6.91	20.54	74 137

5 Conclusions

We proposed a computationally efficient approach to the numerical treatment of problems in which most of the domain can be efficiently discretized with regularly positioned nodes, while scattered nodes are used near irregularly shaped domain boundaries to reduce the discretization-related errors. The computational effectiveness of the spatially-varying approximation method, employing FD-like approximation on regular nodes and RBF-FD on scattered nodes, is demonstrated on a solution to a two-dimensional de Vahl Davis natural convection problem.

We show that the proposed hybrid method, can significantly improve the computational efficiency compared to the pure RBF-FD approximation, while introducing minimal cost to the accuracy of the numerical solution. A convergence analysis from Fig. 7 shows good agreement with the reference de Vahl Davis solutions.

In the continuation, the hybrid method is applied to a more general natural convection problem in two- and three-dimensional irregular domains, where the elegant mathematical formulation of the meshless methods is further exposed by introducing h-refinement towards the irregularly shaped obstacles. In both cases, the hybrid method successfully obtained the numerical solution and proved to be computationally efficient, with execution time gains nearing 50 %.

Nevertheless, the scattered node layer width and the aggressiveness of h-refinement near the irregularly shaped domain boundaries should be further investigated, as both affect the computational efficiency and stability of the solution procedure. In addition, future work should also include more difficult problems, such as mixed convection problems and a detailed analysis of possible surface effects, e.g. scattering, at the transition layer between the scattered and regular domains.

Acknowledgements. The authors would like to acknowledge the financial support of Slovenian Research Agency (ARRS) in the framework of the research core funding No. P2-0095, the Young Researcher program PR-10468 and research project J2-3048.

Conflict of Interest. The authors declare that they have no conflict of interest. All the co-authors have confirmed to know the submission of the manuscript by the corresponding author.

References

1. Bayona, V., Flyer, N., Fornberg, B., Barnett, G.A.: On the role of polynomials in RBF-FD approximations: II. Numerical solution of elliptic PDEs. J. Comput. Phys. **332**, 257–273 (2017)
2. Bourantas, G., et al.: Strong-form approach to elasticity: hybrid finite difference-meshless collocation method (FDMCM). Appl. Math. Model. **57**, 316–338 (2018). https://doi.org/10.1016/j.apm.2017.09.028
3. Chorin, A.J.: Numerical solution of the Navier-Stokes equations. Math. Comput. **22**(104), 745–762 (1968)
4. Wan, D.C., Patnaik, B.S.V., Wei, G.W.: A new benchmark quality solution for the buoyancy-driven cavity by discrete singular convolution. Numer. Heat Transfer Part B: Fundam. **40**(3), 199–228 (2001). https://doi.org/10.1080/104077901752379620
5. Ding, H., Shu, C., Yeo, K., Xu, D.: Simulation of incompressible viscous flows past a circular cylinder by hybrid FD scheme and meshless least square-based finite difference method. Comput. Methods Appl. Mech. Eng. **193**(9–11), 727–744 (2004)
6. El Kadmiri, R., Belaasilia, Y., Timesli, A., Kadiri, M.S.: A hybrid algorithm using the fem-meshless method to solve nonlinear structural problems. Eng. Anal. Boundary Elem. **140**, 531–543 (2022). https://doi.org/10.1016/j.enganabound.2022.04.018

7. Flyer, N., Fornberg, B., Bayona, V., Barnett, G.A.: On the role of polynomials in RBF-FD approximations: I. Interpolation and accuracy. J. Computat. Phys. **321**, 21–38 (2016)
8. Fornberg, B., Flyer, N.: A Primer on Radial Basis Functions with Applications to the Geosciences. SIAM (2015)
9. Javed, A., Djidjeli, K., Xing, J., Cox, S.: A hybrid mesh free local RBF-cartesian FD scheme for incompressible flow around solid bodies. Int. J. Math. Comput. Nat. Phys. Eng. **7**, 957–966 (2013)
10. Kosec, G.: A local numerical solution of a fluid-flow problem on an irregular domain. Adv. Eng. Softw. **120**, 36–44 (2018)
11. Kosec, G., Šarler, B.: Solution of thermo-fluid problems by collocation with local pressure correction. Int. J. Numer. Methods Heat Fluid Flow **18**, 868–882 (2008)
12. Liu, G.R.: Meshfree Methods: Moving Beyond the Finite Element Method. CRC Press (2009)
13. Sadat, H., Couturier, S.: Performance and accuracy of a meshless method for laminar natural convection. Numer. Heat Transfer Part B: Fundam. **37**(4), 455–467 (2000). https://doi.org/10.1080/10407790050051146
14. van der Sande, K., Fornberg, B.: Fast variable density 3-D node generation. SIAM J. Sci. Comput. **43**(1), A242–A257 (2021)
15. Shankar, V., Kirby, R.M., Fogelson, A.L.: Robust node generation for meshfree discretizations on irregular domains and surfaces. SIAM J. Sci. Comput. **40**(4), 2584–2608 (2018). https://doi.org/10.1137/17m114090x
16. Slak, J., Kosec, G.: Refined meshless local strong form solution of Cauchy-Navier equation on an irregular domain. Eng. Anal. Boundary Elem. **100**, 3–13 (2019). https://doi.org/10.1016/j.enganabound.2018.01.001
17. Slak, J., Kosec, G.: Adaptive radial basis function-generated finite differences method for contact problems. Int. J. Numer. Meth. Eng. **119**(7), 661–686 (2019). https://doi.org/10.1002/nme.6067
18. Slak, J., Kosec, G.: On generation of node distributions for meshless PDE discretizations. SIAM J. Sci. Comput. **41**(5), A3202–A3229 (2019)
19. Slak, J., Kosec, G.: Medusa: a C++ library for solving PDEs using strong form mesh-free methods. ACM Trans. Math. Softw. (TOMS) **47**(3), 1–25 (2021)
20. Tolstykh, A., Shirobokov, D.: On using radial basis functions in a "finite difference mode" with applications to elasticity problems. Comput. Mech. **33**(1), 68–79 (2003)
21. Tritton, D.J.: Physical Fluid Dynamics. Oxford Science Publ, Clarendon Press (1988). https://doi.org/10.1007/978-94-009-9992-3
22. de Vahl Davis, G.: Natural convection of air in a square cavity: a bench mark numerical solution. Int. J. Numer. Meth. Fluids **3**(3), 249–264 (1983)
23. Wendland, H.: Scattered Data Approximation, vol. 17. Cambridge University Press (2004)
24. Zamolo, R., Nobile, E.: Solution of incompressible fluid flow problems with heat transfer by means of an efficient RBF-FD meshless approach. Numer. Heat Transf. Part B: Fundam. **75**(1), 19–42 (2019)

Oscillatory Behaviour of the RBF-FD Approximation Accuracy Under Increasing Stencil Size

Andrej Kolar-Požun[1,2](✉), Mitja Jančič[1,3], Miha Rot[1,3], and Gregor Kosec[1]

[1] Parallel and Distributed Systems Laboratory, Jožef Stefan Institute, Ljubljana, Slovenia
{andrej.pozun,mitja.jancic,miha.rot,gregor.kosec}@ijs.si
[2] Faculty of Mathematics and Physics, University of Ljubljana, Ljubljana, Slovenia
[3] Jožef Stefan International Postgraduate School, Ljubljana, Slovenia

Abstract. When solving partial differential equations on scattered nodes using the Radial Basis Function generated Finite Difference (RBF-FD) method, one of the parameters that must be chosen is the stencil size. Focusing on Polyharmonic Spline RBFs with monomial augmentation, we observe that it affects the approximation accuracy in a particularly interesting way - the solution error oscillates under increasing stencil size. We find that we can connect this behaviour with the spatial dependence of the signed approximation error. Based on this observation we are then able to introduce a numerical quantity that indicates whether a given stencil size is locally optimal.

Keywords: Meshless · Stencil · RBF-FD · PHS

1 Introduction

Radial Basis Function generated Finite Differences (RBF-FD) is a method for solving Partial Differential Equations (PDEs) on scattered nodes that has recently been increasing in popularity. It uses Radial Basis Functions (RBFs) to locally approximate a linear differential operator in a chosen neighbourhood, generalising the well known finite difference methods. This neighbourhood used for the approximation is referred to as the stencil of a given point and is commonly chosen to simply consist of its n closest neighbours.

Among the different possible choices of a RBF used, the Polyharmonic Splines (PHS) with appropriate polynomial augmentation stand out due to the fact that they possess no shape parameter, eliminating all the hassle that comes with having to find its optimal value. PHS RBF-FD has been studied extensively and proved to work well in several different contexts [1,2,6–8]. Unlike in the case of RBFs with a shape parameter, where the approximation order is determined by the stencil size [3], in PHS RBF-FD it is determined by the degree of the monomials included in the augmentation [1]. Despite that, the choice of an appropriate stencil size can have a substantional impact on the accuracy. More

J. Mikyška et al. (Eds.): ICCS 2023, LNCS 14076, pp. 515–522, 2023.
https://doi.org/10.1007/978-3-031-36027-5_40

precisely, the accuracy of the method displays an oscillatory behaviour under increasing stencil size.

In the remainder of the paper, we present this observation and our findings. Ideally we would like to be able to predict the stencil sizes that correspond to the accuracy minima or at least provide some indicator on whether a given stencil size is near the minimum.

The following section describes our problem setup along with the numerical solution procedure and in Sect. 3 our results are discussed.

2 Problem Setup

Our analyses are performed on the case of the Poisson equation

$$\nabla^2 u(\mathbf{x}) = f(\mathbf{x}), \tag{1}$$

where the domain is a disc $\Omega = \{\mathbf{x} \in \mathbb{R}^2 : \|x - (0.5, 0.5)\| \leq 0.5\}$. We choose the function $f(\mathbf{x})$ such, that the problem given by Eq. (1) has a known analytic solution. Concretely, we choose

$$u(x, y) = \sin(\pi x)\sin(\pi y), \tag{2}$$

$$f(x, y) = -2\pi^2 \sin(\pi x)\sin(\pi y) \tag{3}$$

with the Dirichlet boundary conditions given by a restriction of $u(\mathbf{x})$ to the boundary $\partial\Omega$.

We discretise the domain with the discretisation distance $h = 0.01$, first discretising the boundary and then the interior using the algorithm proposed in [4]. The Laplacian is then discretised using the RBF-FD algorithm as described in [5], where we choose the radial cubics as our PHS ($\phi(r) = r^3$) augmented with monomials up to degree $m = 3$, inclusive. This requires us to associate to each discretisation point $\mathbf{x}_i$ its stencil, which we take to consist of its n nearest neighbours. We can now convert the PDE (1) into a sparse linear system, which we then solve to obtain an approximate solution $\hat{u}(\mathbf{x})$. The source code is readily available in our git repository[1].

The chosen analytical solution $u(\mathbf{x})$ is displayed in Fig. 1, which additionally serves as a visual representation of how the domain is discretised.

Having both the analytical and approximate solutions, we will be interested in the approximation error. It will turn out to be useful to consider the signed pointwise errors of both the solution and the Laplacian approximation:

$$e^{\pm}_{\text{poiss}}(\mathbf{x}_i) = \hat{u}_i - u_i, \tag{4}$$

$$e^{\pm}_{\text{lap}}(\mathbf{x}_i) = \tilde{\nabla}^2 u_i - f_i, \tag{5}$$

where $\tilde{\nabla}^2$ is the discrete approximation of the Laplacian and we have introduced the notation $u_i = u(\mathbf{x}_i)$. The "poiss" and "lap" subscripts may be omitted in the text when referring to both errors at once.

[1] https://gitlab.com/e62Lab/public/2023_cp_iccs_stencil_size_effect.

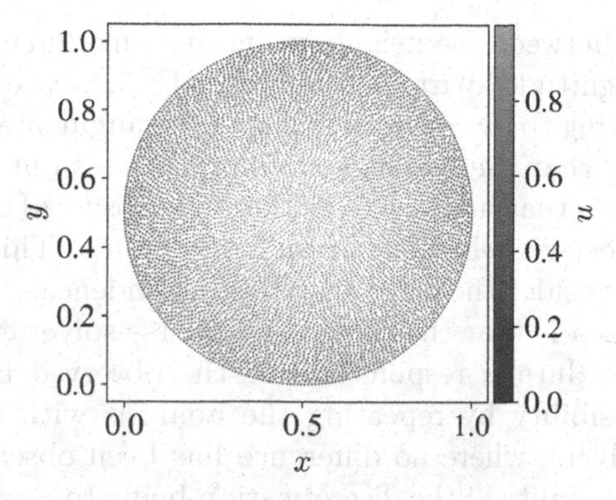

Fig. 1. The analytical solution to the considered Poisson problem.

As a quantitative measure of the approximation quality, we will also look at the average/max absolute value error:

$$e_{\text{poiss}}^{\max} = \max_{\mathbf{x}_i \in \mathring{\Omega}} |e_{\text{poiss}}^{\pm}(\mathbf{x}_i)|, \tag{6}$$

$$e_{\text{poiss}}^{\text{avg}} = \frac{1}{N_{\text{int}}} \sum_{\mathbf{x}_i \in \mathring{\Omega}} |e_{\text{poiss}}^{\pm}(\mathbf{x}_i)| \tag{7}$$

and analogously for $e_{\text{lap}}^{\max}$ and $e_{\text{lap}}^{\text{avg}}$. N_{int} is the number of discretisation points inside the domain interior $\mathring{\Omega}$.

In the next section we will calculate the approximation error for various stencil sizes n and further investigate its (non-trivial) behaviour.

It is worth noting that the setup considered is almost as simple as it can be. The fact that we have decided not to consider a more complicated problem is intentional - there is no need to complicate the analysis by considering a more complex problem if the investigated phenomena already appears in a simpler one. This reinforces the idea that such behaviour arises from the properties of the methods used and not from the complexity of the problem itself.

3 Results

In Fig. 2 we see that $e_{\text{poiss}}^{\max}(n)$ oscillates with several local minima (at stencil sizes $n = 28, 46$) and maxima (stencil sizes $n = 17, 46$). The dependence $e_{\text{poiss}}^{\max}(n)$ seems to mostly resemble a smooth function. This is even more evident in $e_{\text{poiss}}^{\text{avg}}(n)$. The errors of the Laplacian are also plotted and we can observe that $e_{\text{lap}}^{\text{avg}}(n)$ has local minima and maxima at same stencil sizes. Such regularity implies that the existence of the minima in the error is not merely a coincidence, but a consequence of a certain mechanism that could be explained. Further understanding of this mechanism would be beneficial, as it could potentially allow us to predict the location of these local minima a priori. Considering that the error difference between the neighbouring local maxima and minima can be over an order of magnitude apart this could greatly increase the accuracy of the method without having to increase the order of the augmentation or the discretisation density. Note that the behaviour of $e_{\text{lap}}^{\max}$ stands out as it is much more irregular. This implies that in order to explain the observed oscillations, we have to consider the collective behaviour of multiple points. This will be confirmed later on, when we consider the error's spatial dependence.

An immediate idea is that the choice of a sparse solver employed at the end of the solution procedure is responsible for the observed behaviour. We have eliminated this possibility by repeating the analysis with both the SparseLU and BiCGSTAB solvers, where no difference has been observed. The next idea we explore is the possibility of the discretisation being too coarse. Figure 3 shows that under discretisation refinement $e_{\text{poiss}}^{\max}(n)$ maintains the same shape and is just shifted vertically towards a lower error. The latter shift is expected, as we are dealing with a convergent method, for which the solution error behaves as $e \propto h^p$ as $h \to 0$, where p is the order of the method. We also show $e_{\text{poiss}}^{\max}(h)$ in a log-log scale for some different stencil sizes. It can be seen that the slopes and therefore the orders p generally do not change with the stencil size and that the observed oscillations mainly affect the proportionality constant in $e \propto h^p$. The stencil dependence of the error proportionality constant has already been observed in similar methods [3,9].

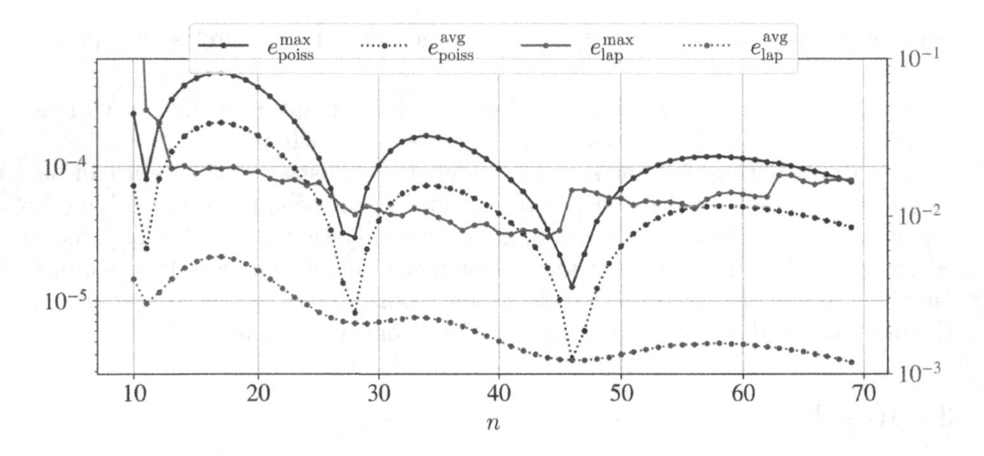

Fig. 2. Dependence of the approximation errors on the stencil size n.

Next we check if boundary stencils are responsible for the observed behaviour as it is known that they can be problematic due to their one-sideness [2]. In Fig. 4 we have split our domain into two regions - the nodes near the boundary $\{\mathbf{x}_i \in \Omega : \|\mathbf{x}_i - (0.5, 0.5)\| > 0.4\}$ are coloured red, while the nodes far from the boundary $\{\mathbf{x}_i \in \Omega : \|\mathbf{x}_i - (0.5, 0.5)\| \leq 0.4\}$ are black. We can see that the dependence of $e_{\text{poiss}}^{\max}(n)$ marginally changes if we keep the stencil size near the

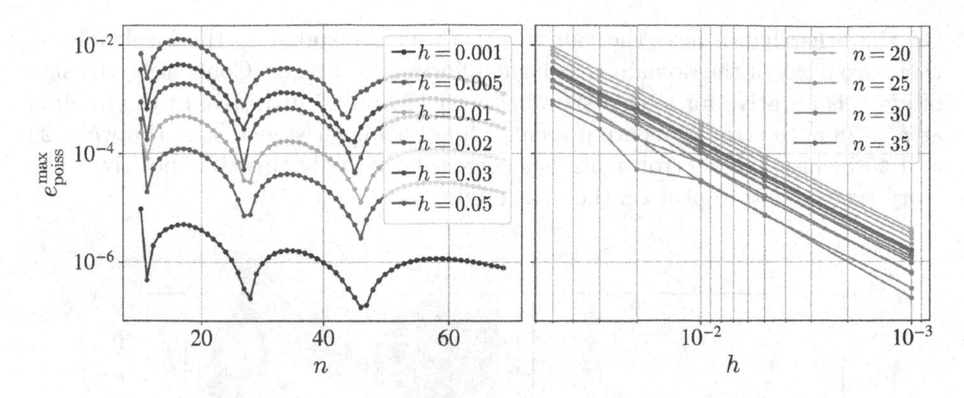

Fig. 3. Behaviour of the approximation errors under a refinement of the discretisation.

boundary fixed at $n = 28$ (corresponding to one of the previously mentioned minima), while only changing the stencil sizes of the remaining nodes. This shows that the observed phenomena is not a consequence of the particularly problematic boundary stencils.

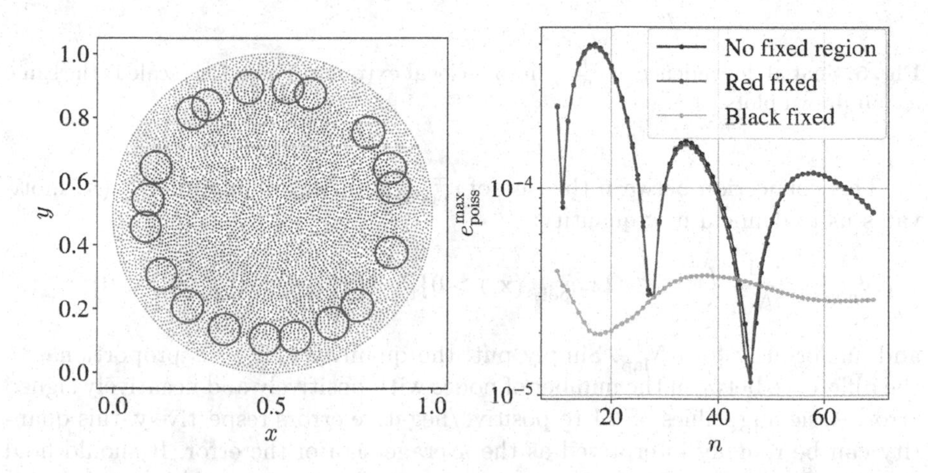

Fig. 4. The seperation of the domain into two regions is seen on the left, where the green circles show the radii of the biggest stencils considered ($n = 69$). The right graph shows the error dependence when either of the regions is at a fixed stencil size $n = 28$. The previous result with no fixed stencil size regions is also shown.

Figure 5 provides some more insight into the mechanism behind the oscillating error. Here we have plotted the spatial dependence of the signed error $e_{\text{poiss}}^{\pm}$ for those stencils that correspond to the marked local extrema. We can observe that in the maxima, the error has the same sign throughout the whole domain.

On the other hand, near the values of n that correspond to the local minima there are parts of the domain that have differing error signs. Concretely, the sign of $e^{\pm}_{\text{poiss}}$ is negative for stencil sizes between 17 and 27 inclusive. In the minima at $n = 28$ both error signs are present, while for bigger stencil sizes (between 29 and 45 inclusive) the error again has constant sign only this time positive. The story turns out to repeat on the next minimum.

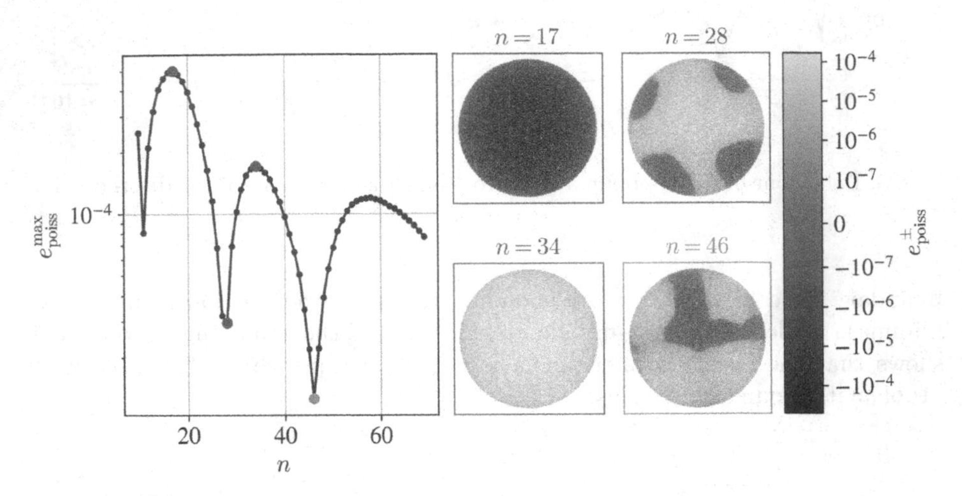

Fig. 5. Spatial dependence of $e^{\pm}_{\text{poiss}}$ in some local extrema. The colour scale is the same for all drawn plots.

This connection between the sign of $e^{\pm}_{\text{poiss}}$ and the minima in $e^{\max}_{\text{poiss}}(n)$ motivates us to define a new quantity:

$$\delta N^{\pm}_{\text{poiss}} = \frac{1}{N_{\text{int}}} \left(|\{\mathbf{x}_i \in \mathring{\Omega} : e^{\pm}_{\text{poiss}}(\mathbf{x}_i) > 0\}| - |\{\mathbf{x}_i \in \mathring{\Omega} : e^{\pm}_{\text{poiss}}(\mathbf{x}_i) < 0\}| \right) \quad (8)$$

and analogously for $\delta N^{\pm}_{\text{lap}}$. Simply put, the quantity $\delta N^{\pm}_{\text{poiss}}$ is proportional to the difference between the number of nodes with positively and negatively signed error. Assigning values of ± 1 to positive/negative errors respectively, this quantity can be roughly interpreted as the average sign of the error. It should hold that $|\delta N^{\pm}_{\text{poiss}}|$ is approximately equal to 1 near the maxima and lowers in magnitude as we approach the minima. Figure 6 confirms this intuition - $\delta N^{\pm}_{\text{poiss}}(n)$ changes its values between ± 1 very abruptly only near the n that correspond to the minima of $e^{\max}_{\text{poiss}}(n)$. A similar conclusion can be made for $\delta N^{\pm}_{\text{lap}}$, which acts as a sort of "smoothed out" version of $\delta N^{\pm}_{\text{poiss}}(n)$.

At a first glance, $N^{\pm}_{\text{lap}}$ seems like a good candidate for an error indicator - it has a well-behaved dependence on n, approaches ± 1 as we get closer to the error maxima and has a root near the error minima. The major downside that completely eliminates its applicability in the current state is the fact that we need access to the analytical solution to be able to compute it.

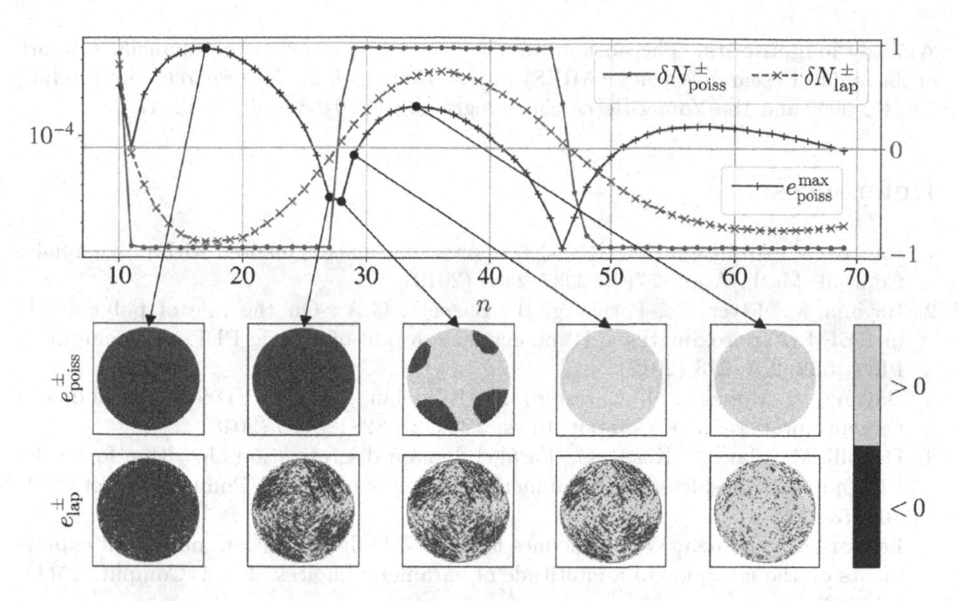

Fig. 6. The quantities $\delta N^{\pm}(n)$ along with the spatial profiles of the signs of $e^{\pm}$ for some chosen stencil sizes. For convenience, $\delta N^{\pm} = 0$ is marked with an orange line.

4 Conclusions

Our study started with a simple observation - when solving a Poisson problem with PHS RBF-FD, the approximation accuracy depends on the stencil size n in a non-trivial manner. In particular, there exist certain stencil sizes where the method is especially accurate. A priori knowledge of these stencil sizes could decrease the solution error without any additional effort and is therefore strongly desirable. We have made a small step towards understanding this phenomena, eliminating various common numerical issues as the cause. Looking at the spatial dependence of the signed solution error, we have noticed that in the stencil sizes corresponding to the local error minima, the signed solution error is not strictly positive or negative. This is unlike the generic stencil sizes, where the error has the same sign throughout the domain. Motivated by this observation, we have introduced a quantity that is roughly the average sign of the pointwise Laplace operator errors and appears to have a root only near the stencils corresponding to the local error minima.

The research presented is a step towards defining a more practically useful indicator, which would reveal the most accurate stencil sizes even without having access to the analytical solution and is a part of our ongoing research. Additional future work includes more rigorous theoretical explanations for the observations presented. Further experimental investigations should also be made, particularly to what extent our observations carry over to different problem setups - other differential equations and domain shapes.

Acknowledgements. The authors would like to acknowledge the financial support of Slovenian Research Agency (ARRS) in the framework of the research core funding No. P2-0095 and the Young Researcher programs PR-10468 and PR-12347.

References

1. Bayona, V.: An insight into RBF-FD approximations augmented with polynomials. Comput. Math. Appl. **77**(9), 2337–2353 (2019)
2. Bayona, V., Flyer, N., Fornberg, B., Barnett, G.A.: On the role of polynomials in RBF-FD approximations: II. numerical solution of elliptic PDEs. J. Computat. Phys. **332**, 257–273 (2017)
3. Bayona, V., Moscoso, M., Carretero, M., Kindelan, M.: RBF-FD formulas and convergence properties. J. Comput. Phys. **229**(22), 8281–8295 (2010)
4. Depolli, M., Slak, J., Kosec, G.: Parallel domain discretization algorithm for RBF-FD and other meshless numerical methods for solving PDEs. Comput. Struct. **264**, 106773 (2022)
5. Le Borne, S., Leinen, W.: Guidelines for RBF-FD discretization: numerical experiments on the interplay of a multitude of parameter choices. J. Sci. Comput. **95**(1), 8 (2023)
6. Oruç, Ö.: A radial basis function finite difference (RBF-FD) method for numerical simulation of interaction of high and low frequency waves: Zakharov-Rubenchik equations. Appl. Math. Comput. **394**, 125787 (2021)
7. Shankar, V., Fogelson, A.L.: Hyperviscosity-based stabilization for radial basis function-finite difference (RBF-FD) discretizations of advection-diffusion equations. J. Comput. Phys. **372**, 616–639 (2018)
8. Strniša, F., Jančič, M., Kosec, G.: A meshless solution of a small-strain plasticity problem. In: 2022 45th Jubilee International Convention on Information, Communication and Electronic Technology (MIPRO), pp. 257–262. IEEE (2022)
9. Tominec, I., Larsson, E., Heryudono, A.: A least squares radial basis function finite difference method with improved stability properties. SIAM J. Sci. Comput. **43**(2), A1441–A1471 (2021)

On the Weak Formulations of the Multipoint Meshless FDM

Irena Jaworska(✉)

Cracow University of Technology, Kraków, Poland
irena@cce.pk.edu.pl

Abstract. The paper discusses various formulations of the recently developed higher order Multipoint Meshless Finite Difference Method. The novel multipoint approach is based on raising the order of approximation of the unknown function by introducing additional degrees of freedom in stencil nodes, taking into account e.g. the right hand side of the considered differential equation. It improves the finite difference solution without increasing the number of nodes in an arbitrary irregular mesh. In general, the standard version of the Meshless (Generalized) FDM is based on the strong problem formulation. The extensions of the multipoint meshless FDM allow for analysis of boundary value problems posed in various weak formulations, including variational ones (Galerkin, Petrov-Galerkin), minimization of the energy functional, and the meshless local Petrov-Galerkin. Several versions of the multipoint method are proposed and examined. The paper is illustrated with some examples of the multipoint numerical tests carried out for the weak formulations and their comparison with those obtained for the strong formulation.

Keywords: Meshless FDM · Higher order approximation · Weak formulations · Multipoint method · Homogenization · Elastic-plastic problem

1 Introduction

The Finite Element Method (FEM) has been the most commonly applied method in the field of engineering computations, especially in computational mechanics, leading to solutions of practical engineering problems, such as fracture mechanics, elasticity, structural and fluid dynamics, many more. However, the problems dealing with for example the mesh distortion and frequent remeshing requirements are not efficient to solve by the FEM discretization. This was the motivation for the development of the alternative, so called meshless methods, prescribed by the set of nodal values only. Compared to the FEM, the arbitrarily irregularly distributed ‘cloud of nodes’ without any imposed structure can be easily modified [21, 22]. Therefore, this technique leads to greater flexibility and is more convenient and attractive to implement the adaptive process.

The local approximation of the unknown function in the meshless methods is performed around the nodes. A growing variety of meshless methods differ from each other

J. Mikyška et al. (Eds.): ICCS 2023, LNCS 10476, pp. 523–531, 2023.
https://doi.org/10.1007/978-3-031-36027-5_41

in the process of construction of this local approximation, such as Moving Least Squares (MLS) [1] based methods: Meshless Finite Difference Method (MFDM/GFDM) [2], Diffuse Element Method [3], Element Free Galerkin [4], Finite Point Method [5], Meshless Local Petrov-Galerkin (MLPG) [6]; kernel methods: Smooth Particle Hydrodynamics [7, 8], Reproducing Kernel Particle Method [9], and Partition of Unity (PU) methods: *hp*-clouds [10], PU FEM [11], and many more.

Meshless methods can be classified in many different ways, among other by division into two categories: methods based on strong formulation and methods based on weak formulations. Many of these methods have also been developed in both forms. One of the oldest, as well as the most well-known meshless strong-form method is the Meshless Finite Difference Method (MFDM) [2, 12] which was later generalized into many meshless variants. Higher order method extensions, such as the Multipoint MFDM [13, 14] and MFDM based on the corrections terms [15] were developed for various formulations of the boundary value problems: strong (local), weak (global), and mixed (local-global) one. The multipoint meshless FDM posed in various weak formulations is briefly presented in this paper.

2 Multipoint Problem Formulation

The solution quality of boundary value problems (BVP), solved using meshless finite difference, may be improved through the application of two mechanisms. The first one is based on the mesh density increase, preferably using an adaptive (*h*-type) solution approach. The second mechanism is provided by rising the approximation order (*p*-type). Several concepts may be used in the last case [14]. In the global formulation]. In this research, a return is being made to the old Collatz idea [16] of the multipoint FDM, which was combined with the MFDM to develop the new higher order multipoint meshless FDM.

The method is based on raising the order of approximation of the unknown function u by introducing additional degrees of freedom in the star nodes and using combinations of values of searched function and the right hand side of the MFD equations taken at all nodes of each MFD star.

Multipoint MFDM includes the following basic modifications and extensions: besides the local (strong), also the global (weak) formulations of BVP may be considered as well; arbitrary irregular meshes may be used; the moving weighted least squares (MWLS) approximation is assumed instead of the interpolation technique.

Let us consider the local (strong) formulations of boundary value problems for the n-th order PDE in the domain Ω

$$\begin{cases} \mathcal{L}u(P) = f(P), & \text{for} \quad P \in \Omega \\ \mathcal{G}u(P) = g(P), & \text{for} \quad P \in \partial\Omega \end{cases}, \tag{1}$$

or an equivalent global (weak) one involving integral with appropriate boundary conditions, where $\mathcal{L}$, $\mathcal{G}$ are respectively the n-th and m-th order differential operators.

In the FDM solution approach, the classical difference operator would be presented in the following form (assuming $u_i = u(P_i)$ and $f_i = f(P_i)$)

$$\mathcal{L}u_i \approx Lu_i \equiv \sum_{j(i)} c_j u_j = f_i \quad \Rightarrow \quad Lu_i = f_i, \quad c_j = c_{j(i)}. \tag{2}$$

In the multipoint formulation, the MFDM difference operator Lu is obtained by the Taylor series expansion of unknown function u, including higher order derivatives, and using additional degrees of freedom at nodes. For this purpose one may use for example a combination of the right hand side values f of the given equation at each node of the stencil using arbitrarily distributed clouds of nodes:

$$\mathcal{L}u_i \approx Lu_i \equiv \sum_{j(i)} c_j u_j = \sum_{j(i)} \alpha_j f_j \quad \Rightarrow \quad Lu_i = Mf_i. \tag{3}$$

Here, j is the index of a node in a selected stencil (FD star), i is the index of the central node of the stencil, Mf_i is a combination of the equation right hand side values at the stencil associated to index i, f may present the value of the whole operator $\mathcal{L}u_i$ or a part of it only, such as a specific derivative $u^{(k)}$. In general, L may be referred to differential eqs, boundary conditions and integrand in the global formulation of BVP.

Two basic versions of the Multipoint MFDM–general and specific are considered [14]. The specific approach, presented above (3), is simpler and easier in implementation, but its application is more restricted, mainly to linear BVP. In the specific formulation, the values of the additional degrees of freedom are known. In the global formulation [13], where $u^{(k)}$ derivative is assumed instead of f

$$\sum_{j(i)} c_j u_j = \sum_{j(i)} \alpha_j u_j^{(k)}, \tag{4}$$

they are sought. In each of these multipoint FD cases, one may usually obtain higher order approximation of the FD operators, using the same stencil, as needed to generate Lu_i in the classical FDM approach based on interpolation.

Such extended higher order MFDM approach also enables analysis of boundary value problems posed in various weak (global) formulations, including different variational formulations (Galerkin, Petrov-Galerkin), minimization of the energy functional, and MLPG. Various versions of the multipoint method in global formulations are briefly discussed.

3 Weak Formulations of the Multipoint MFDM Approach

Besides the development of the multipoint MFDM for analysis of boundary value problems in the strong (local) formulation, the multipoint method was also extended to the weak (global) formulations.

The global formulation may be posed in the domain Ω in general as:

- *a variational principle*

$$b(u,\ v) = l(v), \qquad \forall\ v \in V, \tag{5}$$

where b is a bilinear functional dependent on the test function v and trial function u (solution of the considered BVP), V is the space of test functions, l is a linear operator dependent on v.

- *minimization of the potential energy functional*

$$\min_u I(u), \quad I(u) = 1/2b(u,u) - l(u). \tag{6}$$

In both cases, corresponding boundary conditions have to be satisfied.

In the variational formulations one may deal with the Petrov-Galerkin approach when u and v are different functions from each other, and the Bubnov-Galerkin one, when u and v are the same. Assuming a trial function u locally defined on each subdomain Ω_i within the domain Ω one may obtain a global-local formulation of the Petrov-Galerkin type (MLPG). The test function v may be defined here in various ways. In particular, one may assume it as also given locally in each subdomain Ω_i. Usually, the test function is assumed to be equal to zero elsewhere, though it may be defined in many other ways.

In *the MLPG5 formulation* [17] the Heaviside type test function is assumed. In each subdomain Ω_i around a node P_i, $i = 1, 2, \ldots, N$, in the given domain Ω (e.g. in each Voronoi polygon in 2D) the test function is equal to one ($v = 1$) in Ω_i and is assumed to be zero outside. Hence any derivative of v is also equal to zero in the whole domain Ω. Therefore, relevant expressions in the functional (5) $b(u, v)$ and in $l(v)$ vanish, reducing in this way the amount of calculations involved.

Let us consider the following two dimensional elliptic problem

$$\nabla^2 u = f(P),\ P(x,y) \in \Omega,\ u|_\Gamma = 0, \tag{7}$$

which satisfies the differential equation of the second order with Dirichlet conditions on the boundary Γ of the domain Ω.

When the variational form is derived directly from the (6) by integration over the domain, *the first nonsymmetric variational form* is considered:

$$\int_\Omega (u_{xx} + u_{yy}) v \, d\Omega = \int_\Omega f v \, d\Omega,\ u \in H^2,\ v \in H^0. \tag{8}$$

After differentiation by parts, *the symmetric Galerkin form* is obtained

$$\int_\Omega \left(u_x v_x + u_y v_y\right) d\Omega + \int_\Omega f v \, d\Omega = \int_\Gamma \left(u_x n_x + u_y n_y\right) v \, d\Gamma,\ u \in H^1,\ v \in H^1, \tag{9}$$

where n_x and n_y denotes the normal vectors. For the Heaviside test function, *the MLPG5 formulation* is as follows:

$$\int_{\Omega_i} f v \, d\Omega_i = \int_{\Gamma_i} \left(u_x n_x + u_y n_y\right) v \, d\Gamma_i,\ u \in H^1,\ v \in H^1. \tag{10}$$

All variants of the global (weak) formulation of the multipoint method may be realized by the meshless MFDM using regular or totally irregular meshes like it is in the case of the local formulation of BVP.

In all weak formulations of the multipoint MFDM developed here, the unknown trial function u and its derivatives are always approximated using the multipoint finite difference formulas (3) or (4) (MWLS technique based on the stencil subdomain). However, assumption and discretization of the test function v and its derivatives at Gauss points may be done in many ways–calculated by any type of approximation as well as by the simple interpolation on the integration subdomain, which can be other than the stencil subdomain, e.g. Delaunay triangle or Voronoi polygon. This is a direct result of the assumption that u and v functions may be different, and a choice of the test function v should not influence the BVP solution. This statement was positively tested using benchmark problems (Fig. 2).

4 Numerical Analysis

4.1 Benchmark Tests

Several benchmark tests of the application of the multipoint method to the BVP globally formulated were carried out. The multipoint approach was tested in the Galerkin as well as in the MLPG5 formulations, and minimization of the potential energy functional. The solutions were compared with the corresponding ones obtained for the strong (general and specific) multipoint MFDM formulations. The results of the numerical tests of 2D Poisson's problem done are presented in Figs. 1, 2, and 3 (h is the distance between the nodes).

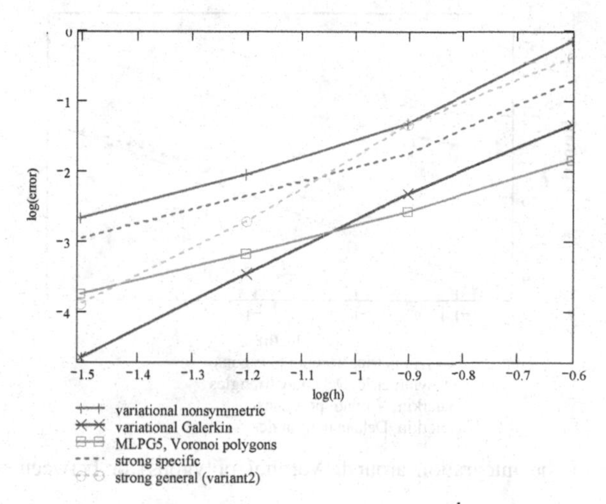

Fig. 1. Convergence of the strong formulations (Multipoint, 3rd approx. Order) and the various weak (nonsymmetric, Galerkin, MLPG5) ones

The type of approximation of the test function does not significantly influence the results (Fig. 2). In the benchmark tests (Fig. 3) the quadrilateral integration subdomains (Voronoi polygons, integration around nodes) give slightly better results than the triangular ones (Delaunay triangles, integration between nodes). The clear advantage of the MLPG5 formulation relies on a significant reduction of numerical operations needed to obtain the final solution of the problems tested, and, as a consequence–the computational time reduction.

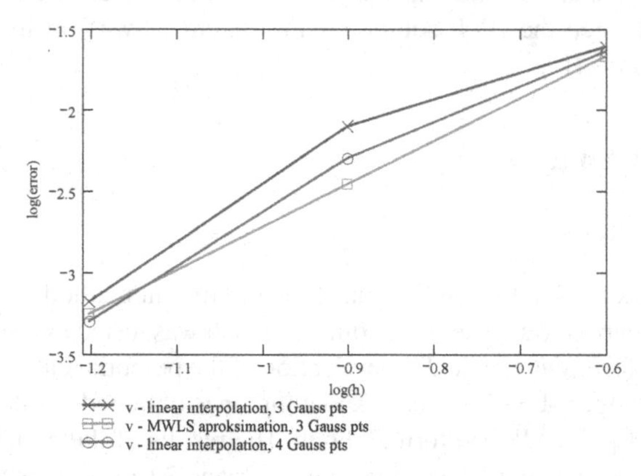

Fig. 2. Influence of the test function interpolation, nonsymmetric variational form

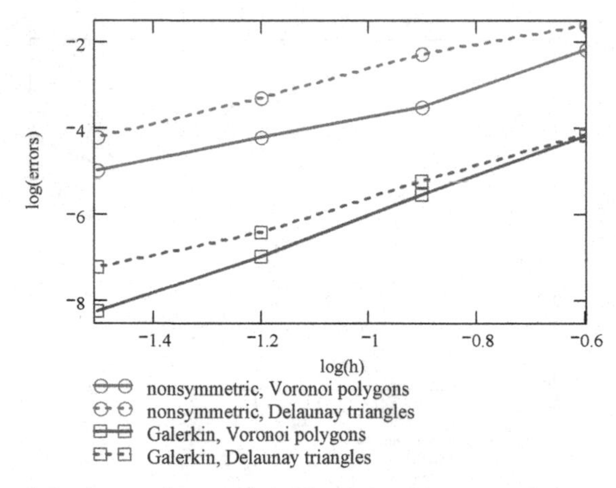

Fig. 3. Influence of the integration around (Voronoi polygons) or between nodes (Delaunay triangles)

4.2 Weak and Strong Formulations in Nonlinear and Multiscale Analyses

The problem formulation may have an influence on the algorithm and the results, especially in the case of more demanding problems, such as multiscale analysis of heterogeneous materials and elastic-plastic problems [18]. The multiscale analysis was carried out by the multipoint method using weak formulations. On the other side, in the elastic-plastic analysis of physically nonlinear problems–the multipoint approach was applied to the strong formulation. In both types of analyses, the problems deal with the jump between types or states of materials.

The oscillations occurred on the interface of the matrix and inclusions of the heterogeneous materials when the nodes of one stencil belong to different material types (Fig. 4). In this situation the only solution was to adjust the stencil to the inclusion distribution [19]. In nonlinear analysis, the similar situation was expected at the elastic-plastic boundary. However, the preliminary numerical results [20] do not demonstrate the oscillations phenomenon (Fig. 5). It seems, that the difference is caused by the assumed type of the problem formulation. Further research is planned.

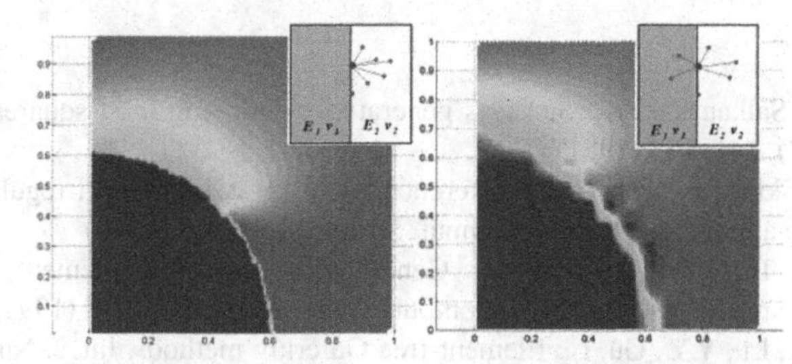

Fig. 4. RVE solution (strain ε_{xx}) obtained on the stencil generated either independently on the inclusion distribution or adjusted to it

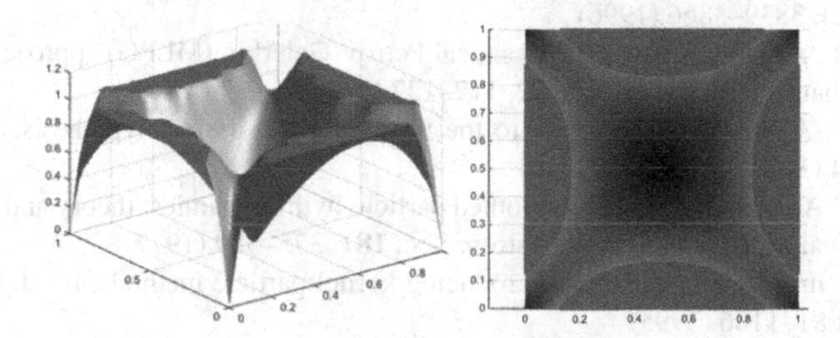

Fig. 5. The total shear stress, coarse and fine mesh (red color depict the plastic zones)

5 Final Remarks

The paper presents an extension of the novel high order multipoint MFDM to the various weak (global) formulations of the boundary value problems. Taken into account were some versions of the weak formulation including the variational Galerkin, local Petrov-Galerkin (MLPG), and minimum of the total potential energy.

The various tests carried out confirm that the multipoint approach may be a useful solution tool for the analysis of boundary value problems given in the global formulation and provide valuable results, close enough to the local (strong) form. Additionally, the advantages of the MLPG5 approach due the Heaviside type test function used–reducing the amount of calculations involved–may be noticed.

The specific or general multipoint MFD operators should be applied to the trial function in the case of the global or global-local formulations of the analyzed problem, and numerical integration is additionally required here. The type of approximation of the test function is not very important here.

Although the results obtained by the multipoint technique are close enough in both strong and weak formulations in general, in the case of more demanding problems, such as nonlinear or multiscale analyses, the problem formulation may influence on the computational algorithm of the method. In the case of the weak formulation, it may be necessary to match the stencils (MFD stars) to the interfaces of the different types or states of the material. Further research is planned.

References

1. Lancaster, P., Salkauskas, K.: Surfaces generated by moving least-squares method. Math. Comp. **155**(37), 141–158 (1981)
2. Liszka, T., Orkisz, J.: The finite difference method at arbitrary irregular grids and its applications in applied mechanics. Comput. Struct. **11**, 83–95 (1980)
3. Nayroles, B., Touzot, G., Villon, P.: Generalizing the finite element method: diffuse approximation and diffuse elements. Comput. Mech. **10**(5), 307–318 (1992)
4. Belytschko, T., Lu, Y.Y., Gu, L.: Element-free Galerkin methods. Int. J. Numer. Meth. Eng. **37**(2), 229–256 (1994)
5. Onate, E., Idelsohn, S.R., Zienkiewicz, O.C., Taylor, R.L.: A finite point method in computational mechanics: applications to convective transport and fluid flow. Int. J. Numer. Meth. Eng. **39**(22), 3839–3866 (1996)
6. Atluri, S.N., Zhu, T.: A new meshless local Petrov-Galerkin (MLPG) approach in computational mechanics. Comput. Mech. **22**, 117–127 (1998)
7. Lucy, L.B.: A numerical approach to the testing of the fission hypothesis. Astron. J. **82**, 1013–1024 (1977)
8. Gingold, R.A., Monaghan, J.J.: Smoothed particle hydrodynamics: theory and application to non-spherical stars. Mon. Not. R. Astron. Soc. **181**, 375–389 (1977)
9. Liu, W.K., Jun, S., Zhang, Y.F.: Reproducing kernel particle methods. Int. J. Numer. Meth. Eng. **20**, 1081–1106 (1995)
10. Duarte, C.A., Oden, J.T.: An hp adaptive method using clouds. Comput. Meth. Appl. Mech. Eng. **139**, 237–262 (1996)
11. Melenk, J.M., Babuska, I.: The partition of unity finite element method: basic theory and applications. Comput. Meth. Appl. Mech. Eng. **139**, 289–314 (1996)
12. Orkisz, J.: Finite Difference Method (Part III). In: Kleiber, M. (ed.) Handbook of Computational Solid Mechanics, pp. 336–432. Springer, Berlin (1998)
13. Jaworska, I.: On the ill-conditioning in the new higher order multipoint method. Comput. Math. Appl. **66**(3), 238–249 (2013)
14. Jaworska, I., Orkisz, J.: Higher order multipoint method – from Collatz to meshless FDM. Eng. Anal. Bound. Elem. **50**, 341–351 (2015)

15. Milewski, S.: Meshless finite difference method with higher order approximation – applications in mechanics. Arch. Comput. Meth. Eng. **19**(1), 1–49 (2012)
16. Collatz, L., Numerische Behandlung von Differential-gleichungen. Springer, Heidelberg (1955). https://doi.org/10.1007/978-3-662-22248-5
17. Atluri, S.N., Shen, S.: The Meshless Local Petrov-Galerkin (MLPG) Method. Tech Science Press (2002)
18. Jaworska, I.: Multipoint meshless FD schemes applied to nonlinear and multiscale analysis. In: Groen, D., de Mulatier, C., Paszynski, M., Krzhizhanovskaya, V.V., Dongarra, J.J., Sloot, P.M.A. (eds.) Computational Science – ICCS 2022. ICCS 2022. Lecture Notes in Computer Science, vol. 13353. pp. 55-68. Springer, Cham (2022). https://doi.org/10.1007/978-3-031-08760-8_5
19. Jaworska, I.: On some aspects of the meshless FDM application for the heterogeneous materials. Int. J. Multiscale Comput. Eng. **15**(4), 359–378 (2017)
20. Jaworska, I.: Generalization of the multipoint meshless FDM application to the nonlinear analysis. Comput. Math. Appl. **87**(3), 1–11 (2021)
21. Chen, J.-S., Hillman, M., Chi, S.-W.: Meshfree methods: progress made after 20 years. J. Eng. Mech. **143**(4), 04017001 (2017)
22. Nguyen, V.P., Rabczuk, T., Bordas, S., Duflot, M.: Meshless methods: a review and computer implementation aspects. Math. Comput. Simul. **79**, 763–813 (2008)

[illegible] Minkowski [illegible] less [illegible] methods [illegible]
[illegible] Arch. Comput. Meth. [illegible]
16. [illegible] Meshfree [illegible]
[illegible] (2000) [illegible]
17. [illegible]
[illegible]
18. [illegible] supported [illegible] analysis [illegible] Mock, [illegible] PMA [illegible] Computational Science [illegible] Lecture Notes in Computer Science [illegible] 2022 [illegible]
19. [illegible] heterogeneous [illegible]
20. [illegible]
21. [illegible]
22. [illegible] Comput. Math. Appl. [illegible] 2009 [illegible]

Multiscale Modelling and Simulation

Convolutional Recurrent Autoencoder for Molecular-Continuum Coupling

Piet Jarmatz(✉), Sebastian Lerdo, and Philipp Neumann

Chair for High Performance Computing,
Helmut Schmidt University, Hamburg, Germany
jarmatz@hsu-hh.de

Abstract. Molecular-continuum coupled flow simulations are used in many applications to build a bridge across spatial or temporal scales. Hence, they allow to investigate effects beyond flow scenarios modeled by any single-scale method alone, such as a discrete particle system or a partial differential equation solver. On the particle side of the coupling, often molecular dynamics (MD) is used to obtain trajectories based on pairwise molecule interaction potentials. However, since MD is computationally expensive and macroscopic flow quantities sampled from MD systems often highly fluctuate due to thermal noise, the applicability of molecular-continuum methods is limited. If machine learning (ML) methods can learn and predict MD based flow data, then this can be used as a noise filter or even to replace MD computations, both of which can generate tremendous speed-up of molecular-continuum simulations, enabling emerging applications on the horizon.

In this paper, we develop an advanced hybrid ML model for MD data in the context of coupled molecular-continuum flow simulations: A convolutional autoencoder deals with the spatial extent of the flow data, while a recurrent neural network is used to capture its temporal correlation. We use the open source coupling tool MaMiCo to generate MD datasets for ML training and implement the hybrid model as a PyTorch-based filtering module for MaMiCo. It is trained with real MD data from different flow scenarios including a Couette flow validation setup and a three-dimensional vortex street. Our results show that the hybrid model is able to learn and predict smooth flow quantities, even for very noisy MD input data. We furthermore demonstrate that also the more complex vortex street flow data can accurately be reproduced by the ML module.

Keywords: Flow Simulation · Machine Learning · Denoising · Data Analytics · Molecular Dynamics · Molecular-Continuum

1 Introduction

One of the fundamental nanofluidics tools in engineering, biochemistry and other fields are molecular dynamics (MD) simulations [5,14]. MD has the potential to assess the properties of novel nanomaterials, for example it has been applied to water desalination, in order to develop highly permeable carbon nanotube

J. Mikyška et al. (Eds.): ICCS 2023, LNCS 14076, pp. 535–549, 2023.
https://doi.org/10.1007/978-3-031-36027-5_42

membranes for reverse osmosis [20]. However, in many such application cases the challenging computational cost of MD imposes a barrier that renders full-domain simulations of larger nanostructures infeasible. Thus, coupled multiscale methods, where MD is restricted to critical locations of interest and combined with a continuum flow model to create a molecular-continuum simulation, are typically used to enable computationally efficient simulations of macroscopic flows. In the water purification example, the MD simulation of the carbon nanotube confined flow can be accompanied by a Hagen-Poiseuille flow solver to yield a multiscale method that still captures molecular physics but also makes water transport predictions for larger laboratory-scale membranes [2].

Here we focus on cases where the flow in the MD domain could be described by the continuum solver, unlike carbon nanotubes, so that we can use the additional continuum information to validate coupling methodology. Since from a software design perspective, implementing a coupling can come with many challenges, many general-purpose frameworks for arbitrary multiphysics simulations are available (e.g. [3,19,21]), however only a few focus on molecular-continuum flow [8,15,18]. In our recent work [8] we have presented MaMiCo 2.0, an open source C++ framework designed to create modular molecular-continuum simulations. Data sampled from MD often suffers from high hydrodynamic fluctuations, i.e. thermal noise, but many coupling schemes and continuum methods depend on smooth flow data. Hence, MaMiCo supports two ways to obtain smoother data: ensemble averaging and noise filtering. For the former, an ensemble of independent MD simulation instances is launched on the same subdomain, so that their results can be averaged [13]. Depending on the temperature, typically 50–200 instances are necessary to obtain a stable transient two-way coupled simulation [23]. For noise filtering, there is a flexible filtering system with several filter module implementations, such as proper orthogonal decomposition (POD) or Non-Local Means (NLM) [6]. NLM can reduce the number of MD instances required for a certain flow result accuracy, approximately by a factor of 10, for details see [6]. MaMiCo 2.0 can change the number of active MD instances dynamically at run time of the coupled simulation for on-the-fly error control [8].

However, also with a combination of ensemble averaging to obtain averaged data for stable coupling and noise filtering to reduce the number of MD instances, a problem of major importance remains to be the computational cost of MD, which restricts the applicability of the methodology to limited scenarios. A promising approach to further tackle this are machine learning (ML) methods which can support the coupling in this case in two ways: First, if ML is used as an advanced noise filter in the filtering system of MaMiCo, then this can help to reduce the number of MD instances. Second, if an ML model is able to learn and predict the behavior of MD, then it can be used to avoid costly computations and to replace the MD simulation, at least for some of the time steps or some of the instances. Note that from ML perspective, both use cases are very similar: in either case, the ML module receives as input data the state of MD, i.e. the data sampled from MD at coupling time step t, as well as the information from the continuum solver at the outer MD boundaries, and the ML module is supposed to output only the filtered MD state, either at t or at $t+1$. Note that while these use cases are the primary motivation for the ML developments presented in this paper, here we will focus on definition, training, validation and analysis of the novel ML module for the purposes of both filtering and prediction of the MD flow description, yet abstaining from replacing MD entirely which we consider a natural next step in the near future, since this is an active field of research. For instance, in [9] a surrogate model of MD simulations is developed, which can accurately predict a small number of key features from MD, based on a fully connected neural network. Here we focus on architectures for grid-based flow data instead, since fully connected networks are not scalable to operate on large inputs.

In recent years, many works have studied various neural network architectures for complex fluid flow and turbulent CFD problems [4,12]. For instance, Wiewel et al. have developed a hybrid model using a convolutional neural network (CNN) based autoencoder (AE) and an LSTM, which is a type of recurrent neural network (RNN) [22]. While they achieve significant speed-ups compared to traditional solvers, they did not investigate the impacts of noisy input data. However, in 2022, Nakamura and Fukagata have performed a robustness analysis, investigating the effects of noise perturbation in the training data in the context of an CNN-AE for turbulent CFD [11], although without employing an RNN. Here, we combine and use insights from both these works by applying a similar hybrid model to a different context, molecular dynamics data, which often contains a higher level of noise.

The goal of this paper is to introduce a convolutional recurrent hybrid model and to demonstrate that it is able to do both: work as an advanced MD noise filter and accurately predict the behavior of the MD simulation over time in a transient three-dimensional molecular-continuum flow.

In Sect. 2 we introduce the coupling methodology, software tools, flow solvers and scenarios used in this paper. In Sect. 2.1 we give details how they are applied to generate the datasets for the ML training. Section 3 develops a convolutional recurrent autoencoder model for MD data. Therefore, first we define an AE in Sect. 3.1, then we introduce an RNN for the CNN latent space in Sect. 3.2, and finally in Sect. 3.3 we combine both of them into a hybrid model. Section 4 gives details about our implementation of this hybrid model, documents the training approach and defines hyper-parameters. We first show results of the hybrid model for a Couette flow validation test case in Sect. 5, and then demonstrate and explain its capabilities for a more complex vortex street in Sect. 6. Finally, Sect. 7 summarizes our insights and provides an outlook to future research that is rendered possible by this work.

2 Molecular-Continuum Coupled Flow

For our molecular-continuum simulations we consider a domain decomposition into a small molecular region placed within a much larger continuum domain, with nested time stepping between the two solvers. For the datasets used in this paper, on the particle side MaMiCo's in-house MD code *SimpleMD* is applied to model a Lennard-Jones fluid and on the continuum side we use the *lbmpy* [1] software package to apply a Lattice-Boltzmann method (LBM).

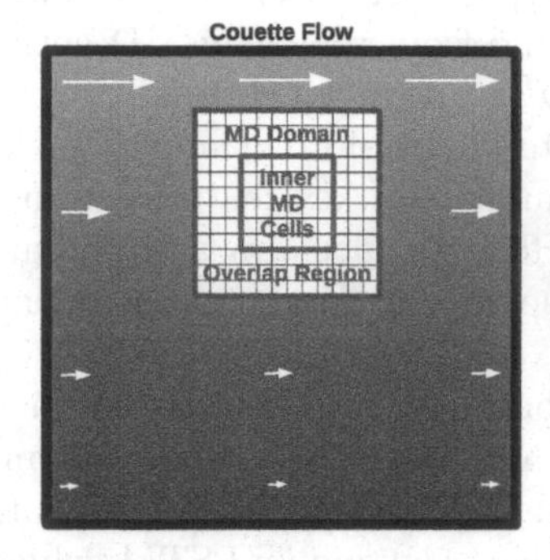

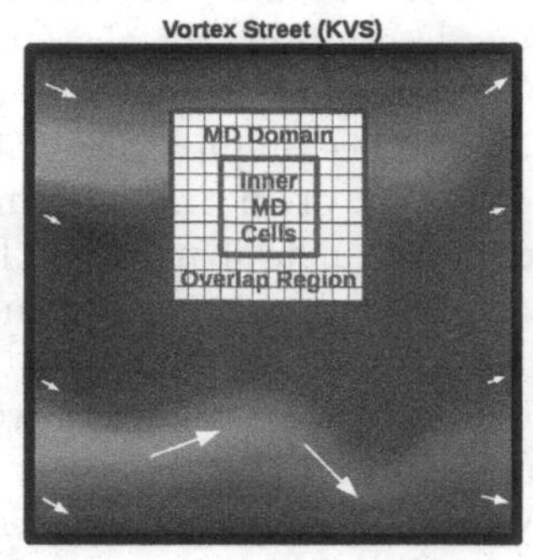

Fig. 1. Transient coupled flow scenarios; MD receives continuum data in overlap region, ML aims to predict MD data in inner cells. Left: flow between two parallel plates. Right: vortex street, obstacle not shown.

In this paper, as illustrated in Fig. 1, we use a Couette flow case similar to *scenario 2A* defined in [6] and a vortex street test case similar to *scenario 1*, corresponding to the benchmark *3D-2Q* by Schäfer et al. [17]. The coupling methodology, software and flow test scenario setup have been described thoroughly in previous publications, thus the reader is referred to [6,7] for more details. The Kármán vortex street (KVS), while not a common nanofluidics scenario, is excellent for investigating and comparing the performance of ML models, since it yields a sufficiently complex multidimensional transient flow with challenging non-linear data. In both scenarios we zero-initialize the flow. We start to couple with MD during a transient start-up phase. Since coupling is already enabled while the continuum solver is establishing the flow, the velocities vary more. This can be used to reveal potential mismatches between MD data and ML predictions. The tests performed in this paper use a single MD instance only, for simplicity and because this is the most challenging test case for an ML module, with the highest possible level of noise in the data sampled from MD. All simulations used in this paper are *one-way* coupled, meaning that flow data is transferred from the continuum solver to the particle system, but not in the other direction.

2.1 Dataset Creation

The filtering system of MaMiCo operates on a regular grid of voxel cells that covers the MD domain, where each cell contains quantities, such as density, temperature or velocity, which are sampled from the MD simulations and averaged over all particles within that cell. Typically each cell contains ca. 10 to 100 particles, depending on the exact simulation configuration. Note that since we train and use ML models as modules in the filtering system of MaMiCo, they do not have any access to individual particles, but only to grid-based quantities. This means that the ML model does not operate on the full state of the MD system, but instead on the information that is exchanged between MD and continuum solver. If ML can predict this information with a sufficient accuracy, then it can replace the MD simulation from the perspective of the continuum solver – even without any knowledge of the internal MD state.

In this paper, the ML training and validation datasets generated by MaMiCo are multichannel volumes in the shape of $[900 \times 3 \times 24 \times 24 \times 24]$. The first dimension refers to 900 coupling cycles, i.e. macroscopic simulation time steps, while the second dimension refers to three flow velocity components per voxel, u_x, u_y and u_z. For simplicity, we disregard the other quantities stored in the cells, but the ML methodology can be applied to them in an analogous way. The remaining dimensions define the volume of interest spanning $24 \times 24 \times 24$ cells covering the entire MD domain. There is an overlap region that covers the three outer cell layers of the MD domain, where data from the continuum solver is received and applied to the MD system as a boundary condition, using momentum imposition and particle insertion algorithms, for details see [6]. On the inner MD domain consisting of $18 \times 18 \times 18$ cells, quantities are sampled from MD for transfer to the continuum solver. The goal of the ML model developed in Sect. 3 is to predict a future filtered state of these quantities in the inner MD domain, excluding the overlap region, while given as input a noisy state from past coupling time steps of the entire domain, including the overlap region. This means that the ML model can access the data that is or would be transferred from the continuum solver to MD, and in turn tries to replace both, MD simulation and noise filters, in order to forecast the data that would finally be transferred back to the continuum solver one coupling cycle later.

Our data stems from 20 KVS and 21 Couette flow MaMiCo simulations. To generate this amount of data, we vary three parameters and choose seven Couette wall velocities at three MD positions, and five KVS init-times at four MD positions. For example, a KVS dataset where the coupling starts after 22000 LBM steps and the MD domain is placed north-west of the KVS center is labeled as '22000_NW' (compare Fig. 6). The most important parameters defining the simulations in this paper are: 175616 molecules per instance, density $\rho \approx 0.813$, Lennard-Jones parameters $\sigma_{LJ} = \epsilon_{LJ} = 1$, cutoff radius $r_{cutoff} = 2.5$, temperature $T = 1.1$, yielding a kinematic viscosity $\nu \approx 2.63$, given in dimensionless MD units (see [6]). In the KVS scenario, we couple with 50 MD steps per cycle to a D3Q19-TRT LBM on 12580620 LB cells, and place the center of MD at $50\% \times 73.2\% \times 50\%$ of domain size in *3D-2Q* setup, plus small variations. All of this yields a data size of 12.2 *GB* (in binary format), we split it into 80% training and 20% validation data sets. More detailed information generating the datasets can be accessed online[1].

3 Convolutional Recurrent Autoencoder

In this section, we introduce our ML model developed to make time-series predictions of microscale flow velocity distributions. Designing such a model must reflect the spatial and temporal dependencies in the underlying physics [4,22]. Advances in computer vision have shown that spatially correlated data, such as

[1] https://github.com/HSU-HPC/MaMiCo_hybrid_ml.

images or volumes, are best dealt with CNNs [10]. Advances in machine translation and speech recognition have shown that sequentially correlated data such as language can be modeled with RNNs [24]. Our models follow a hybrid approach combining their advantages. Our experiments have shown that in this MD data context, a classical AE performs better than a model based on the *U-Net* architecture [16]. Thus, in the following we focus on a concept similar to the approach presented by Nakamura et al. [12].

3.1 Convolutional Autoencoder (AE)

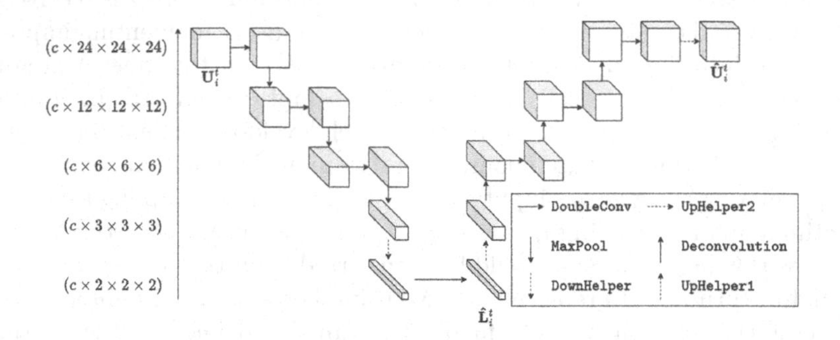

Fig. 2. Schematic of a single CNN AE as used for the triple model approach.

A convolutional AE is a type of CNN that consists of an encoding and a decoding path, as shown in Fig. 2. It aims to learn a dense comprehensive representation of the input. The sizes of our datasets make RNN approaches impractical, thus we use an AE to encode lower-dimensional representations of the input, called latent space [12]. Let $\mathbf{U}$ be the velocity distribution input data and $h(\mathbf{U})$ describe the encoding function. The encoding function $h(\mathbf{U})$ maps the input data $\mathbf{U}$ to a latent space representation $\hat{\mathbf{L}}$, using convolutional and pooling layers such that

$$h(\mathbf{U}) = \hat{\mathbf{L}} = \sigma(\mathbf{W}_h \mathbf{U} + \mathbf{b}_h). \tag{1}$$

Here, $\mathbf{W}_h$ and $\mathbf{b}_h$ are the weights and biases of the encoding function, and σ is a non-linear activation function, such as ReLU. The decoding function $g(\hat{\mathbf{L}})$ maps $\hat{\mathbf{L}}$ back to the original input space $\hat{\mathbf{U}}$ using transposed convolutional and upsampling layers such that

$$g(\hat{\mathbf{L}}) = \hat{\mathbf{U}} = \sigma(\mathbf{W}_g \hat{\mathbf{L}} + \mathbf{b}_g). \tag{2}$$

Here $\mathbf{W}_g$ and $\mathbf{b}_g$ are the weights and biases of the decoding function. The AE is trained by minimizing a reconstruction loss function $\mathcal{L}(\mathbf{U}, \hat{\mathbf{U}})$ which measures the difference between the original input $\mathbf{U}$ and the reconstructed input $\hat{\mathbf{U}}$. The models used in this paper follow a single or triple model approach. The former uses a single AE to operate on the entire velocity distribution $\mathbf{U}$ while the triple model uses three identical instances of an AE designed to operate on a single velocity channel $\mathbf{U}_i$. For each AE, in the encoding path, the first three horizontal groups apply a `DoubleConv` layer consisting of two ReLU-activated 3×3 same convolutions followed by a `MaxPool` layer consisting of a 2×2 max pooling operator. This operator halves each spatial dimension. Next, a `DownHelper` consisting of a ReLU-activated 2×2 valid convolution is applied in order to further reduce dimensionality. Finally, another `DoubleConv` layer is applied thereby yielding the latent space $\hat{\mathbf{L}}_i$. In the decoding path, $\hat{\mathbf{L}}_i$ is first passed to an `UpHelper1` consisting of a 2×2 deconvolution. Next, a `Deconvolution` tailored to double the spatial dimensions followed by a `DoubleConv` is applied. This is repeated until the input spatial dimensionality is restored. Finally, an `UpHelper2` consisting of a 3×3 same convolution tailored to a single channel output is applied. When the output of the AE is used in the hybrid model, only the inner 3×18^3 values are selected and passed on towards the continuum solver, since they correspond to the inner MD cells excluding the overlap layers.

3.2 Recurrent Neural Network

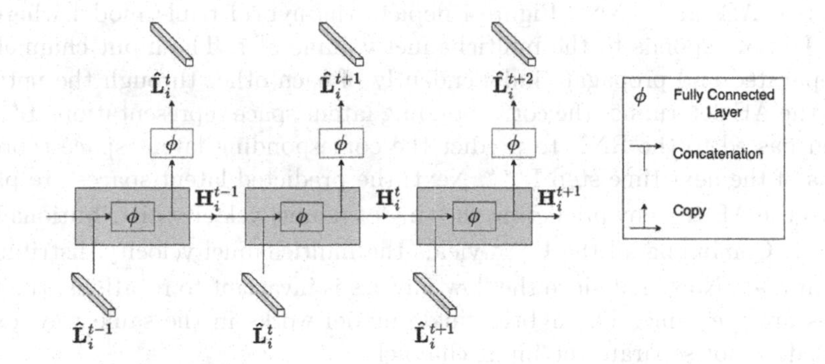

Fig. 3. Schematic of one unrolled RNN as used for the time-series prediction of microscale fluid flow latent spaces.

We account for the temporal dependency via an RNN with a hidden state as depicted in Fig. 3. The hidden state at a given time step is a function of the hidden state of the previous step and the input of the current step. With this, the hidden state is able to preserve historical information from previous time steps of the sequence. Let $\mathbf{H}^t$ be the hidden state at time t, given as [25]

$$\mathbf{H}^t = \sigma(\hat{\mathbf{L}}^t \mathbf{W}_{xh} + \mathbf{H}^{t-1} \mathbf{W}_{hh} + \mathbf{b}_{hh}). \tag{3}$$

Here, $\mathbf{W}_{xh}$ are the input weights, $\mathbf{W}_{hh}$ are the hidden state weights and $\mathbf{b}_{hh}$ are the biases. The RNN latent space output $\hat{\mathbf{L}}^{t+1}$ is then the output of a fully connected layer such that

$$\hat{\mathbf{L}}^{t+1} = \mathbf{H}^t \mathbf{W}_{hq} + \mathbf{b}_{qq}. \tag{4}$$

$\mathbf{W}_{hq}$ and $\mathbf{b}_{qq}$ are output weights and biases. A single or triple AE requires a similar RNN approach. This means one RNN operates on $\hat{\mathbf{L}}$ in a single model, while the triple model requires three RNNs to each operate on one component $\hat{\mathbf{L}}_i$. Figure 3 depicts such an RNN that is a component of a triple model.

3.3 Hybrid Model

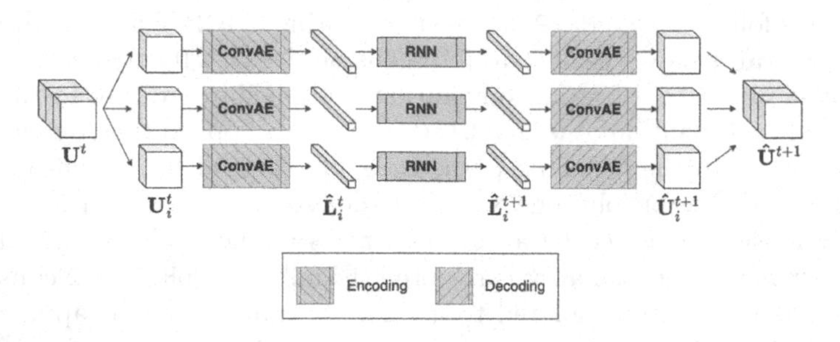

Fig. 4. Schematic of the convolutional recurrent autoencoder as employed in the triple model approach for the task of time-series prediction of cell-wise averaged fluid flow velocities.

Our two hybrid model architectures combine the single and triple model convolutional AEs and RNNs. Figure 4 depicts the hybrid triple model where the input $\mathbf{U}^t$ corresponds to the multichannel volume at t. The input channels $\mathbf{U}_i^t$ are separated and propagate independently of each other through the network. First, the AE determines the corresponding latent space representations $\hat{\mathbf{L}}_i^t$. This is then passed to the RNN to predict the corresponding latent space representations of the next time step $\hat{\mathbf{L}}_i^{t+1}$. Next, the predicted latent spaces are passed back to the AE thereby predicting the single channel velocity distributions $\hat{\mathbf{U}}_i^{t+1}$ for $t+1$. Combining all the $\hat{\mathbf{U}}_i^{t+1}$ yields the multichannel velocity distributions $\hat{\mathbf{U}}^{t+1}$ at $t+1$. Note that since the flow physics is invariant to rotations, the three models are the same. The hybrid single model works in the same way, except that it does not separate the input channels.

4 Implementation and Training Approach

We implement the convolutional recurrent hybrid model for MaMiCo using the open-source machine learning framework PyTorch[2]. PyTorch offers CUDA support for GPU-based deep learning. In the following we briefly describe our training approaches for the AE and the RNN. As [12,22] show, they can be trained separately. Both the single and triple model approaches require to first train the AE by itself. After having successfully trained the AEs, the latent spaces can be generated from the original datasets. Then the RNN can be trained on them. The RNN must be trained on the same AE it is actually used with, i.e. if the AE ever changes, then the RNN has to be trained again. For brevity, the training configurations and hyper-parameters are presented in Table 1. As listed there, we apply the single model to the Couette scenario and the triple model to the KVS scenario, because only the KVS dataset shows a multidimensional transient flow.

[2] https://pytorch.org/docs/stable/index.html.

Table 1. Training Configurations

Model	Conv. AE		RNN	
	Single	Triple	Single	Triple
Dataset	Couette	KVS	Couette	KVS
Loss, Activ. Fn.	MAE, ReLU	MAE, ReLU	MAE, tanh	MAE, tanh
Optimizer	opt.Adam	opt.Adam	opt.Adam	opt.Adam
Batch Size	32	32	32	32
Epochs, Learn. Rate	250, 1e−4	100, 1e−4	250, 1e−4	15, 1e−4
#Layers, Seq. Size	–	–	1, 25	1, 25
Shuffled, Augmented	True, False	True, True	True, False	True, True

Convolutional Autoencoder (AE). The triple models are trained with augmented versions of the velocity resulting from swapping the channels of the original datasets, i.e. the permutations $(\mathbf{U}_0, \mathbf{U}_1, \mathbf{U}_2)$ and $(\mathbf{U}_1, \mathbf{U}_2, \mathbf{U}_0)$ and $(\mathbf{U}_2, \mathbf{U}_0, \mathbf{U}_1)$ are used. This is done so that the models are encouraged to learn a more general mapping by means of a greater variance in the inputs.

RNN. The RNN takes a sequence of the past 25 latent spaces $[\hat{\mathbf{L}}^{t-24}, \ldots, \hat{\mathbf{L}}^{t}]$ and performs the time-series prediction to yield an estimate for $\hat{\mathbf{L}}^{t+1}$. We choose a sequence length of 25 here, because in our experiments that performed best at minimizing validation loss, e.g. it was 34% better compared to a sequence length of 15 latent spaces. A range of about 10–50 is generally reasonable because it should be long compared to the frequency of fluctuations in MD data, but short compared to simulation run time.

In contrast to more common approaches where model prediction is sanctioned by means of loss quantification w.r.t. to the model target output, i.e. $\mathcal{L}(\hat{\mathbf{L}}^{t+1}_{i,\text{targ}}, \hat{\mathbf{L}}^{t+1}_{i,\text{pred}})$, we implement a loss quantification in the velocity space by comparing the decoded latent space prediction to the target single channel velocity distribution, i.e. $\mathcal{L}(\mathbf{U}^{t+1}_i, g(\hat{\mathbf{L}}^{t+1}_{i,\text{pred}}))$. This helps to train the RNN in such a way that the combined hybrid model, including the CNN decoder, minimizes the error in its predicted flow velocities.

5 Results – Couette Flow Scenario

In order to validate the relatively simple single model hybrid ML approach, we choose a Couette flow start-up test scenario as defined in [6]. There is no macroscopic flow in Y and Z directions, thus we focus on the direction of the moving wall, i.e. the X component of the velocity u_x, which is shown in Fig. 5 over 850 simulation time steps, i.e. excluding a 50 step initialization phase. Figure 5a averages u_x over a line of cells and displays its standard deviation over these cells as a lighter shaded area, while Fig. 5b shows the value for one cell only, which is in the center of the MD domain. It can be observed that the raw MD

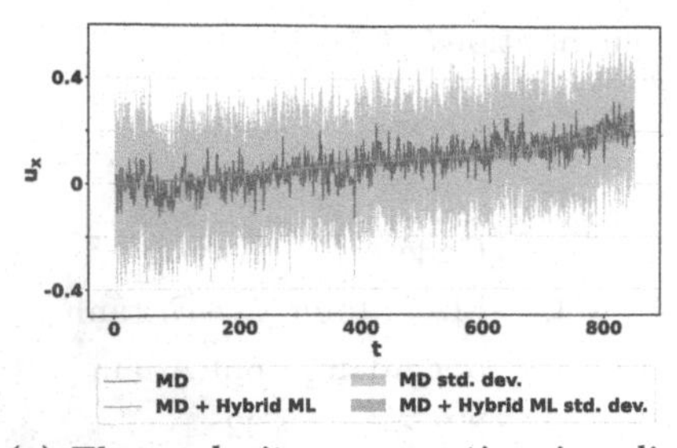

(a) Flow velocity u_x over time in a line of 24 cells through MD domain

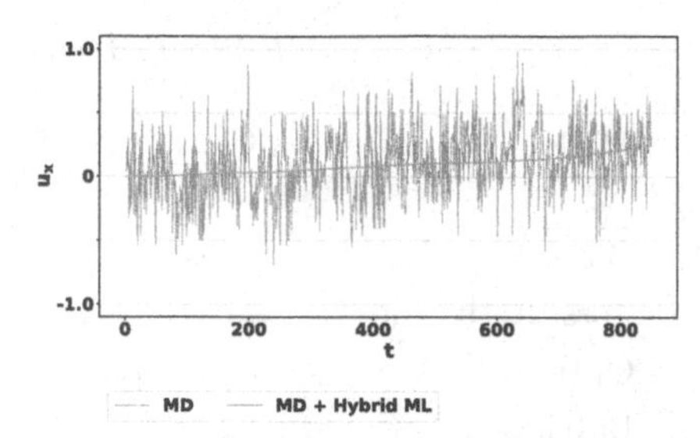

(b) Flow velocity u_x over time in one cell centered in MD domain

Fig. 5. Comparison of CNN+RNN hybrid ML model predictions in orange with raw MD data in blue for x-component of flow velocity u_x in Couette flow scenario. (Color figure online)

data exhibits a high level of random noise, caused by thermal fluctuations. Here we investigate the noise filtering properties of the ML model. In Fig. 5a one can notice that the standard deviation of the ML output grows slightly over time (i.e. the orange shaded area gets wider). However, it can be seen in both Fig. 5a and 5b that the hybrid ML predictions constitute a very stable noiseless signal, that are in good agreement with the mean flow displayed by the fluctuating MD data. Note that the ML model was not trained with information from a continuum flow solver, and also not with any noise filtering algorithm, instead it was trained on noisy raw MD data only (test case 'C_1_5_M' in the online repository). The filtering effect seen in Fig. 5 is obtained in space due to the dimensionality reduction into the latent space of the CNN AE and in time due to the application of the RNN (shown separately below, see Fig. 6 and 7), thus the desired filtering effect is already designed into the architecture of the hybrid model.

6 Results – Kármán Vortex Street Scenario

To predict more complex flow patterns, we set up a vortex street scenario (KVS). It exhibits multidimensional non-steady signals in each of the flow velocity components, so that the ML performance can be investigated adequately. Figure 6 compares the ML predictions with raw MD data, for one of the cells in the center of the MD domain. Both Figs. 6 and 7 evaluate the ML models on test cases from the validation set, i.e. on data on which they have **not** been trained. Figure 6 shows the performance of the AE only, i.e. the result of en- and decoding MD data. This aims to help the reader distinguish the CNN and the RNN impacts on the hybrid model behavior. It can be seen in Fig. 6 that the AE is able to represent all information necessary to capture and preserve the mean flow characteristics, while it does not preserve spatial noise present in the MD data (i.e. visually that the orange curve follows only a sliding mean of the blue curve). However, it can also be seen that the raw AE output is more noisy than the

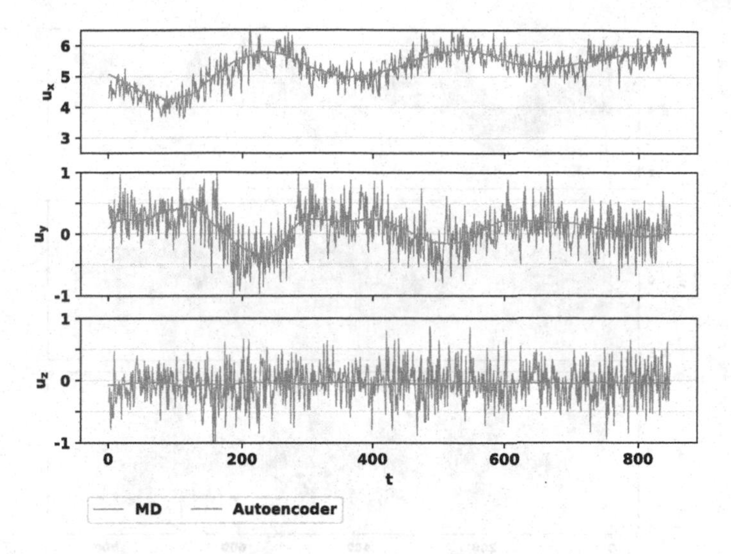

Fig. 6. AE vs. MD, KVS validation '22000_NW', for one cell in MD domain center

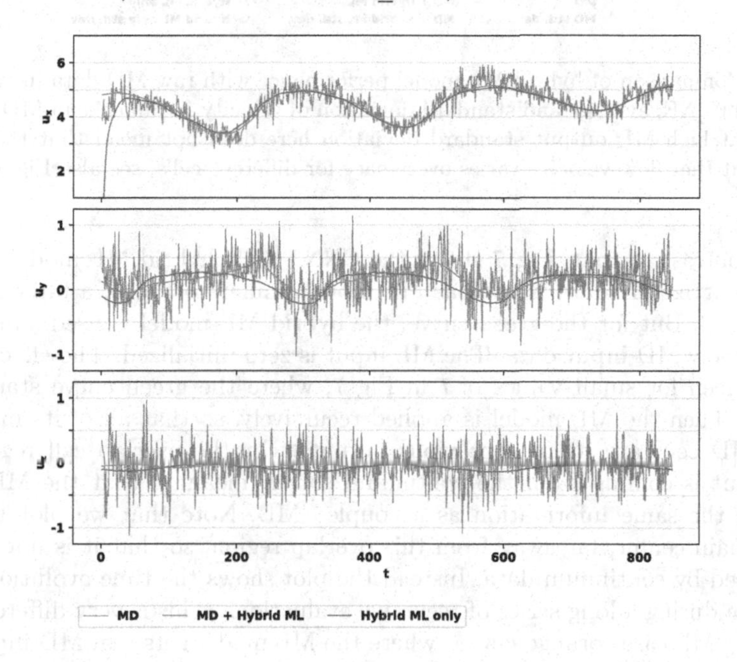

Fig. 7. Raw MD vs. CNN+RNN hybrid ML model performance with and without MD input data, KVS validation case '20000_NE', for one cell in center of MD domain (Color figure online)

hybrid model data (i.e. orange curve less smooth than in Fig. 7), as the temporal smoothing and prediction coming from the RNN is missing here.

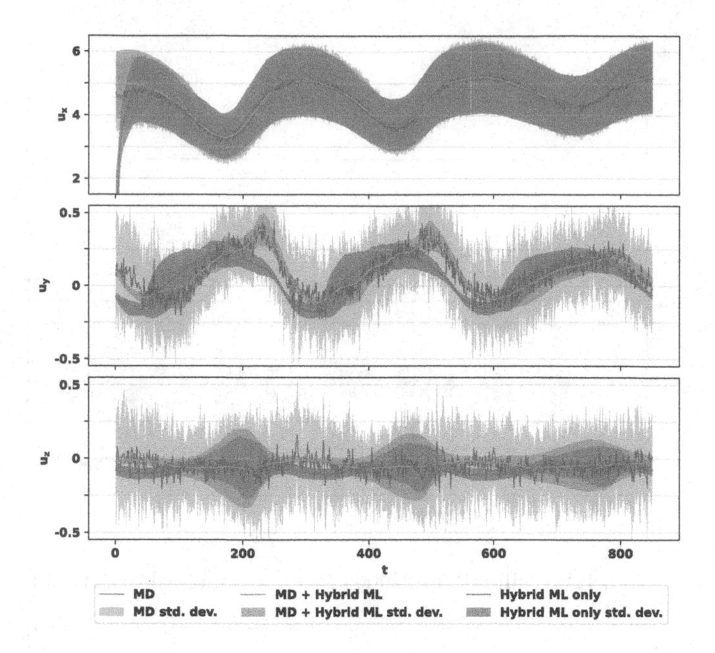

Fig. 8. Comparison of hybrid ML model performance with raw MD data in validation case 20000_NE, average and standard deviation of velocity for all cells in MD domain. Note that high ML output standard deviation here does not mean that the data is noisy, but that flow velocity varies over space for different cells, see also Fig. 1.

In contrast to that, Fig. 7 shows the CNN+RNN hybrid ML model. For the orange curve, MD data is used as ML input, similar to Fig. 5, a filtering effect is obtained. But for the green curve, the hybrid ML model is used standalone, without any MD input data. The ML input is zero-initialized. The effect of this can be seen for small values of t in Fig. 7, where the green curve starts close to zero. Then the ML model is applied recursively, so that all of its inputs for inner MD cells are its own previous outputs. For the overlap cell region, the ML input is coming from the continuum solver, meaning that the ML model receives the same information as a coupled MD. Note that we plot a cell in the domain center, far away from this overlap region, so that it is not directly influenced by continuum data. Instead the plot shows the time evolution of the ML data during a long series of recursive evaluations. This is very different from the MD+ML case (orange curve) where the ML model gets real MD input data in every time step and performs a prediction for only one step into the future. Since no MD data is fed into the ML model here, there is less confidence that the model will stick to physical MD behavior, especially over longer time spans. However, except of just two small offsets for u_y around $t = 250$ and $t = 550$, it is still in an excellent agreement with the correct particle data, revealing the prediction capabilities of the hybrid model.

To give more details about the computational performance of these surrogate model predictions, compared to the original full-scale MD: the AE training in this case took ten hours on a single NVIDIA A100 GPU, the RNN training one hour, yielding eleven hours in total for the hybrid model training. The evaluation of the trained model in our experiments on an Intel Xeon 8360Y CPU costs 81.7 ms, while running the MD model sequentially on the same CPU requires about 19 200 ms per coupling cycle.

Figure 8 presents the same insights as Fig. 7, but in a greater level of detail. It shows the same test case, i.e. flow data on which the ML models have **not** been trained. Here, instead of only one cell, a line of 24 cells in the MD domain is plotted, with a shaded area representing the standard deviations and a continuous line representing the average values of the respective flow velocity components. It is visible that both the MD+ML as well as the ML-only approach deliver results matching the true particle data – except the aforementioned shift in u_y for ML-only, which is however clearly inside the standard deviation of the real MD data. Thus, this can be considered to be an excellent prediction result for a very noisy and challenging complex validation data set.

7 Conclusions

We have pointed out that a hybrid convolutional recurrent autoencoder is a promising approach to learn and predict molecular dynamics data in the context of a molecular-continuum coupled simulation, in order to explore potentials for acceleration of simulation execution. To make our results reproducible, we provided an implementation of the proposed approach based on PyTorch and documented the dataset generation and training processes. Unlike existing approaches, we have introduced a type of loss function for the RNN training that quantifies loss in the velocity space, instead of in the latent space, leading to an enhanced integration of the hybrid model components. We solved problems caused by interdependent flow components by introducing a triple model approach. Our results demonstrate that the hybrid CNN+RNN ML model does not only work as a powerful advanced noise filtering system, but even in a complex flow scenario, it is able to replace the MD system entirely and accurately predict the data that MD would generate for the coupled simulation.

We have considered one-way coupled simulations only. This was reasonable here because we wanted be able to analyze the behavior of the MD system and our ML model operating on particle data, or even analyze the ML module entirely replacing the particle system, independent of any additional effects which would stem from two-way coupling of the two systems. Thus, the impacts and limitations of the ML model in a two-way coupled setup constitute an open question for the future. While we have applied our approach to MD only, we expect it to naturally generalize to other types of particle systems, such as dissipative particle dynamics or smoothed particle hydrodynamics. They tend to expose the macroscopic solver to less noise and thus facilitate AE and RNN training, however their reduced computational cost makes it more difficult to obtain a significant performance benefit with a ML surrogate model. While we have shown that our ML model opens chances for drastic increases of computational efficiency, a further systematic investigation of the speed-ups achievable in practice by this method, including scaling to large-scale simulations on HPC clusters, would be essential. The fact that it can be expensive to re-initialise a MD simulation, after skipping some time steps, might play a crucial role here. So far the hybrid model was only tested for 'toy' scenarios where the average behavior of the MD system can be modeled by a macroscopic solver stand-alone. Thus, future work suggesting itself would be to conduct experiments with scenarios where the MD simulation actually provides differing physical characteristics, such as fluid-structure interaction or flows through carbon nanotube membranes.

Acknowledgments. We thank the projects "hpc.bw" and "MaST" of dtec.bw – Digitalization and Technology Research Center of the Bundeswehr for providing computational resources, as the HPC cluster "HSUper" has been used to train and validate the ML models presented in this paper. We also want to thank Prof. Zhen Li and the MuthComp Group of Clemson University for fruitful discussions and exchange of ideas as well as provision of office space and IT systems.

References

1. Bauer, M., Köstler, H., Rüde, U.: lbmpy: automatic code generation for efficient parallel lattice Boltzmann methods. J. Comput. Sci. **49**, 101269 (2021)
2. Borg, M.K., Lockerby, D.A., Ritos, K., Reese, J.M.: Multiscale simulation of water flow through laboratory-scale nanotube membranes. J. Membrane Sci. **567**, 115–126 (2018). ISSN 0376-7388
3. Bungartz, H.J., et al.: preCICE - a fully parallel library for multi-physics surface coupling. Comput. I& Fluids **141**, 250–258 (2016)
4. Fukami, K., Hasegawa, K., Nakamura, T., Morimoto, M., Fukagata, K.: Model order reduction with neural networks: application to laminar and turbulent flows. SN Comput. Sci. **2**, 1–16 (2021)
5. Grinberg, L.: Proper orthogonal decomposition of atomistic flow simulations. J. Comput. Phys. **231**(16), 5542–5556 (2012)
6. Jarmatz, P., Maurer, F., Wittenberg, H., Neumann, P.: MaMiCo: non-local means and pod filtering with flexible data-flow for two-way coupled molecular-continuum HPC flow simulation. J. Comput. Sci. **61**, 101617 (2022)
7. Jarmatz, P., Neumann, P.: MaMiCo: parallel noise reduction for multi-instance molecular-continuum flow simulation. In: Rodrigues, J.M.F., et al. (eds.) ICCS 2019. LNCS, vol. 11539, pp. 451–464. Springer, Cham (2019). https://doi.org/10.1007/978-3-030-22747-0_34
8. Jarmatz, P., et al.: MaMiCo 2.0: an enhanced open-source framework for high-performance molecular-continuum flow simulation. SoftwareX **20**, 101251 (2022). ISSN 2352-7110
9. Kadupitiya, J., Sun, F., Fox, G., Jadhao, V.: Machine learning surrogates for molecular dynamics simulations of soft materials. J. Comput. Sci. **42**, 101107 (2020)
10. Lecun, Y., Bottou, L., Bengio, Y., Haffner, P.: Gradient-based learning applied to document recognition. Proc. IEEE **86**(11), 2278–2324 (1998). https://doi.org/10.1109/5.726791

11. Nakamura, T., Fukagata, K.: Robust training approach of neural networks for fluid flow state estimations. Int. J. Heat Fluid Flow **96**, 108997 (2022)
12. Nakamura, T., Fukami, K., Hasegawa, K., Nabae, Y., Fukagata, K.: Convolutional neural network and long short-term memory based reduced order surrogate for minimal turbulent channel flow. Phys. Fluids **33**(2), 025116 (2021). https://doi.org/10.1063/5.0039845
13. Neumann, P., Bian, X.: MaMiCo: transient multi-instance molecular-continuum flow simulation on supercomputers. Comput. Phys. Commun. **220**, 390–402 (2017)
14. Niethammer, C., et al.: ls1 mardyn: the massively parallel molecular dynamics code for large systems. J. Chem. Theory Comput. **10**(10), 4455–4464 (2014)
15. Ren, X.G., Wang, Q., Xu, L.Y., Yang, W.J., Xu, X.H.: HACPar: an efficient parallel multiscale framework for hybrid atomistic-continuum simulation at the micro-and nanoscale. Adv. Mech. Eng. **9**(8) (2017)
16. Ronneberger, O., Fischer, P., Brox, T.: U-Net: convolutional networks for biomedical image segmentation. In: Navab, N., Hornegger, J., Wells, W.M., Frangi, A.F. (eds.) MICCAI 2015. LNCS, vol. 9351, pp. 234–241. Springer, Cham (2015). https://doi.org/10.1007/978-3-319-24574-4_28
17. Schäfer, M., Turek, S., Durst, F., Krause, E., Rannacher, R.: Benchmark computations of laminar flow around a cylinder. In: Hirschel, E.H. (ed.) Flow Simulation with High-Performance Computers II. NNFM, pp. 547–566. Springer, Cham (1996). https://doi.org/10.1007/978-3-322-89849-4_39
18. Smith, E.: On the coupling of molecular dynamics to continuum computational fluid dynamics. Sch. Mech. Eng. (2013)
19. Tang, Y.H., Kudo, S., Bian, X., Li, Z., Karniadakis, G.E.: Multiscale universal interface: a concurrent framework for coupling heterogeneous solvers. J. Comput. Phys. **297**, 13–31 (2015)
20. Thomas, M., Corry, B.: A computational assessment of the permeability and salt rejection of carbon nanotube membranes and their application to water desalination. Philos. Trans. R. Soc. A: Math. Phys. Eng. Sci. **374**(2060), 20150020 (2016)
21. Veen, L.E., Hoekstra, A.G.: Easing multiscale model design and coupling with MUSCLE 3. In: Krzhizhanovskaya, V.V., et al. (eds.) ICCS 2020. LNCS, vol. 12142, pp. 425–438. Springer, Cham (2020). https://doi.org/10.1007/978-3-030-50433-5_33
22. Wiewel, S., Becher, M., Thuerey, N.: Latent space physics: towards learning the temporal evolution of fluid flow. In: Computer Graphics Forum, vol. 38, pp. 71–82. Wiley Online Library (2019)
23. Wittenberg, H., Neumann, P.: Transient two-way molecular-continuum coupling with OpenFOAM and MaMiCo: a sensitivity study. Computation **9**(12) (2021). https://doi.org/10.3390/computation9120128. ISSN 2079-3197
24. Yin, W., Kann, K., Yu, M., Schütze, H.: Comparative study of CNN and RNN for natural language processing. arXiv:1702.01923 (2017)
25. Zhang, A., Lipton, Z.C., Li, M., Smola, A.J.: Dive into deep learning. arXiv preprint arXiv:2106.11342 (2021)

Developing an Agent-Based Simulation Model to Forecast Flood-Induced Evacuation and Internally Displaced Persons

Alireza Jahani(✉), Shenene Jess, Derek Groen, Diana Suleimenova, and Yani Xue

Department of Computer Science, Brunel University London, Uxbridge UB8 3PH, UK
alireza.jahani@brunel.ac.uk

Abstract. Each year, natural disasters force millions of people to evacuate their homes and become internally displaced. Mass evacuations following a disaster can make it difficult for humanitarian organizations to respond properly and provide aid. To help predict the number of people who will require shelter, this study uses agent-based modelling to simulate flood-induced evacuations. We modified the Flee modelling toolkit, which was originally developed to simulate conflict-based displacement, to be used for flood-induced displacement. We adjusted the simulation parameters, updated the rule set, and changed the development approach to address the specific requirements of flood-induced displacement. We developed a test model, called DFlee, which includes new features, such as the simulation of internally displaced persons and returnees. We tested the model on a case study of a 2022 flood in Bauchi state, Nigeria, and validated the results against data from the International Organization for Migration's Displacement Tracking Matrix. The model's goal is to help humanitarian organizations prepare and respond more effectively to future flood-induced evacuations.

Keywords: Forced Displacement · Natural Disaster · Flood · Internally Displaced Persons · Agent-based Modelling and Simulation

1 Introduction

The impact of climate change on the world has caused an increase in humanitarian needs [1], particularly in poorer countries. The resulting extreme weather events, such as droughts, floods, wildfires, hurricanes, and tornadoes, are directly linked to rising temperatures and sea levels. Flooding is the leading cause of climate-related displacement, accounting for over 9% of natural disaster displacements [2]. The number of flood-related disasters has been increasing in recent years due to global warming and deforestation. In 2021, more than 10 million people were internally displaced due to 749 flooding events [3]. There is a common misconception that the effects of climate change on displacement are temporary. However, in most cases, people are permanently displaced due to the

J. Mikyška et al. (Eds.): ICCS 2023, LNCS 14076, pp. 550–563, 2023.
https://doi.org/10.1007/978-3-031-36027-5_43

destruction of their homes and communities by floods. For this reason, climate change is now widely considered a humanitarian disaster.

Despite the ongoing devastating effects of flooding worldwide, little research has been done to investigate the displacement of affected people and provide useful information for humanitarian organizations and governments. Therefore, this study focuses on simulating the evacuation caused by floods and the movement of internally displaced people who are seeking a safe shelter. Previous research on flood and disaster-induced evacuation has mainly focused on behavioural models, pre-evacuation behaviour, and risk perception. In addition, computational techniques such as random forests and agent-based modelling have been widely used in flood and disaster management, especially in relation to the reasons for evacuation and predictors of displacement [4]. However, this project has identified a gap in the literature regarding individual movement, distribution of internally displaced persons (IDPs), and evacuation destinations.

To simulate flood-induced evacuation, this study employed the Flee agent-based modelling toolkit, which is typically used to simulate the geographical movement of people fleeing from conflict and violence [5]. The aim was to extend its applications and adapt the Flee toolkit to simulate flood-induced evacuation during a specific flood event, with the intention of providing insights to prepare for future extreme flooding. While Flee has not been used to model flood-induced evacuations before, there are similarities between climate-driven events and conflict-driven events in terms of the mass forced displacement of people. However, there are differences in the locations of shelters and camps between the two events. In this study, some shelters are located within the flooded areas but at lower flood levels, rather than refugees being displaced to neighbouring countries. Moreover, Flee has not been used to model a small region instead of an entire country or state, but this study aimed to repurpose the model to investigate the movement of IDPs during a flood event in small regions.

The paper is structured into six main sections. The second section is a literature review that provides an overview of the current research on flood-induced evacuation and the existing models used to simulate the movement of people during flood events. The third section explains the research problem and the approach taken to develop the simulation and describes the conceptual model. It also discusses the construction of an instance and the necessary parameters and data sets needed to run the agent-based model for flood-induced evacuation. The fourth section contains a case study, while the fifth section discusses and evaluates the model's results. The final section summarizes the study's findings, identifies opportunities for future research, and reflects on the study's limitations.

2 Literature Review

Severe flooding has the potential to displace individuals temporarily or permanently, leading them to seek inadequate accommodations and limited resources [6]. Displacement can be particularly challenging for vulnerable individuals, emphasizing the need to understand the composition and quantity of people who may be evacuated due to floods and displaced to shelters and sites [7]. By comprehending the movement of people in affected areas and meeting their specific needs, there is an opportunity to increase their resilience to flood events. To aid in resource planning for those who are least equipped, a reliable and robust simulation of people's movement is essential.

Previous studies have focused whether on evacuation modelling for smaller scales or on migration modelling for larger scales, especially after conflicts, but little research has been done on the movement of people after natural disasters. While flood-induced evacuation models have considered the behaviour and decisions of people during a flood event, they have neglected the outcomes of these decisions, such as the destinations of evacuees and their distribution across shelters and sites.

2.1 Evacuation Modelling Approaches

Several studies have been conducted to comprehend the evacuation process by considering human behaviour and decision-making. These models are not restricted to flood evacuations; rather, they are developed for various types of disasters such as hurricanes, cyclones, and war-related evacuations. Lim et al. [8] used discrete choice modelling and determined that demographic and household characteristics, as well as the perception and movement of the hazard, play a role in the decision to evacuate. Hasan et al. [9] expanded on this work and examined the effect of these characteristics on departure timing by utilizing a random parameter hazard-based modelling approach. These studies highlighted that evacuation choices are not binary and require interconnectivity of parameters to reproduce human behaviour accurately. This type of modelling is useful in decision-making and parameter significance determination. However, to simulate forced displacement, a more robust modelling approach is required, such as behavioural, spatial, and agent-based modelling approaches, as described below.

Behavioural models aim to understand the required time for people to evacuate during a disaster, how they decide to evacuate, and their actions leading up to the evacuation [10]. Researchers such as Coxhead et al. [11] and Lovreglio et al. [12] have studied displacement decisions in Vietnam and building occupants' behaviour during emergencies, respectively. However, these studies do not focus specifically on flood-induced displacement and natural disasters in general. The primary focus of this method revolves around pre-evacuation modelling and does not delve into the evacuation process itself or the related movements.

Spatial models can be utilized to simulate the movement of evacuees during a flood event by incorporating travel behaviour and traffic networks. Hybrid route choice models [13] and multiple criteria evaluation approaches with traffic microsimulations [14] have proven to be effective in transport and route modelling. These approaches can provide flood and disaster management recommendations by predicting the movement of evacuees in terms of their travel choices across the transportation network. The mode of transport taken by evacuees can greatly influence their movement, and the routes they take may be determined by their pre-trip knowledge of the area and their destinations. However, transport models do not account for the final destination choice of evacuees. It is important to note that during a flood, an effective evacuation requires clear routes and a functioning transport network, which may not always be available.

Agent-based modelling is a widely used technique to simulate forced displacement and the behaviour of agents. It is capable of handling complex decision-making processes, making it suitable for modelling human behaviour and the effects of a flood event [15]. Agents can be assigned various attributes to model flood-induced evacuations, and agent-based modelling is an appropriate approach to consider and simulate human behaviour in the context of flood risk. SiFlo is a model that integrated emotions and social norms into individual behaviour and considered various motivations and priorities people may have when facing flood risk, such as escaping or helping others, or protecting their belongings [16]. The model could predict the decisions that would reduce or increase damage and casualties, but it did not model or describe the journeys and destinations of the evacuated people. Agent-based models can incorporate demographic data, which is known to impact evacuation decisions and risk perception [8,9,13,17]. Characteristics like age and gender can affect an individual's behaviour during a flood event, including their pace of movement and need for assistance. By modelling the interactions between agents of different demographics and the influence of their characteristics on one another [18], agent-based approaches can capture the effects of demographic data on flood-related behaviours.

The objective of this study is to use agent-based modelling to simulate the displacement of people during floods and analyse the behaviours of different demographics. While there is limited literature on quantifiable movement during flood-induced displacement, understanding how evacuees move towards shelters and accommodation sites and the time it takes them to reach their destinations can aid policymakers and authorities in designing better evacuation strategies and allocating resources during prolonged floods. By knowing the routes people are likely to take, authorities can keep these routes open and accessible during emergencies and provide necessary supplies to those affected. Additionally, this simulation can help identify gaps in monitoring infrastructure and forecast the supplies required based on the demographics of displaced individuals.

3 Development Approach

The primary objective of this study is to use the modified version of the Flee modelling toolkit to simulate the movement of IDPs in response to a flood event. To achieve this, several objectives need to be accomplished. First, we must establish the assumptions regarding flood-induced displacement, based on reports from organizations such as the International Organisation for Migration (IOM), to create a general set of rules for an agent-based model and establish the constraints for autonomous decision-making within the model. Next, we need to create a conceptual model based on these assumptions to provide a clear understanding of what the final simulation aims to accomplish. Once we have chosen a case study, we can modify the general rule set to better reflect the specific requirements and processes of the flood-induced displacement scenario. Figure 1 illustrates the different stages of the proposed simulation development approach for DFlee. Regarding the quality of the data used, it is important to note that there are incomplete data sets for some flooding events and in certain situations, there is a high level of uncertainty and bias in the reports that have been published. The authors suggest that collaborating more with the relevant organizations could help reduce these issues.

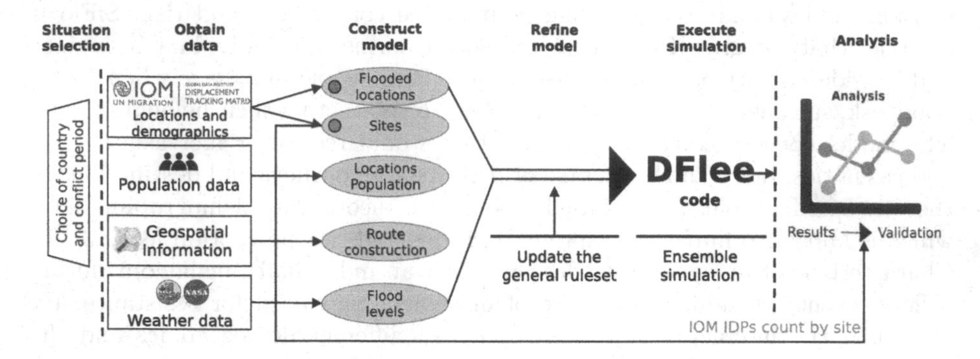

Fig. 1. DFlee simulation development approach

3.1 Assumptions

This particular section outlines the assumptions we utilized to comprehend the design prerequisites of the conceptual model for flood-induced displacement. These assumptions were derived from pertinent literature and will require further confirmation from additional specialists in the future.

Assumption 1 - Floods cause internal displacement: In the event of natural disasters, people usually seek shelter in safe locations such as temporary shelters, with their relatives, or in hotels and other accommodation facilities that are located away from the affected areas. Displacements caused by climate-related disasters are often temporary and local, with individuals returning to their homes once the threat has passed. However, if homes are destroyed and residents are unable to return, the displacement can become permanent. Since floods typically occur in specific regions, individuals are typically displaced within their country's borders, making them IDPs, unless they are close to a border and seek refuge in a neighbouring country.

Assumption 2 - Floods need immediate response and evacuation: In contrast to other types of mass movement and forced displacement, which might take months or years in terms of the time scale of the displacement, natural disasters such as floods require immediate evacuation and demand a prompt response from authorities and humanitarian organizations. Therefore, simulation scenarios for these situations may require shorter time scales, hours, days, or weeks.

Assumption 3 - People might use various ways of transportation (i.e., walking, driving, and river crossing): Floods can rapidly inundate communities, forcing them to evacuate as quickly as possible by any means necessary, such as walking or using boats [19]. This is particularly true when the water level is high and the transportation infrastructure is inundated. Factors related to transportation also play a role in how evacuees reach and choose their evacuation destination, including the distance to available shelters, the method of transportation, and access to private or alternative means of transportation, among other factors that may not be immediately apparent [20]. However, certain individuals may lack means of transportation and consequently become stranded in their current positions. This matter has not been scrutinized in this paper, but we plan to explore it in forthcoming research.

Assumption 4 - Based on the available reports on people displaced by natural disasters, people might want to return to their homes after the flood recession: When floodwaters eventually subside after a period that may range from days to weeks, those displaced by the flood may want to return to their homes, making their displacement temporary. However, if the damage caused by the flood is severe enough, individuals may become permanently displaced and may need to resettle elsewhere because they cannot or do not want to return home.

Assumption 5 - The evacuation decision includes when, how, and where to go questions: Once an individual decides to evacuate, they will also need to consider various factors such as when, how and where to evacuate. The choice of destination may be based on socio-economic and demographic factors. Available destination options include public shelters, hotels, and the homes of friends or relatives. Wealth and ethnicity may influence a person's preference for shelter or non-shelter destinations [13]. However, this research only focuses on the movement of flood-induced IDPs towards official shelters and camps, and private accommodations are not taken into account.

Assumption 6 - Some forms of shelters include higher grounds or even high buildings: It is important to note that unlike in situations of conflict, refugees in flood-prone areas do not necessarily have to leave the area if they can find safety in buildings or homes that are located at higher elevations or have more floors that are closer to their homes. Typically, residents of areas with a high risk of floods, such as Bangladesh, are accustomed to relocating to higher ground during the wet season with their belongings. However, since we could not find any dependable data to support this assumption, we have decided to exclude it from our research.

Assumption 7 - Cities and areas can have different levels of floods: In the event of a flood, the risk level of various locations within a city or area can vary, resulting in differences in the likelihood of people choosing these areas as safe havens or leaving them as high-risk areas. These variations can occur during each simulation period and in every time step.

3.2 Conceptual Model

In this section, we will first create a conceptual agent-based model using a hypothetical flood event for the purpose of refining the model. Then, in the next section, we will develop a model for a selected country and run a simulation. To validate the accuracy and performance of the model, we will compare the simulation results with data obtained from historic flood events and IOM reports and data sets. Figure 2 illustrates a conceptual model that showcases the evolution of a flood event in an imaginary scenario with four locations, each with the potential to flood at three different levels. The flooding level affects the agent's decision to either move towards the nearest safe location or stay in their current location. It is important to note that agents will return home once the situation improves. At each time step, following the update of flood levels, types of locations, population, as well as the number of internally displaced individuals from adjacent areas and those evacuating from the current location, the agents will be spawned to the adjacent safe location.

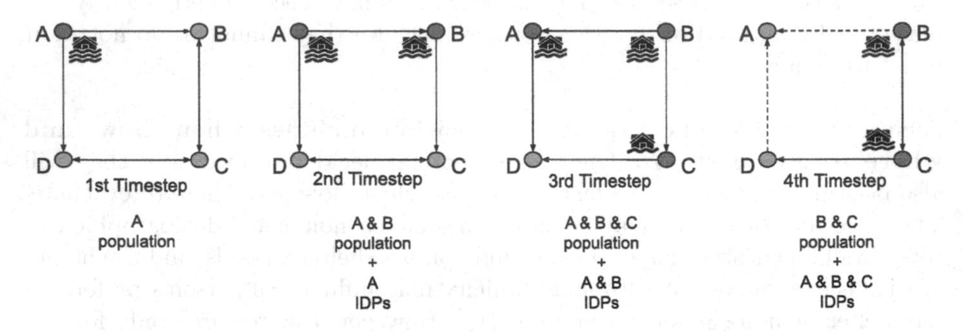

Fig. 2. The conceptual model, represented by a simple location graph and the evolution of a flood event over four timesteps

In this particular case, during the first time step, only location A is inundated with the most severe level of flooding, level three. As a result, the residents of location A will relocate to the nearest safe locations, which are locations B and D. During the second time step, location B is flooded at level two, resulting in its residents fleeing to the safer location, location C. Furthermore, the newly arrived and displaced inhabitants of location A from the previous time step will be combined with the overall number of people seeking refuge. In the third time step, location C will be flooded at level one, while the flood levels of locations A and B will be updated to level two and level three, respectively. This indicates that residents and internally displaced individuals in location B, who have not yet evacuated, may choose whether to go to location A or C since their flood levels are lower. As in previous time steps, the total number of people seeking asylum must be updated to include all three flooded populations and IDPs from the two flooded locations from the previous time step. In this scenario's fourth and last time step, the situation in location A has changed to safe, and its residents can decide whether to stay in their current shelter or return home, which is represented by dashed arrows, depending on how much damage their homes have sustained as a result of the flood. Naturally, the overall number of agents seeking shelter must be updated in the same manner as in the previous time steps.

The proposed model's flowchart is presented in Fig. 3. The flowchart highlights probabilities and variables that can be changed by users to explore different rule sets. These variables are shown by dashed shapes, while the probabilities are shown by dotted shapes. The authors have selected the present parameters for the rule set in accordance with their previous experience and the reports on displacement modeling, to suit the given scenario. The flowchart runs through every time step, during which agents will decide whether to move towards their destination or return home or stay at their current location due to lower flood levels compared to adjacent areas. The model will use the provided data by IOM's DTM from the studied areas to determine the demographic makeup of the agents. This will allow the model to estimate the age and gender of the agents in each location and link at each time step. The number of male and female agents and their ages in the camps are then compared to reference data to validate the model's results.

3.3 Model Inputs

The main input data is provided by IOM's Displacement Tracking Matrix (DTM) reports which contain important details about the number and demographic characteristics of IDPs affected by a flood event and their destinations. The reports also provide insights into the transportation methods used by the affected people, the capacity of shelter sites, and their geographical coordinates. To determine the flood levels at the affected locations during an event, we collected climate and weather-related datasets. For this purpose, we utilized maps of flooded areas provided by the National Aeronautics and Space Administration (NASA) to extract the location graph. This information will be used to update the flood level of affected areas.

4 A Case Study: Nigeria, Bauchi State

We conducted a case study of Bauchi state, located in the northeastern part of Nigeria, which experienced flooding due to heavy rainfall and strong winds. As a result, a total of 2,185 people were affected, with 90% of them being displaced to

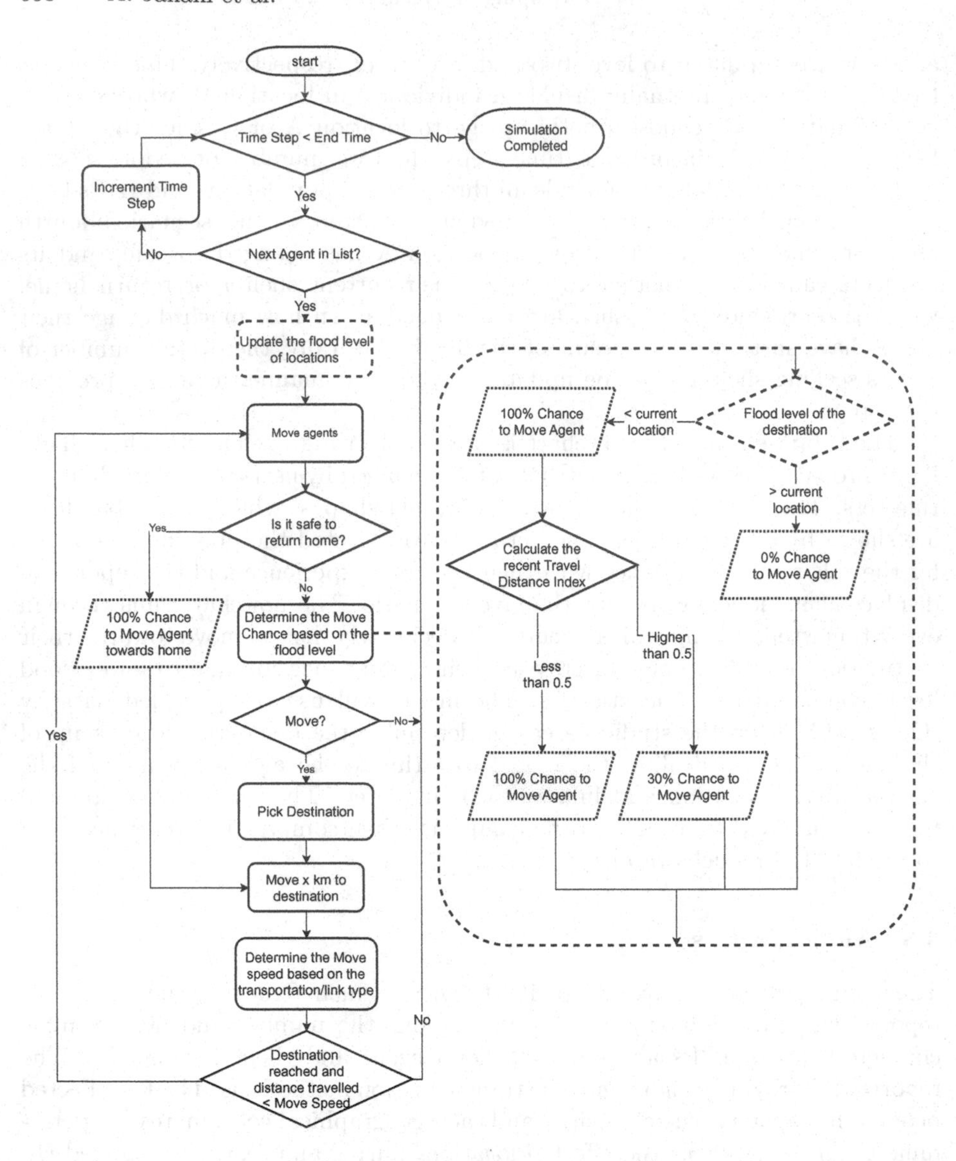

Fig. 3. A flowchart of algorithm assumptions in the DFlee agent-based code to demonstrate the rule set predicting forced displacement destinations updated with flood level data

neighbouring communities in seven Local Government Areas (LGAs): Damban, Gamawa, Ganjuwa, Jama'are, Ningi, Warji, and Zaki. Based on an assessment by the IOM's DTM program conducted between September 1st and 12th, 2022 [21], Filin Shagari, a community in the Kariya ward of Ganjuwa LGA, was the worst-hit site, with 78 houses damaged and an estimated 868 people affected. In Ariya, Kubdiya LGA, 68 individuals were displaced to Kore primary school, and 33 individuals were displaced from Kubdiya in Kubdiya ward of Gamawa LGA to Apda Ward in Zaki LGA. The flooding affected a total of 222 houses, leaving 256 households in need of shelter, repair kits, and non-food items as most houses need re-enforcement with brick blocks. According to the IOM [21], 120 casualties were reported as a result of the Flood. Figure 4(a) depicts the map of the affected areas in the Bauchi state and (b) shows the location graph extracted from the IOM's report.

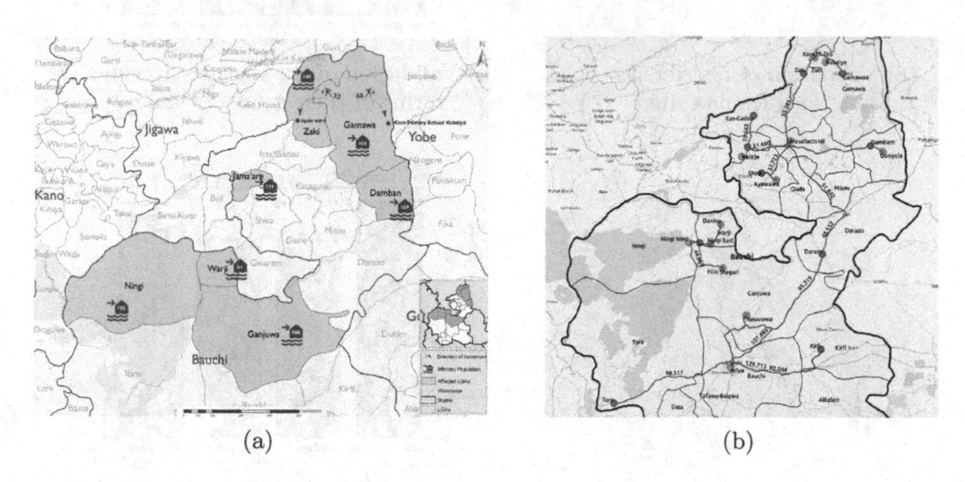

(a) (b)

Fig. 4. Nigeria, Bauchi state location graphs: (a) The map of the affected areas by heavy rainfalls and floods in Bauchi state [21] and (b) The location graph extracted from the IOM's report

Our proposed simulation model aims to forecast the evacuation patterns of people after flood events in the case study and compare the results against the IOM's DTM data. Figure 4(b) depicts the location graph, which includes all the flooded locations and shelters where the agents would go to find asylum. These locations are interconnected with routes found using the ORS tools plugin in QGIS [22]. The agents move through these routes using a weighted probability function based on route length, with distances estimated using driving, walking, and river-crossing routes. The total number of displaced people is extracted from IOM reports using linear interpolation between data points. Agents representing people move with a probability of movechance to different locations to find safety, with a chance of returning to their homes when the flood level decreases, according to the fourth assumption. As mentioned above, we modified the original Flee toolkit in Python to run the simulation and compared the results against IOM-published data, calculating the sum of absolute differences in agent counts for each shelter location as a proportion of the total number of displaced people [5].

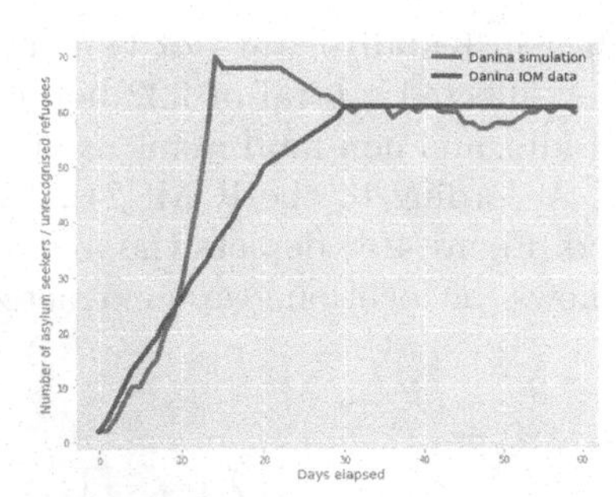

(a) Validation of registered people at a Danina site

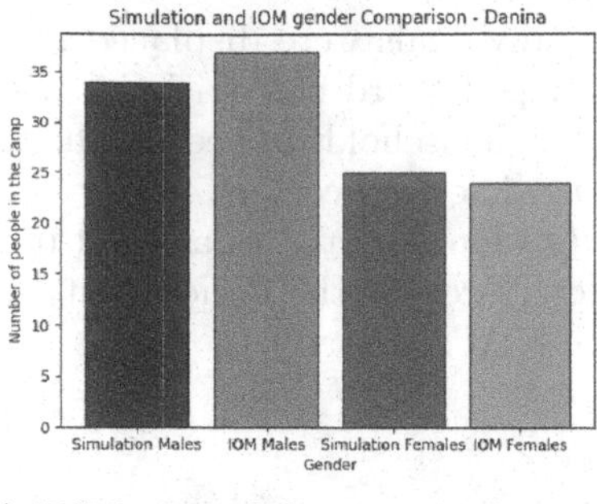

(b) Validation of the registered people in terms of their gender (Danina)

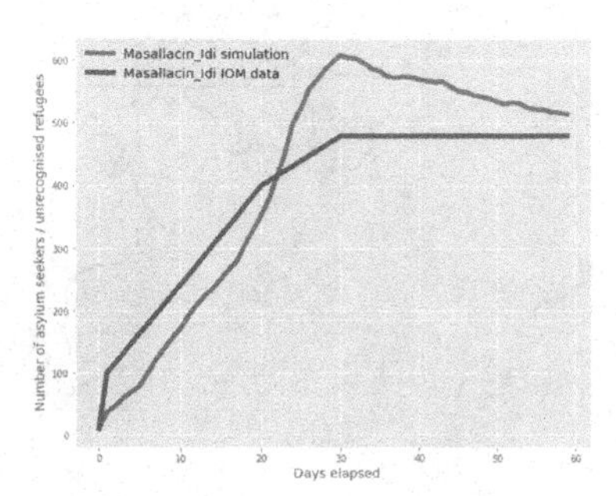

(c) Validation of registered people at a Masallacin-Idi site

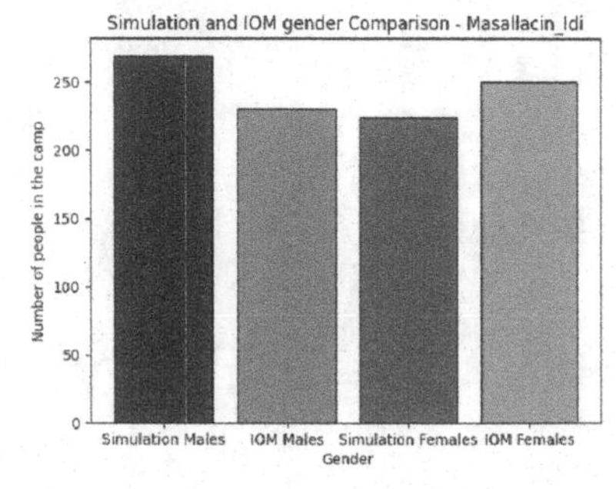

(d) Validation of the registered people in terms of their gender (Masallacin-Idi)

Fig. 5. Sample camp simulation results

5 Results and Discussion

In Fig. 5, we compare simulation results with data from the International Organization for Migration's (IOM) Displacement Tracking Matrix (DTM) for two sample sites: Danina (a) and Masallacin-Idi (c). IDPs arrived at these sites from their hometowns in Bauchi state. The first 30 days of the simulation, with the highest level of floods, showed a steep initial increase in displaced persons, both in data and in the simulation. However, subsiding the flood level in many cities across Bauchi decreased the number of arrivals to these sites after Day 30. Our simulation predicted more arrivals than the IOM data reported, possibly because the IOM report only included major affected towns and cities, which we used as flooded locations in the simulation. The lack of geospatial knowledge about the region caused a preliminary location graph, which likely did not reflect the complete set of locations and routes in this region. Despite these limitations, the average relative difference remained below 0.4 after the 20 days of simulation (see Fig. 6), when most agents have settled in shelters and sites and the flood has subsided. Furthermore, the total number of agents in sites in the simulation was slightly less than reported in the data, as some agents remained in transit within the region. The simulation results for the gender of registered people in the sites were very close to the number of males and females in these sites, as shown in Fig. 5(b) and (d).

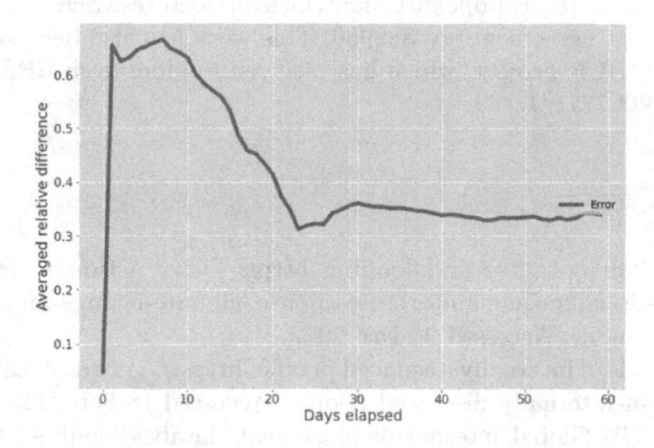

Fig. 6. Overview of the averaged relative differences for the flood event in the Bauchi state in August 2022

6 Conclusion

In conclusion, this study addresses the gap in the literature regarding the simulation of flood-induced evacuations and the movement of internally displaced people. The study employs an agent-based modelling toolkit called Flee, originally used to simulate conflict-based displacement, and modifies it to be used for flood-induced displacement. The resulting model, DFlee, simulates flood-induced evacuations and the movement of internally displaced people seeking safe shelter. The model is tested on a case study of a 2022 flood in Bauchi state, Nigeria, and validated against data from the International Organization for Migration's Displacement Tracking Matrix. The results suggest that the model can help humanitarian organizations prepare and respond more effectively to future flood-induced evacuations. This study contributes to the existing literature by providing a novel approach to simulating flood-induced displacement and highlights the need for further research to address the complexities of natural disasters and displacement. Furthermore, it should be noted that the current study considered a longer period of evacuations after the flood events and did not examine the initial actions taken by evacuees, such as seeking refuge on nearby higher grounds. It may be necessary for future research to take into account such alternative safe locations for temporary shelter. Additional investigations into these aspects could refine the model's precision and effectiveness in forecasting and respond to natural disasters, and utilizing supplementary climate-related data resources such as European Centre for Medium-Range Weather Forecasts (ECMWF) could further augment its capabilities. Moreover, our plan is to expand this research in the future by coupling additional datasets and models using a multiscale methodology.

Acknowledgements. This work is supported by the ITFLOWS project, which has received funding from the European Union Horizon 2020 research and innovation programme under grant agreement nos 882986. This work has also been supported by the SEAVEA ExCALIBUR project, which has received funding from EPSRC under grant agreement EP/W007711/1.

References

1. ActionAid. Climate change and flooding. https://www.actionaid.org.uk/our-work/emergencies-disasters-humanitarian-response/climate-change-and-flooding#footnote1_yg2m8ur. Accessed 20 Dec 2022
2. The rising levels of internally displaced people. https://geographical.co.uk/culture/rising-levels-of-internally-displaced-people. Accessed 18 Feb 2023
3. IDMC (IDMC), Global internal displacement database [online] (2022). https://www.internal-displacement.org/database/displacement-data
4. Best, K.B., et al.: Random forest analysis of two household surveys can identify important predictors of migration in Bangladesh. J. Computat. Soc. Sci. **4**, 77–100 (2021)
5. Groen, D.: Simulating refugee movements: where would you go? Procedia Comput. Sci. **80**, 2251–2255 (2016)
6. Nilsson, C., Riis, T., Sarneel, J.M., Svavarsdóttir, K.: Ecological restoration as a means of managing inland flood hazards. BioScience **68**, 89–99 (2018)
7. IDMC (IDMC), Systematic data collection and monitoring of 1displacement and its impacts at local, national, regional and international level to inform comprehensive needs and risk assessments for the formulation of policy and plans (2018). https://unfccc.int/sites/default/files/resource/WIM%20TFD%20III.1-3%20Output.pdf
8. Lim, M.B.B., Lim, H.R., Piantanakulchai, M., Uy, F.A.: A household-level flood evacuation decision model in Quezon City, Philippines. Nat. Hazards **80**, 1539–1561 (2016)
9. Hasan, S., Mesa-Arango, R., Ukkusuri, S.: A random-parameter hazard-based model to understand household evacuation timing behavior. Transp. Res. Part C: Emerg. Technol. **27**, 108–116 (2013)
10. Kuligowski, E.D., Gwynne, S.M.V.: The need for behavioral theory in evacuation modeling. In: Klingsch, W., Rogsch, C., Schadschneider, A., Schreckenberg, M. (eds.) Pedestrian and Evacuation Dynamics 2008, pp. 721–732. Springer, Heidelberg (2010). https://doi.org/10.1007/978-3-642-04504-2_70
11. Coxhead, I., Nguyen, V.C., Vu, H.L.: Internal migration in Vietnam, 2002–2012. In: Liu, A.Y.C., Meng, X. (eds.) Rural-Urban Migration in Vietnam. PE, pp. 67–96. Springer, Cham (2019). https://doi.org/10.1007/978-3-319-94574-3_3
12. Lovreglio, R., Ronchi, E., Nilsson, D.: A model of the decision-making process during pre-evacuation. Fire Saf. J. **78**, 168–179 (2015)
13. Pel, A.J., Bliemer, M.C., Hoogendoorn, S.P.: A review on travel behaviour modelling in dynamic traffic simulation models for evacuations. Transportation **39**, 97–123 (2012)

14. Alam, M.J., Habib, M.A., Pothier, E.: Shelter locations in evacuation: a multiple criteria evaluation combined with flood risk and traffic microsimulation modeling. Int. J. Disaster Risk Reduction **53**, 102016 (2021)
15. Yin, W., Murray-Tuite, P., Ukkusuri, S.V., Gladwin, H.: An agent-based modeling system for travel demand simulation for hurricane evacuation. Transp. Res. Part C: Emerg. Technol. **42**, 44–59 (2014)
16. Taillandier, F., Di Maiolo, P., Taillandier, P., Jacquenod, C., Rauscher-Lauranceau, L., Mehdizadeh, R.: An agent-based model to simulate inhabitants' behavior during a flood event. Int. J. Disaster Risk Reduction **64**, 102503 (2021)
17. Wang, Z., Wang, H., Huang, J., Kang, J., Han, D.: Analysis of the public flood risk perception in a flood-prone city: the case of Jingdezhen City in China. Water **10**(11), 1577 (2018)
18. Nakanishi, H., Black, J., Suenaga, Y.: Investigating the flood evacuation behaviour of older people: a case study of a rural town in japan. Res. Transp. Bus. Manag. **30**, 100376 (2019)
19. Pregnolato, M., Ford, A., Wilkinson, S.M., Dawson, R.J.: The impact of flooding on road transport: a depth-disruption function. Transp. Res. Part D: Transp. Environ. **55**, 67–81 (2017)
20. Troncoso Parady, G., Hato, E.: Accounting for spatial correlation in Tsunami evacuation destination choice: a case study of the great East Japan earthquake. Nat. Hazards **84**, 797–807 (2016)
21. International Organization for Migration (IOM), Flash Report: Flood Incidents North-East Nigeria - Bauchi State (2022). https://dtm.iom.int/reports/nigeria-flood-flash-report-bauchi-state-12-september-2022
22. QGIS Development Team, QGIS Geographic Information System. QGIS Association (2022)

Towards a Simplified Solution of COVID Spread in Buildings for Use in Coupled Models

Lucas Heighington and Derek Groen(✉)

Brunel University London, Kingston Lane, London UB8 3PH, UK
{2052561,Derek.Groen}@brunel.ac.uk

Abstract. We present a prototype agent-based simulation tool, Flu and Coronavirus Simulation inside Buildings (FACS-iB), for SARS-Cov2 to be used in an enclosed environment such as a supermarket. Our model simulates both the movement and breathing patterns of agents, to better understand the likelihood of infection within a confined space, given agent behaviours and room layout. We provide an overview of the conceptual model, its implementation, and showcase it in a modelled supermarket environment. In addition, we demonstrate how the model can be coupled to the Flu and Coronavirus Simulator (FACS), which is currently used to model the spread of SARS-CoV2 in cities and larger regions.

Keywords: Agent-based Modelling · Covid-19 Transmission · Breathing Modelling

1 Introduction

For the past 3 years, Covid-19 has had a profound effect on society, infecting 678,000,000 people and leading to the deaths of 6,800,000 lives [1]. These numbers are directly related to the policies introduced in each country, as their governments and officials try to determine what their best options are. Computational models played an important role in the decision making for many governmental strategies, from social and behavioural impacts to epidemic forecasting, along with the scale of these simulations from countries to counties [2]. The results these models produce are vital in predicting where hotspots may arise, where extra resources need to be sent and to get an insight into how the virus is spreading throughout a population.

Current simulation models were produced to try and understand the SARS-Cov-2 virus as it spreads, in the hopes of being able to predict where and how it works. As a result, the current models tend to focus on two main functions. Some models focus on large scale areas, modelling the transmission of the virus throughout a population. These areas can range from small towns all the way to countries or globally. The other type of model simulates the fluid dynamics of particulates around an individual, and how those droplets travel.

These models do have limitations in what they are able to simulate. Larger scale population models tend to drastically simplify the scenario they are trying to replicate [3], with the agents usually moving set intervals and infection rates being limited to

J. Mikyška et al. (Eds.): ICCS 2023, LNCS 14076, pp. 564–571, 2023.
https://doi.org/10.1007/978-3-031-36027-5_44

just a simple percentage. On the other hand, fluid dynamic models only focus on the immediate surroundings of 1–2 individuals [4], drastically limiting its ability to provide information on scenarios involving several people, along with the simulation itself being very computationally intensive [5].

In this work, we aim to implement ideas from both large-scale population models and fluid dynamics in a simplified manner, to produce a model that can simulate enclosed environments, whilst being resource friendly, in the hope that it will fill the gap that exists between current models.

2 Conceptual Model

Our model attempts to replicate the movement and interactions that may occur within a room or building by making use of independent agents that move around the environment by randomly selecting a direction to travel in and designing the outcome of potential interactions.

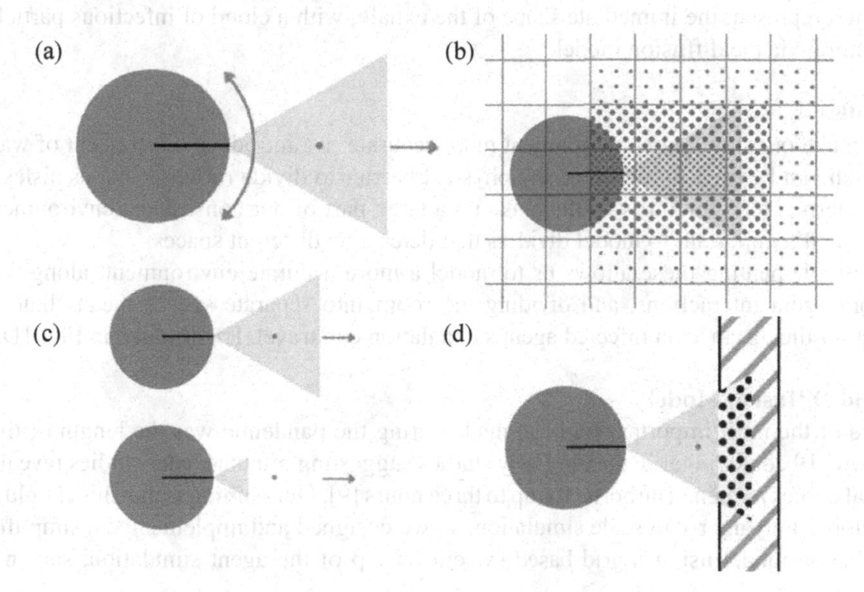

Fig. 1. Agent Model

Agent Movement

As a first step, we decided to use a modified version of the Random Walk algorithm, however, instead of using variables of orientation (N, E, S, W, etc.) we have implemented a randomized degree of rotation. Using the current direction of the agent, it will randomly select to turn either 5° to the left or right before moving forward. This model for the movement of the agent produces, in our opinion, a more fluid and realistic movement, as people tend to turn before they walk.

It is also important for the agent to recognize the boundaries of the simulation, and so if the agent finds itself within a fixed range from the boundary of the simulation (indicated by the circular point placed in front of it shown in Fig. 1A), it will again randomly choose to turn between 45° to 180°, before checking again if its heading towards the boundary. If it is, it will perform the randomization again, if not, it will continue moving forward.

We are aware that this implementation is limited, as humans tend to have destinations and will take the shortest path possible to get to them. It may be possible to improve this using a routing algorithm, which make the agents movements more meaningful.

Breathing, Coughing and Sneezing

To keep the simulation simplified, we represent the volume of air produced by the agent through a cone as this is the most accurate shape for a simplified exhale [6]. To model the action of breathing, we resize this cone every 2 s to represent the rate of breathing [7]. The smaller cone models the act of breathing in, whilst the larger cone represents breathing out, with the breath travelling up to 1 m away.

This can be seen in Fig. 1C, with the two smaller diagrams showing the agent breathing out and the agent breathing in. As for the agent coughing or sneezing, we simply extend the cone a set distance, two metres for coughing and six metres for sneezing [8], which represent the immediate shape of the exhale, with a cloud of infectious particles forming via the diffusion model.

Walls

To make our simulation environment more accurate, we introduce the concept of walls which can be used to represent any physical barrier to divide rooms or act as aisles in a supermarket. Obviously, walls make up a large part of our constructed environment, and so it's important to model dividers that determine different spaces.

Implementing these allows us to model a more accurate environment, along with more agent interactions, and dividing the room into separate spaces means that we restrict the distance an infected agent's exhalation can travel. Highlighted in Fig. 1D.

Grid Diffusion Model

One of the most important findings made during the pandemic was the length of time Covid-19 could linger in the air. Early studies suggesting minutes, later studies revealed viral copies remained airborne for up to three hours [9]. Our opinion is that this should be factored into any room scale simulation, so we designed and implemented a simplified diffusion model using a grid-based system on top of the agent simulation, shown in Fig. 1B.

In this model, we increase the number of droplets per cell in the grid by factoring in the method of exhale and the distance from the agent, with breathing producing fewer droplets than a sneeze or cough and dividing that by the cell's distance to the agent. We then use this value to influence the probability of an agent getting infected, with the simple concept that a higher number of droplets equates to a higher chance of infection, through lingering particles.

It should be noted that the design for our diffusion model is limited by the features we have chosen to simulate, as it does not factor in two variables that may impact the results. Those being aerosol dynamics and varying viral load, which can influence infection rates by the movement of aerosolised particles and the number of particles within the air, respectively, as elaborated on by Clifford K. Ho [10].

3 Establishing a Realistic Base Infection Probability

Before we could simulate anything, it was important for the model to use a more accurate infection probability. To do this, we calculated the given probability of an individual getting infected within a fixed environment. We calculated that the chance of infection for an individual in a two metre by two metre area, over 24 h is around seven percent [Evidence. 1 & 2] Using this information, we can recalculate the probability that an individual gets infected within one hour [8]:

$$(1 - 0.07)^{(1/12)} = \sim 0.994 = (1 - 0.006)$$

With the value of 0.006 calculated, we then needed to try to get as close to that value with the simulation. To do this, we first calculated the scale our simulation would be working at, using the average width of male (41.50 cm) and female (36.50 cm) shoulders [7] and taking the average between the two and then dividing that by the number of pixels in the diameter of our agents.

$$39\ \text{cm}/20\ \text{pixels} = 1.95\ \text{cm/pixels}$$

Using that value, we can calculate the number of pixels needed to represent any object, including the size of the boundaries of our simulation. As mentioned in the research behind the chance of infection in 24 h, we remodelled that space as a two metre by two metre window for the agents to randomly walk around in. With the environment established, we then ran the simulation for 1 h of simulated time and repeated this 100 times, each time altering the infection probability until the results reflected our calculated per-hour chance.

For this model, we landed on a value of 0.0005%, which within an hour simulation gives us only 0 or 1 infected. This value will continue being tested as other parts of the model are added, but we feel that this value is accurate enough for the sake of our current testing.

4 Multiscale Simulation Approach

In previous sections we presented a design and prototype implementation of FACS-iB, and how it can be used to approximate the spread of infectious diseases through the air and droplets (using simplified cone shapes). One of the main motivators to develop FACS-iB is the wish to add additional detail in the disease transmission dynamics of Flu And Coronavirus Simulator (FACS) code [11]. We have previously coupled FACS with the CHARM hospital model [12] to model hospital load resulting from infection patterns generated by FACS. As of now, however, there is no explicit coupling between FACS and actual in-building infection models, and the infection in buildings in FACS itself is only resolved using a single simplified equation.

In Fig. 2 we present a graphical overview of a multiscale simulation approach that extends FACS with infectious predictions using FACSiB.

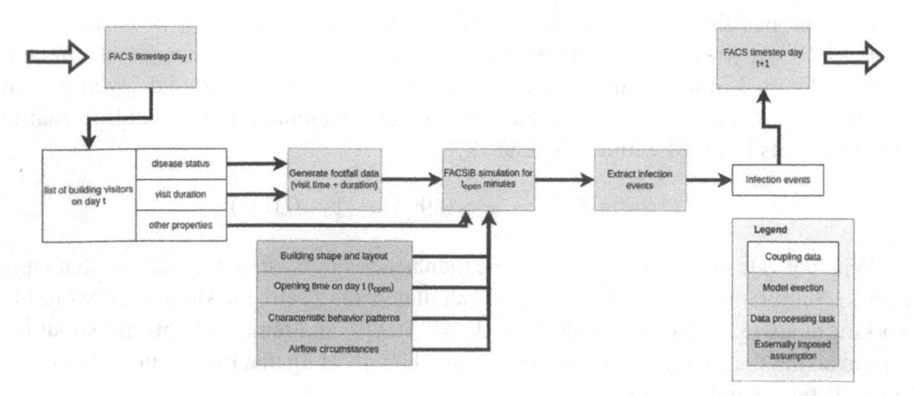

Fig. 2. Schematic overview of a coupled interaction between FACS and FACSiB

Here, the FACSiB model is started every day for each relevant building in the FACS model and run for the full period that the building is open that day. The FACSiB simulation then takes in a footfall profile, which is generated from the visiting information that we can extract from FACS. In addition, FACSiB requires a range of assumptions relating to the properties of the building, the opening times, the typical movement behaviour of people in such buildings, and airflow-related characteristics. After FACSiB has been run, we then use a post-processing script to extract infection events which are then passed back to FACS. The multiscale simulation approach we present here is currently under development, with the aim of obtaining preliminary results later this year.

5 Showcase

After finding an acceptable infection probability we ran some simulations to see what results we could get, using a small population of agents within the room, with 1–2 agents starting off infected.

Figure 3 shows our simulation's seven-point moving average when changing the scale of the environment, and having one agent spawned as infected, over a period of 6 h. The general trend the simulation produced, is that with a smaller room, a higher rate of infection occurs, a trend we would believe to be accurate. The graph shows how in a larger room (10 m × 10 m) the infection rate is slower, taking longer to reach higher levels. Whereas, in the smaller rooms (1 m × 1 m and 3 m × 3 m), the infection rate is much faster. However, we do want to mention that six people in a 1 m × 1 m room is an unrealistic scenario and we would expect its results to be more dramatic. The average produced by the 5 m × 5 m room, however, experiences an anomalous trend where more infections were recorded near the four-hour interval than the five-hour interval, and so the line creates a small wave pattern. In Fig. 4, we change the minimum infectious dose required to trigger the probability function to run, ranging between a diffusion cell value of 0 to 250. Again, the overall trend is to be expected, a lower minimum dosage requirement leads to a higher infection rate. However, the results for dosages of 200 and 250 seem to be faster than predicted, this is potentially due to the way the number of droplets within a cell influences the infection rates, but this may need some further investigating.

6 Discussion

In this paper, we have presented our current work on a simplified model for the spread of COVID-19 in an enclosed environment using agent-based simulation. We have highlighted the design for our model, using individual agents which are able to randomly move around whilst representing their breathing through a visual cone, and how they are able to infect each other. We have also highlighted our plans for implementing this model into a larger simulation, such as FACS, with the aim to produce more accurate results. The model in its current form, publicly available on Github [13], we believe, highlights the potential for agent-based modelling to simulate viral spread within an enclosed environment, and its potential to simulate viruses other than COVID-19 through differing infection rates. We are aware of the current implementation's limitations, which does not include features such as physical dividers, aerosol dynamics or varying viral load and resistance, which we believe can be future additions to the model.

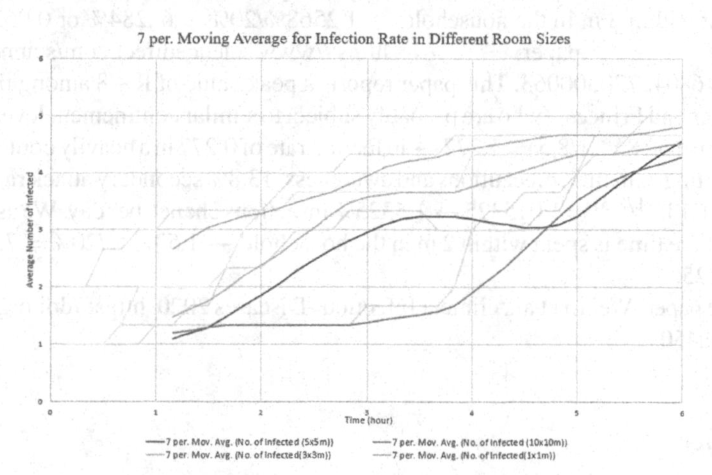

Fig. 3. Comparison of room size on infection rates over 24 h

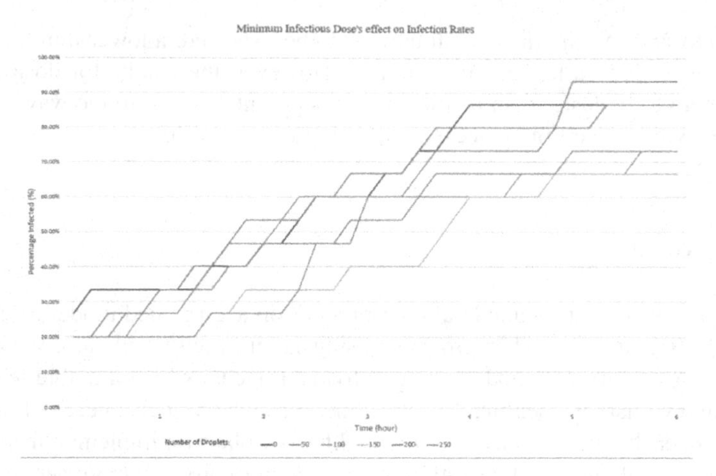

Fig. 4. Comparison of room shape on infection rates over 24 h

Summary of evidence:

1. Source paper: Qifang Bhi et al., Lancet, 2020. DOI: https://doi.org/10.1016/S1473-3099(20)30287-5. 11.2% secondary attack rate in house holds: $1.112^{(1/8.5)} = 1.012568 \rightarrow .2568\%$ infection chance per day. We assume that 20% of the time is spent within 2 m in the household $\rightarrow 1.2568\%/20\% = 6.284\%$ or 0.06284
2. Source paper: https://www.sciencedirect.com/science/article/pii/S2468042720300063. This paper reports a peak value of R ~ 8 among the crew of the Diamond Princess (who are probably subject to similar confinement levels). Deriving from that: $8^{**}(1/8.5) = 1.277 \rightarrow$ in fection rate of 0.277 in a heavily confined cruise ship setting with little precautions and awareness. 13.8% secondary attack rate in house holds: $1.138^{(1/8.5)} = 1.015325 \rightarrow 1.5325\%$ infections chance per day. We assume that 20% of the time is spent within 2 m in the household $\rightarrow 1.5325\%/20\% = 7.6625\%$ or 0.076625.
3. Source paper: Wei Li et al., clinical Infectious Diseases 2020. https://doi.org/10.1093/cid/ciaa450.

References

1. Worldometre. (n.d.): COVID-19 Coronavirus Pandemic. https://www.worldometres.info/coronavirus/. Accessed 4 Mar 2023
2. Adam, D.: "Nature," Special report: The simulations driving the world's response to COVID-19, 3 April 2020. https://www.nature.com/articles/d41586-020-01003-6. Accessed 30 Nov 2022
3. Epstein, J.M., Axtell, R.: Growing Artificial Societies: Social Science from the Bottom Up. Brookings Institution Press, Massachusetts (1996)
4. Rosti, M.E., Olivieri, S., Cavaiola, M., et al.: Fluid dynamics of COVID-19 airborne infection suggests urgent data for a scientific design of social distancing. Sci. Rep. **10**, 22426 (2020). https://doi.org/10.1038/s41598-020-80078-7

5. Tu, J., Yeoh, G.H., Liu, C.: Computational Fluid Dynamics a Practical Approach. Oxford Butterworth-Heinemann, Oxford (2018)
6. Shao, S., et al.: Risk assessment of airborne transmission of COVID-19 by asymptomatic individuals under different practical settings. J. Aerosol. Sci. **151**, 105661 (2021). https://doi.org/10.1016/j.jaerosci.2020.105661
7. respelearning.scot. (n.d.). Normal breathing—RESPe. https://respelearning.scot/topic-1-anatomy-and-physiology/normal-breathing
8. Xie, X., Li, Y., Chwang, A.T., Ho, P.L., Seto, W.H.: How far droplets can move in indoor environments–revisiting the wells evaporation-falling curve. Indoor Air **17**(3), 211–225 (2007). https://doi.org/10.1111/j.1600-0668.2007.00469.x. PMID: 17542834
9. Doremalen, N.V., et al.: Aerosol and surface stability of SARS-CoV-2. New England J. Med. **382**(16), 1564–1567 (2020)
10. Ho, C.K.: Modeling airborne pathogen transport and transmission risks of SARS-CoV-2. Appl. Math. Model. **95**, 297–319 (2021). https://doi.org/10.1016/j.apm.2021.02.018. PMID: 33642664, PMCID: PMC7902220
11. Mahmood, I., et al.: FACS: a geospatial agent-based simulator for analysing COVID-19 spread and public health measures on local regions. J. Simul. **16**(4), 355–373 (2022)
12. Anagnostou, A., et al.: FACS-CHARM: a hybrid agent-based and discrete-event simulation approach for Covid-19 management at regional level. In: 2022 Winter Simulation Conference (WSC), pp. 1223-1234. IEEE (2022)
13. Lucas. Final Year Project: Covid19-AgentSim. GitHub. https://github.com/Foxi-GB/FYP-Covid19-AgentSim. Accessed 23 Apr 2023

Epistemic and Aleatoric Uncertainty Quantification and Surrogate Modelling in High-Performance Multiscale Plasma Physics Simulations

Yehor Yudin[1,2](✉), David Coster[1], Udo von Toussaint[1], and Frank Jenko[1,2]

[1] Max-Planck-Institute for Plasma Physics, 85748 Garching, Germany
{yehor.yudin,david.coster,udo.v.toussaint,frank.jenko}@ipp.mpg.de
[2] CS Department, Technical University of Munich, 85748 Garching, Germany

Abstract. This work suggests several methods of uncertainty treatment in multiscale modelling and describes their application to a system of coupled turbulent transport simulations of a tokamak plasma. We propose a method to quantify the usually aleatoric uncertainty of a system in a quasistationary state, estimating the mean values and their errors for quantities of interest, which is average heat fluxes in the case of turbulence simulations. The method defines the stationarity of the system and suggests a way to balance the computational cost of simulation and the accuracy of estimation. This allows, contrary to many approaches, to incorporate of aleatoric uncertainties in the analysis of the model and to have a quantifiable decision for simulation runtime. Furthermore, the paper describes methods for quantifying the epistemic uncertainty of a model and the results of such a procedure for turbulence simulations, identifying the model's sensitivity to particular input parameters and sensitivity to uncertainties in total. Finally, we introduce a surrogate model approach based on Gaussian Process Regression and present a preliminary result of training and analysing the performance of such a model based on turbulence simulation data. Such an approach shows a potential to significantly decrease the computational cost of the uncertainty propagation for the given model, making it feasible on current HPC systems.

Keywords: Uncertainty Quantification · Sensitivity Analysis · Plasma Physics · Multiscale Modelling · High Performance Computing · Surrogate Modelling · Gaussian Process Regression

1 Introduction

Thermonuclear fusion is a prospective source of clean and abundant renewable energy with a potential for worldwide deployment, independent of geography

Supported by organisation "Munich School of Data Science".

J. Mikyška et al. (Eds.): ICCS 2023, LNCS 14076, pp. 572–586, 2023.
https://doi.org/10.1007/978-3-031-36027-5_45

and politics. To achieve fusion conditions, one needs to bring plasma temperature and density to sufficiently high values and isolate the plasma from the external matter for a sufficiently long time. One of the most studied methods to confine a plasma is to do it in a toroidal magnetic trap. The most developed type of device for confinement is a tokamak, an axisymmetric toroidal chamber with magnetic coils. One of the critical phenomena interesting for designing and analysing such confinement devices is the distribution of heat and density of the plasma over the minor radius of the torus. In the latest decades, it was discovered that transport properties of plasmas are dominated by microscopic dynamics of turbulent nature [7].

One of the challenges of such a study is the multiscale nature of the occurring phenomena, with properties of the processes happening across the size of an entire device, of the order of meters, during the confinement relevant times, of the order of seconds, are influenced and driven by microscopic processes, of the order of millimetres and microseconds.

One of the techniques in computational fields applied to tackle such challenges is coupling single-scale models, each solving a subset of equations describing processes happening on a particular scale and being implemented as a separate computer code. Such codes can, in principle, be run as a standalone program, accepting inputs and producing outputs, adhering to defined data structures.

In this work, we study the simulation of temperature profiles in a tokamak plasma, where plasma properties on the smallest scales define heat transport, and the equations for transport, equilibrium and microscopic turbulence are solved via separate codes, similar to the approach described by Falchetto et al. [7].

The following practical step interesting for scientific purposes is estimating uncertainties in the quantities of interest (QoIs) studied on the largest scale, considering that those uncertainties come from different sources.

To handle uncertainties, a quantity of interest can be treated as a random function of its parameters, also understood as random variables. In order to estimate uncertainties of function values using only a finite number of evaluations, assumptions can be made on the function, and an intermediate model or metamodel can be created based on a given sample of evaluations. This surrogate model describes the function's dependency on a selected subset of parameters and serves as a replacement for evaluations of the original function for practical purposes.

In this work, we apply surrogate modelling to the turbulent transport plasma simulation, replacing the solution of the microscopic turbulence equation with a fast model able to infer mean values and uncertainties of turbulent heat flux.

2 Methodology

2.1 Epistemic Uncertainty

One of the most common concerns in science is the uncertainty of inference or predictions due to a lack of knowledge of actual values of specific parameters or lack of knowledge incorporated into the model, in other words, epistemic uncertainty.

In theoretical fields, particularly in computational areas, mathematical models often involve unknown parameters that can only be estimated probabilistically based on prior knowledge, observations or experiments. The uncertainty propagation problem means finding an approximation of the probability density function of quantities of interest in the model, given a set of uncertain parameters that are understood as random variables with known probability density functions.

As it is often difficult to retrieve a particular parametric form of QoI PDFs, the useful statistics describing the properties of output random variables are the statistical moments of the density.

Polynomial Chaos Expansion. One of the most commonly used methods for uncertainty quantification is Polynomial Chaos Expansion (PCE) [18]. The core idea of the method is to expand the function of output Y of inputs X using the finite basis of orthonormal polynomials $P_i(x)$ up to the order of p, each depending on a subset of input components and being orthogonal to their PDFs:

$$Y \approx \hat{Y}(X) = \sum_{i=0}^{N} c_i P_i(X) \tag{1}$$

For the function $Y = f_X(x)$, the expansion coefficient ω_k and the ordinate values x_k could be determined via a quadrature scheme based on a Spectral Projection method exploiting the orthogonality of the basis polynomials and using normalisation factors H_i for polynomials P_i:

$$c_i = \frac{1}{H_i} \int_{\Omega} Y(x) P_i(x) f_X(x) dx \approx \frac{1}{H_i} \sum_{k=1}^{N} Y(x_k) P_i(x_k) \omega_k \tag{2}$$

The benefit of such an orthonormal approach is that the expansion coefficients are sufficient to calculate the estimates of the moments of the output PDF.

$$\hat{\mathbb{E}}[Y] = c_1, \quad \hat{\mathbb{V}}[Y] = \sum_{i=2}^{N} \mathbb{E}[P_i^2(X)] c_i \tag{3}$$

However, this method suffers from the curse of dimensionality as the number of required samples grows exponentially, like a binomial coefficient, with the number of dimensions $d : N = (d+p)!/p! \cdot d!$.

Sensitivity Analysis. A particular type of statistics over a sample of code's QoI responses aims to define the contribution of a single parameter, or a subset of parameters, to the uncertainty of the output. In this work, we study Sobol indices [17], which is a global variance-based sensitivity metric. The quantity we estimate is the variance of the mathematical expectations of the function output, conditioned on a subset of input components $\{i\}$ and normalised by the total variance of the output. Here we use the first-order Sobol indices, identifying the influence of a single parameter i; higher-order Sobol indices, determining the influence of a subset of interacting parameters $\{i\}$; and total Sobol indices, identifying the total effect of each parameter, using a set of all components $-i$, excluding a single i component:

$$S_{\{i\}} = \frac{\mathbb{V}[\mathbb{E}[Y|X_{\{i\}}]]}{\mathbb{V}[Y]}, \quad S_i^{tot} = 1 - \frac{\mathbb{V}[\mathbb{E}[Y|X_{-i}]]}{\mathbb{V}[Y]} \tag{4}$$

Using such an indicator, one could judge which parameters are most important for the behaviour of the computational model and decide whether a particular parameter's value can be fixed in subsequent studies or whether more values should be tried out.

2.2 Aleatoric Uncertainty

One of the types of uncertainties, different from the epistemic, is the aleatoric uncertainty coming from the inherent properties of the model and its solver. In the first place, it is related to the dynamic stochastic behaviour of the quantities of interests, for which the model is solved.

Several methods for analysing aleatoric uncertainties commonly involve two principles: separating epistemic uncertainties and performing statistical analyses of output quantities. One approach uses replica sampling, in which a certain number of model replicas are solved for each set of input parameters, followed by analysing the distributions of quantities of interest values [19]. Other methods are suitable for analysing the quasistationary nonlinear behaviour of the model in the vicinity of a certain attractor over a considerable time of solution [3].

This work utilises a practical measure for an aleatoric uncertainty based on analysis of the standard error of the mean values of quantities of interest $E = \mathrm{SEM}[y] = \sigma[y]/\sqrt{n_{\mathrm{eff}}}$ for a sample of observations $\mathbf{Y}^{\mathrm{eff}}$, each taken in an autocorrelation time window of length t_{A} apart from the previous one. Such an approach allows judging the requirements of a single solution for a given set of parameters and controlling the level of aleatoric uncertainty. However, such a measure only provides information on the second moments of the quantity of interest distribution. Furthermore, it is not always well incorporable into a single framework of analysis of both epistemic and aleatoric uncertainties.

2.3 Surrogate Modelling

This work primarily focuses on applying Gaussian Process Regression (GPR), a non-parametric probabilistic machine learning model fully defined by the covariance matrix of observed data samples and capable of providing a posterior PDF of output given any input values from the considered support [15].

When the Matern-3/2 kernel k describes covariance between any two parametric points x_i, x_j, determining a PDF $p(f(x_*))$ of a posterior for new data x_* using a standard GPR based on a sample of observed inputs X and outputs y with a Normal likelihood looks like following:

$$p(f(x_*)|\mathbf{X}, \mathbf{y}, x_*) \sim \mathcal{N}(\mu(x_*), \sigma^2(x_*)) \tag{5}$$

$$\sigma^2(x_*) = K(x_*, x_*) - K(x_*, \mathbf{X})^\top \left(K(\mathbf{X}, \mathbf{X}) + \sigma_e^2(I)\right)^{-1} K(x_*, \mathbf{X}) \tag{6}$$

$$K = (k(x_i, x_j))_{i,j=1..N} \tag{7}$$

$$k(x_i, x_j) = \sigma^2 \left(1 + \sqrt{3}\frac{|x_i - x_j|}{l}\right) \exp\left(-\sqrt{3}\frac{|x_i - x_j|}{l}\right) \tag{8}$$

The method allows for capturing uncertainties in the underlying data that come from the lack of information on the parametric dependencies and internal noise. Furthermore, this method has a high prediction uncertainty in the region of the parametric space for which there was not enough training data, as the posterior would have a higher variance for points decorrelated with the observed sample. Such a high uncertainty indicates the untrustworthiness of the regression results and the model's limitations.

One of the ideas behind the surrogate modelling approach is to create flexible and adaptive surrogates based on new data. Online Learning allows adding new data into the training sample to improve the surrogate gradually used as a proxy solver [10]. Active Learning uses the regression model's prediction uncertainty and information about sought dependency to determine new points in the parameter space for the evaluation of the function of interest [14].

In this work, we apply Gaussian Process Regression trained on a data set describing performed turbulence simulations. We trained such a surrogate model for the dependency of heat fluxes on the temperature values and its gradient for a particular configurational location in a tokamak plasma.

3 Numerical Results

3.1 Computational Model

The object of the study is the effects of turbulent processes happening at microscopic time and space scales on the temperature profiles of a nuclear fusion device across the confinement scales. For that, we use the Multiscale Fusion Workflow (MFW) [11] consisting of three main models, each implemented as a separate computer code, serving as an independent workflow component capable of exchanging its solution with others in a black-box fashion. The models are the equilibrium code that describes plasma geometry, the transport code that evolves the plasma profiles and the turbulence code used to compute effective transport coefficients (Fig. 1). The code that solves equilibrium Grad-Shafranov equations for plasma density and the magnetic field is CHEASE [12], a high-resolution fixed-boundary solver. Evolution of temperature and density is performed with ETS [4], a 1D code for energy and particle transport.

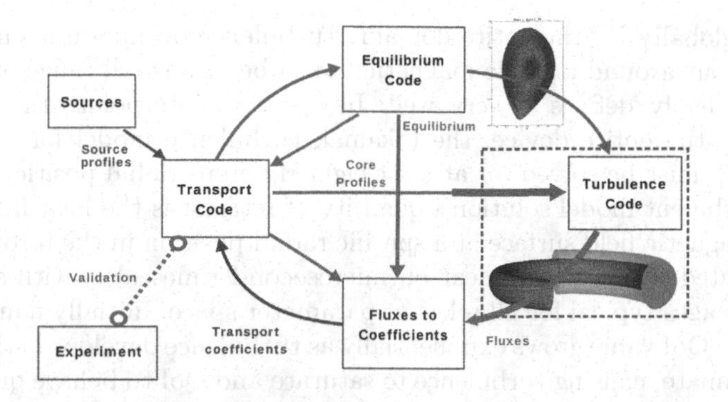

Fig. 1. Main components of the Multiscale Fusion Workflow. The turbulence model is the one that requires the most computational resources to solve, and that exposes the highest sensitivity to its inputs.

In this work, we analyse the turbulence component of the workflow, for which we use the 3D nonlinear gyrofluid turbulence code GEM [16] in its local flux tube approximation. Such turbulence models are very sensitive to input parameters and computationally expensive to solve. Thus, it is a prime point for performing a study of uncertainties in the joint multiscale model and the component for which a fast and easy-to-evaluate surrogate model would be of interest to create. As discussed in [5], performing UQ on the MFW with GEM is currently computationally unreasonable since 100 000 core hours are required for a single run, whereas a surrogate model should be at least 10 000 times cheaper (1 core instead of 1024, and a factor of 10 per cycle around the MFW loop).

Codes store their respective solution in standardised data structures of Consistent Physical Objects (CPO) [8], and the turbulence quantities of interests are stored in a structure describing core transport properties. In principle, the workflow implemented to analyse the flux time traces and prepare a dataset for surrogate training and utilisation can support any code capable of interfacing via CPOs.

The resulting data structures are passed from a code to a code instance in a point-to-point fashion using the MUSCLE2 coupling library [2]. Currently, a version supporting various turbulence models is being implemented using the MUSCLE3 library [20].

3.2 Simulations and Resulting Time Traces

A number of simulations of turbulence code for different input parameter values were performed to study the behaviour of quantities of interest in turbulence. Furthermore, this data was used to analyse what information can be included in a surrogate model and as an initial data set for the model training.

Transport equations describe temperature and density behaviour using a 1D approach, where quantities are functions of radial coordinate and time but must be solved globally for the entire domain. Turbulence occurs on a smaller scale, and behaviour around a single magnetic flux tube or a small range of the radial domain typically defines it very well. In order to interpolate the turbulence solution for the entire device, the tokamak turbulence model for the ASDEX Upgrade [7] must be solved for at least eight different radial positions [5].

The turbulent model solution's quantity of interest is the heat flux averaged across a magnetic field surface at a specific radial position in the toroidal device. The computed result is nonlinear on microsecond timescales, with several features and stages typical for all relevant parameter space. Initially non-turbulent, the effective QoI value grows exponentially as turbulence develops until nonlinear effects dominate, causing turbulence to saturate and QoI to behave quasistationary. Although mean values remain around a particular mean value, the QoI is also locally chaotic and fluctuates substantially over time.

Defining whether the system is saturated, accurately representing the mean output values for a particular case, and decoupling fluctuating behaviour are crucial to creating an effective surrogate model parametrising the transport properties.

Macroscopic kinetic profiles behave similarly to micro-level turbulent fluxes over confinement-relevant times. Starting from initial conditions, they converge towards a stationary level self-consistent with the coupled model. The profiles dynamically and nonlinearly interact with local turbulent behaviour, resulting in a quasistationary solution. In order to capture the dependency of turbulent fluxes in relevant parameter space, a surrogate model must be based on a large sample of code solutions covering the relevant regions.

Creating a sample for surrogate training involves running the same code for multiple input parameter values and managing the simulation and training data, which pose a significant challenge. Using the resulting surrogate as a microscopic code replacement requires ensuring it captures QoI dependency well for all practically possible parameter values. This requires a training dataset covering enough parameter space information; the surrogate should indicate its applicability limits and epistemic uncertainty due to a lack of training data.

Given the practical challenges, this work suggests another novel computational workflow besides one of the multiscale simulations to solve the turbulent transport problem. The new workflow analyses turbulent fluxes time traces on the microscopic temporal scales and defines the stationarity of the QoI as well as its mean values and estimation error.

3.3 Aleatoric Uncertainty

One of the work's challenges was to separate fundamental irreducible uncertainty from epistemic parametric uncertainty in a fluctuating system. It was important to establish the stationary mean values for analysing key parametric dependencies and to use fluctuations only for statistical error estimation. However, solving the model for a long enough time to gather sufficient data is computationally demanding and gathering a large sample would be prohibitively expensive.

Estimating error levels during the course of a simulation run can help decide when to stop the simulation and significantly save computational costs.

The procedure for analysing the QoI time traces $y(t)$, representing a model solution for a single parametric point, constitutes of the following steps:

1. For the time-traces $y(t)$ of length t_n representing a model solution in a scalar quantity of interest at time steps $\{t\}$, we select a part in the saturated phase:
   (a) Define the $y(t)$ ramp-up phase duration $t_{\text{r.u.}}$: here, in practice, for the long term, we chose an initial 15% of readings
   (b) Discard the readings from the ramp-up phase
2. Downsample the readings:
   (a) Calculate the Auto-Correlation Time: $t_{\text{A}} = \text{ACT}[y(t)] = \min t^* : \frac{1}{(t_n - t^*)} \sum_{t=t^*}^{t_n} (y(t) - \bar{y}) \cdot (y(t - t^*) - \bar{y}) < \frac{1}{t_n} \sum_{t=t_1}^{t_n} y^2(t) \cdot e^{-1}$
   (b) Split time series in saturated phase into $n_{\text{eff}} = \lfloor \frac{t_n}{t_{\text{A}}} \rfloor$ windows
   (c) For every autocorrelation time window of length t_{A} choose a downsampled reading as a mean value $y_i^{\text{eff}} = \frac{1}{t_{\text{A}}} \sum_{j=i \cdot t_{\text{A}}}^{(i+1) \cdot t_{\text{A}}} y_j$ for an effective time step $t_i^{\text{eff}} = \frac{1}{t_{\text{A}}} \sum_{j=i \cdot t_{\text{A}}}^{(i+1) \cdot t_{\text{A}}} j$
   (d) Collect downsampled readings into a new set $\mathbf{Y}^{\text{eff}} = \{y_i^{\text{eff}}\}$
3. Test the stationarity of the resulting time series:
   (a) Here: compute an ordinary least-squares linear multivariate regression model of downsampled QoI readings over the time
   (b) Apply Normal Equations to find coefficients: $\hat{\theta} = \left(\mathbf{X}^\top \mathbf{X}\right)^{-1} \mathbf{X}^\top \mathbf{Y}$ where $\mathbf{X}$ consists of effective time steps t_i^{eff}
   (c) Test if the linear regression coefficients are below a chosen relative tolerance: $\hat{\theta} < \epsilon_{\text{tol}}$
      i. If the coefficients are too large, continue running the simulation for another t_{run} time steps
      ii. If the coefficients are small enough, stop the simulation and proceed to the statistics calculation
4. Compute the essential statistics estimates:
   (a) Mean: $\mu[y] = \frac{1}{n_{\text{eff}}} \sum_{y \in \mathbf{Y}^{\text{eff}}} y$
   (b) Standard deviation: $\sigma[y] = \left(\frac{1}{n_{\text{eff}} - 1} \sum_{y \in \mathbf{Y}^{\text{eff}}} (y - \mu[y])^2 \right)^{-\frac{1}{2}}$
   (c) Standard Error of the Mean: $\text{SEM}[y] = \sigma[y] / \sqrt{n_{\text{eff}}}$

We applied this procedure for the time traces produced by turbulence code iteratively, in particular to ion heat flux values over time, sampling after a number of solution time steps, and analysed the behaviour of the mean estimate and its error.

Figure 2 describes mean values calculated with the sequential procedure based on the data of time traces of heat fluxes in a quasistationary phase.

After the relative change of the mean and the relative change of the standard error of the mean converge below a certain threshold, as shown in Fig. 3, we considered the duration of a simulation sufficient enough to have a reasonable estimate of the average flux and its error and added this data to a dataset

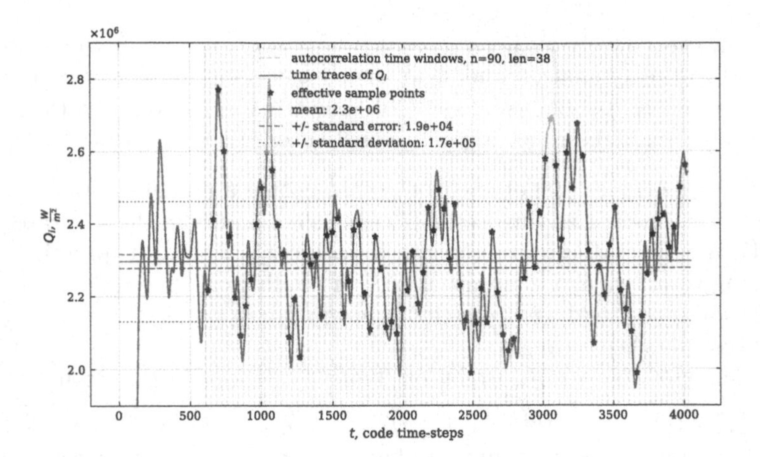

Fig. 2. The standard error of the mean is calculated accounting to discarding 15% of reading that do not belong to stationary phase, calculating the autocorrelation time of the remaining time traces, taken a sub-sample constituent of a single mean reading per autocorrelation window and renormalising mean error using the effective size of the sample.

describing the computational turbulence model behaviour. Subsequently, this dataset is used to draw conclusions on the model's parametric uncertainties and train a surrogate model.

3.4 Epistemic Uncertainty

To explore the epistemic uncertainties of the properties of the turbulence model, we performed an uncertainty propagation, taking a particular input of macroscopic parameter with a value taken from experimental measurement and, assuming that on top of this value, there is uncertainty, described as a normally distributed random variable with a Coefficient of Variation, a ratio of the variance of a quantity to its mean value, $\mathrm{CV}(x) = \sqrt{\mathbb{V}[x]}/\mathbb{E}[x]$ equal to 0.1. The statistics for average ion and electron heat flux $\mathbf{y} = (Q_i(\rho), Q_e(\rho))$ were estimated using a PCE method with Hermite polynomials of order $p = 2$ applied for four input parameters of ion and electron temperatures profile values and gradients: $\mathbf{x} = T_i(\rho), T_e(\rho), \nabla T_i(\rho), \nabla T_e(\rho)$.

The uncertainty propagation was performed using EasyVVUQ [9] library, which manages the definition of input uncertainties, sampling scheme, generation of input data structures and collection of the output ones, as well as calculating statistics. The batches of the turbulence code instances were run on the MPCDF COBRA supercomputer using QCG-PilotJob [13] middle-ware library, which manages HPC resources of a single computational allocation and distributes it to run multiple instances of the code with a particular parallelisation set-up.

One of the issues of uncertainty propagation through the model that exposes chaotic behaviour on some scales is to untangle the epistemic and aleatoric uncer-

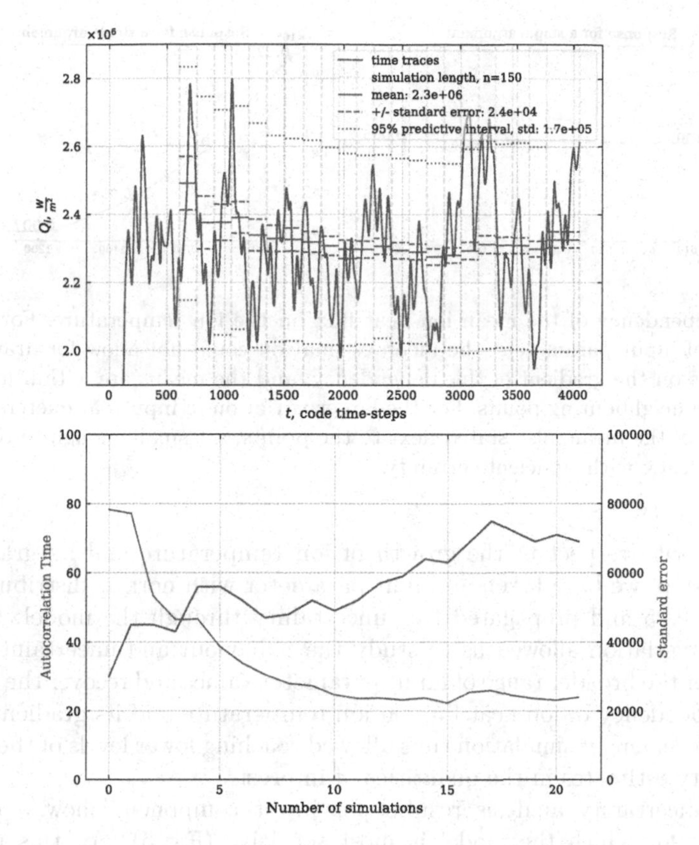

Fig. 3. Sequential analysis of time traces with an interval of 150 time-steps per window, defined by 6 h of computation on a single MPCDF's COBRA node. After each step of time traces computation, we recalculate the autocorrelation time, repopulate an effective data sample, and estimate new values for the mean, standard deviation and standard error of the mean. With the growth of the sample, the mean estimation converges to the true value, and the standard error of the mean decreases to around 1% of the respective mean value after 2300 time steps out of 4000 in total. Here, one can decide to stop the computation using a tolerance threshold on the relative change of the mean estimate of 10^{-3} or by the error of the mean dropping below 1.1% Coefficient of Variation. The plots of the standard error of the mean and autocorrelation time dependency on number simulation steps indicate a convergence of the time traces analysis with simulation time.

tainties. Before analysing the parametric uncertainty propagation results, the time traces produced by the simulations are processed as described in the previous subsection.

The result of this study with an error of $\mathrm{CV}(\mathbf{x}) = 0.1$ was inconclusive, as there was a significant overlap of the error bounds for mean values of quantities of interest for neighbouring points in parametric space (Fig. 4). In this case, it was impossible to conclude with certainty whether the ion heat flux trend was rising

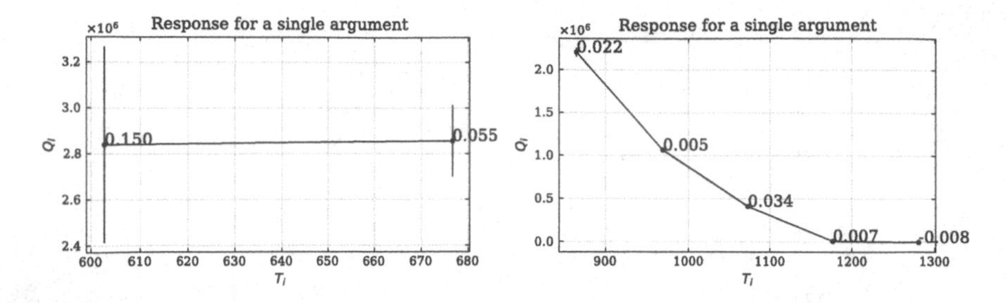

Fig. 4. Dependency of the mean ion heat flux on the ion temperature. For the small variation of input parameters, the mean estimation could not allow for drawing good conclusions on the gradient of this dependency, and the means are within uncertainty intervals of neighbouring points. For the larger variation of input parameters, the relative errors of the mean, also stated next to the points, are small enough to reconstruct the dependency with sufficient certainty.

or falling with respect to the growth of ion temperature and its gradient. As the next step, we have taken an input parameter with normal distribution with $\mathrm{CV}(\mathbf{x}) = 0.25$ and propagated this uncertainty through the model. Enlarging the input variation allowed us to study the behaviour and uncertainties of the model over the broader range of input parameter values and recover the character of the dependency of ion heat flux on ion temperature and its gradient (Fig. 4). Furthermore, longer simulation runs allowed reaching lower levels of the aleatoric uncertainty estimates in the quantities of interest.

The uncertainty analysis results per input component show a dominant parameter for which the model is most sensitive (Fig. 5). For this particular prior PDF over parametric space, the behaviour of the electron heat flux is most sensitive to variation in electron temperature. Still, other parameters and their interactions also play a significant role. Other metrics that were chosen to analyse the model's epistemic uncertainty are the Coefficient of Variation in the quantity of interest and the Uncertainty Amplification Factor [6]. In this work, the coefficient of variation of the model's output ion heat flux is about $\mathrm{CV}(Q_i) \approx 2.96$, showing a significant range of possible values that QoI can take in the study.

The Uncertainty Amplification Factor is the ratio of the Coefficient of Variation of a QoI to the coefficient of variation of the input parameter $\mathrm{UAF}(y|x) = \mathrm{CV}(y)/\mathrm{CV}(x)$. In the case of the studied model and chosen input parameters prior distribution, the Uncertainty Amplification Factor of ion heat flux is $\mathrm{UAF}(Q_i|\mathbf{x}) \approx 11.9$ showing a significant sensitivity of the model to the parametric uncertainty.

3.5 Surrogate Model

Having performed the original epistemic uncertainty propagation, we have collected a substantial sample of the physical model's input-output pairs. This sample was used to fit a surrogate model able to capture the behaviour of the physi-

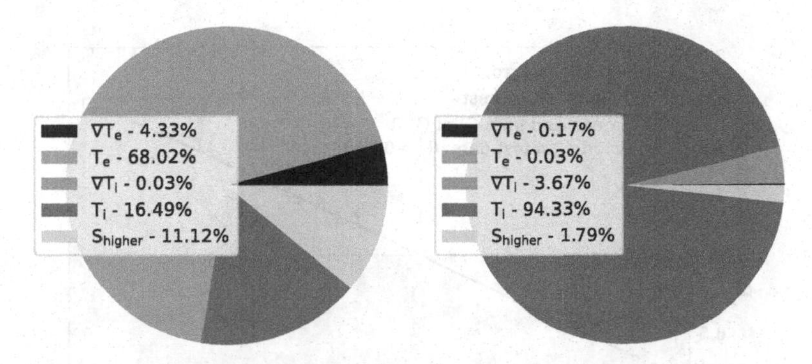

Fig. 5. Sobol indices for electron and ion heat fluxes. Most of the variation in electron heat flux is produced by the variation in the electron temperature. Interactions of the input parameters explain some of the non-negligible variation. For the ion heat flux, the output variation is dominated by a single parameter in this regime. However, a fraction is still explained by the nonlinear interaction of parameters.

cal model QoI means values and uncertainties over the selected input parametric subspace. Here, the QoIs and the input parameters are the same as described in the epistemic uncertainty propagation results section.

The GPR surrogate provides an intermediate model based on the assumptions on the regularity of the actual dependency, primarily encoded in the covariance kernel and the likelihood of the regression model. Unlike the PCE surrogate, which depends on the predefined polynomial order and is difficult to adapt given new information about the sought dependency, the uncertainties of the GPR one produce strong indications that the information on the actual dependency is sufficient for the regression and is easily adaptable through sequential design and active learning algorithms.

Here we present the results of the Gaussian Process Regression surrogate trained on 50% of the collected simulation dataset (Fig. 6). We developed functionality for the training and analysis of surrogates based on EasySurrogate [1] library. After performing a hyper-parameter optimisation, a particular set of categorical parameter values was chosen to judge surrogates by the highest coefficient of determination of QoI values predicted by a surrogate y_i^{surr} compared to the ground truth simulation values $y_i^{\mathrm{g.t.}}$ for the model testing: $R^2 = 1 - \sum_{i=1}^{n_{\mathrm{test}}} \left(y_i^{\mathrm{g.t.}} - y_i^{\mathrm{surr}}\right)^2 / \sum_{i=1}^{n_{\mathrm{test}}} \left(y_i^{\mathrm{g.t.}} - \overline{y^{\mathrm{g.t.}}}\right)^2$.

The future plan is to apply a validation test for the applicability of such a surrogate as a replacement for the solution of the turbulence model for given inputs describing core profiles and inferring the average heat fluxes and effective transport coefficients. The original simulation workflow utilising the turbulence code GEM required around 27.0 s on 1024 cores of Cineca MARCONI supercomputer [11] to predict fluxes for the next iteration of core profiles advancement, whether the EasySurrogate GPR model requires a fraction of a second on a single core for the same procedure. Furthermore, applying a GPR model would allow the detection of points in parameter space where the surrogate uncertainty is too high and should not be applied as a proxy for a high-fidelity solver. When it

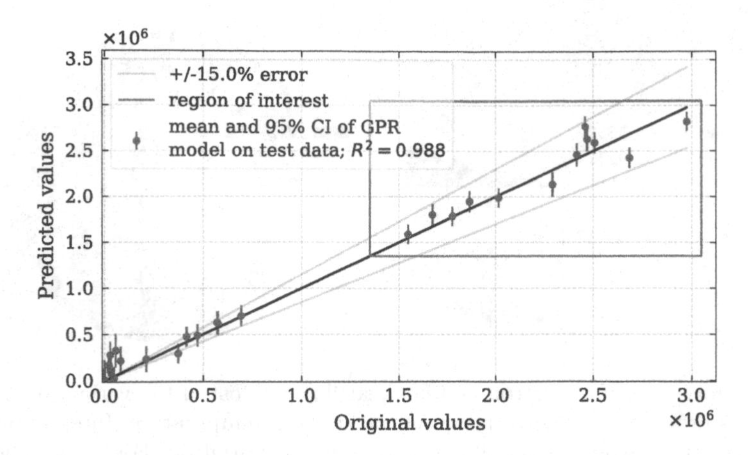

Fig. 6. GPR surrogate model test results with the coefficient of determination $R^2 = 0.988$. Inferred values of ion heat flux compared to the observed values produced by the physical model. Vertical bars denote the ± 1.96 of the standard deviation of the predictive posterior distribution, grey bounds denote 15% relative error, and dotted lines denote the region of interest. For most test samples, the true value is within 95% prediction interval, and the predicted mean is within 15% relative error.

is understood that a GPR surrogate performs sub-par for a region of interest in parametric space, some work was done on designing an Active Learning scheme based on Bayesian Optimisation and Sequential Design of Experiments.

4 Discussion

In this paper, we presented a number of methods to treat uncertainties in multiscale simulations. This includes ways to measure aleatoric and epistemic uncertainties, to control and reduce the amount of computational resources required to quantification of the uncertainties, as well as to identify situations when uncertainties are hindering analysis of the physical model behaviour. The suggested method for aleatoric uncertainty processing is applicable when some model dependency could be decomposed into the mean and fluctuation behaviour and allows for a trade-off between the computational effort for the model solution and the accuracy of the mean behaviour estimation. The surrogate modelling method proposes a way to make epistemic uncertainty propagation cheaper by constructing a data-based machine learning GPR proxy model for an expensive physical model solver that could be used as a replacement in multiscale coupled simulations and for sampling in uncertainty quantification methods.

We applied the described methods for the dependency of average heat fluxes produced by plasma turbulence code GEM based on tokamak core temperature profiles and their gradients. While studying such fluxes' time traces, we identified cases when the simulations converged early enough to allow reducing the duration of runs and saving computational resources. The QoI mean levels were estimated with sufficient certainty to draw conclusions about sought dependency and use solution data as a basis for further analysis and surrogate model training. The subsequently performed uncertainty propagation and sensitivity analysis showed situations where aleatoric uncertainties could not allow concluding on parametric dependencies. Also, we defined cases with a strong indication that a particular set of variables influences output variation dominantly. The surrogate modelling approach of fitting a Gaussian Process Regression model to the dataset of flux computing turbulent simulations showed that it is possible to train a metamodel that allows capturing the mean behaviour of the fluxes well within the posterior predictive bounds for a region of interest in parameter space of the model.

The development of a surrogate for the turbulence code opens up the much-needed possibility of doing detailed UQ for the whole MFW workflow with the accuracy of the underlying turbulence code but without the expense of the turbulence code in the workflow.

Future work would include testing the surrogate as a proxy of the expensive turbulence code, capable of inferring fluxes and subsequently defining the effective transport coefficients. This includes implementing an expanded version of a Multiscale Fusion Workflow, capable of switching between different implementations of the turbulence model and validating the results of the coupled turbulent transport simulations using a turbulence surrogate against a version using the gyrofluid code. Subsequently, such a modified workflow should include capacity for adaptive surrogate training through Online Learning and Active Learning algorithms, as well as methods to select turbulence model implementation in the course of a single simulation.

Acknowledgements. The authors of this paper would like to acknowledge the support of the Poznan Supercomputer and Networking Centre (PSNC) and Max-Planck Computational Data Facility (MPCDF). Research by Yehor Yudin is funded by the Helmholtz Association under the "Munich School for Data Science - MuDS".

References

1. EasySurrogate github. https://github.com/wedeling/EasySurrogate. Accessed 2 Mar 2023
2. Borgdorff, J., et al.: Distributed multiscale computing with MUSCLE 2, the multiscale coupling library and environment. J. Comput. Sci. **5**(5), 719–731 (2014). https://doi.org/10.1016/j.jocs.2014.04.004
3. Brajard, J., et al.: Combining data assimilation and machine learning to infer unresolved scale parametrization. Phil. Trans. R. Soc. A: Math. Phys. Eng. Sci. **379**(2194) (2021). https://doi.org/10.1098/rsta.2020.0086
4. Coster, D., et al.: The European transport solver. IEEE Trans. Plasma Sci. **38**(9), 2085–2092 (2010). https://doi.org/10.1109/TPS.2010.2056707
5. Coster, D., et al.: Building a turbulence-transport workflow incorporating uncertainty quantification for predicting core profiles in a tokamak plasma. Nuclear Fusion (2021)

6. Edeling, W., et al.: The impact of uncertainty on predictions of the CovidSim epidemiological code. Nat. Comput. Sci. **1**(2). https://doi.org/10.1038/s43588-021-00028-9
7. Falchetto, G., et al.: The European Integrated Tokamak Modelling (ITM) effort: achievements and first physics results. Nucl. Fusion **54**(4), 043018 (2014). https://doi.org/10.1088/0029-5515/54/4/043018, 00005
8. Imbeaux, F., et al.: A generic data structure for integrated modelling of tokamak physics and subsystems. Comput. Phys. Commun. **181**(6), 987–998 (2010). https://doi.org/10.1016/j.cpc.2010.02.001
9. Jancauskas, V., Lakhlili, J., Richardson, R., Wright, D.: EasyVVUQ: verification, validation and uncertainty quantification for HPC simulations (2021). https://github.com/UCL-CCS/EasyVVUQ
10. Leiter, K., Barnes, B., Becker, R., Knap, J.: Accelerated scale-bridging through adaptive surrogate model evaluation. J. Comput. Sci. **27**, 91–106 (2018). https://doi.org/10.1016/j.jocs.2018.04.010
11. Luk, O., et al.: ComPat framework for multiscale simulations applied to fusion plasmas. Comput. Phys. Commun. **239**, 126–133 (2019). https://doi.org/10.1016/j.cpc.2018.12.021
12. Lütjens, H., Bondeson, A., Sauter, O.: The CHEASE code for toroidal MHD equilibria. Comput. Phys. Commun. **97**(3), 219–260 (1996)
13. Piontek, T., et al.: Development of science gateways using QCG—lessons learned from the deployment on large scale distributed and HPC infrastructures. J. Grid Comput. **14**(4), 559–573 (2016). https://doi.org/10.1007/s10723-016-9384-9
14. Preuss, R., von Toussaint, U.: Global optimization employing Gaussian process-based Bayesian surrogates. Entropy **20** (2018). https://doi.org/10.3390/e20030201
15. Rasmussen, C.E., W.K.: Gaussian Processes for Machine Learning. The MIT Press (2006). https://doi.org/10.7551/mitpress/3206.001.0001
16. Scott, B.D.: Free-energy conservation in local gyrofluid models. Phys. Plasmas **12**(10), 102307 (2005). https://doi.org/10.1063/1.2064968
17. Sobol, I.: On sensitivity estimation for nonlinear mathematical models. Math. Model. **2**(1), 112–118 (1990)
18. Sullivan, T.: Introduction to Uncertainty Quantification. TAM, vol. 63. Springer, Cham (2015). https://doi.org/10.1007/978-3-319-23395-6
19. Vassaux, M., et al.: Ensembles are required to handle aleatoric and parametric uncertainty in molecular dynamics simulation. J. Chem. Theory Comput. **17**(8) (2021). https://doi.org/10.1021/acs.jctc.1c00526
20. Veen, L.E., Hoekstra, A.G.: Easing multiscale model design and coupling with MUSCLE 3. In: Krzhizhanovskaya, V.V., et al. (eds.) ICCS 2020. LNCS, vol. 12142, pp. 425–438. Springer, Cham (2020). https://doi.org/10.1007/978-3-030-50433-5_33

Network Models and Analysis: From Foundations to Complex Systems

Applying Reinforcement Learning to Ramsey Problems

Tomasz Dzido(✉)

Department of Information Systems, Gdynia Maritime University, 81-225 Gdynia, Poland
t.dzido@wznj.umg.edu.pl

Abstract. The paper presents the use of reinforcement learning in edge coloring of a complete graph, and more specifically in the problem of determining Ramsey numbers. To the best of our knowledge, no one has so far dealt with the use of RL techniques for graph edge coloring.

The paper contains an adaptation of the method of Zhou *et al.* to the problem of finding specific Ramsey colorings. The proposed algorithm was tested by successfully finding critical colorings for selected known Ramsey numbers. The results of proposed algorithm are so promising that we may have a chance to find unknown Ramsey numbers.

Keywords: Reinforcement learning · Ramsey numbers · Learning-based optimization

1 Introduction

One of the popular ways looking at Ramsey's theory is in the context of graph theory, and more specifically edge coloring of graphs. To put it quite simply, we want to answer the following question: If we have a complete graph K_n on n vertices where every edge is arbitrarily colored either blue or red, what is the smallest value of n that guarantees the existence of either a subgraph G_1 which is blue, or a subgraph G_2 which is red? This smallest search n is called a 2-color Ramsey number $R(G_1, G_2)$. Initially, only the case when subgraphs G_1 and G_2 are complete subgraphs was considered. Greenwood and Gleason [8] established the initial values $R(K_3, K_4) = 9$, $R(K_3, K_5) = 14$ and $R(K_4, K_4) = 18$ in 1955. Unfortunately, in the case of exact values, there has been very little progress for many years, and for many Ramsey numbers. For example, note that the most recent exact result for a 2-color Ramsey number for two complete graphs is $R(K_4, K_5) = 25$ and was obtained by McKay and Radziszowski in 1995 [9]. Therefore, Ramsey numbers for subgraphs other than complete became popular very quickly. Many interesting applications of Ramsey theory arose in the field of mathematics and computer science, these include results in number theory, algebra, geometry, topology, set theory, logic, information theory and theoretical computer science. The theory is especially useful in building and analyzing communication nets of various types. Ramsey theory has been applied by Frederickson and Lynch to a problem in distributed computations [7], and by

J. Mikyška et al. (Eds.): ICCS 2023, LNCS 14076, pp. 589–596, 2023.
https://doi.org/10.1007/978-3-031-36027-5_46

Snir [13] to search sorted tables in different parallel computation models. The reader will find more applications in Rosta's summary titled "Ramsey Theory Applications" [11].

In recent years, the use of Machine learning (ML) techniques in solving combinatorial problems has significantly increased. Bengio et al. [1] noted that models that are formed by combining ML techniques and combinatorial optimisation strengthen the training procedures. ML is useful especially in discovering or creating certain desirable patterns in graphs, which will be shown later in this article. In 2022, Kai Siong Yow and Siqiang Luo gave a very interesting survey [14]. They reviewed classic graph problems that have been addressed by using learning-based algorithms, particularly those employ ML techniques.

Common approach that is gaining popularity is reinforcement learning (RL), where an agent interacts with its environment in discrete time steps, and learns an (nearly) optimal policy to maximise the reward over a course of actions. There are three key elements in a RL agent, i.e., states, actions and rewards. At each instant a RL agent observes the current state, and takes an action from the set of its available actions for the current state. Once an action is performed, the RL agent changes to a new state, based on transition probabilities. Correspondingly, a feedback signal is returned to the RL agent to inform it about the quality of its performed action [16].

Grouping problems aim to partition a set of items into a collection of mutually disjoint subsets according to some specific criterion and constraints. Grouping problems naturally arise in numerous domains, including, of course, the problem of graph coloring. Zhou, Hao and Duval in [16] presented the reinforcement learning based local search (RLS) approach for grouping problems, which combines reinforcement learning techniques with a descent-based local search procedure. To evaluate the viability of the proposed RLS method, the authors used the well-known graph vertex coloring problem (GCP) as a case study. To the best of our knowledge, no one has so far dealt with the use of RL techniques for graph edge coloring. All the more, there are no known attempts to use these techniques in estimating the value of Ramsey numbers. The most commonly used heuristics are local search, simulated annealing or tabu search. In this paper, we present how the reinforcement learning based local search (RLS) approach presented in [16] can be used to find lower bounds of some Ramsey numbers. Our proposed application of RLS approach belongs to the category of learning generative models of solutions. This method was used, among others, in [4] for the solution of the flow-shop scheduling problem.

To sum up the introduction, the main method used in the article is an adaptation of the RLS approach presented in [16] and the rest of this paper is organized as follows. Section 2 provides useful notation and definitions. Section 3 describes the application of the RLS method in Ramsey's theory. In Sect. 4, we discussed the results obtained from the computer simulations and present possible improvements. The article ends with a short summary and an indication of the direction of further research.

2 Notation and Definitions Related to Graphs and Ramsey Numbers

Let $G = (V(G), E(G))$ be an undirected graph. K_m denotes the complete graph on m vertices, C_m - the cycle of length m, P_m - the path on m vertices and $K_{1,m}$ - the star of order $m+1$. An edge k-coloring of a graph G is any function $f : E(G) \to \{1, 2, 3, ..., k\}$. In this paper we only consider edge 2-colorings. Since colorings involve only two colors (blue and red) further on will be reffered to as colorings (instead of edge 2-colorings). For graphs G_1, G_2 a coloring f is a $(G_1, G_2; n)$ – *coloring* if and only if f is a 2-coloring of the complete graph K_n and f contains neither a G_1 colored with color 1 nor a G_2 colored with color 2.

Definition 1. *The Ramsey number $R(G_1, G_2)$ for graphs G_1 and G_2 is the smallest positive integer n such that there is no $(G_1, G_2; n)$ – coloring.*

Definition 2. *A coloring $(G_1, G_2; n)$ is said to be critical if $n = R(G_1, G_2) - 1$.*

The following theorem is a well-known result on Ramsey number for two cycles, which was established independently in [6] and [12].

Theorem 1 ([6,12]). *Let m, n be integers, where $3 \le m \le n$.*

$$R(C_m, C_n) = \begin{cases} 6 & (m,n) = (3,3), (4,4), \\ 2n-1 & m \textit{ is odd and } (m,n) \neq (3,3), \\ n-1+\frac{m}{2} & m \textit{ and } n \textit{ are even and } (m,n) \neq (4,4), \\ max\{n-1+\frac{m}{2}, 2m-1\} & m \textit{ is even and } n \textit{ is odd.} \end{cases}$$

As we have seen, in the case of two cycles we know everything, however only partial results for C_m versus stars $K_{1,n}$ are known. The most known general exact result for even cycles is:

Theorem 2 ([15]).

$$R(C_m, K_{1_n}) = \begin{cases} 2n & \textit{for even } m \textit{ with } n < m \le 2n, \\ 2m-1 & \textit{for even } m \textit{ with } 3n/4 + 1 \le m \le n. \end{cases}$$

Besides the exact values many lower and upper bounds for various kinds of graphs have appeared in the literature. Radziszowski in his regularly updated dynamic survey "Small Ramsey Numbers" [10] lists all known nontrivial values and bounds for Ramsey numbers. Lower bounds on Ramsey numbers are mostly proved by giving a witness that doesn't have the desired Ramsey property. Such a witness (called a critical coloring) could be part of a general construction, or found 'at random' by a heuristic algorithm. While lower bounds on Ramsey numbers can be established by giving one coloring which does not have the desired property, to prove an upper bound one must give an argument implying that all colorings of a certain order complete graph have the desired property. Mostly, this is done by using general or specific theorems to vastly reduce the number of possible counter-example colorings. The remaining colorings sometimes must then be enumerated by a computer to verify that none of them is a critical coloring.

3 RLS Applied to Determining Critical Colorings for Some Ramsey Numbers

In the problem of determining the exact value of the Ramsey number, it is often the case that we can find it for large n, but we do not know the value for small cases. The algorithm presented below can successfully fill this gap. This algorithm can be adapted to various types of graphs, but for the purposes of the article, we will only present a case of applying it to two cycles and to the case of a cycle and a star.

We already know everything about Ramsey numbers of the type $R(C_k, C_m)$ (see Theorem 1), so we will be able to easily verify the obtained results. Let us assume that we are looking for the critical $(C_k, C_m; n) - coloring$, where $k \leq m$. To apply the proposed RLS approach to this purpose, we need to specify the search space Ω, the neighborhood, the evaluation function $f(S)$, final acceptance criterium and method of choosing initial state.

First, for a given partition of complete graph K_n into 2 graphs: G_1 and G_2, where $V(G_1) = V(G_2) = V(K_n)$, $E(K_n) = E(G_1) \cup E(G_2)$ and $E(G_1) \cap E(G_2) = \emptyset$, we define space Ω to be the family of all possible edge 2-colorings $S = \{G_1, G_2\}$ such that the subgraph G_i is colored with the color i, where $i \in \{1, 2\}$. The neighborhood of a given coloring is constructed by changing the color of an edge belonging to at least one forbidden cycle.

The objective function and the final acceptance criterium are defined intuitively: $f(S)$ is simply equal to the number of edges that are in at least one forbidden cycle in S. Accordingly, a candidate solution S is the desired critical coloring if $f(S) = 0$.

As the initial state a complete graph K_n is taken. Originally, the RLS procedure starts with a random solution taken from the search space Ω. In order to increase the size and speed up finding the desired critical Ramsey coloring, we can use various known graph-type-specific properties. They can reduce the number of remaining edges to be colored or narrow down the number of edges of a given color. The use of this type of methods requires specialist knowledge of Ramsey properties for a given type of graphs, here we will focus only on possible examples for the studied two-cycle problem.

The length of a shortest cycle and the length of a longest cycle in G are denoted by $g(G)$ and $c(G)$, respectively. A graph G is weakly pancyclic if it contains cycles of every length between $g(G)$ and $c(G)$. $G \cup H$ stands for vertex disjoint union of graphs, and the join $G + H$ is obtained by adding all of the edges between vertices of G and H to $G \cup H$. The following interesting properties are known.

Theorem 3 ([5]). *Let G be a graph of order $n \geq 6$. Then $max\{c(G), c(\overline{G})\} \geq \lceil 2n/3 \rceil$, where $\overline{G}$ is the complement of G.*

Theorem 4 ([2]). *Every nonbipartite graph G of order n with $|E(G)| > \frac{(n-1)^2}{4} + 1$ is weakly pancyclic with $g(G) = 3$.*

Theorem 5 ([3]). *Let G be a graph on n vertices and m edges with $m \geq n$ and $c(G) = k$. Then*

$$m \leq w(n,k) = \frac{1}{2}(n-1)k - \frac{1}{2}r(k-r-1), \text{ where } r = (n-1) \bmod (k-1),$$

and this result is the best possible.

For example, consider the number $R(C_8, C_8)$. Theorem 1 leads to $R(C_8, C_8) = 11$. That means we are looking for critical $(C_8, C_8; 10)$ – *coloring*. On the other hand, we know that $R(B_2, B_2) = 10$, where $B_2 = K_2 + \overline{K_2}$. By combining the value of this number and Theorem 3, we get the property that the remaining number of edges to be colored by adapting the RLS algorithm is at least 9 less (where B_2 and C_7 have the greatest possible intersection). Looking at it from another angle, if the number of edges of one color exceeds 27 (combining Theorems 4 and 5, where $w(10, 7) = 27$), in this color we have cycle C_8, which we avoid. This means that the number of edges of K_{10} in each color belongs to the set $\{18, ..., 27\}$.

We define a probability matrix P of size $n \times 2$ (n is the number of edges and 2 is the number of colors). An element p_{ij} denotes the probability that the i-th edge $e \in E(G)$ is colored with the j-th color. Therefore, the i-th row of the probability matrix defines the probability vector of the i-th edge and is denoted by p_i. At the beginning, all the probability values in the probability matrix are set as $\frac{1}{2}$. It means that all edges are colored with one of the two available colors with equal probability. To achieve a local optimum, the current solution (coloring) S_t at instant t is then enhanced by DB-LS, a descent-based local search algorithm which iteratively improves this solution by a neighboring solution of better quality according to the evaluation function. In our case, we simply change the color of each edge belonging to any cycle to the opposite color, and calculate the objective function. This process stops either when a critical coloring is found (i.e., a solution with f(S) = 0), or no better solution exists among the neighboring solutions (in this later case, a local optimum is reached). It means that for current solution S_t the locally best solution $\overline{S_t}$ is generated (of course, if it exists). Next, for each edge e_i, we compare its colors in S_t and $\overline{S_t}$. If the edge does not change its color (say c_i), we reward the selected color c_i (called correct color) and update its probability vector p_i according to:

$$p_{ij}(t+1) = \begin{cases} \alpha + (1-\alpha)p_{ij}(t) & \text{if } j = u \\ 1 - (\alpha + (1-\alpha)p_{ij}(t)) & \text{otherwise.} \end{cases}$$

where α $(0 < \alpha < 1)$ is a reward factor. When edge e changes its color to the opposite color (say c_v, $v \neq u$), we penalize the discarded color c_v (called incorrect color) and update its probability vector p_i according to:

$$p_{ij}(t+1) = \begin{cases} (1-\beta)p_{ij}(t) & \text{if } j = v \\ 1 - ((1-\beta)p_{ij}(t)) & \text{otherwise.} \end{cases}$$

where β $(0 < \beta < 1)$ is a penalization factor. In the next step, a smoothing technique is applied on the probability vector of each edge $e_i \in E(G)$. For this we first calculate the value $p_{iw} = max\{p_{i1}, p_{i2}\}$. Then we check if $p_{iw} > p_0$ is true, where p_0 is a smoothing probability. If so, then we update probability vector p_i according to:

$$p_{ij}(t) = \begin{cases} \rho \cdot p_{ij}(t) & \text{if } j = w \\ 1 - \rho \cdot p_{ij}(t) & \text{otherwise.} \end{cases}$$

Once the probability of a color in a probability vector achieves a given threshold (i.e., p_0), it is reduced by multiplying a smoothing coefficient (i.e., $\rho < 1$) to forget some earlier decisions.

At each iteration of RLS, each edge e_i needs to select a color c_j from two available colors according to its probability vector p_i. As in [16], we adopted the hybrid selection strategy which combines randomness and greediness and is controlled by the noise probability ω. With a noise probability ω, random selection is applied; with probability $1 - \omega$, greedy selection is applied. The purpose of selecting a color with maximum probability (greedy selection) is to make an attempt to correctly select the color for an edge that is most often falsified at a local optimum. Selecting such a color for this edge may help the search to escape from the current trap. On the other hand, using the noise probability has the advantage of flexibility by switching back and forth between greediness and randomness. Also, this allows the algorithm to occasionally move away from being too greedy [16].

Now consider the case of Ramsey numbers of the type $R(C_m, K_{1,n})$. The procedure is basically the same as above, with the difference that the objective function is calculated in a different way. The objective function is defined as follows: $f(S)$ is equal to the sum of the number of edges that are in at least one forbidden cycle C_m colored with color 1 and the number of vertices of red (color 2) degree at least n, and the number of red edges in the red neighborhood of these vertices. This surprising last number came from observing the behavior of the machine learning algorithm considered above for two cycles. Probably the current geometric similarity speeds up the algorithm.

4 Computational Experiments

The basis of our software framework consisted of the package NetworkX, which includes a graph generator, tool to find all cycles of a given length and several other utilities for graph manipulation. All tests were carried out on a PC under 64-bit operating system Windows 11 Pro Intel(R) Core(TM) i5-1135G7 @ 2.40 GHz 2.42 GHz, RAM 16 GB compiled with aid of Python 3.9.

To obtain the desired colorings, each instance was solved 10 times independently with different random seeds. Each execution was terminated when a Ramsey coloring is found or the number of iterations without improvement reaches its maximum allowable value (500). As a result of the computational experiments, the values of all learning parameters were determined. Table 1 shows the descriptions and setting of the parameters used for our experiments. The considered colorings and the times of receiving the appropriate colorings are presented in Table 2. For small cases, it was from a dozen to several dozen iterations, for larger ones it did not exceed 700. The case of coloring $(C_8, C_8; 10)$ was considered in 3 versions: random coloring or random $(C_8, C_8; 8)$-coloring or random $(C_8, C_8; 9)$-coloring was given at the start, respectively. In addition to the colorings presented in Table 2, a number of calculations lasting several dozen hours were also performed in order to find the $(C_8, K_{1,10}; 15)$ – *coloring*. The calculations were started with various types of $(C_8; K_{1,10}; 14)$ – *colorings*. Each of the calculations stopped at some point and for at least 600 iterations the locally best solution was no longer corrected. Due to this fact, it can be assumed that $R(C_8, K_{1,10}) = 15$ and Conjecture 1 from [15] holds for $m = 8$ and $k = 10$.

Table 1. Parameters of Algorithm RLS.

Parameter	Description	Value
ω	noise probability	0.2
α	reward factor for correct color	0.1
β	penalization factor for incorrect color	0.5
ρ	smoothing coefficient	0.55
p_0	smoothing probability	0.955

Table 2. The times of determining the given colorings.

Coloring	Comp. time(s)	Coloring	Comp. time(s)
$(C_6, C_6; 7)$	<1 s	$(C_{10}, C_{10}; 10)$	<1 s
$(C_6, C_8; 8)$	<1 s	$(C_{10}, C_{10}; 11)$	159–473 s
$(C_6, C_8; 9)$	9–13 s	$(C_6, K_{1,6}; 10)$	<30 s
$(C_8, C_8; 10)$ ver 1	106–372 s	$(C_6, K_{1,7}; 10)$	<30 s
$(C_8, C_8; 10)$ ver 2	101–118 s	$(C_8, K_{1,5}; 9)$	29–552 s
$(C_8, C_8; 10)$ ver 3	12–85 s	$(C_8, K_{1,6}; 11)$	17–23 min
$(C_8, C_{10}; 10)$	45–180 s	$(C_8, K_{1,9}; 14)$	69–236 min
$(C_8, C_{10}; 11)$	6–78 min	$(C_8, K_{1,10}; 14)$	62–189 min

The frequent spread of computation time comes from the random, and therefore unpredictable, pre-coloring of the graph. In order to improve the speed of the algorithm, various pre-coloring can be applied and a number of useful Ramsey properties can be used. Examples of such actions are presented above. The structure of the graphs and the objective function used are also important. It is possible that other objective functions than those presented above in the article can be used. The same is true of machine learning parameters. It is likely that for other classes of graphs they should be adapted to them. The performed calculations indicate that within a dozen or so hours at most, we are able to determine the coloring for a graph with 15 vertices (i.e. having 105 edges), as long as the task is to color all the edges. Python was used for the implementation, but there are faster languages, such as ANSI C.

5 Conclusion

The adaptation of the method of Zhou, Hao and Duval [16] and the obtained results show that reinforcement learning can be considered as another and promising heuristic that can be used in determining Ramsey numbers. Appropriate selection of the graph structure, objective function, learning parameters, pre-coloring or even fixing the colors of certain edges can truly bring measurable results in determining unknown values of Ramsey numbers. Future work can be started by applying this method to the open problems contained in [10].

References

1. Bengio, Y., Lodi, A., Prouvost, A.: Machine learning for combinatorial optimization: a methodological tour d'horizon. Eur. J. Oper. Res. **290**(2), 405–421 (2021)
2. Brandt, S.: A sufficient condition for all short cycles. Discret. Appl. Math. **79**, 63–66 (1997)
3. Caccetta, L., Vijayan, K.: Maximal cycles in graphs. Discret. Math. **98**, 1–7 (1991)
4. Ceberio, J., Mendiburu, A., Lozano, J. A.: The Plackett-Luce ranking model on permutation-based optimization problems. In: Proceedings of the IEEE Congress on Evolutionary Computation (CEC), pp. 494–501. IEEE, Cancun (2013)
5. Faudree, R.J., Lesniak, L., Schiermeyer, I.: On the circumference of a graph and its complement. Discret. Math. **309**, 5891–5893 (2009)
6. Faudree, R.J., Schelp, R.H.: All Ramsey numbers for cycles in graphs. Discret. Math. **8**, 313–329 (1974)
7. Frederickson, G., Lynch, N.: Electing a leader in a synchronous ring. J. Assoc. Comput. Mach. **34**, 98–115 (1987)
8. Greenwood, R.E., Gleason, A.M.: Combinatorial relations and chromatic graphs. Can. J. Math. **7**, 1–7 (1955)
9. McKay, B.D., Radziszowski, S.P.: $R(4,5) = 25$. J. Graph Theory **19**, 309–322 (1995)
10. Radziszowski, S.: Small Ramsey numbers. Electron. J. Combin. (2021). Dynamic Survey 1, revision 16
11. Rosta, V.: Ramsey Theory Applications. Electron. J. Combin. (2004). Dynamic Survey 13
12. Rosta, V.: On a Ramsey type problem of J.A. Bondy and P. Erdoös, I & II. J. Combin. Theory Ser. B **15**, 94–120 (1973)
13. Snir, M.: On parallel searching. SIAM J. Comput. **14**, 688–708 (1985)
14. Yow, K.S., Luo, S.: Learning-based approaches for graph problems: a survey. https://arxiv.org/abs/2204.01057. Accessed 4 Jan 2022
15. Zhang, Y., Broersma, H., Chen, Y.: Narrowing down the gap on cycle-star Ramsey numbers. J. Combin. **7**(2–3), 481–493 (2016)
16. Zhou, Y., Hao, J.-K., Duval, B.: Reinforcement learning based local search for grouping problems: a case study on graph coloring. Expert Syst. Appl. **64**, 412–422 (2016)

Strengthening Structural Baselines for Graph Classification Using Local Topological Profile

Jakub Adamczyk(✉) and Wojciech Czech

AGH University of Science and Technology, Kraków, Poland
{jadamczy,czech}@agh.edu.pl

Abstract. We present the analysis of the topological graph descriptor Local Degree Profile (LDP), which forms a widely used structural baseline for graph classification. Our study focuses on model evaluation in the context of the recently developed fair evaluation framework, which defines rigorous routines for model selection and evaluation for graph classification, ensuring reproducibility and comparability of the results. Based on the obtained insights, we propose a new baseline algorithm called Local Topological Profile (LTP), which extends LDP by using additional centrality measures and local vertex descriptors. The new approach provides the results outperforming or very close to the latest GNNs for all datasets used. Specifically, state-of-the-art results were obtained for 4 out of 9 benchmark datasets. We also consider computational aspects of LDP-based feature extraction and model construction to propose practical improvements affecting execution speed and scalability. This allows for handling modern, large datasets and extends the portfolio of benchmarks used in graph representation learning. As the outcome of our work, we obtained LTP as a simple to understand, fast and scalable, still robust baseline, capable of outcompeting modern graph classification models such as Graph Isomorphism Network (GIN). We provide open-source implementation at GitHub.

Keywords: Graph representation learning · Graph classification · Fair evaluation · Graph descriptors · Baseline models

1 Introduction

Graph classification is an essential variant of supervised learning problems, gaining popularity in many scientific fields due to the growing volume of structured datasets, which encode pairwise relations between modeled objects of different types. The applications of graph classification algorithms range from cheminformatics [11], where high-level properties of molecules such as toxicity or mutagenicity are predicted, to sociometry [29], biology [32] and technology [16], tackling different classes of complex networks, whose non-trivial dynamics can be explained by learning structural patterns.

Graph classification poses the inherent problem of measuring the dissimilarity between objects which do not lie in metric space but have combinatorial nature.

J. Mikyška et al. (Eds.): ICCS 2023, LNCS 14076, pp. 597–611, 2023.
https://doi.org/10.1007/978-3-031-36027-5_47

This challenge is typically addressed by extracting isomorphism-invariant representations in the form of feature vectors [33] (also called graph embeddings, descriptors, fingerprints) or by constructing explicit pairwise similarity measures known as graph kernels [13]. More recently, the graph embedding problem was successfully reformulated within the framework of deep convolutional neural networks. Adopting the concept of convolution to vertex neighborhoods by introducing hierarchical iterative operators on multidimensional states of vertices allowed for building task-specific, low-dimensional representations for vertices, edges, and, after global pooling, the whole graph [27].

Baselines are the crucial elements of the fair comparison frameworks used in machine learning. As deep learning methods become increasingly powerful, the baseline algorithms used for their evaluation should also provide competitive results, forming good reference points for analyzing algorithms' performance. The recent development of Graph Neural Networks (GNNs), which automatically extract task-relevant features via deep learning, increased the number of attempts to solve various graph classification tasks [31,34]. Nevertheless, fair evaluation practices were frequently neglected in reported studies, and only recently the need for more rigorous model evaluation was highlighted [4]. This increased the demand for more powerful yet simple and fast baseline methods.

Motivated by recent findings regarding the discriminative power of Local Degree Profile (LDP) [2], which, together with SVM as the classification model, were proven to be competitive with the newest GNN models, we study the robustness and scalability of the new Local Topological Descriptor (LTP) built using histograms of specific descriptors representing vertex and edge structural features. We also use Random Forest classifier instead of SVM to reduce the sensitivity of baseline to hyperparameter tuning [19]. All experiments are performed in the regime of fair evaluation framework [4] to ensure replicability and to correct inaccuracies present in some earlier works (such as reporting accuracy on validation set). We also propose performance improvements in the implementation of LDP and LTP, resulting in better scalability and enabling the computation on large and dense social network benchmarks.

The key contributions of this work are the thorough analysis of the graph classification baseline composed of LDP and SVM, reporting limitations of this approach, the proposal of a new topological baseline utilizing Random Forest and experimental evaluation showing its robustness compared to the state-of-the-art. In addition, we present a modular software framework for LDP/LTP-based graph classification together with the associated open-source Python code shared on GitHub.

2 Related Works

In the early works related to graph comparison, the concept of graph edit distance (GED) was introduced [1]. It was based on calculating the optimal sequence of elementary operations (adding/removing vertex/edge) required to transform one graph into another. Computational complexity prevented GED algorithms from being widely used for larger graphs (>1000 vertices). Nevertheless, multiple successful attempts at classifying attributed graphs were reported based on benchmark dataset [21]. The IAM graph database [21] formed the first consistent framework for comparing the efficiency of graph classification algorithms based on GED or graph embedding methods.

Graph embedding forms the comprehensive field of methods and applications aimed at the generation of multidimensional graph invariants/descriptors, which can be recognized as graph feature extraction or feature engineering [33]. Graph descriptors can be assigned to the vertex, the edge, or the graph itself. Representing a graph as a vector enables using a multitude of unsupervised and supervised machine learning algorithms suitable for tabular data. The most popular graph invariants come from the field of complex networks and spectral graph theory. They are represented by several graph centrality measures, *clustering coefficient*, *efficiency* and permutation invariants constructed from the eigenpairs of Laplace matrix [26]. Generic-purpose vertex and edge descriptors can be aggregated to form a high-dimensional graph representation such as B-matrix [3] or, after including vertex attributes, even more expressive relation order histograms [15]. Graph descriptors can be also extracted by mining frequent patterns/subgraphs, resulting in the topological fingerprint suitable for structural pattern recognition but also querying graph databases [14]. This approach was further extended by introducing domain-specific representations such as molecular fingerprints [22], which are widely used in the prediction of biochemical properties. As graph embedding algorithm can be adjusted to the domain, graph type (e.g. attributed vs. non-attributed), or even available computing resources, the topological descriptors still represent promising are for graph feature engineering and, as presented in [2], can compete with state-of-the-art graph representation learning techniques.

The concept of graph substructure mining was generalized in the form of graph kernels [13], which assess the structural similarity between two graphs by pairwise comparison of their subcomponents. Most typically, the concept of R-convolution [10] is applied as a generic purpose convolution framework for discrete structures. One of the most interesting representatives of this group are Weisfeiler-Lehman kernels designed for subtrees, edges, shortest paths, and whole graphs [23]. They utilize the concept of Weisfeiler-Lehman test of isomorphism, which was also used in the construction of Graph Isomorphism Networks (GIN). Another family of graph kernels uses the concept of optimal assignment [6] to reduce the number of pairwise sub-kernel computations required to obtain a similarity value. In this work, we focus on explicit graph embedding and graph representation learning, skipping graph kernels as a less feasible solution for scalable graph classification.

In case of graph embedding and graph kernels, the domain-specific knowledge can be incorporated by designing specific substructure descriptors with the help of experts. Such an approach can be treated as an example of feature engineering. The different method, providing automatic, task-specific graph feature extraction, is represented by Graph Representation Learning models exemplified by

Graph Neural Networks (GNNs). They form a modern and extensively studied framework for graph classification, with dozens of available models and specific taxonomy [18]. Graph Isomorphism Networks (GIN) [28] were designed to be as powerful as Weisfeiler-Lehman isomorphism test in discriminating graphs. They were reported to achieve state-of-the-art results on graph classification benchmarks; therefore, they will be used as the main reference point for evaluating our method. GraphSAGE [9] is a general inductive framework for different convolutional GNNs, providing a new neighborhood sampling method, which ensures fixed-size aggregation sets to limit computational overhead related to processing hubs, present, e.g., in social networks. The aggregation function for node states can be treated as a hyperparameter and tuned on the validation set. The newer DiffPool model [30] generates hierarchical graph representations using a differentiable graph pooling layer. It assigns nodes to clusters to achieve coarse-graining of input for the next layer, which reduces computation time. Edge-Conditioned Convolution (ECC) model [24] introduces convolutions over local graph neighborhoods using edge labels and custom coarsening procedure subsampling vertices on pooling layers to reduce graph size. The high classification accuracy was achieved by ECC on molecular datasets. Also, Deep Graph Convolutional Neural Network (DGCNN) model [31] proposes custom localized graph convolution similar to spectral filters and related to Weisfeiler-Lehman subtree kernel. Additionally, the new SortPooling layer is introduced, enabling standard neural network training on graphs. All models mentioned in this paragraph will be used in the evaluation of the new LTP baseline.

3 Methods

Graph classification tasks can be organized into well-defined pipeline. First, the graph is typically represented as a sparse adjacency matrix. Optionally, vertex and edge feature matrices can be used, if they are available. The matrices provide the input to the feature extraction algorithm, which outputs a feature vector representing a graph embedding in a metric space. Next, a tabular classification algorithm is used. For explicit feature extraction methods, such as LDP or graph kernels, the graph invariants are calculated by an algorithm distinct from the classifier. This allows using arbitrary algorithms for both parts. For graph representation learning methods, such as GNNs, those representations are typically learned end-to-end using a differentiable framework and gradient-based optimization, with multilayer perceptron (MLP) as a classifier. This potentially increases flexibility and makes embeddings more task-related, but requires vastly more data and computational resources.

Local Degree Profile (LDP) [2] proposes a feature extraction based on vertex degree statistics, which are calculated for each node in the graph and then aggregated into the embedding vector. Following conventions from [2], we denote the graph as $G(V, E)$, where V is the set of vertices (or nodes) v and E is the set of edges e. Degrees of neighboring nodes form a multiset $DN(v) = \{\text{degree}(v) | (u, v) \in E\}$. For each node, we then calculate the following statistics: degree(v), min(DN), max(DN), mean(DN), std(DN). This way, for each node, we obtain the summary statistics of itself and its 1-hop neighborhood. They are then aggregated for the whole graph by calculating a histogram or empirical distribution function (EDF) over each feature. They are concatenated for all features, forming a final graph embedding. The number of bins used for aggregation and the choice between histogram and EDF are hyperparameters. There are also additional hyperparameters reflecting the method of preprocessing the features before the aggregation. Normalization can be applied: separately per graph, dividing the degrees by the highest value (this results in representing the feature value relative to the rest of the graph), or for the whole dataset, dividing by the highest degree in the dataset. Also, based on the observation that node degrees follow a power law for social networks, one can use log scale for aggregating features.

Any additional node- or edge-based structural descriptors can be included in the LDP in the form of histograms. The authors experiment with multiple ones: neighbors degree sum, lengths of shortest paths, closeness centrality, Fiedler vector and Ricci curvature. They remark that only shortest paths gave visible advantage, but could only be calculated in reasonable time on bioinformatics datasets, which have small molecular graphs. However, the gains using the shortest paths are consistent, about 2%, indicating that incorporating edge-based information can be beneficial. It should be noted that additional descriptors rapidly increase the dimensionality of the resulting embedding, which may result in degraded performance due to the curse of dimensionality, so only a limited number of well-chosen descriptors should be used.

Over the years, a vast number of node and edge descriptors were proposed. Among them, there are three commonly used groups, describing very different structural properties of the graph: centrality measures, link prediction indexes, and sparsification scores. Centrality indicates the importance or the influence of the node in the graph. Different measures focus on, e.g., how much information flows though a given node, or how many walks go through that node. They can also be defined for edges, indicating importance of connections in the graph. Link prediction indexes describe edges and are used to suggest the new edges to be added to the graph. They analyze the neighborhoods of the nodes, assigning high scores to the potential edges between nodes that share a large portion of their neighbors, or which have neighborhoods leading to shorter paths between them. Graph sparsification algorithms aim to eliminate edges, which are the least important for keeping the overall structure of the graph, especially in relation to hub nodes and communities. They aim to locally incorporate more global information about graph topology, assigning higher scores to more important edges.

During preliminary experiments, we surveyed descriptors representing each of those groups and available in the Networkit [25] library. While almost all gave promising results, we selected one from each group: edge betweenness centrality, Jaccard Index and Local Degree Score. The selection was based on intuitions that those particular descriptors will bring more edge-focused or more global information than only node degree statistics, enhancing the discriminative power of the baseline.

Edge betweenness centrality (EBC) [7] is a centrality measure based on shortest paths, which measures how much influence the edge has over the flow of information in the network. It is defined as the fraction of the shortest paths in the graph going through the edge $e = (u, v)$:

$$EBC(e) = \sum_{s,t;(s,t)\neq(u,v)} \frac{\sigma_{st}(e)}{\sigma_{st}}, \tag{1}$$

where σ_{st} is the total number of shortest paths between nodes s and t, and $\sigma_{st}(e)$ is the number of those paths that go through e. This can be computed using Floyd-Warshall algorithm, which will give infinity values for disconnected graphs; we simply omit them in our implementation. We selected this descriptor, since it is based on shortest paths, which gave the good results in [2], and also takes into consideration the cyclic structure of the graph, e.g., it distinguishes molecules with linear scaffolds vs those with more ring-like topology.

Jaccard Index (JI) [12] is a normalized overlap between node neighborhoods $N(u)$ and $N(v)$:

$$JI(u, v) = \frac{|N(u) \cap N(v)|}{|N(u) \cup N(v)|} \tag{2}$$

We calculate it for the existing edges $e = (u, v)$ in the graph, obtaining a descriptor of a 3-hop subgraph. This feature should better discriminate between graph with visible community substructures and those without such node clusters.

Local Degree Score (LDS) [17] was proposed to detect edges between hubs, i.e. nodes with locally high degree, and keep only those edges after graph sparsification. For each node v, the rank of its neighbor u, $\text{rank}(v, u)$ is the number of neighbors of v with degree lower than u. Note that this is asymmetrical, i.e. $\text{rank}(u, v) \neq \text{rank}(v, u)$. For each edge, the Local Degree Score is defined as:

$$LDS(e) = \max\left(1 - \frac{\ln \text{rank}(v, u)}{\ln \text{degree}(v)}, 1 - \frac{\ln \text{rank}(u, v)}{\ln \text{degree}(u)}\right), \tag{3}$$

which is simply taking the higher value from perspective of u or v, since either of them can be the hub node, giving a high LDS value. We selected this feature, since it can indicate the dispersion of nodes in the graph. If there are few edges with high LDS, it indicates that there are a few well separated clusters in the graph, centered around hub nodes, e.g., communities or well-connected functional groups in chemistry.

We propose to use the LDP descriptors together with the additional features described above, creating the **Local Topological Profile**. This method incorporates additional graph topology information in a local fashion, enhancing LDP with more discriminative power. Of course, this increases the computational cost, which we discuss below.

LDP authors remark that shortest path lengths are not used for social network datasets due to unreasonably long computing time. Their implementation, however, uses NetworkX [8] for computing graph descriptors, which is a Python library, performing sequential computations. Instead, we propose to use Networkit [25], a parallelized C++ library. This way, we are able to utilize modern CPUs with multiple cores and compute descriptors in parallel. We do not perform a timing comparison, as we also could not finish computation with NetworkX in any reasonable time. The experiments on subsets of datasets indicate that Networkit is at least a one or two degrees of magnitude faster even on much smaller, molecular graphs. For this reason, our whole implementation of descriptors computation is based on Networkit.

Classification algorithm applied to feature vectors has a direct influence on both accuracy and scalability. LDP used Support Vector Machine (SVM) with a Gaussian kernel, which is a powerful classifier traditionally used with graph kernels, since they work well with small datasets. However, they are not scalable, since kernel calculation alone takes $O(n^2)$ for n graphs in the dataset. Moreover, they are sensitive to hyperparameter choice [19], hence requiring extensive tuning to obtain good results. In addition, they are typically trained with the Sequential Minimal Optimization (SMO) algorithm, which is inherently sequential, not utilizing modern CPUs with multiple cores. Linear SVMs, while faster to compute, typically give worse results, which the authors of LDP also observe.

We propose to change SVM to a Random Forest (RF) classifier. It is bagging ensemble of decision trees, which means that each tree can be trained independently in parallel, increasing scalability. Decision tree induction is also very fast, relying on a greedy top-down algorithm. They typically give good results with default hyperparameters, requiring only a sufficiently high number of trees [20]. In preliminary study, we did not observe any significant effect of hyperparameter tuning, even with large hyperparameter grids, hence we skipped this step.

More importantly, in contrast to LDP [2], we use a fair evaluation protocol with test sets. We use the fair comparison procedure from [4], adapted in the following way. The datasets and their statistics are summarized in Table 1. We use the same test splits, and apply 10-fold CV for testing. However, since our baseline is fast to compute, we can afford to perform inner 5-fold CV for validation and hyperparameter tuning, instead of holdout. Following [4], we report mean and standard deviation of accuracy on test sets.
We tune the following hyperparameters for LDP (the same as the authors of [2]):

- number of bins: [30, 50, 70, 100]
- aggregation: [histogram, EDF]
- normalization: [none, graph, dataset]
- log scale: [false, true]

We tune the following hyperparameters for SVM (the same as authors of [2]):

- C (regularization): $[10^{-3}, 10^{-2}, ..., 10^2, 10^3]$
- γ (Gaussian kernel bandwidth): $[10^{-2}, 10^{-1}, ..., 10^1, 10^2]$

Table 1. Statistics of datasets used, following [4].

Dataset	# Graphs	Avg. # Nodes	Avg. # Edges	# Classes
DD	1178	284.32	715.66	2
NCI1	4110	29.87	32.30	2
PROTEINS	1113	39.06	72.82	2
ENZYMES	600	32.63	64.14	6
IMDB-B	1000	19.77	96.53	2
IMDB-M	1500	13.00	65.94	3
REDDIT-B	2000	429.63	497.75	2
REDDIT-5K	4999	508.82	594.87	5
COLLAB	5000	74.49	2457.78	3

For RF, we do not perform tuning, instead setting the following parameters (based on Scikit-learn defaults) for dataset with n samples and d features: 500 trees, minimizing *Gini impurity*, using $\sqrt{d}$ features, sampling n samples with replacement.

We use PyTorch Geometric [5] for data loading and computing node degree features, Networkit [25] for computing EBC, JI and LDS descriptors, and Scikit-learn to implement SVM and RF. We perform all experiments using 12th Gen Intel Core i7-12700KF 3.61 GHz processor with 32 GB of RAM. Feature extraction processes graphs sequentially, while feature calculation is done in parallel, using all available cores. We use all available cores for RF (`n_jobs=-1`) and for grid search. We performed experiments to answer the following questions:

1. Can we improve training speed and prediction accuracy, using RF instead of kernel SVM? If so, by how much?
2. Is tuning all LDP hyperparameters necessary? Can we eliminate some hyperparameters, or set reasonable defaults, in order to decrease tuning time?
3. Do additional descriptors increase prediction accuracy? Can we use all 3 additional descriptors to get the best average improvement?
4. What is the difference in training speed between the original LDP (using SVM and with hyperparameter tuning) and our proposed LTP (using RF and without tuning)?
5. How does LTP compare against baselines from [4] and GNNs?

4 Results and Discussion

The first experiment concerned comparison of LDP prediction accuracy when using RF (without tuning) instead of SVM (with tuning). We include both linear and kernel SVM results. We used a reasonable default values based on LDP paper [2]: 50 bins, histogram aggregation, normalization per graph, and linear scale. As shown in Table 2, RF always gave better results than both linear and kernel SVM. Interestingly, in some cases linear SVM outperformed kernel SVM, contradicting findings in [2]. The average improvement of RF over SVMs across all datasets is 3.4%, but can be as high as 7.3% on ENZYMES or 5.9% on DD. Additionally, the timings presented in Fig. 1 indicate that RF is about an order of magnitude faster than SVM, being the result of both a more parallelizable algorithm and no need for hyperparameter tuning. Based on this finding, we only use RF in further experiments.

Table 2. Classification accuracy on testing sets using LDP features and three analyzed models. The best result for each dataset is marked in bold.

Dataset	Linear SVM	Kernel SVM	RF
DD	68.2 ± 4.3	68.9 ± 4.0	**74.9 ± 3.4**
NCI1	65.8 ± 2.7	71.5 ± 2.8	**73.8 ± 2.0**
PROTEINS	66.6 ± 3.2	66.0 ± 3.3	**71.1 ± 3.1**
ENZYMES	25.7 ± 6.0	29.5 ± 5.2	**36.8 ± 5.8**
IMDB-B	60.2 ± 4.3	64.3 ± 3.6	**65.9 ± 2.2**
IMDB-M	39.6 ± 3.4	35.3 ± 2.6	**43.9 ± 2.4**
REDDIT-B	78.0 ± 2.7	88.1 ± 2.1	**89.6 ± 1.5**
REDDIT-5K	46.4 ± 2.0	52.3 ± 1.5	**52.8 ± 1.4**
COLLAB	68.0 ± 2.3	71.0 ± 2.1	**73.5 ± 2.2**

To verify the necessity of tuning LDP hyperparameters, we set the default values and vary a single hyperparameter at a time. We used 50 bins, histogram aggregation, normalization per graph, and linear scale. Due to space constraints, we do not include the whole results tables, but they are available on GitHub. For each hyperparameter, we calculate the number of times each value gave the best result. Additionally, for each value, we also calculate the absolute average difference between its result and the best result for a given hyperparameter on each dataset. The lower the absolute difference, the better, since it means that a given hyperparameter value, on average, gives the best results among all its possible values. results are presented in Table 3.

For the *number of bins*, we can clearly select 50 bins as the optimal value. While 30 bins gave the best results the same number of times, on average they performed worse compared to the optimal hyperparameter value. Similarly, for *normalization* it is evident that we do not need to perform any kind of normalization, since using no normalization obtained the best results on majority of datasets and on average. This is somewhat contrary to the results obtained in LDP paper [2], but it is apparently an advantage of RF, since it considers each feature separately, while calculating tree splits. For *aggregation* method, the results are very close, both for number of wins and average difference compared to the best result. In this case, the choice does not matter that much, and we choose the simpler histogram method. The linear *scale* obtained much better

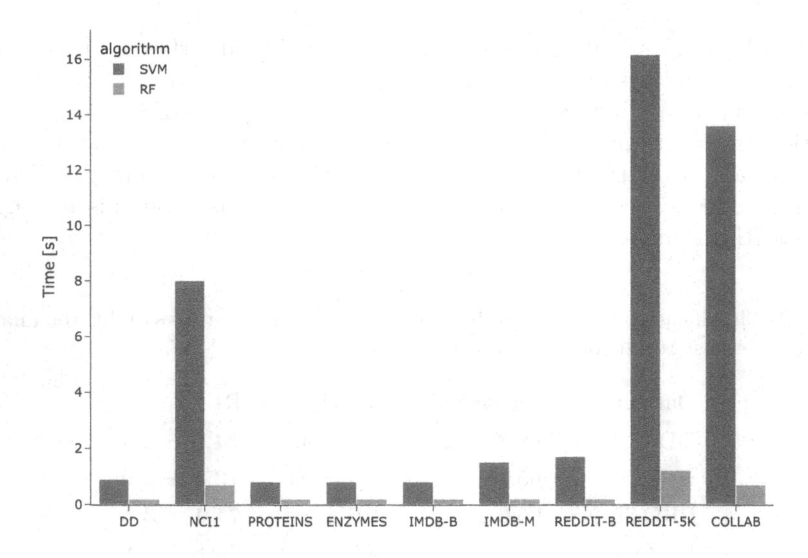

Fig. 1. Training time using LDP features: SVM vs RF clasifier.

results on average than the log scale, so the choice is obvious. Overall, this means that we can confidently recommend default values for all LDP hyperparameters, and tuning them is not particularly helpful. This dramatically decreases the computational cost, while having little effect on accuracy on average, which is a desirable tradeoff in a baseline method.

To assess whether additional descriptors increase accuracy of this method, we performed another set of experiments. We start with basic LDP, and add one additional descriptor at a time: lengths of shortest paths (SP), edge betweenness centrality (EBC), Jaccard Index (JI) and Local Degree Score (LDS). Finally, we check our proposed Local Topological Profile (LTP) method, combining LDP with EBC, JI and LDS descriptors. In all experiments, we keep the same hyperparameters: 50 bins, histogram aggregation, no normalization, and linear scale. As shown in Table 4, in every case the additional descriptors achieved the best result, while LTP was the best on 6 out of 9 datasets. On PROTEINS and IMDB-M it was the second best, being worse than the best by just 0.1% on the latter. It was also the third best on NCI1. Therefore, we can conclude that adding selected descriptors definitely increases the discriminatory power of this method. LTP is a robust method, performing the best on average, and using it eliminates the need to tune descriptor selection.

For performance analysis, we compare the original method from [2], i.e., LDP + SVM + hyperparameter tuning, with the proposed method, i.e. LTP + RF, without any tuning. For the former, we tune feature extraction hyperparameter and SVM hyperparameters separately, since grid search on combined parameters grids would result in approximately 20 times larger number of models to be trained for a single test fold, which is infeasible. We measure the time for the whole experimental procedure, i.e. feature extraction and classifier training

Table 3. Number of wins and absolute average difference between the result for a given hyperparameter value and the best result for any hyperparameter value. Higher number of wins is better, lower absolute average difference is better. For each hyperparameter, the value with the lowest absolute average difference has been marked in bold.

Hyperparameter	Value	# Wins	Abs. avg. difference compared to best
Number of bins	30	4	0.69%
	50	**4**	**0.22%**
	70	1	0.70%
	100	0	0.63%
Normalization	**None**	**5**	**0.23%**
	Graph	2	2.03%
	Dataset	2	0.39%
Aggregation	Histogram	5	0.71%
	EDF	**4**	**0.69%**
Scale	**Linear**	**4**	**0.35%**
	Log	5	0.64%

and tuning for all 10 test folds. This way, we take all characteristics of both approaches into consideration: LDP taking more time due to tuning, and LTP due to extracting more features. As shown in Fig. 2, our proposed LTP approach is vastly superior to LDP in terms of speed, being 1–3 orders of magnitude faster on all datasets. It should be noted that our LDP implementation is nevertheless much faster than the original one, since we use PyTorch Geometric to compute LDP features in parallel with optimized C++ subroutines. The original Python-based, sequential implementation in NetworkX would be additionally 1–2 orders of magnitude slower, based on preliminary experiments. Our method was also very fast on datasets with large number of large graphs (REDDIT datasets and COLLAB), which indicates good scalability. This is especially important considering that graph datasets are getting larger and baselines also have to scale well.

Lastly, we compare accuracy of LTP to GNNs from [4], based on the same fair evaluation framework (compatible settings for model selection and model evaluation). For social networks, we compare against stronger models, using node degree. The outcome is summarized in Table 5. For easier comparison, in Table 6 we also present the average rank of the model across all datasets, i.e. on average which place, from 1st to 8th, it took. Our LTP approach achieves state-of-the-art results on IMDB-B, IMDB-M, REDDIT-B and COLLAB, achieving as much as 3.8% higher accuracy on COLLAB than the previous best method, GIN. Note that our method makes use of graph topology exclusively, ignoring node and edge features. This explains why on bioinformatics datasets we did not get as good results. In fact, on DD, PROTEINS and ENZYMES the best result is achieved

Table 4. Classification accuracy of LDP with additional descriptors and LTP. The best result for each dataset is marked in bold.

Dataset	LDP	LDP + SP	LDP + EBC	LDP + JI	LDP + LDS	LTP
DD	76.0 ± 3.0	76.3 ± 2.8	77.0 ± 3.6	76.0 ± 3.4	75.8 ± 2.6	**77.1 ± 3.7**
NCI1	77.2 ± 1.5	76.1 ± 1.6	76.8 ± 1.7	76.6 ± 1.4	**77.4 ± 1.6**	77.0 ± 1.9
PROTEINS	70.6 ± 1.7	71.9 ± 2.2	**73.0 ± 3.2**	72.6 ± 3.2	71.4 ± 3.0	72.7 ± 4.2
ENZYMES	37.4 ± 4.0	37.2 ± 5.4	40.2 ± 6.5	40.0 ± 6.6	38.7 ± 5.6	**42.5 ± 4.1**
IMDB-B	71.3 ± 3.3	72.2 ± 4.0	72.9 ± 4.6	73.0 ± 4.3	74.2 ± 4.2	**74.5 ± 4.3**
IMDB-M	49.0 ± 4.4	49.2 ± 4.1	49.2 ± 5.0	49.3 ± 4.5	**50.1 ± 4.8**	50.0 ± 4.6
REDDIT-B	89.6 ± 1.2	90.5 ± 2.1	90.1 ± 1.7	89.6 ± 1.3	91.1 ± 1.1	**91.1 ± 1.0**
REDDIT-5K	51.9 ± 1.6	51.9 ± 1.9	51.7 ± 1.8	52.7 ± 2.0	53.1 ± 1.9	**53.3 ± 1.5**
COLLAB	75.7 ± 2.0	76.5 ± 2.2	76.8 ± 1.9	76.8 ± 1.8	78.7 ± 2.4	**79.4 ± 2.5**

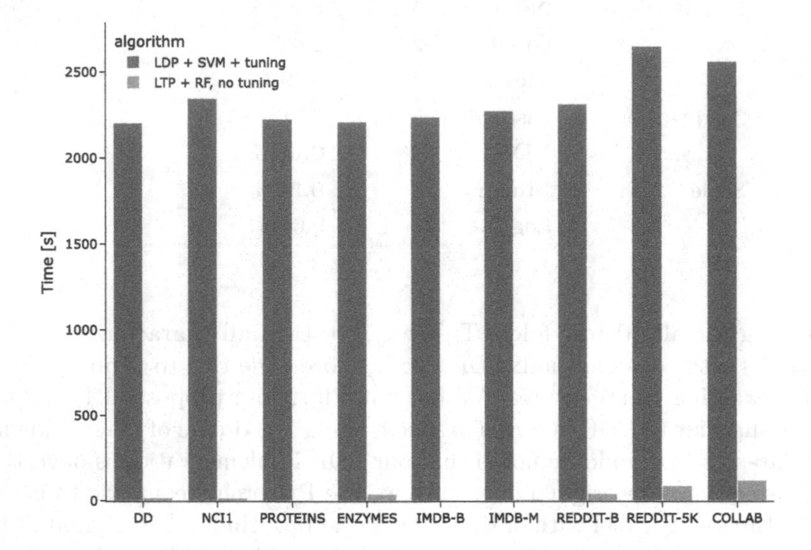

Fig. 2. Experiment time using LDP and LTP approaches.

by exclusively feature-based baseline from [4], which does not use graph topology at all. On average, LTP obtained the best rank among all models, beating even a theoretically very powerful GIN architecture. Additionally, all GNNs require GPUs and many hours of computation, while our method gives results in mere seconds.

Table 5. Comparison of accuracy with fair comparison results from [4]. Higher is better. Best result for each dataset has been marked in bold.

Dataset	Baseline[4]	DGCNN	DiffPool	ECC	GIN	GraphSAGE	LDP	LTP
DD	**78.4 ±4.5**	76.6 ±4.3	75.0 ± 3.5	72.6 ± 4.1	75.3 ± 2.9	72.9 ± 2.0	76.0 ± 3.0	77.1 ± 3.7
NCI1	69.8 ± 2.2	76.4 ± 1.7	76.9 ± 1.9	76.2 ± 1.4	**80.0 ± 1.4**	76.0 ± 1.8	77.2 ± 1.5	77.0 ± 1.9
PROTEINS	**75.8 ± 3.7**	72.9 ± 3.5	73.7 ± 3.5	72.3 ± 3.4	73.3 ± 4.0	73.0 ± 4.5	70.6 ± 1.7	72.7 ± 4.2
ENZYMES	**65.2 ± 6.4**	38.9 ± 5.7	59.5 ± 5.6	29.5 ± 8.2	59.6 ± 4.5	58.2 ± 6.0	37.4 ± 4.0	42.5 ± 4.1
IMDB-B	70.8 ± 5.0	69.2 ± 3.0	68.4 ± 3.3	67.7 ± 2.8	71.2 ± 3.9	68.8 ± 4.5	71.3 ± 3.3	**74.5 ± 4.3**
IMDB-M	49.1 ± 3.5	45.6 ± 3.4	45.6 ± 3.4	43.5 ± 3.1	48.5 ± 3.3	47.6 ± 3.5	49.0 ± 4.4	**50.0 ± 4.6**
REDDIT-B	82.2 ± 3.0	87.8 ± 2.5	89.1 ± 1.6	OOR	89.9 ± 1.9	84.3 ± 1.9	89.6 ± 1.2	**91.1 ± 1.0**
REDDIT-5K	52.2 ± 1.5	49.2 ± 1.2	53.8 ± 1.4	OOR	**56.1 ± 1.7**	50.0 ± 1.3	51.9 ± 1.6	53.3 ± 1.5
COLLAB	70.2 ± 1.5	71.2 ± 1.9	68.9 ± 2.0	OOR	75.6 ± 2.3	73.9 ± 1.7	75.7 ± 2.0	**79.4 ± 2.5**

Table 6. Comparison of average model ranks. The best result is marked in bold.

	Baseline [4]	DGCNN	DiffPool	ECC	GIN	GraphSAGE	LDP	LTP
Average rank	3.8	5.2	4.6	7.6	2.7	5.4	4	**2.6**

5 Conclusions

We presented the new structural baseline for graph classification called Local Topological Profile (LTP). The research questions addressing its efficiency and scalability in comparison to related LDP method and competitive GNN methods were studied in the experimental section, where we conclude that using the Random Forest classifier instead of SVM improved the accuracy and the speed of computation by a large margin and this observation applies to all datasets used. We note that tuning of feature extraction hyperparameters is not necessary therefore, we can use default values for all datasets, decreasing tuning time significantly. More importantly, we observe that introducing additional topological descriptors increases predictive accuracy in LTP method significantly. Using all three proposed descriptors (Edge Betweenness Centrality, Jaccard Index, Local Degree Score) gives very good prediction results across nine benchmark datasets, at the same time LTP is 2–3 orders of magnitude faster than the original LDP approach. Finally, we achieve state-of-the-art results on 4 out of 9 benchmark datasets, and in other cases get very strong accuracy, comparable to or even outcompeting modern GNNs, while using exclusively the graph topology. We share the software package with the research community, hoping that it can be useful in comparing results achieved by state-of-the-art graph classification models.

In our future work, we plan to extend the number of vertex/edge descriptors and enrich the expressive power of LTP towards more global features such as *eccentricity*. We also plan to merge LTP feature extraction and baselines from [4], to strengthen performance on more feature-focused bioinformatics datasets.

Acknowledgements. The research presented in this paper was financed from the funds assigned by Polish Ministry of Science and Higher Education to AGH University of Science and Technology. We would like to thank Alexandra Elbakyan for her work and support for accessibility of science.

References

1. Bunke, H.: On a relation between graph edit distance and maximum common subgraph. Pattern Recogn. Lett. **18**(8), 689–694 (1997)
2. Cai, C., Wang, Y.: A simple yet effective baseline for non-attributed graph classification. arXiv preprint arXiv:1811.03508 (2018)
3. Czech, W.: Invariants of distance k-graphs for graph embedding. Pattern Recogn. Lett. **33**(15), 1968–1979 (2012)
4. Errica, F., Podda, M., Bacciu, D., Micheli, A.: A fair comparison of graph neural networks for graph classification. arXiv preprint arXiv:1912.09893 (2019)
5. Fey, M., Lenssen, J.E.: Fast graph representation learning with PyTorch Geometric. In: ICLR Workshop on Representation Learning on Graphs and Manifolds (2019)
6. Fröhlich, H., Wegner, J.K., Sieker, F., Zell, A.: Optimal assignment kernels for attributed molecular graphs. In: Proceedings of the 22nd International Conference on Machine Learning, pp. 225–232 (2005)
7. Girvan, M., Newman, M.E.: Community structure in social and biological networks. Proc. Natl. Acad. Sci. **99**(12), 7821–7826 (2002)
8. Hagberg, A., Swart, P., S Chult, D.: Exploring network structure, dynamics, and function using networkx. Technical report, Los Alamos National Lab. (LANL), Los Alamos, NM (United States) (2008)
9. Hamilton, W., Ying, Z., Leskovec, J.: Inductive representation learning on large graphs. Adv. Neural Inf. Process. Syst. **30** (2017)
10. Haussler, D., et al.: Convolution kernels on discrete structures. Technical report, Citeseer (1999)
11. Hu, W., et al.: Open graph benchmark: datasets for machine learning on graphs. Adv. Neural Inf. Process. Syst. **33**, 22118–22133 (2020)
12. Jaccard, P.: The distribution of the flora in the alpine zone. 1. New Phytol. **11**(2), 37–50 (1912)
13. Kriege, N.M., Johansson, F.D., Morris, C.: A survey on graph kernels. Appl. Netw. Sci. **5**(1), 1–42 (2020)
14. Kuramochi, M., Karypis, G.: An efficient algorithm for discovering frequent subgraphs. IEEE Trans. Knowl. Data Eng. **16**(9), 1038–1051 (2004)
15. Łazarz, R., Idzik, M.: Relation order histograms as a network embedding tool. In: Paszynski, M., Kranzlmüller, D., Krzhizhanovskaya, V.V., Dongarra, J.J., Sloot, P.M.A. (eds.) ICCS 2021. LNCS, vol. 12743, pp. 224–237. Springer, Cham (2021). https://doi.org/10.1007/978-3-030-77964-1_18
16. Li, Y., Gu, C., Dullien, T., Vinyals, O., Kohli, P.: Graph matching networks for learning the similarity of graph structured objects. In: International Conference on Machine Learning, pp. 3835–3845. PMLR (2019)
17. Lindner, G., Staudt, C.L., Hamann, M., Meyerhenke, H., Wagner, D.: Structure-preserving sparsification of social networks. In: Proceedings of the 2015 IEEE/ACM International Conference on Advances in Social Networks Analysis and Mining 2015, pp. 448–454 (2015)
18. Liu, R., et al.: Taxonomy of benchmarks in graph representation learning. arXiv preprint arXiv:2206.07729 (2022)
19. Probst, P., Boulesteix, A.L., Bischl, B.: Tunability: importance of hyperparameters of machine learning algorithms. J. Mach. Learn. Res. **20**(1), 1934–1965 (2019)

20. Probst, P., Wright, M.N., Boulesteix, A.L.: Hyperparameters and tuning strategies for random forest. Wiley Interdiscip. Rev. Data Min. Knowl. Discov. **9**(3), e1301 (2019)
21. Riesen, K., Bunke, H.: IAM graph database repository for graph based pattern recognition and machine learning. In: da Vitoria Lobo, N., Kasparis, T., Roli, F., Kwok, J.T., Georgiopoulos, M., Anagnostopoulos, G.C., Loog, M. (eds.) SSPR /SPR 2008. LNCS, vol. 5342, pp. 287–297. Springer, Heidelberg (2008). https://doi.org/10.1007/978-3-540-89689-0_33
22. Rogers, D., Hahn, M.: Extended-connectivity fingerprints. J. Chem. Inf. Model. **50**(5), 742–754 (2010)
23. Shervashidze, N., Schweitzer, P., Van Leeuwen, E.J., Mehlhorn, K., Borgwardt, K.M.: Weisfeiler-lehman graph kernels. J. Mach. Learn. Res. **12**(9) (2011)
24. Simonovsky, M., Komodakis, N.: Dynamic edge-conditioned filters in convolutional neural networks on graphs. In: Proceedings of the IEEE Conference on Computer Vision and Pattern Recognition, pp. 3693–3702 (2017)
25. Staudt, C.L., Sazonovs, A., Meyerhenke, H.: NetworKit: a tool suite for large-scale complex network analysis. Netw. Sci. **4**(4), 508–530 (2016)
26. Wilson, R.C., Hancock, E.R., Luo, B.: Pattern vectors from algebraic graph theory. IEEE Trans. Pattern Anal. Mach. Intell. **27**(7), 1112–1124 (2005)
27. Wu, Z., Pan, S., Chen, F., Long, G., Zhang, C., Philip, S.Y.: A comprehensive survey on graph neural networks. IEEE Trans. Neural Netw. Learn. Syst. **32**(1), 4–24 (2020)
28. Xu, K., Hu, W., Leskovec, J., Jegelka, S.: How powerful are graph neural networks? arXiv preprint arXiv:1810.00826 (2018)
29. Yanardag, P., Vishwanathan, S.: Deep graph kernels. In: Proceedings of the 21th ACM SIGKDD International Conference on Knowledge Discovery and Data Mining, pp. 1365–1374 (2015)
30. Ying, Z., You, J., Morris, C., Ren, X., Hamilton, W., Leskovec, J.: Hierarchical graph representation learning with differentiable pooling. Adv. Neural Inf. Process. Syst. **31** (2018)
31. Zhang, M., Cui, Z., Neumann, M., Chen, Y.: An end-to-end deep learning architecture for graph classification. In: Proceedings of the AAAI Conference on Artificial Intelligence, vol. 32 (2018)
32. Zhang, X.M., Liang, L., Liu, L., Tang, M.J.: Graph neural networks and their current applications in bioinformatics. Front. Genet. **12**, 690049 (2021)
33. Zhang, Y.J., Yang, K.C., Radicchi, F.: Systematic comparison of graph embedding methods in practical tasks. Phys. Rev. E **104**(4), 044315 (2021)
34. Zhou, Y., Zheng, H., Huang, X., Hao, S., Li, D., Zhao, J.: Graph neural networks: taxonomy, advances, and trends. ACM Trans. Intell. Syst. Technol. (TIST) **13**(1), 1–54 (2022)

Heuristic Modularity Maximization Algorithms for Community Detection Rarely Return an Optimal Partition or Anything Similar

Samin Aref[1(✉)], Mahdi Mostajabdaveh[2], and Hriday Chheda[3]

[1] Department of Mechanical and Industrial Engineering, University of Toronto, Toronto M5S 3G8, Canada
aref@mie.utoronto.ca
[2] Huawei Technologies Canada, Burnaby V5C 6S7, Canada
[3] Department of Computer Science, University of Toronto, Toronto M5S 2E4, Canada

Abstract. Community detection is a fundamental problem in computational sciences with extensive applications in various fields. The most commonly used methods are the algorithms designed to maximize modularity over different partitions of the network nodes. Using 80 real and random networks from a wide range of contexts, we investigate the extent to which current heuristic modularity maximization algorithms succeed in returning maximum-modularity (optimal) partitions. We evaluate (1) the ratio of the algorithms' output modularity to the maximum modularity for each input graph, and (2) the maximum similarity between their output partition and any optimal partition of that graph. We compare eight existing heuristic algorithms against an exact integer programming method that globally maximizes modularity. The average modularity-based heuristic algorithm returns optimal partitions for only 19.4% of the 80 graphs considered. Additionally, results on adjusted mutual information reveal substantial dissimilarity between the sub-optimal partitions and any optimal partition of the networks in our experiments. More importantly, our results show that near-optimal partitions are often disproportionately dissimilar to any optimal partition. Taken together, our analysis points to a crucial limitation of commonly used modularity-based heuristics for discovering communities: they rarely produce an optimal partition or a partition resembling an optimal partition. If modularity is to be used for detecting communities, exact or approximate optimization algorithms are recommendable for a more methodologically sound usage of modularity within its applicability limits.

Keywords: Community detection · Network science · Modularity maximization · Integer programming · Graph optimization

1 Introduction

Community detection (CD), the process of inductively identifying communities within a network, is a core problem in computational sciences, particularly,

J. Mikyška et al. (Eds.): ICCS 2023, LNCS 10476, pp. 612–626, 2023.
https://doi.org/10.1007/978-3-031-36027-5_48

in physics, computer science, biology, and computational social science [19,50]. Among common approaches for CD are the algorithms which are designed to maximize a utility function, modularity [37], across all possible ways that the nodes of the input network can be partitioned into communities. Modularity measures the fraction of edges within communities minus the expected fraction if the edges were distributed randomly; with the random distribution of the edges being a null model that preserves the node degrees. Despite their name and design philosophy, current modularity maximization algorithms, which are used by no less than tens of thousands of peer-reviewed studies [28], are not guaranteed to maximize modularity [24,35,38]. This has led to uncertainty [20,24] in the extent to which they succeed in returning a maximum-modularity (optimal) partition or something similar.

Modularity is among the first objective functions proposed for optimization-based detection of communities [18,37]. Several limitations [16,18,21,41] of modularity including the resolution limit [17] have led researchers to develop alternative CD methods using stochastic block modeling [23,32,40,45], information theoretic approaches [43,44], and alternative objective functions [2,34,36,47]. Modularity-based algorithms are the most commonly used method for CD [19,46]. Despite the widespread adoption of modularity-based heuristics, there is uncertainty [20,24] in their success in maximizing modularity. This study aims to address this uncertainty by quantifying the extent to which eight commonly used heuristics [7,8,13,30,31,46,48,50] succeed in returning an optimal partition or a partition resembling an optimal partition. After describing the methods and materials, we present the main results followed by a discussion of the methodological ramifications and future directions.

2 Methods and Materials

This study aims to investigate the extent to which eight commonly used heuristic modularity maximization algorithms [7,8,13,30,31,46,48,50] succeed in returning an optimal partition or a partition similar to an optimal partition. To achieve this objective, we quantify the proximity of their results to the globally optimal partition(s), which we obtain using an exact Integer Programming (IP) model for maximizing modularity [1,9,15]. We do not claim that maximum-modularity partitions represent best partitions. Throughout the paper, we use the terms network and graph interchangeably.

2.1 Modularity

Consider the simple graph $G = (V, E)$ with $|V| = n$ nodes, $|E| = m$ edges, adjacency matrix entries a_{ij}, and a partition $X = \{V_1, V_2, \ldots, V_k\}$ of the node set V into k communities. The modularity function $Q_{(G,X)}$ is computed [18,37] according to Eq. (1)

$$Q_{(G,X)} = \frac{1}{2m} \sum_{(i,j)\in V^2} \left(a_{ij} - \gamma \frac{d_i d_j}{2m} \right) \delta(i,j) \tag{1}$$

where d_i represents the degree of node i, γ is the resolution parameter[1], and $\delta(i,j)$ is 1 if nodes i and j are in the same community otherwise 0. The term associated with each pair of nodes (i,j) is alternatively represented as $b_{ij} = a_{ij} - \gamma\frac{d_i d_j}{2m}$ and referred to as the modularity matrix entry for (i,j).

2.2 Modularity Maximization

The modularity maximization problem for input graph $G = (V, E)$ involves finding a partition X^* whose associated $Q_{(G,X^*)}$ is globally maximum over all possible partitions of the node set V.

2.3 Sparse IP Formulation of Modularity Maximization

Consider the simple graph $G = (V, E)$ with modularity matrix entries b_{ij}, obtained using the resolution parameter γ. We use the binary decision variable x_{ij} for each pair of distinct nodes $(i,j), i < j$. Their community membership is either the same (represented by $x_{ij} = 0$) or different (represented by $x_{ij} = 1$). Accordingly, the problem of maximizing the modularity of input graph G can be formulated as an IP model [15] as in Eq. (2).

$$\begin{aligned} \max_{x_{ij}} Q &= \frac{1}{2m}\left(\sum_{(i,j)\in V^2, i<j} 2b_{ij}(1-x_{ij}) + \sum_{(i,i)\in V^2} b_{ii}\right) \\ \text{s.t.}\quad & x_{ik} + x_{jk} \geq x_{ij} \quad \forall (i,j)\in V^2, i<j, k\in K(i,j) \\ & x_{ij} \in \{0,1\} \quad \forall (i,j)\in V^2, i<j \end{aligned} \tag{2}$$

In Eq. (2), the optimal objective function value equals the maximum modularity for the input graph G. An optimal community assignment is characterized by the optimal values of the x_{ij} variables. $K(i,j)$ indicates a minimum-cardinality separating set [15] for the nodes i, j. Using $K(i,j)$ in the IP model of this problem leads to a more efficient formulation with $\mathcal{O}(n^2)$ constraints [15] instead of $\mathcal{O}(n^3)$ constraints as in earlier IP formulations of the problem [1,9]. Solving this optimization problem is NP-complete [9,35]. We use the *Gurobi* solver (version 10.0) [22] to solve it for the small and mid-sized instances as outlined in Subsect. 2.6.

2.4 Reviewing Eight Heuristic Modularity Maximization Algorithms

We evaluate eight modularity maximization heuristics known as Clauset-Newman-Moore (CNM) [13], Louvain [7], Leicht-Newman (LN) [30], Combo [46], Belief [50], Paris [8], Leiden [48], and EdMot-Louvain [31]. We have used the Python implementations of these eight algorithms which are accessible in the Community Discovery library (*CDlib*) version 0.2.6 [42].

[1] Without loss of generality, we set $\gamma = 1$ for all the analysis in this paper.

We briefly describe how these eight algorithms use modularity to discover communities. The CNM algorithm initializes each node as a community by itself. It then follows a greedy scheme of merging two communities that contribute the maximum positive value to modularity [13]. The Louvain algorithm involves two sets of iterative steps: (1) locally moving nodes for increasing modularity and (2) aggregating the communities from the first step [7]. Despite Louvain being the most commonly used modularity-based algorithm [28], it may sometimes lead to disconnected components in the same community [48]. The LN algorithm uses spectral optimization to maximize modularity which also supports directed graphs [30]. The Combo algorithm is a general optimization-based CD method which supports modularity maximization among other tasks. It involves two sets of iterative steps: (1) finding the best merger, split, or recombination of communities to maximize modularity and (2) performing a series of Kernighan-Lin bisections [26] on the communities as long as they increase modularity [46]. The Belief algorithm seeks the consensus of different high-modularity partitions through a message-passing algorithm [50] motivated by the premise that maximizing modularity can lead to many poorly correlated competing partitions. The Paris algorithm is suggested to be a modularity-maximization scheme with a sliding resolution [8]; that is, an algorithm capable of capturing the multi-scale community structure of real networks without a resolution parameter. It generates a hierarchical community structure based on a simple distance between communities using a nearest-neighbour chain [8]. The Leiden algorithm attempts to resolve a defect of the Louvain algorithm in returning badly connected communities. It is suggested to guarantee well-connected communities in which all subsets of all communities are locally optimally assigned [48]. The EdMot-Louvain algorithm (EdMot for short) is developed to overcome the hypergraph fragmentation issue observed in previous motif-based CD methods [31]. It first creates the graph of higher-order motifs (small dense subgraph patterns) and then partitions it using the Louvain method to heuristically maximize modularity using higher-order motifs [31].

To evaluate these eight modularity-based algorithms in maximizing modularity, we quantify (1) the ratio of their output modularity to the maximum modularity for each input graph and (2) the maximum similarity between their output partition and any optimal partition of that graph. We obtain optimal partitions by solving the IP model in Eq. (2) using the Gurobi solver (version 10.0) with a termination criterion ensuring global optimality [22].

2.5 Measures for Evaluating Heuristic Algorithms

For a quantitative measure of proximity to global optimality, we define and use the *Global Optimality Percentage* (GOP) as the fraction of the modularity returned by a heuristic method for a network divided by the globally maximum modularity for that network (obtained by solving the IP model in Eq. (2)). In all cases where the modularity returned by a heuristic method equals the maximum modularity for the input graph, we set GOP $= 1$. In cases where a heuristic algorithm returns a partition with a negative modularity value, we set

GOP = 0 to facilitate easier interpretation of proximity to optimality based on non-negative GOP values.

We also use a quantitative measure for the similarity of a partition to an optimal partition. Normalized Adjusted Mutual Information (AMI) [49] is a measure of similarity between two partitions of the same network. Unlike normalized mutual information [49], AMI adjusts the measurement based on the similarity that two partitions may have by pure chance. AMI for a pair of identical partitions (or permutations of the same partition) equals 1. For two different partitions, however, AMI takes a smaller value (including 0 or negative values close to 0 for two extremely dissimilar partitions).

2.6 Data and Resources

For our computational experiments, we include 60 real networks[2] with no more than 2812 edges as well as 10 Erdős-Rényi graphs and 10 Barabási-Albert graphs with 125–153 edges. These instance sizes were chosen to ensure all algorithms terminate within a reasonable time. The computational experiments were implemented in Python 3.9 using a notebook computer with an Intel Core i7-11800H @ 2.30 GHz CPU and 64 GB of RAM running Windows 10.

3 Results

We present the main results from our experiments in the following four subsections. In Subsect. 3.1, we compare partitions from different algorithms on a single network. In Subsect. 3.2, we examine the multiplicity of optimal partitions and investigate the similarity between multiple optimal partitions of the same networks. In Subsect. 3.3, we evaluate the effectiveness of the heuristic algorithms on 80 networks by measuring the distance of sub-optimal partitions from an optimal partition. Finally, in Subsect. 3.4, we investigate the success rate of the heuristic algorithms in finding an optimal partition.

3.1 Comparing Partitions from Different Algorithms on One Network

Figure 1 shows one graph and its nine partitions returned by nine CD methods. This graph[3] represents an anonymized Facebook *ego network*[4]. Nodes represent Facebook users, and an edge exists between any pair of users who were friends on Facebook in April 2014 [33]. Communities are shown using node colors.

Panel 1a of Fig. 1 shows an optimal partition obtained by solving the IP model in Eq. (2) for the network facebook_friends. It involves $k = 28$ communities, and a maximum modularity value of $Q^* = 0.7157714$. The partitions from the eight

[2] All networks are accessible from the Netzschleuder with the details in the Appendix.
[3] facebook_friends network [33] from the Netzschleuder repository.
[4] A network of one person's social ties to other persons and their ties to each other.

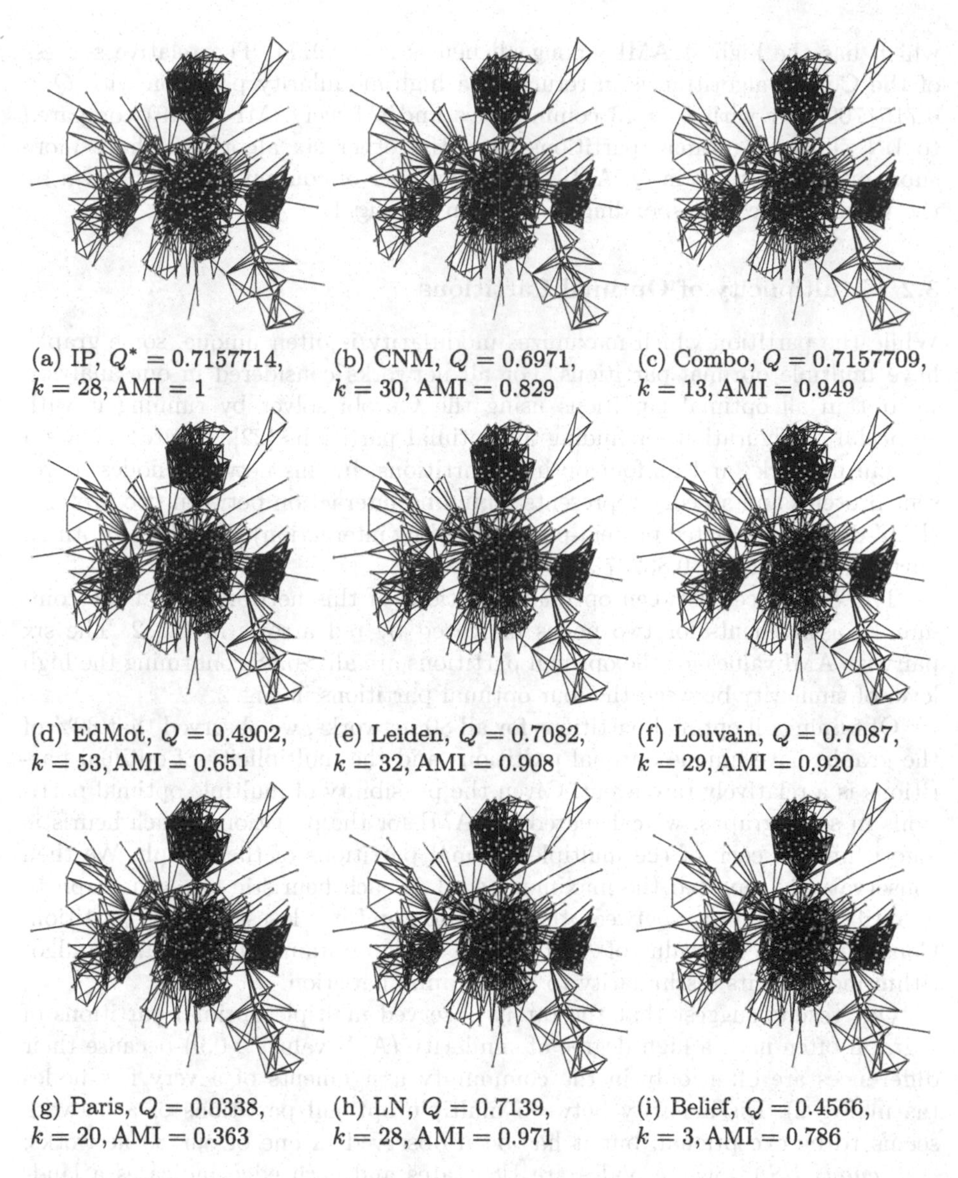

(a) IP, $Q^* = 0.7157714$, $k = 28, \text{AMI} = 1$
(b) CNM, $Q = 0.6971$, $k = 30, \text{AMI} = 0.829$
(c) Combo, $Q = 0.7157709$, $k = 13, \text{AMI} = 0.949$
(d) EdMot, $Q = 0.4902$, $k = 53, \text{AMI} = 0.651$
(e) Leiden, $Q = 0.7082$, $k = 32, \text{AMI} = 0.908$
(f) Louvain, $Q = 0.7087$, $k = 29, \text{AMI} = 0.920$
(g) Paris, $Q = 0.0338$, $k = 20, \text{AMI} = 0.363$
(h) LN, $Q = 0.7139$, $k = 28, \text{AMI} = 0.971$
(i) Belief, $Q = 0.4566$, $k = 3, \text{AMI} = 0.786$

Fig. 1. Modularity maximization for one network using nine methods leading to one optimal partition (panel a) and eight sub-optimal partitions (panels b-i) with different Q, k, and AMI values. (Magnify the high-resolution color figure on screen for more details.)

heuristic modularity maximization algorithms are all sub-optimal as depicted in panels 1b–1i of Fig. 1. Compared to other algorithms, the two algorithms Combo and LN have more success in achieving proximity to an optimal partition. LN returns a partition with $k = 28$ communities and a modularity of $Q = 0.7139$

which has the highest AMI among all heuristics (0.971). The relative success of the Combo algorithm is in returning a high-modularity partition with $Q = 0.7157709$, but with $k = 13$ communities and a lower AMI (0.949) compared to LN. The sub-optimal partitions from the other six algorithms have more substantial variations in Q, AMI, and k (number of communities) as shown by the values in the corresponding subcaptions in Fig. 1.

3.2 Multiplicity of Optimal Partitions

While the partition which maximizes modularity is often unique, some graphs have multiple optimal partitions. For all networks considered in our analysis, we obtain all optimal partitions using the Gurobi solver by running it with a special configuration for finding all optimal partitions [22]. Figure 2 shows a protein network[5] and its four optimal partitions. In this network, nodes represent proteins and an edge represents a binding interaction between two proteins (PDZ-domain-mediated protein-protein binding interaction) [6]. All four optimal partitions have $Q^* = 0.80267$ and $k = 29$.

The differences between optimal partitions of this network are in the community assignments for two nodes indicated by red arrows in Fig. 2. The six pairwise AMI values for the optimal partitions are all >0.98 confirming the high level of similarity between the four optimal partitions in Fig. 2.

Obtaining all optimal partitions for all 80 networks, we observed that 89% of the graphs have unique optimal partitions and the multiplicity of optimal partitions is a relatively rare event. Given the possibility of multiple optimal partitions in some graphs, we calculated the AMI for the partition of each heuristic algorithm and each of the multiple optimal partitions of that graph. We then conservatively reported the maximum AMI of each heuristic for each graph to quantify the similarity between that partition and its closest optimal partition. Consequently, a low value of AMI for a partition obtained by a heuristic algorithm indicates its dissimilarity to any optimal partition.

Our results suggest that the rarely observed multiple optimal partitions of a graph often have a high degree of similarity (AMI values >0.9) because their differences are often only in the community assignments of a very few nodes (as in Fig. 2). Dissimilarity between multiple optimal partitions of a network seems to be exceptional, but it has been observed in one of our 80 networks: *contiguous USA*[6], where nodes are US states and each edge indicates a land-based border between two states. The AMI of the two optimal partitions for this network is exceptionally low (0.34). Upon further investigation, we observed that one optimal partition combines five communities of the other optimal partition together. This makes the two partitions related in terms of belonging to a clustering hierarchy, while they are not similar according to an AMI definition of partition similarity. These exceptional cases are possible due to the mathematical symmetries resulted from the value of γ used in Eq. (1) for defining

[5] interactome_pdz network [6] from the Netzschleuder repository.
[6] contiguous_usa network [27] from the Netzschleuder repository.

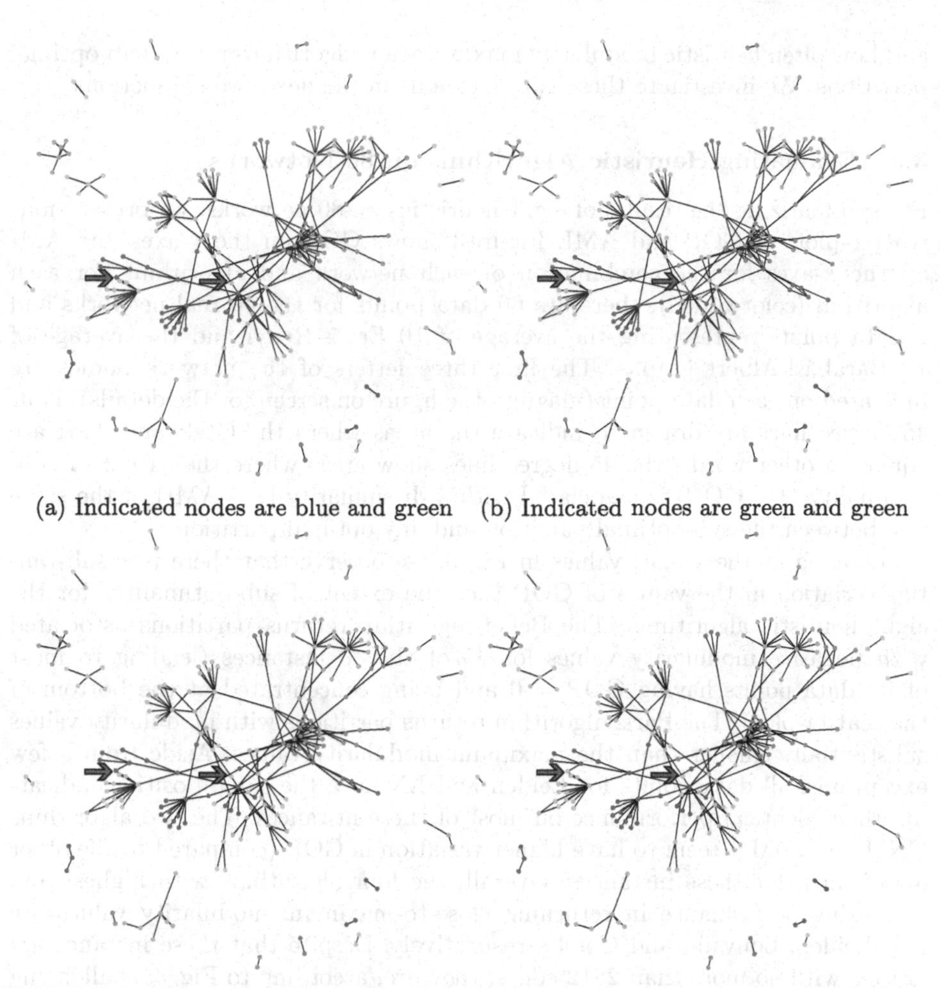

(a) Indicated nodes are blue and green
(b) Indicated nodes are green and green
(c) Indicated nodes are blue and orange
(d) Indicated nodes are green and orange

Fig. 2. A protein network and its four optimal partitions (panels a-d). The red arrows show the differences between optimal partitions. (Magnify the high-resolution color figure on screen for more details.) (Color figure online)

modularity. Our results suggest that there is usually a distinct uniqueness to an optimal partition (or a group of similar optimal partitions) for a given network in comparison to sub-optimal partitions. This new perspective is contrary to the premise that maximizing modularity leads to many competing partitions with almost the same modularity [50] and no clear way of selecting between them [41]. It is the failure to actually maximize modularity that may lead to many poorly correlated competing partitions with unknown distances from the desired objective (both in modularity and in partition similarity). What remains to be analyzed is how different sub-optimal partitions are from an optimal partition and how often heuristic modularity maximization algorithms return sub-optimal partitions. We investigate these two questions in the next two subsections.

3.3 Evaluating Heuristic Algorithms on 80 Networks

For summarizing the results of eight heuristics on 80 networks, we present four scatter-plots of GOP and AMI. Figure 3 shows GOP on the y-axes and AMI on the x-axes for the combination of each network and algorithm. For each algorithm (color-coded), there are 60 data points for the 60 real networks and 2 data points representing the average of 10 Erdős-Rényi and the average of 10 Barabási-Albert graphs. The first three letters of the network names are indicated on each data point (magnify the figure on screen for the details). Four 45-degree lines are drawn to indicate the areas where the GOP and AMI are equal. In other words, the 45-degree lines show areas where the extent of sub-optimality (1 − GOP) is associated with a dissimilarity (1 − AMI) of the same size between the sub-optimal partition and any optimal partition.

Looking at the y-axes values in Fig. 3, we observe that there is a substantial variation in the values of GOP (i.e. the extent of sub-optimality) for the eight heuristic algorithms. The Belief algorithm returns partitions associated with negative modularity values for 45 of the 80 instances (leading to most of its data points having GOP = 0 and being concentrated at the bottom of the scatter-plot). The Paris algorithm returns partitions with modularity values substantially smaller than the maximum modularity values. Aside from a few exceptions, all data points for Leiden and LN have the same position indicating their identical performance on most of these instances. The two algorithms CNM and EdMot seem to have higher variation in GOP (compared to the other algorithms) for these instances. Overall, the four algorithms with highest and increasing performance in returning close-to-maximum modularity values are LN, Leiden, Louvain, and Combo respectively. Despite that these instance are graphs with no more than 2812 edges, they are, according to Fig. 3, challenging instances for these heuristic algorithms to optimize. Given that modularity maximization is an NP-complete problem [9,35], one can argue that the performance of these heuristic methods in term of proximity to an optimal partition does not improve for larger networks.

The x-axes values in Fig. 3 show considerable dissimilarity between the sub-optimal partitions and an optimal partition for these 80 instances. Except for the Combo algorithm, a large number of the sub-optimal partitions obtained by these heuristic algorithms have AMI values smaller than 0.6. This indicates that their sub-optimal partitions are substantially different from any optimal partition. Even for data points concentrated at the top of the scatter-plots which have $0.95 < \text{GOP} < 1$, we see AMI values substantially smaller than 1. Compared to the other seven heuristics, Combo appears to consistently return partitions with large AMIs on a larger number of these 80 instances.

Focusing on the position of data points, we observe that they are mostly located above their corresponding 45-degree line. This indicates that sub-optimal partitions tend to be disproportionately dissimilar to any optimal partition (as

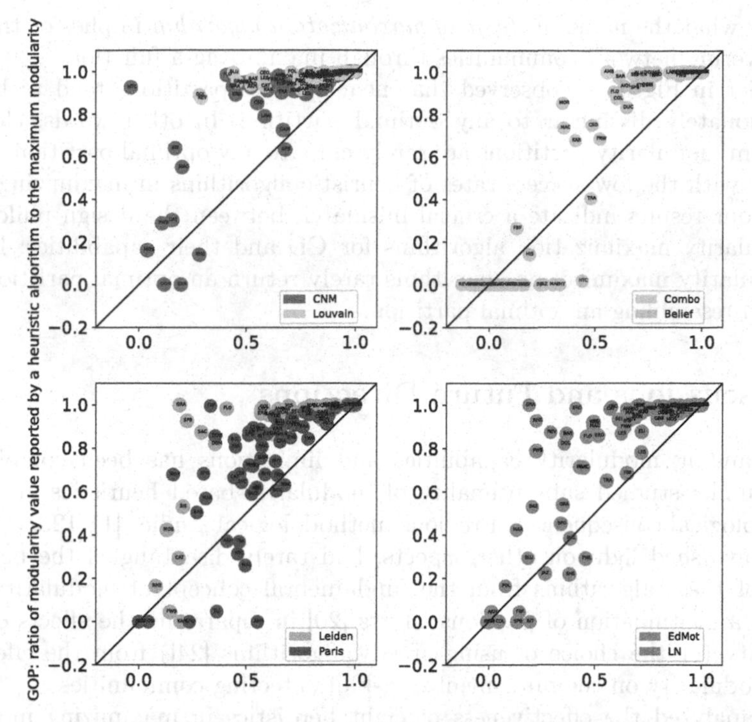

Fig. 3. Global optimality percentage and normalized adjusted mutual information measured for eight modularity maximization heuristics in comparison with (all) globally optimal partitions. (Magnify the high-resolution figure on screen for more details.)

foreshadowed in [14]). This result goes against the naive viewpoint that close-to-maximum modularity partitions are also close to an optimal partition. Our results are aligned with previous concerns that these heuristics may result in degenerate solutions far from the underlying community structure [20] and they have a high risk of algorithmic failure [24].

3.4 Success Rate of Heuristic Algorithms in Maximizing Modularity

Our GOP results for the eight heuristic algorithms allow us to answer a fundamental question about the heuristic modularity maximization algorithms: how often each algorithm returns an optimal (a maximum-modularity) partition? We report the fraction of networks (out of 80) for which a given algorithm returns an optimal partition. Combo [46] has the highest success rate, returning an optimal partition for 55% of the networks. LN [30] and Leiden [48] maximize modularity for 36.2% of the networks considered. Louvain [7] has a success rate of 18.7%. The algorithms CNM [13], EdMot [31], Paris [8], and Belief [50] have success rates of 5%, 2.5%, 1.2%, and 0% respectively. These are arguably low success rates for what the name *modularity maximization algorithm* implies or the idea of discovering network communities through maximizing a function.

Earlier in Fig. 3, we observed that near-optimal partitions tend to be disproportionately dissimilar to any optimal partition. In other words, close-to-maximum modularity partitions are rarely close to any optimal partition. Taken together with the low success rates of heuristic algorithms in maximizing modularity, our results indicate a crucial mismatch between the design philosophy of modularity maximization algorithms for CD and their capabilities: heuristic modularity maximization algorithms rarely return an optimal partition or a partition resembling an optimal partition.

4 Discussions and Future Directions

Understanding modularity capabilities and limitations has been complicated by the under-studied sub-optimality of modularity-based heuristics and their methodological consequences. Previous methodological studies [11,12,29,36,41], which have shed light on other aspects, had rarely disentangled the heuristic aspect of these algorithms from the fundamental concept of modularity. Our study is a continuation of previous efforts [20] in separating the effects of sub-optimality (or the choice of using greedy algorithms [24]) from the effects of using modularity on the fundamental task of detecting communities.

We analyzed the effectiveness of eight heuristics in maximizing modularity. While our findings are limited to a few algorithms, their combined usage by tens of thousands of peer-reviewed studies [28] motivates the importance of this assessment. Most heuristic algorithms for modularity maximization tend to scale well for large networks [51]. They are widely used not only because of their scalability or ease of implementation [24], but also because their high risk of algorithmic failure is not well understood [24]. The scalability of these heuristics comes at a cost: their partitions have no guarantee of proximity to an optimal partition [20] and, as our results showed, they rarely return an optimal partition. Moreover, we showed that their sub-optimal partitions tend to be disproportionately dissimilar to any optimal partition.

Neither using modularity nor succeeding in maximizing it is required for CD at the big-picture level. A recent study suggests modularity maximization is the most problematic CD method and considers it harmful [41]. Another study shows that, given computational feasibility, exact maximization of multiresolution modularity outperforms other CD methods in accurate and stable retrieval of planted communities [4] suggesting the relevance of modularity for CD. For some applications and contexts, *general* CD algorithms [39] which scale to large instance sizes are needed. However, for a "narrow set of tasks" [39, pp.7], involving small and mid-sized networks, *specialized* algorithms which outperform general algorithms are useful.

Our findings suggest that if modularity is to be used for detecting communities, developing approximation [10,14,25] and exact [3,4] algorithms are recommendable for a more methodologically sound usage of modularity within its applicability limits. Exact algorithms can also reveal the formal guarantees of performance [19] for accurate modularity-based algorithms.

A promising path forward could be using the advances in integer programming to develop a specialized accurate algorithm for solving the modularity maximization IP models [1,9,15] for networks of practical relevance within the limits of computational feasibility. New heuristic and approximation algorithms that strike a balance between accurate computations and scalability may also be useful particularly for large-scale networks.

Acknowledgements. We are thankful to the three anonymous referees and Ly Dinh for their helpful comments. We acknowledge Zachary P. Neal for pointing us to this problem and Santo Fortunato for the encouraging correspondence. Accessing CDlib, Netzschleuder, and ICON has been particularly helpful in this study for which we thank Giulio Rossetti, Tiago P. Peixoto, Aaron Clauset, and everyone contributing to these open science initiatives. This study has been supported by the Data Sciences Institute at the University of Toronto.

Author contributions. Conceptualization (SA); data curation (SA, HC); formal analysis (SA, MM); funding acquisition (SA); investigation (SA); methodology (SA,MM); project administration (SA); resources (SA, MM); software (SA, HC, MM); supervision (SA, MM); validation (SA, MM); visualization (SA, MM); writing - original draft preparation (SA); writing - review & editing (SA, MM).

Appendix

The data on 20 random graphs used in this study are available in a *FigShare* data repository [5]. The 60 real networks are loaded as simple undirected graphs. They are available in the publicly accessible network repository Netzschleuder with the 60 names below:

dom, packet_delays, sa_companies, ambassador, florentine_families, rhesus_monkey, kangaroo, internet_top_pop, high_tech_company, moviegalaxies, november17, moreno_taro, sp_baboons, bison, dutch_school, zebras, cattle, moreno_sheep, 7th_graders, college_freshmen, hens, freshmen, karate, dutch_criticism, montreal, ceo_club, windsurfers, elite, macaque_neural, sp_kenyan_households, contiguous_usa, cs_department, dolphins, macaques, terrorists_911, train_terrorists, highschool, law_firm, baseball, blumenau_drug, lesmis, fresh_webs, sp_office, swingers, polbooks, game_thrones, football, football_tsevans, sp_high_school_new, foodweb_baywet, revolution, foodweb_little_rock, student_cooperation, jazz_collab, interactome_pdz, physician_trust, malaria_genes, marvel_partnerships, facebook_friends, netscience.

For more information on each network and its original source, one may check the Netzschleuder website by adding the network name at the end of the url: https://networks.skewed.de/net/. For example, https://networks.skewed.de/net/malaria_genes provides additional information for the *malaria_genes* network. In cases of multiple networks existing with the same name in Netzschleuder, we have used the first network (e.g. we have used the *HVR_1* network from https://networks.skewed.de/net/malaria_genes).

References

1. Agarwal, G., Kempe, D.: Modularity-maximizing graph communities via mathematical programming. Eur. Phys. J. B **66**(3), 409–418 (2008). https://doi.org/10.1140/epjb/e2008-00425-1
2. Aldecoa, R., Marín, I.: Deciphering network community structure by surprise. PLoS ONE **6**(9), 1–8 (2011). https://doi.org/10.1371/journal.pone.0024195
3. Aloise, D., Cafieri, S., Caporossi, G., Hansen, P., Perron, S., Liberti, L.: Column generation algorithms for exact modularity maximization in networks. Phys. Rev. E **82**(4), 046112 (2010). https://doi.org/10.1103/PhysRevE.82.046112
4. Aref, S., Chheda, H., Mostajabdaveh, M.: The Bayan algorithm: detecting communities in networks through exact and approximate optimization of modularity. arXiv preprint arXiv:2209.04562 (2022)
5. Aref, S., Chheda, H., Mostajabdaveh, M.: Dataset of networks used in accessing the Bayan algorithm for community detection (2023). https://doi.org/10.6084/m9.figshare.22442785
6. Beuming, T., Skrabanek, L., Niv, M.Y., Mukherjee, P., Weinstein, H.: PDZBase: a protein-protein interaction database for PDZ-domains. Bioinformatics **21**(6), 827–828 (2005)
7. Blondel, V.D., Guillaume, J.L., Lambiotte, R., Lefebvre, E.: Fast unfolding of communities in large networks. J. Stat. Mech. Theory Exp. **2008**(10), P10008 (2008). https://doi.org/10.1088/1742-5468/2008/10/P10008
8. Bonald, T., Charpentier, B., Galland, A., Hollocou, A.: Hierarchical graph clustering using node pair sampling. In: MLG 2018–14th International Workshop on Mining and Learning with Graphs. London, UK (2018)
9. Brandes, U., et al.: On modularity clustering. IEEE Trans. Knowl. Data Eng. **20**(2), 172–188 (2007)
10. Cafieri, S., Costa, A., Hansen, P.: Reformulation of a model for hierarchical divisive graph modularity maximization. Ann. Oper. Res. **222**, 213–226 (2014)
11. Chen, S., et al.: Global vs local modularity for network community detection. PLoS ONE **13**(10), 1–21 (2018). https://doi.org/10.1371/journal.pone.0205284
12. Chen, T., Singh, P., Bassler, K.E.: Network community detection using modularity density measures. J. Stat. Mech. Theory Exp. **2018**(5), 053406 (2018). https://doi.org/10.1088/1742-5468/aabfc8
13. Clauset, A., Newman, M.E., Moore, C.: Finding community structure in very large networks. Phys. Rev. E **70**(6), 066111 (2004)
14. Dinh, T.N., Li, X., Thai, M.T.: Network clustering via maximizing modularity: approximation algorithms and theoretical limits. In: 2015 IEEE International Conference on Data Mining, pp. 101–110 (2015). https://doi.org/10.1109/ICDM.2015.139
15. Dinh, T.N., Thai, M.T.: Toward optimal community detection: from trees to general weighted networks. Internet Math. **11**(3), 181–200 (2015)
16. Fortunato, S.: Community detection in graphs. Phys. Rep. **486**(3–5), 75–174 (2010)
17. Fortunato, S., Barthélemy, M.: Resolution limit in community detection. Proc. Natl. Acad. Sci. **104**(1), 36–41 (2007)
18. Fortunato, S., Hric, D.: Community detection in networks: a user guide. Phys. Rep. **659**, 1–44 (2016). https://doi.org/10.1016/j.physrep.2016.09.002
19. Fortunato, S., Newman, M.E.: 20 years of network community detection. Nat. Phys. **18**, 848–850 (2022)

20. Good, B.H., De Montjoye, Y.A., Clauset, A.: Performance of modularity maximization in practical contexts. Phys. Rev. E **81**(4), 046106 (2010)
21. Guimerà, R., Sales-Pardo, M., Amaral, L.A.N.: Modularity from fluctuations in random graphs and complex networks. Phys. Rev. E **70**, 025101 (2004)
22. Gurobi Optimization Inc.: Gurobi optimizer reference manual (2023). https://gurobi.com/documentation/10.0/refman/index.html. Accessed 16 Feb 2023
23. Karrer, B., Newman, M.E.J.: Stochastic blockmodels and community structure in networks. Phys. Rev. E **83**, 016107 (2011)
24. Kawamoto, T., Kabashima, Y.: Counting the number of metastable states in the modularity landscape: algorithmic detectability limit of greedy algorithms in community detection. Phys. Rev. E **99**(1), 010301 (2019)
25. Kawase, Y., Matsui, T., Miyauchi, A.: Additive approximation algorithms for modularity maximization. J. Comput. Syst. Sci. **117**, 182–201 (2021). https://doi.org/10.1016/j.jcss.2020.11.005
26. Kernighan, B.W., Lin, S.: An efficient heuristic procedure for partitioning graphs. Bell Syst. Tech. J. **49**(2), 291–307 (1970)
27. Knuth, D.E.: The Stanford GraphBase: A Platform for Combinatorial Computing, vol. 1. ACM Press, New York (1993)
28. Kosowski, A., Saulpic, D., Mallmann-Trenn, F., Cohen-addad, V.P.: On the power of Louvain for graph clustering. In: Larochelle, H., Ranzato, M., Hadsell, R., Balcan, M., Lin, H. (eds.) Advances in Neural Information Processing Systems, vol. 33 (NeurIPS'20) (2020)
29. Lancichinetti, A., Fortunato, S.: Limits of modularity maximization in community detection. Phys. Rev. E **84**(6), 066122 (2011). https://doi.org/10.1103/PhysRevE.84.066122
30. Leicht, E.A., Newman, M.E.J.: Community structure in directed networks. Phys. Rev. Lett. **100**(11), 118703 (2008). https://doi.org/10.1103/PhysRevLett.100.118703
31. Li, P.Z., Huang, L., Wang, C.D., Lai, J.H.: EdMot: an edge enhancement approach for motif-aware community detection. In: Proceedings of the 25th ACM SIGKDD International Conference on Knowledge Discovery & Data Mining, pp. 479–487 (2019)
32. Liu, X., et al.: A scalable redefined stochastic blockmodel. ACM Trans. Knowl. Discov. Data (TKDD) **15**(3), 1–28 (2021)
33. Maier, B.F., Brockmann, D.: Cover time for random walks on arbitrary complex networks. Phys. Rev. E **96**(4), 042307 (2017)
34. Marchese, E., Caldarelli, G., Squartini, T.: Detecting mesoscale structures by surprise. Commun. Phys. **5**(1), 1–16 (2022)
35. Meeks, K., Skerman, F.: The parameterised complexity of computing the maximum modularity of a graph. Algorithmica **82**(8), 2174–2199 (2020)
36. Miasnikof, P., Shestopaloff, A.Y., Bonner, A.J., Lawryshyn, Y., Pardalos, P.M.: A density-based statistical analysis of graph clustering algorithm performance. J. Complex Netw. **8**(3), 1–33 (2020)
37. Newman, M.E.J.: Modularity and community structure in networks. Proc. Natl. Acad. Sci. **103**(23), 8577–8582 (2006). https://doi.org/10.1073/pnas.0601602103
38. Newman, M.E.J.: Equivalence between modularity optimization and maximum likelihood methods for community detection. Phys. Rev. E **94**(5), 052315 (2016). https://doi.org/10.1103/PhysRevE.94.052315
39. Peel, L., Larremore, D.B., Clauset, A.: The ground truth about metadata and community detection in networks. Sci. Adv. **3**(5), e1602548 (2017)

40. Peixoto, T.P.: Efficient Monte Carlo and greedy heuristic for the inference of stochastic block models. Phys. Rev. E **89**(1), 012804 (2014)
41. Peixoto, T.P.: Descriptive vs. Inferential Community Detection in Networks: Pitfalls, Myths and Half-Truths. Elements in the Structure and Dynamics of Complex Networks, Cambridge University Press, Cambridge (2023)
42. Rossetti, G., Milli, L., Cazabet, R.: CDLIB: a Python library to extract, compare and evaluate communities from complex networks. Appl. Netw. Sci. **4**(1), 1–26 (2019)
43. Rosvall, M., Bergstrom, C.T.: An information-theoretic framework for resolving community structure in complex networks. Proc. Natl. Acad. Sci. **104**(18), 7327–7331 (2007). https://doi.org/10.1073/pnas.0611034104
44. Rosvall, M., Bergstrom, C.T.: Maps of random walks on complex networks reveal community structure. Proc. Natl. Acad. Sci. **105**(4), 1118–1123 (2008). https://doi.org/10.1073/pnas.0706851105
45. Serrano, B., Vidal, T.: Community detection in the stochastic block model by mixed integer programming (2021)
46. Sobolevsky, S., Campari, R., Belyi, A., Ratti, C.: General optimization technique for high-quality community detection in complex networks. Phys. Rev. E **90**(1), 012811 (2014)
47. Traag, V.A., Aldecoa, R., Delvenne, J.C.: Detecting communities using asymptotical surprise. Phys. Rev. E **92**, 022816 (2015). https://doi.org/10.1103/PhysRevE.92.022816
48. Traag, V.A., Waltman, L., van Eck, N.J.: From Louvain to Leiden: guaranteeing well-connected communities. Sci. Rep. **9**(1) (2019). https://doi.org/10.1038/s41598-019-41695-z
49. Vinh, N.X., Epps, J., Bailey, J.: Information theoretic measures for clusterings comparison: variants, properties, normalization and correction for chance. J. Mach. Learn. Res. **11**(95), 2837–2854 (2010). http://jmlr.org/papers/v11/vinh10a.html
50. Zhang, P., Moore, C.: Scalable detection of statistically significant communities and hierarchies, using message passing for modularity. Proc. Natl. Acad. Sci. **111**(51), 18144–18149 (2014)
51. Zhao, X., Liang, J., Wang, J.: A community detection algorithm based on graph compression for large-scale social networks. Inf. Sci. **551**, 358–372 (2021)

Analyzing the Attitudes of Consumers to Electric Vehicles Using Bayesian Networks

Karolina Bienias[1(✉)] and David Ramsey[2]

[1] Department of Operations Research and Business Intelligence, Wrocław University of Science and Technology, Wrocław, Poland
karolina.bienias@pwr.edu.pl

[2] Department of Computer Science and Systems Engineering, Wroclaw University of Science and Technology, Wrocław, Poland

Abstract. Road transport, as 'a producer' of carbon dioxide (CO_2), causes high levels of air pollution, especially in cities. A suggested solution to this situation is the effective diffusion of electric vehicles (EVs). Regulations in the European Union aim to encourage consumers to buy electric cars. In addition, car manufacturers are constantly expanding their range of hybrid vehicles (HEVs) and EVs. Nonetheless, consumers have still many doubts regarding adopting an EV. Our survey among social media users investigates the attitudes and readiness of consumers to adopt HEVs and EVs. To investigate the factors underlying consumers' attitudes to such vehicles, Bayesian networks were used as an exploratory tool. This paper presents results of this analysis.

Keywords: electric vehicle · hybrid electric vehicle · innovation diffusion · consumers · willingness to pay · survey · Bayesian network

1 Introduction

The European Union aims to reduce greenhouse gas emissions, produce more energy from renewable sources and improve energy efficiency. The EU has proposed various tools, such as: financial incentives, infrastructure developments, and strategies to encourage people to purchase EVs (Hawkins et al. 2013; Sierzchula et al. 2014; Pasaoglu et al. 2012). Similar strategies have already been introduced in Poland to encourage consumers to buy or rent an EV. However, the current share of EVs in the Polish market is not sufficient to claim the successful diffusion of EVs.

This paper presents results from an online survey conducted among social media users in 2020. The aim of the study is to obtain better knowledge about consumers' opinions regarding EVs and HEVs.

2 Data Collection, the Sample and Methods

The dissemination of innovation through social media channels can bring effective results (Hanna et al. 2011). People who have self-perception as leaders of opinion formation, build strong online networks and have significant effect on users' news sharing intention

J. Mikyška et al. (Eds.): ICCS 2023, LNCS 10476, pp. 627–634, 2023.
https://doi.org/10.1007/978-3-031-36027-5_49

in social media (Ma et al. 2014). Studies show that social media campaigns can easily reach out to consumers in fast and cost efficient way (Chawla and Chodak 2018; Reid 2014). Also, consumers are willing to search for information and exchange opinions with other users through social media channels.

To the best of our knowledge, the awareness and acceptance of EVs/HEVs among social media users have not been checked yet. Our study aims to fill this gap.

The respondents targeted were residents of Poland above the age of 18. Hence, the questionnaire was only conducted in the Polish language. The survey was disseminated on the Facebook platform. Answers from 858 questionnaires were analyzed.

The on-line survey was split into six parts, including a section checking respondents' knowledge and opinion about electric and hybrid vehicles. The definitions of the variables and their coding are presented in Table 1. The sections of the survey cover demographics (M1–M6), information about cars in the households of respondents (H1–H11), questions regarding the evaluation of electric and hybrid cars (OH1–OH8 and OE1–OE10) and further questions about respondents' opinion on electric and hybrid vehicles (F1–F7). Furthermore, respondents were asked about hypothetical situations in which they could use electric vehicles in everyday life (S1–S2), (D1–D8) including their opinion about the prices of electric and hybrid cars, as well as the possibility of enjoying the benefits of EVs (P1–P11). Respondents indicated their degree of acceptance of hybrid and electric vehicles on the basis of a standard five-point Likert scale.

Table 1. Definitions of the variables and coding (N = 858).

Variable	Code	Description
Gender	M1	nominal variable
Age	M2	ordinal variable
Level of education	M3	ordinal variable
Size of home town/city	M4	ordinal variable
Voivodeship (region)	M5	nominal variable
Number of people in the household	M6	ordinal variable
Number of cars in the household	H1	ordinal variable
Number of cars possessed	H2	ordinal variable
Source of the cars used in the household	H3	nominal variable
Price of the most expensive car purchased	H4	ordinal variable
Type of engine	H7–H10	nominal variable
Type of hybrid vehicle (type of engine)	H11	nominal variable
Evaluation of hybrid cars	OH1–OH8	(1) No/ (2) Rather no/ (3) Hard to say/ (4) Rather yes/ (5) Yes

(*continued*)

Table 1. (*continued*)

Variable	Code	Description
Evaluation of electric cars	OE1–OE11	(1) No/ (2) Rather no/ (3) Hard to say/ (4) Rather yes/ (5) Yes
Previous rental of vehicle or electric vehicle	S1–S2	(1) No/ (2) I don't remember/ (3) Yes
General opinion of electric and hybrid vehicles	F1–F2, F6–F7	(1) Negative (2) Rather negative (3) Hart to say (4) Rather positive (5) Positive
Family/friends' ownership of a hybrid/electric vehicle	F3–F4	(1) No/ (2) Rather no/ (3) Hard to say/ (4) Rather yes/ (5) Yes
Occurrence of conversations about hybrid/ electric vehicle	F5	(1) No/ (2) I don't remember/ (3) Yes
Convenience of electric vehicle in everyday life	D1–D7	(1) No/ (2) Rather no/ (3) Hard to say/ (4) Rather yes/ (5) Yes
Usefulness of electric car in the household	D8	nominal variable
Willingness to pay for electric/ hybrid vehicle	P1–P2	ordinal variable
Attractiveness of subsidies for purchasing electric and hybrid cars	P3–P4	(1) No/ (2) Rather no/ (3) Hard to say/ (4) Rather yes/ (5) Yes
Obstacles for buying/using an electric car	P5–P11	(1) No/ (2) Rather no/ (3) Hard to say/ (4) Rather yes/ (5) Yes

3 Analysis Using a Bayesian Network

Bayesian network analysis aims to infer the underlying network of relationships between a set of categorical variables. We used Kendall's test of correlation to analyze the strength and direction of association between pairs of ordinal variables (variables that are ordered with respect to a scale - for testing purposes, all the scales were orientated from the worst to the best condition from the point of view of HEV/EV propagation). When at least one of the variables in a pair was nominal (no natural ordering), then we analyzed the relationship between them using the chi-square test of association. To interpret interesting relations indicated by the Bayesian network, we analysed cross tables for the appropriate pair of variables.

A Bayesian network presents the relationships between variables in graphical form. Variables that are directly linked in such a network are strongly associated with other. Typically, the variable in a pair which influences the other is called the parent variable (Markowska-Przybyła and Ramsey 2015; Borgelt et al. 2009). However, in our data the direction of influence is often unclear, and so we avoid indicating which variable is the parent variable and interpret links as direct relations between variables. The choice of an appropriate network is based on the likelihood of the data under a given model and a penalty function that penalizes the complexity of the model. The most commonly used criteria for choosing such a network are the Bayesian Information Criterion (BIC) and the Akaike Information Criterion (AIC), which uses a weaker penalty on the complexity of a model than the BIC and thus tends to select more complex models. We used the catnet package in the R environment to derive Bayesian networks describing the underlying structure of the data (Balov and Salzman 2017).

The model obtained on the basis of the AIC criterion, under the assumption that any variable can have only one parent (due to the complexity of the networks derived), is illustrated in Fig. 1. Three separate networks of variables were found. In total, the network contains 60 nodes.

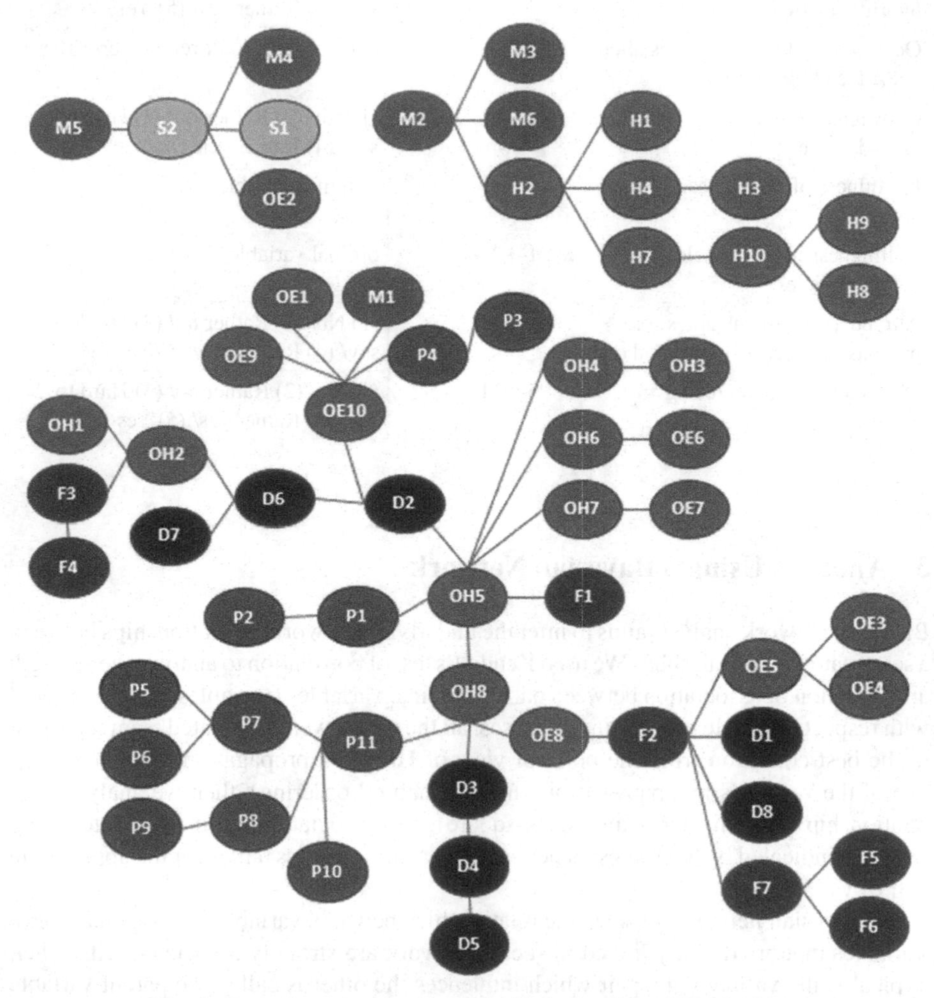

Fig. 1. Bayesian network created on the basis of the AIC criterion.

4 Results and Discussion

The size of a respondent's hometown is positively associated with experience of renting an EV (Kendall's correlation coefficient for M4 and S2 is 0.241, $p < 0.001$). The age of respondents (M2) and the purpose for which cars are used in a household (H2) are clearly associated. As age increases, cars are increasingly used for professional purposes instead of just private ($p < 0.001$).

The relationship between the price of the most expensive car in a household (H4) and whether a household bought new, second hand or both types of car (H3) shows that those who have only bought second hand cars tend to spend less ($p < 0,001$). Almost half of such respondents (48.2% of 525 cases) have always paid less than 25 thousand PLN (just over €5000), while the majority of households that have bought a new car in the past have spent at least 100 thousand PLN (more than €20000). The present cost of EVs is in this range.

The opinion that HEVs should be introduced to the market on a larger scale (OH5) is strongly associated with one's personal opinion about HEVs (F1) (similar relation with EVs), as well with the willingness to pay extra for HEVs (P1) and EVs (P2) compared to petrol-engine cars. Positive opinions about EVs (F2) are associated with respondents having used an EV as a replacement car (D1), as well as with positive opinions among their relatives and friends about both HEVs (F6) and EVs (F7).

Respondents with positive or rather positive opinion about EVs (F2) declare that they could accept an EV as a second car in their household (64.4% of 306 cases and 74.4% of 227 cases accordingly).

The gender of respondents (M1) is associated with knowledge regarding EVs, especially with regard to the possibility of charging an EV at home (OE10). Male respondents mostly agree with statement that an EV can be charged using a household outlet (68,4% of male respondents), while just 22,5% of female respondents gave the same answer.

Opinions regarding possible obstacles to buying/ using an EV (P5–P11) formed one branch of the Bayesian network and are closely related to opinions regarding the relative cost of running a HEV/ EV compared to a vehicle with a combustion engine (OH8/OE8). The relation between OH8 and P11 shows that, even if respondents have difficulties with estimating the operating costs for a HEV (37.1% of respondents/318 cases chose the answer 'Hard to say', a similar proportion of answers to OE8), 53.8% of these respondents (171 cases) think that such difficulty in cost estimation is not or rather not an obstacle to buying/using an EV.

The general opinion of respondents about electric vehicles (F2) is in strong relation with the general opinion of respondents' family/friends about electric vehicles (F7) (Kendall's correlation coefficient is 0.573, $p < 0.001$). The survey confirms that respondents' opinion of innovation (like EV in this case) is strongly influenced by the opinion of their environment (Edwards 1953; Podsakoff et al. 2003).

The knowledge of whether any family members/friends have a HEV (F3) is strongly correlated with information of whether any family members/friends have an EV (F4) (Kendall's correlation coefficient is 0.401, $p < 0.001$). 36.7% of respondents (315 cases) do not know anyone with a HEV/EV. 14.9% of respondents (128 cases) know owners of both HEV and EV vehicles. 21.7% of respondents (186 cases) know someone with a HEV, but no-one with an EV. Just 11 respondents (1.3%) know someone with an EV but no-one with an HEV.

Table 2 presents how many respondents own a hybrid or electric vehicle. Table 3 presents the structure of vehicle types in the households of respondents.

Table 2. Types of engine in the cars owned by a household (N = 858).

Type of engine	Yes		No	
Internal combustion engine	786	91,6%	72	8,4%
Hybrid engine	40	4,7%	818	95,3%
Electric motor	35	4,1%	823	95,9%

Table 3. Composition of car engines in a household (N = 858).

All types of engines in a household	Share	
No car in the household	41	4,8%
ONLY internal combustion engine	745	86,8%
Internal combustion engine AND hybrid engine	19	2,2%
Internal combustion engine AND electric motor	22	2,6%
ONLY hybrid engine	18	2,1%
Hybrid engine AND electric motor	3	0,3%
ONLY electric motor	10	1,2%

Table 4 presents share of passenger cars by fuel type in Poland in 2020 according to the report of the European Automobile Manufacturers' Association (ACEA).

The Bayesian network confirms that the following factors have the highest influence on ownership of a HEV/EV and willingness to pay for HEV/ EV: the price of hybrid/electric cars, the positive opinion of consumers about HEVs/Evs, as well as the positive opinion of friends and relatives about HEVs/EVs, the possibility to enjoy such privileges as free parking in paid parking zones and the assumption that an electric vehicle would work well in the household as the only car or second/additional car.

Table 4. Share of passenger cars by fuel type in Poland and EU in 2020 (ACEA, 2022).

	Petrol	Diesel	Battery electric	Plug-in hybrid	Hybrid electric	LPG
Poland	44,80%	40,2%	0,01%	0,0%	1,0%	13,8%
EU	51,7%	42,8%	0,5%	0,6%	1,2%	2,5%

The factors considered to be most important for the diffusion of EVs on the Polish market will be included in an agent-based model (ABM). This ABM will aid us in checking whether methods that are effective in other countries can also be successfully implemented in Poland. The results of comparative analysis with chosen countries, together with the results from simulations will indicate policies that should be implemented on the Polish market in the near future. Implementing the most efficient policies could lead to a more effective introduction of EVs.

5 Limitations of the Study

The study conducted has some limitations. Firstly, the survey was limited linguistically, as it was only conducted in Polish to investigate the opinion of Polish social media users. In addition, the study was also limited by the way the online survey was disseminated, focusing only on social media users. Hence, consumers who are not active on social media were not included in the sample. Subsequent research could target a more diverse demographic and a wider audience to make the survey group more representative of the Polish population.

6 Conclusions and Future Work

The factors considered to be most important for the diffusion of EVs on the Polish market will be included in an agent-based model (ABM). This ABM will aid us in checking whether methods that are effective in other countries can also be successfully implemented in Poland. The results of comparative analysis with chosen countries, together with the results from simulations will indicate policies that should be implemented on the Polish market in the near future. Implementing the most efficient policies could lead to a more effective introduction of EVs.

Funding. This work was supported by the National Science Center (NCN, Poland) by grant no. 2021/41/N/HS4/04445.

References

ACEA: Vehicles in use, Europe 2022 Report (2022). https://www.acea.auto/files. Accessed 15 Jan 2023

Balov, N., Salzman, P.: CatNet: categorical Bayesian network inference. R package version 1.4 (2012)

Borgelt, C., Steinbrecher, M., Kruse, R.R.: Graphical models: Representations for Learning Reasoning and Data Mining. Wiley, Weinheim (2009)

Chawla, Y., Chodak, G.: Recommendations for Social Media Activities to Positively Influence the Economic Factors. In: Jedlička, P., Marešová, P., Soukal, I., (eds.) Double-Blind Peer-Review Proceedings Part I. of the International Scientific Conference Hradec Economic Days 2018, 30–31 January 2018. University of Hradec Králové: Hradec Králové, Czech Republic, vol. 8, pp. 328–338 (2018)

Edwards, A.L.: The relationship between the judged desirability of a trait and the probability that the trait will be endorsed. J. Appl. Psychol. **37**(2), 90–93 (1953)

Hanna, R., Rohm, A., Crittenden, V.: We're all connected: the power of the social media ecosystem. Bus. Horiz. **54**, 265–273 (2011)

Hawkins, T.R., et al.: Comparative environmental life cycle assessment of conventional and electric vehicles. J. Ind. Ecol. **17**, 53–64 (2013)
Ma, L., Lee, C.S., Goh, D.H.: Understanding news sharing in social media: an explanation from the diffusion of innovations theory. Online Inf. Rev. **38**, 598–615 (2014)
Markowska-Przybyła U., Ramsey D.: A game theoretical study of generalized trust and reciprocation in Poland. II. A description of the study group. Oper. Res. Decis. **2**, 51–73 (2015)
Pasaoglu, G., et al.: Potential vehicle fleet CO2 reductions and cost implications for various vehicle technology deployment scenarios in Europe. Energy Policy **40**, 404–421 (2012)
Podsakoff, P.M., MacKenzie, S.B., Lee, J.-Y., Podsakoff, N.P.: Common method biases in behavioral research: a critical review of the literature and recommended remedies. J. Appl. Psychol. **88**(5), 879–903 (2003)
Reid, C.K.: The state of digital advertising. EContent **37**, 16–17 (2014)
Sierzchula, W., et al.: The influence of financial incentives and other socio-economic factors on electric vehicle adoption. Energy Policy **68**, 183–194 (2014)

Parallel Triangles and Squares Count for Multigraphs Using Vertex Covers

Luca Cappelletti[1](✉), Tommaso Fontana[1], Oded Green[3], and David Bader[2]

[1] University of Milan, Milan, Italy
luca.cappelletti1@unimi.it
[2] New Jersey Institute of Technology, Newark, USA
[3] Georgia Tech, Atlanta, GA, USA

Abstract. Triangles and squares count are widely-used graph analytic metrics providing insights into the connectivity of a graph. While the literature has focused on algorithms for global counts in simple graphs, this paper presents parallel algorithms for global and per-node triangle and square counts in large multigraphs. The algorithms have linear improvements in computational complexity as the number of cores increases. The triangle count algorithm has the same complexity as the best-known algorithm in the literature. The squares count algorithm has a lower execution time than previous methods. The proposed algorithms are evaluated on six real-world graphs and multigraphs, including protein-protein interaction graphs, knowledge graphs and large web graphs.

Keywords: Graph · Multigraph · Triangles · Squares · Count

1 Introduction

The study of complex networks and their properties has been an active area of research in recent years. One of a network's most fundamental and well-studied properties is its clustering coefficient [13], which measures the fraction of triangles in a network, where a triangle is defined as three nodes that are all connected. The computation of the clustering coefficient [7] is a crucial step in many graph analytics tasks, including community detection and link prediction.

The original vertex-cover-based algorithms for counting triangles and squares, as described in [4], used vertex covers to reduce the number of set intersections and avoid unnecessary element comparisons. While these algorithms were shown to be much more efficient than traditional baselines, there are still several areas for improvement.

Self-loops or multiple edges between nodes, i.e., when the graph is a multigraph, are common in real-world and knowledge graphs. Both original algorithms assume that these features were either removed or not present. Other algorithms in the literature that handle multigraphs are approximated and implicitly remove the multi-edges instead of considering them [6,11]. All these algorithms only

J. Mikyška et al. (Eds.): ICCS 2023, LNCS 14076, pp. 635–646, 2023.
https://doi.org/10.1007/978-3-031-36027-5_50

provide the global counts of triangles and squares, respectively, but in many use cases, the per-node count would be more valuable.

This paper presents an updated parallel version of the algorithms presented in [4], addressing the above-mentioned shortcomings. Specifically, our algorithms support graphs containing self-loops and multigraphs and provide the number of triangles and squares per node. The updated algorithms' asymptotic worst-case computational complexities are equal to or lower than the original algorithms in real-world sparse graphs. All algorithms are implemented as part of the GRAPE [1] library, and the experiments are provided as library tutorials.[1]

2 Notation

A graph $G = (V, E)$ is composed of a set of nodes V and a set of edges E. A node $v \in V$ has neighbours $\mathcal{N}(v)$ and has degree $d(v)$ equal to the cardinality of its neighbours, $|\mathcal{N}(v)|$. When we sequentially iterate over a node's neighbors, we assume that they are sorted, as is common in several graph data structures.

In a multigraph, the neighbors of a node $v \in V$, $\mathcal{N}(v)$ may be a multiset, i.e., a set with repeated elements. Given a node $w \in V$ and a multiset $\mathcal{N}(v)$, we denote the multiplicity function $m_{\mathcal{N}(v)}(w) : V \rightarrow \mathbb{N}$ of as the number of times a node w appears in the neighbourhood $\mathcal{N}(v)$.

In the per-node version of the algorithms, we use atomic instructions [5]. Atomic instructions are low-level hardware operations guaranteed to complete without affecting other memory operations. They are helpful in multi-threaded and concurrent programming, allowing multiple threads to access and modify shared memory locations without the risk of race conditions and data corruption. An atomic fetch add is a specific type of atomic operation that retrieves the current value stored in a memory location and adds a specified value to it, returning the original value. This operation is used to increment the value of a shared memory location in a thread-safe manner without the risk of two or more threads interfering with each other. In real-world sparse graphs, the risk of multiple write attempts using atomic fetch add is very low, as the graph is sparse, and thus there are fewer interactions between nodes. We will denote atomic fetch-add operations as $\texttt{+=}_A$.

3 Computation of Vertex Covers

A vertex cover $\hat{V} \subseteq V$ is a subset of vertices in a graph such that each edge has at least one endpoint in the vertex cover. The algorithms use vertex covers to minimize the number of required set intersections. Any vertex cover suffices for the purpose, and there are different heuristics to obtain them. Three heuristics were explored based on different node sorting methods and whether to add one or both nodes of an edge to the vertex cover. Obtaining a vertex cover has a complexity of $O(|E|)$, which is negligible compared to the algorithms' complexity.

[1] https://github.com/AnacletoLAB/grape/tree/main/tutorials.

The paper explores three vertex cover schemas: Natural, Decreasing node degree, and Increasing node degree. The natural schema uses the order of nodes as they are loaded into the graph and adds both the edge source and destination. The Decreasing node degree schema sorts the nodes by decreasing node degree, prioritizing nodes with more edges, and only inserts the source nodes. The Increasing node degree schema sorts the nodes by increasing node degree, prioritizing nodes with fewer edges, and only inserts the source nodes.

4 Counting Triangles

We start by describing the global triangle count (Algorithm 1), which takes as input a graph $G = (V, E)$ and a vertex cover $\hat{V} \subseteq V$.

The counter t is initialized to zero, representing the number of triangles times three. It loops in parallel over all vertices in the cover $v_1 \in \hat{V}$ (Line 2). The key insight is that, by definition, every triangle has at least two nodes in the vertex cover [4]. Requiring the first two nodes to be in cover allows us to reduce the total necessary comparisons in the inner loops. For each vertex v_1, it loops over all of its neighbors in the vertex cover $v_2 \in \mathcal{N}(v_1) \cap \hat{V}$ (Line 3). Since we assume the neighbors are sorted if v_2 is greater than or equal to v_1 (in the case of self-loops), the loop is stopped early (Line 4), and thus halves the computational requirements. For v_2, the algorithm loops over all common neighbors of v_1 and v_2, $v_3 \in \mathcal{N}(v_1) \cap \mathcal{N}(v_2)$, which are the nodes that close the triangle (Line 6). To avoid self-loops, the iteration is skipped if v_3 equals v_1 or v_2, which are excluded from the set. To account for triangles composed by multigraph edges, we compute the multiplicities product of the v_3 node in the neighborhoods of the other two nodes, i.e., $c = m_{\mathcal{N}(v_1)}(v_3) m_{\mathcal{N}(v_2)}(v_3)$ (Line 7). If v_3 is in the cover, the counter t is incremented by c (Line 9) because it will be re-encountered two other times. Conversely, if v_3 is not in the vertex cover $\hat{V}$, the counter t is incremented by $3c$ (Line 11) because the node will not be visited again. The algorithm concludes by returning the number of triangles, i.e., the counter divided by three $t/3$. Since the computation of each outer loop are independent, distributed approaches such as map-reduce are possible.

Time Complexities. The computation of the algorithm can be distributed up to $p = |\hat{V}|$ cores. The two inner loops require $O(d_{cover}^2)$ to iterate over all the in-cover neighbors of v_1, which requires at most d_{cover} to compute. The v_3 loop iterates the intersection of the neighbors of v_1 and v_2, which requires at most d_{cover}. The time complexity of the algorithm is $O(|\hat{V}| d_{cover}^2 / p)$.

Algorithm 1: Triangle counts

```
   Input  : G = (V, E), cover V̂ ⊆ V
   Output: Graph-wide triangles t
 1 t ← 0;
 2 for v1 ∈ V̂ do in parallel
 3 |  for v2 ∈ N(v1) ∩ V̂
 4 |  |  if v2 ≥ v1 then
 5 |  |  |  break;
 6 |  |  for v3 ∈ N(v1) ∩ N(v2) \ {v1, v2}
 7 |  |  |  c=m_N(v1)(v3) · m_N(v2)(v3);
 8 |  |  |  if v3 ∈ V̂ then
 9 |  |  |  |  t+=c;
10 |  |  |  else
11 |  |  |  |  t+=3c;
12 return t / 3;
```

4.1 Per Node Triangle Count

In the per-node count (Algorithm 2) we have a vector of atomic counters $\boldsymbol{t}$, one for each node. The triangle count for v_1 is always incremented by the multiplicity factor c (Line 8). If v_3 is not in the cover $\hat{V}$, the triangle count for v_2 and v_3 is also incremented by c. Using atomic additions ensures that each node's triangle count is updated safely, even with concurrent access from multiple threads. Finally, the algorithm returns the vector $\boldsymbol{t}$ of triangle counts per node.

The time complexity of the per-node algorithm remains $O(|\hat{V}| d^2_{cover}/p)$. However, to achieve perfect parallelization using atomic instructions, the processes should simultaneously modify the same counters as little as possible. This is possible in sparse real-world graphs. Still, the algorithm will behave worse than sequentially in degenerate cases, such as cliques, as simultaneous modification will result in the eviction of cache lines and CPU stalls, adding time overhead.

Algorithm 2: Per node count

```
   Input  : G = (V, E), cover V̂ ⊆ V
   Output: Vector of triangles t per node
 1 t ← vector with |V| atomic zeros;
 2 for v1 ∈ V̂ do in parallel
 3 |  for v2 ∈ N(v1) ∩ V̂
 4 |  |  if v2 ≥ v1 then
 5 |  |  |  break;
 6 |  |  for v3 ∈ N(v1) ∩ N(v2) \ {v1, v2}
 7 |  |  |  c=m_N(v1)(v3) · m_N(v2)(v3);
 8 |  |  |  t[v1]+=_A c;
 9 |  |  |  if v3 ∉ V̂ then
10 |  |  |  |  t[v2]+=_A c;
11 |  |  |  |  t[v3]+=_A c;
12 return t;
```

5 Counting Squares

We describe the global square count (Algorithm 3), for a graph $G = (V, E)$ and a vertex cover $\hat{V} \subseteq V$.

The algorithm from [4] employed a double iteration on the vertex cover to check all the pairs of nodes in the cover and the intersection of their neighbors. We can speed up the square counts on sparse graphs by skipping the pairs of nodes that would produce empty intersections. In our approach, we iterate once $v_1 \in \hat{V}$ on the vertex cover and on the second-order neighbors of v_1 in the vertex cover, i.e., $v_3 \in \hat{V}_{v_1}$ where $\hat{V}_{v_1} = \bigcup_{v_2 \in \mathcal{N}(v_1)} \mathcal{N}(v_2) \cap \hat{V}$. By definition, we will only iterate on a pair of nodes in the cover with at least one common neighbor. We want to efficiently iterate on the set of *unique* second-order neighbors in the cover $\hat{V}_{v_1}$; to do so, we need to keep track of the visited nodes $\bar{V}$ to avoid counting squares multiple times. In our implementation to represent $\bar{V}$, we used a bitmap with $|V|$ bits for each thread which is cleared at the start of each new root node v_1. The algorithm initializes the counter s to zero, representing the number of squares times two. It loops in parallel over all vertices in the vertex cover $v_1 \in \hat{V}$ (Line 2). For each vertex v_1, it loops over all of its neighbors $v_2 \in \mathcal{N}(v_1)$ (Line 3). If v_2 equals v_1, we skip to the next neighbor to avoid self-loops. For each v_2, we iterate on all its neighbor in the vertex cover $v_3 \in \mathcal{N}(v_2) \cap \hat{V}$. Since we assume the neighbors are sorted if v_3 is greater than v_1, the loop is stopped early (Line 6), which is done to avoid checking twice the same node and roughly halves the time requirements. We have to skip self-loops $v_3 = v_2$, backward edges $v_3 = v_1$, and already visited nodes $v_3 \in \bar{V}$. Then, we add v_3 to the visited nodes $\bar{V}$ (Line 8).

We initialize the multiplicity counters of neighbors of v_1 in cover v_{in}, out of cover v_{out}, and the sums of the squared multiplicities v_{in^2}, v_{out^2}. We iterate over each common neighbour of v_1 and v_3 excluding the nodes v_1, v_3 themselves. We compute the product of multiplicities of v_4 in v_1 and v_2. If the node v_4 is in cover $v_4 \in \hat{V}$, this multiplicity and its square are added to the counters v_{in} and v_{in^2}, conversely, they are added to v_{out} and v_{out^2}.

We add to the s counter the four counters to obtain the number of squares involving v_1, v_3, v_4, v_2 is counted as part of v_4 nodes. Since we will not encounter multiple times the nodes outside of the cover forming squares with v_1 and v_3, we need to account for the squares they form with themselves $v_{out}^2 - v_{out^2}$, which are all pairs of *distinct* nodes, the squares they form with the in cover nodes $2v_{out}v_{in}$, and the squares formed by nodes in cover $(v_{in}^2 - v_{in^2})/2$, which will be encountered twice. The algorithm concludes by returning the number of squares, $s/2$.

Time Complexity. The algorithm's three inner loops require $O(d_{cover}^2 d_{graph})$ because the algorithm will iterate over all the in-cover neighbors of v_1, which requires at most d_{cover} to compute. The v_3 loop has to compute the neighbors of v_2, which takes at most d_{graph}. The v_4 loop computes the intersection of the neighbors of v_1 and v_2, which will require at most d_{cover}. Therefore, the time complexity of the algorithm is $O(|\hat{V}| d_{cover}^2 d_{graph}/p)$, for $p \leq |\hat{V}|$. This analysis ignored the costs relative to the set $\bar{V}$ due to its strict dependency on the implementation details and because it was negligible in our experiments. This analysis ignored the costs relative to the set $\bar{V}$ due to its strict dependency on the implementation details. A sensible choice may be to use a bitmap paired with a vector, the bitmap for fast reading and updating, and the vector to keep track of the words of memory to reset. These require $O(1)$ time for reading and updating it. The time needed to reset it is proportional to the number of elements in it. This would add a multiplicative factor to the complexity, which in the worst case is $O(d_{cover}d_{graph})$. In practice, this operation is bottle-necked by the memory bandwidth of RAM, so even for large bitmaps, the resetting is practically negligible compared to loops.

Algorithm 3: Square counts

```
   Input  : G = (V, E), cover V̂ ⊆ V
   Output: Number of squares s
1  s ← 0;
2  for v1 ∈ V̂ do in parallel
3  |  V̄ ← ∅;
4  |  for v2 ∈ N(v1) \ {v1}
5  |  |  for v3 ∈ N(v2) ∩ V̂ \ {v1, v2} \ V̄
6  |  |  |  if v3 > v1 then
7  |  |  |  |  break;
8  |  |  |  V̄ ← {v3} ∪ V̄;
9  |  |  |  v_in, v_out, v_in2, v_out2 ← 0;
10 |  |  |  for v4 ∈ N(v1) ∩ N(v3) \ {v1, v3}
11 |  |  |  |  c=m_N(v1)(v4) · m_N(v3)(v4);
12 |  |  |  |  if v4 ∈ V̂ then
13 |  |  |  |  |  v_in+=c;
14 |  |  |  |  |  v_in2+=c^2;
15 |  |  |  |  else
16 |  |  |  |  |  v_out+=c;
17 |  |  |  |  |  v_out2+=c^2;
18 |  |  |  s+=v_out^2 - v_out2 + (v_in^2 - v_in2)/2 + 2 v_out v_in;
19 return s/2;
```

5.1 Per Node Version

We have a vector of atomic counters $\boldsymbol{s}$, one for each node. Since the number of squares contributed by v_1, v_3 and all $v_4 \in \mathcal{N}(v_1) \cap \mathcal{N}(v_3)$ is obtained from the factor of multiplicities of each $v_4 \in \mathcal{N}(v_1) \cap \mathcal{N}(v_3)$, we need first to compute the counters of the nodes in cover v_{in} and the nodes out of cover v_{out}), and afterward dispense the number of squares among the nodes properly. The necessity to iterate twice on the neighbors $v_4 \in \mathcal{N}(v_1) \cap \mathcal{N}(v_3)$ effectively duplicates the time requirements of the per-node algorithm. The counts of the node v_1 and v_3, which are the root vertex cover nodes, are incremented by the number of squares they form with the in-vertex and out-of-vertex nodes, $v_{out} \cdot v_{in}$. Each node $v_4 \in \mathcal{N}(v_1) \cap \mathcal{N}(v_3)$ count is incremented depending on the multiplicity factor c and whether it is in cover or not. Nodes in the cover will be re-encountered, while nodes outside will be only encountered once alongside the root nodes v_1 and v_3. We double the number of squares deriving from other out-of-cover nodes to account for the latter nodes encountered once. Since in the number of out-of-cover nodes v_{out}, we also count the node's multiplicity factor c, we must subtract that twice. We observe that by summing the obtained square, the total will be four times the total number of squares obtained from the global algorithm. Analogously to the global version, the per-node algorithm is distributable. The time complexity of the per-node algorithm remains $O(|\hat{V}| d_{cover}^2 d_{graph}/p)$.

Algorithm 4: Per node count

```
   Input  : G = (V, E), cover V̂ ⊆ V
   Output: Number of squares s per node
 1 s ← vector with |V| atomic zeros;
 2 for v1 ∈ V̂ do in parallel
 3     V̄ ← ∅;
 4     for v2 ∈ N(v1) \ {v1}
 5         for v3 ∈ N(v2) ∩ V̂ \ {v1, v2} \ V̄
 6             if v3 > v1 then
 7                 break;
 8             V̄ ← {v3} ∪ V̄;
 9             v_in, v_out ← 0;
10             for v4 ∈ N(v1) ∩ N(v3) \ {v1, v3}
11                 c=m_N(v1)(v4) · m_N(v3)(v4);
12                 if v4 ∈ V̂ then
13                     v_in+=c;
14                 else
15                     v_out+=c;
16             s[v1]+=_A v_out v_in;
17             s[v3]+=_A v_out v_in;
18             for v4 ∈ N(v1) ∩ N(v3) \ {v1, v3}
19                 c=m_N(v1)(v4) m_N(v3)(v4);
20                 if v4 ∈ V̂ then
21                     s[v4]+=_A c(v_out + v_in − c);
22                 else
23                     s[v4]+=_A c(2(v_out − c) + v_in);
24 return s;
```

6 Experiments

Experiments were conducted on a computer with an *AMD Ryzen 9 3900x CPU* (12 cores, 24 threads) and 128 GB RAM using six real-world graphs, including protein-protein interaction graphs, knowledge graphs, and web graphs. Table 1 summarizes the datasets, including the graph ID used in all result tables.

Table 1. Summary of the datasets' main characteristics

Graph id	Graph name	Nodes	Edges	d_{graph}
1	Saccharomyces Cerevisiae [12]	7*K*	1*M*	2.7*K*
2	Homo Sapiens [12]	20*K*	6*M*	7.5*K*
3	Mus Musculus [12]	22*K*	7*M*	7.6*K*
4	KGCOVID19 [9]	570*K*	18*M*	122*K*
5	Friendster [10]	65*M*	1.8*G*	5*K*
6	ClueWeb09 [2,10]	1.6*G*	7.8*G*	6.4*M*

6.1 Impact of Vertex Cover Schema

In this section, we present the results of our evaluation of the performance of the triangle and square counting algorithms for various vertex covers. Table 2 provides information on the vertex cover size and time requirements of six different graphs using the three vertex cover schemas described in Sect. 3: natural, decreasing, and increasing. The size of the vertex cover for each graph using each schema is given in the $|\hat{V}|$ column and the maximum degree of each vertex in the graph is given in the d_{cover} column, the percentage of vertices covered by the vertex cover is given in the % column. Finally, the time it took to compute the vertex cover using the given schema is in the Time column. The table indicates that the vertex cover size can vary depending on the schema. The decreasing schema typically produces the smallest vertex cover, and the increasing schema produces the largest. The time it takes to compute the vertex cover also varies depending on the schema used, with the decreasing schema typically being the slowest and the increasing schema typically being the fastest, beating even the **natural** approach, which does not involve any sorting procedures, contrarily to the other two schemas. The table also shows that as the size of the graph increases, the time it takes to compute the vertex cover generally increases as well. Nevertheless, it remains a fraction of the time necessary to compute the same graph's triangles or squares counts.

Table 2. Vertex cover size by vertex cover schema

	Natural				Decreasing				Increasing			
Id	$\|\hat{V}\|$	d_{cover}	%	*Time*	$\|\hat{V}\|$	d_{cover}	%	*Time*	$\|\hat{V}\|$	d_{cover}	%	*Time*
1	6240	2729	93%	77 ms	5720	2729	85%	90 ms	6393	2092	95%	70 ms
2	19200	7507	98%	2 ms	18475	7507	94%	2 ms	19384	6940	99%	1 ms
3	20756	7669	94%	3 ms	19524	7669	88%	3 ms	21300	7296	96%	1 ms
4	217*K*	122*K*	38%	12 ms	180*K*	122*K*	31%	50 ms	540*K*	22*K*	94%	22 ms
5	37*M*	5214	57%	6 s	31*M*	5214	48%	15 s	65*M*	3507	99%	6 s
6	456*M*	6444*K*	27%	52 s	277*M*	6444*K*	16%	171 s	1672*M*	2*M*	99%	106 s

Our experiments revealed that the choice of vertex cover has a significant impact on the performance of the triangle counting algorithm. Table 3 shows the execution time and the number of counted triangles for each vertex cover, both in the global and per-node versions. Notably, the algorithm achieved the fastest performance when using the increasing vertex cover, followed by the natural and decreasing vertex covers. This can be attributed to the fact that the increasing vertex cover, while being the least efficient in terms of the number of nodes covered, effectively excludes the nodes with higher degrees, which can substantially reduce the algorithm's time requirements by a quadratic factor. The choice of vertex cover should therefore be carefully considered when applying our algorithm to real-world graphs, especially those with a high degree of heterogeneity in their node degrees.

Table 3. Triangle counts time by vertex cover

Id	Number of Triangles	Natural		Decreasing		Increasing	
		Global	*Per node*	*Global*	*Per node*	*Global*	*Per node*
1	48834553	231 ms	208 ms	228 ms	207 ms	226 ms	291 ms
2	399408889	2442 ms	2313 ms	2431 ms	2434 ms	2424 ms	2317 ms
3	713495427	3752 ms	3518 ms	3822 ms	3693 ms	3720 ms	3549 ms
4	402950936	3290 ms	3081 ms	3575 ms	3807 ms	2812 ms	2669 ms
5	4173724142	248 s	248 s	250 s	259 s	250 s	244 s
6	31013019123	293 m	301 m	296 m	305 m	43 m	43 m

In Table 4, we present the results of the square counting algorithm using three different vertex covers. The table shows the time taken and the number of squares counted for each strategy. Interestingly, our results suggest that there is no clear optimal vertex cover strategy for this algorithm. This implies that the algorithm's performance is not highly dependent on the choice of vertex cover.

Table 4. Square counts time by vertex cover

Id	Number of Squares	Natural		Decreasing		Increasing	
		Global	*Per node*	*Global*	*Per node*	*Global*	*Per node*
1	17223337716	2 s	6 s	2 s	6 s	2 s	6 s
2	250013165364	40 s	101 s	40 s	99 s	40 s	102
3	659991475347	48 s	126 s	48 s	124 s	49 s	129 s
4	709420799404	104 s	248 s	216 s	516 s	415 s	1058 s
5	465803364346	38.5 h	76 h	37.5 h	35 h	39 h	77 h

6.2 Scalability

To evaluate the scalability of our algorithms, we conducted a series of experiments with varying numbers of threads, including 1, 6, 12 (utilizing all cores), and 24 (using hyper-threading). As shown in Table 5, our algorithms demonstrated linear scaling with the number of cores, confirming their effectiveness in exploiting parallel processing resources. However, we observed some sub-linear scaling when hyper-threading was utilized. Nonetheless, our results demonstrate that our algorithms are highly scalable and capable of achieving significant performance improvements when executed on multi-core systems.

Table 5. Square and triangle count times with natural vertex cover per thread number

	Triangles								Squares							
	Global				*Per node*				*Global*				*Per node*			
Id	1	6	12	24	1	6	12	24	1	6	12	24	1	6	12	24
1	4 s	0.7 s	0.35 s	0.2 s	4 s	0.6 s	0.3 s	0.2 ms	36 s	6 s	3 s	2 s	80 s	16 s	8 s	6 s
2	46 s	8 s	4 s	2 s	42 s	7 s	3.5 s	2 s	12 m	113 s	47 s	40 s	26 m	5 m	147 s	101 s
3	68 s	11 s	6 s	4 s	63 s	10 s	5 s	4 s	14 m	2 m	68 s	48 s	30 m	6 m	3 m	126 s
4	55 s	9 s	5 s	3 s	52 s	9 s	4 s	4 s	27 m	5 m	134 s	104 s	57 m	10 m	5 m	248 s

7 Future Works

This paper presented parallel algorithms for global and per-node triangle and square counts in large multigraphs. While our proposed algorithms have shown improvements in computational complexity, there is still room for future work to optimize further and improve the efficiency of the algorithms.

Firstly, we have identified that the current time complexity of the square count algorithm is $O(|\hat{V}|d_{cover}^2 d_{graph}/p)$, and we have not yet found ways to exploit the vertex cover to reduce the number of checks on 2 of the four vertices of the graph. Future research could explore the design of better algorithms that leverage these two nodes to reduce the computational requirements further.

Secondly, while we focused on developing efficient algorithms for triangle and square counts, we have not explored other algorithms for larger circuits using vertex cover-based acceleration. By searching for efficient algorithms for larger circuits, solutions with lower computational requirements could be discovered that also apply to the count of squares and possibly even triangles.

Thirdly, while our proposed triangle count algorithm can process ClueWeb09 in 40 min, the square count algorithm still cannot process graphs with billions of nodes in reasonable wall times. Future work could investigate the use of GPU-accelerated implementations to close this gap and enable faster execution of the square count on large instances.

In addition to the optimization and improvement of the algorithms, another important avenue for future work is the exploration of the use of these tools in the context of real-world applications, such as graph clustering [8]. While our algorithms provide a fast and efficient way to count triangles and squares and, therefore, to calculate clustering coefficients, we have not yet fully investigated their potential in the analysis of biological graphs such as protein-protein interaction graphs. These graphs are of significant interest in bioinformatics and have important applications in drug discovery and disease diagnosis [3,14]. Future research could explore the application of our proposed algorithms to these types of graphs, and investigate how the resulting triangle and square counts and clustering coefficients could be used to gain insights into the structure and function of large dynamic biological systems. By leveraging the power and efficiency of our algorithms, we believe that our tools could have important implications for the analysis of real-world graphs and the discovery of new insights in a variety of fields.

8 Conclusions

We have presented a set of parallel algorithms for counting triangles and squares in large multigraphs, which have demonstrated significant improvements in computational complexity compared to the best-known algorithms in the literature. Our algorithms achieve linear scaling with the number of available cores and have been evaluated on a range of real-world graphs and multigraphs, including protein-protein interaction graphs, knowledge graphs, and large web graphs. We have also shown that different vertex covers for square counts, both in the global and per-node versions, show no dominant option. In contrast, the increasing vertex covers heuristic is clearly dominant in the triangle counts. These findings could have important implications for optimising and designing future algorithms for counting triangles and squares in large multigraphs.

While our proposed algorithms have demonstrated significant improvements in computational complexity and efficiency, the limited scalability of the squares count algorithm on large instances highlights the need for future studies in high-performance computing settings. These could include exploring the use of GPUs and computing clusters to optimize the efficiency of the algorithm further and enable the processing of larger graphs in reasonable wall times.

Overall, our work contributes to the growing body of research on graph analytics and provides a valuable tool for researchers and practitioners working in a range of fields. By enabling fast and efficient counting of triangles and squares in large multigraphs, our algorithms have the potential to facilitate new insights and discoveries in areas such as bioinformatics, social network analysis, and web mining, among others. We hope that our work will inspire further research in this area and lead to new developments in the field of graph analytics.

References

1. Cappelletti, L., et al.: GRAPE: fast and scalable Graph Processing and Embedding (2021). arXiv: 2110.06196 [cs.LG]
2. Clarke, C.L., Craswell, N., Soboroff, I.: Overview of the trec: web track, p. 2009. Technical report DTIC Document (2009)
3. Friedel, C.C., Zimmer, R.L.: Inferring topology from clustering coefficients in protein-protein interaction networks. BMC Bioinf. **7**(1), 1–15 (2006)
4. Green, O., Bader, D.A.: Faster clustering coefficient using vertex covers. In: 2013 International Conference on Social Computing, pp. 321–330. IEEE (2013)
5. Part Guide: Intel® 64 and IA-32 architectures software developer's manual. In: Volume 3A: System Programming Guide 2.11 (2011)
6. Jha, M., Comandur, S., Pinar, A.: When a graph is not so simple: counting triangles in multigraph streams. Technical report Sandia National Lab. (SNL-CA), Livermore, CA (United States) (2013)
7. Li, Y., Shang, Y., Yang, Y.: Clustering coefficients of large networks. Inf. Sci. **382**, 350–358 (2017)
8. Nascimento, M.C.V., Carvalho, A.C.P.L.F.: A graph clustering algorithm based on a clustering coefficient for weighted graphs. J. Braz. Comput. Soc. **17**, 19–29 (2011)

9. Reese, J.T., et al.: KG-COVID-19: a framework to produce customized knowledge graphs for COVID-19 response. Patterns **2**(1), 100155 (2021). ISSN: 2666–3899. https://doi.org/10.1016/j.patter.2020.100155. https://www.sciencedirect.com/science/article/pii/S2666389920302038
10. Rossi, R.A., Ahmed, N.K.: Networkrepository: a graph data repository with interactive visual analytics. In: 29th AAAI Conference on Artificial Intelligence, Austin, Texas, USA, pp. 25–30 (2015)
11. Stefani, L.D., et al.: Triest: counting local and global triangles in fully dynamic streams with fixed memory size. ACM Trans. Knowl. Disc. Data (TKDD) **11**(4), 1–50 (2017)
12. Szklarczyk, D., et al.: The STRING database in 2021: customizable protein-protein networks, and functional characterization of user-uploaded gene/measurement sets. Nucleic Acids Res. **49**(D1), D605–D612 (2021)
13. Watts, D.J., Strogatz, S.H.: Collective dynamics of 'small- world'networks. Nature **393**(6684), 440–442 (1998)
14. Zaki, N., Efimov, D., Berengueres, J.: Protein complex detection using interaction reliability assessment and weighted clustering coefficient. BMC Bioinf. **14**(1), 1–9 (2013)

Deep Learning Attention Model for Supervised and Unsupervised Network Community Detection

Stanislav Sobolevsky[1,2](✉)

[1] Center for Urban Science and Progress, New York University, Brooklyn, NY, USA
sobolevsky@nyu.edu

[2] Department of Mathematics and Statistics and Institute of Law and Technology, Masaryk University, Brno, Czech Republic

Abstract. Network community detection is a complex problem that has to utilize heuristic approaches. It often relies on optimizing partition quality functions, such as modularity, description length, stochastic block-model likelihood etc. However, direct application of the traditional optimization methods has limited efficiency in finding the global maxima in such tasks. This paper proposes a novel bi-partite attention graph neural network model for supervised and unsupervised network community detection, suitable for unsupervised optimization of arbitrary partition quality functions, as well as for minimization of a loss function against the provided partition in a supervised setting. The model is demonstrated to be helpful in the unsupervised improvement of suboptimal partitions previously obtained by other known methods like Louvain algorithm for some of the classic and synthetic networks. It is also shown to be efficient in supervised learning of the provided community structure for a set of classic and synthetic networks. Furthermore, the paper serves as a proof-of-concept for the broader application of graph neural network models to unsupervised network optimization.

Keywords: Complex networks · Community detection · Deep Learning · Graph Neural Networks

1 Introduction

The network community detection saw a wide range of applications, including social science [36], biology [21], and economics [35]. In particular, partitioning the networks of human mobility and interactions is broadly applied to regional delineation [1,3,7,20,24,37,43,44], and urban zoning [27,28,42].

Community detection is a complex problem, and multiple algorithms have been proposed to address it. Some of them are straightforward, such as hierarchical clustering [23] or the Girvan-Newman [16] algorithm, while the majority rely on optimization techniques for various objective functions. The most well-known partition quality function is modularity [33,34] assessing the relative strength of edges and quantifying the cumulative strength of the intra-community links.

J. Mikyška et al. (Eds.): ICCS 2023, LNCS 14076, pp. 647–654, 2023.
https://doi.org/10.1007/978-3-031-36027-5_51

A large number of modularity optimization strategies have been suggested over the last two decades [8,12,13,19,22,29,31,33,34,41,45]. Comprehensive overviews are presented in [14,15] and later surveys [25,26].

As the problem is known to be NP-hard [9]), there is no efficient algorithm that could guarantee finding the optimal partition. Therefore, optimization has to rely on heuristic algorithms, which often fail to reach the optimal partition, and, therefore, may require further fine-tuning. Although in some cases, an algorithmic optimality proof of the partition is possible [4–6,40].

The rise of deep learning and graph neural networks in particular, offer new opportunities. Recently graph neural networks (GNNs) have become increasingly popular for supervised classifications and unsupervised embedding of the graph nodes with diverse applications in text classification, recommendation systems, traffic prediction, computer vision etc. [47]. And GNNs were already attempted to be applied for community detection, including supervised learning of the ground-truth community structure [11] as well as some unsupervised learning of the node features enabling representation modeling of the network, including stochastic block-model [10] and other probabilistic models with overlapping communities [38] or more complex self-expressive representation [2]. However, existing GNN applications for unsupervised community detection has been limited so far, and largely overlook unsupervised modularity optimization.

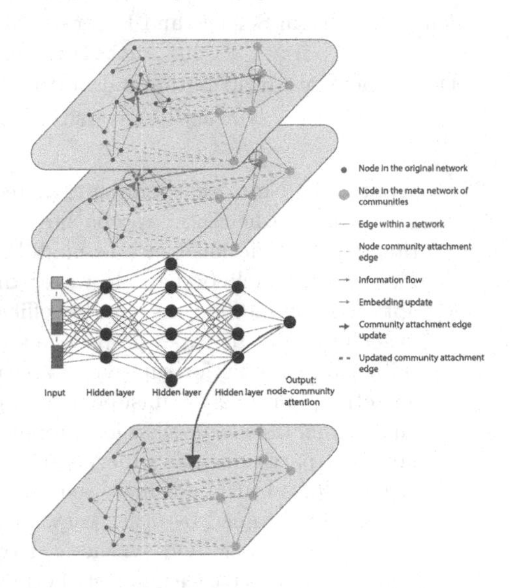

Fig. 1. A deep learning framework for network community detection.

This work proposes a suitable network augmentation with an additional layer of community meta-nodes, and a novel deep learning model over such a network for supervised and unsupervised community detection, capable of optimizing arbitrary quality functions for the network partition, including modularity.

2 Methodology

In our recent paper [39], we proposed a simple recurrent GNN-inspired algorithm to serve as a proof-of-concept for unsupervised modularity optimization. The algorithm tunes node community attachment through an iteration of the GNN-style transformations. It reached the best-known partitions for some of the classical networks, and provided a scheme for fine-tuning the network community structure with a flexible trade-off between quality and speed.

Table 1. Performance on the proposed deep learning algorithm improving partition modularity from Louvain algorithm for some classic and synthetic network examples

Network	Louvain	Improvement
Word adjacency network in David Copperfield [32]	0.305052	0.309279
Amazon co-purchases of political books, orgnet.com	0.527082	0.527236
LFR 500 nodes	0.837450	0.837501
LFR 1000 nodes	0.888029	0.888615
LFR 2000 nodes	0.901624	0.902277

In order to further improve the efficiency of the approach, we propose a new model which consists of a two-layer bi-partite convolutional graph neural network stacked with a fully connected attention vanilla neural network Fig. 1. The model takes certain initial network node embedding as the input, such as the personalized Pagerank probability vectors (for each source node defining the stationary probability distribution of a Markov chain that, with probability $\alpha = 0.15$, randomly transitions following the link structure of the network, and with a probability $1 - \alpha$ teleports to a source node [17]), further reduced in dimensionality using a linear principal component method. The edges between the network nodes and community meta-nodes are initially defined with respect to a certain initial node community attachment (either a known one to be improved if used for partition fine-tuning, or based on clustering initial node embedding), and can be further updated as the model updates the node community attachment. The two layers of a graph neural network propagating the initial node embedding over such a bi-partite network generates the final embedding for both - the original network nodes as well as the community meta-nodes. Finally, the vanilla neural network (in the experiments below, the configuration with five hidden layers and batch normalization has been evaluated) for each pair of the original network node and the community meta-node takes the stacked vectors of those node embedding, generated by the graph neural network, and computes the relative attention score between the two nodes. The resulting node community attachment (a "fuzzy" probabilistic one rather than discrete) is redefined proportional to those attention scores. The weights of both - the final vanilla neural network as well as the graph neural network layers are trained together, within the backpropagation framework, aiming to optimize the final objective function - either the quality function of the resulting network partition in the unsupervised setting (like modularity or stochastic block-model likelihood) or the loss function between this resulting partition and the known partition in the supervised setting (like categorical cross-entropy). In order to improve the convergence stability and final performance of the model in the unsupervised setting, it can be initially pre-trained to reconstruct a previously known community structure prior to the final unsupervised fine-tuning phase.

3 Results

The approach turned out to be efficient in fine-tuning the results of other algorithms, e.g. a popular Louvain algorithm [8]. The Table 1 provides examples of such improvement reached by the Python 3.7 implementation of the proposed model for several classic and Lancichinetti-Fortunato-Radicchi (LFR) synthetic networks (Table 2). And while for the two classic networks - Amazon political books and Word Adjacency network in David Copperfield - other known efficient algorithms like Combo were also capable of improving the partition, for the three provided cases of LFR networks, the fine-tuned partition, provided by the proposed deep learning algorithm, is the best partition known to us.

Table 2. Out-of-sample performance on the proposed deep learning algorithm in supervised learning of the best-known or given partition for some classic network examples (community reconstruction accuracy for the 40% randomly masked nodes) in comparison with the label propagation baseline

Network	Accuracy	Baseline
Amazon.com co-purchases of political books, www.orgnet.com	97.22%	94.44%
Dolphins' Social Network [30]	95.00%	95.00%
Network of Jazz Musicians [18]	93.10%	88.51%
Neural network of C. Elegans [46]	86.96%	75.65%
Metabolic network of C. Elegans [13]	64.33%	57.31%

The proposed approach can also be applied to other quality functions, such as a block-model likelihood or description length.

Furthermore, the model can perform supervised community detection, extrapolating the community structure provided for a certain part of the network nodes to the rest of them. The out-of-sample reconstruction accuracy for the best-known partition often ranges within 90–99% for a number of classic (Table 2) and LFR synthetic networks (Table 3).

For comparison, the out-of-sample accuracy of the label propagation baseline algorithm (nodes with unknown community attachments get attached according to the majority of their neighbors with known attachments) for most of the provided networks falls noticeably short of the accuracy achieved by the proposed deep learning approach.

Those cases represent initial proof-of-concept results, while fine-tuning of the model's configuration could further help improve the performance. Also, evaluation of the approach on a broader range of examples and comparison against known state-of-the-art/baseline supervised community detection approaches remains the subject of future work.

Finally, as the deep learning model configuration does not depend on the dimensionality of the network or the number of network communities but only on the selected dimensionality of the node embedding, it makes it possible to consider transferring the pre-trained model architectures and parameters between

Table 3. Out-of-sample performance on the proposed deep learning algorithm in supervised learning of the best-known or given partition for LFR synthetic networks (community reconstruction accuracy for the 40% randomly masked nodes) in comparison with the label propagation baseline

Network	Size	Accuracy	Baseline
1	500	93.65%	91.53%
2	500	99.47%	93.65%
3	500	94.18%	91.53%
4	500	95.24%	89.42%
5	500	98.94%	96.30%
6	1000	98.61%	93.98%
7	1000	96.99%	91.67%
8	1000	95.83%	89.35%
Average		96.61 ± 2.23%	92.18 ± 2.36%

the networks. And similarly to [39], iterating an ensemble of partition fine-tuning models (pre-trained over select sample networks) over the target network partition may provide the best practical results.

4 Conclusions

To summarize, the novel bi-partite attention graph neural network has been proposed for supervised and unsupervised network community detection. The model augments the original network with the meta-nodes representing the network communities and learns the node embedding as well as the relevance links between the two types of network nodes.

It was proven useful for the supervised reconstruction of the network community structure for both - classic and synthetic networks, consistently outperforming a baseline label propagation algorithm. In an unsupervised setting, we found the model helpful in fine-tuning the suboptimal network partitions obtained for some of the classic and synthetic networks by other known community detection algorithms, like Louvain.

While the presented results serve as a proof of concept of the proposed deep learning model's utility for supervised and unsupervised community detection, its further fine-tuning and extensive evaluation, as well as exploring the potential of transfer learning between the networks, is the subject of our future research.

Acknowledgements. This research was supported by the MUNI Award in Science and Humanities (MASH Belarus) of the Grant Agency of Masaryk University under the Digital City project (MUNI/J/0008/2021). The work of Stanislav Sobolevsky was also partially supported by ERDF "CyberSecurity, CyberCrime and Critical Information Infrastructures Center of Excellence" (No. CZ.02.1.01/0.0/0.0/16_019/0000822). The author thanks Mingyi He for valuable help with visualization, and Dr. Alexander Belyi for insightful discussions.

References

1. Amini, A., Kung, K., Kang, C., Sobolevsky, S., Ratti, C.: The impact of social segregation on human mobility in developing and industrialized regions. EPJ Data Sci. **3**(1), 1–20 (2014). https://doi.org/10.1140/epjds31
2. Bandyopadhyay, S., Peter, V.: Self-expressive graph neural network for unsupervised community detection. arXiv preprint arXiv:2011.14078 (2020)
3. Belyi, A., et al.: Global multi-layer network of human mobility. Int. J. Geogr. Inf. Sci. **31**(7), 1381–1402 (2017)
4. Belyi, A., Sobolevsky, S.: Network size reduction preserving optimal modularity and clique partition. In: Gervasi, O., Murgante, B., Hendrix, E.M.T., Taniar, D., Apduhan, B.O. (eds.) ICCSA 2022. LNCS, pp. 19–33. Springer, Cham (2022). https://doi.org/10.1007/978-3-031-10522-7_2
5. Belyi, A., Sobolevsky, S., Kurbatski, A., Ratti, C.: Improved upper bounds in clique partitioning problem. J. Belarusian State Univ. Math. Inform. **2019**(3), 93–104 (2019). https://doi.org/10.33581/2520-6508-2019-3-93-104
6. Belyi, A., Sobolevsky, S., Kurbatski, A., Ratti, C.: Subnetwork constraints for tighter upper bounds and exact solution of the clique partitioning problem. arXiv preprint arXiv:2110.05627 (2021)
7. Blondel, V., Krings, G., Thomas, I.: Regions and borders of mobile telephony in Belgium and in the brussels metropolitan zone. Brussels Studies. La revue scientifique électronique pour les recherches sur Bruxelles/Het elektronisch wetenschappelijk tijdschrift voor onderzoek over Brussel/The e-journal for academic research on Brussels (2010)
8. Blondel, V.D., Guillaume, J.L., Lambiotte, R., Lefebvre, E.: Fast unfolding of communities in large networks. J. Stat. Mech: Theory Exp. **2008**(10), P10008 (2008)
9. Brandes, U., et al.: Maximizing modularity is hard. arXiv preprint physics/0608255 (2006)
10. Bruna, J., Li, X.: Community detection with graph neural networks. Stat **1050**, 27 (2017)
11. Chen, Z., Li, X., Bruna, J.: Supervised community detection with line graph neural networks. arXiv preprint arXiv:1705.08415 (2017)
12. Clauset, A., Newman, M.E.J., Moore, C.: Finding community structure in very large networks. Phys. Rev. E **70**, 066111 (2004). https://doi.org/10.1103/PhysRevE.70.066111
13. Duch, J., Arenas, A.: Community detection in complex networks using extremal optimization. Phys. Rev. E **72**, 027104 (2005). https://doi.org/10.1103/PhysRevE.72.027104
14. Fortunato, S.: Community detection in graphs. Phys. Rep. **486**, 75–174 (2010)
15. Fortunato, S., Hric, D.: Community detection in networks: a user guide. Phys. Rep. **659**, 1–44 (2016)
16. Girvan, M., Newman, M.: Community structure in social and biological networks. Proc. Natl. Acad. Sci. USA **99**(12), 7821–7826 (2002)
17. Gleich, D.F.: PageRank beyond the web. SIAM Rev. **57**(3), 321–363 (2015)
18. Gleiser, P.M., Danon, L.: Community structure in jazz. Adv. Complex Syst. **06**(04), 565–573 (2003). https://doi.org/10.1142/S0219525903001067

19. Good, B.H., de Montjoye, Y.A., Clauset, A.: Performance of modularity maximization in practical contexts. Phys. Rev. E **81**, 046106 (2010). https://doi.org/10.1103/PhysRevE.81.046106
20. Grauwin, S., et al.: Identifying and modeling the structural discontinuities of human interactions. Sci. Rep. **7** (2017)
21. Guimerà, R., Nunes Amaral, L.A.: Functional cartography of complex metabolic networks. Nature **433**(7028), 895–900 (2005). https://doi.org/10.1038/nature03288
22. Guimera, R., Sales-Pardo, M., Amaral, L.A.N.: Modularity from fluctuations in random graphs and complex networks. Phys. Rev. E **70**(2), 025101 (2004)
23. Hastie, T.: The Elements of Statistical Learning: Data Mining, Inference, and Prediction: With 200 Full-Color Illustrations. Springer, New York (2001)
24. Hawelka, B., Sitko, I., Beinat, E., Sobolevsky, S., Kazakopoulos, P., Ratti, C.: Geo-located Twitter as proxy for global mobility patterns. Cartogr. Geogr. Inf. Sci. **41**(3), 260–271 (2014)
25. Javed, M.A., Younis, M.S., Latif, S., Qadir, J., Baig, A.: Community detection in networks: a multidisciplinary review. J. Netw. Comput. Appl. **108**, 87–111 (2018)
26. Khan, B.S., Niazi, M.A.: Network community detection: a review and visual survey. arXiv preprint arXiv:1708.00977 (2017)
27. Landsman, D., Kats, P., Nenko, A., Kudinov, S., Sobolevsky, S.: Social activity networks shaping St. Petersburg. In: Proceedings of the 54th Hawaii International Conference on System Sciences, p. 1149 (2021)
28. Landsman, D., Kats, P., Nenko, A., Sobolevsky, S.: Zoning of St. Petersburg through the prism of social activity networks. Procedia Comput. Sci. **178**, 125–133 (2020)
29. Lee, J., Gross, S.P., Lee, J.: Modularity optimization by conformational space annealing. Phys. Rev. E **85**, 056702 (2012). https://doi.org/10.1103/PhysRevE.85.056702
30. Lusseau, D., Schneider, K., Boisseau, O.J., Haase, P., Slooten, E., Dawson, S.M.: The bottlenose dolphin community of Doubtful Sound features a large proportion of long-lasting associations. Behav. Ecol. Sociobiol. **54**(4), 396–405 (2003). https://doi.org/10.1007/s00265-003-0651-y
31. Newman, M.E.J.: Fast algorithm for detecting community structure in networks. Phys. Rev. E **69**, 066133 (2004). https://doi.org/10.1103/PhysRevE.69.066133
32. Newman, M.E.: Finding community structure in networks using the eigenvectors of matrices. Phys. Rev. E **74**(3), 036104 (2006)
33. Newman, M.: Modularity and community structure in networks. Proc. Natl. Acad. Sci. **103**(23), 8577–8582 (2006)
34. Newman, M., Girvan, M.: Finding and evaluating community structure in networks. Phys. Rev. E **69**(2), 026113 (2004)
35. Piccardi, C., Tajoli, L.: Existence and significance of communities in the world trade web. Phys. Rev. E **85**, 066119 (2012). https://doi.org/10.1103/PhysRevE.85.066119
36. Plantié, M., Crampes, M.: Survey on social community detection. In: Ramzan, N., van Zwol, R., Lee, J.S., Clüver, K., Hua, X.S. (eds.) Social Media Retrieval. CCN, pp. 65–85. Springer, London (2013). https://doi.org/10.1007/978-1-4471-4555-4_4
37. Ratti, C., et al.: Redrawing the map of Great Britain from a network of human interactions. PLoS ONE **5**(12), e14248 (2010). https://doi.org/10.1371/journal.pone.0014248

38. Shchur, O., Günnemann, S.: Overlapping community detection with graph neural networks. arXiv preprint arXiv:1909.12201 (2019)
39. Sobolevsky, S., Belyi, A.: Graph neural network inspired algorithm for unsupervised network community detection. Appl. Netw. Sci. **7**(1), 1–19 (2022)
40. Sobolevsky, S., Belyi, A., Ratti, C.: Optimality of community structure in complex networks. arXiv preprint arXiv:1712.05110 (2017)
41. Sobolevsky, S., Campari, R., Belyi, A., Ratti, C.: General optimization technique for high-quality community detection in complex networks. Phys. Rev. E **90**(1), 012811 (2014)
42. Sobolevsky, S., Kats, P., Malinchik, S., Hoffman, M., Kettler, B., Kontokosta, C.: Twitter connections shaping New York City. In: Proceedings of the 51st Hawaii International Conference on System Sciences (2018)
43. Sobolevsky, S., Sitko, I., Des Combes, R.T., Hawelka, B., Arias, J.M., Ratti, C.: Money on the move: big data of bank card transactions as the new proxy for human mobility patterns and regional delineation. The case of residents and foreign visitors in Spain. In: 2014 IEEE International Congress on Big Data (BigData Congress), pp. 136–143. IEEE (2014)
44. Sobolevsky, S., Szell, M., Campari, R., Couronné, T., Smoreda, Z., Ratti, C.: Delineating geographical regions with networks of human interactions in an extensive set of countries. PLoS ONE **8**(12), e81707 (2013)
45. Sun, Y., Danila, B., Josić, K., Bassler, K.E.: Improved community structure detection using a modified fine-tuning strategy. EPL (Europhysics Letters) **86**(2), 28004 (2009). http://stacks.iop.org/0295-5075/86/i=2/a=28004
46. White, J.G., Southgate, E., Thomson, J.N., Brenner, S.: The structure of the nervous system of the nematode Caenorhabditis elegans. Philos. Trans. R. Soc. Lond. B, Biol. Sci. **314**(1165), 1–340 (1986). https://doi.org/10.1098/rstb.1986.0056. http://rstb.royalsocietypublishing.org/content/314/1165/1.abstract
47. Wu, Z., Pan, S., Chen, F., Long, G., Zhang, C., Philip, S.Y.: A comprehensive survey on graph neural networks. IEEE Trans. Neural Netw. Learn. Syst. (2020)

On Filtering the Noise in Consensual Communities

Antoine Huchet(✉), Jean-Loup Guillaume, and Yacine Ghamri-Doudane

L3i, La Rochelle University, La Rochelle, France
{antoine.huchet,jean-loup.guillaume,yacine.ghamri}@univ-lr.fr

Abstract. Community detection is a tool to understand how networks are organised. Ranging from social, technological, information or biological networks, many real-world networks exhibit a community structure. *Consensual* community detection fixes some of the issues of classical community detection like non-determinism. This is often done through what is called a *consensus* matrix. We show that this consensus matrix is not filled with relevant information only, it is noisy. We then show how to filter out some of the noise and how it could benefit existing algorithms.

Keywords: Consensual Community Detection · Noise · Complex Networks

1 Introduction

Biological, social, technological and information networks can be represented by graphs [1]. Thus, graph analysis has become an important tool to understand such networks. Graphs representing real-world data exhibit particular features that make them far from regular. Few nodes tend to have a lot of neighbours while many nodes tend to have few neighbours. The distribution of edges is not homogeneous: parts of the graph are densely connected, while between such dense parts, there tend to be only a few edges. Such feature of real-world graphs is called *community structure* [2] and finding such densely connected parts of a graph is called *community detection*. In a network of purchase relationship between customers and products, community detection can identify communities of customers with similar interests, thus improve recommendation systems. In social graphs, it could help identify group of people: families, friends or coworkers.

Many community detection algorithms exists like Walktrap [3], Infomap [4] or Louvain [5]. Some are non-deterministic like Louvain, in which the result is determined by the order in which the nodes are visited. Since the nodes may be visited in any order, such algorithm may produce different partitions of communities. Therefore, we need a way to pick a "good" partition. In order to do so, we

The authors would like to sincerely thank Bivas Mitra (from IIT Kharagpur, India) for the insightful discussions in the early stages of the article. This work has been partially funded by the ANR MITIK project, French National Research Agency (ANR), PRC AAPG2019.

J. Mikyška et al. (Eds.): ICCS 2023, LNCS 14076, pp. 655–670, 2023.
https://doi.org/10.1007/978-3-031-36027-5_52

would need a criterion to sort the partitions and pick the one that is the most representative of the actual community structure of the network.

In the absence of such criterion, we combine the information of different partitions of communities into *consensual* communities [6] (also known in the literature as community *cores* or *constant* communities). Such communities allow working with dynamic graphs as it becomes easier to follow communities in a timestamped network when the computation is deterministic [7].

Our contribution is twofold: we show that the information from the different partitions of communities is noisy. Then when combining the partitions into consensual communities, we show that some of the noise can be avoided.

This article is organised as follows: Sect. 2 formally defines general graphs concepts needed in Sect. 3, which gives an overview of related works in the area of consensual community detection. Then, Sect. 4 shows that the computation carries noise, and Sect. 5 shows a way to filter out some of the noise. Section 6 shows how our ideas can improve some state of the art algorithms, on some synthetic and real data. The last section concludes the article.

2 Definitions

A *graph* $G = (V, E)$ is made of a set of nodes V and a set of edges $E \in V \times V$, where $|V| = n$ and $|E| = m$. Within a graph, *communities* are defined as a partition of the nodes. That is, each node belongs to exactly one community.

To evaluate the quality of a set of communities, the *Modularity* [8] compares the number of intra-community edges to the expected number of intra-community edges in a random graph with the same number of nodes, same number of edges and same degree distribution as the starting graph. It is defined as

$$Q = \frac{1}{2m} \sum_{i,j} \left[A_{ij} - \frac{k_i k_j}{2m} \right] \delta(c_i, c_j),$$

where i and j are nodes of the graph, and A_{ij} is 1 if there is an edge between i and j, 0 otherwise. m is the number of edges of the graph, k_i is the degree of i, and $\delta(c_i, c_j)$ is 1 if nodes i and j are in the same community, 0 otherwise.

If we have ground-truth communities, we can directly compare a partition given by any community detection algorithm to this ground-truth. In this case we can use the Normalised Mutual Information (*NMI*) [9]. It is based on a *confusion matrix* N where N_{ij} is the number of nodes of the ground-truth community i that also belong to the computed community j. It if formally defined as

$$I(A, B) = \frac{-2 \sum_{i=1}^{c_A} \sum_{j=1}^{c_B} N_{ij} \log\left(\frac{N_{ij} N}{N_{i.} N_{.j}}\right)}{\sum_{i=1}^{c_A} N_{i.} \log\left(\frac{N_{i.}}{N}\right) + \sum_{j=1}^{c_B} N_{.j} \log\left(\frac{N_{.j}}{N}\right)},$$

where c_A (resp. c_B) is the number of ground-truth (resp. computed) communities. The sum over row i (resp. column j) of N is denoted $N_{i.}$ (resp. $N_{.j}$). The NMI takes values from 0 (independent partitions) to 1 (same partitions).

Finally it is also possible to use synthetic graphs generated using the Lancichinetti, Fortunato and Radicchi (*LFR*) model [10]. This model creates synthetic graphs with a known community structure. They are generated with a *mixing* parameter μ that ranges from 0 to 1, and denotes the fraction of edges that a node shares with other communities of the graph. Thus, the smaller μ, the stronger the communities are and the easier it will be to detect them.

3 Related Work

Most community detection algorithms are non-deterministic. Thus, running any such algorithm $\mathscr{A}$ multiple times on the same graph G may result in different partitions. Moreover, these partitions cannot be discriminated on the basis of their modularity alone, as it has been shown that there are significantly different partitions with similar modularity [11]. To address this issue, a solution is to search similarities between different partitions to obtain *consensual communities.*

These similarities has also been studied for random graphs. It has been shown that random graphs contain good communities (or at least partitions of high modularity) which is very unnatural. However, different partitions are very dissimilar from each other. The absence of similarities might therefore indicate that the communities are not meaningful [12]. Peixoto et al. showed that when there is too much diversity in several partitions, it is in general not possible to obtain a consistent answer. Therefore, consensual communities are expected to yield good results only when there is a community structure in the graph.

3.1 The Consensus Matrix

To record similarities between partitions, most algorithms rely, in one way or another, on a *consensus matrix* P. First, a (classical) community detection algorithm $\mathscr{A}$ is ran n_p times on a graph G. Since $\mathscr{A}$ is non-deterministic, the n_p results will most likely be different. Then, the consensus matrix P is built as follows: the *consensus coefficient* P_{ij} is the number of times that nodes i and j were placed in the same community by $\mathscr{A}$ during the n_p executions. P_{ij} is thus between 0 and n_p. Then, the weighted *consensus graph* G_P whose adjacency matrix is P is built: nodes i and j are linked in G_P by an edge whose weight is P_{ij}. If $P_{ij} = 0$, then nodes i and j are not connected [6,7].

Rather than using $\mathscr{A}$ multiple times, some authors obtain diversity by using different algorithms to generate the first n_p. The Azar method [13] picks n_p different algorithms and each of them is run only once on G. Then the information from the n_p partitions is aggregated in a consensus matrix as described before, from which they build the associated consensus graph.

Liu et al. [14] chose to avoid local optimum of modularity by applying perturbations to the initial graph, then used a modified version of Louvain [5] that may explores a wider range of possible solutions. Finally, they aggregate the result in a consensus matrix. Burgess et al. [15] focuses on adding intra-community edges to the starting graph G, hoping to increase the efficiency of community detection algorithm. They do so by computing the similarity between pairs of nodes, based on metrics such as Jaccard [16] or the number of common neighbours. The edges are then added in a non-deterministic way to G, and they execute $\mathscr{A}$ on G. The process is repeated n_p times to obtain a consensus matrix.

Rasero et al. [17] use another version of consensus matrix that is not obtained through community detection but from the original data. They study brain connectivity graphs from different persons and aggregate the different graphs into a consensus matrix.

3.2 From Consensus Matrix to Consensual Communities

The first type of approach developed by Seifi et al. [7] does not require additional community detection. It considers that low values of P_{ij} are not significant and therefore keep only the values P_{ij} higher than a given threshold τ and set all the others to 0. This gives a thresholded matrix P_τ and the associated graph G_{P_τ}. The connected components of G_{P_τ} are then directly the consensual communities. Work has been done to get rid of τ and decide whether the P_{ij} entry is statistically significant before keeping it or not [18]. In the same vein, Chakraborty et al. [19] consider that constant communities are groups of nodes that are always in the same community in several executions (i.e., $P_{ij} = 1$).

Another approach consists in executing one final community detection algorithm on the consensus graph. Liang et al. [20] use the Label Propagation algorithm (LPA) [21] n_p times to build the consensus matrix, then perform a final single execution of a weighted version of LPA on the consensus graph. This has been generalized for instance in the LF procedure [6] that runs $\mathscr{A}$ n_p times again on G_P. At this point, if the n_p partition just computed are all the same, they are considered consensual and the algorithm stops. Otherwise, a new consensus matrix is built, and the process is repeated until all n_p partitions are the same.

Wang and Fleury developed an algorithm for overlapping communities that builds a consensus graph, merges nodes when their consensus coefficient is greater than a threshold α and repeat the process [22]. A similar idea using a consensus matrix has been developed by Yang and Leskovec [23].

3.3 Complexity Issues

The consensus matrix P is an $n \times n$ matrix. The computation and storage of such a matrix quickly becomes infeasible on large graphs. This is unfortunately regularly the case with complex networks with up to several billions of nodes [1]. Several studies have therefore worked on improving the computation of the consensus matrix, as in the algorithms of Tandon [24] and ECG [25] where they only compute the entries P_{ij} for the edges $\{i, j\}$ of G. Instead of having n^2 entries in P, they only have m, the number of edges in the graph and all other entries are 0. Tandon chooses to construct a consensus graph G_P on which $\mathscr{A}$ is run n_p times until convergence. ECG, on the other hand, runs $\mathscr{A}$ one last time on the consensus graph G_P and considers the result as final.

The presence of edges between communities limits the efficiency of community detection algorithms. If there were only intra-community edges, the communities would be the connected components of the graph and identifying them would be trivial. Therefore, adding all the entries that correspond to the edges of the graph into the consensus matrix adds noise. The two main contributions of this article are to highlight the presence of noise and to find another criterion that would decrease further the number of entries, hence limit this noise. In addition, decreasing the number of entries in P could also reduce the temporal and spatial complexity of the consensus matrix computation.

4 Information is Noisy

This section shows that the consensus matrix P is noisy. We do this by showing that the quality of the community is maximal when the consensus matrix is only partially filled. This means that there is some noise in P when it is full.

First, we generate a synthetic LFR graph G (along with its ground-truth communities). We use a distance DIST between nodes that will be described in more details in the following subsections. Next, we execute n_p times a community detection algorithm $\mathscr{A}$ on G. Finally, we build a consensus matrix P, but we fill it one entry P_{ij} at a time in increasing order of $\text{DIST}(i, j)$. In case of a tie, we break it by selecting an arbitrary pair of nodes among the tie. After each addition in the incomplete consensus matrix P, we build the associated partial consensus graph G_P and execute $\mathscr{A}$ on G_P. We then compute the NMI between the LFR ground-truth communities and the communities we just computed. We iterate this way until P is completely filled. This method allows us to study how the NMI varies based on the number of entries in P.

Our methodology (see Algorithm 1) uses several parameters: the LFR graph and its mixing parameter μ that defines the proportion of links between communities; the community detection algorithm $\mathscr{A}$, the number n_p of executions of $\mathscr{A}$ and the distance DIST. In the rest of the paper we use $\mathscr{A}$ = Louvain as it is very fast, $n_p = 12$ as it has been shown that the consensus matrix rapidly converges as n_p grows [25] and we will vary μ and use several DIST.

Algorithm 1. ORDERING (an LFR graph G, $\mathscr{A}$, n_p, DIST)

1: Build an ordered list L of the pairs of nodes (i, j), following the order given by DIST
2: Execute n_p times algorithm $\mathscr{A}$ on G
3: **while** L is not empty **do**
4: Pick the pair (i, j) whose $\text{dist}(i, j)$ is the smallest among L
5: Compute the consensus coefficient P_{ij} and add it to P
6: Execute $\mathscr{A}(G_P)$
7: Compute the NMI between the ground truth communities, and those found at the previous step
8: **end while**
9: **return** The plot with the NMI as a function of the number of entries in P

4.1 Graph Distance

First we consider the graph distance (smallest number of consecutive edges) between two nodes as DIST, and apply Algorithm 1 for different values of μ. We use the graph distance since in Tandon [24] and ECG [25] the authors restrict themselves to connected pairs of nodes, i.e., nodes at distance 1.

Figure 1 shows the NMI as a function of the number of entries in P, for LFR graphs with 10 000 nodes and 3 different values of the mixing parameter μ. The graphs were generated with the implementation available online. The parameters used for LFR are: number of nodes $n = 10.000$, average degree $k = 20$, maximum degree $\max_k = 50$, degree sequence exponent $t_1 = 2$, community size distribution exponent $t_2 = 3$, minimum community size $\min_c = 10$, maximum community size $\max_c = 50$, weight mixing parameter $\mu_w = 0.6$.

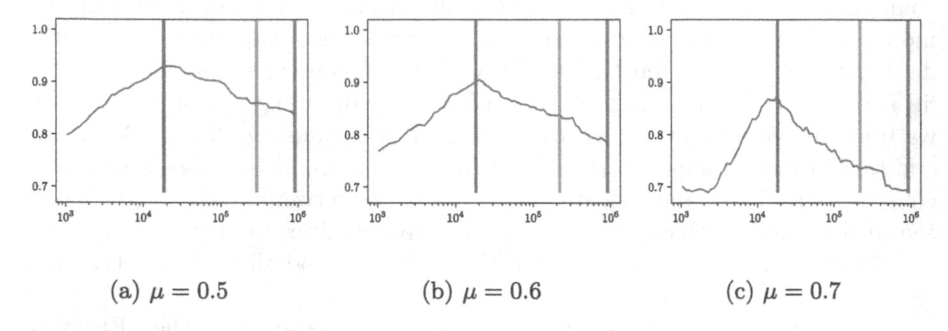

Fig. 1. NMI vs number of entries of P filled using the graph distance ordering. LFR graphs with 10.000 nodes, values of the mixing parameter μ from .5 to .7. The vertical bars represent change of distance: entries that are added before the blue bar correspond to pairs of nodes at distance 1 (using a random order as a tie-breaker). Between the blue and orange bars are pairs at distance 2... (Color figure online)

Figures 1 provide us with three main information. First of all, higher values of μ correspond to lower values of NMI. Since higher values of μ creates less pronounced communities, they are harder to find. The maximum values obtained are respectively .93, .904 and .874 respectively for μ from .5 to .7.

Then, the shape of the curves indicates that entries need to be filled. Indeed, after a certain number of entries, the NMI decreases. Intuitively, community detection algorithms would work better if there were no inter-community edges, and if all intra-community edges were present. The non-determinism of $\mathscr{A}$ makes the n_p partitions potentially different and thus, some non-zero consensus coefficients correspond to pairs of nodes that are in different communities in the ground-truth. The more we add such values, the harder it is for the algorithm to find the real communities, even though these values are very close to 0.

Finally, we can observe that, using the graph distance, the NMI is maximised when all entries corresponding to pairs of nodes at distance 1 are present in the consensus matrix, plus some of the entries at distance 2 (the peak is located a little after the vertical blue bar). Even if Tandon [24] and EGC [25] do not use exactly the same method as we do, it might be possible to do a little better than these algorithms by adding a few more entries in the consensus matrix.

To improve these algorithms we would therefore need to select some at distance 2. However, since the graph distance can only take integer values, we are limited by its precision and we therefore need another criterion that would help discriminate such entries.

4.2 Edge Clustering Coefficient

Intuitively we want to use a distance that will first add entries corresponding to intra-community pairs of nodes. Several criteria have been used to identify such pairs and they are often used to find communities. We can cite, among others, the Jaccard coefficient [16], the Cosine similarity, the Hamming distance [26] or the Edge Clustering Coefficient (*ECC*) [27] that measures the similarity between nodes i and j, and is defined as

$$ECC(i,j) = \frac{nb\_common\_neigh}{\min(\deg(i), \deg(j))},$$

where nb_common_neigh is the number of neighbours that nodes i and j have in common, while deg() is the degree of a node.

Most of these similarities give good results, even though we ruled out Cosine because of its higher computing time. We present here the results for the ECC.

We proceed as before: we generate an LFR graph G, but time we use the ECC as DIST. We compute the ECC of all pairs of nodes that are at distance one or two. Nodes at distance three or higher cannot have any neighbour in common, thus their ECC will be zero. They should be avoided to maximise the NMI anyway, as shown in Figs. 1. Finally, we use Algorithm 1 but insert values in decreasing order of the ECC.

Figure 2 shows the NMI as a function of the number of entries in P, for 3 different values of μ of LFR graphs with 10 000 nodes. The graphs were generated with the same implementation and the same parameters as Fig. 1. The ECC value being a real number, we can now precisely choose which entry to add in the consensus matrix, based on their ECC. We also notice that the maximum NMI is much greater for the ECC than graph distance. They are respectively .9994, .9910 and .8864 for μ from .5 to .7 (compared to .93, .904 and .874).

In real scenarios, we do not know the ground truth communities, so we cannot compute the NMI. We therefore need to decide beforehand how many entries need to be inserted in P to get as close as possible to the maximum NMI. The next section focuses on finding such an optimum number of entries.

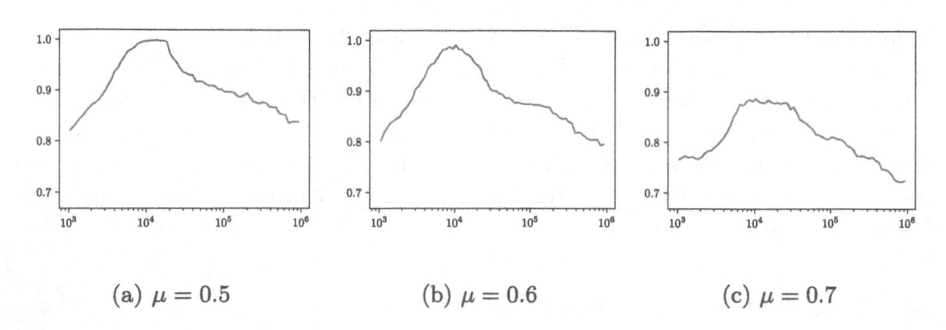

(a) $\mu = 0.5$ (b) $\mu = 0.6$ (c) $\mu = 0.7$

Fig. 2. NMI vs number of entries of P using the ECC in decreasing order. LFR graphs with 10.000 nodes and different values of μ.

5 Finding an Optimum Number of Entries – Filtering Out The noise

This subsection is devoted to showing how, based on some correlations, we can estimate the number of entries to insert in the consensus matrix.

Figure 3a shows that the average modularity obtained across multiple executions of $\mathscr{A}$ is linearly correlated with the mixing parameters μ on LFR graphs. We use the correlation:

$$Q = 1 - \mu$$

Kaminski et al. studied such phenomenon on the ABCD graph model, which is very close to LFR [28]. They shown that the modularity of the ground truth partition of an ABCD graph with a mixing parameter μ asymptotically reaches $1 - \mu$ as the number of nodes n grows. They also observe that the modularity is smaller for smaller graphs, and converges as n grows.

Figure 3b shows that the mixing parameter μ is correlated to the ECC value τ above which pairs of nodes should be added to the consensus matrix to maximise the NMI. That is, all the pairs of nodes $\{i, j\}$ whose ECC is greater than τ should have their consensus coefficient put in the consensus matrix.

$$\tau = -1.414\mu + 0.991$$

Combining these two correlations it is therefore possible to deduce the optimum number of entries in the consensus matrix from the first n_p executions of $\mathscr{A}$. We compute several partitions (which is anyway the first step to compute the consensus matrix), we compute the average modularity and deduce the value of μ from which we deduce the lower limit on ECC. These correlations are made on LFR synthetic graphs and may not be valid for real graphs. However, we will see in Sect. 6 that they give good results even in these situations.

Another potential limitation to these correlations is that the average modularity over several executions increases with the size of the graph and stabilizes once the number of nodes of a graph reaches 20 000 (see Fig. 3c). A solution

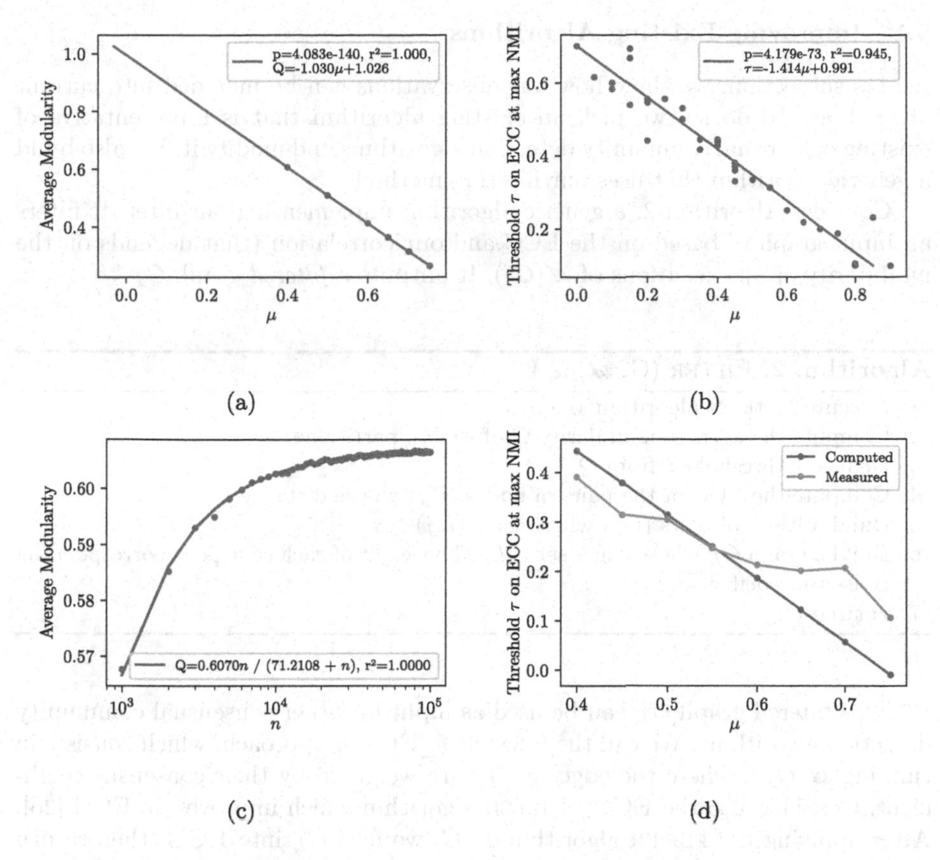

Fig. 3. Average modularity as a function of μ (top left). Threshold τ as a function of μ (top right). Q vs the number of nodes n (bottom left). τ vs μ at max NMI, computed with our filter and measured (bottom right)

would be to artificially increase the average modularity for graphs with less than 20 000 nodes. Figure 3c shows the modularity Q as a function of n, for LFR graphs with $\mu = 0.4$. We repeat the operation for different values of μ. We get a fit in the general case $\frac{Q_{\max} \times n}{(K+n)}$, where $Q_{\max}$ is the modularity for a sufficiently large graph, and $K = 71.21$. Therefore:

$$Q_{\max} = \frac{KQ}{n} + Q$$

The threshold τ should therefore be deduced from $Q_{\max}$ if the graph has less than 20 000 nodes.

Finally, Fig. 3d shows the *computed* threshold τ along with the *measured* threshold, with which the NMI would have been maximised, for different values of μ.

5.1 Improving Existing Algorithms

In this subsection, we show how our observations can be included into current algorithms. To do so, we pick an existing algorithm that is representative of existing consensual community detection algorithms and modify it. We also build a generic algorithm that uses our filtering method.

Consider Algorithm 2, a generic algorithm implementing our filter. It filters an input graph G based on the ECC and our correlation (that depends on the modularity of n_p executions of $\mathscr{A}(G)$). It outputs a *filtered* graph G_L.

Algorithm 2. FILTER (G, $\mathscr{A}$, n_p)

1: Execute n_p times algorithm $\mathscr{A}$ on G
2: Compute the average modularity Q of the n_p partitions
3: Deduce a threshold τ from Q
4: Compute the ECC of the pairs of nodes (i, j) whose distance is ≤ 2
5: Build a list L of pairs (i, j) whose $ECC(i, j) > \tau$
6: Build a graph G_L whose edges set is L. The weight of each edge is its corresponding consensus coefficient
7: **return** G_L

This filtered graph G_L can be used as input for other consensual community detection algorithms. We call the GENERIC_FILTER approach, which consists in running $\mathscr{A}(G_L)$, where the edges of G_L are weighted by their consensus coefficient. Consider also the ECG_FILTER algorithm which improves on ECG [25]. After applying the FILTER algorithm on G, we feed G_L into ECG, then return the consensual communities found by ECG. We also study the TANDON_FILTER algorithm, which builds G_L in the same way, then feeds it to TANDON. Note that since TANDON end ECG work on unweighted graphs, we don't bother weighting them. The next section focuses on studying the performance of both algorithms.

6 Experiments

First we run our algorithm on several LFR graphs and compare the NMI along with the running time compared to other state of the art algorithms, then we repeat the experiment on real-world graphs. In this section, we chose $\mathscr{A}$ = Louvain and $n_p = 64$ as we chose to parallelize the executions of $\mathscr{A}$ and our machine has got 64 cores.[1]

[1] Implementation details with correlations and correction for small graphs can be found on Software Heritage. The ordering criteria (based on the ECC) along with such correlations allows to improve on some current consensual community detection algorithms. The implementation behind Figs. 1 and 2 is available on Software Heritage.

6.1 Synthetic Graphs

First, we generate LFR graphs with 10 000 nodes, using the parameters from Sect. 4.1 and μ varies from 0.4 to 0.7. Figure 4a shows the NMI as a function of μ, and Fig. 4b shows the running time as a function of the number of nodes. We compare the filtered version of Tandon and ECG against their regular version. We also display our generic filter, and Louvain and Infomap as baselines. For low values of μ, the filtered version of ECG provides a higher NMI, at the cost of a higher running time compared the regular ECG algorithm. For high values of μ, the NMI rapidly decreases, providing an NMI of 0.42 and 0.30 for $\mu = 0.7$ and 0.75 respectively, which is a lot lower than the regular ECG. Our generic filter algorithm gives a higher NMI compared to ECG, but is outperformed by Tandon's algorithm. However, it uses a lot less time and memory, allowing it to work on bigger graphs. We also observe a sharp decrease in NMI for high values of μ. For Tandon's algorithm, the filtered version provides a slim NMI improvement at the cost of a slightly higher running time.

Note that since a fair amount of the running time is spent filling the consensus matrix, the memory consumption grows with the running time.

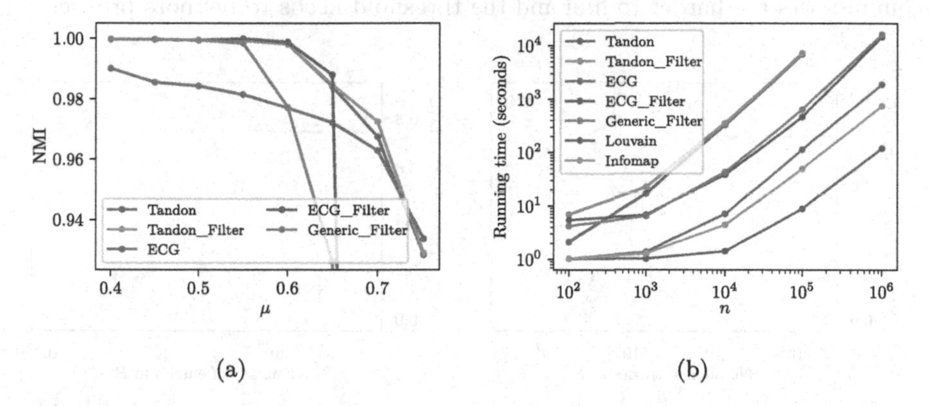

Fig. 4. NMI as a function of μ, LFR graphs with 10 000 nodes, with different algorithms (left). Running time as a function of n, LFR graphs, $\mu = 0.6$ (right)

Even though classical community detection algorithms suffer from the limits described before, we also applied this protocol to Louvain and Infomap. The NMI is naturraly much lower with these two algorithms: Louvain (resp. Infomap) goes from 0.84 to 0.76 (resp. from 0.81 to 0.78) for μ from 0.4 to 0.7.

6.2 Limitations

Overall, Fig. 4a shows that the filtered algorithms tend to perform better than their non-filtered counterparts for low values of μ, and perform worse on high values of μ. We believe that this is due to a combination of several effects.

As μ grows, the number of edges inside communities decreases, while the number of inter-community edges increases. This means that the ECC of pairs of nodes located in different communities will also tend to increase and our filter is more likely to add noise in the consensus matrix, which will amplify the problem, if μ is sufficiently large. Second, as μ grows, the modularity of the ground-truth partition tends to decrease. Therefore, as observed by Aynaud et al. [29] a community detection algorithm that focuses on modularity maximisation may find a partition with a higher modularity than the ground-truth. In this case it might be interesting to use non-modularity based algorithms.

Last, Fig. 5a shows the NMI as a function of the number of entries added in the consensus matrix, along with the value of the ECC pair added to the matrix. For low values of μ, we observe a sharp decrease in the ECC values, which corresponds to the maximum NMI value. This allows some imprecision in our threshold on the ECC value. As long as the threshold falls within the sharp decrease, we should be close to the maximum NMI possible. However, as μ grows, the sharp decrease tends to happen after the maximum NMI (see Fig. 5b). The decrease tends to be less sharp, thus requiring the threshold to be more precise. This makes the ECC threshold harder to set as a small imprecision might make a big change in the number of entries added to the consensus matrix. All in all, communities are harder to find and the threshold needs to be more precise.

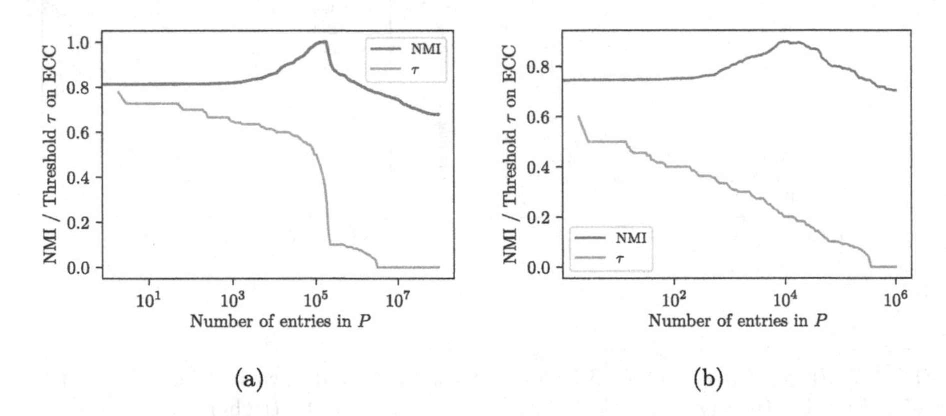

Fig. 5. NMI and ECC threshold as a function of the number of entries in the consensus matrix, for $\mu = 0.4$ (left). NMI and ECC threshold as a function of the number of entries in the consensus matrix, for $\mu = 0.7$ (right)

According to Orman et al [30], in LFR graphs, when μ is greater than 0.5, the communities are less well defined, and as μ increases we are left with a scale-free network with little to no community structure. At one extreme, if μ is very low, the partition is very easy to find. At the other extreme, there is no more community structure in the graph. Therefore, the consensus methods are particularly useful in intermediate ranges of μ.

6.3 Real Graphs

Football Dataset. In the football dataset [2], nodes represent US football teams, and the edges represent games played between the teams. The communities are the conference in which the team play. There are 115 nodes, 613 games and 12 communities in this dataset. The results are summarized in Table 1. We see that our generic filter algorithm provides the highest NMI along with the lowest running time, while our filtered version of ECG provides a higher NMI, but a higher running time than the classical ECG algorithm. We compare the performance against two classical community detection algorithms: the Louvain method [5], which gives a lower NMI than the other algorithms we tested, and the Infomap algorithm [3], which gives the second highest NMI on this dataset.

DBLP Dataset. The Digital Bibliography & Library Project (DBLP) dataset [31] is a co-authorship network where nodes are authors and two authors are linked by an edge if they co-published a paper. It is a larger graph than football, with 317 thousands nodes and about 1 million edges. The ground truth communities are the publication venues. Results are summarized in Table 1. First, we note a big difference in running time between the different algorithms, ranging from 12 s to 35 min. The *filtered* algorithms are the longest because of the computation of the ECC and the generation of the consensus graph. The highest NMI is provided by Infomap, at the cost of a high running time. The filtered version of the ECG algorithm provides the second highest NMI, but also has a high running time. Our generic filter algorithm performs poorly on this dataset, while Louvain gives the lowest NMI but is the fastest by far. We could not run Tandon's algorithm because the running time was too high.

Table 1. NMI and running time of several algorithms for football and DBLP

Algorithm	Football		DBLP	
	NMI	Running Time	NMI	Running Time
ECG	0.9079	1.10 s	0.6030	171 s
Tandon	0.8976	1.29 s	–	–
Tandon_Filter	0.8981	2.54 s	–	–
Generic_Filter	0.6823	1.21 s	0.4613	2236 s
ECG_Filter	0.9349	1.99 s	0.6288	1278 s
Louvain	0.8879	1.02 s	0,4769	12 s
Infomap	0.9242	1.12 s	0.7738	1305 s

7 Conclusion

Most community detection algorithms are non-deterministic and it is often necessary to aggregate the results of multiple runs to find a consensus, summarized in a consensus matrix. However, in real data, community are not perfectly defined. There are inter-community edges that carry noise. We have shown that the information in the consensus matrix is therefore noisy but that it is possible to filter out some of this noise. We then used these observations to improve existing algorithms, and verified the effectiveness of our approach on synthetic and real graphs. ECG_FILTER tends to yield a higher NMI than ECG at the cost of a higher running time. Our generic filter, GENERIC_FILTER, also provides a high NMI in most of our tests, while being more scalable than Tandon's algorithm.

A closer look at existing consensus community detection algorithms would also be relevant. Indeed, we believe that our general observations would be useful for most algorithms that use a consensus matrix and that could therefore benefit from noise filtering. Moreover, some algorithms like the LF procedure perform community detection in several steps, it could be worthwhile to filter the graph at each step. Finally, the correlations of Fig. 3 would be worth proving, the same way it was done for Fig. 3a [28].

References

1. Fortunato, S.: Community detection in graphs. Phys. Rep. **486**(3–5), 75–174 (2010)
2. Girvan, M., Newman, M.E.: Community structure in social and biological networks. Proc. Natl. Acad. Sci. **99**(12), 7821–7826 (2002)
3. Pons, P., Latapy, M.: Computing communities in large networks using random walks. In: Yolum, I., Güngör, T., Gürgen, F., Özturan, C. (eds.) ISCIS 2005. LNCS, vol. 3733, pp. 284–293. Springer, Heidelberg (2005). https://doi.org/10.1007/11569596_31
4. Rosvall, M., Axelsson, D., Bergstrom, C.T.: The map equation. Eur. Phys. J. Spec. Top. **178**(1), 13–23 (2009)
5. Blondel, V.D., Guillaume, J.-L., Lambiotte, R., Lefebvre, E.: Fast unfolding of communities in large networks. J. Stat. Mech. Theory Exp. **2008**(10), P10008 (2008)
6. Lancichinetti, A., Fortunato, S.: Consensus clustering in complex networks. Sci. Rep. **2**(1), 1–7 (2012)
7. Seifi, M., Guillaume, J.-L.: Community cores in evolving networks. In: Proceedings of the 21st International Conference on World Wide Web, 2012, pp. 1173–1180 (2012)
8. Newman, M.E.: Modularity and community structure in networks. Proc. Natl. Acad. Sci. **103**(23), 8577–8582 (2006)
9. Danon, L., Diaz-Guilera, A., Duch, J., Arenas, A.: Comparing community structure identification. J. Stat. Mech. Theory Exp. **2005**(09), P09008 (2005)
10. Lancichinetti, A., Fortunato, S., Radicchi, F.: Benchmark graphs for testing community detection algorithms. Phys. Rev. E **78**(4), 046110 (2008)
11. Good, B., de Montjoye, Y.-A., Clauset, A.: Performance of modularity maximization in practical contexts. Phys. Rev. E **81**, 046106 (2010). https://link.aps.org/doi/10.1103/PhysRevE.81.046106
12. Campigotto, R., Guillaume, J.-L., Seifi, M.: The power of consensus: random graphs have no communities. In: Proceedings of the 2013 IEEE/ACM International Conference on Advances in Social Networks Analysis and Mining, 2013, pp. 272–276 (2013)

13. Kheirkhahzadeh, M., Analoui, M.: A consensus clustering method for clustering social networks. Stat. Optim. Inf. Comput. **8**(1), 254–271 (2020)
14. Liu, Q., Hou, Z., Yang, J.: Detecting spatial communities in vehicle movements by combining multi-level merging and consensus clustering. Remote Sens. **14**(17), 4144 (2022)
15. Burgess, M., Adar, E., Cafarella, M.: Link-prediction enhanced consensus clustering for complex networks. PLoS ONE **11**(5), e0153384 (2016)
16. Jaccard, P.: Distribution de la flore alpine dans le bassin des dranses et dans quelques régions voisines. Bull. Soc. Vaud. Sci. Nat. **37**, 241–272 (1901)
17. Rasero, J., Pellicoro, M., Angelini, L., Cortes, J.M., Marinazzo, D., Stramaglia, S.: Consensus clustering approach to group brain connectivity matrices. Netw. Neurosci. **1**(3), 242–253 (2017)
18. Mandaglio, D., Amelio, A., Tagarelli, A.: Consensus community detection in multilayer networks using parameter-free graph pruning. In: Phung, D., Tseng, V.S., Webb, G.I., Ho, B., Ganji, M., Rashidi, L. (eds.) PAKDD 2018. LNCS (LNAI), vol. 10939, pp. 193–205. Springer, Cham (2018). https://doi.org/10.1007/978-3-319-93040-4_16
19. Chakraborty, T., Srinivasan, S., Ganguly, N., Bhowmick, S., Mukherjee, A.: Constant communities in complex networks. Sci. Rep. **3**(1), 1–9 (2013)
20. Liang, Z.-W., Li, J.-P., Yang, F., Petropulu, A.: Detecting community structure using label propagation with consensus weight in complex network. Chin. Phys. B **23**(9), 098902 (2014)
21. Raghavan, U.N., Albert, R., Kumara, S.: Near linear time algorithm to detect community structures in large-scale networks. Phys. Rev. E **76**(3), 036106 (2007)
22. Wang, Q., Fleury, E.: Detecting overlapping communities in graphs. In: European Conference on Complex Systems 2009 (ECCS 2009) (2009)
23. Yang, L., Yu, Z., Qian, J., Liu, S.: Overlapping community detection using weighted consensus clustering. Pramana **87**(4), 1–6 (2016). https://doi.org/10.1007/s12043-016-1270-2
24. Tandon, A., Albeshri, A., Thayananthan, V., Alhalabi, W., Fortunato, S.: Fast consensus clustering in complex networks. Phys. Rev. E **99**(4), 042301 (2019)
25. Poulin, V., Théberge, F.: Ensemble clustering for graphs. In: Aiello, L.M., Cherifi, C., Cherifi, H., Lambiotte, R., Lió, P., Rocha, L.M. (eds.) COMPLEX NETWORKS 2018. SCI, vol. 812, pp. 231–243. Springer, Cham (2019). https://doi.org/10.1007/978-3-030-05411-3_19
26. Hamming, R.W.: Error detecting and error correcting codes. Bell Syst. Tech. J. **29**(2), 147–160 (1950)
27. Radicchi, F., Castellano, C., Cecconi, F., Loreto, V., Parisi, D.: Defining and identifying communities in networks. Proc. Natl. Acad. Sci. **101**(9), 2658–2663 (2004)
28. Kamiński, B., Pankratz, B., Prałat, P., Théberge, F.: Modularity of the ABCD random graph model with community structure. J. Complex Netw. **10**(6), cnac050 (2022)
29. Aynaud, T., Blondel, V.D., Guillaume, J.-L., Lambiotte, R.: Multilevel local optimization of modularity. In: Graph Partitioning, chap. 13, pp. 315–345. John Wiley and Sons, Ltd. (2013). https://doi.org/10.1002/9781118601181.ch13. ISBN: 9781118601181

30. Orman, G.K., Labatut, V.: A comparison of community detection algorithms on artificial networks. In: Gama, J., Costa, V.S., Jorge, A.M., Brazdil, P.B. (eds.) DS 2009. LNCS (LNAI), vol. 5808, pp. 242–256. Springer, Heidelberg (2009). https://doi.org/10.1007/978-3-642-04747-3_20
31. Yang, J., Leskovec, J.: Defining and evaluating network communities based on ground-truth. In: Proceedings of the ACM SIGKDD Workshop on Mining Data Semantics, 2012, pp. 1–8 (2012)

Author Index

J. Mikyška et al. (Eds.): ICCS 2023, LNCS 14076, pp. 671–673, 2023.
https://doi.org/10.1007/978-3-031-36027-5

Printed in the United States
by Baker & Taylor Publisher Services